# Lecture Notes in Computer Science    11644

More information about this series at http://www.springer.com/series/7409

De-Shuang Huang · Kang-Hyun Jo ·
Zhi-Kai Huang (Eds.)

# Intelligent Computing Theories and Application

15th International Conference, ICIC 2019
Nanchang, China, August 3–6, 2019
Proceedings, Part II

 Springer

*Editors*
De-Shuang Huang
Tongji University
Shanghai, China

Kang-Hyun Jo
University of Ulsan
Ulsan, South Korea

Zhi-Kai Huang
Nanchang Institute of Technology
Nanchang, China

ISSN 0302-9743                ISSN 1611-3349   (electronic)
Lecture Notes in Computer Science
ISBN 978-3-030-26968-5        ISBN 978-3-030-26969-2   (eBook)
https://doi.org/10.1007/978-3-030-26969-2

LNCS Sublibrary: SL3 – Information Systems and Applications, incl. Internet/Web, and HCI

This Springer imprint is published by the registered company Springer Nature Switzerland AG
The registered company address is: Gewerbestrasse 11, 6330 Cham, Switzerland

# Preface

The International Conference on Intelligent Computing (ICIC) was started to provide an annual forum dedicated to the emerging and challenging topics in artificial intelligence, machine learning, pattern recognition, bioinformatics, and computational biology. It aims to bring together researchers and practitioners from both academia and industry to share ideas, problems, and solutions related to the multifaceted aspects of intelligent computing.

ICIC 2019, held in Nanchang, China, during August 3–6, 2019, constituted the 15th International Conference on Intelligent Computing. It built upon the success of previous ICIC conferences held in Wuhan, China (2018), Liverpool, UK (2017), Lanzhou, China (2016), Fuzhou, China (2015), Taiyuan, China (2014), Nanning, China (2013), Huangshan, China (2012), Zhengzhou, China (2011), Changsha, China (2010), Ulsan, Republic of Korea (2009), Shanghai, China (2008), Qingdao, China (2007), Kunming, China (2006), and Hefei, China (2005).

This year, the conference concentrated mainly on the theories and methodologies as well as the emerging applications of intelligent computing. Its aim was to unify the picture of contemporary intelligent computing techniques as an integral concept that highlights the trends in advanced computational intelligence and bridges theoretical research with applications. Therefore, the theme for this conference was "Advanced Intelligent Computing Technology and Applications." Papers focusing on this theme were solicited, addressing theories, methodologies, and applications in science and technology.

ICIC 2019 received 609 submissions from 22 countries and regions. All papers went through a rigorous peer-review procedure and each paper received at least three review reports. Based on the review reports, the Program Committee finally selected 217 high-quality papers for presentation at ICIC 2019, included in three volumes of proceedings published by Springer: two volumes of *Lecture Notes in Computer Science* (LNCS), and one volume of *Lecture Notes in Artificial Intelligence* (LNAI). This volume of *Lecture Notes in Computer Science* (LNCS) includes 73 papers.

The organizers of ICIC 2019, including Tongji University, Nanchang Institute of Technology, and East China Institute of Technology, as well as Shandong University at Weihai, made an enormous effort to ensure the success of the conference. We hereby would like to thank the members of the Program Committee and the referees for their collective effort in reviewing and soliciting the papers. We would like to thank Alfred Hofmann, executive editor from Springer, for his frank and helpful advice and guidance throughout, and for his continuous support in publishing the proceedings. In particular, we would like to thank all the authors for contributing their papers. Without the high-quality submissions from the authors, the success of the conference would not have been possible. Finally, we are especially grateful to the International Neural Network Society, and the National Science Foundation of China for their sponsorship.

June 2019

De-Shuang Huang
Kang-Hyun Jo
Zhi-Kai Huang

# ICIC 2019 Organization

## General Co-chairs

De-Shuang Huang, China
Shengqian Wang, China

## Program Committee Co-chairs

Kang-Hyun Jo, South Korea
Phalguni Gupta, India

## Organizing Committee Co-chairs

Chengzhi Deng, China
Zhikai Huang, China
Yusen Zhang, China

## Organizing Committee Members

Shumin Zhou, China
Wei Tian, China
Yan Li, China
Keming Liu, China
Shaoquan Zhang, China
Liling Zhang, China

## Award Committee Chair

Vitoantonio Bevilacqua, Italy

## Tutorial Chair

M. Michael Gromiha, India

## Publication Chair

Ling Wang, China

## Special Session Chair

Abir Hussain, UK

## Special Issue Chair

Kyungsook Han, South Korea

## International Liaison Chair

Prashan Premaratne, Australia

## Workshop Co-chairs

Jair Cervantes Canales, Mexico
Michal Choras, Poland

## Publicity Co-chairs

Valeriya Gribova, Russia
Laurent Heutte, France
Chun-Hou Zheng, China

## Exhibition Contact Chair

Di Wu, Tongji University, China

## Program Committee Members

| | | |
|---|---|---|
| Abir Hussain | Dah-Jing Jwo | Tianyong Hao |
| Khalid Aamir | Shaoyi Du | Mohd Helmy Abd Wahab |
| Kang-Hyun Jo | Dunwei Gong | Hao Lin |
| Angelo Ciaramella | Xiaoheng Deng | Hongmin Cai |
| Wenzheng Bao | Meng Joo Er | Xinguo Lu |
| Binhua Tang | Eros Pasero | Hongjie Wu |
| Bin Qian | Evi Syukur | Jianbo Fan |
| Bingqiang Liu | Fengfeng Zhou | Jair Cervantes |
| Bin Liu | Francesco Pappalardo | Junfeng Xia |
| Li Chai | Gai-Ge Wang | Juan Carlos |
| Chin-Chih Chang | LJ Gong | Figueroa-Jiangning Song |
| Wen-Sheng Chen | Valeriya Gribova | Joo M. C. Sousa |
| Michal Choras | Michael Gromiha Maria | Ju Liu |
| Xiyuan Chen | Naijie Gu | Ka-Chun Wong |
| Jieren Cheng | Guoliang Li | Kyungsook Han |
| Chengzhi Liang | Fei Han | Seeja K. R. |

Yoshinori Kuno
Laurent Heutte
Xinyi Le
Bo Li
Yunxia Liu
Zhendong Liu
Hu Lu
Fei Luo
Haiying Ma
Mingon Kang
Marzio Pennisi
Gaoxiang Ouyang
Seiichi Ozawa
Shaoliang Peng
Prashan Premaratne
Boyang Qu
Rui Wang
Wei-Chiang Hong
Xiangwei Zheng
Shen Yin
Sungshin Kim

Surya Prakash
TarVeli Mumcu
Vasily Aristarkhov
Vitoantonio Bevilacqua
Ling Wang
Xuesong Wang
Waqas Haider Khan
  Bangyal
Bing Wang
Wenbin Liu
Weidong Chen
Wei Jiang
Wei Wei
Weining Qian
Takashi Kuremoto
Shitong Wang
Xiao-Hua Yu
Jing Xiao
Xin Yin
Xingwen Liu
Xiujuan Lei

Xiaoke Ma
Xiaoping Liu
Xiwei Liu
Yonggang Lu
Yongquan Zhou
Zu-Guo Yu
Yuan-Nong Ye
Jianyang Zeng
Tao Zeng
Junqi Zhang
Le Zhang
Wen Zhang
Qi Zhao
Chunhou Zheng
Zhan-Li Sun
Zhongming Zhao
Shanfeng Zhu
Quan Zou
Zhenran Jiang

## Additional Reviewers

Huijuan Zhu
Yizhong Zhou
Lixiang Hong
Yuan Wang
Mao Xiaodan
Ke Zeng
Xiongtao Zhang
Ning Lai
Shan Gao
Jia Liu
Ye Tang
Weiwei Cai
Yan Zhang
Yuanpeng Zhang
Han Zhu
Wei Jiang
Hong Peng
Wenyan Wang
Xiaodan Deng
Hongguan Liu
Hai-Tao Li

Jialing Li
Kai Qian
Huichao Zhong
Huiyan Jiang
Lei Wang
Yuanyuan Wang
Biao Zhang
Ta Zhou
Wei Liao
Bin Qin
Jiazhou Chen
Mengze Du
Sheng Ding
Dongliang Qin
Syed Sadaf Ali
Zheng Chenc
Shang Xiang
Xia Lin
Yang Wu
Xiaoming Liu
Jing Lv

Lin Weizhong
Jun Li
Li Peng
Hongfei Bao
Zhaoqiang Chen
Ru Yang
Jiayao Wu
Dadong Dai
Guangdi Liu
Jiajia Miao
Xiuhong Yang
Xiwen Cai
Fan Li
Aysel Ersoy Yilmaz
Agata Giełczyk
Akila Ranjith
Xiao Yang
Cheng Liang
Alessio Ferone
José Alfredo Costa
Ambuj Srivastava

Mohamed Abdel-Basset

Angelo Ciaramella

Anthony Chefles

Antonino Staiano

Antonio Brunetti

Antonio Maratea

Antony Lam

Alfredo Pulvirenti

Areesha Anjum

Athar Ali Moinuddin

Mohd Ayyub Khan

Alfonso Zarco

Azis Ciayadi

Brendan Halloran

Bin Qian

Wenbin Song

Benjamin J. Lang

Bo Liu

Bin Liu

Bin Xin

Guanya Cai

Casey P. Shannon

Chao Dai

Chaowang Lan

Chaoyang Zhang

Chuanchao Zhang

Jair Cervantes

Bo Chen

Yueshan Cheng

Chen He

Zhen Chen

Chen Zhang

Li Cao

Claudio Loconsole

Cláudio R. M. Silva

Chunmei Liu

Yan Jiang

Claus Scholz

Yi Chen

Dhiya AL-Jumeily

Ling-Yun Dai

Dongbo Bu

Deming Lei

Deepak Ranjan Nayak

Dong Han

Xiaojun Ding

Domenico Buongiorno

Haizhou Wu

Pingjian Ding

Dongqing Wei

Yonghao Du

Yi Yao

Ekram Khan

Miao Jiajia

Ziqing Liu

Sergio Santos

Tomasz Andrysiak

Fengyi Song

Xiaomeng Fang

Farzana Bibi

Fatih Adıgüzel

Fang-Xiang Wu

Dongyi Fan

Chunmei Feng

Fengfeng Zhou

Pengmian Feng

Feng Wang

Feng Ye

Farid Garcia-Lamont

Frank Shi

Chien-Yuan Lai

Francesco Fontanella

Lei Shi

Francesca Nardone

Francesco Camastra

Francesco Pappalardo

Dongjie Fu

Fuhai Li

Hisato Fukuda

Fuyi Li

Gai-Ge Wang

Bo Gao

Fei Gao

Hongyun Gao

Jianzhao Gao

Jianzhao Gao

Gaoyuan Liang

Geethan Mendiz

Geethan Mendiz

Guanghui Li

Giacomo Donato
Cascarano

Giorgio Valle

Giovanni Dimauro

Giulia Russo

Linting Guan

Ping Gong

Yanhui Gu

Gunjan Singh

Guohua Wu

Guohui Zhang

Guo-Sheng Hao

Surendra M. Gupta

Sandesh Gupta

Gang Wang

Hafizul Fahri Hanafi

Haiming Tang

Fei Han

Hao Ge

Kai Zhao

Hangbin Wu

Hui Ding

Kan He

Bifang He

Xin He

Huajuan Huang

Jian Huang

Hao Lin

Ling Han

Qiu Xiao

Yefeng Li

Hongjie Wu

Hongjun Bai

Hongtao Lei

Haitao Zhang

Huakang Li

Jixia Huang

Pu Huang

Sheng-Jun Huang

Hailin Hu

Xuan Huo

Wan Hussain Wan Ishak

Haiying Wang

Il-Hwan Kim

Kamlesh Tiwari

M. IkramUllah Lali

Ilaria Bortone

H. M. Imran

Ingemar Bengtsson
Izharuddin Izharuddin
Jackson Gomes
Wu Zhang
Jiansheng Wu
Yu Hu
Jaya Sudha
Jianbo Fan
Jiancheng Zhong
Enda Jiang
Jianfeng Pei
Jiao Zhang
Jie An
Jieyi Zhao
Jie Zhang
Jin Lu
Jing Li
Jingyu Hou
Joe Song
Jose Sergio Ruiz
Jiang Shu
Juntao Liu
Jiawen Lu
Jinzhi Lei
Kanoksak
    Wattanachote
Juanjuan Kang
Kunikazu Kobayashi
Takashi Komuro
Xiangzhen Kong
Kulandaisamy A.
Kunkun Peng
Vivek Kanhangad
Kang Xu
Kai Zheng
Kun Zhan
Wei Lan
Laura Yadira
Domínguez Jalili
Xiangtao Chen
Leandro Pasa
Erchao Li
Guozheng Li
Liangfang Zhao
Jing Liang
Bo Li

Feng Li
Jianqiang Li
Lijun Quan
Junqing Li
Min Li
Liming Xie
Ping Li
Qingyang Li
Lisbeth Rodríguez
Shaohua Li
Shiyong Liu
Yang Li
Yixin Li
Zhe Li
Zepeng Li
Lulu Zuo
Fei Luo
Panpan Lu
Liangxu Liu
Weizhong Lu
Xiong Li
Junming Zhang
Shingo Mabu
Yasushi Mae
Malik Jahan Khan
Mansi Desai
Guoyong Mao
Marcial Guerra de
    Medeiros
Ma Wubin
Xiaomin Ma
Medha Pandey
Meng Ding
Muhammad Fahad
Haiying Ma
Mingzhang Yang
Wenwen Min
Mi-Xiao Hou
Mengjun Ming
Makoto Motoki
Naixia Mu
Marzio Pennisi
Yong Wang
Muhammad Asghar
    Nadeem
Nadir Subasi

Nagarajan Raju
Davide Nardone
Nathan R. Cannon
Nicole Yunger Halpern
Ning Bao
Akio Nakamura
Zhichao Shi
Ruxin Zhao
Mohd Norzali Hj Mohd
Nor Surayahani Suriani
Wataru Ohyama
Kazunori Onoguchi
Aijia Ouyang
Paul Ross McWhirter
Jie Pan
Binbin Pan
Pengfei Cui
Pu-Feng Du
Kunkun Peng
Syed Sadaf Ali
Iyyakutti Iyappan
    Ganapathi
Piyush Joshi
Prashan Premaratne
Peng Gang Sun
Puneet Gupta
Qinghua Jiang
Wangren Qiu
Qiuwei Li
Shi Qianqian
Zhi-Xian Liu
Raghad AL-Shabandar
Rafał Kozik
Raffaele Montella
Woong-Hee Shin
Renjie Tan
Rodrigo A. Gutiérrez
Rozaida Ghazali
Prabakaran
Jue Ruan
Rui Wang
Ruoyao Ding
Ryuzo Okada
Kalpana Shankhwar
Liang Zhao
Sajjad Ahmed

| | | |
|---|---|---|
| Sakthivel Ramasamy | Vitoantonio Bevilacqua | Yaolai Wang |
| Shao-Lun Lee | Valeriya Gribova | Yaping Yang |
| Wei-Chiang Hong | Guangchen Wang | Yue Chen |
| Hongyan Sang | Hong Wang | Yongchun Zuo |
| Jinhui Liu | Haiyan Wang | Bei Ye |
| Stephen Brierley | Jingjing Wang | Yifei Qi |
| Haozhen Situ | Ran Wang | Yifei Sun |
| Sonja Sonja | Waqas Haider Bangyal | Yinglei Song |
| Jin-Xing Liu | Pi-Jing Wei | Ying Ling |
| Haoxiang Zhang | Wei Lan | Ying Shen |
| Sebastian Laskawiec | Fangping Wan | Yingying Qu |
| Shailendra Kumar | Jue Wang | Lvjiang Yin |
| Junliang Shang | Minghua Wan | Yiping Liu |
| Wei-Feng Guo | Qiaoyan Wen | Wenjie Yi |
| Yu-Bo Sheng | Takashi Kuremoto | Jianwei Yang |
| Hongbo Shi | Chuge Wu | Yu-Jun Zheng |
| Nobutaka Shimada | Jibing Wu | Yonggang Lu |
| Syeda Shira Moin | Jinglong Wu | Yan Li |
| Xingjia Lu | Wei Wu | Yuannong Ye |
| Shoaib Malik | Xiuli Wu | Yong Chen |
| Feng Shu | Yahui Wu | Yongquan zhou |
| Siqi Qiu | Wenyin Gong | Yong Zhang |
| Boyu Zhou | Wu Zhang | Yuan Lin |
| Stefan Weigert | Zhanjun Wang | Yuansheng Liu |
| Sameena Naaz | Xiaobing Tang | Bin Yu |
| Sobia Pervaiz | Xiangfu Zou | Fang Yu |
| Somnath Dey | Xuefeng Cui | Kumar Yugandhar |
| Sotanto Sotanto | Lin Xia | Liang Yu |
| Chao Wu | Taihong Xiao | Yumin Nie |
| Yang Lei | Xing Chen | Xu Yu |
| Surya Prakash | Lining Xing | Yuyan Han |
| Wei Su | Jian Xiong | Yikuan Yu |
| Qi Li | Yi Xiong | Yong Wang |
| Hotaka Takizawa | Xiaoke Ma | Ying Wu |
| FuZhou Tang | Guoliang Xu | Ying Xu |
| Xiwei Tang | Bingxiang Xu | Zhiyong Wang |
| Li-Na Chen | Jianhua Xu | Shaofei Zang |
| Yao Tuozhong | Xin Xu | Chengxin Zhang |
| Qing Tian | Xuan Xiao | Zehui Cao |
| Tianyi Zhou | Takayoshi Yamashita | Tao Zeng |
| Junbin Fang | Atsushi Yamashita | Shuaifang Zhang |
| Wei Xie | Yang Yang | Yan Zhang |
| Shikui Tu | Zhengyu Yang | Liye Zhang |
| Umarani Jayaraman | Ronggen Yang | Zhang Qinhu |
| Vahid Karimipour | Xiao Yang | Sai Zhang |
| Vasily Aristarkhov | Zhengyu Yang | Sen Zhang |

Shan Zhang
Shao Ling Zhang
Wen Zhang
Wei Zhao
Bao Zhao
Zheng Tian
Sijia Zheng
Zhenyu Xuan
Fangqing Zhao
Zhao Fangqing

Zhipeng Cai
Xing Zhou
Xiong-Hui Zhou
Lida Zhu
Ping Zhu
Qi Zhu
Zhong-Yuan Zhang
Ziding Zhang
Junfei Zhao
Zhe Li

Juan Zou
Quan Zou
Qian Zhu
Zunyan Xiong
Zeya Wang
Yatong Zhou
Shuyi Zhang
Zhongyi Zhou

# Contents – Part II

# Particle Swarm Optimization-Based Power Allocation Scheme for Secrecy Sum Rate Maximization in NOMA with Cooperative Relaying

Carla E. Garcia[1] ⓘ, Mario R. Camana[1] ⓘ, Insoo Koo[1(✉)] ⓘ,
and Md Arifur Rahman[2]

[1] University of Ulsan, Ulsan, South Korea
iskoo@ulsan.ac.kr
[2] Campus of Rennes, Cesson Sevigne, France

**Abstract.** In this paper, we study a particle swarm optimization (PSO)-based power allocation scheme for physical-layer security under downlink non-orthogonal multiple-access (NOMA) with cooperative relaying where a source node communicates directly with a nearby user and with a distant user via decode-and-forward relay. First, we formulate the optimization problem to maximize the secrecy sum rate (SSR) under the constraints of a minimum rate at each user and maximum transmission power at the source and the relay. Then, we stablish the lower and upper boundaries of the variables that need to be optimized, and we derive a solution via the proposed low computational-complexity scheme based on PSO algorithm. Next, we compare the SSR performance of the proposed scheme with two baseline schemes: orthogonal multiple access (OMA) and NOMA without cooperative relaying. Finally, simulation results show that NOMA with cooperative relaying can increase the SSR, compared to OMA and NOMA without cooperative relaying.

**Keywords:** Particle swarm optimization · Non-orthogonal multiple-access · Physical-layer security · Secrecy sum rate · Orthogonal multiple access · Cooperative relaying

## 1 Introduction

The rapid advancement of wireless communications networks has led to tremendous increases in the data traffic of the networks. Fifth-generation (5G) technology needs to support massive connectivity of users and Internet of Things devices to meet the demand for high data rates, high-energy efficiency, low latency, and improved quality of service (QoS). Multiple access techniques in wireless communications can be broadly categorized into two different approaches: orthogonal multiple access (OMA) and non-orthogonal multiple access (NOMA) [1]. In OMA, the same frequency channel is shared by all users at different times, so that it is not possible to use the same resource elements (e.g., frequency bands, time slots) in order to perform multiple access among several users. In contrast to OMA, NOMA allows allocation of one

© Springer Nature Switzerland AG 2019
D.-S. Huang et al. (Eds.): ICIC 2019, LNCS 11644, pp. 1–12, 2019.
https://doi.org/10.1007/978-3-030-26969-2_1

frequency channel to multiple users within the same cell at the same time in order to obtain advantages like improved spectral efficiency, higher cell-edge throughput, relaxed channel feedback, and low transmission latency [2].

In power-domain NOMA, different levels of transmission power are assigned to different users according to their channel conditions in order to obtain high system performance. In downlink NOMA, multiple users' information signals are overlaid at the transmitter, and successive interference cancellation (SIC) is applied at the receiver to decode the signals one by one until the desired users' information is obtained.

Cooperative communications (CC) uses a cooperative user to transmit a base station signal to the intended user, and it can also enhance the coverage of the network. To improve the transmission reliability, different cooperative NOMA schemes have been proposed in [3, 4] to evaluate the performance in outage probability and data rates of the systems. With the concept of CC in NOMA, a dynamic power allocation (PA) scheme was proposed in [5], where a closed form approximation is derived for outage probability, and a dynamic PA scheme is proposed to improve the throughput of the network. In wireless communications, due to the broadcasting nature, secure information exchange between the sender and the receiver is critical when an eavesdropper coexists in the network to intercept the legitimate users' communications. CC is a very useful method to improve the physical-layer security (PLS) of wireless communications [6–8], in which the secrecy rate (SR) of the networks is maximized based on solving the relay selection and the PA problems of the networks. Similarly, the concept of PLS enhancement can be applied in NOMA to recognize robust secure transmission.

SSR maximization in NOMA was studied in the literature [9, 10] through a closed-form solution of an optimal PA policy that maximizes the SSR. The results validated significant SSR improvement by NOMA, compared to OMA. However, neither study considered a cooperative relaying system. SR performance in a downlink NOMA system with relay was analyzed in [11, 12] based on evaluation of the secrecy outage probability. Simulation results demonstrated that cooperative relay in a NOMA system can be very useful in improving the secrecy performance of NOMA users.

In [13], the authors proposed maximizing the achievable SSR under NOMA with an amplify-and-forward (AF) relay system. The solution consists of decoupling the optimization problem into two subproblems and solving PA and relay beamforming in an iterative manner, which has high computational complexity. That is unlike our paper, where we use a decode-and-forward (DF) relay and an optimization solution based on the low computational-complexity particle swarm optimization (PSO) method [14]. A low complexity PA scheme is essentially required for SSR maximization in NOMA with cooperative relaying, which motivates us to propose a PSO-based PA scheme which can converge faster than the exhaustive search-based (ES) method. Exhaustive search is useful to obtain the optimal optimization solution. However, its computational complexity is very high when the number of users is increased in the network.

Our main contributions are summarized as follows.

- We provide a solution to maximize the SSR of NOMA system with cooperative relaying by applying a low-complexity PA scheme based on the PSO algorithm, which greatly decreases the complexity of the optimization problem.
- Maximization of the SSR under NOMA with cooperative relaying system is analyzed by satisfying the minimum rate for users and the maximum transmission power requirements of the network.
- Simulation results reveal that NOMA achieves significant SSR improvement in comparison with conventional OMA. Moreover, the proposed scheme can reach near-optimal performance, compared with an (ES)-based scheme with much lower complexity.
- We also verify through simulation results that the proposed PSO-based scheme with cooperative relaying significantly improves SSR performance compared to a scheme without cooperative relaying in NOMA systems.

The remainder of this paper is organized as follows. The system model is briefly discussed in Sect. 2. The problem formulation of NOMA with cooperative relaying and the PSO-based PA scheme are described in Sect. 3. Simulation results of the proposed scheme and a comparison with the OMA scheme are presented in Sect. 4. Finally, the conclusion is presented in Sect. 5.

## 2   System Model

We consider a downlink NOMA with cooperative relaying system comprised of one source, $S$, two users, one half duplex relay, $R$, employing a DF protocol, and an eavesdropper, $E$, as shown in Fig. 1. The nearby user is denoted as $u_1$ and the distant user is denoted $u_2$. Assume that each node only has a single antenna and there is no direct transmission link between the source and the distant user due to obstacles or shadowing effect. The channel coefficients of S to R, S to $u_1$, S to E, R to $u_2$, R to $u_1$, and R to E are denote as $h_{SR}$, $h_{Su_1}$, $h_{SE}$, $h_{Ru_2}$, $h_{Ru_1}$, $h_{RE}$. The Additive white Gaussian noise (AWGN) at all nodes is denoted as $z_j$ and is assumed to be independent, with variances of $\sigma_j^2$, $j \in \{u_1, u_2, R \text{ and } E\}$. In this paper, we also assume that the variance in the noise of the users, the eavesdropper, and the relays is the same as the variance for $\sigma^2$, where $\sigma^2 = \sigma_E^2 = \sigma_R^2 = \sigma_{u_1}^2 = \sigma_{u_2}^2$.

NOMA with a cooperative relaying system involves two stages: Stage A and Stage B. In Stage A, the source transmits nearby- and distant-users' messages, and then relay, nearby user, and eavesdropper receive them. According to NOMA principles, the nearby user performs SIC to remove the distant-user message. In Stage B, the relay transmits the distant-user message, and the eavesdropper, nearby user, and distant user receive it. In the following, we explain the operations in stage A and B in more details. In Stage A, the source sends the superimposed signal, $x = \sqrt{p_1}x_1 + \sqrt{p_2}x_2$ where $x_1$, $x_2 \in \mathbb{C}$, are the information-bearing messages for the nearby user and the distant user, respectively, with corresponding transmit power $p_1$ and $p_2$.

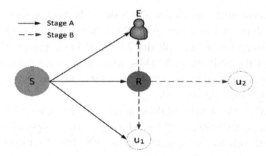

**Fig. 1.** Transmission under NOMA with cooperative relaying.

The received signal at nearby user $u_1$, the relay, and the eavesdropper are given by

$$y_j^A = \sqrt{p_1}x_1 h_{Sj} + \sqrt{p_2}x_2 h_{Sj} + z_j, \tag{1}$$

where $j \in \{u_1, R, E\}$. The rate of the nearby-user data at $u_1$ and at eavesdropper can be written as (2) and (3), respectively.

$$R_{x_1, Su_1} = \frac{1}{2}\log_2(1 + \frac{p_1 h_{Su_1}^2}{\sigma^2}). \tag{2}$$

$$R_{x_1, SE} = \frac{1}{2}\log_2(1 + \frac{p_1 h_{SE}^2}{p_2 h_{SE}^2 + \sigma^2}). \tag{3}$$

Hence, the achievable secrecy rate of the nearby user is the rate of nearby-user data at $u_1$ minus the rate of nearby-user data at eavesdropper [10, 12], given by

$$R_{u_1} = \left[R_{x_1, Su_1} - R_{x_1, SE}\right]^+, \tag{4}$$

where $[x]^+ = \max(0, x)$.

In Stage B, the relay decodes $x_2$ by treating $x_1$ as noise, and then, the relay sends $x_2$, which is received by the eavesdropper, nearby user $u_1$, and distant user $u_2$. Accordingly, the received signal in Stage B at $u_1$, $u_2$, and the eavesdropper is given by

$$y_j^B = \sqrt{p_r}x_2 h_{Rj} + n_j, \tag{5}$$

where $p_r$ is the transmission power at relay and $j \in \{u_1, u_2, E\}$. The rate of the distant-user data at $u_1$ and at the eavesdropper can be written as

$$R_{x_2, j} = \frac{1}{2}\log_2(1 + \frac{p_2 h_{Sj}^2}{p_1 h_{Sj}^2 + \sigma^2} + \frac{p_r h_{Rj}^2}{\sigma^2}), \tag{6}$$

where $j \in \{u_1, E\}$. The rate of the distant-user data at $u_2$ and at the relay is given by (7) and (8), respectively.

$$R_{x_2,Ru_2} = \frac{1}{2}\log_2(1 + \frac{p_r h_{Ru_2}^2}{\sigma^2}). \tag{7}$$

$$R_{x_2,SR} = \frac{1}{2}\log_2(1 + \frac{p_2 h_{SR}^2}{p_1 h_{SR}^2 + \sigma^2}). \tag{8}$$

Since the relay executes the DF protocol, and nearby user $u_1$ applies SIC, the achievable rate and the achievable secrecy rate for distant user $u_2$ are given by (9) and (10), respectively.

$$R_{u_2,min} = \min(R_{x_2,u_1}, R_{x_2,Ru_2}, R_{x_2,SR}). \tag{9}$$

$$R_{u_2} = \left[R_{u_2,min} - R_{x_2,E}\right]^+. \tag{10}$$

## 3   Problem Formulation and PSO-Based PA Scheme for Secrecy Sum Rate Maximization

In this paper, we focus on finding a low-complexity PA scheme for secrecy sum rate maximization in downlink NOMA with cooperative relaying, which is equivalent to maximizing $R_{u_1} + R_{u_2}$ under the constraint of a minimum rate at each user and the maximum transmission power at the source and the relay, as follows

$$P1: \max_{p_1,p_2,p_r} R_{u_1} + R_{u_2} \tag{11a}$$

$$R_{x_1,Su_1} \geq \gamma_1, \tag{11b}$$

$$R_{u_2,min} \geq \gamma_2, \tag{11c}$$

$$p_1 + p_2 \leq P_T^{max}, \tag{11d}$$

$$p_r \leq P_R^{max}, \tag{11e}$$

where $\gamma_1$ is the minimum rate at nearby user $u_1$, $\gamma_2$ is the minimum rate at distant user $u_2$, $P_T^{max}$ is the maximum transmission power at the source, and $P_R^{max}$ is the maximum transmission power at the relay.

### 3.1   PSO-Based PA Scheme for Secrecy Sum Rate Maximization in NOMA with Cooperative Relaying

In this paper, we propose a low-complexity PA scheme based on a PSO algorithm [14], which tries to find the approximately optimal solution in the search space base on a swarm of particles that updates from iteration to iteration.

We first establish the lower and upper boundaries of the variables $p_1$, $p_2$ and $p_r$, which is needed to be optimized for SSR maximization. The minimum transmission power of the nearby user is calculated based on constraint (11b), the minimum transmission power of the distant user and the minimum transmission power of relay is calculated based on the constraint (11c) and are given by (12), (13) and (14), respectively.

$$P_{1min} = \left(2^{2\gamma_1} - 1\right)\frac{\sigma^2}{h^2_{Su_1}}. \tag{12}$$

$$P_{2min} = \max(0, \theta_1, \theta_2), \tag{13}$$

$$P_{rmin} = \max(0, \varphi_1, \varphi_2), \tag{14}$$

where  $\theta_1 = (2^{2\gamma_2} - 1)\frac{P_{1min}h^2_{SR} + \sigma^2}{h^2_{SR}}$,   $\theta_2 = \left(2^{2\gamma_2} - 1 - \frac{P^{max}_R h^2_{Ru_1}}{\sigma^2}\right)\frac{P_{1min}h^2_{Su_1} + \sigma^2}{h^2_{su_1}}$.   $\varphi_1 = (2^{2\gamma_2} - 1)\frac{\sigma^2}{h^2_{Ru_2}}$, and  $\varphi_2 = \left(2^{2\gamma_2} - 1 - \frac{P^{max}_T h^2_{Su_1}}{P_{1min}h^2_{Su_1} + \sigma^2}\right)\frac{\sigma^2}{h^2_{Ru_1}}$.

The procedure of the proposed PSO-based algorithm in solving SSR maximization problem $P_1$ is described in Table 1, where the input parameters of PSO correspond to the maximum iteration number, the number of particles in a swarm, the inertia weight for velocity update, and the cognitive and social parameters, denoted as $M_I$, $N_P$, $I_W$, $c_1$, and $c_2$, respectively. Each of the $m$-th particle's position is a vector of three elements, $p_1$, $p_2$ and $p_r$, the feasible search region is denoted as $\left[p_{1\,min}, P^{max}_T\right]$, $\left[p_{2\,min}, P^{max}_T\right]$ and $\left[p_{r\,min}, P^{max}_R\right]$, respectively. The position of the $m$-th particle ($m = 1, 2, 3, \ldots, N_P$) has the elements $x_{p_1,m}$, $x_{p_2,m}$, and $x_{p_r,m}$ as follows:

$$\mathbf{x}_m = \left\{x_{p_1,m}, x_{p_2,m}, x_{p_r,m}\right\} \tag{15}$$

The initial global best position is assigned by selecting the maximum objective value from among all the initial positions of the particles. The velocities of the particles are updated, based on [15], can be given by (16) and the positions of the particles are accelerated can be given by (17).

$$\mathbf{v}^{(k+1)}_m = I^{(k)}_W \mathbf{v}^{(k)}_m + j_1 c_1 \left(\mathbf{pb}_m - \mathbf{x}^{(k)}_m\right) + j_2 c_2 \left(\mathbf{gb} - \mathbf{x}^{(k)}_m\right), \tag{16}$$

$$\mathbf{x}^{(k+1)}_m = \mathbf{x}^{(k)}_m + \mathbf{v}^{(k+1)}_m. \tag{17}$$

where $v^{(k)}_m$ is the velocity of the $m$-th particle in the $k$-th iteration, and $j_1$ and $j_2$ are random numbers between 0 and 1. Moreover, $\mathbf{x}^{(k)}_m$ is the position of the $m$-th particle in the $k$-th iteration. The individual best position of the $m$-th particle is denoted by $\mathbf{pb}_m$ and the global best position of all particles in the $k$-th iteration is denoted by $\mathbf{gb}$.

In order to obtain the global optimum values of the $p_1$, $p_2$ and $p_r$ variables, each particle is recalculated by updating the velocity and position of each particle and by evaluating the global best position of all particles until finding the best $m$-th particle to achieve the maximum value of SSR. The process is repeated until meeting the stopping criteria.

**Table 1.** The proposed algorithm based on PSO to solve the problem P1.

---

1:  Input parameters of PSO

2:  Set iteration count $k=1$

3:  Assign the initial particles' positions.

$$\mathbf{x}^{(k)}=\left\{\left(x_{p_1,1}^{(k)},x_{p_2,1}^{(k)},x_{p_r,1}^{(k)}\right)...\left(x_{p_1,m}^{(k)},x_{p_2,m}^{(k)},x_{p_r,m}^{(k)}\right)\right\}$$

4:  Assign the initial particles' velocities.

$$\mathbf{v}^{(k)}=\left(\mathbf{v}_1^{(k)},\mathbf{v}_2^{(k)},...,\mathbf{v}_m^{(k)}\right)$$

5:  Calculate SSR by computing $R_{u_1}+R_{u_2}$ for each $m$-th particle,

and denoting by $SSR\left(\mathbf{x}_m^k\right),\forall m$.

6:  Set the initial best particles' positions, $\forall m$: $\mathbf{pb}_m=\mathbf{x}_m^{(k)}$

7:  Set the initial global best position. $\mathbf{gb}=\arg\max_{1\leq m\leq NP} SSR\left(\mathbf{x}_m^{(k)}\right)$

8:  Increase the set iteration count $k=k+1$.

9:  **For** each particle $m$ **do**

10:    From (16), update the particles' velocities

11:    From (17), update the particles' positions

12:    Update the best particles' positions.

if $SSR\left(\mathbf{x}_m^k\right)>SSR(\mathbf{pb}_m)$ then

$\mathbf{pb}_m=\mathbf{x}_m^{(k)}$

end if

13:    Update the global best particles' positions, $\forall m$.

if $SSR\left(\mathbf{x}_m^k\right)>SSR(\mathbf{gb})$ then

$\mathbf{gb}=\mathbf{x}_m^{(k)}$

end if

14: **end for**

15: **if** $k<M_I$ **then**

$k=k+1$ and go to step 9

**else**

go to step 16

**end if**

16: **Return** the best values $\left\{p_1^*,p_2^*,p_r^*\right\}=\mathbf{gb}$ to obtain the maximum

value of SSR indicated in the problem *P1*.

---

## 3.2   Baseline Scheme to Solve the Secrecy Sum Rate Maximization Problem in OMA

In this subsection, we describe the baseline scheme to find a low-complexity PA scheme for secrecy sum rate maximization problem in OMA with cooperative relaying. In OMA, in the first time slot, the message $x_1$ is sent from the source to the nearby user $u_1$. Then, in the second time slot, the message $x_2$ is sent from the source to the relay. After the relay receives the message $x_2$, in the third time slot, the relay forwards this message to the distant user $u_2$. Accordingly, the rate of nearby-user data at $u_1$ and at eavesdropper can be written as (18) and (19), respectively.

$$R_{x_1, S_{u_1}\_OMA} = \frac{1}{3}\log_2\left(1 + \frac{p_1 h_{Su_1}^2}{\sigma^2}\right) \tag{18}$$

$$R_{x_1, SE\_OMA} = \frac{1}{3}\log_2\left(1 + \frac{p_1 h_{SE}^2}{\sigma^2}\right) \tag{19}$$

Consequently, the achievable secrecy rate of the nearby user is given by

$$R_{u_1\_OMA} = \left[R_{x_1, S_{u_1}\_OMA} - R_{x_1, SE\_OMA}\right]^+. \tag{20}$$

The rate of the distant-user data at $u_2$ and at relay are given by (21) and (22), respectively:

$$R_{x_2, S_{u_2}\_OMA} = \frac{1}{3}\log_2\left(1 + \frac{p_r h_{Ru_2}^2}{\sigma^2}\right) \tag{21}$$

$$R_{x_2, SR\_OMA} = \frac{1}{3}\log_2\left(1 + \frac{p_2 h_{SR}^2}{\sigma^2}\right) \tag{22}$$

The achievable rate of distant user $u_2$ is expressed as follows:

$$R_{u_2, min\_OMA} = \min\left(R_{x_2, S_{u_2}\_OMA}, R_{x_2, SR\_OMA}\right) \tag{23}$$

The rate of the distant-user data at the eavesdropper is given by

$$R_{x_2, E\_OMA} = \frac{1}{3}\log_2\left(1 + \frac{p_2 h_{Sj}^2}{\sigma^2} + \frac{p_r h_{Rj}^2}{\sigma^2}\right) \tag{24}$$

Consequently, the achievable secrecy rate of the distant user is expressed as follows:

$$R_{u_2\_OMA} = \left[R_{u_2, min\_OMA} - R_{x_2, E\_OMA}\right]^+. \tag{25}$$

In the same manner as $P1$, our aim is to maximize the secrecy sum rate, subject to the minimum rate requirements of the users and by satisfying the transmission power constraints of the source and the relay. Therefore, the optimization problem considering OMA with cooperative relaying is formulated as follows:

$$P2: \quad \max_{p_1,p_2,p_r} R_{u_1\_OMA} + R_{u_2\_OMA} \tag{26a}$$

$$R_{x_1,S_{u_1\_OMA}} \geq \gamma_1, \tag{26b}$$

$$R_{u_2,\min\_OMA} \geq \gamma_2, \tag{26c}$$

$$p_1 \leq P_T^{\max}, \tag{26d}$$

$$p_2 \leq P_T^{\max}, \tag{26e}$$

$$p_r \leq P_R^{\max}, \tag{26f}$$

Similar to the previous case, we propose a low-complexity PA scheme based on a PSO method to solve secrecy sum rate maximization problem $P2$.

## 4   Simulation Results

In this section, we present the simulation results of the proposed PA scheme for secrecy sum rate maximization in downlink NOMA with cooperative relaying programmed in the MATLAB software. Moreover, we provide a performance comparison among the proposed scheme, conventional OMA, and NOMA without cooperative relaying.

In the simulations, the channel coefficients are modeled as $h_{S_{u_1}} : CN\left(0, d_{S_{u_1}}^{-v}\right)$, $h_{SE} : CN\left(0, d_{SE}^{-v}\right)$, $h_{Ru_2} : CN\left(0, d_{Ru_2}^{-v}\right)$, $h_{Ru_1} : CN\left(0, d_{Ru_1}^{-v}\right)$, $h_{RE} : CN\left(0, d_{RE}^{-v}\right)$, where $d_{ij}$ denotes the distance between node $i$ and $j$, and $v$ is the path-loss exponent. We used the value of the path-loss exponent at $v = 4$. The noise power was $\sigma_E^2 = \sigma_R^2 = \sigma_{u_1}^2 = \sigma_{u_2}^2 = \sigma^2 = -60$ dBm. The distances between the nodes were as follows: $d_{SR} = 50$, $d_{S_{u_1}} = 30$, $d_{SE} = 50$, $d_{Ru_2} = 50$, $d_{Ru_1} = 30$, and $d_{RE} = 30$ in meters.

Figure 2 shows the convergence of the proposed low-complexity PSO-based algorithm given in Table 1 in terms of the iteration index. We see that the SSR improves as the iteration index is increased. Furthermore, we can observe that the SSR can converge in less than 100 iterations. Consequently, we set the maximum number of iterations, $M_I$ to be 60 for further simulations. The rest of the simulation parameters for the PSO method were chosen, based on [16], as follows: number of particles $N_P = 15$, inertia weight for velocity update $I_W = 0.7$, and scaling factors $c_1 = 1.494$ and $c_2 = 1.494$.

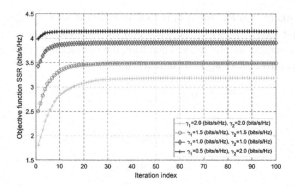

**Fig. 2.** Convergence of the PSO-based algorithm with different data rate requirements at user 1 $(\gamma_1)$ and user 2 $(\gamma_2)$.

Figure 3(a) shows the SSR performance according to the minimum data rate at nearby user 1, $\gamma_1$, and at distant user 2, $\gamma_2$, when the maximum transmission power at the source and the maximum transmission power at the relay are limited to $P_T^{\max} = P_R^{\max} = 40$ dBm. The SSR performance was obtained under the proposed scheme and the OMA scheme, and we can observe that the proposed scheme is superior to the traditional OMA. This is because in OMA, the users share the same frequency channel at different times. By contrast, NOMA users can utilize the same frequency channel at the same time. This fact improves the spectral efficiency in the NOMA system. Moreover, we see that the simulation result of the proposed PSO-based algorithm is very close to that obtained by the ES method, which is useful for obtaining the optimal optimization solution. However, the computational complexity of ES is very high when the number of users in the network is increased. Therefore, we use PSO-based PA scheme, which permits obtaining an approximately optimal solution and can converge faster than the ES-based method. To show the performance of the ES scheme, we use 10 channel realizations because of the long computational time. However, we use 1000 channel realizations for the rest of the simulations.

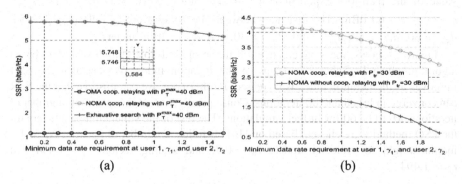

(a)                                                                 (b)

**Fig. 3.** SSR performance comparison according to the minimum data rate requirements at user 1 $(\gamma_1)$ and user 2 $(\gamma_2)$. (a) for the exhaustive search, NOMA and OMA (b) between NOMA with cooperative relaying and NOMA without cooperative relaying.

Based on the minimum data rates at nearby user 1 $(\gamma_1)$ and distant user 2 $(\gamma_2)$, Fig. 3(b) shows a comparison of SSR performance between our scheme using cooperative relaying and a NOMA scheme without employing comparative relaying when the maximum transmission power at the source and the maximum transmission power at the relay are limited to $P_T^{\max} = P_R^{\max} = 30$ dBm. The results show that the proposed scheme achieves higher SSR by using NOMA with cooperative relaying than by using NOMA without cooperative relaying. This behavior is because the distant user can easily get the QoS requirement by utilizing the relaying, especially when this user is far from the source. In the NOMA scheme without cooperative relaying, the source provides more transmission power to satisfy the QoS requirement of the distant user, causing that the SINR increases at the eavesdropper, and hence, the SSR decreases.

## 5 Conclusion

In this paper, we studied a low-complexity PA scheme based on a PSO algorithm to maximize the SSR of NOMA cooperative relaying. SSR is maximized based on the minimum rate requirements of the users and by satisfying the transmission power constraints of the source and the relay in a downlink NOMA system. We compared our proposed scheme with the optimal ES-based scheme and showed that the proposed scheme can achieve near-optimal performance with much lower computational complexity. In addition, we compared the performance of NOMA with a baseline OMA scheme to show the effectiveness of NOMA at improving the SSR of the system. We also showed that NOMA with cooperative relaying provides better performance than NOMA without cooperative relaying.

**Acknowledgment.** This work was supported by the National Research Foundation of Korea (NRF) grant through the Korean Government (MSIT) under Grant NRF-2018R1A2B6001714.

## References

1. Wang, P., Xiao, J., Ping, L.: Comparison of orthogonal and nonorthogonal approaches to future wireless cellular systems. IEEE Vehi. Techn. **1**(3), 4–11 (2006)
2. Wan, D., Wen, M., Ji, F., Yu, H., Chen, F.: Non-orthogonal multiple access for cooperative communications: challenges, opportunities, and trends. IEEE Wireless Commun. **25**(2), 109–117 (2018)
3. Choi, J.: Non-orthogonal multiple access in downlink coordinated two-point systems. IEEE Commun. **18**(2), 313–316 (2014)
4. Ding, Z., Peng, M., Poor, H.: Cooperative non-orthogonal multiple access in 5G systems. IEEE Commun. Lett. **19**(8), 1462–1465 (2015)
5. Li, Y., Li, Y., Chen, Y., Ye, Y., Zhang, H.: Performance analysis of cooperative NOMA with a shared AF relay. IET Commun. **12**(19), 2438–2447 (2018)
6. Wang, D., Bai, B., Chen, W., Han, Z.: Secure green communication via untrusted two-way relaying: a physical layer approach. IEEE Trans. Commun. **64**(5), 1861–1874 (2016)
7. Wang, D., Bai, B., Chen, W., Han, Z.: Achieving high energy efficiency and physical-layer security in AF relaying. IEEE Trans. Wireless Commun. **15**(1), 740–752 (2016)

8. Wang, D., Bai, B., Chen, W., Han, Z.: Energy efficient secure communication over decode-and-forward relay channels. IEEE Trans. Commun. **63**(3), 892–905 (2015)
9. Zhang, Y., Wang, H., Yang, Q., Ding, Z.: Secrecy sum rate maximization in non-orthogonal multiple access. IEEE Commun. Lett. **20**(5), 930–933 (2016)
10. Tang, J., et al.: Optimization for maximizing sum secrecy rate in SWIPT-enabled NOMA systems. IEEE Access **6**, 43440–43449 (2018)
11. Liu, C., Zhang, L., Xiao, M., Chen, Z., Li, S.: Secrecy performance analysis in downlink NOMA systems with cooperative full-duplex relaying. In: 2018 IEEE International Conference on Communications Workshops (ICC Workshops), Kansas City, USA, pp. 1–6 (2018)
12. Abbasi, O., Ebrahimi, A.: Secrecy analysis of a NOMA system with full duplex and half duplex relay. In: 2017 Iran Workshop on Communication and Information Theory (IWCIT), Tehran, pp. 1–6 (2017)
13. Tian, M., Zhao, S., Li, Q., Qin, J.: Secrecy sum rate optimization in nonorthogonal multiple access AF relay networks. IEEE Syst. J., 1–4 (2018). https://doi.org/10.1109/JSYST.2018.2867365
14. Zhang, Y., Wang, S., Ji, G.: A comprehensive survey on particle swarm optimization algorithm and its applications. Math. Prob. Eng. **501**, 931256 (2015)
15. Rahman, M., Lee, Y., Koo, I.: Energy-efficient power allocation and relay selection schemes for relay-assisted D2D communications in 5G wireless networks. Sensors **18**(9), 2865 (2018)
16. Tuan, P., Koo, I.: Optimal multiuser MISO beamforming for power-splitting SWIPT cognitive radio networks. IEEE Access **5**, 14141–14153 (2017)

# A Novel Binary Particle Swarm Optimization for Multiple Sequence Alignment

Yanlian Du$^{(\boxtimes)}$, Jingxuan He, and Changcun Du

Hainan University, Haikou 570228, Hainan, China
181551@hainanu.edu.cn

**Abstract.** Multiple sequence alignment (MSA) is a basic operation in bioinformatics, which is known to be an NP-complete problem. With the advent of the post genetic era, the number of biological sequences is increasing exponentially, it is urgent to develop high-efficiency heuristic parallel algorithms to solve MSA. In this paper, a novel binary particle swarm optimization (NBPSO) is proposed for MSA. Firstly, a novel binary code is designed to encode particle to adapt PSO to MSA, velocity and position updating formulas are refined accordingly. Regarding of the 'illegal' phenomenon of MSA after positions updating, based on probability theory, an auxiliary particle re-initialization strategy is proposed. Multi-swarm PSO is used to make the searching more parallel, diversity index $dt$ is proposed based on Hamming Distance, mutation takes place when $dt$ falls below threshold. The proposed NBPSO-MSA is operated on two benchmarks, i.e., OX-Bench and BAliBASE. Experiments show that, NBPSO-MSA outperforms other 12 state-of-the art aligners in accuracy, the trajectories of particles show that, diversity index and mutation help multi-swarm NBPSO jump out of local optima and lead to rapid convergence.

**Keywords:** Multiple sequence alignment · Particle swarm optimization · Novel binary particle swarm optimization · Diversity index

## 1 Introduction

Multiple sequence alignment (MSA) is a basic operation in Bioinformatics. It plays an important role in discovering sequence motifs, analyzing phylogeny and predicting structure. As a basic subject of Bioinformatics, MSA algorithms have been around for 40 years. So far, many aligners have been proposed, either based on Dynamic Programming (DP), or based on heuristic algorithm. DP is an accurate aligner, the cost is high complexity, with $O(n^s)$ time and space [1]. Most of the aligners may have high precision in alignment effect, but either with high complexity or with numerous parameters. When changing to other sequences, new parameters need to be reset. With the advent of the post-genetic era, the number of biological sequences is increasing exponentially, it is urgent to develop high-efficiency heuristic algorithms to solve it.

In recent years, Genetic Algorithm, Ant Colony Algorithm, Particle Swarm Optimization (PSO) and other algorithms have been extended to MSA, trying to solve the problems of accuracy, inherent complexity and parameters selection. PSO, which was

© Springer Nature Switzerland AG 2019
D.-S. Huang et al. (Eds.): ICIC 2019, LNCS 11644, pp. 13–25, 2019.
https://doi.org/10.1007/978-3-030-26969-2_2

proposed in 1995 [2], has been successfully applied to various fields, but there is little research on MSA using binary parallel PSO. In [3], PSO is designed to use binary code to encode the biological sequences, and a probability model is introduced. However, in the iteration process, all particles share the same probability matrix, which does not conform to the idea of PSO that, particles fly according to their velocity, personal best (*pbest*), and global best (*gbest*), leading to problems such as local optima, premature convergence etc. In [4], Sun Jun et al. use PSO to train parameters of HMM, then HMM is used to resolve MSA. In [5], Zhan et al. first optimize the parameters of HMM by PSO, then HMM is combined with partition function to resolve MSA. In both [4, 5], PSO is used to resolve the parameters of HMM instead of resolving MSA directly, this reduces the feasibility of the algorithm, in the way that different models have to be trained again according to different sequences and different objective functions, what's more, learning HMM is itself a difficult task, which inherently increases the propagation of error. In [6], bi-level discrete PSO based on set-theory is proposed to solve MSA, level one optimizes optimal sequence length, whereas level two optimizes optimum gap positions for best alignment. In [7, 8], a characteristic-based framework for MSA is proposed, PSO was used to train the parameter configuration of the framework.

In this paper, inspired by the idea of PSO tracking the optimal particle in an iterative process [2, 9], the problem of finding the optimal MSA is transformed into a problem of finding the optimal particle. The particle is encoded as a 0/1 position matrix, the components of position are "1" or "0", "1" represents a gap inserted, while "0" represents a base (or residue), and all "0" components from the $i$th row corresponds to the $i$th sequence to be aligned, in this way, each particle represents a MSA. The goal is to find the significant locations for "1" in the optimal particle. Velocity is designed to estimate the probability that, the value of the corresponding component in position matrix will change, it is quantified by the distance between the individual and the optimal, which represents the pace of the individual and the optimal. The proposed algorithm gives a new definition of velocity and position, designs a novel binary coding according to the problem of MSA, while the idea of NBPSO is exactly the same with the original PSO, and shares the same parameter settings, so all variants of PSO can be used to solve MSA.

The main contributions of this paper are as follows:

(1) Based on the binary code, we design a novel binary particle swarm optimization for multiple sequence alignment (NBPSO-MSA).
(2) According to the 'illegal' phenomenon, an auxiliary position re-initialization strategy is proposed, to make the search more effective.
(3) Multi-swarm PSO is used to make the search parallel, diversity index $dt$ is proposed based on Hamming Distance. To preserve the diversity of the population, when $dt$ falls below threshold, mutation takes place in each sub-swarm accordingly.
(4) NBPSO-MSA is operated on OX-Bench and BAliBASE benchmarks. The experimental results of NBPSO-MSA are compared with 12 well-known aligners.

The rest of the paper is organized as follows. Section 2 introduces the conventional PSO and its variants. Section 3.1 gives a mathematical description of MSA. Section 3.2 presents the proposed NBPSO-MSA, including the representation scheme, related operators, an auxiliary position re-initialization strategy and diversity index.

Section 3.3 presents the experimental results and the comparisons between NBPSO-MSA and other aligners. Section 5 concludes the paper and outlines future work.

## 2 Particle Swarm Optimization

### 2.1 Continuous Particle Swarm Optimization

Particle swarm optimization (PSO) was first proposed by Kennedy and Eberhart in 1995 [2], as a swarm-based heuristic random search algorithm, inspired by behavior patterns such as bird flocking and fish foraging. It was originally designed to find optimal solutions in a continuous space. In the original PSO, $m$ particles cooperate to search for the global optimum in the $n$-dimensional search space. Every particle maintains a position. In the iterative process, each particle uses its own search experience (self-cognitive) and the whole swarm's search experience (social-influence) to update the velocity and fly to a new position. The updating rules are as follows:

$$v_{i+1} = v_i + c_1 r_1 (pbest_i - x_i) + c_2 r_2 (gbest - x_i) \tag{1}$$

$$x_{i+1} = x_i + v_{i+1} \tag{2}$$

where $pbest_i$ is the best solution yielded by the $i$th particle and $gbest$ is the best-so-far solution obtained by the whole swarm. $c_1$ and $c_2$ are two parameters to weigh the importance of self-cognitive and social-influence, respectively. $r_1$ and $r_2$ are random numbers uniformly distributed in $[0, 1]$.

As the original PSO is easy to implement and effective, an amount of research effort has been made to improve its performance in various ways. The first successor is introduced by updating the inertia weight $\omega$ according to the following rule (3) [10]:

$$v_{i+1} = wv_i + c_1 r_1 (pbest_i - x_i) + c_2 r_2 (gbest - x_i) \tag{3}$$

This work indicates that, a relatively large inertia weight is better for global search, while a small one enhances the ability of local search. A scheme to decrease $\omega$ linearly from 0.9 to 0.4 was proposed accordingly. Constriction-factor variant was proposed, it is basically the same as inertia-weight variant, except the velocity updating formula (4) [11]. In [12], Clerc and Kennedy analyze the convergence behavior of PSO in detail and introduce a constriction factor, to guarantee the convergence of the algorithm.

$$v_{i+1} = \chi(v_i + c_1 r_1 (pbest_i - x_i) + c_2 r_2 (gbest - x_i)) \tag{4}$$

Another active research area is to develop different topologies. Kennedy and Mendes suggest using the local best position $lbest$ of a neighborhood and modify the velocity updating rule into formula (5) [13],

$$v_{i+1} = wv_i + c_1 r_1 (pbest_i - x_i) + c_2 r_2 (lbest - x_i) \tag{5}$$

The neighborhood can be defined by different topologies, such as Ring, URing, von Neumann and so on. In general, a large neighborhood is better for simple problems, and a small neighborhood is more suitable for complex problems.

Meanwhile, many researchers have developed variants of PSO to prevent premature convergence by modifying the learning strategies of particles or combining PSO with other search techniques. A representative PSO variant is the CLPSO proposed by Liang et al. [14]. The algorithm updates velocity using formula (6),

$$v_{i+1}^j = w v_i^j + c r^j \left( pbest_{f_i(j)}^j - x_i^j \right) \tag{6}$$

where $c$ is a parameter, $r^j$ is a random number in [0, 1], and $pbest_{f_i(j)}^j$ means the $j$th dimension of the $pbest_{f_i(j)}^j$ position of the particle $f_i(j)$, $pbest_{f_i(j)}$ can be $pbest_i$ or any other particles' $pbest$ position. CLPSO has been shown to be excellent for complex multimodal function optimization problems.

The third group uses hybrid strategies and the fourth uses multi-swarm techniques.

## 2.2 Discrete Particle Swarm Optimization (DPSO)

PSO can also be applied to some discrete optimization problems of integer variables. Many DPSO algorithms have been proposed [9, 15, 16]. Chen et al. classify the existing DPSO algorithms into four types. Swap operator based PSO uses a permutation of numbers as position and a set of swaps as velocity. Space transformation based PSO uses various techniques to transform the position, defined as a vector of real numbers, to the corresponding solution in discrete space. Fuzzy matrix based PSO defines the position and velocity as a fuzzy matrix, and encode it to a feasible solution. Incorporating based PSO consists of hybrid methods incorporating some heuristic or problem dependent techniques. Chen et al. propose a discrete set-based PSO (S-PSO) with probabilities, velocity is defined as a set with probabilities, operators are replaced by procedures defined on the set. Experiments on TSP and MKP show that S-PSO fits for combinatorial optimization problems very well [15]. Wu et al. then adapt S-PSO to generate covering arrays, and propose two auxiliary strategies to enhance the performance of S-PSO [16].

## 2.3 Mathematical Description of Multiple Sequence Alignment

A sequence $S$ is a string consisting of a number of characters from an alphabet, e.g., $S = s_1 s_2 \cdots s_l, s_i \in A, i = 1, 2, \cdots, l$, where $A$ is a set with finite number of characters, $l$ is the length of the sequence $S$. For DNA sequence, $n = 4$, $A = \{A, C, G, T\}$, which representing four distinct nucleotides; for protein sequences, $n = 20$, each character represents a unique amino acid in $A = \{A, R, N, D, C, Q, E, G, H, I, L, K, M, F, P, S, T, W, Y, V\}$,

A sequence set is a set consisting of a given number of sequences, e.g., $U = (S_1, S_2, \cdots, S_m)$, where $S_i = s_{i1} s_{i2} \cdots s_{il_i}, i = 1, 2, \cdots, m$, $s_{ij} \in A$, $l_i$ is the length of the $i$th sequence $S_i$ in sequence set $U$. Then the sequence alignment of $U$ can be defined as a matrix $Y = (y_{ij})_{m \times n}$ satisfying the following conditions:

1. $y_{ij} \in \Lambda \cup \{-\}$, where " $-$ " indicates a gap inserted;
2. If delete all " $-$ " in the $i$th row of $Y$, the left $y_{ij}$ of $Y$ corresponds to $S_i$;
3. No column in $Y$ consisting only of " $-$ ", i.e., $\sum_i^m x_{ij} < m, \forall j = 1, 2, \cdots, n$.
4. When $m = 2$, it is called pairwise sequence alignment (PSA), else it is called MSA while $m > 2$.

Based on the above discussion, MSA can be described as a Maximum problem (7):

$$Max f(U) \tag{7}$$

$$s.t. \ U \oplus V = Y$$

$$y_{1j} + y_{2j} + \cdots + y_{mj} < m, \forall j = 1, 2, \cdots, n$$

$$l_i + r_i = n, \forall i = 1, 2, \cdots, m$$

$$u_{ik} \in \Lambda, \ y_{ij} \in \Lambda \cup \{-\}, \ v_{it} \in N$$

$$i = 1, 2, \cdots, m; k = 1, 2, \cdots, l_i; j = 1, 2, \cdots, n; t = 1, 2, \cdots, r_i.$$

where $V$ records the indexes of gaps inserted, the operator $\oplus$ describes how to insert gaps into the sequences in $U$, e.g., 'ATGATC' $\oplus$ (0, 3, 6) = '-ATG-ATC-'. $f$ is the objective function, i.e., the scoring function of MSA. $r_i$ is the number of gaps inserted in the $i$th sequence $S_i$. To assess the quality of a test alignment with respect to a reference alignment, in this paper, we take into account two well-known scoring methods [7, 8]:

*Q-Score* (also known as *Sum-of-Pairs Score, QS*): Defines $p_{ijk}$ as 1 if $y_{ji} = y_{ki}$, 0 else; the score for the $i$th column $p_i$ is defined as (8), the $Q$-score for the MSA is equal to (9),

$$p_i = \sum_{j=1}^{n} \sum_{k=1, k \neq j}^{n} p_{ijk} \tag{8}$$

$$QS = \frac{\sum_{i=1}^{n} p_i}{\sum_{i=1}^{rn} r p_i} \tag{9}$$

where $rn$ is the number of columns in the reference alignment, $rp_i$ is the score for the $i$th column in the reference alignment.

*Total Column Score* (*TC-Score*): Defines a score $c_i$ for the $i$th column in the alignment: $c_i = 1$ if all the residues are aligned with the reference alignment; otherwise, $c_i = 0$. The *TC-score* for the MSA is then calculated as formula (10),

$$TC = \frac{\sum_{i=1}^{rn} c_i}{n} \tag{10}$$

## 3   The Novel Binary PSO for MSA (NBPSO-MSA)

In this section, the NBPSO-MSA is described. To adapt the idea of PSO to solve MSA, NBPSO-MSA applies a novel binary representation scheme for MSA, and redefines the term "position" and "velocity" and all related operators for the discrete space according to MSA. In order to overcome the disadvantages of PSO, diversity index is proposed based on Hamming Distance, mutation is activated when the diversity index $d_t$ falls below the threshold $d_{low}$. The pseudo code of NBPSO-MSA is showed in Algorithm 1.

### 3.1   Representation Scheme

For MSA, two problems are significant: (1). How many gaps should be inserted; (2). Which locations of particle should be gaps.

According to problem (1), in this paper, OX-bench and BAliBASE benchmarks are used, which provided with reference alignments, the lengths of the aligned sequences are consistent with those in reference alignments. For problem (2), the key problem of MSA is how to detect locations of gaps that satisfies conditions (1)–(3). Exhaustive alignment covers exponential schemes of the system. With exponential time and space complexities, alignment becomes prohibitive as the number of sequences and the lengths of the sequences increase. Moreover, only a few locations are significant, in other words, exploring key locations can lead to rapid convergence. From this point of view, in this paper, each particle represents a candidate optimal MSA, a population of $m$ particles search in parallel for the optimal MSA.

**Table 1.** State of $pbest_{ij} - x_{ij}$ & $gbest_{ij} - x_{ij}$

| X | pbest or gbest | |
|---|---|---|
|   | 0 | 1 |
| 0 | 0 | 1 |
| 1 | −1 | 0 |

**Table 2.** State of $|pbest_{ij} - x_{ij}| + |gbest_{ij} - x_{ij}|$

| | | X | | | |
|---|---|---|---|---|---|
| | | 1 | 0 | | |
| pbest | 1 | 0 | 2 | 1 | gbest |
| | | 1 | 1 | 0 | |
| | 0 | 1 | 1 | 1 | gbest |
| | | 2 | 0 | 0 | |

The position of each particle is encoded as a matrix $X_{s \times n}$, where $s$ is the number of sequences to be aligned, $n$ is the length of the aligned sequences, its components $x_{ij}$, indicates whether a gap is inserted or not at the $j$th location of the $i$th sequence, with a range of $\{0, 1\}$. The velocity of each particle is encoded as a matrix $V_{s \times n}$ of the same type as the position matrix, and its components $v_{ij}$ estimates the probability of the change of the corresponding component in position matrix, with a range of $[0,1]$. The range of the $pbest_{ij} - x_{ij}$ and the $gbest_{ij} - x_{ij}$ both are $\{-1,0,1\}$, as shown in Table 1, "−1" indicates that the $pbest_{ij}$ or $gbest_{ij}$ is not gap while $x_{ij}$ do, in this case, $v_{ij}$ should be large to increase the probability that $x_{ij}$ switch to 0; "1" indicates that the $pbest_{ij}$ or $gbest_{ij}$ is gap while $x_{ij}$ do not, in this case, $x_{ij}$ should switch to "1", $v_{ij}$ should also be increased to make itself consistent with the optimal particle; differently, "0" indicates that the $pbest_{ij}$ or $gbest_{ij}$ and $x_{ij}$ are both gap or both not, in that case, $x_{ij}$ has been consistent with $pbest_{ij}$ and $gbest_{ij}$, then $v_{ij}$ should be decreased, preventing it from going away from the optimal.

## 3.2 Velocity Updating

In NBPSO, velocity is a matrix $V_{s \times n}$, whose components represent the probabilities that, the corresponding components in $X_{s \times n}$ switch from "1" to "0", or "0" to "1".

As is presented in Table 1, the states of $pbest_{ij} - x_{ij}$ or $gbest_{ij} - x_{ij}$ is $\pm 1$ or 0. Furthermore, the states of $|pbest_{ij} - x_{ij}| + |gbest_{ij} - x_{ij}|$ is presented in Table 2. 0 conveys the fact that, the component of the current particle is consistent with both the *pbest* and *gbest*, change is not needed; 1 indicates that the current particle is consistent with only one of the *pbest* and *gbest*, change is needed; while 2 represents the fact that the current particle is inconsistent with *pbest* or *gbest*, urgent change is needed. Consequently, velocity is designed to represent probability of the change of the particle, it is firstly quantified by the distance between the individual and the optimal, then, calculated by formula (11), finally serves as a probability that stimulates the particle, guides the swarm to fly to *gbest*. Velocity and position updating rules are as followed (11) and (12).

$$v_{t+1} = w_t v_t + c_1 r_1 |pbest_t - x_t| + c_2 r_2 |gbest - x_t| \tag{11}$$

$$\begin{cases} if(rand() < S(v_t)) \ then \ x_{t+1} = 1 - x_t \\ else \ x_{t+1} = x_t \end{cases} \tag{12}$$

where *rand()* is a quasi-random number uniformly distributed in [0,1], $S(v)$ is as formula (13), $w_t$ linearly decreases from 0.9 to 0.4, $c_1 = 2$, $c_2 = 2$, $v_{max} = 6$.

$$S(v) = Sigmoid(v) = \frac{1}{1 + e^{-v}} \tag{13}$$

---

**Algorithm 1** NBPSO-MSA

---

1: Initialize
2: **while** the termination condition is not met, **do**
3:     Each sub-swarm operates PSO independently
4:     Mutation
5:     Migration
6:     $t = t + 1$
7: **end while**
8: **return** *gbest*

---

## 3.3 Position Updating

As velocity is designed to represent the probability of the change of the current particle, acts as a compass to guide the swarms. After velocity updating, particles update their position value by formula (12). The larger the velocity, the more probably it will switch, either from "1" to "0" or "0" to "1"; the smaller the velocity, the more probably it remains the same. After all components are updated, the particle probably becomes illegal. That is, some particles fail to satisfy conditions (1)–(3) in three ways, the number of "1" components from the same row exceeds maximum, the number of "1" components from the same row is less than needed, the third is, some columns are all "1". In order to overcome these, an auxiliary particle re-initialization strategy based on probability theory is proposed for MSA.

As the populations evolve, *pbest* and *gbest* become more and more optimal. There must be some significant locations that frequently contribute to high fitness, e.g., type *A* locations which frequently keep "1" in *gbest*. For this sake, type *A* location is defined as definition 1. Position re-initialization is designed as Algorithm 2.

---

**Algorithm 2** Position re-initialization

---
1: **for** each particle $x_i$ **do**
2:     **for** the $j$th sequence $x_{ij}$ of particle $x_i$ **do**
3:         $R = $ the indexes of $v_{ij}$ in descending order
4:         **for** each base or gap $k$
5:             **if** rand()$<$s$(v_{ijR(k)})$ **then**
6:                 $x_{ijR(k)} = 1 - x_{ijR(k)}$
7:             **else** $x_{ijR(k)} = x_{ijR(k)}$ **end if**
8:         **end for**
9:         **if** sum$(x_{ij}) > r_j$ , **then**
10:             $d = $ sum$(x_{ij})-r_j$
11:             $Set=\{$locations of 1 in $x_{ij} - A\}$
12:             $q=$size$(Set)$
13:             Randomly generate a permutation $p = P_q^d$
14:             **for** each dimension $k$ $(k=1,2,\cdots,d)$ **do**
15:                 $x_{ijSet(p(k))} = 0$
16:             **end for**
17:         **end if**
18:         **if** sum$(x_{ij}) < r_j$ , **then**
19:             $d = r_j - $ sum$(x_{ij})$
20:             $Set=\{$locations of 0 in $x_{ij}\}$
21:             $q=$size$(Set)$
22:             Randomly generate a permutation $p = P_q^d$
23:             $k=1,kk=1$
24:             **while** $k<q$ and $kk<d+1$ $(k=1,2,\cdots,q)$
25:                 $x_{ijSet(p(k))}=1$
26:                 **if** sum$(x_{i \cdot Set(p(k))})=s$ **then**
27:                     $x_{ijSet(p(k))} = 0$
28:                 **else** $kk = kk +1$
29:                 **end if**
30:             **end while**
31:         **end if**
32:     **end for**
33: **end for**

---

**Definition 1** (*Type A location*): Let $A_i$ be a set of locations of "1" which appear in the *gbest* before the *i*th iteration, $gb_i$ be a set of locations of "1" which appear in the *gbest* in the *i*th iteration, then,

$$A_{i+1} = A_i \cap gb_{i+1} \tag{14}$$

From formulas (11–12), it can be inferred and preliminary tests also show that, particles rapidly converge to the optimal in each iteration, which is a double-edged sword, resulting in strong search ability but along with loss of diversity and premature convergence. One way to overcome these disadvantages is to employ a diversity strategy. In NBPSO, Hamming Distance is used to design diversity index as followed.

**Definition 2** (*Diversity index*): The diversity index of the swarm in the *t*th iteration is,

$$d_t = \sum_{k=1}^{m} \sum_{i=1}^{s} \sum_{j=1}^{n} \left( x_{ijk}^t + gbest_{ij}^t - 2x_{ijk}^t gbest_{ij}^t \right) \tag{15}$$

where $m$ is the population size of the swarm, $s$ is the number of sequences to be aligned, $x_{ijk}^t$ is the $(i,j)$ element of particle $X_k$ in the *t*th iteration, $gbest_{ij}^t$ is the $(i,j)$ element of the *gbest* of the swarm in the *t*th iteration.

For a population of $m$ particles, each with $\sum_{i=1}^{s} r_i$ "1" elements. If all particles are all the same, the diversity index $d_t$ achieves its minimum, i.e., $d_{min} = 0$. On the other hand, if all $m$ particles are different from each other, $d_t$ reaches its maximum, i.e., $d_{max} = 2(m-1) \sum_{i=1}^{s} r_i$. The less the $d_t$, the worse the diversity; the more the $d_t$, the better the diversity. Only when diversity index declines below threshold $d_{low}$, should mutation take place to help the *gbest* out of local optimum. In our preliminary experimental results, $d_{low} = 0.5d_{max}$ may lead to good performance.

## 4    Experiments and Results

To prove the effectiveness of NBPSO-MSA, we use an ordinary computer to test. Intel (R) Pentium (R) 3.30 GHz CPU, 4 GB memory, windows 10 operating system. In the experiments, the parameter settings are the same: 6 sub-swarms, each containing 20 particles, the *gbest* of each sub-swarm is migrated to its neighbor sub-swarm in clockwise direction, max iteration No. is 1000. The fitness function is Q-score.

To test NBPSO-MSA, we compare the experimental results of NBPSO-MSA with 12 state-of-the art aligners [7] based on two MSA alignment benchmarks: OX-Bench and BAliBASE, which can be downloaded from websites http://www.compbio.dudee. ac.uk/, http://www.lbgi.fr/balibase/BalibaseDownload/ respectively. Both are for benchmarking MSA methods which include reference alignment datasets. OX-Bench consists of 395 datasets which are divided into four groups by identity (ID), i.e., [0,0. 15), [0.15,0.3), [0.3,0.5), [0.5,1), each groups involve different numbers of datasets, which is made up by different number of sequences, and sequences are of different lengths. For each dataset of each benchmark, we record the mean of the Q-score, the TC-score and the hitting time of optima in 30 independent runs. For each group, we calculate the average of the Q-score, the TC-score and the running time of all the datasets in the group, the results of NBPSO-MSA and comparisons with 12 state-of-the art aligners are presented in Table 3 and Table 5 respectively.

NBPSO-MSA achieves the optimal Q-score among all aligners in all groups in OX-Bench, TC-score achieves the optimal 3 times except for Group Four. The running time is among the middle in each group, ranks 4 in average. An interesting phenomenon is that, the Q-score of NBPSO-MSA exceeds the one of reference alignment, as showed in Table 4, NBPSO-MSA identifies all the base pairs in the reference alignment i.e., the base pairs in shadow fonts; moreover, NBPSO-MSA may identify more base pairs than biological experts do, i.e., the base pairs in shadow bold fonts.

**Table 3.** Comparisons between NBPSO-MSA and 11 aligners on OX-Bench (*T*: second)

| Group | | One: [0,0.15) | | | Two: [0.15,0.3) | | | Three: [0.3,0.45) | | | Four: [0.5,1) | | |
|---|---|---|---|---|---|---|---|---|---|---|---|---|---|
| Aligner | *MT* | *QS* | *TC* | *T* | *QS* | *TC* | *T* | *QS* | *TC* | *T* | *QS* | *TC* | *T* |
| NBPSO-MSA | .37 | **.52** | **.16** | .22 | **.71** | **.41** | .35 | **.93** | **.81** | 0.53 | **1.08** | .94 | 0.38 |
| Clustal W | .07 | .29 | .11 | .03 | .64 | .38 | .06 | .88 | .75 | .12 | .98 | **.95** | .070 |
| DIALIGN | 1.07 | .14 | .02 | .16 | .52 | .24 | .51 | .84 | .68 | 2.89 | .97 | .93 | .73 |
| Kalign2 | **.03** | .27 | .11 | **.01** | .62 | .37 | **.02** | .89 | .77 | **.06** | .98 | **.95** | **.02** |
| MAFFT | .39 | .27 | .12 | .28 | .65 | **.41** | .35 | .89 | .77 | .57 | .98 | **.95** | .34 |
| MUSCLE | .16 | .31 | **.16** | .06 | .64 | **.41** | .10 | .88 | .76 | .32 | .98 | .94 | .06 |
| FSA | 1.71 | .11 | 0 | .35 | .48 | .21 | 1.70 | .83 | .65 | 3.38 | .97 | .93 | 1.39 |
| MSAProbs | .95 | .25 | .14 | .11 | .62 | .37 | .52 | .88 | .75 | 2.68 | .98 | **.95** | .49 |
| MUMMALS | .66 | .26 | .07 | .36 | .65 | .40 | .97 | .88 | .75 | .925 | .98 | **.95** | .39 |
| ProbAlign | .61 | .27 | .13 | .08 | .61 | .36 | .31 | .88 | .75 | 1.75 | .98 | **.95** | .31 |
| ProbCons | .93 | .26 | .12 | .13 | .61 | .36 | .51 | .87 | .74 | 2.58 | .98 | **.95** | .50 |
| T-Coffee | 2.62 | .21 | .05 | .36 | .62 | .37 | 1.45 | .87 | .74 | 7.30 | .98 | **.95** | 1.37 |

*MT*: the mean of time among 4 groups; *QS*: mean of Q-Score; *TC*: mean of TC-Score; *T*: mean of time hitting the optima.

**Table 4.** Reference alignment and best alignment of NBPSO-MSA

| Refenrence alignment |
|---|
| LDSPTGIDFSDITANSFTVHWIA-PRATITGYRIRHH-PEHFSGRPREDRVPHSRNSITLTNLTPGTEYVVSIVALNGREESPLLIGQQSTVS LDAPSQIEVKDVTDTTALITWFK-PLAEIDGIELTYGIKDVPG-DRTTIDLTEDENQYSIGNLKPDTEYEVSLISRRGDMSSNPAKETFTT— VPPPTDLRFTNIGPDTMRVTWAPPPSIDLTNFLVRYSPVKNEE-DVAELSISPSDNAVVLTNLLPGTEYVVSVSSVYEQHESTPLRGRQKTG- |
| Best alignment of NBPSO-MSA |
| LDSPTGIDFSDITANSFTVHWI-APRATITGYRIRHHPE-HFSGRPREDRVPHSRNSITLTNLTPGTEYVVSIVALNGREESPLLIGQQSTVS LDAPSQIEVKDVTDTTALITW-FKPLAEIDGIELTYG-IKDVPGDRTTIDLTEDENQYSIGNLKPDTEYEVSLISRRGDMSSNPAKETFTT— VPPPTDLRFTNIGPDTMRVTWAPPPSIDLTNFLVRYSPVKNE-EDVAELSISPSDNAVVLTNLLPGTEYVVSVSSVYEQHESTPLRGRQKTG- |

**Table 5.** Comparisons between NBPSO-MSA and 7 aligners on BAliBASE (*Time*: second)

| Aligners | Average | QS | TC | Time |
|----------|---------|-----|-----|------|
| NBPSO-MSA | **0.7920** | **0.9123** | 0.6717 | 1.036 |
| Clustal W | 0.5942 | 0.6963 | 0.4921 | – |
| DIALIGN | 0.5843 | 0.6863 | 0.4822 | – |
| ProbPFP | 0.6154 | 0.8250 | 0.6703 | – |
| ProbAlign | 0.7490 | 0.8253 | **0.6727** | – |
| MUSCLE | 0.6694 | 0.7560 | 0.5827 | – |
| ProbCons | 0.7339 | 0.8155 | 0.6522 | – |
| T-Coffee | 0.7288 | 0.8082 | 0.6493 | – |

*Average*: average of QS and TC; *QS*: Q-Score; *TC*: TC-Score

We can see the trajectory of particles from Fig. 1 that, in the process of NBPSO, particles fly to the optimal in a state of chaos without getting stuck in local optima. Figure 2 shows that, in NBPSO particles jump out of local optima frequently and finally converges to the global optima in an average of 430 iterations, the higher the identity of the dataset, the more rapidly NBPSO may converges. From this point of view, NBPSO is more suitable for alignment of datasets with high identities, which is consistent with reality, since regions with high identity may be visually identified.

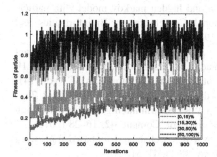

**Fig. 1.** Trajectory of particle

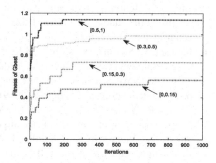

**Fig. 2.** Convergence of NBPSO-MSA

NBPSO-MSA is also operated on BAliBASE benchmark, the results are showed in Table 5. NBPSO-MSA achieves the best Q-score again, though the TC-score is very close to the optima, NBPSO-MSA still achieves the best average of Q-score and TC-score. This may due to the fact that, the fitness function of NBPSO-MSA is Q-score.

From the above, NBPSO-MSA is proved to outperform state-of-the art aligners in accuracy. Figures 1 and 2 show that, diversity index and mutation help NBPSO jump out of local optima and lead to rapid convergence. This reflects the fact that NBPSO is a greatly efficient parallel heuristic algorithm which can be further developed.

# 5   Conclusions

NBPSO-MSA is proposed. Firstly, binary code is designed to encode particle to adapt PSO to MSA, an auxiliary particle re-initialization strategy was proposed, diversity index $d_t$ is proposed, mutation takes place when $d_t$ falls below threshold. To test the proposed NBPSO-MSA, we compare the experimental results with 12 state-of-the art aligners based on two MSA alignment benchmarks: OX-Bench and BAliBASE. Experimental results show that, NBPSO-MSA achieves and even exceeds the optimal in terms of Q-score. Experiments show that, the proposed NBPSO-MSA is promising.

# References

1. Edgar, R.C.: MUSCLE: multiple sequence alignment with high accuracy and high throughput. Nucleic Acids Res. **32**(5), 1792–1797 (2004)
2. Kennedy, J., Eberhart, R.: Particle swarm optimization. In: IEEE International Conference on Neural Networks, Conference Proceedings, no. 4, pp. 1942–1948 (1995)
3. Yan, C.: Study on biological multi-sequence alignment based on probabilistic particle swarm optimization. Yanshan University (2010)
4. Sun, J., Palade, V., Wu, X., Fang, W.: Multiple sequence alignment with hidden markov models learned by random drift particle swarm optimization. IEEE/ACM Trans. Comput. Biol. Bioinf. **11**(1), 243–257 (2014)
5. Zhan, Q., Wang, N., Jin, S., Tan, R., Jiang, Q., Wang, Y.: ProbPFP: a multiple sequence alignment algorithm combining partition function and hidden markov model with particle swarm optimization. In: 2018 IEEE International Conference on Bioinformatics and Biomedicine, vol. 2, no. 21, pp. 1290–1295 (2019)
6. Lalwani, S., Sharma, H., Mohan, M.K., Deep, K.: An efficient bi-level discrete PSO variant for multiple sequence alignment. In: Yadav, N., Yadav, A., Bansal, J.C., Deep, K., Kim, J.H. (eds.) Harmony Search and Nature Inspired Optimization Algorithms. AISC, vol. 741, pp. 797–807. Springer, Singapore (2019). https://doi.org/10.1007/978-981-13-0761-4_76
7. Álvaro, R., Vanneschi, L., Castelli, M., Miguel, A.V.R.: Swarm intelligence for optimizing the parameters of multiple sequence aligners. Swarm Evol. Comput. **42**, 16–28 (2018)
8. Álvaro, R., Vanneschi, L., Castelli, M., Miguel, A.V.R.: A characteristic-based framework for multiple sequence aligners. IEEE Trans. Cybern. **48**(1), 41–51 (2016)
9. Kennedy, J., Eberhart, R.C.: A discrete binary version of the particle swarm algorithm. In: Proceedings IEEE International Conference System Man Cybernetics, pp. 4104–4109 (1997)
10. Bansal, J.C., Singh, P.K., Saraswat, M., Verma, A., Jadon, S.S., Abraham, A.: Inertia weight strategies in particle swarm optimization. In: Proceedings 2011 3rd World Congress on Nature and Biologically Inspired Computing (NaBIC2011), pp. 633–640 (2011)
11. Iwamatsu, M.: Locating all the global minima using multi-species particle swarm optimizer: the inertia weight and the constriction factor variants. In: IEEE Congress on Evolutionary Computation, pp. 816–822 (2006)
12. Clerc, M., Kennedy, J.: The particle swarm-explosion, stability, and convergence in a multidimensional complex space. IEEE Trans. Evol. Comput. **6**(1), 58–73 (2002)
13. Mendes, R., Kennedy, J., Neves, J.: Watch thy neighbor or how the swarm can learn from its environment. In: IEEE Swarm Intelligence Symposium, pp. 88–94 (2013)

14. Liang, J.J., Qin, A.K., Suganthan, P.N., Baskar, S.: Comprehensive learning particle swarm optimizer for global optimization of multimodal functions. IEEE Trans. Evol. Comput. **10** (3), 281–295 (2006)
15. Chen, W.N., Zhang, J., Chung, H.S., Zhong, W.L., Wu, W.G., Shi, Y.H.: A novel set-based particle swarm optimization method for discrete optimization problems. IEEE Trans. Evol. Comput. **14**(2), 278–300 (2010)
16. Wu, H., Nie, C., Kuo, F., Leung, H., Colbourn, C.J.: A discrete particle swarm optimization for covering array generation. IEEE Trans. Evol. Comput. **19**(4), 575–591 (2015)

# A Diversity Based Competitive Multi-objective PSO for Feature Selection

Jianfeng Qiu[1,2], Fan Cheng[1,2], Lei Zhang[1,2], and Yi Xu[1(✉)]

[1] School of Computer Science and Technology,
Anhui University, Hefei 230039, China
noverfitting@gmail.com
[2] Key Lab of Intelligent Computing and Signal Processing of Ministry
of Education, Anhui University, Hefei 230039, China

**Abstract.** Multi-Objective Particle Swarm Optimization (MOPSO) for feature selection has attracted increasing attention of researchers recently. However, in the existing methods, quick convergence usually degrades the diversity of the population, especially when many irrelevant and redundant features involved in them. To this end, a diversity based competitive multi-objective particle swarm optimization for feature selection problem (named D-CMOPSO) is proposed. In D-CMOPSO, a diversified competition based learning mechanism is proposed to improve the quality of found feature subset, which consists of three parts: exemplar particle construction, pairwise competition, and diversified learning strategy. The proposed competition mechanism utilizes the above three parts to boost the diversity in the following generations. Moreover, in order to guide the initial population to evolve the promising area, a maximal information coefficient based initialization strategy is also suggested. The experimental results demonstrate that the proposed D-CMOPSO is competitive for feature selection problem.

**Keywords:** Competitive mechanism · Diversity enhancement ·
Feature selection · Multi-objective optimization · Particle swarm algorithm

## 1 Introduction

Recently, classification has been widely used in many supervised learning problems [1, 2]. Since a large number of irrelevant or redundant features involved, the performance of classification suffers from the curse of dimensionality [1]. Feature selection can be naturally modeled as an optimization problem and usually as a NP-hard problem [1]. For this reason, evolutionary algorithms (EAs) have been used for feature selection [3, 5]. However, it is more suitable to consider several conflicting objectives simultaneously, e.g. accuracy and feature number. Thus, multi-objective evolutionary algorithms (MOEAs) can find their niche in this area.

An early work proposed by Lac et al. [5] utilized multi-objective genetic algorithm to find a set of Pareto optimal feature subsets. After that, several multi-objective

D.-S. Huang et al. (Eds.): ICIC 2019, LNCS 11644, pp. 26–37, 2019.
https://doi.org/10.1007/978-3-030-26969-2_3

evolutionary feature selection approaches, such as CMDPSOFS [3] and HMPSOFS [4], have been proposed recently, which has proven MOEA can tackle NP-hard problems effectively.

In this paper, we continue this research line by proposing a Diversity enhancement based Competitive Multi-Objective PSO, termed D-CMOPSO, to further improve the performance of MOPSO. The main contributions can be summarized as follows.

- A diversity enhancement based competitive mechanism is proposed for feature selection, where exemplar particles and competitor particles are suggested to update current population.
- Based on the suggested diversity enhancement competitive mechanism, a multi-objective particle swarm optimization algorithm, named D-CMOPSO, is proposed for identifying a set of informative feature subsets. Moreover, a problem-specific initialization approach is introduced to generate the initial population.
- We evaluate the effectiveness of proposed D-CMOPSO on six benchmark data sets with different sizes compared with several baseline algorithms for feature selection. Experimental results show that our method performs better in finding trade-off solutions, which indicates that D-CMOPSO is competitive and promising.

The rest of this paper is organized as follows. The preliminaries about multi-objective feature selection and related work are presented in Sect. 2. Section 3 gives the detail of the proposed algorithm and the empirical results on the benchmark data sets are reported in Sect. 4. Finally, Sect. 5 gives the conclusions and future work.

## 2 Preliminaries and Related Work

In this section, we first give some preliminaries about feature selection problem, and then present the related work about MOEAs for feature selection.

### 2.1 Feature Selection Problem

The feature selection algorithms are usually classified into two categories: filter based methods and wrapper based methods [1]. In this paper, we focus on wrapper method and formulate the feature selection problem as multi-objective problem (MOP) by maximizing accuracy and minimizing feature number simultaneously, which can be defined as follows.

$$Minimize \begin{cases} f_1(x) = |x| \\ f_2(x) = 1 - accuracy(x) \end{cases} \tag{1}$$

where $x$ is a candidate feature subset, $f_1(x)$ is the number of the selected features in $x$ and $f_2(x)$ denotes the error rate of the learned classification model.

## 2.2    Related Work on MOEAs Based Feature Selection

Due to the wide applications of MOEAs, much efforts have been devoted to develop MOEA based feature selection algorithms [3, 4, 6].

An early research on MOEA based feature selection algorithm, was proposed by Lac et al. [5]. Since the promising performance of MOEA, a variety of competitive MOEAs have been suggested to explore their potential in feature selection. Xue et al. [3] developed a particle swarm based MOEA, named CMDPSOFS, to solve multi-objective feature selection problem. The performance of CMDPSOFS was superior over the state-of-the-arts MOEAs. In recognizing the superiority of PSO in solving multi-objective feature selection, a series of MOPSO based feature selection approaches have been suggested [4, 8]. A several recent surveys can be found in [7, 9].

In our previous work in [10], a variant of MOPSO, CMOPSO, was suggested to improve the performance of MOPSO. However, we find it is hard to directly extend to solve discrete combination optimization problem. Therefore, in this paper, we have proposed a diversity enhancement based competitive multi-objective particle swarm optimization, namely D-CMOPSO, for feature selection.

## 3    The Proposed Algorithm

### 3.1    The General Framework of D-CMOPSO

The main framework of D-CMOPSO consists of three main phases. Firstly, a prior information based initialization strategy is suggested to achieve the competitive particles. Secondly, a diversity enhancement competitive mechanism is proposed to guide the evolution of the particles, which is the main component of the proposed algorithm. Thirdly, the $NP$ better particles are to be selected from the combination of population $P$ and updated population $P'$. Similarly, the environmental selection mechanism in CMOPSO [10] is directly adopted in our proposed algorithm.

From the above explanation of Algorithm 1, we can find that there are two important components in the proposed D-CMOPSO, which are the proposed initial strategy and the diversity enhancement competition mechanism. In the following, we will elaborate them in detail.

---

**Algorithm 1: The General Framework of D-CMOPSO**

**Input:** $Data$: the given data set;
  $Dim$: the dimensionality of the $Data$;
  $NP$: the number of particles in population $P$;
  $Maxiter$: the maximum number of iteration;
**Output:** $PF$: the set of particles in Pareto front;
  $P \leftarrow$ InitializePopulation($Data$, $Data.Label$, $NP$);
  $V \leftarrow$ RandomVelocityVector($NP$, $Dim$);
  **while** $iter < Maxiter$ **do**
    $Para \leftarrow \{P, V, iter, Maxiter\}$;
    $ES \leftarrow$ non_dominated_sorted($P$); // Elite set
    $P', V' \leftarrow$ DiversCompete($Para$, $ES$);
    $V \leftarrow V'$;
    $P \leftarrow$ EnvironmentalSelection ($P'$, $P$);

---

## 3.2    The Prior Information Based Initialization Strategy

To identify the wide correlation of between feature and label, we adopt a novel metric, namely, maximal information coefficient (*MIC*) [11], to identify the correlation between each feature vector $f_i$ and label vector $c$, which is defined as follows.

$$R = \{MIC(f_i, c)\}_{i=1,2,\cdots,D} \tag{2}$$

where $R$ is the *MIC* vector carrying the information of any pair of feature and label, $f_i$ is the *ith* feature with $M$ sample values on feature $i$ and denoted as $f_i = (f_{i,1}, f_{i,2}, \cdots, f_{i,M}), i = 1, 2, \cdots, D$, $D$ is the dimensionality of the data set. The whole procedure of population initialization based on MIC is presented in Algorithm 2.

Firstly, calculating the *MIC* of every feature according to formula (2), seeing in Line 2–4. Then, for each dimensionality of particle $i$, e.g., *Pop(i)*, two features $m$ and $n$ are randomly selected. Feature $m$ or feature $n$ is chosen and set as 1 in *Pop(i, m)* or *Pop(i, n)* by comparing the value of *MIC*.

## 3.3    The Diversity Enhancement Competition Mechanism

The proposed diversity enhancement competition mechanism mainly consists of two steps. Firstly, to obtain good exemplar information, a new exemplar vector construction method is proposed, which utilizes the correlation between features and labels. Moreover, a diversity based competition mechanism is also suggested, which can find the competitors to enhance the diversity of the population. Finally, the exemplar and competitor based update process are proposed. The whole procedure above is presented in Algorithm 3. The details of each component are depicted respectively as follows.

### Exemplar Vector Construction
Comprehensive Learning PSO (CLPSO) proposed by Liang et al. [12] synthesized the advantages from different dimensionality in different particles. Inspired by the fact, in this paper, the suggested constructed exemplar vector utilizes the prior information between features and labels. The procedure of constructing exemplar vectors is performed as follows.

First, symmetric uncertainty (*SU*) [13], is employed for evaluating the correlation between the feature and the class and ranking the features. $SU(f_i, C), i \in 1, 2, \cdots, D$, represents a correlation between the *i-th* feature and the class label C. $Mean(SU(F, C))$ denotes the mean value of all. $SU_i$ To maintain randomness, we choose the exemplar randomly for the remaining dimensionality, which are the indexes of the chosen particles. For the remaining dimension $d$, we select randomly two particles, namely, $m$ and $n$, from the population. If the level of particle m, $P_n$, obtained by non-dominated sorting, is smaller than $Level(P_m)$ of the particle $n$, the $d$ dimension of the particle $m$ is chosen into the corresponding dimension of the exemplar vector, otherwise, the particle $n$ is chosen. The whole construction procedure is presented in Algorithm 4.

---

**Algorithm 2:** InitializePopulation

**Input:** *Data*: the data set with $M$ samples and $D$ features;
    *Data.Label*: the set of class label;
    $NP$: the number of particles in population $P$;
**Output:** *Pop*: the initial population;
    $(f_1, f_2, \cdots, f_D)$ ← extract the feature vectors in *Data*;
    $c$ ← *Data.Label*;
1:   $R$ ← zeros$(1, D)$;
2:   **for** $j = 1$ to $D$ **do**
3:     $R(j) = MIC(f_j, c)$;
4:   **end for**
5:   $Pop$=zeros$(NP, D)$;
6:   **for** $i = 1$ to $NP$ **do**
7:     **for** $j = 1$ to $\lfloor rand * D \rfloor$ **do**
8:       $[m, n]$ ← randomly select two features $m$ and $n$;
9:       **if** $MIC(f_m, c) \geq MIC(f_n, c)$ **then**
10:         $Pop(i, j)$ ← the $m$th feature is selected;
11:       **end if**
12:       **if** $MIC(f_m, c) < MIC(f_n, c)$ **then**
13:         $Pop(i, j)$ ← the $n$th feature is selected;
14:       **end if**
15:     **end for**
16:   **end for**

---

## Diversity Based Competition Mechanism

To overcome the population diversity degradation, we propose a diversity based competitive mechanism to select a competitor to update the particle in current population. Assuming the current particle is $P_i$, the procedure of identifying competitor is performed and shown in Algorithm 5.

---

**Algorithm 3:** DiversCompete

**Input:** *Para*: the set of variables;
    $ES$: the elite particle set;
**Output:** $P'$: the updated population;
    $V'$: the updated velocity matrix;
1:   $P$ ← *Para.P*;
2:   $NP, D$ ← sizeof$(P)$;
    /* exemplar vector construction*/
3:   $Exemplar$ ← zeros(NP,D);
4:   **for** $i = 1$ to $NP$ **do**
5:     $Exemplar(i)$ ← ExemplarVector (P);
6:   **end for**
    /* diversity based competition mechanism */
7:   $P_{win}$=zeros$(NP, D)$;
8:   $P_{win}$ ← DiversityCompetition$(Para, ES)$;
9:   $V'$ ← Updating the velocity of the population using 4;
10:   $P'$ ← Updating the position of the population using 5;

---

**Algorithm 4:** ExemplarVector

**Input:** $P$: the population;
**Output:** $Exemplar_p$: the exemplar vector corresponding to the particle $p$;
1:   $D$ ← Length$(p)$;
2:   $Examplar_p$=zeros$(1, D)$;
3:   **for** $j = 1$ to $D$ **do**
4:     **if** $SU(p_j) \geq Mean(SU)$ **then**
5:       $Examplar_p(j)$ ← $p_i(j)$;
6:     **end if**
7:     randomly choose $m$ and $n$ from $P$;
8:     **if** $Level(P_m) \leq Level(P_n)$ **then**
9:       $Examplar_p(j)$ ← $p_m(j)$;
10:     **else**
11:       $Examplar_p(j)$ ← $p_n(j)$;
12:     **end if**
13:   **end for**

Firstly, two particles $P_m$ and $P_n$ are randomly selected from the elite particle set. If both $P_m$ and $P_n$ dominate the $P_i$ then one of them will be randomly selected as competitor. If the two particles $P_m$, $P_n$ and the current particle $P_i$ are non-dominated among them, we choose the dimensions from $P_m$ or $P_n$, which corresponds the larger velocity value in velocity vector $V_m$ or $V_n$, to construct a virtual competitor. Moreover, since the elite particle set is updated iteratively, in early iterations, the chosen elite particles may lie in different levels, even not in the Pareto front, therefore, to further boost diversity, we propose an angle based adaptive competition mechanism for generate the competitor. To be specific, first, a number of $k \in (0, 1)$ is randomly generated. Secondly, if $k \leq \left(1 - \frac{iter}{Maxiter}\right)$, the particle $P_m$ will become competitor if the angle $\theta_1$ between $P_m$ and $P_i$ is smaller than the $\theta_2$ between $P_n$ and $P_i$. Otherwise, the elite particle with greater angle is chosen as competitor in later stage for updating the particle $P_i$. Algorithm 5 gives the detailed procedure.

**The Proposed Updating Strategy**
The update process for the particle $P$ is performed via learning not only from the exemplar but also from the competitor. Let $P_i$ and $V_i$ be the velocity and position of the *i-th* particle respectively, $1 \leq i \leq NP$. The updating equations are adopted to calculate the updated $P_i$ and $V_i$:

$$V'_i = r_1 * V_i + r_2 * \left(P_{comp}(i) - P_i\right) + r_3 * \left(P_{exemplar}(i) - P_i\right) \tag{3}$$

$$P'_i = P_i + V'_i \tag{4}$$

where $r_1, r_2, r_3 \in (0, 1)$ are three random numbers, $P_{comp}(i)$ and $P_{exemplar}(i)$ are the position of the competitor and exemplar of the particle $i$ respectively.

---

**Algorithm 5: DiversityCompetition**

**Input:** *Para*: the set of variables;
    *EP*: the elite particle set;
**Output:** $P_{win}$: the competitor population of the current population $p$;
 1: $P \leftarrow Para.P$; //the current population
 2: $V \leftarrow Para.V$; //the velocity matrix of the current population
 3: $iter \leftarrow Para.iter$;
 4: $Maxiter \leftarrow Para.Maxiter$;
 5: $[NP, D] \leftarrow$ sizeof($P$);
 6: randomly choose $P_m$ and $P_n$ from $EP$;
 7: $P_{comp}$=zeros($NP, D$);
 8: **for** $i = 1$ to $NP$ **do**
 9:    **if** $P_m \prec P_i$ and $P_n \prec P_i$ **then**
10:      $P_{comp}(i) \leftarrow$ randomly choose one particle from $P_m$ or $P_n$;
11:    **else if** $P_m \nprec P_i$ and $P_n \nprec P_i$ **then**
12:      $J1 = Find(V_m \geq V_n)$;
       $P_{comp}(i, J1) \leftarrow P_m(J1)$;
       $J2 = Find(V_n \geq V_m)$;
       $P_{comp}(i, J2) \leftarrow P_n(J2)$;
13:    **else**
14:      calculate the angle $\theta_1$ between $P_m$ and $P_i$, and $\theta_2$ between $P_n$ and $P_i$;
15:      **if** $rand() \leq (1 - \frac{iter}{Maxiter})$ **then**
16:        $P_{comp}(i) \leftarrow P_{argmin\{\theta_1,\theta_2\}}$;
17:      **else**
18:        $P_{comp}(i) \leftarrow P_{argmax\{\theta_1,\theta_2\}}$;
19:      **end if**
20:    **end if**
21: **end for**

## 4    Experimental Results

In this section, a series of experiments are conducted to empirically validate the performance of the proposed algorithm D-CMOPSO on six benchmark data sets with different size by comparing it with state-of-the-arts.

### 4.1    Comparison Algorithms

The performance of the D-CMOPSO is compared with several representative feature selection algorithms, namely, LFS [14], 2SFS [15], NSGA-II [16], and CMDPSO [3]. Among them, LFS is a typical non-EA feature selection approach. The remaining algorithms are EA based feature selection approaches, which can be broadly classified into two categories: one is single objective based algorithms, e.g. 2SFS, the other is representative MOEAs, such as NSGA-II, CMDPSO, which have been used to address multi-objective feature selection problem.

### 4.2    Benchmark Data Set

The compared algorithms are tested on six benchmark data sets, which can be available from UCI machine learning repository [18]. Table 1 depicts the detailed characteristics of the adopted data sets.

In Table 1, #Instances and #Features represent the number of samples and features respectively. For each data set, all samples are randomly partitioned into two sets: 70% as the training set and 30% as the testing set. The classification accuracy is used to evaluate the quality of the selected feature subset by 10-fold cross-validation on the training set, where K Nearest Neighbor (KNN) classifier is employed to evaluate the performance of the selected feature subset ($K = 5$).

**Table 1.** The detailed information of six data sets used in the experiments

| No | Date set | #Features | #Instances | #Classes |
|----|----------|-----------|------------|----------|
| 1 | Climate | 18 | 540 | 2 |
| 2 | German | 24 | 1000 | 2 |
| 3 | Lungcancer | 56 | 32 | 3 |
| 4 | Sonar | 60 | 208 | 2 |
| 5 | Musk1 | 166 | 476 | 2 |
| 6 | LSVT | 309 | 126 | 2 |

### 4.3    Parameter Setting

For the comparison algorithms, we adopt the recommended parameter values suggested in their original papers. Moreover, in all the following experiments, the population size is set to 50, the maximum iteration is fixed to 100. Except for LFS, each algorithm is performed 30 runs independently on each data set.

## 4.4   Performance Evaluation

In this paper, we adopt several common used evaluation measures to compare the performance of the algorithms [3, 9]. To be specific, in LFS, since it is a deterministic algorithm, only one run is need for this algorithm. For single-objective based EAs, e.g., 2SFS, in 30 independent runs, 30 solutions which denote different feature subsets are obtained. For MOEAs, since a set of Pareto optimal solutions are obtained in each run, two metrics, "-PF" and "-AVG", are used to evaluate the performance of the MOEAs in feature selection, which are formally defined in [3, 9].

## 4.5   The Comparison Results Between D-CMOPSO and Baselines

### D-CMOPSO vs. Non-EA and Single-Objective Based EA

Figure 1 shows the compared results of the proposed algorithm and the other two feature selection algorithms, one is non-EA approach and the other is the single-objective based EA. From Fig. 1, we can find that on almost of the experimental sets, EA perform much better than non-EA approach, LFS, which indicates that it is a feasible way to solve feature selection problem. In addition, it can be also observed that D-CMOPSO can provide more accurate and diverse solutions than single objective EA.

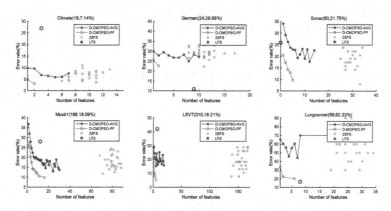

**Fig. 1.**   Results obtained by the proposed algorithm, LFS and 2SFS on six benchmark data sets

### D-CMOPSO vs. Multi-objective Evolutionary Algorithms

Table 2 gives the experimental results on the two metrics of "-PF" and "-AVG". Due to the space limitation, we only plot the compared results of "-PF" on benchmark data sets in Fig. 2 and the results of "-PF" are similar. In Fig. 2, D-CMOPSO-PF can achieve better classification performance with smaller features than that using all features in all data sets.

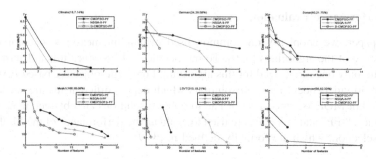

**Fig. 2.** Non-dominated solutions on the union set of 30 different Pareto fronts obtained by D-CMOPSO, CMDPSO and NSGA-II on six benchmark data sets

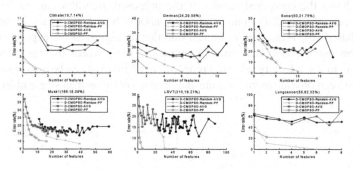

**Fig. 3.** The comparison results obtained by D-CMOPSO and D-CMOPSO with random initialization adopted on six benchmark data sets

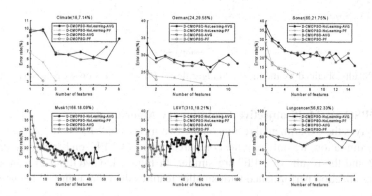

**Fig. 4.** The comparison results obtained by D-CMOPSO and D-CMOPSO with no learning strategy adopted in objective space on six benchmark data sets

D-CMOPSO is proposed to solve feature selection problems using the framework of CMOPSO, an efficient MOEA for tackling with continuous problem. An empirically comparison about its computation complexity can be found in [10]. Compared with CMOPSO, the evaluation of classifier on the selected feature subset is added into D-CMOPSO. Therefore, the computational complexity is a little higher than CMOPSO and comparable to the other wrapper based MOEAs for feature selection.

**Strategy Effectiveness Analysis in D-CMOPSO**
In this section, we investigate the influence of the two suggested strategies on six benchmark data sets. Figure 3 has shown the results between the proposed algorithm with D-CMOPSO-Random, where D-CMOPSO-Random is the same algorithm as our proposed D-CMOPSO, except that the random way is adopted to initialize population. Figure 4 presents the results of D-CMOPSO-NoLearning, which has the only difference is that the proposed updating strategy adopted in our algorithm. D-CMOPSO-NoLearning employed the original learning strategy used in CMOPSO.

From the results, we can find that D-CMOPSO outperforms D-CMOPSO-random and D-CMOPSO-NoLearning on most of data sets, since more and better solution information is integrated into the learning process.

**Table 2.** The results of "-PF" and "-AVG" on all non-dominated solutions obtained by MOEAs. "Size" denotes the number of the selected feature, "Error Rate" and "Std" are the classification using the selected feature subset and standard deviation, respectively.

| Data | Method | AVG | | | PF | | |
|------|--------|------|------------|------|------|------------|------|
| | | Size | Error Rate | Std | Size | Error Rate | Std |
| Climate | CMDPSO | 5.2 | 7.33 | 1.24 | 3.3 | 4.53 | 1.98 |
| | NSGA-II | 4.5 | **7.23** | 1.47 | 2.0 | 4.63 | 2.18 |
| | D-CMOPSO | **4.0** | 7.46 | 1.63 | **1.5** | **4.01** | 1.31 |
| German | CMDPSO | 6.5 | 29.29 | 2.52 | 4.3 | 23.75 | 0.92 |
| | NSGA-II | 7.0 | 28.37 | 1.84 | 3.5 | **23.33** | 2.58 |
| | D-CMOPSO | **5.5** | **27.76** | 1.41 | **1.5** | 24.00 | 1.89 |
| Lungcancer | CMDPSO | 7.5 | 58.82 | 8.41 | 1.5 | 35.00 | 7.07 |
| | NSGA-II | 6.3 | 52.80 | 9.94 | **1.0** | **20.00** | 0.00 |
| | D-CMOPSO | **6.2** | **52.60** | 9.44 | 3.0 | 25.19 | 7.14 |
| Sonar | CMDPSO | 13.8 | 23.07 | 6.47 | 4.8 | 16.82 | 7.58 |
| | NSGA-II | 11.5 | 22.16 | 4.77 | **2.5** | 16.67 | 6.67 |
| | D-CMOPSO | **7.1** | **22.04** | 4.40 | 3.0 | **14.99** | 4.21 |
| Musk 1 | CMDPSO | 30.9 | 18.50 | 3.44 | 15.5 | 15.21 | 4.02 |
| | NSGA-II | 40.8 | **16.42** | 2.86 | 19.2 | **10.77** | 1.69 |
| | D-CMOPSO | **14.5** | 19.50 | 5.40 | **6.5** | 16.40 | 6.16 |
| LSVT | CMDPSO | 41.5 | 22.04 | 4.59 | 18.5 | 14.47 | 9.31 |
| | NSGA-II | 76.6 | **20.05** | 5.15 | 55.2 | 11.58 | 6.34 |
| | D-CMOPSO | **9.4** | 20.24 | 3.82 | **2.8** | **11.17** | 6.90 |

## 5 Conclusion and Future Work

In this paper, we have proposed a diversity enhancement based competitive MOPSO algorithm named D-CMOPSO for feature selection. Experimental results have demonstrated the superiority of the proposed D-CMOPSO. In the future, we would like to further combine this competitive mechanism with other frameworks of MOEA, such as LMEA [17], to improve the performance for high-dimensional problem.

**Acknowledgment.** This work is supported by the Anhui University College Students Innovation and Entrepreneurship Training Program, Natural Science Foundation of China (Grant No. 61876184 and 61822301), the Key Program of Natural Science Project of Educational Commission of Anhui Province (Grant No. KJ2017A013) and the Academic and Technology Leader Imported Project of Anhui University (No. J01006057). This work was also supported in part by the Natural Science Foundation of Anhui Province (Grant No. 1708085MF166, No. 1908085MF219, and No. 1908085QF271), Humanities and Social Sciences Project of Chinese Ministry of Education (Grant No. 18YJC870004). Provincial Quality Engineering Project of Anhui Province (Grant No. 2017jyxm0086).

## References

1. Agor, J., Özaltın, O.Y.: Feature selection for classification models via bilevel optimization. Comput. Oper. Res. **106**, 156–168 (2019)
2. Khammassi, C., Krichen, S.: A GA-LR wrapper approach for feature selection in network intrusion detection. Comput. Secur. **70**, 255–277 (2017)
3. Xue, B., Zhang, M., Browne, W.N.: Particle swarm optimization for feature selection in classification: a multi-objective approach. IEEE Trans. Cybern. **43**(6), 1656–1671 (2013)
4. Zhang, Y., Gong, D.W., Cheng, J.: Multi-objective particle swarm optimization approach for cost-based feature selection in classification. IEEE/ACM Trans. Comput. Biol. Bioinf. **14**(1), 64–75 (2017)
5. Lac, H.C., Stacey, D.A.: Feature subset selection via multi-objective genetic algorithm. In: Proceedings. 2005 IEEE International Joint Conference on Neural Networks, vol. 3, pp. 1349–1354. IEEE (2005)
6. Karaboga, D., Gorkemli, B., Ozturk, C., Karaboga, N.: A comprehensive survey: artificial bee colony (ABC) algorithm and applications. Artif. Intell. Rev. **42**(1), 21–57 (2014)
7. Brezočnik, L., Fister, I., Podgorelec, V.: Swarm intelligence algorithms for feature selection: a review. Appl. Sci. **8**(9), 1521 (2018)
8. Nguyen, H.B., Xue, B., Liu, I., Andreae, P., Zhang, M.: New mechanism for archive maintenance in PSO-based multi-objective feature selection. Soft. Comput. **20**(10), 3927–3946 (2016)
9. Xue, B., Zhang, M., Browne, W.N., Yao, X.: A survey on evolutionary computation approaches to feature selection. IEEE Trans. Evol. Comput. **20**(4), 606–626 (2015)
10. Zhang, X., Zheng, X., Ran, C., Qiu, J., Jin, Y.: A competitive mechanism based multi-objective particle swarm optimizer with fast convergence. Inf. Sci. **427**, 63–76 (2018)
11. Reshef, D.N., et al.: Detecting novel associations in large data sets. Science **334**(6062), 1518–1524 (2011)

12. Liang, J.J., Qin, A.K., Suganthan, P.N., Baskar, S.: Comprehensive learning particle swarm optimizer for global optimization of multimodal functions. IEEE Trans. Evol. Comput. **10**(3), 281–295 (2006)
13. Press, W.H., Flannery, B.P., Teukolsky, S.A., Vetterling, W.T.: Numerical recipes in C. Art Sci. Comput. **10**(1), 176–177 (1995)
14. Gutlein, M., Frank, E., Hall, M., Karwath, A.: Large-scale attribute selection using wrappers. In: IEEE Symposium on Computational Intelligence and Data Mining, pp. 332–339 (2009)
15. Xue, B., Zhang, M., Browne, W.N.: New fitness functions in binary particle swarm optimisation for feature selection. In: 2012 IEEE Congress on Evolutionary Computation, pp. 1–8 (2012)
16. Deb, K., Pratap, A., Agarwal, S., Meyarivan, T.: A fast elitist multiobjective genetic algorithm: NSGA-II. IEEE Trans. Evol. Comput. **6**(2), 182–197 (2000)
17. Zhang, X., Tian, Y., Cheng, R., Jin, Y.: A decision variable clustering based evolutionary algorithm for large-scale many-objective optimization. IEEE Trans. Evol. Comput. **22**(1), 97–112 (2018)
18. UCI machine learning repository. https://archive.ics.uci.edu/ml/. Accessed 30 Mar 2019

# A Decomposition-Based Hybrid Estimation of Distribution Algorithm for Practical Mean-CVaR Portfolio Optimization

Yihua Wang$^{(\boxtimes)}$ and Wanyi Chen

College of Artificial Intelligence, Nankai University, Tianjin 300350,
People's Republic of China
wangyihua@mail.nankai.edu.cn, wychen@nankai.edu.cn

**Abstract.** This paper addresses a practical mean-CVaR portfolio optimization problem, which maximizes mean return and minimizes CVaR. Since the practical constraints are considered, the problem is proved to be NP-hard. To solve this complex problem, we decompose it into an asset selection problem and a proportion allocation problem. For the asset selection problem, an estimation of distribution algorithm (EDA) is developed to determine which assets are included in the portfolio. Once the asset selection is fixed in each generation of the EDA, the proportion of each asset is determined by solving the proportion allocation problem using the linear programming. To guarantee the diversity of the obtained solutions, the probability model (PM) is divided into a set of sub-PMs according to the decomposition of the objective space. A knowledge-based initialization and a cooperation-based local search are designed to improve the solutions obtained in the initialization stage and the search process, respectively. The proposed decomposition-based hybrid EDA (DHEDA) is tested on real-world datasets and compared with an existing algorithm. Numerical results demonstrate the effectiveness and efficiency of the DHEDA.

**Keywords:** Portfolio optimization · EDA · Decomposition · CVaR

## 1 Introduction

The portfolio optimization (PO) is a multi-objective optimization problem (MOP) aiming to maximize return and minimize risk. The PO was first formulated as a mean-variance (M-V) model by Markowitz in 1952 [1]. In the M-V model, the mean is used to measure the expected return and the variance is used to measure the investment risk. Although this theory is seen as the beginning of the modern portfolio theory, the M-V model is criticized for lack of practicality. Therefore, many other models are developed to describe the problem based on the classical model.

To improve the efficiency of portfolio assessment, many different risk measures are adopted to replace the variance, such as Value-at-Risk (VaR), Conditional Value-at-Risk (CVaR), Mean-Absolute Deviation (MAD), and Minimax (MM) [2]. Among these risk measures, VaR and CVaR are the most widely-used. VaR reflects the maximum loss of a portfolio that will not be exceeded with a given probability [3]. While CVaR is described as the average expected loss which exceeds VaR [4].

© Springer Nature Switzerland AG 2019
D.-S. Huang et al. (Eds.): ICIC 2019, LNCS 11644, pp. 38–50, 2019.
https://doi.org/10.1007/978-3-030-26969-2_4

Compared to VaR, CVaR has the following advantages. (1) The optimization of CVaR is a convex optimization problem, which can be easily solved by many optimization techniques [5]. (2) CVaR is a "coherent" risk measurement satisfying the subadditivity property [6]. (3) Minimizing CVaR typically leads to a portfolio with a small VaR [7].

Moreover, some practical constraints are added to the PO model to enlarge the application range, such as boundary constraint (BC), cardinality constraint (CC), transaction costs (TC) and transaction lots (TL). The details of these constraints can be found in [2]. Many papers addressed the mean-CVaR (M-CVaR) PO considering different practical constraints. In [8], a single-stage M-CVaR PO with TC was formulated as a mixed integer linear programming (MILP). To solve the MILP, a two-phase cutting plane algorithm was proposed and tested on large-scale problems. While in [9], the TC were considered in a multi-stage mean-semivariance-CVaR PO, which was formulated by stochastic programming and solved by a hybrid algorithm combining genetic algorithm (GA) and particle swarm optimization (PSO). Another hybrid algorithm based on PSO and artificial bee algorithm was proposed in [10] to solve a realistic PO which simultaneously considers TC and TL. The CC was considered in [11, 12]. In [11], a M-CVaR PO with CC was solved by a reweighted $l_1$-norm method. In [12], the CC was included in both the mean-VaR and M-CVaR PO, and different multi-objective evolutionary algorithms (MOEAs) were developed to solve these problems.

From the brief review, it can be seen that since some practical PO problems are nonconvex and NP-hard, it is significant to develop more advanced and efficient optimization techniques to solve these problems. In this paper, a decomposition-based hybrid estimation of distribution algorithm (DHEDA) is proposed to solve the practical M-CVaR PO with cardinality and class constraints. To reduce the solving complexity, the problem is decomposed into an asset selection problem and a proportion allocation problem. The proposed DHEDA fuses the estimation of distribution algorithm (EDA) [13] and linear programming-based (LP-based) approach together to solve the decomposed problem. The probability model of the EDA is specially designed according to the decomposition of the objective space to guarantee the diversity of the obtained solutions. Problem-specific knowledge is used to design the initialization and local search methods. The numerical results demonstrate the effectiveness of the proposed DHEDA.

The rest of the paper is organized as follows. In Sect. 2, the practical M-CVaR PO is formulated. In Sect. 3, the proposed DHEDA and the decomposition strategy are described. The computational results and comparisons are given in Sect. 4. Finally, the paper ends with some conclusions and future work in Sect. 5.

## 2   Practical Mean-CVaR Portfolio Optimization

### 2.1   Formulation of Practical M-CVaR Portfolio Optimization

The practical M-CVaR portfolio optimization is formulated based on the standard M-V model [1] with the notations listed in Table 1. The main difference is that the practical

M-CVaR portfolio optimization adopts CVaR as the risk measurement instead of variance. In this paper, a scenario-based method is used to calculate CVaR as follows.

$$z_t(\boldsymbol{x}) = \sum_{i=1}^{N} r_{it} \cdot x_i, \forall t \tag{1}$$

$$VaR_\alpha(\boldsymbol{x}) = -\inf\{z_{(t_\alpha)} | \sum_{j=1}^{t_\alpha} w_j \geq \alpha\} \tag{2}$$

$$CVaR_\alpha(\boldsymbol{x}) = -E\{z_t(\boldsymbol{x}) | z_t(\boldsymbol{x}) < -VaR_\alpha(\boldsymbol{x})\} \tag{3}$$

where formula (1), (2), and (3) calculate the expected return under scenario $t$, VaR of portfolio $\boldsymbol{x}$, and CVaR of portfolio $\boldsymbol{x}$, respectively. In formula (2), the returns $z_{(j)}$ are arranged by increasing order such that $z_{(1)} \leq z_{(2)} \leq \ldots \leq z_{(T)}$ and $w_j$ denotes the realization probability of scenario $(j)$. The formula (2) indicates the VaR at confidence level $\alpha$ to be the opposite number of the $t_\alpha$-th biggest expected return $z_{(t_\alpha)}$, i.e. the maximum expected loss that the portfolio cannot exceed with probability $\alpha$. In formula (3), the CVaR of portfolio $\boldsymbol{x}$ equals to the average expected loss exceeding the VaR, which can be calculated as: $\sum_{t=1}^{T} (z_t(\boldsymbol{x}) - VaR_\alpha(\boldsymbol{x}))^+ / \sum_{t=1}^{T} sgn(z_t(\boldsymbol{x}) - VaR_\alpha(\boldsymbol{x}))^+$, where $(z_t(\boldsymbol{x}) - VaR_\alpha(\boldsymbol{x}))^+ = \max\{z_t(\boldsymbol{x}) - VaR_\alpha(\boldsymbol{x}), 0\}$, $sgn(\cdot)$ returns 1 if the expression $\cdot$ is positive and 0 otherwise.

**Table 1.** Notation descriptions.

| Parameters | | | |
|---|---|---|---|
| $i$ | Index of assets, $i \in \{1, \ldots, N\}$ | $\alpha$ | Confidence level |
| $N$ | Number of assets | $w_t$ | Realization probability of scenario $t$ |
| $t$ | Index of scenarios, $t \in \{1, \ldots, A\}$ | $l_i$ | Minimum proportion of asset $i$ |
| $T$ | Number of scenarios | $u_i$ | Maximum proportion of asset $i$ |
| $m$ | Index of classes, $m \in \{1, \ldots, M\}$ | $L_m$ | Minimum proportion of class $m$ |
| $M$ | Number of classes | $U_m$ | Maximum proportion of class $m$ |
| $J$ | Number of assets in each class | $r_{it}$ | Historical return of asset $i$ under scenario $t$ |
| $K$ | Number of assets in the portfolio | | |
| Decision variables | | | |
| $x_i$ | Investment proportion of asset $i$ | $\delta_i$ | 1, if asset $i$ is in the portfolio; 0, otherwise |

Moreover, several practical constraints are added into the standard model to adapt to realistic situations, including cardinality constraint, boundary constraints, and class constraints. Specific descriptions of these practical constraints are as follows.

- Cardinality constraint: restricts the number of assets in a portfolio.
- Asset boundary constraints: specify the range of each asset proportion.
- Class constraints: divide the assets into several exclusive classes according to the industry classification, such as energy and finance.
- Class boundary constraints: specify the range of each class proportion.

The practical M-CVaR portfolio optimization is formulated as follows.

$$\max. \mu(\boldsymbol{x}) = \sum_{t=1}^{T} w_t \cdot z_t(\boldsymbol{x}) \tag{4}$$

$$\min. \rho(\boldsymbol{x}) = CVaR_{\alpha}(\boldsymbol{x}) \tag{5}$$

s.t.:

$$\sum_{i=1}^{N} x_i = 1 \tag{6}$$

$$\sum_{i=1}^{N} \delta_i = K \tag{7}$$

$$l_i \cdot \delta_i \le x_i \le u_i \cdot \delta_i, \forall i \tag{8}$$

$$L_m \le \sum_{i \in C_m} x_i \le U_m, \forall m \tag{9}$$

$$\delta_i \in \{0, 1\}, \forall i \tag{10}$$

The objective (4) is to maximize the expected return, where $z_t(\boldsymbol{x})$ is calculated by (1). In this paper, the realization probability of each scenario is assumed to be the same such that $w_t = 1/T, \forall t$. The objective (5) is to minimize the CVaR at confidence level $\alpha$ which is calculated by (2) and (3). The constraint (6) is the budget constraint which ensures all the money should be invested. The constraint (7) and (8) are the cardinality constraint and asset boundary constraints, respectively. The constraint (9) defines the class constraints and the class boundary constraints. The constraint (10) indicates $\delta_i$ to be binary variables.

## 3  DHEDA for Practical M-CVaR Portfolio Optimization

In this section, we propose a decomposition-based hybrid estimation of distribution algorithm to solve the practical M-CVaR PO. As the decomposition techniques are used, including both the decomposition of the problem and the decomposition of the objective space, the algorithm is called "decomposition-based". Since the LP-based approach is fused into the classical EDA, the algorithm is called a "decomposition-based hybrid EDA". The details of the proposed DHEDA are described in the following sections.

### 3.1  Outline of the DHEDA

To reduce the complexity of solving the practical M-CVaR PO, we decompose the problem into an upper-level problem to determine the asset selection and a lower-level problem to determine the proportion allocation. For solving the upper-level problem, a search based algorithm is used to determine the asset selection for each search generation. Once the asset selection of the portfolio is fixed, the problem reduces to an asset proportion allocation problem which can be transformed to an equivalent multi-objective linear programming (MOLP) problem according to [14]. The equivalent MOLP is formulated as follows.

$$\text{max. (4)}$$
$$\min. \rho(x, \beta) = \beta + \left(\sum_{t}^{T} u_t\right) / (T \cdot (1 - \alpha)) \tag{11}$$

s.t.: (6), (9), and the followings.

$$l_i \cdot \delta_i' \leq x_i \leq u_i \cdot \delta_i', \ \forall i \tag{12}$$

$$u_t \geq 0, \ \forall t \tag{13}$$

$$u_t \geq z_t(x) - \beta, \ \forall t \tag{14}$$

where $x$ and $\beta$ are decision variables; $\delta_i'$ are fixed parameters determined by the upper-level problem; $u_t$ are slack variables. Since all the objectives and constraints are linear functions with respect to the decision variables $x$ and $\beta$, the equivalent problem is a MOLP which can be solved by LP-based approaches.

In this paper, the EDA is employed to work as the search framework to solve the upper-level problem owing to its success in solving many other discrete optimization problems [15–17]. To guarantee the diversity of the obtained solutions, the probability model (PM) of the EDA consists of several sub-PMs which are designed for different regions of the objective space. The non-dominated solutions found in the search process are reserved in an archive, which is initialized by a knowledge-based heuristic in the initialization stage. In each generation of the search process, a sampling method is developed to generate the new population, which is decoded into feasible solutions by solving the lower-level problem using Gurobi. Then the non-dominated solutions are selected as the elite solutions, which are used to update the PM. To generate more competitive solutions, a cooperation-based local search is implemented using both the elite solutions and the solutions in the archive. Finally, the archive is updated by the non-dominated solutions found in the generation.

The flowchart of the proposed DHEDA is illustrated in Fig. 1.

**Fig. 1.** Flowchart of the DHEDA

## 3.2   Encoding and Decoding Scheme

To solve the upper-level problem using EDA, the asset selection of a portfolio is encoded as an individual containing $K$ integers $\Pi = \{\pi_1, \pi_2, \ldots, \pi_M, \ldots, \pi_K\}$, where each integer denotes the index of a selected asset. To ensure that each class of assets is included in the portfolio, the first $M$ assets are selected from each class, while the remaining $K - M$ assets are selected without class limitations.

To determine the proportion of each asset, the lower-level problem is solved in the decoding stage. As the lower-level problem is formulated as a MOLP, the weighted-sum method is used to convert the problem into $R_n$ single-objective linear programing (SOLP) problems, where $R_n$ is a user-determined parameter. Then Gurobi is used to obtain the optimal solution of each SOLP. Accordingly, each individual can be decoded into $R_n$ feasible solutions $S_r = (\Pi, \Phi_r), r = 1, \ldots, R_n$, containing the same asset selection $\Pi = \{\pi_1, \pi_2, \ldots, \pi_K\}$ and different proportion allocations $\Phi_r = \{\phi_{\pi_1}, \phi_{\pi_2}, \ldots, \phi_{\pi_K}\}, r = 1, \ldots, R_n$, where $\phi_{\pi_k}$ denotes the proportion of the asset $\pi_K$. Specific procedures of the decoding are as follows.

**Step 1.** Normalize the objective functions. Let $x^{\mu*}$ and $x^{\rho*}$ be the optimal solutions of optimizing $\mu(x)$ and $\rho(x)$, respectively. Then the normalized objective functions are: $\overline{\mu(x)} = \frac{\mu(x) - \mu(x^{\mu*})}{\mu(x^{\rho*}) - \mu(x^{\mu*})}$, $\overline{\rho(x)} = \frac{\rho(x) - \rho(x^{\rho*})}{\rho(x^{\mu*}) - \rho(x^{\rho*})}$.

**Step 2.** Calculate the objective weights $\lambda_1, \lambda_2, \ldots, \lambda_{R_n}$: $\lambda_1 = 0, \lambda_2 = 1/(R_n - 1)$, $\lambda_3 = 2/(R_n - 1), \ldots, \lambda_{R_n} = 1$.

**Step 3.** Obtain the asset proportions using Gurobi by solving $R_n$ SOLPs, whose objectives are to minimize $\overline{\rho(x)} - \lambda_w \cdot \overline{\mu(x)}$, $w = 1, \ldots, R_n$.

## 3.3   Probability Model and Sampling Method

The probability model (PM) is critical to the EDA since it describes the distribution of solutions in the search space and is used for generating new individuals. A good PM should reflect the characteristics of the problem and help the algorithm to find better tradeoff solutions of the MOP. For this purpose, we decompose the normalized objective space into $R_n$ regions $\Omega_1, \Omega_2, \ldots, \Omega_{R_n}$ as illustrated in Fig. 2(a), where the angles $\gamma_1 = \ldots = \gamma_{R_n} = \pi/(2 \cdot R_n)$. For each region $\Omega$, a sub-PM $P^r$ is developed to estimate the distribution of solutions in $\Omega_r$ so that $R_n$ sub-PMs are developed for the whole objective space. Each sub-PM is designed as a vector containing $N$ real numbers which represent the selection probability of $N$ assets. The $R_n$ sub-PMs constitute the complete PM as illustrated in Fig. 2(b), where $p_{ri}(g) \in [0, 1]$ denotes the selection probability of asset $i$ in $P^r$ at generation $g$.

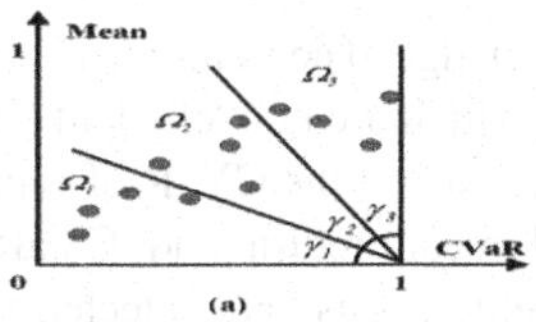

$$P(g) = \begin{bmatrix} P^1(g) \\ P^2(g) \\ \vdots \\ P^{R_n}(g) \end{bmatrix} = \begin{bmatrix} p_{11}(g) & p_{12}(g) & \cdots & p_{1N}(g) \\ p_{21}(g) & p_{22}(g) & \cdots & p_{2N}(g) \\ \vdots & \vdots & \ddots & \vdots \\ p_{R_n 1}(g) & p_{R_n 2}(g) & \cdots & p_{R_n N}(g) \end{bmatrix}$$

(a)                                                                (b)

**Fig. 2.**   (a) An illustration of objective space decomposition (b) An illustration of PM

In each generation of the search progress, a population composed of $N_p$ individuals is generated by sampling the PM. As the number of solutions in the archive varies in different regions, the numbers of individuals generated by different sub-PMs are also different. To make the Pareto front evenly distributed, the number of individuals which are obtained by sampling the sub-PM $P^r$ is calculated as $SN_r = N_p \cdot (N_A - N_{A_r})/((R_n - 1) \cdot N_A)$, where $N_A$ denotes the number of solutions in the archive, $N_{A_r}$ denotes the number of solutions which belong to both the archive and region $\Omega_r$. In this way, more individuals are generated by sampling the sub-PMs whose corresponding regions contain less solutions in the archive. The procedures of generating an individual $\Pi^N$ by sampling the sub-PM $P^r$ are shown in Algorithm 1.

---

**Algorithm 1** Procedure of the sampling method

**Input:** Sub-PM $P^r = [p_{r1} \, p_{r2} \cdots p_{rN}]$

1:   Calculate the index of the $j$-th asset in class $m$: $j' = (m - 1) \cdot J + j$, $j = 1, ..., J$. Initialize the probability array (PA) for class $m$: $PA_m[j] \leftarrow p_{rj'}, \forall m, j = 1, ..., J$, the PA for all assets: $PA[i] \leftarrow p_{ri}, \forall i$
2:   **for** $m = 1$ **to** $M$ **do**
3:     $sum \leftarrow \sum_{j=1}^{J} PA_m[j], PA_m[j] \leftarrow PA_m[j]/sum, j = 1, ..., J$
4:     $\pi_m \leftarrow (m - 1) \cdot J + RouletteWheelMethod(PA_m), PA[\pi_m] \leftarrow 0$
5:   **end for**
6:   **for** $k = m + 1$ **to** $K$ **do**
7:     $sum = \sum_{i=1}^{N} PA[i], PA[i] \leftarrow PA[i]/sum, \forall i$
8:     $\pi_k \leftarrow RouletteWheelMethod(PA), PA[\pi_k] \leftarrow 0$
9:   **end for**

**Output:** Newly generated individual $\Pi^N = \{\pi_1, \pi_2, ..., \pi_K\}$

---

In the procedure, the temporary probability array $PA_m$ is built to sample the asset that belongs to class $m$. The sampled asset index is assigned to $\pi_m$. After the first $M$ elements of the individual are assigned, it is ensured that each class has a "representative" in the generated individual. Then, another temporary probability array $PA$ is built to sample the remaining assets. The function $RouletteWheelMethod(PA)$ returns an asset index which is generated by the roulette wheel method based on the probability array $PA$. The designed sampling method guarantees the feasibility of the generated individual.

## 3.4   Initialization

The initialization stage of the DHEDA includes two parts: the initialization of the probability model and the initialization of the archive. The probability model is initialized uniformly as shown in (15).

$$p_{ri}(0) = 1/N, r = 1, \ldots, R_n, \; \forall i \tag{15}$$

The archive is initialized based on problem-specific knowledge. The main idea is to generate good initial individuals by filtering out some assets which are not "efficient" in terms of both two objectives. First, we calculate the mean return and the CVaR of each single asset. Then the assets are sorted by their related mean return and CVaR, respectively. According to the sorted asset sequences, $P_s$ "elite assets" are selected to construct an "asset pool", where $P_s = (\lfloor K/M \rfloor + 1) \cdot 2 \cdot M$, i.e. $(\lfloor K/M \rfloor + 1) \cdot 2$ assets in each class. Finally, $N_p$ individuals are generated by selecting the assets from the constructed asset pool randomly. The generated individuals are decoded into feasible solutions and the non-dominated solutions constitute the initial archive. Specific procedures of generating an initial individual $\Pi^I$ are shown in Algorithm 2.

---

**Algorithm 2** Procedure of generating an initial individual

1:     Calculate mean return $M_i$ and CVaR $C_i$ of each asset $i$.
2:     Rank the assets by $M_i$ in descending order such that $M_{r_{M1}} \geq M_{r_{M2}} \geq \cdots \geq M_{r_{MN}}$; rank the assets
        by $C_i$ in ascending order such that $C_{r_{C1}} \leq C_{r_{C2}} \leq \cdots \leq C_{r_{CN}}$.
3:     Construct the "asset pool" as a matrix: $AP[m][j] \leftarrow r_{Mj}, AP[m][j+J'] \leftarrow r_{Cj}, \forall m, j = 1, ..., J'$,
        $J' = \lfloor K/M \rfloor + 1$, initialize the asset number of each class in the asset pool: $J_m \leftarrow J' \cdot 2, \forall m$.
4:     **for** $m = 1$ to $M$ **do**
5:         Generate a random number $\bar{j} \in [1, J_m], \pi_m \leftarrow AP[m][\bar{j}]$, erase $\pi_m$ from $AP, J_m \leftarrow J_m - 1$
6:     **end for**
7:     **for** $k = m + 1$ to $K$ **do**
8:         Generate a random number $\bar{m} \in [1, M]$, Generate a random number $\bar{j} \in [1, J_m], \pi_k \leftarrow AP[\bar{m}, \bar{j}]$,
9:         erase $\pi_k$ from $AP, J_m \leftarrow J_m - 1$
10:    **end for**
**Output:** Generated initial individual $\Pi^I = \{\pi_1, \pi_2, ..., \pi_K\}$

---

## 3.5   Elite Selection and Probability Model Update

For each generation, after the population is generated, the individuals in the population are decoded into a set of feasible solutions. As an individual is decoded into $R_n$ solutions using the weighted-sum method, there are $N_p \cdot R_n$ newly generated solutions for each generation. The non-dominated solutions are included in the elite set $E(g)$, which is used to update the probability model $P(g)$.

As $P(g)$ is composed of $R_n$ sub-PMs, the elite set $E(g)$ is also divided into $R_n$ subsets $E_r(g), r = 1, \ldots, R_n$, where $E_r(g)$ contains elite solutions whose objectives locate in $\Omega_r$. Then the population based incremental learning method (PBIL) [18] is used to update the PM $P(g)$ as follows.

$$p_{ri}(g+1) = (1 - \alpha) \cdot p_{ri}(g) + \alpha_l \cdot 1/N_{E_r} \cdot \sum_n^{N_{E_r}} \chi_{\Pi_n^r}(i), r = 1, \ldots, R_n, \forall i \quad (16)$$

where $\alpha_l \in (0, 1)$ denotes the learning rate, $\Pi_n^r$ denotes an individual in subset $E_r(g)$, $N_{E_r}$ denotes the size of $E_r(g)$, and $\chi_{\Pi_n^r}(i)$ is the characteristic function as follows.

$$\chi_{\Pi_n^r}(i) = \begin{cases} 1, i \in \Pi_n^r \\ 0, i \notin \Pi_n^r \end{cases} \quad (17)$$

## 3.6   Local Search and Archive Update

Generally, the local search strategy is often integrated into the EDA to accelerate its speed of convergence. In the DHEDA, a cooperation-based local search (CLS) is designed. The crucial idea of the CLS is to generate neighborhood solutions of the archive by cooperating with the elite solutions. First, the archive $A(g)$ and the elite set $E(g)$ are respectively divided into $R_n$ subsets $A_r(g)$ and $E_r(g)$, $r = 1, \ldots, R_n$ according to the decomposition of the objective space. Then, each solution $S^E = (\Pi^E, \Phi^E)$ in $E_r(g)$ cooperates with each solution $S^A = (\Pi^A, \Phi^A)$ in $A_r(g)$ to generate a neighborhood solution $S^N$. An illustration is shown in Fig. 3, where the green points and the red points represent the elite solutions and archive solutions, respectively. The solutions connected by the dotted lines cooperate with each other to generate neighborhood solutions. Specific procedures of the cooperation operator are as follows.

**Step 1.** If $\Pi^A = \Pi^E$, return $S^E$; otherwise, select the asset $\pi^E$ from $\Pi^E - \Pi^A$ with the biggest investment proportion, i.e. $\phi_{\pi^E} = \max\{\phi_\pi | \pi \in \Pi^E - \Pi^A, \phi_\pi \in \Phi^E\}$.

**Step 2.** Select the asset $\pi^A$ from $\left\{\pi | \pi \in \Pi^A, \pi \in C_m^{\pi^E}\right\}$ with the smallest proportion, where $C_m^{\pi^E}$ is the asset class that $\pi^E$ belongs to.

**Step 3.** Return the neighborhood solution $S^N = (\Pi^{A'}, \Phi^A)$ where $\Pi^{A'} = \Pi^A$ with $\pi^A$ replaced by $\pi^E$.

Accordingly, the complete procedures of the CLS are shown in Algorithm 3.

---

**Algorithm 3** Procedure of the CLS

**Input:** Elite set $E_r(g)$ and archive $A_r(g)$, $r = 1, \ldots, R_n$
1:  **for** $r = 1$ to $R_n$ **do**
2:      **for** $S^E$ in $E_r(g)$ **do**
3:          **for** $S^A$ in $A_r(g)$ **do**
4:              $S^N \leftarrow Cooperate(S^E, S^A)$
5:              Save $S^N$ in set $N(g)$
6:          **end for**
7:      **end for**
8:  **end for**
**Output:** Generated neighborhood solution set $N(g)$

---

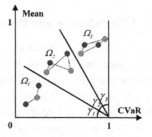

**Fig. 3.** Illustration of the cooperation

After the CLS is performed, the elite set $E(g)$ and the neighborhood solution set $N(g)$ are both used to update the archive $A(g)$. Non-dominated solutions of $E(g)$, $N(g)$, and $A(g)$ are selected to constitute the archive of next generation $A(g+1)$.

## 4    Computational Experiments

### 4.1    Dataset and Experiment Settings

To test the performance of the proposed DHEDA for the practical M-CVaR PO, a set of computational experiments are performed on real-world historical financial market data. Five datasets are constructed based on five famous stock indexes including FTSE100 (UK), N225 (Japan), CSI300 (China), CSI500 (China), and S&P500 (USA). Daily data from 01/01/2018 to 31/12/2018 is considered for all the datasets, however the trading days of different stock markets may be different. By excluding the missing data, the number of assets in the five datasets ranges from 99 to 500. The assets are classified into different classes based on the industry classification. The number of classes in the five datasets (FTSE100 to S&P500) is 9, 10, 10, 19, and 20, respectively. For each dataset, three levels of cardinalities $K_1, K_2, K_3$ are considered. For each portfolio, the minimum proportion of each asset and class is set as $l_i = 0.01$ and $L_m = 0.03$. No upper limits of the asset and class proportion are specified ($u_i = 1, U_m = 1$). The confidence level $\alpha$ is set as 0.01 for all instances. Specific descriptions of the datasets can be found in Table 2.

**Table 2.** Descriptions of the datasets

| Stock index | $N$ | $M$ | $J$ | $T$ | $K_1, K_2, K_3$ |
|---|---|---|---|---|---|
| FTSE100 | 99 | 9 | 11 | 253 | 15, 20, 25 |
| N225 | 220 | 10 | 22 | 245 | 15, 20, 25 |
| CSI300 | 250 | 10 | 25 | 243 | 15, 20, 25 |
| CSI500 | 380 | 19 | 20 | 243 | 30, 35, 40 |
| S&P500 | 500 | 20 | 25 | 251 | 30, 35, 40 |

The DHEDA is implemented in C++ and run on a personal computer with a 3.30 GHz Intel Xeon E3-1230 v2 CPU and 8 GB RAM under Microsoft Windows 7. To demonstrate the effectiveness of the proposed DHEDA, it is compared with the NSGA-II proposed in [12]. For fair comparison, the NSGA-II is also implemented in C++ and run on the same computer with the DHEDA, and the sopping criterion is set as $25 \cdot K$ second CPU time for all the following numerical tests. To evaluate the performance of the algorithms in solving the M-CVaR PO, the following two widely used performance evaluation metrics are adopted in this paper: the inverted generational distance (IGD) [19] and the hyper-volume metric (HV) [20].

## 4.2 Parameter Settings

There are three user-determined parameters in the proposed DHEDA, i.e. the population size $N_p$, the learning rate $\alpha_l$, and the number of decomposed regions of the objective space $R_n$. To investigate the effect of these parameter, the Taguchi method of design-of-experiments [21] is used. Four factor levels are employed for each parameter as listed in Table 3. Accordingly, an orthogonal array $L_{16}(4^3)$ including 16 different ($N_p, \alpha_l, R_n$) combinations are selected. For each combination, the DHEDA runs 10 times independently and the average IGD is calculated as the response value (RV). The trend of the average RV (ARV) are illustrated in Fig. 4.

**Table 3.** Factor levels of parameters

| | Factor level | | | |
|---|---|---|---|---|
| | 1 | 2 | 3 | 4 |
| $N_p$ | 70 | 140 | 210 | 280 |
| $\alpha_l$ | 0.003 | 0.007 | 0.011 | 0.015 |
| $R_n$ | 5 | 10 | 15 | 20 |

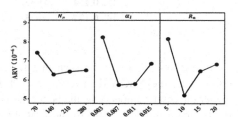

**Fig. 4.** The influence trend of each parameter

From Fig. 4, it can be seen that the ARV values are small, indicating that the performance of the algorithm is not sensitive to the parameters. For the population size $N_p$, a bigger $N_p$ may improve the diversity of the population, however a small $N_p$ is able to increase the search generations. For the learning rate $\alpha_l$, a bigger $\alpha_l$ may speed up the convergence of the algorithm, however a smaller $\alpha_l$ can strengthen the algorithm's exploration ability. The region number $R_n$ is the most significant parameter. A bigger $R_n$ is able to improve the accuracy of the estimation of the probability model, however increase the algorithm complexity as well. Based on the results of the DOE, the parameters of the algorithm are set as: $N_p = 140, \alpha_l = 0.007, R_n = 10$ for the following tests and comparisons.

### 4.3    Comparisons of the Algorithms

The proposed DHEDA is compared with an existing NSGA-II proposed in [12]. The parameters of the NSGA-II are set as the recommended values given in [12]. For each instance, both algorithms run 10 times and the average IGD and HV values are considered as the performance measurement for the comparison of the two algorithms. Specific results are given in Table 4.

**Table 4.** Comparison of the algorithms

| Dataset | $K$ | IGD ($10^{-4}$) | | HV | | CPU time |
|---|---|---|---|---|---|---|
| | | DHEDA | NSGA-II | DHEDA | NSGA-II | |
| FTSE100 | 15 | **0.268** | 0.357 | **0.496** | 0.455 | 380.0 |
| | 20 | **0.193** | 0.287 | **0.500** | 0.451 | 506.8 |
| | 25 | **0.215** | 0.345 | **0.500** | 0.452 | 631.6 |
| N225 | 15 | **0.160** | 0.421 | **0.500** | 0.308 | 376.1 |
| | 20 | **0.250** | 0.470 | **0.483** | 0.282 | 500.5 |
| | 25 | **0.253** | 0.433 | **0.496** | 0.334 | 625.8 |
| CSI300 | 15 | **0.189** | 1.350 | **0.500** | 0.222 | 376.0 |
| | 20 | **0.369** | 1.320 | **0.500** | 0.263 | 500.9 |
| | 15 | **0.357** | 1.180 | **0.497** | 0.264 | 631.2 |
| CSI500 | 30 | **0.101** | 0.818 | **0.499** | 0.255 | 751.8 |
| | 35 | **0.121** | 0.869 | **0.496** | 0.274 | 875.5 |
| | 40 | **0.172** | 0.958 | **0.500** | 0.238 | 1002.6 |
| S&P500 | 30 | **0.115** | 1.050 | **0.499** | 0.255 | 750.8 |
| | 35 | **0.134** | 0.953 | **0.498** | 0.320 | 875.7 |
| | 40 | **0.177** | 1.075 | **0.501** | 0.289 | 1002.2 |

From Table 4, it can be seen that for all instances, the DHEDA achieves better results than the NSGA-II in terms of both the IGD and HV metrics. Moreover, the bigger the asset number $N$ is, the better the DHEDA performs in contrast to the NSGA-II. It demonstrates the effectiveness of the proposed DHEDA, especially in solving complex datasets. To further investigate the performance of the DHEDA and the

NSGA-II, the Pareto fronts obtained by the two algorithms in solving the dataset S&P500 with $K = 30$ are plotted in Fig. 5. It can be seen from the figure that the Pareto front of the DHEDA is more evenly-distributed in the objective space, while the Pareto front of the NSGA-II only covers the middle part of the whole Pareto front. Furthermore, in contrast to the NSGA-II, the DHEDA achieves much more non-dominated solutions in the right region of the objective space than the left, which indicates that the DHEDA is better at finding portfolios with good expected return. Based on the above analysis, it can be concluded that the DHEDA can obtain better non-dominated solutions than the existing algorithm in terms of both proximity and diversity.

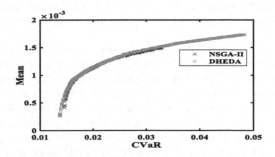

**Fig. 5.** Pareto front of DHEDA and NSGA-II on dataset S&P500 with $K = 30$

## 5  Conclusion

This paper addresses the practical M-CVaR PO, which maximizes the mean return and minimizes the CVaR under a set of practical constraints. A DHEDA is proposed to solve the problem. The computational results show that the proposed DHEDA performs better than the existing NSGA-II. The advantages of the DHEDA can be concluded into the following aspects. (1) The complexity of solving the problem is reduced by decomposing the original problem into an asset selection problem and a proportion allocation problem. (2) The DHEDA combined the EDA and the LP-based approach to solve the asset selection and the proportion allocation, respectively. (3) The decomposition of the objective space and its related sub-PM method guarantee the diversity of the obtained solutions. (4) The proposed knowledge-based initialization and CLS effectively improve the archive in the initialization stage and in the search process, respectively.

The future work is to study more efficient algorithms to solve the practical PO. It is important to consider other practical constraints like the transaction costs in the M-CVaR PO. Other risk measurements such as VaR and MAD are also worth studying.

## References

1. Markowitz, H.: Portfolio selection. J. Financ. **7**, 77–91 (1952)
2. Ertenlice, O., Kalayci, C.B.: A survey of swarm intelligence for portfolio optimization: algorithms and applications. Swarm Evol. Comput. **39**, 36–52 (2018)

3. Lwin, K.T., Qu, R., MacCarthy, B.L.: Mean-VaR portfolio optimization: a nonparametric approach. Eur. J. Oper. Res. **260**, 751–766 (2017)
4. Righi, M.B., Borenstein, D.: A simulation comparison of risk measures for portfolio optimization. Financ. Res. Lett. **24**, 105–112 (2018)
5. Alexander, G.J., Baptista, A.M.: A comparison of VaR and CVaR constraints on portfolio selection with the mean-variance model. Manage. Sci. **50**, 1261–1273 (2004)
6. Pflug, G.C.: Some remarks on the value-at-risk and the conditional value-at-risk. In: Uryasev, S.P. (ed.) Probabilistic Constrained Optimization. NOIA, vol. 49, pp. 272–281. Springer, Boston, MA (2000). https://doi.org/10.1007/978-1-4757-3150-7_15
7. Alexander, S., Coleman, T.F., Li, Y.: Minimizing CVaR and VaR for a portfolio of derivatives. J. Bank Financ. **30**, 583–605 (2006)
8. Takano, Y., Nanjo, K., Sukegawa, N., Mizuno, S.: Cutting plane algorithms for mean-CVaR portfolio optimization with nonconvex transaction costs. Comput. Manag. Sci. **12**, 319–340 (2015)
9. Najafi, A.A., Mushakhian, S.: Multi-stage stochastic mean–semivariance–CVaR portfolio optimization under transaction costs. Appl. Math. Comput. **256**, 445–458 (2015)
10. Qin, Q., Li, L., Cheng, S.: A novel hybrid algorithm for Mean-CVaR portfolio selection with real-world constraints. In: Tan, Y., Shi, Y., Coello, C.A.C. (eds.) ICSI 2014. LNCS, vol. 8795, pp. 319–327. Springer, Cham (2014). https://doi.org/10.1007/978-3-319-11897-0_38
11. Cheng, R., Gao, J.: On cardinality constrained mean-CVaR portfolio optimization. In: The 27th Chinese Control and Decision Conference, pp. 1074–1079. IEEE, Qingdao (2015)
12. Anagnostopoulos, K.P., Mamanis, G.: Multiobjective evolutionary algorithms for complex portfolio optimization problems. Comput. Manag. Sci. **8**, 259–279 (2011)
13. Larrañaga, P., Lozano, J.A.: Estimation of Distribution Algorithms: A New Tool for Evolutionary Computation. Springer, Boston (2001). https://doi.org/10.1007/978-1-4615-1539-5
14. Rockafellar, R.T., Uryasev, S.: Optimization of conditional value-at-risk. J. Risk **2**, 21–42 (2000)
15. Wang, S.Y., Ling, W., Min, L., Ye, X.: An effective estimation of distribution algorithm for solving the distributed permutation flow-shop scheduling problem. Int. J. Prod. Econ. **145**, 387–396 (2013)
16. Fang, C., Kolisch, R., Wang, L., Mu, C.: An estimation of distribution algorithm and new computational results for the stochastic resource-constrained project scheduling problem. Flex. Serv. Manuf. J. **27**, 585–605 (2015)
17. Wu, C., Li, W., Wang, L., Zomaya, A.: Hybrid evolutionary scheduling for energy-efficient fog-enhanced internet of things. IEEE Trans. Cloud Comput. (2018). https://doi.org/10.1109/tcc.2018.2889482
18. Baluja, S.: Population-based incremental learning: a method for integrating genetic search based function optimization and competitive learning. Carnegie-Mellon University, Department of Computer Science, pp. 1–20 (1994)
19. Sierra, M.R., Coello Coello, C.A.: Improving PSO-based multi-objective optimization using crowding, mutation and ∈-dominance. In: Coello Coello, C.A., Hernández Aguirre, A., Zitzler, E. (eds.) EMO 2005. LNCS, vol. 3410, pp. 505–519. Springer, Heidelberg (2005). https://doi.org/10.1007/978-3-540-31880-4_35
20. Zitzler, E., Thiele, L.: Multiobjective evolutionary algorithms: a comparative case study and the strength Pareto approach. IEEE Trans. Evol. Comput. **3**, 257–271 (1999)
21. Montgomery, D.C.: Design and Analysis of Experiments. Wiley, New York (2017)

# CBLNER: A Multi-models Biomedical Named Entity Recognition System Based on Machine Learning

Gong Lejun[1(✉)], Liu Xiaolin[1], Yang Xuemin[1], Zhang Lipeng[1],
Jia Yao[1], and Yang Ronggen[2]

[1] Jiangsu Key Lab of Big Data Security and Intelligent Processing,
School of Computer Science, Nanjing University of Posts and
Telecommunications, Nanjing 210023, China
glj98226@163.com
[2] Faculty Intelligent Science and Control Engineering,
Jinling Institute of Technology, Nanjing 211169, China

**Abstract.** Biomedical named entities is fundamental recognition task in biomedical text mining. This paper developed a system for identifying biomedical entities with four models including CRF, LSTM, Bi-LSTM and BiLSTM-CRF. The system achieved the following performance in test data Genia V3.02: CRF with an F score of 75.91%, LSTM with an F score of 71.69%, BiLSTM with a F score of 74.37%, BiLSTM-CRF with a F score of 76.81%. Experimental results show the performance of BiLSTM-CRF model is better than other three models. Compared with CRF model, Bi-LSTM-CRF model has better recognition effect for biological entities in long text and entities that modified by modifiers. Therefore, CBLNER system lays a foundation for further relationship and event extraction, and could also provide reference for entity recognition research in other fields.

**Keywords:** Named entity recognition · CRF · LSTM · Bi-LSTM ·
BiLSTM-CRF

## 1 Introduction

With the increasing number of published biological literature, the biological knowledge carried in the literature also increases exponentially. Therefore, in order to make full use of published biomedical literature, biomedical text mining has become an important tool for mining new knowledge and theories. Named entities recognition is the fundamental step in biomedical text mining. It acts as the element of text mining involving with all of components including information extraction, text summarization and knowledge discovery. In the early stage, text mining mainly extracted the fact presented in text based on rules. BioIE [1] is a rule-based system for extracting informative sentences. Considering the precise extraction performance, there are some practical platform used the dictionary-based approach. PolySearch [2] is a system using a dictionary approach for extracting relationships between human diseases, genes, mutations, drugs and metabolites based on nine different thesauruses including human

© Springer Nature Switzerland AG 2019
D.-S. Huang et al. (Eds.): ICIC 2019, LNCS 11644, pp. 51–60, 2019.
https://doi.org/10.1007/978-3-030-26969-2_5

genes, human proteins, human disease, approved drugs, endogenous metabolites, protein/gene pathways, human tissues, human organs and sub-cellular localizations. Whatizit [3] is also a system which link protein names to their UniProt-ID using a dictionary generated from the UniProt database. There are some works using machine learning techniques to identify biomedical named entities. These techniques involved with CRFs [4], SVM [5], and HMM [6]. ABNER [7] is a biomedical named entity recognizer tool based on machine learning using CRFs with 70.5% overall F1 score in NLPBA corpus. NLProt [8] is an extracting protein names by combining dictionary and rule-based filtering with support vector machines. Currently, some deep learning approach were also used in this research. Gridach [9] proposed a character-level neural network for identifying biomedical named entities. There are some similar works using deep learning. The latest biomedical naming recognition system is based on the hidden Markov model [10] of transfer learning with an F score of 65.41%. It differs from the traditional biomedical naming entity recognition in that the traditional method requires a lot of target domain tagging data, which requires a lot of manpower to tag information. Hidden Markov model based on transfer learning reduces the requirement of named entity recognition for target domain annotation data in biomedical texts. In this work, we develop a novel biomedical named entities recognition system combining several machine learning models containing Conditional Random Fields (CRF), Long Short-Term Memory (LSTM) [11] and Bidirectional Long Short-Term Memory joint CRF (BiLSTM-CRF) [12] to recognize multi-class entities. The remaining sections are structured as follows: Sect. 2 described the data set used in the paper. Section 3 describes our methods. Section 4 presents implementation and results. Section 5 concludes this paper.

## 2   Dataset

In this paper, we used GENIA [13] corpus to validate the several models' performances. GENIA corpus is annotated by linguistic and informationist with a collection of biomedical literature from the GENIA project which is used to develop text mining systems for the domain of molecular biology. The corpus provide a reference for the development of biomedical text mining system with authority annotated. The dataset used the GENIA 3.02 version which is formed from a controlled search on MEDLINE using MeSH terms containing the keywords: human, transcription factors and blood cells including 2000 biomedical literature with semantically annotated data involving with 18546 sentences and 400000 tokens download at the URL (http://www-tsujii.is.s. u-tokyoac.jp/~genia/topics/Corpus/).

## 3   Methods

Named entity recognition extracts important named entities for instance protein, gene, and diseases, and is a challenging task. It has received considerable attention over the past years. In this study, we develop a multi-view biomedical named entities recognition which called CBLNER. Our system aims at the recognition of the biomedical entities including five classes: Protein, RNA, DNA, Cell-line, Cell-type using machine learning approach. The overview of our system for automatically identifying named entities is shown in Fig. 1.

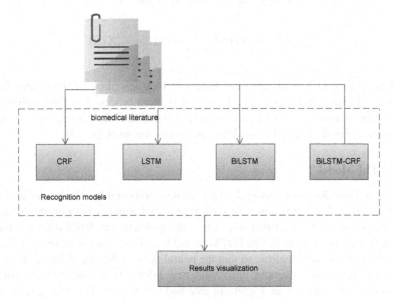

**Fig. 1.** CBLNER's overview

CBLNER contains four recognition modules. Raw texts are put into CBLNER which could call one of four models to process the texts, and produce identified results. The following describes the four recognition models in details.

### 3.1 CRF Model

Named entities recognition could be looked as a sequence segmentation problem which each word in text is a token to be assigned a label. CRF is an undirected graphical model with a linear chain which is well suited to sequence tagging.

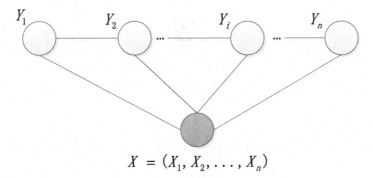

$$X = (X_1, X_2, \ldots, X_n)$$

**Fig. 2.** CRF graphical structure

Let $X = (X_1, X_2, \ldots, X_n)$ be an observed sequence related to words with length $n$. Let $Y = (Y_1, Y_2, \ldots, Y_n)$ be a label sequence which could form a first-order chain as shown in Fig. 2. The line chain conditional field define the conditional probability of a state sequence given an observation sequence to be Eq. (1):

$$P(Y|X) = \frac{1}{Z_0} \exp(\sum_{i=1}^{n} \sum_{j=1}^{m} \lambda_j f_j(Y_{i-1}, Y_i, X, i)) \tag{1}$$

Where $f_j(Y_{i-1}, Y_i, X, i)$ is one of $m$ functions that describes a feature, and $\lambda_j$ is a learned weight for feature function. The aim of training process is for the weights that maximize the likelihood of instances in the training data. $Z_0$ is a normalization factor. The complete detail about CRF model is present in the work [4, 14].

### 3.2    LSTM Model

Long Short Term Memory Network (LSTM) was proposed by Hochreiter and Sch-midhuber [15] in 1997. It is a new type of deep neural network built on the Recurrent Neural Network (RNN) [16] that overcomes the problem that RNN does not handle long-range dependencies well. The LSTM model introduces a set of memory units, that is, records historical information, so that the neural network has the ability to learn past information and update memory units with new information. The memory unit has three control gates: input gate, forgetting gate and output gate. The three gates control the input, read and output of LSTM unit respectively. The model structure diagram of LSTM is shown in Fig. 3.

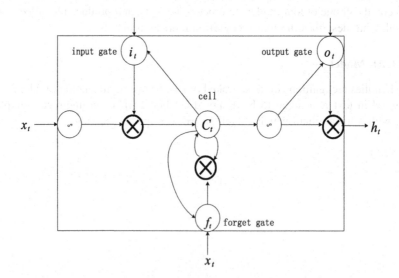

**Fig. 3.** LSTM model graphical structure

The LSTM memory cell is achieved as the following:

$$\sigma(x) = \frac{1}{1 + e^{-x}} \tag{2}$$

$$i_t = \sigma(W_{xi}x_t + W_{hi}h_{t-1} + W_{ci}c_{t-1} + b_i) \tag{3}$$

$$f_t = \sigma(W_f h_{t-1} + U_f x_t + b_f) \tag{4}$$

$$\tilde{c}_t = \tanh(W_c h_{t-1} + U_c x_t + b_c) \tag{5}$$

$$c_t = f_t \odot c_{t-1} + i_t \odot \tilde{c}_t \tag{6}$$

$$o_t = \sigma(W_o h_{t-1} + U_o x_t + b_o) \tag{7}$$

$$h_t = o_t \odot \tanh(c_t) \tag{8}$$

Where $\sigma$ is the activation function sigmod, tanh is the hyperbolic tangent function, and $i_t, f_t, o_t$ represent the input gate, forget gate and output gate at time t, respectively. $w_i, w_f, w_c, w_o$ is the weight matrix of the hidden state. $h_t$, $u_i, u_f, u_c, u_o$ represent the weight matrix of the different gates for input $x_t$ and $b_i, b_f, b_c, b_o$ represent the bias vector. $x_t$ is the input vector at time t, and $h_t$ is the output vector storing all available information at time t.

### 3.3  Bi-LSTM Model

In more granular tasks, LSTM is not enough to use only the front-to-back coding information. The bidirectional long-term and short-term memory network (Bi-LSTM) [17] is composed of forward LSTM and backward LSTM. Therefore, Bi-LSTM can better obtain the two-way semantic dependence of sentences, and process each sequence forward and backward to capture two separate hidden states of past and future information. Bi-LSTM actually trains two LSTM networks, one using the forward sequence, another using the backward order, and finally we contact the two hidden states as the input vector into softmax [18] function. The specific model structure diagram is shown in Fig. 4.

### 3.4  Bi-LSTM-CRF Model

Considering the constraints of state transitions when capturing contextual relationships, and adding CRF to Bi-LSTM, the paper constructs the Bi-LSTM joint CRF model (BiLSTM-CRF) [19]. In this model, CRF could consider the relationship between entity annotations and obtain the overall optimal annotation sequence. Therefore, in this paper, we introduce CRF model into Bi-LSTM model. CRF algorithm uses maximum likelihood estimation to calculate probability, and Viterbi algorithm to decode and annotate the sequence in order to get the optimal annotation sequence. The structure diagram of BiLSTM-CRF model is shown in Fig. 5.

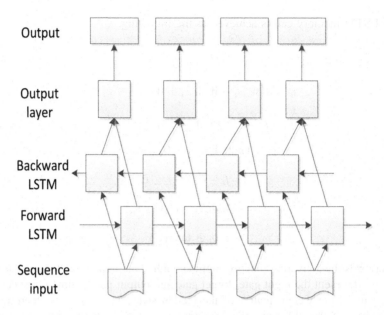

**Fig. 4.** Bi-LSTM model graphical structure

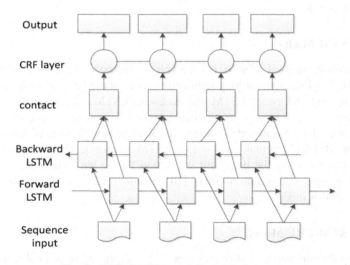

**Fig. 5.** BiLSTM-CRF model graphical structure

## 4 Results and Discussions

In this paper, we developed a biomedical named entity recognition system called CBLNER, which contains the four models including CRF, LSTM, Bi-LSTM and BiLSTM-CRF models using Python programming based on tensorflow and Flask which is lightweight web framework. The three popular measurements are used to evaluate the system performances, i.e. Precision, Recall and F-score. It is also used to the work [20] which describes their definitions in details. The performances are shown in Table 1 for the test dataset Genia version 3.02. From the Table 1, CBLNER achieved 79.04% of precision, 74.72% of recall, and 76.81% of F-score in BiLSTM-CRF which is more superior performance in F-score compared to other three models. Compared with F-score of the work [10], the above four models perform better with state-of the-art.

**Table 1.** CBLNER's performances

| Model | Precise (%) | Recall (%) | F-score (%) |
|-----------|-------------|------------|-------------|
| CRF | 76.00% | 77.98 | 75.91 |
| LSTM | 76.25 | 67.64 | 71.69 |
| Bi-LSTM | 80.07 | 69.43 | 74.37 |
| BiLSTM-CRF | 79.04 | 74.72 | 76.81 |

The corresponding visual effect graph is shown in Figs. 6, 7, 8 and 9 using biomedical literature as an instance (PMID: 90169371), In these figures, different colors indicates different biomedical entities.

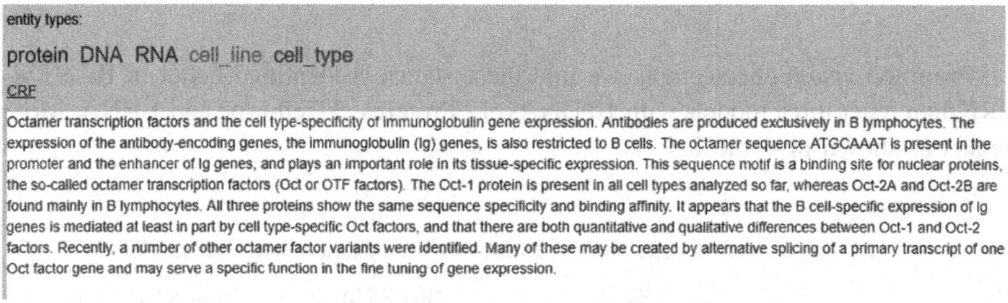

**Fig. 6.** Identified visualization with CRF model

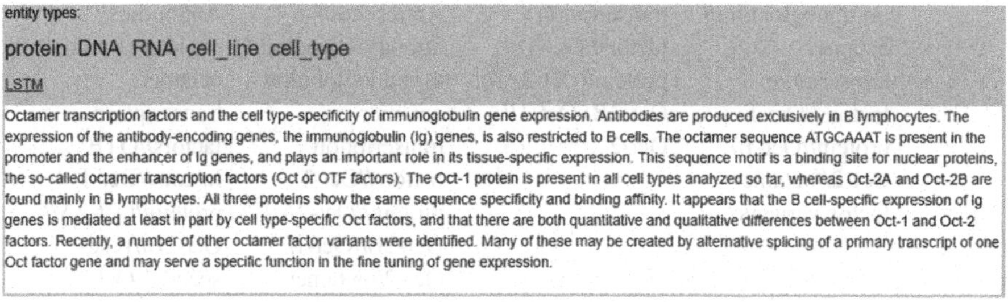

**Fig. 7.** Identified visualization with LSTM model

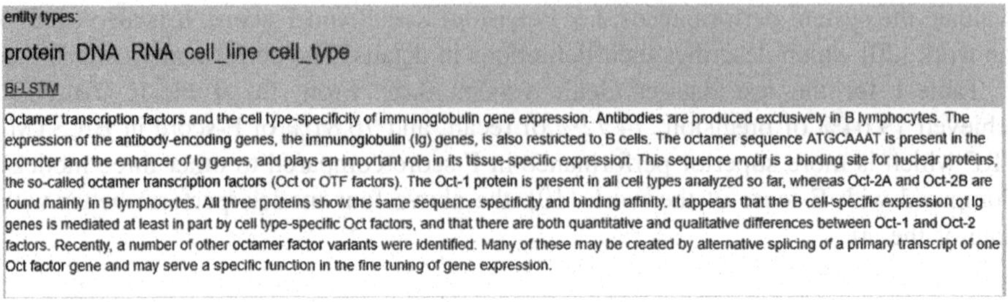

entity types:

protein  DNA  RNA  cell_line  cell_type

Bi-LSTM

Octamer transcription factors and the cell type-specificity of immunoglobulin gene expression. Antibodies are produced exclusively in B lymphocytes. The expression of the antibody-encoding genes, the immunoglobulin (Ig) genes, is also restricted to B cells. The octamer sequence ATGCAAAT is present in the promoter and the enhancer of Ig genes, and plays an important role in its tissue-specific expression. This sequence motif is a binding site for nuclear proteins, the so-called octamer transcription factors (Oct or OTF factors). The Oct-1 protein is present in all cell types analyzed so far, whereas Oct-2A and Oct-2B are found mainly in B lymphocytes. All three proteins show the same sequence specificity and binding affinity. It appears that the B cell-specific expression of Ig genes is mediated at least in part by cell type-specific Oct factors, and that there are both quantitative and qualitative differences between Oct-1 and Oct-2 factors. Recently, a number of other octamer factor variants were identified. Many of these may be created by alternative splicing of a primary transcript of one Oct factor gene and may serve a specific function in the fine tuning of gene expression.

**Fig. 8.** Identified visualization with Bi-LSTM model

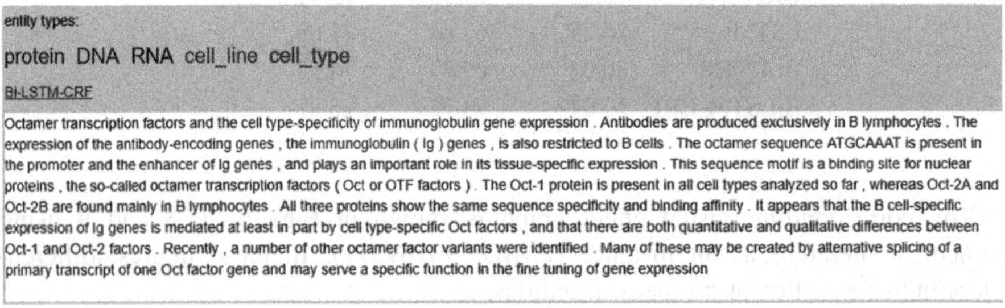

entity types:

protein  DNA  RNA  cell_line  cell_type

Bi-LSTM-CRF

Octamer transcription factors and the cell type-specificity of immunoglobulin gene expression . Antibodies are produced exclusively in B lymphocytes . The expression of the antibody-encoding genes , the immunoglobulin ( Ig ) genes , is also restricted to B cells . The octamer sequence ATGCAAAT is present in the promoter and the enhancer of Ig genes , and plays an important role in its tissue-specific expression . This sequence motif is a binding site for nuclear proteins , the so-called octamer transcription factors ( Oct or OTF factors ) . The Oct-1 protein is present in all cell types analyzed so far , whereas Oct-2A and Oct-2B are found mainly in B lymphocytes . All three proteins show the same sequence specificity and binding affinity . It appears that the B cell-specific expression of Ig genes is mediated at least in part by cell type-specific Oct factors , and that there are both quantitative and qualitative differences between Oct-1 and Oct-2 factors . Recently , a number of other octamer factor variants were identified . Many of these may be created by alternative splicing of a primary transcript of one Oct factor gene and may serve a specific function in the fine tuning of gene expression

**Fig. 9.** Identified visualization with BiLSTM-CRF model

From the visual effect graph, we find the instance is identified better in BiLSTM-CRF than other three models with 10 DNAs, 12 Proteins, 3 Cell-type and 1 RNA. More details as shown in Table 2.

**Table 2.** Identified each kind of entities of the instance with four models

| Type | CRF | LSTM | Bi-LSTM | BiLSTM-CRF |
|------|-----|------|---------|------------|
| Protein | Octamer transcription factors// immunoglobulin// octamer transcription factors//Oct-1 protein//Oct-1// Oct-2//octamer factor variants | Octamer transcription factors//octamer transcription factors//Oct-1 protein//Oct-2A// Oct-2B//Oct-1// Oct-2 | Octamer transcription factors// Antibodies// ntibody-encoding// immunoglobulin// octamer transcription factors//Oct-1 protein//Oct-2A// Oct-2B//Oct-1// Oct-2//octamer factor variants//Oct factor | Octamer transcription factors// Antibodies// nuclear proteins// octamer transcription factors//OTF factors//Oct-1 protein//Oct-2A// Oct-2B//type-specific Oct factors//Oc1-1// Oct-2 factors// octamer factor variants |

(*continued*)

**Table 2.**  (*continued*)

| Type | CRF | LSTM | Bi-LSTM | BiLSTM-CRF |
|---|---|---|---|---|
| DNA | cell type-specificity of immunoglobulin gene//octamer sequence// sequence motif// binding site//Ig genes | cell type-specificity of immunoglobulin gene//octamer sequence// promoter//Ig genes sequence motif// | cell type-specificity of immunoglobulin gene//octamer sequence// promoter// sequence motif//Ig genes | cell type-specificity of immunoglobulin gene//antibody-encoding genes// immunoglobulin (Ig) genes//octamer sequence// promoter// enhancer//Ig genes//sequence motif//binding site//Oct//Ig genes |
| RNA | primary transcript | primary transcript | primary transcript | primary transcript |
| Cell_line | _ | _ | _ | _ |
| Cell_type | _ | _ | _ | B lymphocytes//B cells//B lymphocytes |

Note: "//" is used to split identified entity.

## 5  Conclusions

In this paper, we proposed a system named CBLNER which is a biomedical text named entity recognition system with four models containing CRF, LSTM, Bi-LSTM, BiLSTM-CRF. By the test data, we also evaluated each model's performance. Experimental results show the BiLSTM-CRF is better than other three models with 79.04% precise, 74.72% recall and 76.81% F-score. Compared with CRF model, Bi-LSTM-CRF model has better recognition effect for biological named entities with longer length and modifiers. Compared the latest work [10], CBLNER is the state-of-the-art in named entity recognition. Therefore, CBLNER system lays a foundation for further relationship and event extraction, and could also provide reference for entity recognition research in other fields.

**Acknowledgment.** This work is supported by the Natural Science Foundation of the Higher Education Institutions of Jiangsu Province in China (16KJD520003), National Natural Science Foundation of China (61502243, 61502247, 61572263), China Postdoctoral Science Foundation (2018M632349), Zhejiang Engineering Research Center of Intelligent Medicine under 2016E10011.

# References

1. Divoli, A., Attwood, T.K.: BioIE: extracting informative sentences from the biomedical literature. Bioinformatics **21**(9), 2138–2139 (2005). Epub 2005 Feb 2
2. Cheng, D., Knox, C., Young, N., Stothard, P., Damaraju, S., Wishart, D.S.: PolySearch: a web-based text mining system for extracting relationships between human diseases, genes, mutations, drugs and metabolites. Nucleic Acids Res. **36**(Web Server issue), W399–405 (2008). https://doi.org/10.1093/nar//gkn296
3. Rebholz-Schuhmann, D., Kirsch, H., Couto, F.: Facts from text–is text mining ready to deliver? PLoS Biol. **3**(2), e65 (2005)
4. Yang, Z., Lin, H., Li, Y.: Exploiting the contextual cues for bio-entity name recognition in biomedical literature. J. Biomed. Inform. **41**(4), 580–587 (2008). https://doi.org/10.1016//j.jbi.2008.01.002
5. Mitsumori, T., Fation, S., Murata, M., Doi, K., Doi, H.: Gene/protein name recognition based on support vector machine using dictionary as features. BMC Bioinform. **6**(Suppl 1), S8 (2005)
6. Zhang, J., Shen, D., Zhou, G., Su, J., Tan, C.L.: Enhancing HMM-based biomedical named entity recognition by studying special phenomena. J. Biomed. Inform. **37**(6), 411–422 (2004)
7. Settles, B.: ABNER: an open source tool for automatically tagging genes, proteins and other entity names in text. Bioinformatics **21**(14), 3191–3192 (2005)
8. Mika, S., Rost, B.: NLProt: extracting protein names and sequences from papers. Nucleic Acids Res. **32**(Web Server issue), W634–637 (2004)
9. Gridach, M.: Character-level neural network for biomedical named entity recognition. J. Biomed. Inform. **70**, 85–91 (2017). https://doi.org/10.1016//j.jbi.2017.05.002
10. Gao, B.T., Zhang, Y., Liu, B.: BioTrHMM: biomedical named entity recognition based on transfer learning. Comput. Appl. Res. **36**(01), 51–54 (2019)
11. Lyu, C., Chen, B., Ren, Y., Ji, D.: Long short-term memory RNN for biomedical named entity recognition. BMC Bioinform. **18**(1), 462 (2017). https://doi.org/10.1186//s12859-017-1868-5
12. Dang, T.H., Le, H.Q., Nguyen, T.M., Vu, S.T.: D3NER: biomedical named entity recognition using CRF-biLSTM improved with fine-tuned embeddings of various linguistic information. Bioinformatics **34**(20), 3539–3546 (2018). https://doi.org/10.1093//bioinformatics//bty356
13. Kim, J.D., Ohta, T., Tateisi, Y., et al.: GENIA corpus–semantically annotated corpus for bio-textmining. Bioinformatics **19**(suppl_1), i180 (2003)
14. Avery, I.T.: Conditional random field. Comput. Vis. **3**(2), 637–640 (2012)
15. Hochreiter, S., Schmidhuber, J.: Long short-term memory. Neural Comput. **9**(8), 1735–1780 (2014)
16. Campos, V., Jou, B., Giro-I-Nieto, X., et al.: Skip RNN: learning to skip state updates in recurrent neural networks (2017)
17. Zeng, J., Che, J., Xing, C., et al.: A two-stage Bi-LSTM model for chinese company name recognition. In: International Conference on Artificial Intelligence: Methodology (2018)
18. Milakov, M., Gimelshein, N.: Online normalizer calculation for softmax (2018)
19. Anh, L.T., Arkhipov, M.Y., Burtsev, M.S.: Application of a hybrid Bi-LSTM-CRF model to the task of Russian Named Entity Recognition (2017)
20. Yang, X., Zhang, Z., Yang, R., et al.: Using deep learning to recognize biomedical entities. In: International Conference on Intelligent Systems & Knowledge Engineering. IEEE (2018)

# Dice Loss in Siamese Network for Visual Object Tracking

Zhao Wei[1,2], Changhao Zhang[1,2], Kaiming Gu[1,2], and Fei Wang[1,2(✉)]

[1] National Engineering Laboratory for Visual Information Processing
and Application, XJTU, 99 Yanxiang Road, Xi'an 710054, Shaanxi, China
wfx@mail.xjtu.edu.cn
[2] School of Electronics and Information Engineering, XJTU,
28 West Xianning Road, Xi'an 710049, Shaanxi, China

**Abstract.** The problem of visual object tracking has evolved over the years. Traditionally, it is solved by a model that only learns the appearance of an object online, using the video itself as the only training data. The target in a single object tracking task is a relatively small object in most cases, and the deformation is more serious, referring to the dice loss used in the semantic segmentation problem, we introduced a new objective function to optimize during training based on the Dice coefficient. In this way, we can handle the strong imbalance between foreground and background patches. To cope with the limited amount of annotations available for training, we use random nonlinear transformations and histogram matching to increase the data. We have demonstrated in our experimental evaluation that our method has achieved good performance in challenging test data, while only requiring a small amount of processing time required by other previous methods.

**Keywords:** Object tracking · Dice loss · Siamese network

## 1 Introduction

Visual object tracking aims at locating an unknown object in a video sequence with the object's position initialized in the first frame. Although visual object tracking plays a vital role in a large range of applications, including video analysis, surveillance, and automatic driving, it is still a challenging task. Tracking arbitrary target suffers from numerous factors such as deformation, occlusion as well as cluttered background.

Many methods [6, 29, 2, 5, 11] have been proposed to deal with deformation for better tracking performance with convolutional neural networks (CNNs). ADNet [5] generates actions through reinforcement learning to find locations and sizes of targets in a new frame, but this method needs to fine-tune during tracking for online adaptation to target and background changes, which is a little time-consuming. SA-Siam [29] utilizes a channel attention mechanism to achieve a minimum degree of target adaptation. SiamRPN [2] and the methods improved based on it [6, 29, 2], employ the

---

Z. Wei, C. Zhang and K. Gu—Contributed equally.

D.-S. Huang et al. (Eds.): ICIC 2019, LNCS 11644, pp. 61–70, 2019.
https://doi.org/10.1007/978-3-030-26969-2_6

region proposal subnetwork (RPN) [11] to extract precise proposals which can solve the deformation to some extent. Specifically, SA-Siam and SiamRPN based methods all adopt the CNNs of Siamese structure which can learn the similarities between the template and the instance branches with competitive performance and speed. And they all produce deformation in the template branch to adapt to the target spatial transformation. However, CNNs are inherently limited to model geometric transformations due to the fixed geometric structures in their building modules. Besides, the object in the first frame is given and fixed while the target in the following instance frames may be deformed. Therefore, this paper proposes spatial transform method in the instance branch instead of the template to deal with target deformation, which is more reasonable. According to our knowledge, it is the first time that the spatial transform is utilized in the tracking task.

In this work, compared to SiamRPN based methods which generate anchors to refine the shape, we propose the adaptive spatial transform model in the instance branch to make the learned features robust to the deformation. To be specific, during the correlation operation, Siamese network uses template features as convolution kernels to perform the convolution operation on the instance feature with stride of 1. The convolution stride is 1 on the feature map, which corresponds to several pixels of the original image. It is likely that the precise target location has been omitted during the correlation. And the target in instance branch may not have the same scale as the target in template. Therefore, we add a spatial transform model in the instance branch before correlation, which can learn the offsets in each position of the feature map. The scale and position of the feature map are transformed adaptively by the learned offsets. That is, correlation operation enables free form of the sampling grid in the instance feature map so that it can generate the deformation in the target. Moreover, we achieve regression without adding additional regression losses.

As for the training data, most Siamese network based trackers only discriminate foreground from the non-semantic background, which have poor performance when the backgrounds are cluttered. DaSiamRPN [6] focuses on the balance of non-semantic background and semantic distractors by generating diverse semantic positive and negative pairs. However, semantic distractors extracted from the images may not get a higher score in the response map. Decreasing the score of these distractors may cause the tracker bias to the wrong result. In order to solve this problem, we propose a positive and negative sample mining method which selects the samples based on the classification score of the response map rather than based on the input image pairs.

To sum up, this paper presents a tracking model which adds a target spatial transform network to solve the deformation problem. The main contributions of this paper are as follows:

(1) To our best knowledge, it is the first time to apply the dice loss on the tracking task so that the features learnt by the network can be robust to the variance of multi-scale features.

(2) This paper focuses on hard samples from score map with off-line training, which can make the network distinguish the target with hard sample in the different videos of the same category by providing more elements for training without adding extra inputs.

(3) In-depth experiments are operated to analyze various aspects of our method. Moreover, this paper yields competitive accuracy in visual object tracking in the public benchmark tracking datasets, i.e. OTB2013, OTB2015.

## 2   Related Works

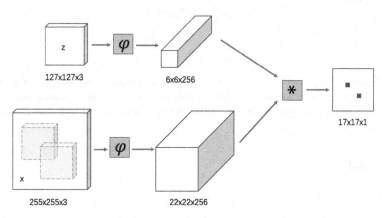

**Fig. 1.** The Siamese-FC framework is constituted by a shared feature extractor and correlation matching. Network input is a pair of images, including target image Z, search image X. Network output is the score map which judges the similarity of input image pairs by correlation operation. In this example, the red and green pixels in the score map contain the similarity of the corresponding sub-windows in the search image X. Best viewed in color. (Color figure online)

A Siamese network consists of two branches which can learn the similarity between the template frame and the instance frame (Fig. 1). The visual tracking algorithm based on Siamese network has been widely adopted because of its outstanding performance and the fast speed. StructSiam [14] network adds three local structures in each branch, including local pattern detection, context modeling and integration module to obtain local features. RASNet [17] employs three attentions mechanism to weigh the space and channel of the SiamFC features for improving the discrimination of the features. EDCF [1] adopts self-coding network to make the network focus more on the detailed description of the target, with combining context-aware correlation filtering to suppress the surrounding interference. SA-Siam [29] utilizes two sets of independent SiamFC to train semantics branch and appearance branch respectively.

The loss function used in above trackers are relatively not accurate enough. The Alex backbone used in SiameseFC will gradually enlarge the input image as the number of network layers deepens. The problem is that the target in the original image will shift in the position of the last layer of the feature, so we It is considered that it is necessary and efficient to compare dice loss between each layer of feature maps, which can reduce the interference of target scale and size caused by deep networks.

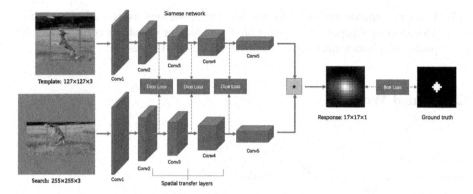

**Fig. 2.** Training framework of the dice loss in Siamese network. We also give the original binary cross-entropy with logistic loss for comparison. Given the same feature extraction in baselines [6, 11], we can apply the dice loss to the feature maps extracted from conv2, conv3 and conv4 layers.

## 3  Method

### 3.1  Overview

The procedure of the proposed method is depicted in Fig. 2. The template image contains the target tightly, which is denoted as Z. Search image includes the target with a larger range background, which is denoted as X. Firstly, features $F_z$ and $F_x$ are extracted from template image Z and search image X by using shared feature extractor $\varphi$. Secondly, a spatial transform is applied to the feature $F_x$ to learn the scale variation of the object in the search image compared with that in the target image. The new feature $FT_x$ which considers the scale variation is obtained by adding the variation to the original feature $F_x$. Then $F_z$ is used as the convolution kernel to perform the convolution operation on $FT_x$, the score map which contains the similarity of template and the each position on search image is obtained. Next, the hard mining is performed based on the score map in a batch, which can mine the both hard positive and negative score maps. Finally, the hard score map is fed into the binary cross entropy (BCE) loss function to calculate the classification loss.

### 3.2  Siamese Architecture

In order to learn the representation of the images, our framework employs the shared backbone $\varphi$ to extract the features from the target Z and search image X, obtaining $F_z \in R^{6 \times 6 \times 256}$, $F_x \in R^{22 \times 22 \times 256}$, respectively.

The advantage of a fully-convolutional network is that, instead of a candidate image of the same size, we can provide as input to the network a much larger search image and it will compute the similarity at all translated sub-windows on a dense grid in a single evaluation. To achieve this, we use a convolutional embedding function and combine the resulting feature maps using a cross-correlation layer.

## 3.3    Cross-Entropy

Cross-Entropy. Also known as log-loss, cross-entropy is the most widely used loss function for classification CNN. When applied to a tracking task, cross-entropy measures the divergence of the predicted probability from the ground truth label for each pair separately and then averages the value over all patches in the mini-batch:

$$L_{CE} = -\frac{1}{N} \sum_{c=1}^{N} \sum_{l=1}^{L} g_l^c log(p_l^c) \tag{1}$$

This loss function tends to under-estimate the prediction probabilities for classes that are under-represented in the mini-batch which is inevitable in our training data.

Weighted Cross-Entropy. The tendency to under-estimate can be mitigated by assigning higher weights to loss contributions from patches with underrepresented class labels:

$$L_{WCE} = \frac{1}{N} \sum_{c=1}^{N} \frac{1}{w_c} \sum_{l=1}^{L} g_l^c log(p_l^c) \tag{2}$$

where $w_c$ is a weight assigned to pixel c computed as a prior probability of ground truth label $r_l^c$ in the given mini-batch.

## 3.4    Dice Loss

The network predictions, which consist of two volumes having the same resolution as the original input data, are processed through a soft-max layer which outputs the probability of each voxel to belong to foreground and to background. In medical volumes such as the ones we are processing in this work, it is not uncommon that the anatomy of interest occupies only a very small region of the scan. This often causes the learning process to get trapped in local minima of the loss function yielding a network whose predictions are strongly biased towards background. As a result the foreground region is often missing or only partially detected. Several previous approaches resorted to loss functions based on sample re-weighting where foreground regions are given more importance than background ones during learning. In this work we propose a novel objective function based on dice coefficient, which is a quantity ranging between 0 and 1 which we aim to maximise. The dice coefficient $L_D$ between two binary volumes can be written as

$$L_D = \frac{2 \sum_i^N p_i g_i}{\sum_i^N p_i^2 + \sum_i^N g_i^2} \tag{3}$$

where the sums run over the N voxels, of the predicted binary segmentation volume pi $\in$ P and the ground truth binary volume gi $\in$ G. This formulation of Dice can be differentiated yielding the gradient

$$\frac{\partial D}{\partial p_j} = 2 \left[ \frac{g_j \left( \sum_i^N p_i^2 + \sum_i^N g_i^2 \right) - 2p_j \left( \sum_i^N p_i g_i \right)}{\left( \sum_i^N p_i^2 + \sum_i^N g_i^2 \right)^2} \right] \tag{4}$$

computed with respect to the j-th patch of the prediction. Using this formulation we do not need to assign weights to samples of different classes to establish the right balance between foreground and background patches, and we obtain results that we experimentally observed are much better than the ones computed through the same network trained optimising a multinomial logistic loss with sample re-weighting.

Further more, Inspired by the Dice coefficient [11] often used to evaluate binary segmentation accuracy, the differentiable soft Dice loss was introduced by Milletari et al. [2] to tackle the class imbalance issue without the need for explicit weighting. One possible formulation is

$$L_{SD} = \frac{1}{I} \sum_{i=1}^{I} 1 - \frac{2 \sum_{l=1}^{L} \sum_{c=1}^{C} p_l^c g_l^c}{\sum_{l=1}^{L} \sum_{c=1}^{C} p_l^c + g_l^c} \tag{5}$$

This allows easy generalization to measuring all patches similarity where $L$ larger than 2 by treating each image as 17 * 17 pairs of target image and sub-window from search image.

### 3.5   Loss Function

The hard score maps H which include the positive and negative score maps are fed into binary cross entropy (BCE) loss to get the hard mining loss, as shown in the Eq. 6.

$$L_{WBC} = -\frac{1}{Q} \sum_{i=1}^{Q} [T_i \times \log(H_i) + (1 - T_i) \times \log(1 - H_i))] \tag{6}$$

Where Q is the number of the hard score map and $T$ is ground truth. We crop the image centering on the target which is descried in Sect. 4.1. The elements of the score map are considered to belong to a positive example if they are within radius $R$ of the centre (accounting for the stride $k$ of the network). So, for the hard positive score maps, $T$ equals to 1 if $k||u - c|| < R$, otherwise it equals to 0. For the hard negative score maps, the ground truth equals to 0 in each position.

Our final form of loss is the sum of binary cross-entropy loss and batch dice loss, the expression is as follows:

$$L = L_{WBC} + L_{SD} \tag{7}$$

# 4 Experiment

## 4.1 Implementation Details

We crop the images from ILSVRC [3] and Youtube-BB [13] centering on the target and resize them to 255 * 255. Compared to ILSVRC which consists of about 4000 videos annotated frame-by-frame, Youtube-BB consists of more than 100,000 videos annotated once in every 30 frames. Then an example image is selected randomly and is copped in the center to 127 * 127. The instance image is selected with equal probability within 100 frames of the example image.

In this paper, we try Alexnet as backbone, which is pretrained from ImageNet [4]. When training with the Alexnet is performed 30 epochs, SGD is adopted with the learning rate of 1e-2 and batchsize of 8. Alexnet processes the crop layer to remove the influence of the padding. During training phase, we freeze the weights of the first 7 * 7 convolution layer, and gradually fine-tune other layers from back to front. We unfreeze the weights of the layers after every five training epochs. There are 50 epochs in total.

## 4.2 Results on OTB2013

OTB-2013 contains 51 real-world sequences, with 11 interference attributes, and two metrics, i.e. bounding box overlap ratio and center location error. By setting a success threshold for each metric, we can get the precision and success plots, which quantitatively measure the performance of different trackers on this dataset. The precision plot shows the percentage of frames that the tracking results are within 20 pixels from the target. The success plot shows the ratios of successful frames when the threshold varies from 0 to 1, where a successful frame means its overlap is larger than given threshold.

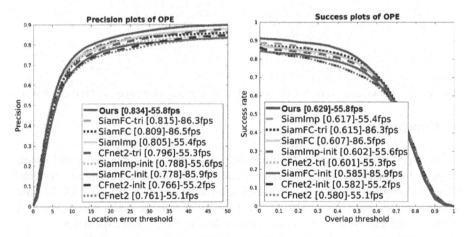

**Fig. 3.** Self-comparisons with variants of baseline trackers. The plots show precision and overlap success rate with AUC on OTB-2013 [11] in terms of OPE.

We compare our method with state of the art trackers, including the baseline SiamFC [11], CFnet [2], GOTURN [2], DSST, SINT, Staple. The experimental results in Fig. 3 show that our method achieves the best performance in success plot, with the improvements of 1.9% than the second best tracker SiamFC-tri. Moreover, compared to the baseline SiamFC, our method improves 2.2% and 2.5% in the success and precision plots.

### 4.3    Results on OTB2015

OTB2015 contains 100 sequences that are collected from commonly used tracking sequences. The evaluation is also based on two metrics: precision and success plot. The area under curve (AUC) of success plot is used to rank tracking algorithm.

We evaluate the proposed algorithms with numerous fast and state-of-the-art trackers including, MEEM, DLSSVM, CFNet, Siam-tri [4], Staple, SAMF, PTAV, and the baseline tracker SiamFC. It is shown in the Fig. 4 that our method is the rank1 both in the success and precision plots. And the proposed method outperforms the baseline 1% in success plot.

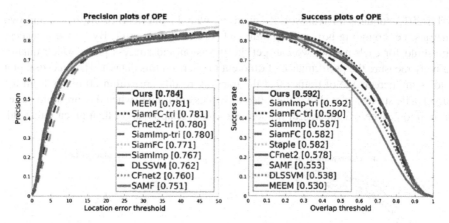

**Fig. 4.** Overlap success plots of OPE with AUC for tracking challenges on OTB-2015 [11].

## 5    Conclusion

In this paper, we have proposed a novel dice loss to achieve a more powerful feature extractor for object tracking by applying it into Siamese network. In contrast to original binary cross-entropy logistic loss, our dice loss can further mine potential relationships among samples and utilize more elements for better training performance. We have shown the effectiveness of the proposed dice loss in theory and experiments. In theoretical analysis, we found that when the network outputs wrong similarity scores, it gives more absolute gradients for feedback in back-propagation. We added this dice loss into three baseline trackers based on Siamese network for experiments. The results on popular tracking benchmarks show that our dice loss can improve the performance without reducing speed for these baselines.

# References

1. Dong, X., Shen, J.: Triplet loss in siamese network for object tracking. In: Ferrari, V., Hebert, M., Sminchisescu, C., Weiss, Y. (eds.) ECCV 2018. LNCS, vol. 11217, pp. 472–488. Springer, Cham (2018). https://doi.org/10.1007/978-3-030-01261-8_28
2. Zhu, Z., Wang, Q., Li, B., Wu, W., Yan, J., Hu, W.: Distractor-aware siamese networks for visual object tracking. In: Ferrari, V., Hebert, M., Sminchisescu, C., Weiss, Y. (eds.) ECCV 2018. LNCS, vol. 11213, pp. 103–119. Springer, Cham (2018). https://doi.org/10.1007/978-3-030-01240-3_7
3. Wang, Q., Teng, Z., Xing, J., Gao, J., Hu, W., Maybank, S.: Learning attentions: residual attentional siamese network for high performance online visual tracking. In: Proceedings of the IEEE Conference on Computer Vision and Pattern Recognition, p. 4854 (2018)
4. Wang, Q., Zhang, M., Xing, J., Gao, J., Hu, W., Maybank, S.: Do not lose the details: reinforced representation learning for high performance visual tracking. In: 27th International Joint Conference on Artificial Intelligence (2018)
5. Li, B., Yan, J., Wu, W., Zhu, Z., Hu, X.: High performance visual tracking with siamese region proposal network. In: Proceedings of the IEEE Conference on Computer Vision and Pattern Recognition, p. 8971 (2018)
6. Zhang, Y., Wang, L., Qi, J., Wang, D., Feng, M., Lu, H.: Structured siamese network for real-time visual tracking. In: Ferrari, V., Hebert, M., Sminchisescu, C., Weiss, Y. (eds.) ECCV 2018. LNCS, vol. 11213, pp. 355–370. Springer, Cham (2018). https://doi.org/10.1007/978-3-030-01240-3_22
7. He, A., Luo, C., Tian, X., Zeng, W.: A two fold siamese network for real-time object tracking. In: Proceedings of the IEEE Conference on Computer Vision and Pattern Recognition (2018)
8. Bertinetto, L., Valmadre, J., Henriques, J.F., Vedaldi, A., Torr, P.H.S.: Fully-convolutional siamese networks for object tracking. In: Hua, G., Jégou, H. (eds.) ECCV 2016. LNCS, vol. 9914, pp. 850–865. Springer, Cham (2016). https://doi.org/10.1007/978-3-319-48881-3_56
9. Held, D., Thrun, S., Savarese, S.: Learning to track at 100 FPS with deep regression networks. In: Leibe, B., Matas, J., Sebe, N., Welling, M. (eds.) ECCV 2016. LNCS, vol. 9905, pp. 749–765. Springer, Cham (2016). https://doi.org/10.1007/978-3-319-46448-0_45
10. Fan, H., Ling, H.: Siamese cascaded region proposal networks for real-time visual tracking. arXiv preprint arXiv:1812.06148 (2018)
11. Yun, S., Choi, J., Yoo, Y., Yun, K., Young Choi, J.: Action-decision networks for visual tracking with deep reinforcement learning. In: Proceedings of the IEEE Conference on Computer Vision and Pattern Recognition, p. 2711 (2017)
12. Lukezic, A., Vojir, T., Cehovin Zajc, L., Matas, J., Kristan, M.: Discriminative correlation filter with channel and spatial reliability. In: Proceedings of the IEEE Conference on Computer Vision and Pattern Recognition, p. 6309 (2017)
13. Real, E., Shlens, J., Mazzocchi, S., Pan, X., Vanhoucke, V.: YouTube-BoundingBoxes: a large high-precision human-annotated data set for object detection in video. In: Proceedings of the IEEE Conference on Computer Vision and Pattern Recognition, p. 5296 (2017)
14. Song, Y., et al.: VITAL: visual tracking via adversarial learning. In: Proceedings of the IEEE Conference on Computer Vision and Pattern Recognition, p. 8990 (2018)
15. Ren, S., He, K., Girshick, R., Sun, J.: Faster R-CNN: towards real-time object detection with region proposal networks. In: Advances in Neural Information Processing Systems, p. 91 (2015)
16. Gordon, D., Farhadi, A., Fox, D.: Real-time recurrent regression networks for visual tracking of generic objects. IEEE Rob. Autom. Lett. 3(2), 788 (2018)

17. Danelljan, M., Häger, G., Khan, F., Felsberg, M.: Accurate scale estimation for robust visual tracking. In: British Machine Vision Conference, 1 September 2014, Nottingham. BMVA Press (2014)
18. Szegedy, C., et al.: Going deeper with convolutions. In: Proceedings of the IEEE Conference on Computer Vision and Pattern Recognition, p. 1 (2015)
19. Tao, R., Gavves, E., Smeulders, A.W.: Siamese instance search for tracking. In: Proceedings of the IEEE Conference on Computer Vision and Pattern Recognition, p. 1420 (2016)
20. Valmadre, J., Bertinetto, L., Henriques, J., Vedaldi, A., Torr, P.H.: End-to-end representation learning for correlation filter based tracking. In: Proceedings of the IEEE Conference on Computer Vision and Pattern Recognition, p. 2805 (2017)
21. Bertinetto, L., Valmadre, J., Golodetz, S., Miksik, O., Torr, P.H.: Staple: complementary learners for real-time tracking. In: Proceedings of the IEEE Conference on Computer Vision and Pattern Recognition, p. 1401 (2016)
22. Wang, Q., Zhang, L., Bertinetto, L., Hu, W., Torr, P.H.: Fast online object tracking and segmentation: a unifying approach. arXiv preprint arXiv:1812.05050 (2018)
23. Russakovsky, O., et al.: Imagenet large scale visual recognition challenge. Int. J. Comput. Vis. 115(3), 211 (2015)
24. Wu, Y., Lim, J., Yang, M.-H.: Online object tracking: a benchmark. In: Proceedings of the IEEE Conference on Computer Vision and Pattern Recognition, p. 2411 (2013)
25. Wu, Y., Lim, J., Yang, M.-H.: Object tracking benchmark. IEEE Trans. Pattern Anal. Mach. Intell. 37(9), 1834 (2015)
26. He, K., Zhang, X., Ren, S., Sun, J.: Deep residual learning for image recognition. In: Proceedings of the IEEE Conference on Computer Vision and Pattern Recognition, p. 770 (2016)
27. Krizhevsky, A., Sutskever, I., Hinton, G.E.: ImageNet classification with deep convolutional neural networks. In: Advances in Neural Information Processing Systems, p. 1097 (2012)
28. Zhipeng, Z., Houwen, P., Qiang, W.: Deeper and wider siamese networks for real-time visual tracking. arXiv preprint arXiv:1901.01660 (2019)
29. Guo, Q., Feng, W., Zhou, C., Huang, R., Wan, L., Wang, S.: Learning dynamic siamese network for visual object tracking. In: Proceedings of the IEEE International Conference on Computer Vision, p. 1763 (2017)

# Fuzzy PID Controller for Accurate Power Sharing in DC Microgrid

Duy-Long Nguyen[1] and Hong-Hee Lee[2(✉)]

[1] Graduate School of Electrical Engineering, University of Ulsan,
Ulsan, South Korea
longnd@tlu.edu.vn
[2] School of Electrical Engineering, University of Ulsan, Ulsan, South Korea
hhlee@mail.ulsan.ac.kr

**Abstract.** In this paper, an intelligent control scheme based on Fuzzy PID controller is proposed for accurate power sharing in DC Microgrid. The proposed Fuzzy PID controller is designed with the aid of a closed loop control based on per unit power of each distributed generator (DG), and accurate power sharing is successfully realized in proportional to each DG's power rating regardless of the line resistance difference or the load change. Thanks to Fuzzy PID controller, the dynamic response becomes faster and the stability of the microgrid system is improved in comparison to conventional PID controller. The superiority of the proposed method is analyzed and verified by simulation in Matlab and Simulink.

**Keywords:** DC microgrid · Power sharing · Droop control · Fuzzy logic control

## 1 Introduction

In order to deal with the problems related to environmental pollution and exhausted fossil fuel supplies, renewable energy sources (RESs) such as solar cell, wind turbines, and hydrogen power have been widely utilized. To exploit these energy source with high efficiency and good performance, power electronic converters are commonly adopted with RESs and energy storage systems (ESSs) to form distributed generators (DGs) [1], and microgrid (MG) concept has been introduced as a promising solution to integrate DGs and loads, to supply the load power as well as support main grid effectively [2, 3]. The MG is classified into DC MG and AC MG. Since there is no reactive power, transformer inrush current, or harmonic issues in DC MG, the transmission and distribution efficiencies in DC MG are higher than those in AC MG [4, 5]. Moreover, the control system in DC MG is simpler in comparison with AC MG because there is no harmonic problems as well as reactive power sharing in DC MG. Besides, DC RESs such as solar power directly provide DC power from the source to DC load, which can reduce number of ac-dc and dc-ac conversion stages. Therefore, in these days, the researches on DC MG are increasing with many projects [6, 7].

Figure 1 shows the basic configuration of a DC microgrid where all units including DGs, battery, supercapacitor, and loads are connected to a common dc bus. In order to

© Springer Nature Switzerland AG 2019
D.-S. Huang et al. (Eds.): ICIC 2019, LNCS 11644, pp. 71–80, 2019.
https://doi.org/10.1007/978-3-030-26969-2_7

achieve power sharing between each DG, droop control method has been generally used without communication network [8, 9]. In this method, the voltage of each DG is regulated based on droop gain. The droop gain is calculated from the rated capacity of the source, and determines the power sharing between DGs [10]. However, because of the line impedance, this method is hard to achieve accurate power sharing.

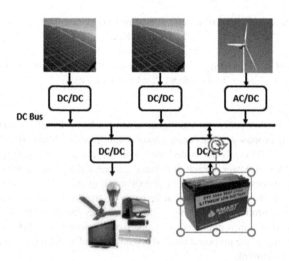

**Fig. 1.** Typical configuration of DC microgrid

To overcome these problems, a variety of researches based on a secondary controller have been conducted [11]. From the communication perspective, there are two types of secondary control scheme: centralized control and distributed control [1, 5]. In centralized approach, by a digital communication network, a central controller requires all local information such as output voltage, output current of DGs, and transmits command values as well as control parameters to the local controller in order to achieve advanced power management such as supervision control, global optimization, prioritizations for charging or discharging of batteries with a different SoC, power balancing, load shedding, and operation mode transitions [1].

Although the centralized control provides a good technical solution to implement the advanced control functionalities, a single point of failure invokes serious problem to the control scheme; if the central controller or key communication link fails, the connection to local controller will be disabled and control objectives cannot be attained [1, 12, 13].

This problem can be solved by means of the distributed control scheme [1, 5]. In this approach, local controllers are connected all together through communication network without the central controller. Even though some communication link failure occurs, the system can be maintained with full functionality [1]. Therefore, the distributed control is useful and robust to control the load power sharing due to its advantages of low cost communication systems and the lower risk of single point failures [14].

In this paper, based on the distributed control scheme, we propose an enhanced fuzzy proportional-integral-derivative (Fuzzy PID) controller in order to achieve accurate power sharing between DGs in DC MG. The proposed Fuzzy PID controller is combined with the advantage of the nonlinear control fuzzy logic control and the zero steady state error of PID controller. Although the Fuzzy PID controller has already introduced in many literatures [15, 16], the application of Fuzzy PID for accurate power sharing in DC MC has not been presented up to now. The proposed regulator provides outstanding performance and guarantees system stability in spite of the load change. The effectiveness of the proposed FPID controller is verified by simulation in Matlab and Simulink.

## 2   Droop and Distributed Controls for Power Sharing in DC MG

### 2.1   Droop Control

Figure 2 shows droop control scheme in DC MG, where only two DGs are considered for simple analysis. From Fig. 2, the equivalent circuit can be obtained simply without considering line resistance as shown in Fig. 3, where $R_{d1}$ and $R_{d2}$ are droop coefficient of DG1 and DG 2, respectively.

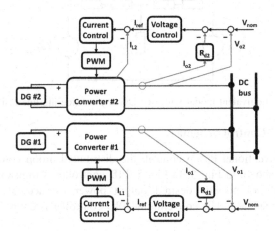

**Fig. 2.** Droop control scheme for DC MG.

From Fig. 3,

$$V_{dc1} - I_1 R_{d1} = V_{dc2} - I_2 R_{d2}. \tag{1}$$

Without loss of generality, assume that two DGs have the same power rating $P_1 = P_2$ and the same output voltage $V_{dc1} = V_{dc2}$. From (1), $I_1 = I_2$, if $R_{d1} = R_{d2}$. Then, the load power is shared equivalently between two DGs. In droop control scheme, the condition $R_{d1} = R_{d2}$ can be achieved easily by adjusting the droop gain or virtual impedances in droop control scheme in Fig. 2.

When the line resistance is considered, the equivalent circuit in Fig. 3 is modified by inserting the line resistance as shown in Fig. 4. Then, (2) is obtained from Fig. 4:

$$V_{dc1} - I_1 \left( R_{d1} + R_{l1} \right) = V_{dc2} - I_2 \left( R_{d2} + R_{l2} \right) \tag{2}$$

In practical applications, the line resistances are different, i.e., $R_{l1} \neq R_{l2}$. Therefore, load currents are not equal ($I_1 \neq I_2$), and the load power is not equally shared between two DGs.

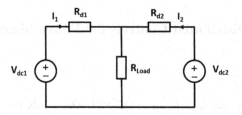

**Fig. 3.** Simplified model of two DGs

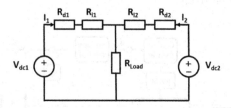

**Fig. 4.** Equivalent model with two DGs by considering line resistance

## 2.2   Distributed Control Scheme

Based on the conventional droop control, the distributed droop control for accurate power sharing is shown in Fig. 5. In Fig. 5, PID controller is used to compensate the voltage magnitude to provide desired output power for accurate power sharing regardless line impedance. The conventional PID controller is defined as following:

$$\Delta V = (k_p + \frac{k_i}{s} + k_d s) \, (p_{pu\_avg} - p_{pu[i]}), \tag{3}$$

where, $p_{pu\_avg}$, $p_{pu[i]}$ are average per unit power and per unit power of $i^{th}$ DG, respectively.

Power per unit is calculated from voltage and current of DG:

$$p_{pu[i]} = \frac{V_{o[i]} \, I_{o[i]}}{P_{rated[i]}}, \tag{4}$$

where $V_{o[i]}$, $I_{o[i]}$, $P_{rated[i]}$ are output voltage, output current, and rated power of $i^{th}$ DG, respectively. The measurement signals of voltage and current are obtained through low bandwidth communication link.

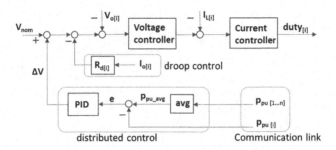

**Fig. 5.** Distributed control scheme.

In the conventional PID controller, the gains of PID controller for power sharing can be tuned by following Ziegler Nichols procedure. However, the performance of the system highly depends on the load condition. This is simply because the PID gains are tuned under the given load condition such as the maximum load. When the load changes, the performance also changes. Meanwhile, fuzzy PID controller can overcome this problem, and show good system performance over the conventional PID controller. As we know, DC MG is basically nonlinear system with a lot of uncertainty, unpredictability due to noise and disturbance. Therefore, Fuzzy PID controller with nonlinear characteristic is suitable and useful to enhance DG performance to achieve accurate power sharing in DC MG.

## 3 Proposed Fuzzy PID Controller for Accurate Power Sharing in DC Microgrid

In a period of thirty years, Fuzzy Logic controllers have been widely used for industrial application. In comparison with the conventional PID controllers, the Fuzzy PID controllers have higher control gains when the system is away from equilibrium state, which enables better performance [15]. Up to now, many different types of Fuzzy PID controllers have been proposed [16–18], and they can be classified into two major categories according to their constructions [15].

One category of fuzzy PID controllers is composed of the traditional PID controller with a set of fuzzy rules and fuzzy logic mechanism. In this type, the PID gain is tuned according to the knowledge base and fuzzy inference, and the PID controller generates the control signal. Due to the nonlinearity of the fuzzy knowledge, it is hard to analyze the stability and performance of this structure.

In second category of fuzzy PID controllers, the control signal is directly derived from knowledge base, a set of control rules and the fuzzy inference with a typical fuzzy logic. This structure is analogous to that of the PID controller, so it is easy to analyze the performance. Moreover, it is easy to borrow the well known tuning method of the conventional PID controller to design the Fuzzy PID controller.

In this paper, a typical Fuzzy PID controller of second category is selected as shown in Fig. 6.

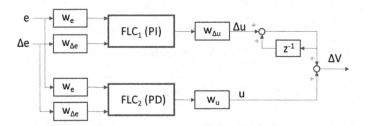

**Fig. 6.** The overall structure of Fuzzy PID controller.

In Fig. 6, e is the error between average power per unit of all DGs and power per unit of $i^{th}$ DG, $\Delta e$ is the change of error:

$$
\begin{aligned}
e(k) &= p_{pu\_avg}(k) - p_{pu[i]}(k), \\
\Delta e(k) &= e(k) - e(k-1)
\end{aligned}
\tag{5}
$$

and $w_e$, $w_{\Delta e}$ are weighting factors, while $w_{\Delta u}$ and $w_u$ are gains of output.

The membership function is given simply to easily analysis, and in addition, the nonlinearity of the simplest fuzzy controller is the strongest [19]. It is shown in Fig. 7.

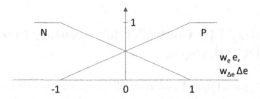

**Fig. 7.** Membership functions of $w_e$ e & $w_{\Delta e}$ $\Delta e$

The fuzzy outputs are singletons defined as P = 1, Z = 0, N = −1. The control outputs of each fuzzy logic controller are calculated as following [15, 17]:

$$
\begin{aligned}
\Delta u &= \frac{w_{\Delta u}}{4 - 2\max(w_e|e|, w_{\Delta e}|\Delta e|)} (w_e e + w_{\Delta e}\Delta e) \\
\Delta u &= \frac{w_{\Delta u}}{4 - 2\alpha} (w_e e + w_{\Delta e}\Delta e)
\end{aligned}
\tag{6}
$$

$$
\begin{aligned}
u &= \frac{w_u}{4 - 2\max(w_e|e|, w_{\Delta e}|\Delta e|)} (w_e e + w_{\Delta e}\Delta e) \\
u &= \frac{w_u}{4 - 2\alpha} (w_e e + w_{\Delta e}\Delta e)
\end{aligned}
\tag{7}
$$

where $\alpha = \max(w_e |e|, w_{\Delta e} |\Delta e|)$.

The total fuzzy control output becomes

$$\Delta V = \sum_0^k \Delta u + u = \sum_0^k \frac{w_{\Delta u} w_{\Delta e}}{4 - 2\alpha} (\Delta e + \frac{\Delta t}{\frac{w_{\Delta e}}{w_e} \Delta t} e) + \frac{w_u w_e}{4 - 2\alpha} (e + \frac{w_{\Delta e} \Delta t}{w_e} \frac{\Delta e}{\Delta t}). \tag{8}$$

If we define the parameters in (8) as following:

$$K_c^{(F)} = \frac{w_{\Delta u} w_{\Delta e}}{4 - 2\alpha}; \ T_i^{(F)} = \frac{w_{\Delta e}}{w_e} \Delta t; \ K_c^{(F)} \frac{T_d^{(F)}}{T_i^{(F)}} = \frac{w_u w_e}{4 - 2\alpha}, \tag{9}$$

the proposed Fuzzy PID control output is finally given in (10) from (8):

$$\Delta V = \sum_0^k K_c^{(F)} (\Delta e + \frac{\Delta t}{T_i^{(F)}} e) + K_c^{(F)} \frac{T_d^{(F)}}{T_i^{(F)}} (e + T_i^{(F)} \frac{\Delta e}{\Delta t}) \tag{10}$$

In order to verify the proposed Fuzzy PID controller relevance to the conventional PID controller, the Fuzzy PID controller in (10) is investigated by analogy with the conventional PID controller. By substituting $\frac{\Delta e}{\Delta t}$ with $\frac{de}{dt}$ into (10), the total control output of fuzzy controller becomes

$$\Delta V \approx \int_0^{k.\Delta t} K_c^{(F)} \dot{e} \, dt + \int_0^{k.\Delta t} \frac{K_c^{(F)}}{T_i^{(F)}} e \, dt + \frac{K_c^{(F)} T_d^{(F)}}{T_i^{(F)}} (e + T_i^{(F)} \dot{e}). \tag{11}$$

(11) has similar shape as that of the conventional PID controller in (12):

$$u_{PID} = \int_0^t K_c \dot{e} \, dt + \int_0^t \frac{K_c}{T_i} e \, dt + \frac{K_c T_d}{T_i} (e + T_i \dot{e}) \tag{12}$$

Therefore, we can say the Fuzzy PID controller in (10) is designed reasonably, and its design process is summarized as following: firstly, tune the parameter of PID controller: $K_c$, $T_i$, $T_d$ by means of well known method such as the Ziegler - Nichol rules, secondly, weighting factors and gains of control output: $w_e$, $w_{\Delta e}$, $w_u$, $w_{\Delta u}$ are calculated from (9) based on the selected control parameters $K_c$, $T_i$ and $T_d$. The overall block diagram for distributed control with Fuzzy PID controller is shown in Fig. 8 respect to Fig. 5.

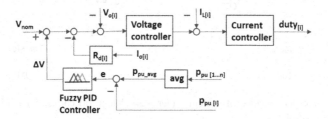

**Fig. 8.** Overall block diagram with Fuzzy PID controller.

## 4 Simulation Results

To verify the effectiveness of the proposed Fuzzy PID controller, DC MG with 3 DGs is simulated by means of Matlab and Simulink. Each DG consists of a buck converter with the parameters given as: $V_{in} = 200$ V; $V_{nom} = 100$ V; $L_1 = L_2 = L_3 = 0.5$ mH; $C_{in} = C_{out} = 2200$ uF; $R_{line1} = 0.1$ $\Omega$; $R_{line2} = 0.2$ $\Omega$; $R_{line3} = 0.15$ $\Omega$; $P_{rated1} = P_{rated2} = P_{rated3} = 1$ kW; $f_{sw} = 20$ kHz; $T_{sample} = 50$ $\mu$s; $K_c = 9.965$; $T_i = 0.285$; $T_d = 0.0001$; $w_e = 0.2$; $w_{\Delta e} = 1140$; $w_u = 0.07$; $w_{\Delta u} = 0.035$.

Figure 9 show the dynamic performance of the system with the conventional PID and Fuzzy PID controllers. From 0 to 1.5 s, because the power sharing controller is not active, the per unit power of each DG is different due to the different line resistance. At 1.5 s, accurate power sharing scheme becomes active. Comparing to the conventional PID controller, the Fuzzy PID has better performance with faster response and shorter settling time.

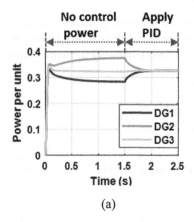
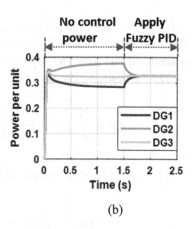

(a)                                                                                    (b)

**Fig. 9.** Dynamic response of system with (a) conventional PID controller (b) Fuzzy PID controller.

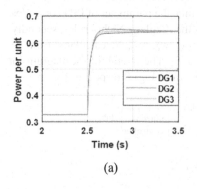
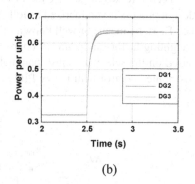

(a)                                                                                    (b)

**Fig. 10.** Dynamic performance of system under load change (a) conventional PID controller (b) Fuzzy PID controller.

Figure 10 shows the dynamic response of system when the load changes from 1 kW to 2 kW at 2.5 s. In comparision with the conventional PID controller, the Fuzzy PID controller has smaller overshoot, and faster response. As a result, it is clear that the proposed Fuzzy PID shows very good dynamic performance compared with the conventional PID controller.

## 5 Conclusion

This paper has presented an intelligent control approach based on the fuzzy logic controller to achieve accurate power sharing in DC microgrids. The nonlinear properties of Fuzzy PID controller with variable control gains brings enhanced power sharing performance in comparison with the traditional PID. The validity of the proposed Fuzzy PID controller is confirmed through simulation in Matlab & Simulink, and simulation results have shown the effectiveness of the method.

**Acknowledgments.** This work was supported in part by the National Research Foundation of Korea Grant funded by the Korean Government under Grant NRF-2018R1D1A1A09081779 and in part by the Korea Institute of Energy Technology Evaluation and Planning and the Ministry of Trade, Industry and Energy under Grant 20194030202310.

## References

1. Dragicevic, T., Xiaonan, L., Vasquez, J.C., Guerrero, J.M.: DC Microgrids—Part I: a review of control strategies and stabilization techniques. IEEE Trans. Power Electron. **31**(7), 4876–4891 (2016)
2. Samanta, B., Al-Balushi, K.: Artificial neural network based fault diagnostics of rolling element bearings using time-domain features. Mech. Syst. Signal Process. **17**(2), 317–328 (2003)
3. Ipakehi, A., Albuyeh, F.: Grid of the future. IEEE Power Energ. Mag. **7**(2), 52–62 (2009)
4. Han, R., Wang, H., Jin, Z., Meng, L., Guerrero, J.M.: Compromised controller design for current sharing and voltage regulation in DC microgrid. IEEE. Tran. Power. Electron. https://doi.org/10.1109/TPEL.2018.2878084
5. Dam, D., Lee, H.: A power distributed control method for proportional load power sharing and bus voltage restoration in a DC microgrid. IEEE. Tran. Ind. Appl. **54**(4), 3616–3625 (2018)
6. Guerrero, J.M., Vasquez, J.C., Matas, J., De Vicuna, L.G., Castilla, M.: Hierarchical control of droop-controlled ac and dc microgrids - a general approach toward standardization. IEEE Trans. Ind. Electron. **58**(1), 158–172 (2011)
7. Vandoorn, T.L., Meersman, B., Degroote, L., Renders, B., Vandevelde, L.: A control strategy for islanded microgrids with DC-link voltage control. IEEE Trans. Power Deliv. **26**(2), 703–713 (2011)
8. Chen, D., Xu, L.: Autonomous dc voltage control of a dc microgrid with multiple slack terminals. IEEE Trans. Power Syst. **27**(4), 1897–1905 (2012)
9. Iravani, R., Khorsandi, A., Ashourloo, M., Mokhtari, H.: Automatic droop control for a low voltage dc microgrid. IET Gener. Transm. Distrib. **10**(1), 41–47 (2016)

10. Ito, Y., Zhongquing, Y., Akagi, H.: DC microgrid based distribution power generation system. In: The 4th International Power Electronics and Motion Control Conference, IPEMC (2004)
11. Dragicevic, T., Vasquez, J.C., Guerrero, J.M., Skrlec, D.: Advanced LVDC electrical power architectures and microgrids: a step toward a new generation of power distribution networks. IEEE Electrif. Mag. 2(1), 54–65 (2014)
12. Lu, X., Guerrero, J., Sun, K., et al.: An improved droop control method for dc microgrids based on low bandwidth communication with dc bus voltage restoration and enhanced current sharing accuracy. IEEE Trans. Power Electron. 29(4), 1800–1812 (2014)
13. Wang, P., Lu, X., Yang, X., Wang, W., Xu, D.: An improved distributed secondary control method for DC microgrids with enhanced dynamic current sharing performance. IEEE Trans. Power Electron. 31(9), 6658–6673 (2016)
14. Anand, S., Fernandes, B.G., Guerrero, J.M.: Distributed control to ensure proportional load sharing and improve voltage regulation in low-voltage dc microgrids. IEEE Trans. Power Electron. 28(4), 1900–1913 (2013)
15. Jian-Xin, X., Hang, C.-C., Liu, C.: Parallel structure and tuning of a fuzzy PID controller. Automica 36, 673–684 (2000)
16. Zhao, Z.Y., Tomizuka, M., Isaka, S.: Fuzzy gain scheduling of PID controllers. IEEE Trans. SMC 23(5), 1392–1398 (1993)
17. Ying, H.: The simplest fuzzy controllers using different inference methods are different nonlinear proportional-integral controllers with variable gains. Automica 29(6), 1579–1589 (1993)
18. Malki, H.A., Li, H.D., Chen, G.R.: New design and stability analysis of fuzzy proportional-derivative control system. IEEE Trans. Fuzzy Syst. 2(4), 245–254 (1994)
19. Buckley, J.J., Ying, H.: Fuzzy controller theory: limit theorem for linear fuzzy control rules. Automica 25(3), 469–472 (1989)

# Precipitation Modeling and Prediction Based on Fuzzy-Control Multi-cellular Gene Expression Programming and Wavelet Transform

YuZhong Peng[1,2], ChuYan Deng[1], HongYa Li[3], DaoQing Gong[1], Xiao Qin[1(✉)], and Li Cai[2]

[1] School of Computer and Information Engineering,
Nanning Normal University, Nanning 530001, China
qhyihui@163.com
[2] School of Computer Science, Fudan University, Shanghai 200433, China
[3] Department of Science, Shangqiu University Applied Science
and Technology College, Kaifeng 475000, China

**Abstract.** Accurate and timely precipitation prediction is very important to development and management of regional water resources, flood disaster prevention/control and people's daily activities and production plans. However, the prediction accuracy is greatly affected by nonlinear and non-stationary features of precipitation data and noise. Many researches show that Multi-cellular Gene Expression Programming (MC-GEP) algorithm has strong function mining ability. This paper designed a Fuzzy-control Multi-cellular Gene Expression Programming algorithm (FMC-GEP) based on Multi-cellular Gene Expression Programming, which used fuzzy control theory to dynamically adjust the probability of genetic manipulation. Then we coupled Fuzzy-control Multi-cellular Gene Expression Programming algorithm with wavelet transform to develop a novel algorithm for precipitation prediction (abbreviated as WT_FMC-GEP). To verify the prediction performance, we conducted experiments using RMSE and MAE as evaluation metrics on the real precipitation data sets in three regions of different continents where climate vary widely. The results show that the WT_FMC-GEP algorithm outperforms other existing prediction algorithms including BP neural network, support vector regression and Gene Expression Programming. It also outperforms such algorithm based on MC-GEP and wavelet transform, thus having a good application prospect.

**Keywords:** Gene Expression Programming · Fuzzy control ·
Adaptive genetic manipulation · Wavelet transform · Precipitation prediction ·
Time series analysis

## 1 Introduction

Precipitation is an important climatic phenomenon in nature. Being an important part of the atmospheric cycle, it is formed under combined action of various complex factors. Heavy precipitation within a short term may easily bring about flood disaster and hence

© Springer Nature Switzerland AG 2019
D.-S. Huang et al. (Eds.): ICIC 2019, LNCS 11644, pp. 81–92, 2019.
https://doi.org/10.1007/978-3-030-26969-2_8

affect the national economy and the people's livelihood. Therefore, accurately predicting future precipitation can provide basis for daily activities and production planning.

Traditional models of precipitation prediction are not easy to accurately mine key information when analyzing and processing large-scale precipitation time series data. Moreover, noise or redundant information may directly affect the prediction quality with low accuracy. Along with development of artificial intelligence (AI) and data mining technology, intelligent computing and data mining technology are applied in regional precipitation prediction. They provide a new and effective method for finding the inherent law of large-scale precipitation time series data. This has become one of the hot research topics recently. In recent years, neural network and SVM algorithm, which can effectively describe the complicated relationships among influencing factors in precipitation time series data, have been widely applied to precipitation modeling and prediction. For example, Devi [1] et al. compared prediction abilities of neural network models, including BPN, CBPN, DTDNN and NARX, and proved the superior performance of the NARX model. Shamshirband [2] et al. proposed a precipitation prediction method combining ANFIS and SVR. Monthly precipitation data from 29 meteorological stations in Serbia were used in a simulation experiment, and the result sufficiently demonstrated the algorithm's effectiveness. Mishra [3] designed a model based on Back Propagation Algorithm and Levenberg-Marquardt training function, using Regression Analysis, MSE and MRE as evaluation indicators, which has a good result. Zainudin [4] studied Naive Bayes, Decision Tree, Neural Network and Support Vector Machine (SVM) and Random Forest for predicting precipitation in Malaysia, and the results showed that a set of random forest classifier could collectively beat a single classifier. Xiang et al. [5] proposed a novel combined model based on the information extracted with EEMD for rainfall prediction.

In general, neural network and SVM algorithm can effectively describe the complicated relationships among various influencing factors in precipitation time series data. However, it is very hard to determine unified structure and parameters of the algorithm itself [6, 7]. Besides, excessive computational complexity makes the algorithm not conducive to large-capacity sample training and learning [8].

Gene Expression Programming (GEP) algorithm is a notable evolutionary algorithm integrating the advantages of individual organization methods of genetic algorithm and genetic programming. It is superior to genetic algorithm and genetic programming, because it can simply encode any complex problem, search in the global space, and has strong ability to search optimum and discover rules and formulas. The existing researches show that GEP algorithm has very strong ability in regression analysis and problems solving, especially in complex problems of nonlinearity that are difficult to solve by traditional methods and require global optimization [9–11]. Nonetheless, current researches use GEP-related algorithms to directly mine the precipitation time series data and make prediction. They do not consider the complex time-frequency components and time-frequency differences. Since the key information contained in the data could not be completely mined, the prediction performance cannot satisfy the practical requirements [12].

Wavelet transform can effectively analyze the time and scale of signals. Its window area is constant and the time and frequency windows are variable, which has the characteristics of multi-resolution analysis. It can solve problems that are difficult to be expressed due to non-stationary time-frequency localization in the data, and can easily separate the effective data from the noise to obtain good denoising effect [13].

Therefore, this work aims at developing a precipitation modeling prediction method based on Fuzzy-control Multi-cellular Gene Expression Programming algorithm and wavelet transform. The effectiveness of the proposed algorithm is verified by experiments with three real data sets.

## 2  Preliminaries

**Wavelet Transform and Reconstruction**
Wavelet is mainly constructed for multi-frequency analysis and realized with Mallat pyramidal algorithm. It focuses on decomposing signals into detail signal and approximation signal. The specific decomposition is shown in Fig. 1:

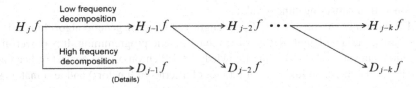

**Fig. 1.** Data signal different frequency decomposition diagram.

Where, limited energy signal $f \in L^2(R)$ is approximate to $H_j f$ at the resolution of $2^j$, and $H_j f$ can be decomposed to the sum of $H_{j-1} f$ and $D_{j-1} f$. $H_{j-1} f$ is the approximation signal obtained by $H_j f$ from low-pass filter and $D_{j-1} f$ is the detail signal from high-pass filter by $H_j f$. Only the approximation signal is decomposed again into detail signal and approximation signal until satisfying the decomposition requirement. Decomposition of Mallat algorithm is shown in Eq. (1):

$$\begin{cases} H_{j+1,k} = \sum_m h_0(m - 2k)H_{j,m} \\ D_{j+1,k} = \sum_m h_1(m - 2k)H_{j,m} \end{cases} \tag{1}$$

Where, $H_j$ and $D_j$ are column vector forms of wavelet coefficient, $h_0$ is the impact response series of low-pass filter, and $h_1$ is the impact response series of high-pass filter. Reconstruction is an inverse process of decomposition, with the formula as shown in Eq. (2):

$$C_{j-1,k} = \sum_m h_0(k - 2m)C_{j,m} + \sum_m h_1(k - 2m)D_{j,m} \tag{2}$$

The refactoring process is shown in Fig. 2:

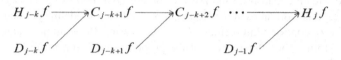

**Fig. 2.** Wavelet reconstruction process

## Multi-cellular Gene Expression Programming

**Gene Expression Programming (GEP).** GEP heuristically searches the optimal solution in a solving space of the to-be-solved problem with chromosome expression. Chromosomes in the population may be regarded as a description for a certain solution of a target problem and are evaluated by fitness value. The more optimal the fitness value is, the better the solution mapped by the chromosome is. The chromosomes will gradually evolve after a series of genetic manipulation and selection until obtaining a solution with a satisfying fitness value.

GEP integrates the linearity & fixed length encoding of genetic algorithm and the nonlinearity, variable length & tree structure of genetic programming. It is powerful in regression analysis and discovering knowledge. GEP chromosome consist of head and tail. The head consists of functors set F (a set of a series of functors) and terminators set T (containing a series of decision variables and constants), and the tail only consists of terminators set T. Assuming that the gene head length is h, the tail length must satisfy Eq. (3), where M is the largest number of function arguments. In terms of decoding way, GEP inherits GP's tree decoding way by transforming the linear genotype of chromosome to corresponding tree structure. Its genetic manipulation is similar to that of GA, and includes such common operations as selection, mutation, crossing and revert sequence.

$$t = h \times (M - 1) + 1 \tag{3}$$

**Multi-cellular Gene Expression Programming (MC-GEP).** Multi-cellular Gene Expression Programming is an improved algorithm based on GEP that introduces homologous gene and cell system [14]. MC-GEP chromosome consists of multiple common genes, DC domain and homologous gene. In terminators set and functors set of the homologous gene, each character represents a gene and symbols connecting the genes. This complex individuality with multiple genes greatly simplifies the process of constructing a powerful genotype/phenotype system. Figure 3 shows a MC-GEP chromosome consisted of two common genes and one homologous gene. Head lengths of a common gene and a homologous gene are 8 and 4 respectively. The terminators set T of the common gene and the homologous gene is $\{?, a, b, c, d, e, f\}$ and $\{0, 1\}$ respectively. '0' and '1' herein represent the first and the second common gene respectively. The functors set F of both the common gene and the homologous gene is

{+,−,*,/,S,sqrt}, wherein S mean the sin functor. DC domain is {A,B,C,D,E,F,G,H,I,J} and the constants corresponding to each character are randomly generated.

```
0 1 2 3 4 5 6 7 8 9 0 1 2 3 4 5 6 7 8 9 0 1 2 3 4 5|0 1 2 3 4 5 6 7 8 9 0 1 2 3 4 5 6 7 8 9 0 1 2 3 4 5|0 1 2 3 4 5 6 7 8 9
+ - * c d a ? b a b a f d e c b d G G D F D B D C G|/ S b + ? d e b f d f a b c d d a B H C E D I B C F|+ * 1 0 1 1 1 1 1 0
```

**Fig. 3.** Multi-cellular chromosome coding structure

The expression tree (ET) of the chromosome subject to decoding is as shown in Fig. 4, and the mathematical expression subject to final decoding is Eq. (4).

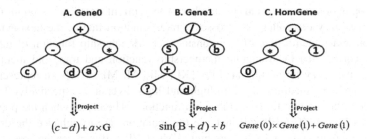

**Fig. 4.** Expression tree of multi-cellular chromosomes. S refer to sin function.

$$((c - d) + a \times G + 1) \times (\sin(B + d) \div b) \qquad (4)$$

# 3  Precipitation Modeling and Prediction Based on Fuzzy Multi-cellular Gene Expression Programming and Wavelet Transform

**Fuzzy-Control Multi-cellular Gene Expression Programming**

Crossover rate of evolutionary algorithms will influence their convergence efficiency to a large extent. And mutation rate determines whether the algorithms can find globally optimal solution out of local optimal solution or not [15]. Nevertheless, similar to other evolutionary algorithms, both GEP and MC-GEP fix initial parameters and adopt single order of genetic manipulation. As the evolution is ongoing, it is not so easy for the algorithms to jump out of local optimal solution because of loss of population diversity. Therefore, fuzzy control theory is introduced to MC-GEP, and Fuzzy-control Multi-cellular Gene Expression Programming (FMC-GEP) is proposed to dynamically and automatically adjust the crossover rate and the mutation rate. In this way, it is easier for GEP to find globally optimal solution out of local optimal solution.

Firstly, population diversity is measured according to dispersion degree of individual fitness in the population. That is to calculate the ratio d of optimal fitness (Fbest) and average fitness (Fave) of the current population. When calculating the minimum of an optimization function, Eq. (5) is used to determine the population diversity. On the contrary, Eq. (6) is used. As the population converges, d gradually approaches 1.

$$d = \frac{F_{\min}}{F_{ave}}, F_{best} < F_{ave} \tag{5}$$

$$d = \frac{F_{ave}}{F_{\max}}, F_{best} > F_{ave} \tag{6}$$

We design three different fuzzy controllers, to describe the size of population diversity and dynamically adjust the crossover rate, mutation rate and constant set mutation rate combined with fuzzy mathematics. These three fuzzy controllers use the current population diversity and the number of the current iterations as input. Outputs of the three fuzzy controllers are crossover rate, mutation rate of the next-generation population and mutation rate of real constant set. Membership function of input and output is constructed by triangular membership function and trapezoid membership function. Five fuzzy linguistic variables {XL, ML, M, MH, XH} are represented by low, low-medium, medium, medium-high and high diversity, respectively. They are used to describe the five fuzzy membership functions. Therefore, when the population diversity is low, FMC-GEP will increase the mutation rate to enhance the diversity. When the population diversity is quite high, FMC-GEP will increase the crossover rate and reduce the mutation rate.

## Precipitation Modeling and Prediction

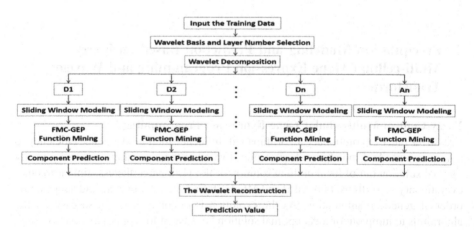

**Fig. 5.** The framework of our algorithm

**Basic Thinking.** The volume of precipitation is jointly affected by various complex factors. The ever-changing fluctuation range and frequency make the precipitation time series data nonlinear and non-stationary. It is difficult to analyze the detailed variation of data and accurately predict precipitation by direct regression analysis on precipitation time series with computing model. Besides, some noises are unavoidable during collection of precipitation data. They will directly interfere in the data fitting and hence affect the prediction result. Therefore, a data preprocessing method is needed to decompose and reconstruct the data, in order to reduce deviation generated from direct modeling and prediction to precipitation data. The needed method should process nonlinear and non-stationary signals and effectively eliminate noise. The components decomposed from orthogonal wavelet basis are orthogonal in pairs, without linear correlation, and the information is not redundant. Hence subtle changes of the data at each level can be further analyzed, changes in signals can be known better, and extracted coefficients of each level can be reconstructed. So the key features of precipitation time series can be captured [16]. Therefore, this paper applies WT_FMC-GEP for precipitation data modeling and prediction. Figure 5 shows the flowchart of WT_FMC-GEP.

Firstly, the high-frequency component $D_1, D_2, \cdots D_n$ and the low-frequency component $A_n$ are obtained subject to wavelet decomposition of precipitation time series according to the selected wavelet basis and the number of decomposing layers. Then FMC-GEP algorithm is evolved to get the fitting model after building models for training samples of each high-frequency and low-frequency component by sliding window algorithm, respectively. Finally, the fitting model solves prediction result of each component, and the prediction results are subject to wavelet reconstruction to get the results of test samples, i.e. the prediction data.

**Wavelet Transform and Function Mining.** The high-frequency and the low-frequency components subject to wavelet decomposition are time series data. When modeling and predicting time series data with GEP, the time series data is usually converted into a delay matrix by sliding window, and the mathematical description is:

$$X(t) = \begin{vmatrix} x_1 & x_2 & \cdots & x_{t-n+1} \\ x_2 & x_3 & \cdots & x_{t-n+2} \\ \cdots & \cdots & \cdots & \cdots \\ x_n & x_{n+1} & \cdots & x_t \end{vmatrix} \tag{7}$$

Where, t refers to time, $x_t$ is the value x at the time $t$, and the size of sliding window is $t - n + 1$. The delay matrix in Eq. (8) shows the relationship between the value after $t - n$ moments and that at the current moment $t - n + 1$. Using function to describe the relationship, time series prediction can be described as finding a function f that is approximate to the time series.

$$x_{t-n+1} = f(x_1, x_2, \cdots, x_{t-n}) \tag{8}$$

Then GEP algorithm is used to find the specific constitution of function f, that is, to find out the fitting model of time series data with the best fitness. In this study, wavelet decomposition is conducted to training sample data of precipitation time series to obtain components, and FMC-GEP is used for function mining of each component, thus obtaining sub-model corresponding to each component.

**Wavelet Reconstruction and Precipitation Prediction.** The value in the corresponding year of each testing sample is computed in combination with the sub-model by function mining with FMC-GEP and the testing sample data. Wavelet reconstruction is conducted in the principle of adding the values in corresponding same year of predicting sample to obtain the final prediction value.

## 4    Experiment Result and Analysis

### Data and Experimental Setup

The experimental data sets herein include the total monthly precipitation data over 55 years in China's Nanning (22°48′N and 108°22′E), Australia's Melbourne (37°50′S and 144°58′E) and America's Seattle (47°37′N and 122°19′W). The climate varies widely in these three regions. Each region has 660 data samples respectively. Their average precipitation per month is 108.78 mm, 85.53 mm and 3.36 mm respectively. Figure 6 shows the specific data distribution of the three regions.

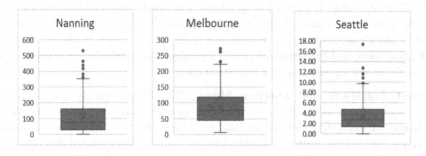

**Fig. 6.** Box-plot of original datasets

**Table 1.** The parameters of GEP correlation algorithm

|                        | GEP          | MC-GEP       | FMC-GEP |
|------------------------|--------------|--------------|---------|
| Number of gene         | 1            | 3            | 3       |
| Homo gene head length  | Null         | 4            | 4       |
| Crossover rate         | 0.3          | 0.3          | Null    |
| Mutation rate          | 0.3          | 0.3          | Null    |
| DC mutation rate       | Null         | 0.25         | Null    |
| Bound of constant      | [−10,10]     |              |         |
| Gene head length       | 15           |              |         |
| Population size        | 100          |              |         |
| Iteration              | 2000         |              |         |
| Terminal set           | {a,b,c,d,e,f,?} |           |         |
| Function set           | {+,−,*,/,sin,cos,sqrt} |      |         |
| Selection mode         | Championships select |        |         |
| Fitness function       | RMSE/MAE     |              |         |

To evaluate the precipitation prediction performance of the proposed algorithm, we use BP, SVM, GEP, WT_MC-GEP and the proposed WT_FMC-GEP algorithm to conduct compared experiments with the same datasets. In addition, we use MAE and RMSE as evaluation metrics. Among them, GEP, BP and SVM algorithms are widely used for precipitation prediction modeling, and are often used to solve other symbol regression problems. WT_MC-GEP algorithm is the state-of-the-art method of precipitation modeling and prediction [17]. Five independent experiments are conducted with each algorithm in order to avoid occasionality of the experimental results. In the experiments, 60% of data sample in each data set is used as training data and 40% as test data, so that the test data may contain certain abnormal values. 4-layer decomposition of wavelet basis Db10 is selected as wavelet transform setup, and the parameter setting of BP, SVM and WT_MC-GEP refer to References [17]. The main parameters of GEP-related algorithms are shown in Table 1.

**Experimental Results and Analysis**

Firstly, we verify changes in the performance of precipitation prediction by MC-GEP with or without fuzzy control in combination with wavelet transform, respectively. That is to compare the performance difference of the proposed WT_FMC-GEP and WT_MC-GEP, as shown in Fig. 7. It can be seen that WT_FMC-GEP evidently outperforms WT_MC-GEP by 48 out of 60 groups of endpoint-compared results. Especially when using MAE as evaluation index of model performance, WT_FMC-GEP outperforms WT_MC-GEP in fitting and prediction for five times on the two data sets of Melbourne and Seattle. When using RMSE as evaluation index, WT_FMC-GEP also outperforms WT_MC-GEP in the prediction effect for five times. Though

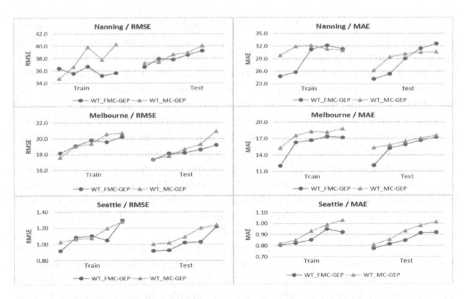

**Fig. 7.** Compared results in 5 times independent experiments

WT_MC-GEP slightly outperforms WT_FMC-GEP in fitting based on the precipitation datasets of Nanning and Melbourne, this indicates that WT_MC-GEP has the same over-fitting issue as most intelligent algorithms. We believe that over-fitting is unavoidable in MC-GEP, since it, like the traditional GEP, can't jump out of local optimal solution. This also demonstrates the proposed WT_FMC-GEP has stronger generalization and better robustness.

Secondly, the proposed algorithm is compared with BP, SVM, GEP and other classical time series prediction algorithms in precipitation prediction, and the result is shown in Table 2.

**Table 2.** FMCGEP-DNN parameters

| Index | | RMSE | | | | | | MAE | | | | | |
|---|---|---|---|---|---|---|---|---|---|---|---|---|---|
| Algorithm dataset | | Nanning | | Melbourne | | Seattle | | Nanning | | Melbourne | | Seattle | |
| | | Fit | Pred | Fit | Pred | Fit | Pred | Fit | Pred | Fit | Pred | Fit | Pred |
| BP | Optimum | 63.200 | 91.054 | 38.791 | 43.079 | 1.881 | 1.948 | 45.849 | 62.560 | 29.833 | 32.648 | 1.383 | 1.474 |
| | Average | 59.801 | 94.726 | 36.469 | 45.498 | 1.724 | 2.108 | 44.031 | 67.555 | 28.188 | 35.094 | 1.303 | 1.553 |
| SVM | Optimum | 74.992 | 77.642 | 41.853 | 39.816 | 2.053 | 2.008 | 53.616 | 55.816 | 31.806 | 29.603 | 1.477 | 1.459 |
| | Average | 74.992 | 77.642 | 41.853 | 39.816 | 2.053 | 2.008 | 53.616 | 55.816 | 31.806 | 29.603 | 1.477 | 1.459 |
| GEP | Optimum | 84.315 | 87.510 | 42.319 | 40.861 | 2.033 | 1.937 | 55.685 | 57.700 | 34.084 | 32.090 | 1.545 | 1.503 |
| | Average | 86.880 | 91.862 | 44.537 | 43.769 | 2.078 | 1.985 | 58.142 | 59.964 | 34.807 | 33.049 | 1.560 | 1.581 |
| WT_MC-GEP | Optimum | **34.686** | 37.216 | **17.590** | 17.370 | 1.025 | 1.000 | 29.740 | 26.167 | 15.249 | 15.299 | 0.816 | 0.807 |
| | Average | 37.817 | 38.428 | 19.422 | 18.791 | 1.126 | 1.106 | 31.192 | 29.347 | 17.597 | 16.429 | 0.921 | 0.917 |
| WT_FMC-GEP | Optimum | 36.346 | **36.588** | 18.123 | **17.314** | 0.913 | 0.915 | 24.740 | 24.111 | 11.995 | 12.092 | 0.804 | 0.773 |
| | Average | **35.841** | 38.010 | **19.335** | 18.266 | **1.086** | 1.019 | **29.011** | 28.478 | **15.880** | 15.399 | 0.868 | 0.852 |

Note: The column names Fit and Pred mean fitting and prediction respectively.

According to the results in Table 2, WT_FMC-GEP obtained the best performance in 10 items out of the 12 compared results of the two evaluation metrics on the three data sets. Wherein, it predicted with the best result when comparing prediction performance in all data sets across all evaluation metrics. WT_FMC-GEP and WT_MC-GEP algorithms in combination with wavelet transform significantly outperform other algorithms in precipitation prediction. Their performance in all data sets across all evaluation metrics was improved by 41.6% at least. This indicates that the deeply hidden information of the data can be indeed mined by the method combining GEP and wavelet transform. It overcomes the neglect of time-frequency components and time-frequency differences by the GEP algorithm in the direct prediction of rainfall time series, and obtains better performance. When using MAE as the evaluation index, WT_FMC-GEP achieves the best performance on all precipitation data sets, no matter in fitting or prediction. Using RMSE as evaluation metrics, WT_FMC-GEP has the optimal prediction performance on all data sets.

To sum up, precipitation prediction algorithm combined with wavelet transform significantly outperforms other algorithms. WT_FMC-GEP has a strong ability to predict time series and it is not prone to over-fitting. It is not only suitable for data modeling and prediction with stably fluctuated variables, but also for that with large fluctuation range and extreme points.

# 5 Conclusion

This paper designed a method to improve algorithm by fuzzy control theory to adjust the genetic manipulation probability of Multi-cellular Gene Expression Programming, then combined the improved algorithm and wavelet transform for precipitation modeling and prediction. Using RMSE and MAE as evaluation metrics, experiments were conducted on three different precipitation data sets. The experimental results showed that the proposed algorithm outperformed BP neural network, support vector regression, GEP and other prediction algorithms in precipitation prediction. It also outperformed the precipitation prediction algorithm based on Multi-cellular Gene Expression Programming and wavelet transform. This sufficiently verifies the effectiveness of the proposed algorithm. Since wavelet function needs to be selected through trials in wavelet transform, the algorithm takes longer time to implement because of the high computational complexity.

However, in the actual application for precipitation prediction, accuracy is emphasized while time efficiency is nearly negligible. Thus WT_FMC-GEP algorithm will have a good prospect in meteorological application and can be transferred to other time series prediction.

**Acknowledgments.** This work was supported in part by the National Natural Science Foundation of China Grant #61562008, #41575051 and #61663047, the Natural Science Foundation of Guangxi Province under Grant No. #2017GXNSFAA198228 and No. #2017GXNSFBA198153; The Project of Scientific Research and Technology Development in Guangxi under Grant No. #AA18118047 and No. #AD18126015. Thanks to the support by the BAGUI Scholar Program of Guangxi Zhuang Autonomous Region of China. Xiao Qin is the corresponding author.

# References

1. Devi, S.R., Arulmozhivarman, P., Venkatesh, C., et al.: Performance comparison of artificial neural network models for daily rainfall prediction. Int. J. Autom. Comput. **13**(5), 1–11 (2016)
2. Shamshirband, S., Gocic, M., Petkovi, D., et al.: Soft-computing methodologies for precipitation estimation: a case study. IEEE J. Sel. Top. Appl. Earth Obs. Remote Sens. **8**(3), 1353–1358 (2015)
3. Mishra, N., Soni, H.K., Sharma, S., et al.: Development and analysis of artificial neural network models for rainfall prediction by using time-series data. Int. J. Intell. Syst. Appl. **10**(1), 16–23 (2018)
4. Zainudin, S., Jasim, D.S., Bakar, A.A.: Comparative analysis of data mining techniques for Malaysian rainfall prediction. Int. J. Adv. Sci. Eng. Inf. Technol. **6**(6), 1148 (2016)
5. Xiang, Y., Gou, L., He, L., et al.: A SVR-ANN combined model based on ensemble EMD for rainfall prediction. Appl. Soft Comput. J. **73**, 874–883 (2018)
6. Huang, D.S., Jiang, W.: A general CPL-AdS methodology for fixing dynamic parameters in dual environments. IEEE Trans. Syst. Man Cybern. - Part B **42**(5), 1489–1500 (2012)
7. Huang, D.S., Du, J.-X.: A constructive hybrid structure optimization methodology for radial basis probabilistic neural networks. IEEE Trans. Neural Netw. **19**(12), 2099–2115 (2008)

8. Hamidi, O., Poorolajal, J., Sadeghifar, M., et al.: A comparative study of support vector machines and artificial neural networks for predicting precipitation in Iran. Theor. Appl. Climatol. **119**(3–4), 723–731 (2015)
9. Peng, Y.Z., Yuan, C.A., Qin, X., et al.: An improved gene expression programming approach for symbolic regression problems. Neurocomputing **137**(15), 293–301 (2014)
10. Zhong, J., Ong, Y.S., Cai, W.: Self-learning gene expression programming. IEEE Trans. Evol. Comput. **20**(1), 65–80 (2016)
11. Samadianfard, S., Asadi, E., Jarhan, S., et al.: Wavelet neural networks and gene expression programming models to predict short-term soil temperature at different depths. Soil Tillage Res. **175**, 37–50 (2018)
12. Li, H., Peng, Y., Deng, C., Pan, Y., Gong, D., Zhang, H.: Multicellular gene expression programming-based hybrid model for precipitation prediction coupled with EMD. In: Huang, D.-S., Bevilacqua, V., Premaratne, P., Gupta, P. (eds.) ICIC 2018. LNCS, vol. 10954, pp. 207–218. Springer, Cham (2018). https://doi.org/10.1007/978-3-319-95930-6_20
13. Daubechies, I.: The wavelet transform, time-frequency localization and signal analysis. J. Renew. Sustain. Energy **36**(5), 961–1005 (2015)
14. Peng, Y.-Z., Yuan, C.-A., Chen, J.-W., Wu, X.-D., Wang, R.-L.: Multicellular gene expression programming algorithm for function optimization. Control Theory Appl. **27**(11), 1585–1589 (2010)
15. Yang, C., Qian, Q., Wang, F., et al.: An improved adaptive genetic algorithm for function optimization. In: IEEE International Conference on Information and Automation, pp. 675–680. IEEE (2017)
16. Nourani, V., Farboudfam, N.: Rainfall time series disaggregation in mountainous regions using hybrid wavelet-artificial intelligence methods. Environ. Res. **168**, 306–318 (2019)
17. Li, H., Peng, Y., Deng, C., Pan, Y., Gong, D., Zhang, H.: A hybrid precipitation prediction method based on multicellular gene expression programming. arXiv:2634993 (2019)

# Integrative Enrichment Analysis
# of Intra- and Inter- Tissues' Differentially
# Expressed Genes Based on Perceptron

Xue Jiang[1,2,3] (iD), Weihao Pan[1] (iD), Miao Chen[1] (iD), Weidi Wang[1,2,3] (iD),
Weichen Song[1,2,3] (iD), and Guan Ning Lin[1,2,3(✉)] (iD)

[1] School of Biomedical Engineering, Shanghai Jiao Tong University,
Shanghai 20030, China
{jiangxue_s,chenmiao95}@sjtu.edu.cn,
wwd-swxx@foxmail.com, song628196@gmail.com,
nickgnl@sjtu.edu.cn
[2] Shanghai Key Laboratory of Psychotic Disorders, Shanghai 20030, China
[3] Brain Science of Technology Research Center, Shanghai Jiao Tong University,
Shanghai 200030, China

**Abstract.** Recent researches in biomedicine have indicated that the molecular basis of chronic complex diseases can also exist outside the disease-impacted tissues. Differentially expressed genes between different tissues can help revealing the molecular basis of chronic diseases. Therefore, it is essential to develop new computational methods for the integrative analysis of multi-tissues gene expression data, exploring the association between differentially expressed genes in different tissues and chronic disease. To analysis of intra- and inter-tissues' differentially expressed genes, we designed an integrative enrichment analysis method based on perceptron (IEAP). Firstly, we calculated the differential expression scores of genes using fold-change approach with intra- and inter- tissues' gene expression data. The differential expression scores are seen as features of the corresponding gene. Next, we integrated all differential expression scores of gene to get differential expression enrichment score. Finally, we ranked the genes according to the enrichment score. Top ranking genes are candidate disease-risk genes. Computational experiments shown that genes differentially expressed between striatum and liver of normal samples are more likely to be Huntington's disease-associated genes, and the prediction precision could be further improved by IEAP. We finally obtained five disease-associated genes, two of which have been reported to be related with Huntington's disease.

**Keywords:** Multi-tissues · Integrative enrichment analysis ·
Differentially expressed gene

D.-S. Huang et al. (Eds.): ICIC 2019, LNCS 11644, pp. 93–104, 2019.
https://doi.org/10.1007/978-3-030-26969-2_9

# 1  Background

Huntington's disease (HD) is a type of neurodegenerative disease severely affecting ones' behavioral and mental health. It is caused by a triplet (CAG) repeat elongation in huntingtin (HTT) gene in chromosome 4 that codes for polyglutamine in the huntingtin protein [1]. The mutant protein can enter the nucleus and alter gene transcription and has many deleterious effects in cells [2]. With the accumulation of the mutant protein, numerous molecular interactions and molecular pathways can be affected, resulting in neuronal dysfunction and degeneration [3, 4]. To date, the complex nature of molecular pathogenesis of HD is still not fully understood.

With the rapid and encouraging development of high-throughput sequencing technologies, large amounts of omics data have been accumulated. At the same time, along with the accumulation of RNA-seq data, many computational methods for investigating the molecular basis candidate targets have also been developed in recent years. The computational methods can be roughly divided into three main categories: network-based methods [5–9], statistics-based enrichment analysis methods [10, 11], and machine learning based methods [12, 13]. Network-based methods can well describe the interactions between molecules, and they often need large amounts of computational time. Machine learning based methods, such as matrix factorization based methods [14, 15] and deep learning methods [16, 17], have been widely used and studied, but they may lead to instable results due to random initialization.

The complex nature of molecular mechanisms and complicated phenotypes in neurodegenerative diseases pose great challenges in identifying disease risk genes. And there is still a large gap between the interpretation of pathological mechanisms and the disease genes screened by various computational methods. Besides, the consistency of the candidate disease gene sets obtained by different methods is poor [16]. It is also very difficult to distinguish disease genes from non-disease genes [14]. At present, it is urgent to develop effective computational methods to improve the accuracy of disease-associated gene prediction and the robustness of the candidate disease gene sets, promoting the understanding of the pathological molecular mechanisms under complex phenotypes.

Recent studies in the biomedical research field have shown that the molecular basis of chronic complex diseases can exist outside the diseased-impacted tissues. And the differentially expressed genes between different tissues are expected to reveal the molecular origin of complex chronic diseases [18–21]. Researcher often screen genes that significantly differential expressed between normal and case samples of different individuals as disease-associated ones. However, large amounts of genes' expression have been affected during the disease development, making it quite difficult to accurately distinguish disease genes from non-disease genes. Moreover, the differentially expressed genes selected with normal and case samples may not helpful for the personalized medicine due to the individual differences. Nevertheless, exploring the differentially expressed genes between different tissues of same individuals may reveal the endogenous answers for the disease development.

According to the above analysis, in this study, we conducted large amounts of experiments to screen Huntington's disease genes using the-sated- of-art methods, including t-test [22], fold change method (FC) [22], flexible non- negative matrix factorization method (FNMF) [14], and joint non-negative matrix factorization meta-analysis method (jNMFMA) [15], to explore the relationship of differentially expressed genes and disease.

To further improve the disease gene prediction accuracy, we integrated the differential expression score of gene between different tissues, namely intra-tissues' differential expression score, and that between normal and case samples of one tissue, namely inter-tissues' differential expression score. Hence, we proposed an integrative enrichment analysis of intra- and inter- tissues' differentially expressed genes based on perceptron model (IEAP). Firstly, we calculated the differential expression scores of genes using FC [22]. The differential expression scores are seen as features of the genes. Next, we integrated the differential expression scores to get an enrichment score of the corresponding gene using perceptron. Top ranking genes were selected as disease ones. Experiments on gene expression data of Huntington's disease shown that the prediction accuracy of IEAP could be as precise as that of the-state-of-art methods, while the gene rankings in ranked list of IEAP are more stable than that of other methods. More importantly, IEAP is much helpful for understanding the mechanisms under complex disease phenotypes.

## 2  Method

In this section, first, we present the key idea of the integrative enrichment analysis. Next, the perceptron model and the learning process are described. Finally, we present the parameter setting of the IEAP.

### Integrative Enrichment Analysis

Enrichment analysis aims to select a set of genes which are significantly differentially expressed between different conditions. The gene set is considered to be strongly correlated to the accuracy of distinguishing one condition from the other. Traditional enrichment analysis methods evaluate the significance of gene set using statistical-based strategy, and then give the corresponding gene set an enrichment score. Usually, the enrichment score is used to measure the importance of the gene set. The higher the enrichment score is, the stronger the biological meaning the gene set is.

Intuitively, the interpretability and biological meaning could be further improved if we integrate all the differential expression scores of one gene. According to the above analysis, we designed integrative enrichment analysis based on an artificial neuron.

### Integrative Enrichment Analysis of Intra- and Inter- Tissues' Differentially Expressed Genes Based on Perceptron

**Model.** We computed the gene intra-tissues' differential expression score and inter-tissues' differential expression score based on FC. The fold change of any two samples must be not less than 1. If the fold change is less than 1, we need to use its reciprocal. The differential expression score of gene is the average fold change of the two type samples. The greater the differential expression score is, the more significant the differential expression of the gene is.

Symbol $x_g = (p_{g1}, \cdots, p_{gn})$ represents the differential expression scores of gene $g$. In this model, the differential expression score $p_{gi}$, $i = 1, \cdots, n$ are seen as features of gene $g$. We train the artificial neuron with the genes in the training set. The labels of genes are denoted as $Y = (y_1, \cdots, y_g)$. The value of $y_g$ is defined by

$$y_g = \begin{cases} 1, & \text{if } g \text{ is disease} - \text{associated}, \\ -1, & \text{if } g \text{ is non} - \text{disease} - \text{associated}. \end{cases} \tag{1}$$

In the artificial neuron model (Fig. 1), $p_i$ represents the differential expression score, and $w_i$ and $b$ are the parameters of the model. The details for the model are shown below.

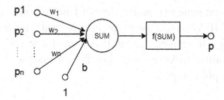

**Fig. 1.** Schematic illustration of IEAP.

In this model, we use sigmoid function to integrate all the features of a gene. The sigmoid function is

$$f_\theta(x_g) = \frac{1}{1 + e^{-\left(\sum_{i=1}^{n} w_i p_{gi} + b\right)}} \tag{2}$$

where $\theta = (W, b)$ represents the parameter setting of the model.

Let $\overline{y_g} = f_\theta(x_g)$ represent the evaluated label of gene $g$, then $\overline{Y} = (\overline{y_1}, \cdots, \overline{y_g})$. The loss function is

$$L(Y, \overline{Y}) = \sum_{g=1}^{N} l(y_g, \overline{y_g}) \tag{3}$$

where $N$ represents the number of genes. The loss function for gene $g$ is defined by

$$l(y_g, \overline{y_g}) = \frac{1}{2}(y_g - \overline{y_g})^2 \tag{4}$$

**Learning.** In this study, cumulative error background propagation algorithm is used to learn the parameters. The learning process is completed until the global error converges. The gradient descent algorithm is used to calculate the gradient in each iteration. The calculation of gradient is given by

$$\Delta w_i = \frac{\partial L}{\partial w_i}$$

$$= \sum_{g=1}^{N} \frac{\partial l_g}{\partial w_i} \tag{5}$$

$$= -\sum_{g=1}^{N} \left(y_g - \overline{y}_g\right) \frac{e^{-\left(\sum_{i=1}^{n} w_i p_{gi} + b\right)}}{1 + e^{-\left(\sum_{i=1}^{n} w_i p_{gi} + b\right)}}$$

$$\Delta b = \frac{\partial L}{\partial b}$$

$$= \sum_{g=1}^{N} \frac{\partial l_g}{\partial b} \tag{6}$$

$$= -\sum_{g=1}^{N} \left(y_g - \overline{y}_g\right) \frac{e^{-\left(\sum_{i=1}^{n} w_i p_{gi} + b\right)}}{1 + e^{-\left(\sum_{i=1}^{n} w_i p_{gi} + b\right)}}$$

$$W(k+1) = W(k) - \eta \Delta W(k) \tag{7}$$

$$b(k+1) = b(k) - \eta \Delta W(k) \tag{8}$$

where $W(k)$ is the parameters in the $k$-th interactions. $\eta$ is the learning rate, which controls the convergent speed.

**Integrative Enrichment Score**

To reduce the computational complexity, we use the Eq. (9) to calculate the integrative enrichment score. Finally, we rank the genes in descending order according to the integrative enrichment scores. Top ranking genes are more likely to be disease-associated genes.

$$E_g = \sum_{i=1}^{n} w_i p_{gi} \tag{9}$$

By training the model, we can clearly know which differential expression score contribute more to the finally integrative enrichment score according to the weights, providing a better understanding of association between disease-phenotypes and the differentially expressed genes.

**Parameter Setting**

Here, we initialize parameters in the model as follows: (1) if the area under the receiver operating characteristic curve (AUC) [23] of the differential expression score is larger than 0.5, the corresponding $w$ is preset to be the AUC, (2) the $b$ is preset to be 0, and (3) the other $w$ is also preset to be 0.

In summary, the detailed training process of the model is shown below.

| Algorithm 1: Training for IEAAN |
| --- |
| 1: Initialize $W$ and $b$ |
| 2: Input training set $S$ |
| 3: For $i = 1,2,L,k$ |
| 4:      Compute $\Delta w_i$ according to Eq. (5) |
| 5:      Compute $\Delta b$ according to Eq. (6) |
| 6:      Update $w_i$ according to Eq. (7) |
| 7:      Update $b$ according to Eq. (8) |
| 8: End |
| 9: Compute the integrative enrichment score for each gene according to Eq. (9) |
| 10: Rank the genes in descending order according to the integrative enrichment scores, and return the ranked list. |

## 3   Results and Discussion

### Gene Expression Data

Gene expression data used in this study were downloaded from http://www.hdinhd.org, which were obtained from Huntington's disease mice through RNA-seq technology. The age of the experimental mice was 6-month-old. The dataset contained three tissues, including striatum, cortex, and liver. There were 8 kinds of genotypes, including ploy Q20, ploy Q80, ploy Q92, ploy Q111, ploy Q140, and ploy Q175. And there were 8 samples for each genotype. The genotype ploy Q20 was normal one, while the rest genotypes were disease ones. The detailed information of the dataset was illustrated in Table 1.

**Table 1.** Experimental data description.

| Age | 6-month-old |
| --- | --- |
| Tissue | Striatum Cortex Liver |
| Genotype | poly Q20 poly Q80 poly Q92 poly Q111 poly Q140 poly Q175 |

There were 23,351 genes in the dataset. Since most of the computational methods screened significant differentially expressed genes as disease ones, the genes whose expression changed slightly during the disease development were difficult to be selected out. Therefore, we conducted a filter step to reduce computational complexity. First, we filtered out the genes with any 0 expression level according to $l0$-norm. Then, we normalized the gene expression of each sample, and ranked the gene in descending order according to the variance of gene. Figure 2 shows the variance of normalized gene expression in striatum, cortex, and liver, respectively. Due to computational methods have no discriminative ability for the genes with small variances, we selected top ranked 4000 genes of each tissue, and then integrated the top ranked 4000 genes of the three tissues. Finally, 6,723 genes were selected out from the whole genome for the analysis of next step. The modifier genes were from [24, 25], including 89 disease-associated genes, and 431 non-disease-associated genes.

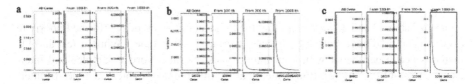

**Fig. 2.** The variance of normalized gene expression in striatum, cortex, and liver, respectively.

### Prediction Performance of t-test, FC, FNMF, and jNMFMA

Large amounts of genes' expression, as well as the interactions between genes, are affected during the Huntington's disease progression. The pathological molecular mechanisms of HD are still unclear. In this study, we conducted many experiments using t-test, FC, FNMF, and jNMFMA, to explore the characteristics of intra-tissues' differentially expressed genes and inter-tissues' differentially expressed genes, respectively. We denoted the experiments using normal samples (gene expression data with genotype Q20) versus normal samples as Normal-Normal, the experiments using normal samples versus case samples (gene expression data with genotype Q80, Q92, Q111, Q140, Q175) as Normal-Case, and the experiments using case samples versus case samples as Case-Case.

For the non-parameter methods, i.e., t-test and FC, we conducted the experiment once to obtain stable gene ranking results. Due to the instability of parametric methods, i.e., FNMF and jNMFMA, we conducted 10 times experiments. Mean and standard deviation of the 10 experimental results were calculated as the final assessment. The experimental results for the four methods were shown in Tables 2, 3, 4 and 5, respectively .

**Table 2.** Performance of the t-test method.

| t-test | Normal_Case | | | Normal_Normal | | | Case_Case | | |
|---|---|---|---|---|---|---|---|---|---|
| | Str | Cor | Liv | Str_Cor | Str_Liv | Cor_Liv | Str_Cor | Str_Liv | Cor_Liv |
| AUC | 0.48 | 0.52 | 0.47 | 0.545 | 0.529 | 0.512 | 0.521 | 0.514 | 0.523 |
| AUPR | 0.16 | 0.17 | 0.15 | 0.200 | 0.186 | 0.178 | 0.178 | 0.178 | 0.183 |

**Table 3.** Performance of the FC method.

| FC | Normal_Case | | | Normal_Normal | | | Case_Case | | |
|---|---|---|---|---|---|---|---|---|---|
| | Str | Cor | Liv | Str_Cor | Str_Liv | Cor_Liv | Str_Cor | Str_Liv | Cor_Liv |
| AUC | 0.55 | 0.51 | 0.56 | 0.472 | 0.582 | 0.584 | 0.485 | 0.583 | 0.583 |
| AUPR | 0.18 | 0.19 | 0.23 | 0.168 | 0.218 | 0.216 | 0.174 | 0.219 | 0.216 |

From Tables 2 and 3, we known that the t-test and FC methods performed poorly in disease gene prediction with intra-tissues' Normal-Case samples. While, t-test and FC methods performed better with both inter-tissues' Normal-Normal samples and inter-tissues' Case-Case samples. It indicated that the differentially expressed genes of inter-tissues are more likely to be disease-associated genes. Besides, the performance of the two methods with inter-tissues' Normal-Normal samples was comparable to that with inter-tissues' Case-Case samples. It demonstrated that the differentially expressed genes between different tissues in health individuals could be likely to be disease-associated genes.

Comparing Tables 2 and 3, the prediction performance of FC method are much better than that of t-test method. Since t-test method used the average information of gene expression, ignoring lots of useful information, it finally leaded to poor results.

Tables 4 and 5 shown the experimental results using jNMFMA and FNMF. The two methods have similar performance in screening disease genes under various conditions.

Through comprehensive comparison of Tables 2, 3, 4 and 5, we can know that the performance of jNMFMA and FNMF methods were better than that of the two statistical-based methods when we screen differentially expressed genes with intra-tissues' gene expression data. However, there were no statistical significance among the results of the 4 methods with gene expression data of inter-tissues. These results indicate that genes differentially expressed among inter-tissues in healthy individuals have great relationship with the disease. It provides a new perspective to understand and screen disease genes.

### Comparison of the Performance IEAP with Other Four Methods

We further analyzed the performance of IEAP and the other four methods. For jNMFMA and FNMF method, we selected the best performed experiment to conduct comparison analysis.

Through comparison of the ROC curves and PR curves of t-test, FC, jNMFMA, and FNMF with Normal-Case samples of striatum, cortex, and liver, respectively, we known the prediction precision of jNMFMA and FNMF for top ranking genes were better. Besides, through comparison of ROC curves and PR curves of the 4 methods with inter-tissues' Normal-Normal samples, and Case-Case samples, respectively, we known that FC, jNMFMA, and FNMF have similar performances, which were better than t-test method. Since t-test used the average expression of gene, some useful information were missed. Besides, due to the random initialization, the ranking of gene in the final ranked lists of jNMFMA and FNMF were unstable. Moreover, the computational complexity of jNMFMA and FNMF was high. Nevertheless, FC as a parameterless method, was simple, effective and stable. If the method was considered

**Table 4.** Performance of the jNMFMA method.

| jNMFMA | Normal_Case | | | Normal_Normal | | | Case_Case | | |
|---|---|---|---|---|---|---|---|---|---|
| | Str | Cor | Liv | Str_Cor | Str_Liv | Cor_Liv | Str_Cor | Str_Liv | Cor_Liv |
| AUC | 0.567 ± 0.02 | 0.554 ± 0.01 | 0.585 ± 0.01 | 0.527 ± 0.01 | 0.534 ± 0.01 | 0.548 ± 0.03 | 0.537 ± 0.02 | 0.581 ± 0.01 | 0.563 ± 0.01 |
| AUPR | 0.207 ± 0.02 | 0.194 ± 0.01 | 0.216 ± 0.01 | 0.181 ± 0.01 | 0.191 ± 0.01 | 0.196 ± 0.01 | 0.187 ± 0.019 | 0.221 ± 0.01 | 0.206 ± 0.01 |

**Table 5.** Performance of the FNMF method.

| FNMF | Normal_Case | | | Normal_Normal | | | Case_Case | | |
|---|---|---|---|---|---|---|---|---|---|
| | Str | Cor | Liv | Str_Cor | Str_Liv | Cor_Liv | Str_Cor | Str_Liv | Cor_Liv |
| AUC | 0.554 ± 0.02 | 0.556 ± 0.01 | 0.569 ± 0.01 | 0.542 ± 0.02 | 0.566 ± 0.02 | 0.537 ± 0.03 | 0.540 ± 0.02 | 0.549 ± 0.03 | 0.545 ± 0.03 |
| AUPR | 0.199 ± 0.02 | 0.197 ± 0.01 | 0.194 ± 0.02 | 0.194 ± 0.01 | 0.192 ± 0.01 | 0.198 ± 0.02 | 0.188 ± 0.01 | 0.195 ± 0.01 | 0.191 ± 0.01 |

**Table 6.** The overlap degree of the top 800 genes in any two ranked lists obtained by FC.

| | | Normal_Case | | | Normal_Normal | | | Case_Case | | |
|---|---|---|---|---|---|---|---|---|---|---|
| | | Str | Cor | Liv | Str_Cor | Str_Liv | Cor_Liv | Str_Cor | Str_Liv | Cor_Liv |
| Normal_Case | Cor | 0.38 | | | | | | | | |
| | Liv | 0.22 | 0.19 | | | | | | | |
| Normal_Normal | Str_Cor | 0.48 | 0.26 | 0.21 | | | | | | |
| | Str_Liv | 0.25 | 0.14 | 0.44 | 0.27 | | | | | |
| | Cor_Liv | 0.21 | 0.13 | 0.45 | 0.21 | 0.92 | | | | |
| Case_Case | Str_Cor | 0.39 | 0.29 | 0.41 | 0.87 | 0.26 | 0.20 | | | |
| | Str_Liv | 0.25 | 0.14 | 0.44 | 0.27 | 0.96 | 0.91 | 0.26 | | |
| | Cor_Liv | 0.21 | 0.12 | 0.45 | 0.20 | 0.91 | 0.96 | 0.19 | 0.92 | |
| IEAP | | 0.23 | 0.14 | 0.46 | 0.23 | 0.93 | 0.95 | 0.23 | 0.93 | 0.95 |

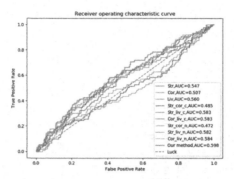

**Fig. 3.** The ROCs of FC-based methods.

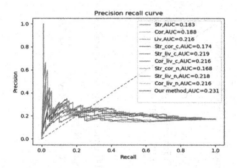

**Fig. 4.** The PR curves of FC-based methods.

in detail, it may yet a better results. Therefore, we conducted experiments using IEAP based on FC to integrate differential expression scores of intra-tissues and inter-tissues. The results of IEAP and FC are show in Figs. 3 and 4.

The AUC of IEAP was 0.598, which was improved by 2.6% compared with the best result of FC. The AUPR was 0.231, which was improved by 5.4% compared with the best result of FC. It was more exciting to see the stability of the results of IEAP.

To verify the consistency of the gene ranked lists, which were obtained from the differentially expression scores using FC with intra-tissues' gene expression data and inter-tissues' gene expression data, we statistics the overlap degree of top ranking genes between any two ranked lists. Since the prediction precision of disease genes was very high when the recall rate was no more than 0.1, we checked the rankings of top 9.

(89 * 0.1 = 8.9) genes in the ranked lists, and found that they were ranked in top 800 of the final ranked lists. So, we statistics the overlap degree of the top 800 genes. From Table 6, we can know that the overlap degrees were larger than 0.20. The overlap degree between the result of IEAP and that of FC with better performance was higher. And the overlap degree between the result of IEAP and that of FC with poor performance was lower. The above analysis results indicate that genes that were differentially expressed between tissues were more susceptible to be affected and differentially expressed during the disease progression.

Integrating the top 800 gene sets of the eight ranked lists, we finally obtained 5 genes simultaneously presented in the top 800 of all ranked list. They were Arpp21 (cAMP-regulated phosphoprotein 21), Rgs4 (regulator of G-protein signaling 4), Rasd2 (RASD family member 2), Gabrd (gamma-aminobutyric acid type A receptor delta subunit), and Tmod1 (tropomodulin 1). These five genes were differentially expressed intra-tissues and inter-tissues.

**Functional Annotation of the 5 Genes**
We conducted functional annotation of these five genes using DAVID [11]. Arpp21 is related to cellular response to heat and nucleic acid binding, Rgs4 is related to inactivation of MAPK activity and GTPase activator activity, Rasd2 is related to synaptic transmission and GTP binding, Gabrd is related to cell junction and GABA-A receptor complex, and Tmod1 is related to muscle contraction and pointed-end action filament capping. The five gene may play a key role during HD progression. It is important to note that Arrpp21 and Rasd2 have also been reported in the article [31], suggesting the effectiveness of IEAP and the significance of the five genes for the disease development.

# 4   Conclusions

In this study, we screen disease genes by prioritizing the differentially expressed genes. To better understand molecular mechanisms under complicated phenotypes, we designed IEAP to integrate the differential expression scores of intra-tissues' and inter-tissues'. We conducted extensive experiments to analyze the performance of different methods with different samples. The best performance of AUC is ~0.6, and AUPR is ~0.23, indicating that the-state-of-art methods cannot effectively distinguish the disease genes from non-disease genes. We also found that differentially expressed genes between different tissues of healthy individuals are likely to be disease genes. So, in the further study, we may develop computational methods from a new perspective to mine the differentially expressed genes between different tissues of healthy individuals. It is also very helpful for understanding the endogenous reasons of disease.

We finally screened five genes, including Arpp21, Rgs4, Rasd2, Gabrd, and Tmod1, two (Arpp21 and Rasd2) of which have been reported to be related with Huntington's disease.

# References

1. Ross, C.A., et al.: Huntington disease: natural history, biomarkers and prospects for therapeutics. Nat. Rev. Neurol. **10**(4), 204 (2014)
2. Seredenina, T., Luthicarter, R.: What have we learned from gene expression profiles in huntington's disease? Neurobiol. Dis. **45**(1), 83 (2012)
3. Wang, X., Huang, T., Bu, G., Xu, H.: Dysregulation of protein trafficking in neurodegeneration. Mol. Neurodegener. **9**(1), 31 (2014)
4. Difiglia, M., et al.: Aggregation of huntingtin in neuronal intranuclear inclusions and dystrophic neurites in brain. Science **277**(5334), 1990 (1997)

5. Kugler, K.G., Mueller, L.A.J., Graber, A., Dehmer, M.: Integrative network biology: graph prototyping for co-expression cancer networks. PLoS ONE 6(7), 22843 (2011)
6. Liu, Z.P.: Identifying network-based biomarkers of complex diseases from high- throughput data. Biomark. Med. 10(6), 633–650 (2016)
7. Xulvibrunet, R., Li, H.: Co-expression networks: graph properties and topological comparisons. Bioinformatics 26(2), 205–214 (2010)
8. Ray, M., Zhang, W.: Analysis of alzheimer's disease severity across brain regions by topological analysis of gene co-expression networks. BMC Syst. Biol. 4(1), 136 (2010)
9. Ideker, T., Krogan, N.J.: Differential network biology. Mol. Syst. Biol. 8(1), 565 (2012)
10. Gwinner, F., et al.: Network-based analysis of omics data: the lean method. Bioinformatics 33(5), 701–709 (2017)
11. Huang, D.W., Sherman, B.T., Lempicki, R.A.: Systematic and integrative analysis of large gene lists using david bioinformatics resources. Nat. Protoc. 4(1), 44 (2009)
12. Bevilacqua, V., Pannarale, P., Abbrescia, M., Cava, C., Paradiso, A., Tommasi, S.: Comparison of data-merging methods with SVM attribute selection and classification in breast cancer gene expression. BMC Bioinformatics 13(S7), 9 (2012)
13. Maulik, U., Mukhopadhyay, A., Chakraborty, D.: Gene-expression-based cancer subtypes prediction through feature selection and transductive SVM. IEEE Trans. Bio-Med. Eng. 60(4), 1111–1117 (2013)
14. Jiang, X., Zhang, H., Zhang, Z., Quan, X.: Flexible non-negative matrix factorization to unravel disease-related genes. IEEE Trans. Comput. Biol. Bioinf. 1(1), 1–11 (2018)
15. Wang, H.Q., Zheng, C.H., Zhao, X.M.: jNMFMA: a joint non-negative matrix factorization meta-analysis of transcriptomics data. Bioinformatics 31(4), 572 (2015)
16. Jiang, X., Zhang, H., Duan, F., Quan, X.: Identify huntington's disease associated genes based on restricted boltzmann machine with RNA-seq data. BMC Bioinformatics 18(1), 447 (2017)
17. Liang, M., Li, Z., Chen, T., Zeng, J.: Integrative data analysis of multi-platform cancer data with a multimodal deep learning approach. IEEE/ACM Trans. Comput. Biol. Bioinf. 12(4), 928–937 (2015)
18. Battle, A., Brown, C.D., Engelhardt, B.E., Montgomery, S.B.: Genetic effects on gene expression across human tissues. Nature 550(7675), 204–213 (2017)
19. Tan, M.H., et al.: Dynamic landscape and regulation of RNA editing in mammals. Nature 550(7675), 249–254 (2017)
20. Tukiainen, T., et al.: Landscape of X chromosome inactivation across human tissues. Nature 550(7675), 244 (2017)
21. Li, X., et al.: The impact of rare variation on gene expression across tissues. Nature 550(7675), 239–243 (2016)
22. Hong, F., Breitling, R.: A comparison of meta-analysis methods for detecting differentially expressed genes in microarray experiments. Bioinformatics 24(3), 374 (2008)
23. Hanley, J.A., Mcneil, B.J.: The meaning and use of the area under a receiver operating characteristic (ROC) curve. Radiology 143(1), 29 (1982)
24. Langfelder, P., et al.: Integrated genomics and proteomics de ne huntingtin cag length-dependent networks in mice. Nat. Neurosci. 19(4), 623 (2016)
25. Yamamoto, S., et al.: A drosophila genetic resource of mutants to study mechanisms underlying human genetic diseases. Cell 159(1), 200–214 (2014)

# Identifying Differentially Expressed Genes Based on Differentially Expressed Edges

Bolin Chen[1,2], Li Gao[1], and Xuequn Shang[1,2(✉)]

[1] School of Computer Science, Northwestern Polytechnical University,
Xi'an, China
`npu_bioinf@hotmail.com`
[2] Key Laboratory of Big Data Storage and Management,
Ministry of Industry and Information Technology,
Northwestern Polytechnical University, Xi'an, China

**Abstract.** Identification of differentially expressed (DE) genes under different experimental conditions is an important task in many microarray-based studies. There are many methods developed to detect DE genes based on either fold-change (FC) strategy or statistical test. However, majority of those methods identify DE genes by calculating the expression values of individual genes, without taking interactions between genes into consideration. In this study, we consider the interaction and importance of genes in the network and believe that the edges in the network also contribute a lot to DE genes. Therefore, we propose three new ideas for calculating the expression values of edges by considering mean expression, minimal expression and partial expression, respectively. Those methods were implemented and evaluated on the microarray data and were compared with existing methods. The results show that the proposed edge-based methods can identify more biologically relevant genes and have high computational efficiency. More importantly, the Min-Edge method outperforms the other methods when feasibility and specificity are considered simultaneously.

**Keywords:** Gene expression · Differentially expressed genes ·
Differentially expressed edges · Microarray data

## 1 Introduction

Detecting differentially expressed (DE) genes under two biological conditions is an essential step and is one of the most common reasons for statistical analysis of Microarray data. Differential expression analysis targets genes with differential expression values in different phenotypes [1], and the identification of DE genes helps us to study disease-related, cell-specific gene expression patterns. Numerous methods have been done for this. A prevalent approach called Fold Change (FC) [2] calculate a ratio of average from the expression values between control and test samples. Levels of cutoffs with 2-fold, such as 0.5 for down-regulated and 2 for up-regulated are usually used, and genes under or above the threshold are selected as DE genes [3]. Because of the cutoffs do not guarantee the reproducibility and take the variability into account [4],

© Springer Nature Switzerland AG 2019
D.-S. Huang et al. (Eds.): ICIC 2019, LNCS 11644, pp. 105–115, 2019.
https://doi.org/10.1007/978-3-030-26969-2_10

so the statistical tests such as *t-test* then become popular. The significance analysis of microarrays (SAM) [5] method impose a restriction on the variability of the genes by adding a value to the denominator of the t-statistic, excluding the genes that do not change or with very high *p-value*. Another popularly used method called Moderated t-statistic (ModT) [6] aims to obtain a *p-value* for each gene for choosing a feasible false discovery rate (FDR). It is an empirical Bayes modification of the *t-test* for detecting DE genes. However, although these methods perform well in identifying DE genes and are widely used, they do not consider the edges in various biomolecular networks, where the changes of edges may result in significant changes of cellular functions. Moreover, traditional DE genes identification methods often obtain a set of isolated genes in a biomolecular network, which is hard to analyze those gene functions from a system biology point of view.

In this study, we present a new idea for detecting DE genes based on the evaluation of DE edges. This idea of detecting DE genes can be applied to any kind of network. Here, we take protein-protein interaction (PPI) [7] network as an example. The method can be regarded as three steps. The first step is to calculate the expression value of edges in the network. Then obtain the DE edges based on the FC method. Finally, the genes at both sides of the DE edges are determined as DE genes. By doing this, DE genes can be obtained by the proposed DE edge-based methods, and those genes tend to be closer to biological functions and have more functional interactions between them.

The rest of the paper is organized as follows. Section 2 describes the methods proposed in this paper. Section 3 addresses the validation results and discussions. Section 4 draws some conclusions.

## 2    Methods

The interactions between genes in the network are reflected by individual edges. Considering the edge of the network also carries certain biological information, we argue that edges should also have corresponding expression values in a biomolecular network. Taking a PPI network for example, it is a set of genes connected by edges representing functional relationships among these genes [8]. The expression value may derive from the genes at both of sides or its neighbors, which provides us with a new idea for identifying DE genes by DE edges instead of by individual genes.

Giving two sets of *control* and *test* samples and a set of genes $g_1, g_2, \cdots, g_n$, let $e_i^c$ and $e_i^t$ be the mean expression value for the gene $g_i (i = 1, 2, \cdots, n)$ in the *control* and *test* samples, respectively, and let $e_{ij}$ be the edge between gene $g_i$ and $g_j$ in the PPI network. We introduce three methods to measure the edge expression value as follows.

### 2.1    Mean-Edge Method

The "Mean-Edge" method is designed to measure the edge expression value by taking the mean expression of two incidence genes. Assuming that the edge $e_{ij}$ has an expression value in the PPI network, it can only be measured by the expression value of the genes at both sides. Therefore, we take the average expression values of genes at both sides as the expression value of the edge $e_{ij}$. Then, the obtained expression value of the edge is taken as an initial value to find the DE edges based on the FC method. Here, we use 2-fold as threshold to select the DE genes. The formula is defined as

$$V_{mean} = log_2 \frac{e_i^t + e_j^t}{e_i^c + e_j^c} \tag{1}$$

Each edge corresponds to a $|V_{mean}|$ value, and an edge with a value greater than 1 is considered to be a DE edge, then the genes at both sides are determined as DE genes.

## 2.2 Min-Edge Method

Based on the idea of method above, we propose another method called "Min-Edge". Since the interaction between genes in the PPI network reflects the functional relationship between genes, we consider the expression value of the edge depends on the genes at both sides, and the expression value of an edge should not larger than the gene expression value of either side. Thus, the minimum of the gene expression value from both sides is taken as the expression value of each edge $e_{ij}$. Then, FC method with 2-fold helps us to calculate the DE edges through selected edges expression values. The formula is defined as

$$V_{min} = log_2 \frac{\min\left\{e_i^t, e_j^t\right\}}{\min\left\{e_i^c, e_j^c\right\}} \tag{2}$$

Using the formula above, we take $|V_{min}| \geq 1$ to find the DE edges $e_{ij}$ set, and then consider that the genes at both sides of the DE edges are DE genes.

## 2.3 β-Edge Method

From the perspective of the edges, the above methods consider that the expression value of each edge is only related to the genes at both sides, ignoring the neighbors of genes. From the perspective of the genes in the network, it is believed that part of the expression of each gene contributes to its neighbors. By taking this scenario into consideration, we present another parameter method called "β-Edge". We assume that the gene expression value $e_{i_t}$ obtained after processing the original data consists of two parts. One part of which is its own expression value $e_{i_s}$, and the other is the expression value $e_{i_n}$ contributed to its neighbors. To be more specific, let

$$e_{i_t} = e_{i_s} + e_{i_n} \tag{3}$$

where $e_{i_s} = \beta * e_{i_t}$, $e_{i_n} = (1 - \beta) * e_{i_t}$, $0 < \beta < 1$.

Based on this idea, the expression value of the edge $e_{ij}$ is calculated as the sum of the expression value of $g_i$ contributes to $g_j$ and the expression value of $g_j$ contributes to

$g_i$, where the expression value contributed by each gene to each of its neighbors is equal, which is defined as

$$e_{ij}^0 = \frac{e_{i_n}}{d_i} + \frac{e_{j_n}}{d_j}$$

(4)

where $d_i$, $d_j$ represents the number of neighbors of $g_i$ and $g_j$ in the network, respectively.

After calculating the expression value of each edge, the DE genes are calculated based on the formula shown below

$$V_\beta = log_2 \frac{e_{ij}^{t_0}}{e_{ij}^{c_0}}$$

(5)

where $e_{ij}^{t_0}$ and $e_{ij}^{c_0}$ represent the expression values of the edge $e_{ij}$ for test and control samples, respectively. An edge with a value $|V_\beta| \geq 1$ is considered to be a DE edge, and the genes at both sides are DE genes.

# 3  Results and Discussions

## 3.1  Data Sources

The microarray data is downloaded from the National Center for Biotechnology Information Gene Expression Omnibus (GEO) (accession number: GSE41089; available at: https://www.ncbi.nlm.nih.gov/geo/query/acc.cgi?acc=GSE41089) [9]. This microarray data contains 22,690 probe sets, 3 samples from uninfected mice (control) and 3 samples from infected mice (test). We use Affymetrix Microarray Suite version 5.0 (MAS) algorithm [10] to preprocess the probe-level data. Following this normalization step, 8089 genes are filtered for further analysis.

The PPI dataset of the mouse is derived from the database of BioGRID [11] consists of 12867 genes and 45140 edges. After deleting the duplicated edges between the same pair of nodes and the edges connecting to itself, there are 12849 genes and 38683 edges for the next analysis.

## 3.2  Results of Microarray Data

We compare three proposed edge-based methods with three existing individual genes-based methods (FC, SAM and ModT) in terms of the performance for differential expression [12] by a two-step analysis based on the DE genes. We use the KEGG [13] pathway enrichment to analyze the scale of the enriched DE genes corresponding to significant pathways, and functional interaction network to illustrate the important functionally related DE genes.

**Pathway Enrichment Analysis.** By conducting six methods on genes, the ModT, SAM, FC, Mean-Edge, β-Edge, and Min-Edge methods obtain 1565, 1636, 1687, 608,

592 and 811 DE genes, respectively. Among these methods, there are 192, 196, 204, 608, 592, and 811 DE genes with PPI relationship, respectively. Apparently, ModT method obtains the least number of DE genes, while the Min-Edge method obtains the most. However, this is not enough to explain whether the Min-Edge method is better for identifying DE genes or not. To evaluate our method more rationally, we carry out pathway enrichment analysis on DE gene sets obtained from each method and investigate the significant biological pathways of DE genes [14].

**Table 1.** Number of genes and pathways obtained by each method.

| Method | Pathway no. | Top 25 | Top 50 | All |
|--------|-------------|--------|--------|-----|
| β-Edge | 116 | 216 | 265 | 340 |
| Min-Edge | 123 | **276** | **354** | **444** |
| Mean-Edge | 114 | 194 | 261 | 335 |
| ModT | 75 | 100 | 111 | 119 |
| SAM | 73 | 96 | 108 | 117 |
| FC | 75 | 98 | 112 | 122 |

Firstly, we conduct PPI correlations and pathway enrichment on the obtained DE genes sets. Table 1 shows the number of DE genes and pathways detected by all methods. "Pathway No." represents the number of pathways for DE genes identified by each method. "Top #" represents the number of DE genes contained in the significant pathway of top #. "All" represents the number of DE genes contained in all significant pathway. Since the significant pathways reveal the particular function of genes present in organisms, we take some more significant pathways for analysis. When the top 50 most significant pathways are taken, the Mean-Edge method obtain 261 DE genes in pathways, and the parametric β-Edge method obtain 265 DE genes. The ModT, SAM and FC have the smaller number of DE genes, while the Min-Edge method has the most. Obviously, no matter how many significant pathways are taken that are sorted by p-value previously, the number of DE genes enriched by the Min-Edge method is the highest.

We then use the Venn [15] diagram to search for common and specific DE genes identified by six methods. When the top 25 significant pathways are taken, the Min-Edge method shares more genes with the β-Edge method and Mean-Edge than with the ModT method. As expected, the edge-based methods (Mean-Edge, β-Edge and Min-Edge) are slightly more sensitive than the *t-statistic* based method as they detect most of genes that detected by the ModT method shown in Fig. 1(a). Figure 1(b) shows that when take the top 25 significant pathways, 78 DE genes are detected by these four methods. The Min-Edge method also shares 89 DE genes with the β-Edge methods, which are not detected by SAM method. And the DE genes enriched by the edge-based method contains most of the DE genes detected by the FC, ModT and SAM method. Surprisingly, the Min-Edge method detect the largest number of DE genes that are not detected by others. We further select all significant pathways to evaluate the evidence of six methods, which are shown in the two figures below. Figures 1(c) and (d) show that the DE genes detected by edge-based methods contain the most of DE genes detected by the FC, SAM and ModT method, of which the Min-Edge can detect the most DE genes. This comparison reveals that the edge-based methods outperforms the previous methods.

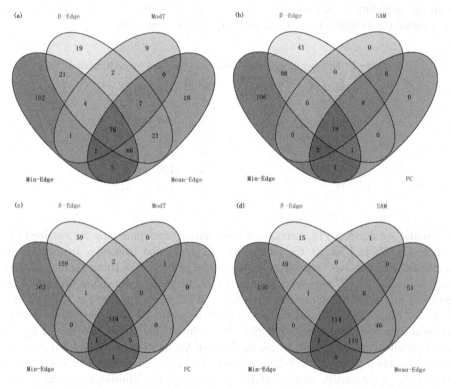

**Fig. 1.** Comparison of the performance of methods with the microarray data.

Secondly, we perform simple pathway enrichment analysis on six DE genes sets and compare the unions of each method to see what biological features DE genes may connote. Since the number of obtained pathways is too large to visualize, we take the union of the top 25 significant pathways to analyze, as Fig. 2 shows. From the perspective of enriched pathways, DE genes obtained by edge-based methods are contained in more significant pathways, while other three-method cannot. Among which the significant pathways obtained by Min-Edge method have very low *p-value*, and these pathways are likely to play an important role in the development of organisms [16]. From the number of DE genes contained in each pathway, the number of DE genes enriched by edge-based methods is more than other methods in the same pathway, and it can be clearly seen that the dot color of the Min-Edge method in figure is darker than other methods.

The above two aspects indicate that the proposed edge-based methods work better in terms of DE genes identification in pathways.

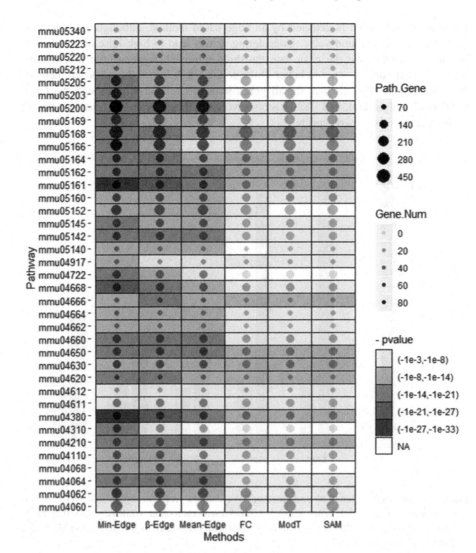

**Fig. 2.** Simple pathway enrichment analysis of DE genes. The *p-value* of the pathway is displayed in different alpha, and NA indicate that this method does not enrich the DE genes into the pathway, which means the darker the color, the smaller the *p-value* of the pathway. The size of the dot (Path.Gene) is proportional to the number of genes originally in the pathway, and the color of the dot (Gene.Num) indicates the number of DE genes enriched in the pathway. Dark color represents a large of DE genes are contained in pathway.

**Functional Interaction Network of DE Genes.** The above experiments show that comparing with FC, ModT and SAM method, a large number of DE genes are detected by DE edges, and most DE genes are enriched in significant pathways, especially by Min-Edge method. The result may be caused by the interaction between genes or the value of edge depends on the genes at both sides, assuming that two DE genes are found once a DE edge is detected by an edge-based method, although the gene may have very low expression value. Considering it may lead to high false discovery rate because we don't take the error rate threshold removed, we construct a functional interaction network of genes to better illustrate the feasibility of the proposed methods [17]. We obtain 2072 clusters for all genes with PPI relationship through the ClusterONE [18] algorithm, and 782 clusters involved DE genes are selected for further useful analyses from two aspects.

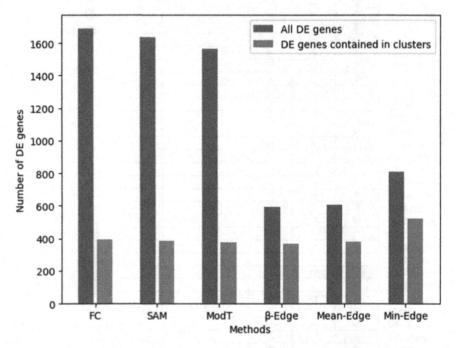

**Fig. 3.** Results of DE genes of each method. We can see from the figure that, vertex-based methods tend to detected DE genes which are isolated with each other, while edge-based methods tend to detect DE genes which are functional related with each other, especially for the Min-Edge method. (Color figure online)

From the perspective of the number of DE genes in each method, as Fig. 3 shows. Blue represents the number of DE genes obtained by each method, while yellow represents the number of DE genes contained in all clusters by each method. Obviously, although the number of DE genes found by traditional methods is much larger than that of the edge, only a small part of them eventually participate in biological functions. On the contrary, the DE genes found by the edge method have high coverage in functional clusters. For example, there are 523 genes among 811 DE genes obtained by the Min-Edge method that ultimately participate in the function, while only 393 among the 1687 DE genes in the FC method. This shows that more functionally related DE genes can be obtained based on the edge methods, of which Min-Edge method is the most, which helps us to better analyze biological mechanisms.

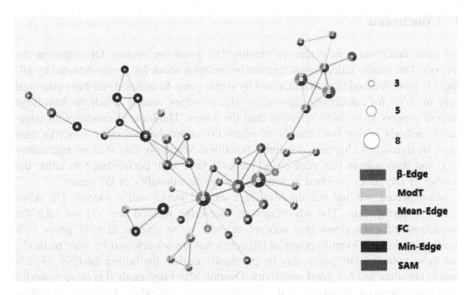

**Fig. 4.** Functional interaction network of DE genes with six methods. Each node represents a functional cluster, the size of which is proportional to the number of genes bearing cluster. The thickness of the edge represents the correlations between clusters. The proportion of color represents the ratio of the DE genes to total genes in this cluster. (Color figure online)

From the perspective of the coverage of DE genes in each cluster shown in Fig. 4. We construct a functional interaction network for each cluster based on the results of clustering. Then we regard each cluster as a new node, and the PPI correlations of genes between clusters as the weights for edges. Weights here indicate the strength of associations between informative related clusters in the network [19]. The wider the edges, the more closely biology functionality between clusters. We combine the biological significance of clustering to assess the feasibility of proposed methods by observing the connectivity of the network. Considering the complexity of the functional network relationship, we extract the strongly interacting clusters and finally select the largest connected component with 48 nodes. This functional interaction network of genes depicts the number of DE genes involved in each method, as well as the correlation between functional clusters [20], represented in ring charts in Fig. 4. We can conclude that more functionally related DE genes can be found by edge-based method in most clusters. Surprisingly, out of six methods, Min-Edge method can find some functionally related DE genes while others cannot, which have the possibility of strongly interacting among the clusters.

According to the analyses of above two aspects, we can conclude that the edge-based method can find more functionally related genes. As expected, this kind of method performs better in the functional interaction network, of which the Min-Edge method works best.

## 4  Conclusion

We have described a new idea to identify DE genes by finding DE edges in the network. Our results indeed show that the information about DE genes detected by DE edges is more focused than that detected by single gene. In addition, we have proposed three methods for calculating expression value of edges, among which the Min-Edge method appears to be more sensitive than the others. The basic assumption for edge-based methods is that the changes of edges in various biomolecular networks may result in significant changes of cellular functions. We apply this idea on microarray data, and demonstrate that edge-based methods have high performance in either the number of DE genes enriched in pathways or the functionality of DE genes.

As expected, we find that the Min-Edge method works well to identify DE genes from microarray data. The advantages of Min-Edge method are: (1) we take the interactions between genes into account in order to be able to identify genes with similar function. (2) identify potential DE genes that are not detected by other methods, and these potential DE genes may be potentially critical for further analysis. (3) it's easy to calculate and has good sensitivity. Overall, Min-Edge method is more powerful than other methods in terms of the sensitivity and feasibility. Given its excellent performance, we believe that Min-Edge method should be another new way to identify the DE genes. The proposed method could be further evaluated by using different protein interaction networks and also a large dataset in the future.

**Acknowledgement.** This work was supported by the National Natural Science Foundation of China under [Grant No. 61602386, 61772426 and 61332014]; the Natural Science Foundation of Shaanxi Province under [Grant No. 2017JQ6008]; the Fundamental Research Funds for the Central Universities, and the Top International University Visiting Program for Outstanding Young scholars of Northwestern Polytechnical University.

## References

1. Dembélé, D., Kastner, P.: Fold change rank ordering statistics: a new method for detecting differentially expressed genes. BMC Bioinformatics **15**, 14 (2014). https://doi.org/10.1186/1471-2105-15-14
2. Shi, L., Tong, W., Fang, H., et al.: Cross-platform comparability of microarray technology: intra-platform consistency and appropriate data analysis procedures are essential. BMC Bioinformatics **6**(2), 1–14 (2005)
3. Lockhart, D.J., Brown, E.L., Wong, G.G., et al.: Expression monitoring by hybridization to high density oligonucleotide arrays. Nat. Biotechnol. **14**(13), 1675–1680 (1996)
4. Mccarthy, D.J., Smyth, G.K.: Testing significance relative to a fold-change threshold is a TREAT. Bioinformatics **25**(6), 765–771 (2009)
5. Tusher, V.G., Tibshirani, R., Chu, G., et al.: Significance analysis of microarrays applied to the ionizing radiation response. Proc. Nat. Acad. Sci. U.S.A. **98**(9), 5116–5121 (2001)
6. Smyth, G.K.: Linear models and empirical bayes methods for assessing differential expression in microarray experiments. Stat. Appl. Genet. Mol. Biol. **3**(1), 1–28 (2004)

7. De Las Rivas, J., Fontanillo, C.: Protein-protein interactions essentials: key concepts to building and analyzing interactome networks. PLoS Comput. Biol. **6**(6), e1000807 (2010). https://doi.org/10.1371/journal.pcbi.1000807

8. Bebek, G.: Identifying gene interaction networks. Methods Mol. Biol. **850**, 483–494 (2012). https://doi.org/10.1007/978-1-61779-555-8_26

9. Silva, G.K., Costa, R.S., Silveira, T.N., Caetano, B.C., et al.: Apoptosis-associated speck-like protein containing a caspase recruitment domain inflammasomes mediate IL-1β response and host resistance to Trypanosoma cruzi infection. J. Immunol. **191**(6), 3373–3383 (2013)

10. Pepper, S.D., Saunders, E.K., Edwards, L.E., Wilson, C.L., Miller, C.J.: The utility of MAS5 expression summary and detection call algorithms. BMC Bioinformatics **30**(8), 273 (2007)

11. Stark, C., Breitkreutz, B.J., Reguly, T., Boucher, L., Breitkreutz, A., Tyers, M.: BioGRID: a general repository for interaction datasets. Nucleic Acids Res. **34**, D535–D539 (2006)

12. Hong, F., Breitling, R.: A comparison of meta-analysis methods for detecting differentially expressed genes in microarray experiments. Bioinformatics **24**(3), 374–382 (2008)

13. Kanehisa, M., Goto, S.: KEGG: kyoto encyclopedia of genes and genomes. Nucleic Acids Res. **28**, 27–30 (2000)

14. Zhang, Q.: A powerful nonparametric method for detecting differentially co-expressed genes: distance correlation screening and edge-count test. BMC Syst. Biol. **12**(1), 58 (2018)

15. Cai, H., et al.: VennPlex–a novel venn diagram program for comparing and visualizing datasets with differentially regulated datapoints. PLoS One **8**, e53388 (2013)

16. Farztdinov, V., Mcdyer, F.A.: Distributional fold change test – a statistical approach for detecting differential expression in microarray experiments. Algorithms Mol. Biol. **7**(1), 29 (2012)

17. Aouiche, C., Chen, B., Shang, X.: Predicting stage-specific cancer related genes and their dynamic modules by integrating multiple datasets. BMC Bioinformatics **20**(S7), 194 (2019)

18. Nepusz, T., Yu, H., Paccanaro, A., et al.: Detecting overlapping protein complexes in protein-protein interaction networks. Nat. Methods **9**(5), 471–472 (2012)

19. Chen, B., Shang, X., Li, M., Wang, J., Wu, F.: Identifying individual-cancer-related genes by rebalancing the training samples. IEEE Trans. NanoBiosci. **15**(4), 309–315 (2016)

20. Shi, G., Wang, Y., Zhang, Ch.: Identification of genes involved in the four stages of colorectal cancer: gene expression profiling. Mol. Cell. Probes **37**, 39–47 (2018)

# Gene Functional Module Discovery via Integrating Gene Expression and PPI Network Data

Fangfang Zhu[1,2], Juan Liu[3], and Wenwen Min[1,3(✉)]

[1] School of Mathematics and Computer Science,
Jiangxi Science and Technology Normal University, Nanchang 330038, China
wenwen.min@qq.com
[2] Library of Jiangxi Science and Technology Normal University,
Nanchang 330038, China
[3] School of Computer, Wuhan University, Wuhan 430072, China

**Abstract.** Sparse Singular Value Decomposition (SSVD) model has been proposed to bicluster gene expression data to identify gene modules. However, traditional SSVD model can only handle the gene expression data where no gene-gene interaction information is integrated. Here, we develop a Sparse Network-regularized SVD (SNSVD) method, which can integrate the gene-gene interaction information into the SSVD model, to identify the underlying gene functional modules from the gene expression data. The simulation results on synthetic data show that SNSVD is more effective than the traditional SVD-based methods.

**Keywords:** Sparse network-regularized SVD ·
Biclustering gene expression data

## 1 Introduction

With the rapid development of (single-cell) RNA-Seq and microarray technologies, huge number of gene expression data have been generated. These data provide new opportunities to study on the gene cooperative mechanisms [1–3]. However, traditional clustering techniques face with the limitation that some genes can co-regulate across some samples rather than all samples in the real biological systems [4]. Therefore, many biclustering methods [1, 5–7] are proposed to discover some sets of genes that co-regulate across some samples.

Recently, some based Sparse Singular Value Decomposition (SSVD) model methods have also been proposed for biclustering gene expression data to discover gene functional modules (biclusters) [8], such as ALSVD [1], L0SVD [9], and so on. However, most of them ignore the prior knowledge from cellular pathway and molecular network, whereas such information (e.g. PPI network) is very useful to improve biological interpretability [10–13].

To address the limitation, we integrate the PPI network in the SSVD model for biclustering gene expression. Although PPI network has been used in many applications

© Springer Nature Switzerland AG 2019
D.-S. Huang et al. (Eds.): ICIC 2019, LNCS 11644, pp. 116–126, 2019.
https://doi.org/10.1007/978-3-030-26969-2_11

for accurate discovery or better biological interpretability [10, 11, 14, 15]. In this paper, we obtain the gene-gene interaction information from the PPI network, and propose a sparse network-regularized SVD (SNSVD) model to identify gene modules by integrating the gene expression data and PPI network. In order to ensure the discovered gene modules in which genes are co-expressed and densely connected in PPI network, we introduce a sparse network-regularized penalty in the model. Compared to traditional regularized penalties, the sparse network-regularized penalty can make the biclustering process tend to select correlated and interacted genes. To evaluate SNSVD, we compare its performance with other representative SSVD models on the artificial data. In the meanwhile, we apply SNSVD to the gene expression data from CGP [16] with the PPI network from Pathway Commons [17] as the gene-gene interaction constraint, and investigate the functionality of the identified modules by SNSVD from multiple perspectives.

## 2 Sparse Network-Regularized SVD Model

Let $\mathbf{X} \in R^{p \times n}$ ($p$ genes and $n$ samples) be the gene expression data, the general Sparse SVD (SSVD) model for biclustering is described as the following:

$$
\begin{aligned}
&\underset{\mathbf{u},\mathbf{v},d}{\text{minimize}} \|\mathbf{X} - d\mathbf{u}\mathbf{v}^T\|_F^2 \\
&\text{subject to } \|\mathbf{u}\|_2^2 \leq 1, \ \|\mathbf{v}\|_2^2 \leq 1, \ P_1(\mathbf{u}) \leq c_1, \ P_2(\mathbf{v}) \leq c_2
\end{aligned}
\tag{1}
$$

where $\mathbf{u}$ and $\mathbf{v}$ are two singular vectors (corresponding to genes and samples respectively) to be determined, $d$ is the singular value, $\| \ \|_F$ is the Frobenius norm, and $\| \ \|_2$ is Euclidean norm. $P_1(\mathbf{u})$ and $P_2(\mathbf{v})$ are two constraint functions to promise the sparsity of $\mathbf{u}$ and $\mathbf{v}$ respectively, $c_1$ and $c_2$ are two parameters to control the sparse levels. Based on the SSVD model, one can get ALSVD or L0SVD methods if using $L_1$-norm or $L_0$-norm as the constraint functions respectively. According to the non-zero elements of $\mathbf{u}$ and $\mathbf{v}$, one can identify a gene module.

As it known to all, if two genes are adjacent in the PPI network, then they may be functionally related and be very likely in the same module. So, we define $P_1(\mathbf{u})$ as the combination of a $L_1$-norm and a sparse network-regularized norm. Suppose $\mathbf{A} \in R^{p \times p}$ is the adjacency matrix of the PPI network. And $\mathbf{A}_{ij} = 1$ if vertex $i$ and vertex $j$ is connected and $\mathbf{A}_{ij} = 0$ otherwise. The degree of vertex $i$ is defined by $d_i = \sum_{j=1}^{p} \mathbf{A}_{ij}$. Thus, the normalized Laplacian matrix $\mathbf{L} = (\mathbf{L}_{ij})$ encoding the PPI network can be defined as:

$$
\mathbf{L}_{ij} = \begin{cases} 1, & \text{if } i = j \text{ and } d_i \neq 0, \\ -\frac{\mathbf{A}_{ij}}{\sqrt{d_i d_j}}, & \text{if } i \text{ and } j \text{ are adjacent}, \\ 0, & \text{otherwise}. \end{cases}
\tag{2}
$$

Correspondingly, we have $\mathbf{u}^T\mathbf{L}\mathbf{u} = \frac{1}{2}\sum_i \sum_j \mathbf{A}_{ij}\left(\frac{\mathbf{u}_i}{\sqrt{d_i}} - \frac{\mathbf{u}_j}{\sqrt{d_j}}\right)^2$ which reaches its minimal if genes within the same module interact with each other [15]. To integrate the PPI network information into the SSVD framework, we introduce the following constraint function on $\mathbf{u}$:

$$P_1(\mathbf{u}) = \lambda\|\mathbf{u}\|_1 + \sigma\mathbf{u}^T\mathbf{L}\mathbf{u} \tag{3}$$

where $\lambda$ and $\sigma$ are two parameters. The $L_1$-norm induces the sparsity constraint of the elements in $\mathbf{u}$ and the quadratic Laplacian norm ($\mathbf{u}^T\mathbf{L}\mathbf{u}$), is to let the biclustering process tend to find genes which are connected with each other in the PPI network.

As for the samples, we only require the biclustering process to promise the sparsity. Therefore, we simply apply a $L_0$-regularized constraint on sample variables $\mathbf{v}$, i.e., $P_2(\mathbf{v}) = \|\mathbf{v}\|_0$. $L_0$-norm is known as the most essential sparsity measure [9].

Thus, our SNSVD model can be represented as follows:

$$\begin{aligned}
&\underset{\mathbf{u},\mathbf{v},d}{\text{minimize}}\|\mathbf{X} - d\mathbf{u}\mathbf{v}^T\|_F^2 \\
&\text{subject to } \|\mathbf{u}\|_2^2 \leq 1,\ \lambda\|\mathbf{u}\|_1 + \sigma\mathbf{u}^T\mathbf{L}\mathbf{u} \leq c_1, \\
&\quad\quad\quad \|\mathbf{v}\|_2^2 \leq 1,\ \|\mathbf{v}\|_0 \leq k_v,
\end{aligned} \tag{4}$$

where $c_1$ and $k_v$ are two super-parameters to control the sparsity levels.

## 3    Solving SNSVD Model

Although there are three parameters to be optimized in Eq. (4). It is notable that once $\mathbf{u}$ and $\mathbf{v}$ are fixed, then $d$ can be determined $d = \mathbf{u}^T\mathbf{X}\mathbf{v}$. Thus, to solve Eq. (4), we just need to optimize $\mathbf{u}$ and $\mathbf{v}$. Inspired by Ref. [18], we propose an alternating iterative strategy to learn $\mathbf{u}$ (left singular vector) and $\mathbf{v}$ (right singular vector), i.e., fixing $\mathbf{v}$ to update $\mathbf{u}$ and fixing $\mathbf{u}$ to update $\mathbf{v}$.

### 3.1    Learning the Left Singular Vector

Fixing $\mathbf{v}$, Eq. (4) can be rewritten as:

$$\begin{aligned}
&\underset{\mathbf{u}}{\text{minimize}}\|\mathbf{X} - d\mathbf{u}\mathbf{v}^T\|_F^2 \\
&\text{subject to } \|\mathbf{u}\|_2^2 \leq 1,\ \lambda\|\mathbf{u}\|_1 + \sigma\mathbf{u}^T\mathbf{L}\mathbf{u} \leq c_1,
\end{aligned} \tag{5}$$

$\|\mathbf{X} - d\mathbf{u}\mathbf{v}^T\|_F^2 = tr(\mathbf{X}\mathbf{X}^T) + d^2 tr(\mathbf{u}\mathbf{v}^T\mathbf{v}\mathbf{u}^T) - 2d\mathbf{u}^T\mathbf{X}\mathbf{v}$, where $tr()$ denotes the trace of a matrix. Since the singular vectors $\mathbf{u}$ and $\mathbf{v}$ are guaranteed to be two unit vectors, $tr(\mathbf{u}\mathbf{v}^T\mathbf{v}\mathbf{u}^T) = 1$. Therefore minimizing $\|\mathbf{X} - d\mathbf{u}\mathbf{v}^T\|_F^2$ is equivalent to minimizing $-\mathbf{u}^T\mathbf{X}\mathbf{v}$. Let $\mathbf{z} = \mathbf{X}\mathbf{v}$, the optimizing problem in (5) can be redefined as follows:

$$\text{minimize} \ -\mathbf{u}^T\mathbf{z}$$
$$\text{subject to} \ \|\mathbf{u}\|_2^2 \le 1, \ \lambda\|\mathbf{u}\|_1 + \sigma\mathbf{u}^T\mathbf{L}\mathbf{u} \le c_1. \tag{6}$$

In order to solve above constrained optimization problem, we define its Lagrangian form as follows:

$$L(\mathbf{u}) = -\mathbf{u}^T\mathbf{z} + \eta\mathbf{u}^T\mathbf{u} + \lambda\|\mathbf{u}\|_1 + \sigma\mathbf{u}^T\mathbf{L}\mathbf{u}, \tag{7}$$

where $\lambda \ge 0$, $\eta \ge 0$, $\sigma \ge 0$ are Lagrangian multipliers. In order to facilitate the calculation without loss of generality, we use $\frac{1}{2}\eta$ instead of $\eta$, $\frac{1}{2}\sigma$ instead of $\sigma$, then (7) can be rewritten as:

$$L(\mathbf{u}) = -\mathbf{u}^T\mathbf{z} + \frac{1}{2}\eta\mathbf{u}^T\mathbf{u} + \lambda\|\mathbf{u}\|_1 + \frac{1}{2}\sigma\mathbf{u}^T\mathbf{L}\mathbf{u}, \tag{8}$$

It can be proven that function (8) is a convex function with respect to $\mathbf{u}$, therefore the optimal solution can be characterized by its sub-gradient equations.

Let $\mathbf{W} = \mathbf{I} - \mathbf{L} = \mathbf{D}^{-1/2}\mathbf{A}\mathbf{D}^{-1/2}$ ($\mathbf{D}$ is a degree diagonal matrix of $\mathbf{A}$), then we have the sub-gradient equations of Eq. (8) as:

$$\frac{\partial L}{\partial \mathbf{u}_j} = -\mathbf{z}_j + \eta\mathbf{u}_j + \lambda s_j + \sigma\mathbf{u}_j - \sigma\mathbf{W}_j\mathbf{u} = 0, \ j = 1, \cdots, p \tag{9}$$

where $s_j = sign(\mathbf{u}_j)$ if $\mathbf{u}_j \ne 0$ and $s_j \in \{t, |t| \le 1\}$ otherwise; and $\mathbf{W}_j$ is the $j$-th row of matrix $\mathbf{W}$.

Let the solution of (9) be $\hat{\mathbf{u}} = (\hat{\mathbf{u}}_1, \hat{\mathbf{u}}_2, \cdots, \hat{\mathbf{u}}_p)$. By using the cycle coordinate descent method [19], we obtain the following coordinate update rule for $\mathbf{u}_j$:

$$\hat{\mathbf{u}}_j = \begin{cases} 0, & \text{if } |\mathbf{z}_j + \sigma\mathbf{W}_j\hat{\mathbf{u}}| \le \lambda \\ \frac{\mathbf{z}_j + \sigma\mathbf{W}_j\hat{\mathbf{u}} - \lambda sign(\hat{\mathbf{u}}_j)}{\eta + \sigma}, & \text{otherwise.} \end{cases} \tag{10}$$

After defining a soft thresholding operator $S(x, \lambda) = sign(x)(|x| - \lambda)_+$, we have $\hat{\mathbf{u}}_j = S(\mathbf{z}_j + \sigma\mathbf{W}_j\hat{\mathbf{u}}, \lambda)/(\eta + \sigma)$. So, we can obtain the normalized solution $\mathbf{u} = \frac{\hat{\mathbf{u}}}{\|\hat{\mathbf{u}}\|_2}$.

### 3.2  Learning the Right Singular Vector

Similarly, fixing $\mathbf{u}$ in Eq. (4), we can get the following constrained optimization problem:

$$\text{minimize} \ \|\mathbf{X} - d\mathbf{u}\mathbf{v}^T\|_F^2$$
$$\text{subject to} \ \|\mathbf{v}\|_2^2 \le 1, \ \|\mathbf{v}\|_0 \le k_v, \tag{11}$$

Let $\mathbf{z} = \mathbf{X}^T\mathbf{u}$, $\hat{\mathbf{v}} = d\mathbf{v}$, we thus have $\|\mathbf{X} - d\mathbf{u}\mathbf{v}^T\|_F^2 = \|\mathbf{z} - \hat{\mathbf{v}}\|_2^2 + c$ where $c = tr(\mathbf{X}^T\mathbf{X}) - \mathbf{u}^T\mathbf{X}\mathbf{X}\mathbf{u}$. Obviously $c$ is a constant value with respect to $\mathbf{v}$. Thus problem (11) is equivalent to:

$$\underset{\hat{\mathbf{v}}}{\text{minimize}}\|\mathbf{z} - \hat{\mathbf{v}}\|_2^2, \ \|\hat{\mathbf{v}}\|_0 \leq k_v. \tag{12}$$

Obviously the optimal solution is $\hat{\mathbf{v}} = \mathbf{z} \cdot I(|\mathbf{z}| \geq |\mathbf{z}|_{(k_v)})$ where $I(\cdot)$ is the indicator function and $\cdot$ is point multiplication function, and $|\mathbf{z}|_{(i)}$ denotes the $i$-th order statistic of $|\mathbf{z}|$, i.e. $|\mathbf{z}|_{(1)} \geq |\mathbf{z}|_{(2)} \geq \cdots \geq |\mathbf{z}|_{(n)}$. That is to way, we only keep the $k_v$ variables of $\mathbf{z}$ corresponding to its $k_v$ largest absolute values. So, the normalized optimal solution of (11) is $\mathbf{v} = \frac{\hat{\mathbf{v}}}{\|\hat{\mathbf{v}}\|_2}$.

### 3.3  SNSVD Algorithm

For a specific pair of $\mathbf{u}$ and $\mathbf{v}$, there is $d = \mathbf{u}^T\mathbf{X}\mathbf{v}$. Finally, we obtain the SNSVD algorithm to solve SNSVD model by iteratively updating $\mathbf{u}$ and $\mathbf{v}$ (See Sects. 3.1 and 3.2), and calculating the corresponding $d$ until the change of $d$ is smaller than a given threshold. SNSVD algorithm is an alternating iterative algorithm. Obviously, the algorithm is convergent and its time complexity is $O(Tnp + Tp^2 + Tn^2)$, where $T$ is the number of iterations.

It is notable that every run of SNSVD algorithm can only obtain a pair of sparse singular vectors ($\mathbf{u}$ and $\mathbf{v}$), which represents one gene module of genes and samples. In order to identify multiple gene modules, we can repeat running the algorithm. After each turn of the iteration, we use the obtained $\mathbf{u}$, $\mathbf{v}$ and $d$ to modify the gene expression data $\mathbf{X}$ (i.e., $\mathbf{X} := \mathbf{X} - d\mathbf{u}\mathbf{v}^T$), the modified $\mathbf{X}$ is then used as the new input data for the next run to obtain the next pair of singular vectors. Moreover, we notice that the algorithm may get different local optima with different initials, therefore in every iteration, we run the algorithm five times with different initials which are generated according to the multivariate standard normal distribution, and choose the best one as the final solution of each turn.

## 4  Comparing with Representative Sparse SVD Methods

### 4.1  Simulation Study

We evaluate the performance of SNSVD with simulated data by comparing it with other sparse SVD based methods L0SVD [9], ALSVD [1] and SCADSVD [20]. Without loss of generality, we define a rank-one true signal matrix as $\mathbf{u}\mathbf{v}^T$ where $\mathbf{u}$ and $\mathbf{v}$ are vectors of size $p \times 1$ and $n \times 1$, respectively. The observed matrix is defined as $\mathbf{X} = \mathbf{u}\mathbf{v}^T + \gamma\boldsymbol{\varepsilon}$, where $\boldsymbol{\varepsilon}$ is a noise matrix each element in which is randomly sampled from a standard normal distribution and $\gamma$ is a nonnegative parameter to control the signal-to-noise ratio. To generate the simulated data, we first generate two sparse singular vectors $\mathbf{u}$ and $\mathbf{v}$ with $p = 200$, $n = 100$ whose first 50 elements equal to 1 (non-zeros), and the remaining ones are zeros. Then we create a series of observation matrices $X$ for each $\gamma$ ranging from 0.02 to 0.06 in steps of 0.005. In addition, we create a simulated prior PPI network for row variables of $X$, where any two nodes in first 50 vertices are connected with probability $p_{11} = 0.3$, and remaining ones are connected with probability $p_{12} = 0.1$.

For each $\gamma$, we generate 50 different noise matrices $\varepsilon$ to get 50 observed matrices $X$ for testing. The average sensitivity and specificity of $u$ (or $v$) on the 50 matrices $X$ are calculated. Moreover, we set $\sigma = 0.5$ according to 5-fold cross validation test, and force the singular vectors to contain 50 non-zero elements with same sparsity level for each method by tuning the parameters so that the results of different methods are comparable. The average sensitivities and specificities of $u$ (or $v$) with different $\gamma$ are compared in Fig. 1, from which we can see that the performance of our proposed method (SNSVD) is superior to that of other methods, which illustrating that SNSVD model can enhance the power of variable selection by integrating the prior network knowledge.

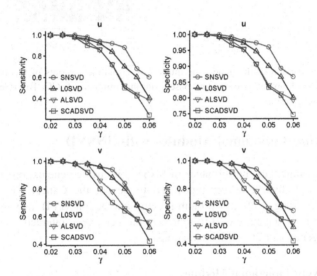

**Fig. 1.** Evaluation of different methods in simulated data. Sensitivity denotes the percentage of true non-zero entries identified, and specificity denotes the percentage of true zero entries identified.

## 4.2  Comparing with L0SVD on the Real Data

Now that L0SVD have shown good performances in other applications [9], we also compare it with our method to further illustrate the importance of integrating PPI network. Therefore, we identify 40 gene modules on CGP data via SNSVD and L0SVD respectively (See Sect. 5 for more details). Figure 2 shows the comparing results of interaction enrichment scores and functional enrichment of the modules. We find that the gene-gene interaction edges of the identified modules in the prior PPI network by SNSVD are significantly higher than that by L0SVD (one-sided Wilcoxon signed rank test $p$-value $< 0.001$) (Fig. 2A), which indicates that SNSVD can find more tightly connected genes than L0SVD. Furthermore, SNSVD obtains a higher number of significant GO BP terms at different levels than L0SVD (one-sided Wilcoxon signed rank test $p$-value < 0.001) (Fig. 2B), showing that incorporating the PPI network does help SNSVD to discover more biological interpretable modules.

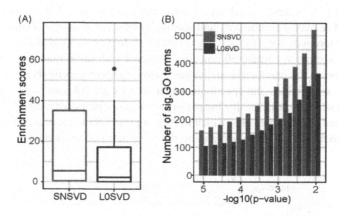

**Fig. 2.** (A) Comparison of the gene-gene interaction enrichment scores of identified modules. (B) Functional enrichment comparison based on the number of GO BP (Gene Ontology Biological Process) terms.

## 5   Identifying Functional Modules with SNSVD

We further investigate the performance of SNSVD by using gene expression data with 641 cell lines of diverse cancer types and tissues in the Cancer Genome Project (CGP) [16], as long with the protein-protein interaction (PPI) network from the Pathway-Commons database [17] to represent the gene-gene interactions. In total, we have 13,321 genes and 262,462 interactions.

### 5.1   Identifying Functional Modules

We set $\sigma = 100$ according to 5-fold cross validation test, and set $k_v = 50$ (control the sample sparsity); we also select a suitable $\lambda$ to force the estimated $\mathbf{u}$ only containing 200 nonzero elements (control the gene sparsity). Using SNSVD, we identify the first 40 pairs singular vectors $\{(\mathbf{u}_1, \mathbf{v}_1), (\mathbf{u}_2, \mathbf{v}_2), \cdots, (\mathbf{u}_{40}, \mathbf{v}_{40})\}$. Let $\mathbf{U} = [\mathbf{u}_1, \mathbf{u}_2, \cdots, \mathbf{u}_{40}]$ and $\mathbf{V} = [\mathbf{v}_1, \mathbf{v}_2, \cdots, \mathbf{v}_{40}]$, where the $i$-th columns of $\mathbf{U}$ and $\mathbf{V}$ correspond to the $i$-th pair sparse singular vectors. To reduce the false positive cases, we first calculate an absolute z-score for each none-zero element of $\mathbf{U}$ (or $\mathbf{V}$) according to:

$$z_{ij} = \frac{\left| |x_{ij}| - \mu_j \right|}{\sigma_j}, \tag{13}$$

where $x_{ij}$ is the value of $i$-th element in $\mathbf{u}_j$ (or $\mathbf{v}_j$). $\mu_j$ is the average value of non-zero elements in $\mathbf{u}_j$ (or $\mathbf{v}_j$), and $\sigma_j$ is the standard deviation. If $z_{ij}$ is greater than a given threshold, the corresponding gene (or sample) is then selected into the module $j$. In this work, we set the given threshold as the top 20 z-score value of non-zero $\mathbf{U}$ (or $\mathbf{V}$) elements. Accordingly, we get 40 functional modules, with 160 genes and 40 samples in average.

## 5.2  Functional Analysis of the Genes in Modules

Firstly, we investigate whether the genes within the same modules are connected with each other in the prior PPI network. To evaluate whether the genes within the same module are tightly connected in the prior PPI network, we use the right tailed hypergeometric test to compute a significance level of each module. The result show that 57% of the 40 modules are significantly inter-connected with each other in PPI network, illustrating that our method tends to cluster genes interacting with each other.

In addition, we also check the biological relevance of all the identified gene modules using gene functional enrichment analysis via DAVID online web tool [21]. By selecting the GO BP (Gene Ontology Biological Process) and KEGG pathways with Benjamini-Hochberg adjusted $p$ - value $<0.05$ as significant ones, we get 766 significant GO BP pathways and 70 significant KEGG pathways enriched. By statistically, 62.5% modules are significantly related with at least one GO BP pathways and 42.5% modules are significantly related with at least one KEGG pathways.

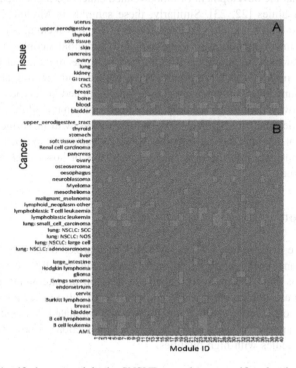

**Fig. 3.** These identified gene modules by SNSVD are subtype-specific related to some tissues or cancer types. (A) is a heatmap in term of tissue. (B) is a heatmap in term of cancer type. Note that each blue square in the two heatmaps corresponds to a significance overlapping relationship with $p$-value $< 0.05$.

### 5.3  Functional Analysis of the Samples in Modules

To evaluate the subtype-specific of samples in modules, we compute the overlapping significance level of between module-samples and cancer/tissue specific samples. There are 641 samples (cell lines) in CGP gene expression data. The 641 samples are from 16 tissues and a tissue type contains about 40 samples. In addition, the 641 samples are also divided into 52 cancer types and a cancer type contains about 12 samples. For each gene module, we first collect a sample set from the module and then compute the overlapping significance levels between its sample set and any one tissue-sample set using the right hypergeometric test (Fig. 3A), and we also compute the overlapping significance levels between its sample set and any one cancer-sample set (Fig. 3B). The results show most of the identified gene modules can be seen subtype-specific gene functional modules which provide insights into the mechanisms of the relationship between different tissues and cancers.

Furthermore, we also find that the cancer/tissue types of some modules are consistent with their corresponding functional pathways. For example, Module 1 contains 47 cell lines significantly overlapping with blood tissue and some blood related cancers (e.g., AML, B cell leukemia, B cell lymphoma, lymphoblastic leukemia, lymphoblastic T cell leukaemia, lymphoid_neoplasm other), while the top enriched GO/KEGG pathways of 174 genes in Module 1 are related the immune system. Some literatures have reported that the development of blood-related cancers is associated with immune pathway abnormalities [22, 23]. Similarly, these samples in Module 2 are also significantly related with some blood related cancers (B cell leukemia, B cell lymphoma, Burkitt lymphoma, lymphoblastic leukemia, and lymphoid_neoplasm other), while many genes in which are significantly enriched in some immune-related pathways. These samples in Module 4 are significantly related with central nervous system (CNS), while many genes in which are significantly enriched in nervous system related GO/KEGG pathways. Although we only show results of some modules, the most other gene modules identified by SNSVD have similar results with these modules. All the results show that our method can identify some cancer/tissue-specific gene modules which provide insights into the molecular mechanisms of how genes act in different cancers and tissues.

## 6  Conclusion

In this paper, we first propose a Sparse Network regularized SVD (SNSVD) model to identify gene functional modules from gene expression data by incorporating gene-gene interactions from PPI network. Then we develop an alternating iterative algorithm, based on a cyclic coordinate descent strategy, to solve the proposed model. By comparing with other representative methods on the simulated data and the real data, the results illustrate that integrating gene-gene interaction information can make SNSVD find modules with high qualities.

**Acknowledgement.** This work has been supported by the National Natural Science Foundation of China 61802157, Youth Project from the Education Department of Jiangxi Province of China GJJ180626.

# References

1. Lee, M.S., et al.: Biclustering via sparse singular value decomposition. Biometrics **66**(4), 1087–1095 (2010)
2. Liquet, B., et al.: Group and sparse group partial least square approaches applied in genomics context. Bioinformatics **32**(1), 35–42 (2015)
3. Min, W., et al.: A two-stage method to identify joint modules from matched MicroRNA and mRNA expression data. IEEE Trans. Nanobiosci. **15**(4), 362–370 (2016)
4. Eren, K., et al.: A comparative analysis of biclustering algorithms for gene expression data. Brief. Bioinform. **14**(3), 279–292 (2013)
5. Sill, M.K., et al.: Robust biclustering by sparse singular value decomposition incorporating stability selection. Bioinformatics **27**(15), 2089–2097 (2011)
6. Oghabian, A., et al.: Biclustering methods: biological relevance and application in gene expression analysis. PLOS One **9**(3), e90801 (2014)
7. Chen, S., Liu, J., Zeng, T.: Measuring the quality of linear patterns in biclusters. Methods **83**, 18–27 (2015)
8. Yang, D., Ma, Z., Buja, A.: Rate optimal denoising of simultaneously sparse and low rank matrices. J. Mach. Learn. Res. **17**(1), 3163–3189 (2016)
9. Asteris, M.K., et al.: A simple and provable algorithm for sparse diagonal CCA. In: 33rd International Conference on Machine Learning, New York, NY, USA, pp. 1148–1157 (2016)
10. Chuang, H., et al.: Network-based classification of breast cancer metastasis. Mol. Syst. Biol. **3**, 140 (2007)
11. Sokolov, A., et al.: Pathway-based genomics prediction using generalized elastic net. PLoS Comput. Biol. **12**, 3 (2016)
12. Hill, S.M., et al.: Inferring causal molecular networks: empirical assessment through a community-based effort. Nat. Methods **13**(4), 310–318 (2016)
13. Glaab, E.: Using prior knowledge from cellular pathways and molecular networks for diagnostic specimen classification. Brief. Bioinform. **17**(3), 440–452 (2016)
14. Lee, E., et al.: Inferring pathway activity toward precise disease classification. PLoS Comput. Biol. **4**, 11 (2008)
15. Li, C., et al.: Network-constrained regularization and variable selection for analysis of genomic data. Bioinformatics **24**(9), 1175–1182 (2008)
16. Iorio, F., et al.: A landscape of pharmacogenomic interactions in cancer. Cell **166**(3), 740–754 (2016)
17. Cerami, E.G., et al.: Pathway commons, a web resource for biological pathway data. Nucleic Acids Res. **39**, D685–D690 (2011)
18. Bolte, J., et al.: Proximal alternating linearized minimization for nonconvex and nonsmooth problems. Math. Program. **146**(1), 459–494 (2014)
19. Friedman, J.H., et al.: Pathwise coordinate optimization. Ann. Appl. Stat. **1**(2), 302–332 (2007)
20. Fan, J., et al.: Variable selection via nonconcave penalized likelihood and its oracle properties. J. Am. stat. Assoc. **96**(456), 1348–1360 (2001)

21. Huang, D., et al.: Systematic and integrative analysis of large gene lists using DAVID bioinformatics resources. Nat. Protoc. **4**(1), 44–57 (2009)
22. Leeksma, O.C., et al.: Germline mutations predisposing to diffuse large B-cell lymphoma. Blood Cancer J. **7**(2), e532 (2017)
23. Disis, M.L.: Immune Regulation of Cancer. J. Clin. Oncol. **28**(29), 4531–4538 (2010)

# A Novel Framework for Improving the Prediction of Disease-Associated MicroRNAs

Wenhe Zhao[1,2], Jiawei Luo[1,2(✉)], and Nguyen Hoang Tu[3]

[1] College of Computer Science and Electronic Engineering,
Hunan University, Changsha 410082, Hunan, China
luojiawei@hnu.edu.cn
[2] Collaboration and Innovation Center for Digital Chinese Medicine
in Hunan Province, Changsha 410082, Hunan, China
[3] Faculty of Information and Technology,
Hanoi University of Industry, Hanoi 100803, Vietnam

**Abstract.** Increasing evidences have shown that human complex diseases associate with plenty of miRNAs. Identifying potential associations between miRNAs and diseases provides great insight into studying the pathogenesis of complex diseases and improving drugs. However, most proposed prediction methods may not consider the existence of some impossible interactions in these unknown interactions which can be regard as negative interactions. In this paper, we proposed a framework to improve the prediction for some existing algorithms. The framework mainly consists of three steps, the first we cluster miRNAs and diseases from the given dataset by using k-medoids in order to find the weakly related interactions from unknown interactions as negative interactions. Secondly, we use existing algorithms to calculate the associated score matrix for miRNAs and diseases based on the given dataset. Finally, we combine the calculated scores with the potential negative interactions to get the final correlation scores. We conduct comprehensive experiments including 5-fold cross validation (5-fold CV) and leave-one-out cross validation (LOOCV) to indicate that our framework has some advantages including improving performance and universal applicability over several of prediction methods.

**Keywords:** miRNA-disease associations · Framework · Disease similarity · miRNA similarity

## 1 Introduction

MicroRNA (miRNA), a class of new noncoding RNAs, is encoded as short inverted repeats in the genomes of animals and plants [1]. The miRNA target genes are transcribed from introns (or host genes) and processed into primary miRNAs (pri-miRNAs) by RNA polymerase II. Plenty of researches conducted by different scholars have testified to the significant influence that miRNAs exert in complex biological

© Springer Nature Switzerland AG 2019
D.-S. Huang et al. (Eds.): ICIC 2019, LNCS 11644, pp. 127–137, 2019.
https://doi.org/10.1007/978-3-030-26969-2_12

procedures [2], involving cell growth and proliferation, apoptosis, and cell cycle [3, 4]. In consideration of the multiple functions of MiRNAs [5, 6], many researchers have designed various computational methods aiming to identify the relationships between miRNAs and genes [7]. For example, there is mounting evidence that lots of diseases, incorporating cancers, nervous system diseases, inherited diseases, etc., may be attributed to mutations or mis-expression of miRNAs [8]. Therefore, identifying potential disease-miRNA associations can help researchers to study the pathogenesis of complex diseases as well as provide effective and appropriate treatments [9, 10].

Aiming to forecast or prioritize miRNAs which has the potential to result in diseases, a lot of scholars have put forward a multitude of computational approaches to find miRNA-disease associations with the support of the hypothesis that miRNAs performing similar functions have the tendency to be linked to phenotypically similar diseases displaying similar phenotypical features [11, 12]. Based on the hypergeometric distribution, Jiang [13] integrated the miRNA function similarity, disease phenotype similarity as well as experimentally supported miRNA-disease interactions to discover miRNAs that potentialy lead to diseases. The function similarity mentioned above contains the similarity of information content of disease terms and phenotype similarity between diseases. Another computational method named miRPD has been proposed by Mørk et al. [14], which can be adopted to identify relationships between miRNAs and diseases through combining the known and predicted miRNA-protein interactions with protein-disease interactions. Gu et al. [15] designed a nonparametric algorithm based on network, namely NCPMDA that simultaneously predicts miRNA-disease associations that can be used in all diseases with no known sample. Random walk is a global network similarity algorithm that prove to be useful in inferring drug-disease interactions [16], predicting the function of protein [17] and discovering miRNA-disease pairs. For instance, Xuan et al. [18] adopted an innovative method to seek miRNAs that are related to diseases with the help of random walk. Taking the interactions with diseases into consideration, they divided miRNA network nodes into unlabeled and labeled ones and then are allocated diverse transition weights. Huang et al. [19] are the first who presented a way to predict disease-associated miRNAs using miRNA expression profiles. Chen et al. [20] invented a novel predicting method which is known as RWRMDA targeting to predict disease-miRNA associations by operating random walks on functional similarity miRNA network. In addition, machine learning-based model was also an important branch for predicting disease-miRNA candidates. Xu et al. [21] also innovated an approach which is able to identify disease-related miRNA through building a support vector machine with which prostate cancer and non-prostate cancer related miRNAs can be easily distinguished. Chen et al. [22, 23] invented two different machine learning models, namely RLSMDA and RBMMMDA, which are applicable in the process of discovering potential miRNA-disease associations. On basis of social network analysis, two methods called KATZ and CATAPULT were put forward by Zou et al. [24] aiming at forecasting possible miRNA-disease associations. Keum et al. [25] categorized unlabeled interactions and negative interactions among unknown interactions based on clustering methods and they used the BLM method and self-training SVM to find out potential drug-target interactions.

Although great achievements have been made in miRNA-disease association inference by using these methods mentioned above, there are still several limitations. Some of these methods may not take into account the existence of some impossible interactions in the unknown interactions. Nevertheless, unknown interactions may cover some impossible interactions which also can be validated as positive, so the possibility of obtaining false-positive results may be much higher. Here, we proposed a novel framework which can improve the predictive performance for some existing methods. In the framework, we first use similarity to cluster miRNAs and diseases respectively and classify unknown associations into unlabeled associations and impossible association based on known associations. Secondly, we calculate the calculated score for a given data set using an existing algorithm. Finally, we combine calculated and impossible associations to get the final score. If the association is impossible, its calculated score is set to the minimum value, and if the association is possible, its calculated score remains the original value. To further demonstrate the powerful ability and universality of our framework in predicting disease-related miRNAs, we implemented 5-Fold Cross Validation and Leave-One-Out Cross Validation on three common algorithms including EPMDA, NCPMDA and RWRMDA to estimate prediction performance of our framework. The experimental results indicate that our framework can improve performance of prediction and apply to most existing algorithms.

## 2 Method

In this chapter, we formally propose a new framework to improve prediction of disease-associated microRNAs for some existing methods. The framework can be divided into three sequential steps: (1) Categorize low correlation from unknown associations; (2) Calculate the scores of disease-associated microRNAs using existing method; (3) Combing calculated scores and classified unknown associations. The framework is described in Fig. 1.

### 2.1 Date Sources

**Human miRNA-Disease Association Network**
The experimentally verified interactions between miRNA and disease are derived from HMDD [11], which collected 5430 entries involving in 495 miRNAs and 383 diseases. Then, the adjacency matrix A record the verified associations between miRNAs and diseases.

**Disease Similarity Network**
Medical Subject Headings (MeSH) [26] is a strict dataset classification system to classify different diseases, which enables the researchers to find the relationships between various diseases in a scientific and effective way. The disease semantic

similarity reference method proposed by Wang et al. [27]. MeSH's hierarchical structure can be regarded as a $DAG_d = (d, T_d, E_d)$, in which $T_d$ refers to the node set which consists node d and its ancestor nodes and $E_d$ refers to the edge set between different diseases.

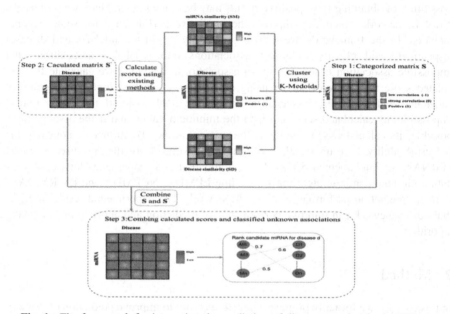

**Fig. 1.** The framework for improving the prediction of disease-associated microRNAs.

**MicroRNA Similarity Network**

Under this assumption that miRNAs with similar function are often associated with similar diseases, Wang et al. calculated the functional similarity of miRNA based on the semantic similarity between the two groups of diseases [27]. The adjacency matrix of miRNA similarity network is a symmetric matrix SM. The entity SM(i, j) in row i and column j is the similarity score between miRNA i and j. The similarity values of the two miRNAs range from 0–1.

## 2.2 Categorizing Low Correlation from Unknown Associations

As we all know, interactions between miRNAs and diseases include positive ones and unknown ones. If there is a miRNA interacted with a disease, the interaction between them will be considered as positive. However, those unknown interactions can be further categorized into two kinds, including unlabeled ones and negative ones [25]. These unlabeled ones represent strong correlations and negative ones represent low correlation or even impossible interactions. In our frame, miRNAs and diseases firstly are clustered by k-medoids clustering respectively. Then, the miRNAs in a cluster will be reckoned to be having a negative interaction with the diseases, if any of the miRNAs in the cluster does not interact with the cluster of diseases. The rest of those unknown interactions will be regarded as unlabeled ones, which have the potential to be positive and may be predicted by using existing algorithm.

In k-medoids, the number of clusters is critical. In our framework, the number of clusters of disease and miRNA can be defined based the resulting number when we divided the overall number of entity by integer $N_m$ and $N_d$ for miRNA and disease respectively (Fig. 2).

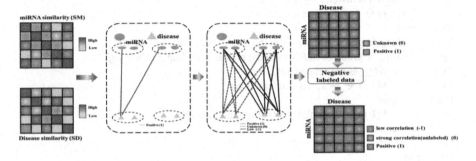

**Fig. 2.** The flowchart of categorizing low correlation from unknown associations.

### 2.3    Calculating the Final Score for the Existing Algorithm

Through the existing method, we can calculate an associated score matrix S. S is the score matrix for all the miRNAs and disease, the higher the score $S(i, j)$, the stronger the correlation between miRNA M$i$ and disease D$j$. At the same time, the above negative sample algorithm can generate an associative matrix S', S' contains only three values $-1, 0, 1$, and we assume that if $S'(i, j) = 1$, it means that M$i$ and disease D$j$ are known to be related. If $S'(i, j) = 0$, there is a possible relationship between miRNA M$i$ and disease D$j$, if $S'(i, j) = -1$ means that miRNA M$i$ and disease D$j$ are not related. Then, we combine the two matrices to get the final score by the following rules:

$$Result\ (i,j) = \begin{cases} S(i,j) & if\ S'_{ij} \geq 0 \\ 0 & if\ S'_{ij} < 0 \end{cases} \tag{1}$$

Finally, on basis of the relevance scores, all the miRNAs with unknown related diseases are ranked and a higher score will be considered potential candidate miRNA for further verification. The detail steps of our frame are described in Algorithm1.

---

**Algorithm 1 Applying Frame for existing algorithm**

**Input:** miRNA similarity **SM**, disease similarity **SD**, miRNA-disease association **Interaction**, existing algorithm **Algorithm()**
**Output:** Final association score matrix **Result.**

1:    →**Step 1: Classify unknown associations**
2:    $k_d$ ← $|D|$ / $N_d$                    // **D** : set of disease    $k_d$ : the num of diseases cluster
3:    $k_m$ ← $|M|$ / $N_m$                    // **M** : set of miRNA    $k_m$ : the num of miRNA cluster
4:    $C_d$ ← k-medodids($k_d$, **SD**);
5:    $C_m$ ← k-medodids($k_m$, **SM**);
6:    for each $i \in |D|$ do
7:     for each $j \in |M|$ do
8:       if **Interaction$(i,j)$** == **1** then
9:         $S(i,j)$ ← **1**; //positive interactions
10:    else
11:        $SD_{di}$ ← group of diseases in the cluster including disease $di$;
12:        $SM_{mj}$ ← group of miRNAs in the cluster including miRNA $mj$;
13:       if $SD_{di}$ is not related $SM_{mj}$ then
14:          $S(i,j)$ ← − **1**; //low interactions
15:       else
16:          $S(i,j)$ ← **0**; //unlabeled(strong) interactions
17:       end if
18:      end if
19:     end
20:   end
21:    →**Step 2: Calculate scores using existing methods**
22:    Initialization: prepare parameters for **Algorithm()**
23:    **S`** ← **Algorithm(SM,SD,interaction)**
24:    →**Step 3: Combine calculated scores with classified unknown associations**
25:    for each $i \in |D|$ do
26:     for each $j \in |M|$ do
27:       if $S(i,j)$ == −**1** then
28:         $Result_{ij}$ ← 0
29:       else
30:         $Result_{ij}$ ← $S`(i,j)$
31:       end if
32:     end
33:   end

---

# 3   Results and Discussion

## 3.1   Related Work

5-fold cross validation (5-fold CV) a common experimental method in machine learning model. For an exact disease $d$ of which the related miRNAs are known, the $d$-related miRNAs (labeled nodes) are split into five subsets. Four of them are utilized to train a prediction model, and the rest one is used as positive samples in testing set. The left miRNAs (unlabeled nodes), which have not been considered to connect interaction with disease d, are also included in the testing dataset. In order to keep independent between the training dataset and test dataset, miRNA similarity and negative sample relationships are recalculated based on the training dataset at each experiment. The testing miRNAs will be ranked according to their association scores after the estimation.

Proposing a threshold £, a labeled node will be regarded to be a precisely identified positive sample if its score is higher than £, and an unlabeled node will be considered as an accurately identified negative sample if its score is lower than £. in order to attain a ROC curve, the False Positive Rates (FPRs) and the True Positive Rates (TPRs) at different £ values are calculated as follow

$$TPR = \frac{TP}{TP+FN}, \ FPR = \frac{FP}{TN+FP} \tag{2}$$

where TP represents the number of successfully identified positive samples, TN demonstrates the number of successfully identified negative samples, FP refers to the total number of misidentified positive samples, and FN means the total number of misidentified negative samples. The global predictive accuracy of a prediction method is measured based on the AUC value which is computed based on varying the threshold.

## 3.2  5-Fold Cross-Validation

To systematically evaluate the performance of our framework, we apply it on the following existing methods: RWRMDA, NCPMDA, EPMDA, and perform 5-fold cross validation experiments using the respective optimal parameter settings (i.e. $\alpha = 0.5, \beta = 0.5$ for EPMDA, NPCMDA without parameters, and Rate $= 0.9$ for RWRMDA) and calculate the performance value. Two parameters $N_m$ and $N_d$ are involving in our framework. The parameters $N_m$ and $N_d$ are number of nodes in each cluster. After investigation of those two parameters, the following values are used in our experiments: $N_m = 8, N_d = 5$. In above existing methods, RWRMDA uses the restart random walk strategy to identify the association of disease miRNAs, starting from some given seed nodes and randomly transferring them to neighbors in the network, while the walker has a certain probability to return to the seed node to restart random transfer. The RWRMDA model mainly includes the following three steps: (1) determining the initial probability of each miRNA; (2) randomly walking on the miRNA similarity network; (3) obtaining the random walk stabilization probability and ranking the miRNA according to the transition probability score. The NCPMDA model primarily uses network-consistent projection scores to identify the associations of miRNA and diseases. Network-consistency refers to the similarity of two eigenvectors in different expressions. The model calculates the projection scores of the miRNA and disease space separately and combines the two scores to ultimately predict disease-associated miRNAs. The EPMDA is a model of miRNA-disease interactions prediction by using two-way diffusion. The model uses the miRNA similarity, disease similarity and miRNA disease association network to calculate the weighted miRNA disease association network.

The performance improvement was shown in Fig. 3. As the result shows, the AUC value of RWRMDA algorithm is 0.725. When the frame is applied to the RWRMDA algorithm, the AUC increases to 0.788, which is 6.3% higher than the previous value. As to the NCPMDA algorithm, the AUC of it reaches 0.899 with the help of our frame, showing a 4.5% increase. In the part of the EPMDA algorithm, the AUC mounts up to 0.912 after applying our frame to it and increased by 2.5% compared with the previous value. Therefore, the use and combination of framework is very meaningful, and the performance improvement of these algorithms can be contributed to the fact that we categorized these unverified interactions as negative and unlabeled ones.

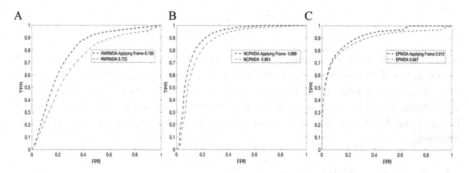

**Fig. 3.** (A) ROC curves for RWRMDA and RWRMDA applying our framework in 5-fold cross validation (B) ROC curves for NCPMDA and NCPMDA applying our framework in 5-fold cross validation (C) ROC curves for EPMDA and EPMDA applying our framework in 5-fold cross validation

In order to achieve a feasible and objective comparison, we test 15 common diseases which associated with at least 80 experimentally verified miRNAs in the context of CV setting as in Xuan et al. [18]. Table 1 are the AUC values of the 15 testing diseases, which indicates our framework has a strong versatility for the dataset and can improve the accuracy of the algorithm prediction whether it is global or local.

What's more, the higher recall value within top K ranking list illustrates the more positive testing samples are identified successfully. Figure 4. lists the recall values of the above 15 given diseases for the top K ranking list. The average recall values of RWRMDA in Top20, Top60 and Top100 are 8.1%, 24.3% and 40.5% respectively. When applying our framework, the average recall of RWRMDA reaches 8.5%, 26.1% and 43.3%. As for NCPMDA, with the help of our framework it obtains higher promotion with the recall rates of 13.3% in Top20, 35% in Top60 and 54% in Top100 compared with the previous value. The performance of NCPMDA increases to 13.6% in Top20, 36.1% in Top60 and 55.8% in Top100.

**Table 1.** AUC of comparison before and after applying our framework for the 15 diseases

| Disease name | AUC | | | | | |
|---|---|---|---|---|---|---|
| | RWRMDA | | NCPMDA | | EPMDA | |
| | Before | After | Before | After | Before | After |
| Breast neoplasms | 0.796 | 0.827 | 0.862 | 0.883 | 0.835 | 0.863 |
| Carcinoma, hepatocellular | 0.726 | 0.738 | 0.818 | 0.830 | 0.799 | 0.808 |
| Carcinoma, non-small-cell lung | 0.795 | 0.799 | 0.872 | 0.886 | 0.866 | 0.870 |
| Carcinoma, renal cell | 0.772 | 0.786 | 0.841 | 0.856 | 0.808 | 0.819 |
| Colorectal neoplasms | 0.778 | 0.785 | 0.851 | 0.856 | 0.847 | 0.851 |
| Glioblastoma | 0.734 | 0.738 | 0.858 | 0.860 | 0.817 | 0.820 |
| Heart failure | 0.744 | 0.880 | 0.804 | 0.895 | 0.797 | 0.892 |
| Lung neoplasms | 0.832 | 0.838 | 0.927 | 0.930 | 0.910 | 0.914 |
| Melanoma | 0.780 | 0.786 | 0.848 | 0.850 | 0.834 | 0.839 |
| Neoplasms | 0.844 | 0.851 | 0.922 | 0.925 | 0.919 | 0.922 |
| Ovarian neoplasms | 0.850 | 0.865 | 0.906 | 0.912 | 0.898 | 0.907 |
| Pancreatic neoplasms | 0.823 | 0.830 | 0.911 | 0.921 | 0.901 | 0.905 |
| Prostatic neoplasms | 0.775 | 0.820 | 0.891 | 0.904 | 0.828 | 0.866 |
| Stomach neoplasms | 0.742 | 0.757 | 0.847 | 0.855 | 0.787 | 0.803 |
| Urinary bladder neoplasms | 0.748 | 0.761 | 0.884 | 0.899 | 0.831 | 0.841 |

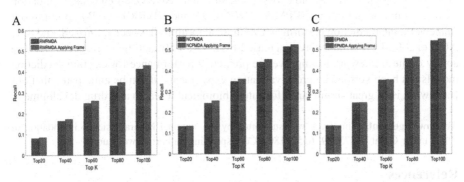

**Fig. 4.** (A) The average recalls for RWRMDA and RWRMDA applying our framework in miRNA-disease association prediction (B) The average recalls for NCPMDA and NCPMDA applying our framework in miRNA-disease association prediction (C) The average recalls for EPMDA and EPMDA applying our framework in miRNA-disease association prediction

### 3.3 Leave-One-Out Cross-Validation

In Leave-One-Out Cross-Validation, for a specific disease d, each miRNA associated with disease d is left as a test miRNA, and other known d-related miRNAs are known as information to predict association and calculate the AUC value. As the result shows, the AUC value of the RWRMDA algorithm is 0.752. When the frame is applied to the RWRMDA algorithm, the AUC increases to 0.811, which is 5.9% higher than the previous value. As to the NCPMDA algorithm, the AUC of it reaches 0.901 with the help of our frame, showing a 3.2% increase. In the part of the EPMDA algorithm, the AUC mounts up to 0.921 after applying our frame to it and increased by 2.9% compared with the previous value (Table 2).

**Table 2.** AUC of comparison before and after applying our framework for LOOCV

| Method | RWRMDA | NCPMDA | EPMDA |
|---|---|---|---|
| AUC applying frame | 0.811 | 0.901 | 0.921 |
| AUC without applying frame | 0.752 | 0.869 | 0.892 |

## 4   Conclusion

Accumulating studies indicate that miRNAs have important effects on biological processes and play crucial roles in various complicated diseases. Consequently, identifying miRNA-disease associations is important for studying the underlying mechanisms of human complex diseases. In this paper, we presented a novel framework that uses k-medoids clustering to classify these unknown interactions into negative and unlabeled ones. The main contribution of our framework is the classification for unverified associations, which can raise score of potential associations and then improve the prediction performance for some existing methods. In order to illustrate the universality of our frame, we perform 5-Fold Cross-Validation and Leave-One-Out Cross-Validation on three common algorithms (EPMDA, NCPMDA and RWRMDA). By applying our framework, the prediction performance of these algorithms has been improved, and their AUC of 5-fold CV values have increased by 2.5%, 4.5% and 6.3% respectively. The ability of the framework to improve the predictive performance for existing prediction models has been verified by cross-validation experiments. It can be anticipated that our framework is of great significance for future biological research and drug development.

**Acknowledgement.** This work was supported by the National Natural Science Foundation of China under Grant 61873089, 61572180. (Corresponding author: Jiawei Luo.)

## References

1. Ambros, V.: The functions of animal microRNAs. Nature **431**(7006), 350–355 (2004)
2. Ding, P., Luo, J., Liang, C., et al.: Human disease MiRNA inference by combining target information based on heterogeneous manifolds. J. Biomed. Inform. **80**, 26–36 (2018)
3. Luo, J., Ding, L., Liang, C., et al.: An efficient network motif discovery approach for co-regulatory networks. IEEE Access **6**, 14151–14158 (2018)
4. Calin, G.A., Croce, C.M.: MicroRNA signatures in human cancers. Nat. Rev. Cancer **6**(11), 857–866 (2006)
5. Li, M., Zheng, R.Q., Li, Q., Wang, J.X., Wu, F.X., Zhang, Z.H.: Prioritizing disease genes by using search engine algorithm. Curr. Bioinform. **11**(2), 195–202 (2016)
6. Chen, B.L., Li, M., Wang, J.X., Shang, X.Q., Wu, F.X.: A fast and high performance multiple data integration algorithm for identifying human disease genes. BMC Med. Genom. **8**(3), S2 (2015)

7. Liu, Y., Luo, J., Ding, P.: Inferring microRNA targets based on restricted Boltzmann machines. IEEE J. Biomed. Health Inform. **23**(1), 427–436 (2018)
8. Ha, M., Kim, V.N.: Regulation of microRNA biogenesis. Nat. Rev. Mol. Cell Biol. **15**(8), 509–524 (2014)
9. Xu, G., et al.: MicroRNA-21 promotes hepatocellular carcinoma HepG2 cell proliferation through repression of mitogen-activated protein kinase 3. BMC Cancer **13**(1), 469 (2013)
10. Luo, J., Ding, P., Liang, C., et al.: Collective prediction of disease-associated miRNAs based on transduction learning. IEEE/ACM Trans. Comput. Biol. Bioinf. **14**(6), 1468–1475 (2017)
11. Li, Y., et al.: HMDD v2.0: a database for experimentally supported human microRNA and disease associations. Nucleic Acids Res. **42**(D1), D1070–D1074 (2014)
12. Zou, Q., Li, J., Song, L., et al.: Similarity computation strategies in the microRNA-disease network: a survey. Brief. Funct. Genom. **15**(1), 55–64 (2016)
13. Jiang, Q., et al.: Prioritization of disease microRNAs through a human phenome-microRNAome network. BMC Syst. Biol. **4**(1), S2 (2010)
14. Mørk, S., Pletscher-Frankild, S., Palleja Caro, A., Gorodkin, J., Jensen, L.J.: Protein-driven inference of miRNA-disease associations. Bioinformatics **30**(3), 392–397 (2014)
15. Gu, C., Liao, B., Li, X., et al.: Network consistency projection for human miRNA-disease associations inference. Sci. Rep. **6**, 36054 (2016)
16. Luo, H.M., et al.: Drug repositioning based on comprehensive similarity measures and Bi-Random walk algorithm. Bioinformatics **32**(17), 2664–2671 (2016)
17. Peng, W., Li, M., Chen, L., Wang, L.S.: Predicting protein functions by using unbalanced random walk algorithm on three biological networks. IEEE/ACM Trans. Comput. Biol. Bioinform. **14**(2), 360–369 (2015)
18. Xuan, P., et al.: Prediction of potential disease-associated microRNAs based on random walk. Bioinformatics **31**(11), 1805–1815 (2015)
19. Huang, Y.A., You, Z.H., Li, L.P., et al.: EPMDA: an expression-profile based computational model for microRNA-disease association prediction. Oncotarget **8**(50), 87033–87043 (2017)
20. Chen, X., Liu, M.X., Yan, G.Y.: RWRMDA: predicting no vel human microRNA–disease associations. Mol. BioSyst. **8**(10), 2792–2798 (2012)
21. Xu, J., Li, C.X., Lv, J.Y., et al.: Prioritizing candidate disease miRNAs by topological features in the miRNA target-dysregulated network: case study of prostate cancer. Mol. Cancer Ther. **10**(10), 1857–1866 (2011)
22. Chen, X., Yan, G.Y.: Semi-supervised learning for potential human microRNA-disease associations inference. Sci. Rep. **4**(1), 5501 (2014)
23. Chen, X., et al.: RBMMMDA: predicting multiple types of disease microRNA associations. Sci. Rep. **8**(5), 13877 (2015)
24. Quan, Z., Jinjin, L., Qingqi, H., et al.: Prediction of MicroRNA-disease associations based on social network analysis methods. Biomed. Res. Int. **2015**, 1–9 (2015)
25. Keum, J., Nam, H.: SELF-BLM: prediction of drug-target interactions via self-training SVM. PLoS One **12**(2), e0171839 (2017)
26. Lipscomb, C.E.: Medical subject headings (MeSH). Bull. Med. Libr. Assoc. **88**(3), 265–266 (2000)
27. Wang, D., Wang, J., Lu, M., et al.: Inferring the human microRNA functional similarity and functional network based on microRNA-associated diseases. Bioinformatics **26**(13), 1644–1650 (2010)

# Precise Prediction of Pathogenic Microorganisms Using 16S rRNA Gene Sequences

Yu-An Huang[1], Zhi-An Huang[2], Zhu-Hong You[3(✉)], Pengwei Hu[4], Li-Ping Li[3], Zheng-Wei Li[5], and Lei Wang[3]

[1] Department of Computing, Hong Kong Polytechnic University, Hung Hom 999077, Hong Kong
[2] Department of Computer Science, City University of Hong Kong, Kowloon 999077, Hong Kong
[3] Xinjiang Technical Institute of Physics and Chemistry, Chinese Academy of Sciences, Ürümqi, China
zhuhongyou@ms.xjb.ac.cn
[4] IBM Research, Beijing 100000, China
[5] School of Computer Science and Technology, China University of Mining and Technology, Xuzhou 221116, China

**Abstract.** Clinical observations show that human microorganisms get involved in various human biological processes. The disruption of a symbiotic balance for host-microbiota relationship is found to cause different types of human complex diseases. Discoverying the associations between microbes and the host health statuses that they affect could provide great insights into understanding the mechanisms of diseases caused by microbes. However, experimental approaches are time-consuming and expensive. Little effort has been done to develop computational models for predicting pathogenic microbes on a large scale. The prediction results yielded by such models are anticipated to boost the identification and characterization of potential human pathogenic microbes. Based on the assumption that microbes of similar characters tend to get involved in diseases of similar symptoms forming functional clusters, in this paper, we develop a group based computational model of Bayesian disease-oriented ranking for inferring the most potential microbes associated with human diseases. It is the first attempt to predict this kind of associations by using 16S rRNA gene sequences. Based on the sequence information of genes, we use two computational approaches (BLAST+ and MEGA 7) to measure how similar each pairs of microbes are from different aspects. On the other hand, the similarity of diseases is computed based on MeSH descriptors. Using the data collected from HMDAD database, the proposed model achieved AUCs of 0.9456, 0.8266, 0.8866 and 0.8926 in leave-one-out, 2-fold, 5-fold and 10-fold cross validations, respectively. Besides, we conducted a case study on colorectal carcinoma and found that 16 out of top-20 predicted microbes can be confirmed by the published literatures. The prediction result is publicly released and anticipated to help researchers to preferentially validate these promising pathogenic microbe candidates via biological experiments.

© Springer Nature Switzerland AG 2019
D.-S. Huang et al. (Eds.): ICIC 2019, LNCS 11644, pp. 138–150, 2019.
https://doi.org/10.1007/978-3-030-26969-2_13

**Keywords:** Microbe–disease associations · 16S rRNA sequence analysis · Computational prediction model · Microflora · Pathogenic microorganisms

# 1 Introduction

In the last decade, the important role of the microorganisms playing in life activities has been realized by the development of metagenomics, functional genomics, metabolomics and proteomics based on high-throughput sequencing techniques, especially 16S ribosomal RNA (rRNA) sequencing. Microorganisms, also known as microbes, mainly include bacteria, archaea, viruses, as well as eukaryotes (protozoa and fungi) [1]. They can inhabit and thrive in nearly all kinds of natural environments, including the human body. And the human microbial communities vary in composition and locate in different parts of human body, including both the external (e.g. skin) and the internal (e.g. the mucosal epithelia of vagina and intestine). It is estimated that up to $10^{13}$ microbes (basically bacteria in the intestine) approximate human cells close to a ratio of one to one [2]. Therefore, human can be regarded as a "meta-organism", which implies a potential relationship between human cells and microorganisms. There have been numerous attempts to investigate into the important function of the human microbiota, which is associated with human diseases or disorders.

The human microbiota and its host maintain coexist in a homeostatic system involving multiple biological activities. Both of them are mutually affected by various factors. The human microbiota varies with multiple factors of its host, such as: age, the genetics, host lifestyle, body site, health status and others (e.g. antibiotics and smoking) [3–6]. The resident microbial communities can provide multiple microbial genome-encoded metabolic functions, which are absent in human cells and therefore can extend the host's metabolic capacity. Their genes and related metabolic processes play a significant role in defense against pathogens, enhance the immune system, access to nutrients, as well as the degradation of toxic compounds [3]. Although diverse pathogenic microorganisms can result in disease, most human microbiota is beneficial. The disruption of the microbial communities could cause particular diseases. The understanding of the fundamental function of the latent host-microbe relationship, could help to explore the complex mechanism of diseases. For example, butyrate is not only the main energy source of intestinal epithelial cell but is also able to prevent signal transduction pathway of the expression of proinflammatory cytokine. The population decline of butyrate-producing microbes was found in subjects with inflammatory bowel disease (IBD), such as *Clostridium leptum* and *Clostridium coccoides* groups [7]. This situation could also decrease butyrate utilization [8]. In other words, the restoration of host-microbe equilibrium can maintain a symbiotic balance for human health.

Technological and molecular developments accelerate the identification and characterization of human microbiota. Ongoing international efforts have been made to expand our knowledge, such as the International Human Microbiome Consortium (during 2011–2015) and the Metagenomics of the Human Intestinal Tract project–MetaHIT (during 2008–2012). These valuable studies revealed how the human microbiota can interfere the development of or susceptibility to a particular disease, steering people's understanding of pathology toward microbial involvement. As high throughput sequencing platforms produce great volumes of sequencing data, some bioinformatics tools or databases are used for downstream analysis and data management [9]. A 16S rDNA analysis toolkit named W.A.T.E.R.S. [10] can be used for sequence alignment, operational taxonomic units (OTUs) determination, phylogenetic tree construction, and etc. Hundreds of microbe-disease associations are publicly available in the Human Microbe-Disease Association Database (HMDAD) [11]. However, there is still much work to be done and progress to be made despite the substantial efforts of the scientific community. The known knowledge is only the tip of the iceberg, and that the interactions of microorganisms with human pathology still remain unclear. Most culture-independent methods based on DNA are still time-consuming and expensive because of limitations like nutritional requirements, *in vitro* growth condition or low survival rate. In recent years, researchers have attempted to develop computational models for refining potential biomarkers based on the heterogeneous biological datasets. Some remarkable achievements have greatly advanced the development in many fields, such as: noncoding RNA-disease association prediction [12, 13], protein-protein interaction (PPI) prediction [14, 15] and drug-target interaction inference [16]. This idea inspires us to devise a computational model for prioritizing the most possible pathogenic microbes, which could be served as ideal biomarkers.

16S rRNA is the component of the 30S small subunit of a prokaryotic ribosome. Because of its slow rates of evolution, 16S rRNA is highly conserved between diverse species of microbes [17]. 16S rRNA gene sequences contain approximately 1500 base pairs long and 9 hypervariable regions (V1-V9), which can vary dramatically between bacteria and make the alignments easier. Thus, 16S rRNA gene sequencing has become a prevalent way in medical microbiology providing natural species-specific signatures for identification and characterization of microbes. We are inspired to firstly measure the microbe similarities based on the 16S rRNA gene sequences.

Two computational models called KATZHMDA [18] and PBHMDA [19] were presented for microbe-disease association prediction based on the known microbe-disease association network and inferred similarity matrices. KATZHMDA is based on a social network prediction algorithm called Katz [20] by calculating the number of walks within the network and their own lengths served as an efficient measure index. PBHMDA is a path-based prediction model by introducing a special depth-first search algorithm traversing all possible paths between microbes and diseases. Despite the reliable prediction performance achieved in both models, some limitations indeed restrict their performance, such as: inevitable bias brought by the predicted similarity matrices and incompatibility of new diseases and new microbes. Here, we present a Group based computational model of Bayesian Disease-oriented Ranking for predicting most potential microbial biomarkers of various diseases (GBDR) based on the known microbe-disease association network derived from HMDAD database. Unlike the previous work, we introduce the heterogeneous biological information to measure diverse similarity matrices, including disease semantic similarity, microbe sequence similarity and microbe evolutionary distance-based similarity. In this way, the proposed computational model can accurately predict the most potential microbes with respect to a specific disease. Compared with the two previous models and other classical algorithms in recommender system, the proposed model obtained the best prediction accuracy based on leave-one-out cross validation (LOOCV) and *k*-fold cross validation (*k*-fold CV). Furthermore, introducing group-based collaboration filtering and diverse similarity matrices was demonstrated to improve prediction accuracy. Finally, we selected an important disease as a case study to manually certificate those predicted microbe candidates in the top-20 list based on the published literatures. In conclusion, the result fully demonstrates the reliability of the proposed model. It is anticipated that GBDR can be an effective tool to accelerate the identification of pathogenic microorganisms.

## 2  Method

### Materials
Three types of biological information are utilized in the proposed model, i.e., the known microbe-disease associations derived from HMDAD database (http://www. cuilab.cn/hmdad) [11], 16S rRNA partial or complete gene sequences downloaded from the Nucleotide Database of the National Centre for Biotechnology Information (NCBI) [21] and Medical Subject Headings (MeSH) descriptors provided by the Nation Library of Medicine (NLM) [22]. It should be noted that, Ma *et al.* [11] searched articles regarding human microbiome-related research published before July 2014. HMDAD database provides 450 non-repetitive microbe-disease entries including 292 microbes and 39 human diseases. The numbers of microbes and diseases are denoted as *nm* and *nd*, respectively. For a better calculation, all these microbe-disease associations are converted into an adjacency binary matrix as variable $\mathcal{R}$ of size nm × nd representing their association relationships. Namely, $\mathcal{R}(m_i, d_j) = 1$ indicates microbe *mi* is known to be associated with disease dj, and vice versa.

### Disease Semantic Similarity
The MeSH descriptors are arranged in a hierarchy for effective classification of various human diseases. For example, the MeSH ID of overnutrition is C18.654.726 and obesity's is C18.654.726.500, which means obesity is categorized into a subtype of overnutrition. Accordingly, all disease relationships can be constructed into respective Directed Acyclic Graphs (DAGs) based on the hierarchy of MeSH IDs [23]. Each disease corresponds to at least one MeSH IDs which numerically index its location(s) in DAGs. Empirically, the shorter distance between the ancestor node *d* and the targeted node *t* it is, the higher weight it presents. It can be formulated as follows:

$$V_d(t) = \begin{cases} 1 & \text{if } t = d \\ \frac{1}{dist(d,t)+1} & \text{if } t \in \{\text{the descendant node of } d\} \\ 0 & \text{otherwise} \end{cases} \tag{1}$$

here $dist(d,t)$ is the shortest distance between the ancestor node *d* and one of its descendant nodes *t*. In this way, the contribution of feature vectors to the investigated diseases can be computed. Then, we further calculated the semantic similarity of any two diseases $d_i$ and $d_j$ by a simple yet effective way, cosine similarity:

$$S_d(\text{di, dj}) = \text{CS}(\text{di, dj}) = \frac{V_{di} * V_{dj}^T}{\|V_{di}\| \|V_{dj}\|} \tag{2}$$

where $V_{di}$ and $V_{dj}$ are the feature vectors of $d_i$ and $d_j$. Figure 1 illustrates the calculation process of disease semantic similarity.

## Microbe Similarity Based on BLAST+ Scores

Basic Local Alignment Search Tool (BLAST) is a specific sequence similarity search program (http://www.ncbi.nlm.nih.gov/blast) [24]. We used its variant BLAST+ [25] to compare a targeted 16S rRNA gene sequence as a nucleotide query sequence against the rest sequences as a nucleotide sequence database in turn. Identity is an important glossary of BLAST+ to measure the extent to which two (nucleotide or amino acid) sequences are invariant in an alignment, often expressed as a percentage. In the case of nucleotide sequences, identity can be used for the purpose of assessing the degree of similarity. In this way, we define a matrix as *Iden* of size $nm \times nm$ recording the identity values yielded from the alignment. Then we calculated the microbe sequence similarity denoted as *MSS* by empirically normalizing *Iden* matrix as follows:

$$\text{MSS}(\text{mi, mj}) = \frac{Iden(mi, mj) - Min.(Iden)}{Max.(Iden) - Min.(Iden)} \tag{3}$$

It should be noted that, among 292 investigated microbes, 5 microbes could not find their corresponding 16S rRNA gene sequences from NCBI. We simply set their sequence similarities as the average value of the similarities of the rest available ones.

## Microbe Similarity Based on MEGA7 Evolutionary Distance Scores

Microbe evolutionary distance-based similarity is also calculated via MEGA 7 [26] (http://www.megasoftware.net/), a molecular evolutionary genetics analysis software. The number of nucleotide substitutions occurring between a pair of sequences is utilized to represent their evolutionary distance, as a fundamental measurement in the study of molecular evolution. It is one type of estimation of divergence times and phylogenetic reconstructions. First, we built a nucleotide sequences alignment by Clustal W, a multiple sequence alignment approach [27]. To eliminate the disturbance caused by gaps, all sequences were subject to the shortest one by removing terminal redundancy at 5' and 3' terminus. Then we used the complete-deletion option for handling gaps and missing data. We selected $p$-distance model to calculate the evolutionary distances via substitutions (including transitions and transversions). $p$-distance [28] for nucleotide sequences is formulated as:

$$\hat{p} = \frac{n_d}{n} \tag{4}$$

where $n_d$ refers to the number of different nucleotides between two tested sequences and $n$ is the total number of nucleotides examined. The higher values of evolutionary distances denote the more evolutionary diversity. We subtracted the evolutionary distances from 1 and the result was denoted by a matrix as *ED* of size $nm \times nm$.

Similarly, the microbe evolutionary distance-based similarity (*MES*) is defined as follows:

$$\text{MES}(\text{mi}, \text{mj}) = \frac{ED(mi, mj) - Min.(ED)}{Max.(ED) - Min.(ED)} \tag{5}$$

In this way, entity $\text{MES}(m_i, m_j)$ denotes the microbe evolutionary distance-based similarity between microbe $m_i$ and microbe $m_j$ from 0 to 1. For those microbes without available 16S rRNA gene sequences, we also set their evolutionary distance-based similarities to the overall mean of the rest available ones.

Finally, we simply combined both microbe sequence similarity *MSS* and microbe evolutionary distance-based similarity *MES* as follows:

$$S_m(m_i, m_j) = \frac{MSS(m_i, m_j) + MES(m_i, m_j)}{2} \tag{6}$$

### Group Preference-Based Bayesian Disease-Oriented Ranking

Based on the previous work [29, 30] in recommender system, the pointwise preference assumption on items, i.e., treating all observed feedbacks as "likes" and unobserved ones as "dislikes", may mislead the learning process. However, the pairwise preference assumption over two items could relax the pointwise preference assumption by treating that a user $u$ is likely to prefer an item $i$ to an item $j$ represented as $\hat{r}_{ui} > \hat{r}_{uj}$ where item $i$ belongs to the observed feedbacks whereas $j$ does not. Empirically, this assumption generates better recommendation results than the pointwise assumption. Inspired by this idea [31, 32], we present group pairwise preference based Bayesian disease-oriented ranking for refining the most potential pathogenic microbes (see Fig. 2 and Table 1).

Based on the known microbe-disease associations for a typical disease $d$, we first define the overall likelihood of pairwise preferences (LPP) among the whole set of microbe denoted as $\mathcal{M}$:

$$\begin{aligned}
\textbf{LPP}(d) &= \prod_{i,j \in \mathcal{M}} \textbf{Pr}(\hat{r}_{di} > \hat{r}_{dj})^{\delta((d,i) \succ (d,j))} \times \left[1 - \textbf{Pr}(\hat{r}_{di} > \hat{r}_{dj})\right]^{[1 - \delta((d,i) \succ (d,j))]} \\
&= \prod_{(d,i) \succ (d,j)} \textbf{Pr}(\hat{r}_{di} > \hat{r}_{dj})\left[1 - \textbf{Pr}(\hat{r}_{di} > \hat{r}_{dj})\right]
\end{aligned} \tag{7}$$

where $(d, i) \succ (d, j)$ means that disease $d$ is more potentially associated with microbe $i$ than microbe $j$. And $\delta((d, i) \succ (d, j))$ is Bernoulli distribution over the binary random variable. Based on the pairwise preference assumption over two items, $\textbf{LPP}(d)$ can be simplified to $\textbf{BDR}(d)$ as follows [31]:

$$\textbf{BDR}(d) = \prod_{i \in \mathcal{M}_d^{tr}} \prod_{j \in \mathcal{M}^{tr} \setminus \mathcal{M}_d^{tr}} Pr(\hat{r}_{di} > \hat{r}_{dj})\left[1 - Pr(\hat{r}_{di} > \hat{r}_{dj})\right] \tag{8}$$

where $i \in \mathcal{M}_d^{tr}$ indicates the known microbe-disease association pair $(i, d)$ in training data and $j \in \mathcal{M}^{tr} \setminus \mathcal{M}_d^{tr}$ represents the microbe-disease association pair $(j, d)$ is unknown.

We assume that the group preference is an overall preference score of a microbe group on a disease. If microbe-disease pair $(i, d)$ is a known association but $(i, b)$ is not, the group preference can be represented as:

$$(\mathcal{G}, d) \succ (\mathcal{G}, b), \text{where } i \in \mathcal{G} \text{ and } \mathcal{G} \subseteq \mathcal{M}_d^{tr} \tag{9}$$

It can assume that the group preference of $\mathcal{G} \subseteq \mathcal{M}_d^{tr}$ on a disease $d$ is probably stronger than the individual preference of microbe $i$ on disease $b$. To learn the unified effect of both individual preference and group preference, we linearly combine them as follows:

$$(\mathcal{G}, d) + (i, d) \succ (i, b) \text{ or } \hat{r}_{\mathcal{G}id} > \hat{r}_{ib} \tag{10}$$

where $\hat{r}_{\mathcal{G}id} = \rho \hat{r}_{\mathcal{G}d} + (1 - \rho)\hat{r}_{id}$ is the combined preference of individual preference $\hat{r}_{id}$ and group preference $\hat{r}_{\mathcal{G}d}$ and parameter $\rho$ controls the weight of two preferences from 0 to 1. We set $\rho$ to 0.5 in this study. Similarly, group Bayesian disease-oriented ranking (GBDR) based on microbe $i$ can be represented as follows:

$$\mathbf{GBDR}(i) = \prod_{d \in \mathcal{D}_i^{tr}} \prod_{b \in \mathcal{D}^{tr} \setminus \mathcal{D}_i^{tr}} \mathbf{Pr}(\hat{r}_{\mathcal{G}id} > \hat{r}_{ib})[1 - \mathbf{Pr}(\hat{r}_{ib} > \hat{r}_{\mathcal{G}id})] \tag{11}$$

where $\mathcal{G} \subseteq \mathcal{M}_d^{tr}, d \in \mathcal{D}_i^{tr}$ means $d$ is verifiably associated with microbe $i$ in training data and $b \in \mathcal{D}^{tr} \setminus \mathcal{D}_i^{tr}$ means disease $b$ has not been confirmed to be related to microbe $i$. For any two microbes $i$ and $j$, the joint likelihood is simply approximated via multiplication like $\mathbf{GBDR}(i, j) \approx \mathbf{GBDR}(i)\mathbf{GBDR}(j)$. Therefore, the overall likelihood for all microbes and all diseases can be formulated as:

$$\mathbf{GBDR} = \prod_{i \in \mathcal{M}^{tr}} \prod_{d \in \mathcal{D}_i^{tr}} \prod_{b \in \mathcal{D}^{tr} \setminus \mathcal{D}_i^{tr}} \mathbf{Pr}(\hat{r}_{\mathcal{G}id} > \hat{r}_{ib})[1 - \mathbf{Pr}(\hat{r}_{ib} > \hat{r}_{\mathcal{G}id})] \tag{12}$$

where $\mathcal{G} \subseteq \mathcal{M}_d^{tr}$. The objective function of GBDR is defined as follows,

$$\min_{\Theta} -\frac{1}{2}\ln GBDR + \frac{1}{2}\mathcal{R}(\Theta) \tag{13}$$

where $\Theta = \{U_i \in \mathbb{R}^{1 \times d}, V_d \in \mathbb{R}^{1 \times d}, b_d \in \mathbb{R}, i \in \mathcal{M}^{tr}, d \in \mathcal{D}^{tr}\}$ is a set of model parameters needed to be learned. We used stochastic gradient descent (SGD) algorithm to optimize the object function in Eq. (13). Before carrying out SGD, a subset of microbes is randomly sampled as the microbe group $\mathcal{G}$. In this way, for each random sampling, it includes a microbe $i$, a disease $d$, a disease $b$ and a microbe group $\mathcal{G}$ where $i \in \mathcal{G}$.

The objective function in Eq. (13) can be written as:

$$
\begin{aligned}
\mathcal{F}(\mathcal{G}, i, d, b) = & -\ln(\hat{r}_{\mathcal{G}id} - \hat{r}_{ib}) + \frac{\alpha_u}{2}\sum_{j \in G}\|U_j\|^2 + \frac{\alpha_v}{2}\|V_d\|^2 \\
& + \frac{\alpha_v}{2}\|V_b\|^2 + \frac{\beta_v}{2}\|b_d\|^2 + \frac{\beta_v}{2}\|b_b\|^2 \\
= & \ln\left[1 + \exp\left(-\hat{r}_{\mathcal{G}id;ib}\right)\right] + \frac{\alpha_u}{2}\sum_{j \in G}\|U_j\|^2 + \frac{\alpha_v}{2}\|V_d\|^2 \\
& + \frac{\alpha_v}{2}\|V_b\|^2 + \frac{\beta_v}{2}\|b_d\|^2 + \frac{\beta_v}{2}\|b_b\|^2
\end{aligned}
\tag{14}
$$

where $\hat{r}_{gid;ib} = \hat{r}_{gid} - \hat{r}_{ib}$, and $\alpha_u$, $\alpha_v$ and $\beta_v$ are regularization weights for latent feature vector matrices $U_j$ (microbe-specific), $V_d$ (disease-specific), $V_b$ (disease-specific) and disease bias $b_d$ and $b_b$ respectively from 0.001 to 0.1. We can then update the model parameters $\Theta$ as:

$$\Theta = \Theta - \gamma \frac{\partial \mathcal{F}(\mathcal{G}, i, d, b)}{\partial \Theta} \tag{15}$$

where $\gamma$ is the learning rate and we fixed $\gamma = 0.01$ in this study. After the learning process reaches the maximum iterations (default: 100), the predicted score of microbe $i$ on disease $d$ is calculated via $\hat{r}_{di} = V_d^T U_i + b_d$. The above process is described by pseudo-code in Algorithm 1.

Then we aggregate $\hat{r}_{di}$ with the integrated microbe similarity $S_m$ and disease semantic similarity $S_d$. For an unknown disease-microbe pair $(d_i, m_j)$, $d' \in \mathcal{D}^{tr}_{m_j}$ means a set of disease verifiably associated with $m_j$ in training data and $m' \in \mathcal{M}^{tr}_{d_i}$ means a set of microbes verifiably associated with $d_i$. Finally, the final prediction score of $d_i$ on $m_j$ could be calculated by adding the mean values as follows:

$$\hat{r}_{d_i m_j} + = \frac{\alpha_d}{|d'|} \sum_{d' \in \mathcal{D}^{tr}_{m_j}} S_d(d_i, d') + \frac{\alpha_m}{|m'|} \sum_{m' \in \mathcal{M}^{tr}_{d_i}} S_m(m_j, m') \tag{16}$$

where parameters $\alpha_d$ and $\alpha_m$ control the weights of $S_m$ and $S_d$ respectively (we set $\alpha_d = \alpha_m = 0.1$). $\hat{r}_{d_i m_j}$ represents the predicted score of the unknown disease-microbe pair $(d_i, m_j)$. The higher value $\hat{r}_{d_i m_j}$ it is, the more possibly related they are. In this way, we can generate a disease-oriented ranking list for a specific disease via picking up the top-$k$ microbes based on the highest predicted scores $\hat{r}$.

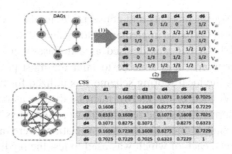

**Fig. 1.** The flowchart of how to calculate the disease semantic similarity.

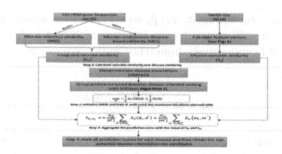

**Fig. 2.** The flowchart of GBDR.

**Table 1.** Computational process of GBDR for predicting microbe-disease associations.

| |
|---|
| **Algorithm: GBDR with the SGD algorithm** |
| **Input:** Training data $\mathcal{R} = \{(m, d)\}$ parameters: $\gamma$, $\alpha_u$, $\alpha_v$, $\alpha_d$, $\alpha_m$ and $\beta_v$, the size of microbe group $|\mathcal{G}|$ ($|\mathcal{G}| = 5$) |
| 1: Initialize the model parameters $\Theta$ |
| 2: for $t_1 = 1,...,T$ do //T=100←maximum iterations |
| 3:    for $t_2 = 1,...,nm$ do |
| 4:       Randomly pick a microbe $i \in \mathcal{M}^{tr}$, a disease $d \in \mathcal{D}_i^{tr}$ and a disease $b \in \mathcal{D}^{tr} \backslash \mathcal{D}_i^{tr}$. |
| 5:       Randomly pick $|\mathcal{G}| - 1$ microbes from $\mathcal{M}_d^{tr} \backslash \{i\}$ as microbe group $\mathcal{G}$. |
| 6:       Calculate $\frac{\partial \mathcal{F}(\mathcal{G}, i, d, b)}{\partial \hat{r}_{\mathcal{G}id; ib}}$ and $\bar{U}_{\mathcal{G}}$ ($\bar{U}_{\mathcal{G}} = \sum_{j \in \mathcal{G}} U_j / |\mathcal{G}|$). |
| 7:       Update $U_j, j \in \mathcal{G}, V_d, V_b, b_d, b_b$, via Eq.(13). |
| 8:    End |
| 9: End |
| 10: Calculate $\hat{r}_{di} = V_d^T U_i + b_d$ |

## 3   Results and Discussion

We prefer the use of square brackets and consecutive numbers. Citations using labels Both LOOCV and $k$-fold CV were implemented to evaluate the prediction performance of the proposed model. GBDR was devised as a disease-oriented ranking computational model. In other words, it is suitable to infer the most possible microbes for a specific disease. Thus, the local scoring scheme was utilized for the assessment. In the framework of LOOCV, each of the known microbe-disease associations was selected as the test sample in turn whereas the remainders were served as the training samples. As for $k$-fold CV, all known microbe-disease associations were randomly divided into $k$ uncrossed sub-datasets, each of which was used to test the prediction performance in turn and others were used to train the model. $K$-fold CV needed to be repeated 100 times to alleviate the bias of the random division. Then the average value was calculated as the final result representing the performance of $k$-fold CV.

The microbe-disease association prediction is actually a binary classification problem mainly depending on whether a microbe has an association with a disease. The receiver operating characteristic (ROC) curve is extensively used to evaluate the binary classification approaches. The ROC curve is determined by sensitivity and specificity with the varying thresholds. In this problem, sensitivity measures what proportion of a test successfully makes a positive microbe-disease association prediction. And specificity measures what proportion of a test successfully makes a negative one. In this way, the ROC curve is plotted accordingly based on the true positive rate (sensitivity) versus false positive rate (1-specificity). Beside the visual evaluation of ROC curve, the area under ROC curve (AUC) is an evaluation coefficient ranging from 0 to 1. In this work, the AUC value of a disease $d$ can be defined as:

$$\mathbf{AUC}_d = \frac{1}{|\mathcal{R}^{te}(d)|} \sum_{(i,j) \in \mathcal{R}^{te}(d)} \delta(\hat{r}_{di} > \hat{r}_{dj}) \tag{17}$$

where $\mathcal{R}^{te}(d) = \{(i,j)|(d,i) \in \mathcal{R}^{te}, (d,j) \notin \mathcal{R} \cup \mathcal{R}^{te}\}$. $\mathcal{R}^{te}(d)$ is a test dataset of $d$, $\hat{r}_{di}$ and $\hat{r}_{dj}$ are predicted values, $\delta()$ is a binary indicator. When the equation within the brackets is true, $\delta() = 1$, otherwise 0. The final AUC value can be averaged as follows:

$$\mathbf{AUC} = \frac{\sum_{u \in D^{te}} AUC_d}{|D^{te}|} \tag{18}$$

Here, $D^{te}$ is a disease set in test data. Normally, AUC = 1 means a perfect prediction and AUC = 0.5 means a completely random one.

In addition, we also implemented $k$-fold CV to further assess the prediction performance of GBDR (see Table 2). As a result, the proposed model yielded average AUCs of 0.8266, 0.8866 and 0.8926 with standard deviations of 0.0805, 0.0270 and 0.0167 with respect to 2-fold, 5-fold and 10-fold CV, respectively. In conclusion, both LOOCV and 5-fold CV fully demonstrated the reliable prediction performance of GBDR. Finally, we selected colorectal carcinoma (CRC) as an important human disease in our case study. As a result, 9 out of top-10 and 16 out of top-20 predicted microbes have been validated by literature evidences. Related information is provided in Table S1.

# 4   Conclusion

In recent years, the human microbiota has obtained the increasing attention due to its key role in human biological processes. Therefore, it has even been considered as "forgotten organ" with human host. Accumulating evidences show that disruption of the host-microbiota homeostatic interactions can increase the risk of various diseases in human body. The development of technology, especially PCR amplification and next-generation sequencing, helps to access massive amounts of sequences, which provides multiple biological datasets for the follow-up downstream analysis. In this work, we leverage 16S rRNA gene to measure microbe sequence similarity and microbe evolutionary distance-based similarity, which have been confirmed to improve the prediction accuracy of the proposed model. Then a group-based computational model of Bayesian disease-oriented ranking (GBDR) is presented to refine the most potential pathogenic microorganisms on a large scale. By combining the known microbe-disease associations, disease semantic similarity, microbe sequence similarity and microbe evolutionary distance-based similarity, the proposed model is evaluated to outperform two state-of-the-art models and other classical algorithms in recommender system via LOOCV. The reliable prediction of GBDR was further demonstrated via k-fold CV and a case study. In conclusion, GBDR could provide insights into identifying the potential ideal biomarkers for human complex diseases. The prediction list of the most possible pathogenic microbes has been published in Table S2 sorted by various specific diseases.

The reliable prediction of GBDR could be attributed to several factors. First, the integrated microbe similarity is effective to reveal the implicit biological information among microbes. Disease semantic similarity can measure the hierarchical correlations between various human diseases. Second, the known microbe-disease association matrix is extremely sparse, which brings about the challenges in prediction. Group-based CF can take advantage of the calculated similarity matrices to extract the additional information of data slices particularly. Third, Bayesian modelling approach of the disease-oriented ranking is more suitable compared with other traditional recommendation algorithms. Finally, the previous computational models were developed to globally predict the most latent microbe-disease associations through rating prediction values in terms of all diseases. The learning samples are disproportionate to many investigated diseases, which makes the previous models degrade performance. The local scoring scheme can alleviate such kind of bias caused by disproportion. Of course, the proposed model also exists several limitations. For example, too many parameters need to be adjusted. And as a disease-oriented model, GBDR cannot be well applied to the microbe-disease association global prediction.

# References

1. Clemente, J.C., Ursell, L.K., Parfrey, L.W., Knight, R.: The impact of the gut microbiota on human health: an integrative view. Cell **148**, 1258–1270 (2012)
2. Sender, R., Fuchs, S., Milo, R.: Are we really vastly outnumbered? Revisiting the ratio of bacterial to host cells in humans. Cell **164**, 337–340 (2016)
3. Savitz, L.D.: The human microbiota: the role of microbial communities in health and disease. Acta Biol. Colomb. **21**, 5–15 (2016)
4. Donia, M.S., et al.: A systematic analysis of biosynthetic gene clusters in the human microbiome reveals a common family of antibiotics. Cell **158**, 1402–1414 (2014)
5. Davenport, E.R., Mizrahi-Man, O., Michelini, K., Barreiro, L.B., Ober, C., Gilad, Y.: Seasonal variation in human gut microbiome composition. PLoS One **9**, e90731 (2014)
6. Mason, M.R., Preshaw, P.M., Nagaraja, H.N., Dabdoub, S.M., Rahman, A., Kumar, P.S.: The subgingival microbiome of clinically healthy current and never smokers. ISME J. **9**, 268–272 (2015)
7. Manichanh, C., et al.: Reduced diversity of faecal microbiota in Crohn's disease revealed by a metagenomic approach. Gut **55**, 205–211 (2006)
8. Thibault, R., Blachier, F., Darcy-Vrillon, B., de Coppet, P., Bourreille, A., Segain, J.P.: Butyrate utilization by the colonic mucosa in inflammatory bowel diseases: a transport deficiency. Inflamm. Bowel Dis. **16**, 684–695 (2010)
9. Huang, Z.A., Wen, Z., Deng, Q., Chu, Y., Sun, Y., Zhu, Z.: LW-FQZip 2: a parallelized reference-based compression of FASTQ files. BMC Bioinform. **18**, 179 (2017)

10. Hartman, A.L., Riddle, S., McPhillips, T., Ludascher, B., Eisen, J.A.: Introducing W.A.T.E. R.S.: a workflow for the alignment, taxonomy, and ecology of ribosomal sequences. BMC Bioinform. **11**, 317 (2010)

11. Ma, W., et al.: An analysis of human microbe-disease associations. Brief. Bioinform. **18**, 85–97 (2017)

12. You, Z.H., et al.: PBMDA: a novel and effective path-based computational model for miRNA-disease association prediction. PLoS Comput. Biol. **13**, e1005455 (2017)

13. Mork, S., Pletscher-Frankild, S., Palleja Caro, A., Gorodkin, J., Jensen, L.J.: Protein-driven inference of miRNA-disease associations. Bioinformatics **30**, 392–397 (2014)

14. Huang, Y.A., You, Z.H., Gao, X., Wong, L., Wang, L.: Using weighted sparse representation model combined with discrete cosine transformation to predict protein-protein interactions from protein sequence. Biomed. Res. Int. **2015**, 902198 (2015)

15. Huang, Y.A., You, Z.H., Chen, X., Yan, G.Y.: Improved protein-protein interactions prediction via weighted sparse representation model combining continuous wavelet descriptor and PseAA composition. BMC Syst. Biol. **10**, 120 (2016)

16. Y.A. Huang, Z.H. You, X. Chen: A systematic prediction of drug-target interactions using molecular fingerprints and protein sequences. Curr. Protein Peptide Sci. (2016)

17. Coenye, T., Vandamme, P.: Intragenomic heterogeneity between multiple 16S ribosomal RNA operons in sequenced bacterial genomes. FEMS Microbiol. Lett. **228**, 45–49 (2003)

18. Chen, X., Huang, Y.A., You, Z.H., Yan, G.Y., Wang, X.S.: A novel approach based on KATZ measure to predict associations of human microbiota with non-infectious diseases. Bioinformatics **33**, 733–739 (2017)

19. Huang, Z.A., et al.: PBHMDA: path-based human microbe-disease association prediction. Front. Microbiol. **8**, 233 (2017)

20. Katz, L.: A new status index derived from sociometric analysis. Psychometrika **18**, 39–43 (1953)

21. Pruitt, K.D., Tatusova, T., Maglott, D.R.: NCBI reference sequences (RefSeq): a curated non-redundant sequence database of genomes, transcripts and proteins. Nucleic Acids Res. **35**, D61–D65 (2007)

22. Lipscomb, C.E.: Medical subject headings (MeSH). Bull. Med. Libr. Assoc. **88**, 265–266 (2000)

23. Wang, D., Wang, J., Lu, M., Song, F., Cui, Q.: Inferring the human microRNA functional similarity and functional network based on microRNA-associated diseases. Bioinformatics **26**, 1644–1650 (2010). (Oxford, England)

24. Johnson, M., Zaretskaya, I., Raytselis, Y., Merezhuk, Y., McGinnis, S., Madden, T.L.: NCBI BLAST: a better web interface. Nucleic Acids Res. **36**, W5–W9 (2008)

25. Camacho, C., et al.: BLAST+: architecture and applications. BMC Bioinform. **10**, 421 (2009)

26. Kumar, S., Stecher, G., Tamura, K.: MEGA7: molecular evolutionary genetics analysis version 7.0 for bigger datasets. Mol. Biol. Evol. **33**, 1870–1874 (2016)

27. Larkin, M.A., et al.: Clustal W and Clustal X version 2.0. Bioinformatics **23**, 2947–2948 (2007)

28. Thomas, R.H.: Molecular evolution and phylogenetics. Heredity **86**, 385 (2001)

29. Hu, Y., Koren, Y., Volinsky, C.: Collaborative filtering for implicit feedback datasets. In: Eighth IEEE International Conference on Data Mining, pp. 263–272 (2009)

30. Pan, R., et al.: One-class collaborative filtering. In: Eighth IEEE International Conference on Data Mining, pp. 502–511 (2008)

31. Rendle, S., Freudenthaler, C., Gantner, Z., Schmidt-Thieme, L.: BPR: Bayesian personalized ranking from implicit feedback. In: Conference on Uncertainty in Artificial Intelligence, pp. 452–461 (2009)
32. Pan, W., Chen, L.: GBPR: group preference based Bayesian personalized ranking for one-class collaborative filtering. In: International Joint Conference on Artificial Intelligence, pp. 2691–2697 (2013)

# Learning from Deep Representations of Multiple Networks for Predicting Drug–Target Interactions

Pengwei Hu[1,2], Yu-an Huang[1], Zhuhong You[3(✉)], Shaochun Li[2],
Keith C. C. Chan[1], Henry Leung[4], and Lun Hu[5(✉)]

[1] Department of Computing, The Hong Kong Polytechnic University,
Hung Hom, Hong Kong
{csyahuang, cskcchan}@comp.polyu.edu.hk
[2] IBM Research, Beijing, China
{hupwei, lishaoc}@cn.ibm.com
[3] Technical Institute of Physics and Chemistry,
Chinese Academy of Sciences, Urumqi, China
zhuhongyou@ms.xjb.ac.cn
[4] Electrical and Computer Engineering,
University of Calgary, Calgary, Canada
leungh@ucalgary.ca
[5] School of Computer Science and Technology,
Wuhan University of Technology, Wuhan, China
hulun@whut.edu.cn

**Abstract.** Many computational approaches have been developed to predict drug-target interactions (DTIs) based on the use of different similarity networks that connect drugs and targets. However, such approaches do not fully exploit all the information available in all similarity networks which can be considered as multiple domain representations of DTIs. As more comprehensive understanding of the latent knowledge underlying the DTI networks requires combining insights obtained from multiple, diverse networks, there is a need for a computational approach to be developed to learn hidden patterns from multiple DTI networks simultaneously for more complete understanding of DTIs. In this paper, we propose such an approached based on a deep multiple DTI network fusion algorithm, called DDTF. to take into consideration all relevant DTI networks. With this DDTF, the identification of DTIs can be made more effective. The DDTF performs its tasks in several steps. Given a set of complex heterogeneous networks, DDTF first uses the network completion algorithm to reconstruct the data representation information to obtain the best network description. To do so, a matrix factorization technique is first used. Based on this approach, the networks obtained from multiple domains are first represented by several similarity matrices and the feature vectors of each pair of drug and protein of the DTI networks are obtained. With these features and representations, we introduce here a novel approach based on non-negative matrix factorization to rescale similarity networks to ensure that the data are reliable. DDTF algorithm constructs a new network to represent the similarity between two vertices. The new similarity network is calculated from the heterogeneous

D.-S. Huang et al. (Eds.): ICIC 2019, LNCS 11644, pp. 151–161, 2019.
https://doi.org/10.1007/978-3-030-26969-2_14

information embedded by a new fusion algorithm. As a final step, DDTF finds a deep representation of each drug or protein in the fused network and use such information for the inference of DTIs. Given the fused deep representations, DDTF can discover optimal projection from a drug network onto a target network. The DDTF algorithm has been tested with real data and experimental results show that DDTF outperforms sophisticated network integration approaches and others significantly. Based on the experiments, it is discovered that the network representation inferred by DDTF has a higher correlation than those yielded by previous work. Moreover, it is noted that completing similarity network based on known networks is a promising direction for drug-target predictions.

**Keywords:** Drug-target interaction prediction ·
Non-negative Matrix Factorization · Network fusion · Deep learning

# 1    Introduction

Drug-target interactions (DTIs) are very important in various processes of drug discovery and form the basis of biological mechanisms. However, the unintended occurrence of drug target interactions has been a major factor in the failure of modern drug development. Despite data manufacturing technologies are accelerating faster than Moore's Law, drug discovery is still a costly and inefficient process. The cost of discovering a new FDA approved drug has doubled every 9 years since 1950, with costs for each new drug estimated at $2.6 billion from a 2013 estimation [1, 2]. The identification of DTIs can, therefore, brings a better understanding of how these interactions function in different biological systems. Therefore, the topic of DTIs analysis and prediction has attracted extensive attention, and researchers have proposed many biotechnology and computational models in this aspect. Given the broad variety of high-throughput biological datasets now being generated, current DTIs prediction is increasingly focusing on methods for their integration.

Many DTIs related information like sequence, structure, side-effects and function of proteins have been collected to public databases. For example, there are hundreds of thousands of human proteins are recorded in Uni-ProtKB database [3]. On the other hand, there are around thousands known drug compounds are deposited in Drug Bank [4]. Other databases such as Super Target and Matador [5], Therapeutic Target Database (TTD) [6], Comparative Toxicogenomics Database [7] and SIDER database [8] have been designed as resources for drug and protein functions. The list of recorded proteins and drugs in these new public databases is growing. Therefore, a considerable number of unexplored compounds and human proteins make it impossible to evaluate drug-target interactions effectively by biological experiment. Normal drug discovery processing may generate products different from the original treatment. In the whole development stage of drug, the instability of drug and target needs to be controlled by various means. To reduce the huge investment of medical experiments and researchers' time, many computational models have been built to elucidate interesting drug-target relationships of most promising candidates for further experimental validation. Various methods care for drug similarity and drug-target nature representations respectively [9].

A kind of popular solutions is feature vector-based methods that face drug and protein features straightforward [10]. They can uncover the description of the hidden knowledge in terms of meaningful features and then generate rules to repeat the experts' decision process. These methods provide biological representations for learning interest patterns such as compound subset and protein subspace [11]. But, current feature-based methods are difficult to handle incomplete knowledge. For example, some protein expression like protein-protein interactome mapping hasn't been fully discovered [12]. Network-based solutions are developed to identify bio-logical interaction by including the similarity matrices of related entities. Lots of computational approaches have been proposed to discover DTIs based on their compound and protein sequence similarity [13–15]. An attractive alternative approach is to integrate various descriptions of drug-target from multiple sources in a statistical learning framework. The integrated prediction of drug-target interaction (DTIs) is both necessary for the development of new drugs and for the research of basic biological functions, and the prospect of further extracting useful knowledge from large-scale data leads to the problem of multi-similarity network fusion. Many network learning methods have proved very effective in solving this problem [16]. However, these methods did not perform well in both cases. One is incomplete putative related net-works, and the other is each domain have multiple network representations.

There have also been several works of network fusion from multiple domains [17, 18]. These approaches adopt multiple data types of same sample to generate similarity networks and then fusing them into one single network. This fusion method has strong robustness against noise and acquisition bias. The specialty of this fusion method is to eliminate irrelevant weak similarity, enhance and retain strong similarity, and has strong robustness against noise and bias. Some existing studies have demonstrated the feasibility of this approach. In [17], genome-wide data were used to identify cancer subtypes after using similarity network fusion. In [18], the author proposed a new algorithm to connect topological network and attribute network for PPI complex clustering.

In this study, we develop a new network-based method called DDTF (a deep multiple DTI networks fusion algorithm) that motivated by the success of network fusion approaches to the problem of constructing networks into one network. DDTF predicts DTIs from multiple network's deep representations with inductive network completion aiming at training a robust training model to rely on known interactions. Firstly, we introduce non-negative matrix factorization (NMF) to complement the unreliable similarity network. Several drug and protein functions about nature features have been used to construct a similarity network as well. Secondly, to solve the difficult problem of multiple networks understanding, we iteratively update all known networks and perform a nonlinear method after their similarity reaches a certain level. Here to calculate the closest converged network nonlinear input data expression, we take spectral theory, it is sent to stacked Auto-Encoder. Finally, we used the matrix completion method to calculate the new expression to predict DTIs. Experimental results show that DDTF is better able to identify DTIs more accurately when compared with the state-of-the-art network fusion algorithm due to it considering of deep fused information.

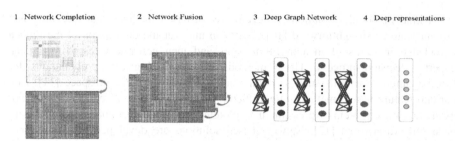

**Fig. 1.** The procedure of DDTF. (1) Example similarity matrix factorization for the network completion. (2) Fused network that constructed by network fusion step (3) Deep autoencoder eliminate edges by generating deep graph representations. (4) The final deep representations.

## 2 Related Work

To supplement and extend biological experiment, scientists have come up with a variety of solutions for predicting drug-target interactions, which can be broken down into categories based on the type of model input. Drug target prediction based on feature model is a good starting point for drug design. They proposed to use direct essential information instead of using matrix representations of another dimension of original DTIs. Cao et al. first try to address compound structure and protein sequence as structure–activity relationship [19]. In [20], the authors only used positive sample DTIs as the training sample and used a special support vector description method to achieve a very good prediction effect. In [11], this work takes advantage of deep learning to encode and decode features for the classifier. One vital problem of current feature-based methods is they cannot project DTIs along the drug and protein respectively. All of the feature-based methods are learning both sides of the feature together. It may make the experiment result cannot be applied to help real drug development. Currently, the latest research direction of DTIs is based on the similarity network of drugs and targets. Compared with the traditional prediction technology of drug target interaction, this method based on similarity is more innovative in the computational model, and their basic research is based on the exploration of drug-drug and target-target similarity. These latest methods have gradually replaced the original methods and been applied in the task of screening potential drug candidates. A general supervised inference method was developed to predict DTI and used drug-based similarity inference, target-based similarity inference [16]. These studies usually use SIMCOMP method to test the similarity of drug molecules [21]. In [22], Yu et al. used the similarity network of drugs and proteins as the input of SVM to carry out two DTIs classification, and fused the experimental results, providing a new solution for drug target prediction. KBMF2K first extracted the original space of drugs and proteins, and then calculated their low-dimensional expression similarity network for DTIs prediction [23]. In [24], they took unknown DTIs as unlabeled samples, used PU learning and similarity information, and adopted three methods to extract reliable negative cases and possible negative cases. As mentioned above, more and more similarity-based works enjoyed similarity matrix representation which helps them generate very high prediction accuracy. However, most approaches based on similarity use single drug and protein networks to infer DTIs. Even some studies integrated multi-domain features like drug side-effects [25, 26], they also ignored the various network level knowledge. The state-of-the-art work first tries to address multiple networks integration by combining the network diffusion algorithm and a dimensionality reduction scheme [27]. They use a subtle fusion method to fuse multiple similarity networks for DTIs prediction. Despite the performance of this work is good, the progress is still unsatisfactory. Because the edge values of the network fused by multiple networks may also lose information and lead to poor prediction results of the fused network, the success of the algorithm also needs to avoid the dilemma that the edge values tend to be the same in the fusion.

## 3 Methodology

Suppose that we are given several drug similarity networks and protein similarity networks constructed from multiple domain information and we want to predict the unknown interactions between the compound and the target protein: (i) introduce NMF to recruit a similarity network reflecting the connections among involved drugs and proteins, thereby generating the reliable network. (ii) the network fusion step, which introduces a fusion method fuses the similarity network obtained from multiple drug or protein information. (iii) the deep representation step, using the principle of spectral clustering, the depth network is built by using the self-encoder with cascade structure, to obtain the depth expression of the fusion network. Besides, we infer the unknown drug–target interactions based on the projection distance of their new representations. Figure 1 shows a procedure of the proposed DDTF.

### 3.1 Constructing High Reliable Drug and Protein Networks

Let $G = \{V, E\}$ represent a network with a two-element tuple, where $V = \{v_i\}(1 \leq i \leq n_v)$ is a set of $n_v$ vertices which means each node represent a drug or a protein in a network, and the edges $E = \{e_{ij}\}$ are weighted by how similar the nodes are. In our work, edge weights of drug or protein network can be represented by an $n \times n$ similarity matrix $M$ with $M(i,j)$ represents the similarity between nodes. For some of drug and protein related similarity network, a common situation is that biological and chemical expression is often very sparse, and limited by experimental conditions, information is incomplete, such as protein-related diseases have not been fully found. Since matrix factorization is efficient to complement matrix, we use the NMF to complement a similar network to obtain more optimized network [28, 29]. We can then take an estimate of our original similarity network and treat it as a more complete network representation.

## 3.2  Network Learning

Proteins have many forms of data expression, and so do drugs. Each form of expression has its own focus, but many other forms of expression are missing. If expressed in the form of similarity network, then each network is interrelated internally. Therefore, an effective way is to merge many forms of similarity network into an ideal network. The fused network should be the most similar representation between networks in the existing domain. Here, we introduce SNF model to solve the problem of fusion network learning [17]. First, a normalized weight matrix $K = D^{-1}L$ is calculated as the full kernel of the vertices. Then, $D$ is a diagonal matrix $D(i, i) = \sum_j L(i, j)$, let $\sum_j K(i, j) = 1$ to avoid being suppressed by self-similarity on $L$ diagonal elements, so as to get a better normalized result:

$$K(i, j) = \frac{L(i, j)}{2 \sum_{k \neq i} L(i, k)} \tag{1}$$

subject to the constraints: $i \neq j$, otherwise $K(i, j) = \frac{1}{2}$. We use KNN to calculate the local affinities, where the domain set of $v_i$ in G is expressed by $N_i$:

$$Q(i, j) = \frac{L(i, j)}{\sum_{k \in N_i} L(i, k)} \tag{2}$$

subject to the constraints: $j \in N_i$, otherwise $Q(i, j) = 0$. Next, the key to SNF is to iteratively update the similarity matrix corresponding to each kind of data expression. Here $K_{t=0}^{(v)}$ is the initial $v$ state matrix, and $Q_{t=0}^{(v)}$ is the kernel matrix:

$$K^{(v)} = Q^{(v)} \times \left( \frac{\sum_{k \neq v} K^{(k)}}{m - 1} \right) \times \left( Q^{(v)} \right)^T, v = 1, 2, 3, \ldots, m \tag{3}$$

The final state fusion matrix is obtained after t step:

$$K^{(v)} = \frac{\sum_{v=1}^{m} K^v}{m} \tag{4}$$

## 3.3  Deep Fusion

Let $G_f = \left( V_f, E_f \right)$ be the new graph after the fusion of multiple expressions. The vertices on the graph represent drugs or proteins $V_f = \{v_{f1}, \ldots, v_{fn}\}$, and the edges represent the fusion similarity between two drugs or proteins. To obtain the deep representation of the fusion network, we introduce a classical deep learning framework: stacked Auto-Encoder [30, 31]. The fusion expression is regarded as an n node expression and sent to the Auto-Encoder, whose goal is to minimize the reconstruction error of input $D^{-1}S$ and output $h_{W,b}(x) \approx x^{(i)}$, to obtain the nonlinear embedding of the fusion network and eliminate the redundant edges. In the lower layer of a stacked

Auto-Encoder, we access an inductive matrix completion method, which is characterized by the output protein and drug deep representation [27, 32] to predict DTIs.

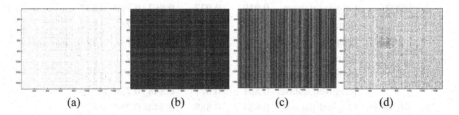

(a)　　　　　　　(b)　　　　　　　(c)　　　　　　　(d)

**Fig. 2.** Protein similarity degree matrix (a) Similarity degree matrix calculated based on protein-protein interactions; (b) Similarity degree matrix calculated based on protein related disease; (c) Similarity degree matrix calculated based on protein sequence; (d) Similarity degree matrix calculated based on above networks fusion.

## 4    Evaluation

### 4.1    Experimental Datasets

In this study, the drug and protein networks which we used to predict DTIs come from [27]. There are two different scale datasets: original dataset and removed dataset. The fewer number one is reconstructed by removing the homologous proteins with high sequence identity scores from the larger one. In the removed dataset, only the DTIs of the original dataset which proteins conditioned the low sequence identity scores can be retained. There are 708 drugs and 1512 protein in both datasets and we observed the number of known DTIs in original dataset and removed dataset are 1923 and 1332 respectively. DrugBank database is our main database for drug information [4], drug-related disease data are taken from CTD database (Comparative Toxicogenomics Database) [7] and SIDER database is our side effect data source [8]. All the proteins include the protein-protein interactions were extracted from the HPRD database [33] and the protein related disease data are taken from CTD [7]. The data which are used to represent 708 drugs include four types of expressions: drug interactions, disease associations, side-effect expressions and their chemical structures. That is to say, we would construct four 708 × 708 similarity network for drugs. The data which are used to represent 1512 proteins include three types of expressions: protein-protein interactions, protein-disease associations and their sequences. That is to say, we would construct three 1512 × 1512 similarity network for proteins. To show the difference between the original similarity networks and fused network, the matrix representations of similarity network based on protein-protein interactions, similarity network based on protein-disease associations, similarity network based on protein sequence and the fused network are shown in Fig. 2.

**Table 1.** Comparison for the Auroc, Auprc, Mcc and F-measure values of the DDTF versus DTINet on the two datasets

| Measures | | AUROC | AUPRC | ACC | MCC | F-measure |
|---|---|---|---|---|---|---|
| DDTF | Original dataset | **0.919** | **0.927** | **0.862** | **0.790** | **0.877** |
| | Removed dataset | **0.901** | **0.908** | **0.860** | **0.785** | **0.862** |
| DTINet | Original dataset | 0.908 | 0.925 | 0.855 | 0.748 | 0.844 |
| | Removed dataset | 0.880 | 0.902 | 0.840 | 0.731 | 0.829 |
| MFDR | Original dataset | 0.902 | 0.893 | 0.855 | 0.734 | 0.819 |
| | Removed dataset | 0.881 | 0.877 | 0.848 | 0.726 | 0.813 |
| KBMF2K | Original dataset | 0.884 | 0.889 | 0.844 | 0.789 | 0.871 |
| | Removed dataset | 0.861 | 0.870 | 0.836 | 0.783 | 0.859 |

## 4.2 Evaluation

Regarding imbalance prediction performance evaluation, it has argued for using ranking measures like AUC (area under the ROC curve) [34]. To further evaluate the performance of the proposed method, we also use the several more measurements like: the prediction accuracy, precision, recall, Matthews's correlation coefficient (MCC) and F-measure. To evaluate the performance of the proposed prediction method, we also applied the DTINet [27], MFDR [11] and KBMF2 K [23] to predict DTIs from multiple similarity networks. To the best of our knowledge, DTINet is only one network fusion method for DTIs prediction so far. In terms of comparison theory, we examined the learned deep fusion representations in the same inductive matrix method with DTINet. Hence, the experimental result just reflects the effect of the proposed method. Due to our DTIs dataset is a high-class imbalance, we use 10-fold cross-validation.

Table 1 reports the accuracies of different algorithms on both original dataset and removed dataset. The scores of AUROCM AUPRC, ACC, MCC and F-MEASURE also be listed in Table 1. As expected, among all measures, DDTF achieves the higher score. This result shows that DDTF extracted more meaningful representations to drug and protein from the fused network and improved the prediction performance.

**Table 2.** The newly confirmed drug-target interactions by public database with high scores

| Drug ID | Protein ID | Evidence |
|---|---|---|
| Levodopa | hsa:1814 | KEGG |
| Levodopa | hsa:1816 | KEGG |
| Isoetharine | hsa:154 | KEGG |
| Metoprolol | hsa:1814 | Drugbank |
| Dipivefrin | hsa:2099 | Drugbank |

### 4.3    Case Study

In this subsection, to prove the DDTF is practical in actual application, we conducted a practical test predict potential DTIs all experimental data set. These data sets were collected from several large public databases several years ago. Therefore, we further verified the newly added DTIs in four public databases including KEGG and Drug-bank. In fact, we considered the top drug-target pairs in DDTF prediction results as highly potential DTIs and cross-verified with some newly added samples in the data-base. Examples of predicted potential drug-target interactions can be obtained in Table 2. As a result, several new drug-target interactions are finally confirmed. In addition, we need to understand that these databases continue to add records, which means that many predicted drug-target pairs are just not laboratory proven. The above case study further confirms that DDTF can be used to predict the interaction of new drug targets.

## 5    Conclusion

In this paper, a network-based fusion approach for DTIs prediction is proposed. DDTF solves two problems in DTI prediction, one is the incomplete biological network and the other is the multi-network fusion. One is fusing the diversity of heterogeneous information embedded in the network data, the other is reducing the incompleteness brought by the vertex features in the heterogeneous network data not fully discovered e.g. Side-effects usually found slowly even if the drug has been listed. DDTF intro-duces NMF, which first characterizes the higher reliable information of some individual network. And then, we were applying an interchanging diffusion algorithm to multiple networks. In addition, we use stacked Auto-Encoder compute deep representations for each node in the networks to approximate the fused network. Deep neural network subtly replaces the step of find k largest eigenvalues of the normalized graph similarity matrix in a spectral procedure. These low-dimensional feature vectors encode nodes in the networks and are readily incorporable for the downstream predictive models. Given the fused deep graph representations, we used an inductive matrix completion for predicting unknown DTIs. We have demonstrated that DDTF can display excellent ability in network integration for accurate DTIs inferring and achieve substantial improvement over the advanced approach. Moreover, experimental results on two real world networks dataset demonstrate that DDTF able to achieves a good detecting performance.

## References

1. Mullard, A.: New drugs cost US 2.6 billion to develop. Nat. Rev. Drug Discov. **13**(12), 877 (2014)
2. Scannell, J., et al.: Diagnosing the decline in pharmaceutical R&D efficiency. Nat. Rev. Drug Discov. **11**, 191–200 (2012)
3. Breuza, L., et al.: UniProt consortium. The UniProtKB guide to the human proteome. Database. (2016)

4. Law, V., et al.: DrugBank 4.0: shedding new light on drug metabolism. Nucleic Acids Res. **42**(D1), D1091–D1097 (2014)
5. Günther, S., et al.: Super target and matador: resources for exploring drug-target relationships. Nucleic Acids Res. **36**(suppl 1), D919–D922 (2008)
6. Chen, X., Ji, Z.L., Chen, Y.Z.: TTD: therapeutic target database. Nucleic Acids Res. **30**(1), 412–415 (2002)
7. Davis, A.P., et al.: The comparative toxicogenomics database: update 2013. Nucleic Acids Res. **41**(D1), D1104–D1114 (2012)
8. Kuhn, M., Campillos, M., Letunic, I., Jensen, L.J., Bork, P.: A side effect resource to capture phenotypic effects of drugs. Mol. Syst. Biol. **6**(1), 343 (2010)
9. Ding, H., Takigawa, I., Mamitsuka, H., Zhu, S.: Similarity-based machine learning methods for predicting drug–target interactions: a brief review. Brief. Bioinform. bbt056 (2013)
10. Wang, L., You, Z.H., Chen, X., Yan, X., Liu, G., Zhang, W.: RFDT: a rotation forest-based predictor for predicting drug-target interactions using drug structure and protein sequence information. Curr. Protein Pept. Sci. **19**(5), 445–454 (2018)
11. Hu, P.W., Chan, K.C.C., You, Z.H.: Large-scale prediction of drug-target interactions from deep representations. In: Proceedings of the International Joint Conference on Neural Networks (IJCNN), Vancouver, Canada, 24–29 July 2016
12. Luo, X., Ming, Z., You, Z., Li, S., Xia, Y., Leung, H.: Improving network topology-based protein interactome mapping via collaborative filtering. Knowl.-Based Syst. **90**, 23–32 (2015)
13. Chen, X., et al.: Drug–target interaction prediction: databases, web servers and computational models. Brief. Bioinform. **17**, bbv066 (2015)
14. Yamanishi, Y., Araki, M., Gutteridge, A., et al.: Prediction of drug-target interaction networks from the integration of chemical and genomic spaces. Bioinformatics **24**(13), i232–i240 (2008)
15. Chen, X., Liu, M.X., Yan, G.Y.: Drug–target interaction prediction by random walk on the heterogeneous network. Mol. BioSyst. **8**(7), 1970–1978 (2012)
16. Cheng, F., et al.: Prediction of drug-target interactions and drug repositioning via network-based inference. PLoS Comput. Biol. **8**(5), e1002503 (2012)
17. Wang, B., et al.: Similarity network fusion for aggregating data types on a genomic scale. Nat. Methods **11**(3), 333–337 (2014)
18. Hu, A.L., Chan, K.C.C.: Utilizing both topological and attribute information for protein complex identification in PPI networks. IEEE/ACM Trans. Comput. Biol. Bioinform. **10**(3), 780–792 (2013)
19. Cao, D.-S., et al.: Large-scale prediction of drug–target interactions using protein sequences and drug topological structures. Anal. Chim. Acta **752**, 1–10 (2012)
20. Hu, P., Chan, K.C., Hu, Y.: Predicting drug-target interactions based on small positive samples. Curr. Protein Pept. Sci. **19**(5), 479–487 (2018)
21. Hattori, M., Okuno, Y., Goto, S., Kanehisa, M.: Heuristics for chemical compound matching. Genome Inform. Ser. **14**, 144–153 (2003)
22. Yu, H., et al.: A systematic prediction of multiple drug-target interactions from chemical, genomic, and pharmacological data. PLoS One **7**(5), e37608 (2012)
23. Gönen, M.: Predicting drug–target interactions from chemical and genomic kernels using Bayesian matrix factorization. Bioinformatics **28**(18), 2304–2310 (2012)
24. Lan, W., et al.: Predicting drug-target interaction using positive-unlabeled learning. Neurocomputing **206**, 50–57 (2016). https://doi.org/10.1016/j.neucom.2016.03.080
25. Yamanishi, Y., Kotera, M., Moriya, Y., Sawada, R., Kanehisa, M., Goto, S.: DINIES: drug–target interaction network inference engine based on supervised analysis. Nucleic Acids Res. **42**(W1), W39–W45 (2014)

26. Takarabe, M., Kotera, M., Nishimura, Y., Goto, S., Yamanishi, Y.: Drug target prediction using adverse event report systems: a pharmacogenomic approach. Bioinformatics **28**(18), i611–i618 (2012)
27. Luo, Y., et al.: A network integration approach for drug-target interaction prediction and computational drug repositioning from heterogeneous information. Nat. Commun. **8**, 573 (2017)
28. Lee, D.D., Seung, H.S.: Learning the parts of objects by non-negative matrix factorization. Nature **401**(6755), 788 (1999)
29. Lee, D.D., Seung, H.S.: Algorithms for non-negative matrix factorization. In: Advances in Neural Information Processing Systems, pp. 556–562 (2001)
30. Tian, F., Gao, B., Cui, Q., Chen, E., Liu, T.-Y.: Learning Deep Representations for Graph Clustering. In: AAAI, pp. 1293–1299 (2014)
31. Bengio, Y., et al.: Greedy layer-wise training of deep networks. In: Advances in Neural Information Processing Systems, vol. 19, p. 153 (2007)
32. Natarajan, N., Dhillon, I.S.: Inductive matrix completion for predicting gene–disease associations. Bioinformatics **30**(12), i60–i68 (2014)
33. Keshava Prasad, T.S., et al.: Human protein reference database—2009 update. Nucleic Acids Res. **37**(suppl_1), 767–772 (2008)
34. Huang, J., Ling, C.X.: Using AUC and accuracy in evaluating learning algorithms. IEEE Trans. Knowl. Data Eng. **17**(3), 299–310 (2005)

# Simulation of Complex Neural Firing Patterns Based on Improved Deterministic Chay Model

Zhongting Jiang[1], Dong Wang[1,2($\boxtimes$)], Huijie Shang[1],
and Yuehui Chen[1,2]

[1] School of Information Science and Engineering,
University of Jinan, Jinan 250022, China
ise_wangd@ujn.edu.cn
[2] Shandong Provincial Key Laboratory of Network Based Intelligent
Computing, Jinan 250022, China

**Abstract.** In this study, the deterministic Chay model is improved considering the $K^+$ channel opening probability during the generation of the generation mechanism of action potential. It can not only simulate the periodic firing, chaos and periodic-adding bifurcation that the original Chay model can simulate, but also simulate the rhythm that the original model cannot simulate, which enhances the simulation ability of the model. The simulation results show that the unification of certainty and randomness in the deterministic system for the first time.

**Keywords:** Deterministic Chay model · Neural firing pattern ·
Action potential · Biological systems · Simulation

## 1 Introduction

The biological nervous system shows strong non-linearity from nerve unit to neural network such as cerebral cortex. Using the non-linearity theory to analyze its complex behavior can deepen our understanding of some phenomena of the nervous system [1–5]. It is also a kind of enrichment and supplement for the non-linearity science itself. Neurodynamics is an interdisciplinary subject formed by the combination of neuroscience and dynamics. Its basic content is to adopt the viewpoint and method of dynamics, especially non-linear dynamics, to study the nervous system as a dynamic system at different disciplines and levels.

With the acceptance of the view that the information encoding mode of nervous system is rhythm rather than frequency, the rhythm mode of nerve discharge has become the focus of neurodynamics research at all levels [6, 7]. In terms of the discharge rhythm of neurons alone, there are very complex rhythm patterns and transitions between them [7–10]. The transmembrane ionic current, which is caused by the opening and closing of ion channels, is the material basis to the generation and change of the action potential, as shown in Fig. 1 [11].

D.-S. Huang et al. (Eds.): ICIC 2019, LNCS 11644, pp. 162–169, 2019.
https://doi.org/10.1007/978-3-030-26969-2_15

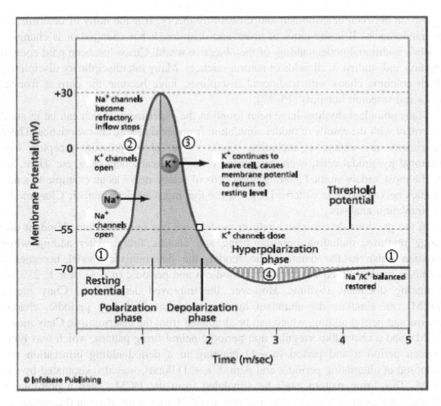

**Fig. 1.** The transmembrane ionic current in different phases of action potential [11]

Certainty and randomness are two aspects of discharge modes at this level. In the early related studies, due to the lack of necessary theoretical knowledge and experimental means, the non-periodic and irregular complex discharge rhythms in the experiments are often considered abnormal rhythms. With the combination of non-linear theory and neuroscience, some of the non-periodic and irregular discharges considered as random rhythms are identified as chaotic rhythms, and the dynamic mechanism of the real random rhythms caused by noise is gradually clear. Among these two kinds of rhythm modes, the most representative one is the chaotic rhythm derived from deterministic mechanism and the random rhythm induced by noise through stochastic self-resonance effect [12–14]. Both chaos and noise induced by stochastic self-resonance emphasize certainty and randomness. Chaos emphasizes stochastic behavior in deterministic system, while stochastic (self-resonance) emphasizes synergistic effect between noise and system.

Chaos is a kind of random behavior generated by a deterministic non-linear system without additional random factors. In short, it is a random behavior generated by a deterministic system. It is a unique form of motion of non-linear dynamics, which is generated and developed with the development of non-linear science. It can be said that the main content of non-linear science is the study of chaos. Chaos eliminates the opposition between determinism and probability theory. It is the unity of determinism and randomness. It is the unity of order and disorder. It has changed or is changing people's traditional understanding of the objective world. Chaos has been paid enough attention and studied in all fields of natural sciences. Many interdisciplinary disciplines, which combine chaos with traditional disciplines, have become the current frontier topics and research hotspots [15–17].

Many abundant rhythms have been found in the experiment, which can be in good agreement with the results of model simulation from the dynamic characteristics. These experiments are extensive, including sciatic nerve, blood pressure receptor, hippocampal pyramidal cells, even myocardial cells, pancreatic beta cells, etc. Here, we use the most widely studied starting point cells of sciatic nerve as an example to study complex nerve discharge patterns [18]. We use the improved deterministic Chay model for simulation analysis.

In the previous deterministic neuron models, we can only simulate abundant discharge rhythms, including periodic firing and chaotic firing. After adding white Gaussian noise to the deterministic model, the deterministic model becomes a stochastic model, which can simulate periodic n and periodic (n + 1) (n = 1, 2, 3, 4) alternating stochastic rhythms. However, the improved deterministic Chay model (IDCM) can simulate the abundant discharge rhythm, including periodic, chaotic patterns and period-adding, which can be simulated from the deterministic Chay model (DCM), and a chaos-like irregular non-periodic neural firing pattern, which was lying between period n and period (n + 1) bursting in a period-adding bifurcation and composed of alternating period n and period (n + 1) bursts, was also simulated by this IDCM. This firing pattern can't be simulated from the DCM. Then a preliminary explanation was tried to give to the analysis results. By the work done in this paper, we hope to enrich the theoretical connotation for the research of nonlinear dynamics and neuroscience, and provide some practical methods.

## 2  Theoretical Chay Model

Chay model [19] is a typical realistic model which describes the firing behavior of neurons based on the dynamic behavior of ion channels. We made a reasonable hypothesis considering the mechanism of action potential generation. When the voltage of action potential reaches the spike peak, that is the joint point between phase 2 and 3 in Fig. 1, the $K^+$ channel will open instantaneously and completely accompanied by the end of depolarization and the beginning of repolarization. The Chay model is improved, a more detailed introduction we have a more detailed introduction in another paper, and the improved formula is as follows:

$$\frac{dV}{dt} = g_I m_\infty^3 h_\infty (V_I - V) + g_{K,V} n^4 (V_K - V) + g_{K,C} \frac{C}{1+C}(V_K - V) + g_L(V_L - V) \quad (1)$$

$$\frac{dn}{dt} = w_K \frac{n_\infty - n}{\tau_n} \quad (2)$$

$$\frac{dC}{dt} = \rho\left(m_\infty^3 h_\infty (V_C - V) - K_C C\right) \tag{3}$$

$V$, $n$ and $C$ represent cell membrane potential, the probability of $K^+$ channel activation and intracellular $Ca^{2+}$ concentration, respectively. $\tau n$ is the steady value of the opening probability of $K^+$ channel. $Vc$ is the equilibrium potential of $Ca^{2+}$ channel. $m\infty$ and $h\infty$ are the steady values of the activation and inactivation probability of $Na^+$-$Ca^{2+}$ channel and $n\infty$ is the steady value of the opening probability of $K^+$ channel. It should be noted that this improved model was a deterministic neuron model. The parameter settings in this paper can be referred to in [20].

## 3   Numerical Simulation Results and Analysis

A series of results suggested that the IDCM could well numerically simulate abundant neural firing patterns and bifurcations as same as those in previous studies, including periodic, chaotic patterns and period-adding.

In the DCM and the IDCM based on peak constraints, when the value of $\lambda n$ is different and $Vc$ from 486 to 0 mV, we can simulate different bifurcation processes with periods, as shown in Fig. 2.

When $\lambda n = 225.8$, the DCM can simulate periodic-adding bifurcation with chaos. With the decrease of $Vc$, the periodic- adding bifurcation of periodic 1 and 2 bursting to chaos, the period- adding bifurcation of periodic 3 bursting to chaos, the period- adding 4 bursting, to spiking via shrinkage, and through inverse period- adding bifurcation to

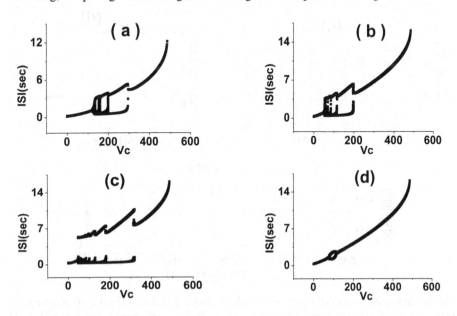

**Fig. 2.** (a) Periodic-adding bifurcation with chaos in DCM ($\lambda n = 225.8$) (b) Periodic-adding bifurcation with chaos in IDCM ($\lambda n = 210$) (c) Periodic-adding bifurcation with irregular firing patterns among neighbouring burstings in the IDCM ($\lambda n = 225.8$) (d) The bifurcation scenario from period 1 bursting to period 1 spiking via period 2 spiking in IDCM ($\lambda n = 190$)

the period-1 spiking, as shown in Fig. 2(a). Considering the generation mechanism of action potential, this paper improves the DCM. When we choose $\lambda n = 210$, we can still simulate the rhythmic transition process with chaotic periodic bifurcation, as shown in Fig. 2(b). When $\lambda n = 225.8$, the irregular firing between period n and period (n + 1) bursting (n = 1, 2, 3, 4) in Periodic-adding bifurcation was numerically simulated, which illustrated in Fig. 2(c). This pattern could be considered as the transitions between period n and period (n + 1) bursts (n = 1, 2, 3, 4), which illustrated in Fig. 3(e). It was intuitively different from chaos because of no other bursts. Obvious single burst was indicated by oblique arrows. When $\lambda n = 190$, the firing changes from periodic 1 bursting to period 1 spiking through period-doubling bifurcation and inverse period -doubling bifurcation. There is no complicated bifurcation process in the process, as shown in Fig. 2(d).

Here, much more attention will be paid to periodic patterns and chaotic patterns, as shown in Fig. 3:

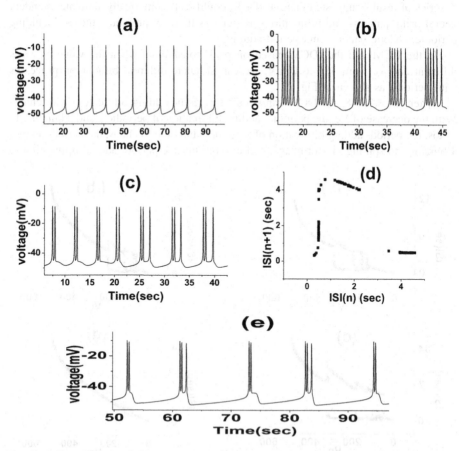

**Fig. 3.** Spike trains of periodic patterns and chaotic patterns in IDCM ($\lambda n = 210$) (a) periodic 1 bursting (b) periodic 6 bursting (c) chaotic rhythm discharge trajectory between period 2 and period 3. The oblique arrow is a rhythm other than period 2 and period 3 (d) First return map of interspike intervals (ISI) (e) irregular non-periodic bursting among period 2 and period 3 in the improved DCM

When $\lambda n = 210$, the IDCM can simulate periodic n bursting (n = 1, 2, 3, 4, 5, 6) and chaotic patterns. $Vc = 300$ mV, we can simulate periodic 1 bursting and $Vc = 56$ mV, we can simulate periodic 6 bursting, which illustrated in Fig. 3(a) and (b). The firing mode among period 2 and period 3 bursting when $Vc = 115$ mV in the IDCM was taken for example. Chaotic patterns is simulated between period 2 and period 3, as shown in Fig. 3(c). The ISI time series transformed from the spike trains are usually used for further analysis. Chaotic multiple firing patterns still have specific structure in the first return map of ISI. From Fig. 3(d), we can see that the first return map of ISI in Fig. 3(c) has an obvious structure, which also verifies that the rhythm is chaotic discharge. The irregular firing between period 2 and period 3 bursting in Periodic-adding bifurcation was numerically simulated, which illustrated in Fig. 3(e). This pattern could be considered as the transitions between period 2 and period 3 bursts. It was intuitively different from chaos because of no other bursts. Obvious single burst was indicated by oblique arrows. This cannot be described in the DCM according to references results. All these familiar and similar activities between two models will be particularly introduced in another article.

## 4 Discussion and Conclusion

In this paper, the IDCM was established considering the $K^+$ channel opening probability during the generation of action potential, and had a powerful simulation capability compared to the original model. The IDCM can simulate the rich rhythm of firing, including periodic, chaotic patterns and period-adding, etc., in addition, the simulation of firing rhythm appeared randomness, it may be only a random rhythm for the appearance of the chaotic system, different from the DCM.

Considering the mechanism of action potential, the introduction of $w_K$ turn the IDCM into discrete with significant sensitive to the $K^+$ channel opening situation. Meanwhile, another factor about the sensory systems of higher animals is that it is very important to define the sensitive intensity of self-perceived information.

The conclusion that the chaos generated in the system is the unity of order and disorder, certainty and randomness has been generally accepted, and the random alternating rhythm induced by peak constraints described in this paper seems to be the unity of certainty and randomness. Only some discharge activities exhibit more deterministic features like periodic firing patterns, while others exhibit more stochastic characteristics like integer multiple firing patterns. Even so, periodic firing patterns still have small fluctuation of the same ISI, integer multiple firing patterns still have specific structure in the First return map of ISI.

The IDCM has stronger simulation ability than the DCM. It can simulate a chaos-like irregular non-periodic neural firing pattern. This rhythm can not be simulated in the DCM. In many previous researches, a noise term was always joined in the first formula to make up the stochastic Chay model (SCM), which was used to study the influence of noise in neural activities. However, the addition of noise will bring some troubles. Noise only works in the characteristic parameter range, and the size of noise will also affect the discharge. The size and range of noise are controlled artificially, and the value taken into account will be incomplete and inefficient. The IDCM in this paper does not need to add noise, which can avoid these problems and improve the efficiency and accuracy of simulation.

Certainly, the hypothesis mentioned above still needs further experimental, stimulation and analytical verification.

**Acknowledgment.** This research was supported by the Shandong Provincial Natural Science Foundation, China (No. ZR2018LF005), the National Key Research and Development Program of China (No. 2016YFC0106000), the Natural Science Foundation of China (Grant No. 61302128), the Youth Science and Technology Star Program of Jinan City (201406003), the Nature Science Research Fund of Jiangsu Province of China (No. BK20161165), and the applied fundamental research Foundation of Xuzhou of China (No. KC17072).

# References

1. Atencio, C.A., Sharpee, T.O., Schreiner, C.E.: Cooperative nonlinearities in auditory cortical neurons. Neuron **58**(6), 956–966 (2008)
2. Li, C.H., Yang, S.Y.: Eventual dissipativeness and synchronization of nonlinearly coupled dynamical network of Hindmarsh-Rose neurons. Appl. Math. Model. **39**(21), 6631–6644 (2015)
3. Shi, R., Hu, G., Wang, S.: Reconstructing nonlinear networks subject to fast-varying noises by using linearization with expanded variables. Commun. Nonlinear Sci. Numer. Simul. **72**, 407–416 (2019)
4. Yang, Y., Solis-Escalante, T., van der Helm, F., Schouten, A.: A generalized coherence framework for detecting and characterizing nonlinear interactions in the nervous system. IEEE Trans. Biomed. Eng. **25**(4), 401–410 (2008)
5. Zhao, Z., Gu, H.: Identifying time delay-induced multiple synchronous behaviours in inhibitory coupled bursting neurons with nonlinear dynamics of single neuron. Proc. IUTAM **22**, 160–167 (2017)
6. Ren, W., Hu, S.J., Zhang, B.J., Wang, F.Z., Gong, Y.F., Xu, J.: Period-adding bifurcation with chaos in the interspike intervals generated by an experimental neural pacemaker. Int. J. Bifurcat. Chaos. **7**(08), 1867–1872 (1997)
7. Yang, M., An, S., Gu, H., Liu, Z., Ren, W.: Understanding of physiological neural firing patterns through dynamical bifurcation machineries. NeuroReport **17**(10), 995–999 (2006)
8. Huang, S., Zhang, J., Wang, M., Hu, C.: Firing patterns transition and desynchronization induced by time delay in neural networks. Phys. A **499**, 88–97 (2018)
9. Jia, B., Gu, H., Xue, L.: A basic bifurcation structure from bursting to spiking of injured nerve fibers in a two-dimensional parameter space. Cogn. Neurodyn. **11**(2), 1–12 (2017)
10. Jia, B., Gu, H.: Dynamics and physiological roles of stochastic firing patterns near bifurcation points. Int. J. Bifurcat. Chaos. **27**(7), 1750113 (2017)
11. Evans-Martin, F.F.: The Nervous System. Infobase Publishing (2009)
12. Bao, B.C., Wu, P.Y., Bao, H., Xu, Q., Chen, M.: Numerical and experimental confirmations of quasi-periodic behavior and chaotic bursting in third-order autonomous memristive oscillator. Chaos. Soliton. Fract. **106**, 161–170 (2018)
13. Gu, H., Zhang, H., Wei, C., Yang, M., Liu, Z., Ren, W.: Coherence resonance–induced stochastic neural firing at a saddle-node bifurcation. Int. J. Mod. Phys. B **25**(29), 3977–3986 (2011)

14. Li, D., Hu, B., Wang, J., Jing, Y., Hou, F.: Coherence resonance in the two-dimensional neural map driven by non-Gaussian colored noise. Int. J. Mod. Phys. B **30**(5), 1650012 (2016)
15. Liu, J., Mao, J., Huang, B., Liu, P.: Chaos and reverse transitions in stochastic resonance. Phys. Lett. A **382**(42), 3071–3078 (2018)
16. Shaw, P.K., Chaubey, N., Mukherjee, S., Janaki, M.S., Iyengar, A.N.: A continuous transition from chaotic bursting to chaotic spiking in a glow discharge plasma and its associated long range correlation to anti correlation behaviour. Phys. A **513**, 126–134 (2019)
17. Zlatkovic, B.M., Samardzic, B.: Multiple spatial limit sets and chaos analysis in MIMO cascade nonlinear systems. Chaos. Soliton. Fract. **119**, 86–93 (2019)
18. Shang, H., Xu, R., Wang, D.: Dynamic analysis and simulation for two different chaos-like stochastic neural firing patterns observed in real biological system. In: International Conference on Intelligent Computing, pp. 749–757 (2017)
19. Chay, T.R.: Chaos in a three-variable model of an excitable cell. Phys. D **16**(2), 233–242 (1985)
20. Shang, H., Jiang, Z., Xu, R., Wang, D., Wu, P., Chen, Y.: The dynamic mechanism of a novel stochastic neural firing pattern observed in a real biological system. Cogn. Syst. Res. **53**, 123–136 (2019)

# Knowledge Based Helix Angle and Residue Distance Restraint Free Energy Terms of GPCRs

Huajing Ling, Hongjie Wu$^{(\boxtimes)}$, Jiayan Han, Jiwen Ding,
Weizhong Lu, and Qiming Fu

School of Electronic and Information Engineering,
Suzhou University of Science and Technology, Suzhou 215009, China
Hongjie.wu@qq.com

**Abstract.** The function of G-protein-coupled receptor (GPCR) in organisms are directly related to its tertiary structure. Protein free energy can well reflect the state stability of protein tertiary structure. Therefore, the research on the free energy of GPCR is of great significance. At present, there is a lack of goldenfree energy constraint GPCR structure in helix domain level in the current researches, which affects GPCR's three-dimensional structure stability. In this paper, a knowledge-based free energy term with the residue distance and angle of the helix of the GPCR as a constraint is established. The energy term is based on the gaussian distribution model, which accurately expresses the free energy of the GPCR. Compared with other energy functions, the experimental data and the result of this model is more accurate.

**Keywords:** GPCR · Free energy term · Gaussian distribution

## 1 Introduction

G Protein-Coupled Receptor (GPCR) is a class of receptor proteins with seven trans-membrane helices. It is responsible for the information between cells and the external environment. It is generally believed that the role and function of a protein in a living organism is directly related to the conformation of its three-dimensional structure [1]. The study found that when the protein has the lowest free energy the three-dimensional structure state is more stable. That is, the native protein conformation is at the lowest energy state in thermodynamics [2]. To this end, the three-dimensional structure of the GPCR can be predicted by studying the magnitude of the free energy value for better application in biomedicine [3]. Yarov-Yarovoy et al. divided the membrane protein and the embedded phospholipid bilayer into 7 layers, quantified the hydrophobicity of the conformation and the polarity of the bilayer, and established their respective energy terms [4]. The energy term is linearly combined to guide the folding and optimization of membrane proteins [5]. There are currently very few energy functions for GPCR, and there is an energy function for Rosetta. Due to the complexity of the relationship between organic molecules and their internal particles, the current energy function does not accurately reflect the energy state of the molecular system, but is approximated on a

© Springer Nature Switzerland AG 2019
D.-S. Huang et al. (Eds.): ICIC 2019, LNCS 11644, pp. 170–176, 2019.
https://doi.org/10.1007/978-3-030-26969-2_16

certain side [6]. In this paper, the method of constructing the energy function makes full use of the three-dimensional structural characteristics of the GPCR. The distance and angle between the helices are used as constraints, and the Gaussian distribution function can be used to obtain more accurate energy terms.

## 2 Data Sources and Data Set

### 2.1 Data Sources

The experimental data source of this paper is from GPCRDB, which is the most abundant database of GPCR protein molecules [7]. Website: http://www.gpcrdb.org/.

### 2.2 Data Set

28 structured GPCRs are used in this paper and we selected 10 representative GPCRs with structural characteristics listed in Table 1.

**Table 1.** Part of the GPCRs used in this paper.

| # | PDB | Ch | Res | Protein name | TM-helix definition |
|---|-----|-----|-----|--------------|---------------------|
| 1 | 2HPY | B | 2.8 | Rhodopsin | 34−64, 71−100, 107−140, 150−172, 200−226, 246−277, 285−309 |
| 2 | 2LNL | A | NA | C-X-C chemokine receptor type 1 | 10−38, 46−73, 80−110, 121−145, 171−200, 210−239, 248−280 |
| 3 | 2ZIY | A | 3.7 | Rhodopsin | 28−58, 65−95, 102−135, 146−168, 192−224, 253−283, 291−314 |
| 4 | 3EML | A | 2.6 | Adenosine receptor A2a | 6−30, 39−63, 75−99, 116−135, 166−186, 209−233, 244−267 |
| 5 | 3ODU | A | 2.5 | C-X-C chemokine receptor type 4 | 10−40, 46−74, 80−113, 118−148, 167−199, 213−243, 250−278 |
| 6 | 3PBL | A | 2.9 | D(3) dopamine receptor | 3−25, 32−60, 69−102, 116−138, 155−185, 194−225, 234−258 |
| 7 | 3RZE | A | 3.1 | Histamine H1 receptor | 4−27, 34−62, 70−103, 114−136, 154−182, 195−224, 232−254 |
| 8 | 3UON | A | 3.0 | Muscarinic acetylcholine receptor M2 | 4−31, 38−67, 74−107, 118−147, 165−194, 206−234, 241−265 |
| 9 | 3V2Y | A | 2.8 | Sphingosine 1-phosphate receptor 1 | 30−57, 64−89, 99−130, 135−157, 178−207, 218−245, 258−279 |
| 10 | 3VW7 | A | 2.2 | Proteinase-activated receptor 1 | 9−43, 46−74, 82−115, 121−144, 171−203, 209−244, 251−279 |

## 3   Methods

### 3.1   Derivation of a Three-Dimensional Fitted Straight Line Formula for Helix

In this paper, the angle between any two helices and the distance of the two endpoint residues are taken as the free energy term [8]. The space rectangular coordinate system is established to get the three-dimensional space coordinates of each point of the helix and a straight line is fitted thereto [9]. In order to get the fitting straight line of GPCR helix better, the three-dimensional straight line fitting formula is further deduced based on least squares (LS) [10].

Simplification of the space straight line as is shown in Eq. 1 [11]:

$$\frac{x - x_0}{m} = \frac{y - y_0}{n} = \frac{z}{1} \tag{1}$$

Equation of n points [12]:

$$\begin{bmatrix} m & x_0 \\ n & y_0 \end{bmatrix} \begin{bmatrix} z_1 \cdots\cdots\cdots z_n \\ 1 \cdots\cdots\cdots 1 \end{bmatrix} = \begin{bmatrix} x_1 \cdots\cdots\cdots x_n \\ y_1 \cdots\cdots\cdots y_n \end{bmatrix} \tag{2}$$

Least squares:

$$\begin{bmatrix} m & x_0 \\ n & y_0 \end{bmatrix} \begin{bmatrix} z_1 \cdots\cdots\cdots z_n \\ 1 \cdots\cdots\cdots 1 \end{bmatrix} \begin{bmatrix} z_1 & 1 \\ \vdots & \vdots \\ z_n & 1 \end{bmatrix} = \begin{bmatrix} x_1 \cdots\cdots\cdots x_n \\ y_1 \cdots\cdots\cdots y_n \end{bmatrix} \begin{bmatrix} z_1 & 1 \\ \vdots & \vdots \\ z_n & 1 \end{bmatrix} \tag{3}$$

LS simplification formula:

$$\begin{bmatrix} m & x_0 \\ n & y_0 \end{bmatrix} = \begin{bmatrix} \sum x_i z_i & \sum x_i \\ \sum y_i z_i & \sum y_i \end{bmatrix} = \begin{bmatrix} \sum z_i^2 & \sum z_i \\ \sum z_i & n \end{bmatrix}^{-1} \tag{4}$$

The equation of the space straight line is obtained if four parameters m, n, x0, y0 are required (Fig. 1).

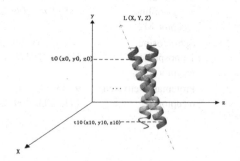

**Fig. 1.** Fitted straight line

Take 11 points on the helix: t0(x0, y0, z0), t1(x1, y1, z1)......t10(x10, y10, z10). The y-axis coordinates are selected according to the 10 equal division of the height of the entire helix y-axis. Then we need to get a straight line vector that the sum of the square of all the points to the line is minimum [13]. This line vector is the line to be fitted.

## 3.2  Helix Angle and Residue Distance Model Based on Gaussian Distribution

Space vector angle formula [14]:

$$\cos(\theta_k) = \cos < \vec{a}, \vec{b} > = \frac{a_1b_1 + a_2b_2 + a_3b_3}{\sqrt{a_1^2 + a_2^2 + a_3^2} \cdot \sqrt{b_1^2 + b_2^2 + b_3^2}}; \tag{5}$$

Distance formula between two points in space [15]:

$$d_k = \left| \overrightarrow{AB} \right| = \sqrt{(x_2 - x_1)^2 + (y_2 - y_1)^2 + (z_2 - z_1)^2}; \tag{6}$$

The energy terms of helix angle and residue distance model are both based on Gaussian distribution [16]:

$$E_{\text{conn}} = -\Sigma_k^N \frac{1}{\sigma_k \sqrt{2\pi}} e^{-\frac{(d_k - \langle d_k \rangle)^2}{2\sigma_k^2}}; \tag{7}$$

$$E_{\text{angle}} = -\Sigma_k^N \frac{1}{\sigma_k \sqrt{2\pi}} e^{-\frac{(\cos(\vartheta_k) - \langle \cos(\vartheta_k) \rangle)^2}{2\sigma_k^2}}; \tag{8}$$

N is the number of connected helix pairs, k is the index of helix pair (Fig. 2).

**Fig. 2.** Helix angle and residue distance

## 3.3   Algorithm Based on Helix Angle and Residue Distance for Energy Term

The pseudo-code is shown in Algorithm 1.

---

**Algorithm 1:**  algorithm based on helix angle and residue distance for energy term

---
Input :

$t_0(x_0,y_0,z_0)...t_{10}(x_{10},y_{10},z_{10})$: Three-dimensional coordinates from $t_0$ to $t_{10}$ of helix i

$s_0(x_0,y_0,z_0)...s_{10}(x_{10},y_{10},z_{10})$: Three-dimensional coordinates from $t_0$ to $t_{10}$ of helix j

$m_0(x_0,y_0,z_0)$: Three-dimensional coordinates of the endpoint residue on helix i

$n_0(x_0,y_0,z_0)$: Three-dimensional coordinates of the endpoint residue on helix j

$d(k)$: the normalized distance of the kth helix pair

$\langle d_k \rangle$: the average of the normalized distance of the quiet helix pair

$\theta_k$: angle of the axis of two helices   $\sigma_k$: deviation of the kth helix paircooling

*N: the number of connected helix pairs   k: the index of helix pair*

1.     Substitute the three-dimensional coordinates of all points of the two helices into the spatial straight line equation
2.     Get the space straight line equation of two helices
3.     Calculate the cos value of the angle between the two vectors according to the vector angle formula
4.     Substitute Three-dimensional coordinates of the endpoint residue space vector distance formula to get the distance between them
5.     Substitute cos $(\theta_k)$ 、 $d_k$、 $\sigma_k$ into free energy term formula
6.     Output : value of energy term

---

In terms of time complexity, the Gaussian distribution model in Algorithm 1 is a polynomial level, and the execution time of the algorithm in this paper is constant, so its complexity is $O(1)$ [17]. In terms of space complexity, the space complexity of Algorithm 1 mainly depends on the three-dimensional coordinate values of the helix and its fitted straight line, so the complexity is $O(n)$.

# 4   Experiments and Discussion

28 GPCRs of known structure are performed in this paper [18]. Finally, the data of the angle of the helix and the distance of the residue and their mean values are obtained, as shown in Table 2.

**Table 2.**  The Mean and deviation of residue distance and helix angle on 28 solved GPCRs

| Helix | I–II | II–III | III–IV | IV–V | V–VI | VI–VII |
|---|---|---|---|---|---|---|
| $d(\text{Å})$ | $1.70 \pm 0.07$ | $1.75 \pm 0.22$ | $1.42 \pm 0.23$ | $0.65 \pm 0.023$ | $0.94 \pm 0.078$ | $1.56 \pm 0.27$ |
| $\cos(\theta)$ | II | III | IV | V | VI | VII |
| I | $0.9 \pm 0.0015$ | $0.69 \pm 0.0059$ | $0.87 \pm 0.0019$ | $0.65 \pm 0.0057$ | $0.81 \pm 0.0044$ | $0.89 \pm 0.0015$ |
| II | | $0.90 \pm 0.0010$ | $0.91 \pm 0.0020$ | $0.83 \pm 0.0037$ | $0.81 \pm 0.0023$ | $0.84 \pm 0.0024$ |
| III | | | $0.85 \pm 0.0007$ | $0.93 \pm 0.0006$ | $0.75 \pm 0.0052$ | $0.72 \pm 0.0063$ |
| IV | | | | $0.91 \pm 0.0018$ | $0.96 \pm 0.0005$ | $0.96 \pm 0.0011$ |
| V | | | | | $0.89 \pm 0.0014$ | $0.83 \pm 0.0056$ |
| VI | | | | | | $0.98 \pm 0.0005$ |

In Table 3, we get the average of the six groups of residue distances and the cos average of the angle between one of the seven helices and the remaining six helices. These two sets of data fully demonstrate that the structural state is very stable. Compared with the average free energy result with other methods', our energy term has a much lower value and much higher accuracy of calculation. The model can better reflect the free energy [19].

**Table 3.** The Mean and deviation of all groups of residue distance and helix angle

| Helix | d(Å) |
|---|---|
| I–II, II–III, III–IV, IV–V, V–VI, VI–VII | $1.34 \pm 0.1485$ |
| Helix | $\cos(\theta)$ |
| I-II, III, IV, V, VI, VII | $0.80 \pm 0.0035$ |
| II-I, III, IV, V, VI, VII | $0.87 \pm 0.0022$ |
| III-I, II, IV, V, VI, VII | $0.81 \pm 0.0033$ |
| IV-I, II, III, V, VI, VII | $0.91 \pm 0.0013$ |
| V-I, II, III, IV, VI, VII | $0.84 \pm 0.0031$ |
| VI-I, II, III, IV, V, VII | $0.87 \pm 0.0024$ |
| VII-I, II, III, IV, V, VI | $0.87 \pm 0.0029$ |

## 5  Conclusion

In this paper, a free energy term formula based on the Gaussian distribution model, using the GPCR helix angle and the same-end residue distance as constraints is constructed [20]. In this paper, 28 known structural GPCRs were tested and a large amount of experimental data was obtained. The results show that the free energy term expression can accurately reflect the free energy of the GPCR.

**Acknowledgement.** This paper is supported by the National Natural Science Foundation of China (61772357, 61502329, 61672371, and 61876217), Jiangsu Province 333 Talent Project, Top Talent Project (DZXX-010), Suzhou Foresight Research Project (SYG201704, SNG201610, and SZS201609)

## References

1. Hauser, A.S., Attwood, M.M., Rask-Andersen, M., et al.: Trends in GPCR drug discovery: new agents, targets and indications. Nat. Rev. Drug Discov. **16**, 829–842 (2017)
2. Flock, T., Hauser, A.S., Lund, N., et al.: Selectivity determinants of GPCR–G-protein binding. Nature **545**(7654), 317–322 (2017)
3. Hilger, D., Masureel, M., Kobilka, B.K.: Structure and dynamics of GPCR signaling complexes. Nat. Struct. Mol. Biol. **25**(1), 4–12 (2018)
4. Kang, Y., Zhou, X.E., Xiang, G., et al.: Crystal structure of rhodopsin bound to arrestin by femtosecond X-ray laser. Nature **523**(7562), 561–567 (2015)

5. Moritsugu, K., Terada, T., Kidera, A.: Free-energy landscape of protein-ligand interactions coupled with protein structural changes. J. Phys. Chem. B **121**(4), 731–740 (2017)

6. Takemura, K., Matubayasi, N., Kitao, A.: Binding free energy analysis of protein-protein docking model structures by evERdock. J. Chem. Phys. **148**(10), 105101 (2018)

7. Gohlke, H., Kiel, C., Case, D.A.: Insights into protein-protein binding by binding free energy calculation and free energy decomposition for the Ras-Raf and Ras-RalGDS complexes. J. Mol. Biol. **330**(4), 891–913 (2003)

8. Lee, H.S., Seok, C., Im, W.: Potential application of alchemical free energy simulations to discriminate GPCR ligand efficacy. J. Chem. Theory Comput. **11**(3), 1255–1266 (2015)

9. Lenselink, E.B., Louvel, J., Forti, A.F., et al.: Predicting binding affinities for GPCR ligands using free-energy perturbation. ACS Omega **1**(2), 293–304 (2016)

10. Advances in free-energy-based simulations of protein folding and ligand binding. Curr. Opin. Struct. Biol. **36**, 25–31 (2016)

11. Suofu, Y., Li, W., Jeanalphonse, F.G., et al.: Dual role of mitochondria in producing melatonin and driving GPCR signaling to block cytochrome c release. Proc. Natl. Acad. Sci. U.S.A. **114**(38), E7997–E8006 (2017)

12. Pavlos, N.J., Friedman, P.A.: GPCR signaling and trafficking: the long and short of it. Trends Endocrinol. Metab. **28**(3), 213–226 (2017)

13. Irannejad, R., Pessino, V., Mika, D., et al.: Functional selectivity of GPCR-directed drug action through location bias. Nat. Chem. Biol. **13**(7), 799 (2017)

14. Eichel, K., Jullié, D., Barsirhyne, B., et al.: Catalytic activation of β-arrestin by GPCRs. Nature **557**, 381–386 (2018)

15. Jean-Charles, P.Y., Kaur, S., Shenoy, S.K.: GPCR signaling via β-arrestin-dependent mechanisms. J. Cardiovasc. Pharmacol. **70**(3), 142–158 (2017)

16. Kumar, B.A., Kumari, P., Sona, C., et al.: GloSensor assay for discovery of GPCR-selective ligands. Methods Cell Biol. **142**, 27–50 (2017)

17. Pándy-Szekeres, G., Munk, C., Tsonkov, T.M., et al.: GPCRdb in 2018: adding GPCR structure models and ligands. Nucleic Acids Res. **46**(Database issue), 440–446 (2017)

18. Mcgregor, K.M., Bécamel, C., Marin, P., et al.: Using melanopsin to study G protein signaling in cortical neurons. J. Neurophysiol. **116**(3), 1082–1092 (2016)

19. Huang, Y., Todd, N., Thathiah, A.: The role of GPCRs in neurodegenerative diseases: avenues for therapeutic intervention. Curr. Opin. Pharmacol. **32**, 96–110 (2017)

20. Gupta, A., Singh, V.: GPCR Signaling in C. Elegans and its implications in immune response. Adv. Immunol. **136**, 203–226 (2017)

# Improved Spectral Clustering Method for Identifying Cell Types from Single-Cell Data

Yuanyuan Li[1,2], Ping Luo[1], Yi Lu[1], and Fang-Xiang Wu[1,3,4(✉)]

[1] Division of Biomedical Engineering,
University of Saskatchewan, Saskatoon, Canada
faw341@mail.usask.ca
[2] School of Mathematics and Physics,
Wuhan Institute of Technology, Wuhan, China
[3] Department of Mechanical Engineering,
University of Saskatchewan, Saskatoon, Canada
[4] Department of Computer Science,
University of Saskatchewan, Saskatoon, Canada

**Abstract.** With the development of single-cell RNA sequencing (scRNA-seq) technology, characterizing heterogeneity at the cellular level has become a new area of computational biology research. However, the infiltration of different types of cells and the high variability in gene expression complicate classification of cell types. In this study, we propose an improved spectral clustering method for clustering single-cell data that avoid the overfitting issue and consider both similarity and dissimilarity, motivated by the observation that same type cells have similar gene expression patterns, but different types of cells produce dissimilar gene expression patterns. To evaluate the performance of the proposed spectral clustering method, we compare it with the traditional spectral clustering method in recognizing cell types on various real scRNA-seq data. The results show that taking intercellular dissimilarity into account can effectively achieve high accuracy and robustness and that our method outperforms the traditional spectral clustering methods in grouping cells that belong to the same cell types.

**Keywords:** Single-cell RNA sequencing data · Spectral clustering · Similarity/dissimilarity matrix · Cell types identification

## 1 Introduction

Recent development in single-cell sequencing technologies has extended our understanding of a series of complex biological phenomena at the single cell level [1]. Datasets arising from these technologies can be used for the reveal of gene expressional differences between individual cells, the discovery of cell types, and the detection of heterogeneity in cell line [2]. However, different types of cells are always heavily infiltrated with each other in the traditional biological experiments [3]. A flexible approach to this problem would be to divide the cells into distinct groups via clustering methods, in order that cells in the same cluster exhibit strikingly similar gene expression patterns.

© Springer Nature Switzerland AG 2019
D.-S. Huang et al. (Eds.): ICIC 2019, LNCS 11644, pp. 177–189, 2019.
https://doi.org/10.1007/978-3-030-26969-2_17

The identification of cell types based on single-cell data is an unsupervised classification problem, and a series of computational approaches have been developed to solve this problem. However, both biological and technical issues give rise to a range of problems. In addition, the number of features (genes) is much larger than the number of cells for classification, which may lead to the overfitting problem. Accordingly, most clustering methods are not effective enough to partition the cells into well-separated groups.

In recent years, to circumvent these problems, many approaches have been proposed that based on single cell gene expression patterns to identify groups of cells. For example, Buettner et al. used latent variable models to remove variation before taking downstream analytic strategies [4]. Xu and Su combined a quasi-clique-based clustering algorithm with the shared nearest neighbor based on similarity measure, which could generate desirable solutions with high accuracy and sensitivity [5]. Shao and Höfer adapted the nonnegative matrix factorization [6, 7] for the unsupervised learning of subpopulations from single-cell gene expression data [8]. Kiselev et al. proposed a novel algorithm named SC3 that achieved better accuracy and robustness by integrating multiple clustering solutions through a consensus approach [9]. For the identification of putative cell types, most of these approaches perform well for some situations by relying on feature selection or dimensionality reduction to reduce the noise and speed up calculations [10].

Spectral clustering (SC), as one of the most popular clustering algorithms, uses the eigenvalues of the similarity matrix derived from the data to perform dimensionality reduction for clustering [11]. Generally, there are three ways to construct similarity matrix: ε-neighborhood, k-nearest neighbor, or fully connected. All ways are based on the distances calculated by several different choices available. The quality of the clustering outcome is sensitive to the computation of eigenvectors of the normalized Laplace matrix and the selection of similarity measurement. Recently, a series of computational approach have been developed to improve SC performance. When a data affinity matrix is fixed, Lu et al. proposed the sparse SC method, which extended the traditional SC with a sparse regularization [12]. Wang et al. made effort to obtain an appropriate cell-to-cell similarity metric by combining multiple kernels, and the similarities could be efficiently adapted [13]. Park and Zhao used multiple doubly stochastic affinity matrices to construct a suitable similarity matrix and imposed sparse structure on the target matrix [14].

While these methods can work well for identifying cell types, they only account for the positive similarities between cells, and ignore the impact of negative similarities on clustering results. In other words, only the similarities are taken into consideration, but the dissimilarities are ignored. This methodology may limit the effectiveness of spectral clustering-based strategies for identifying clusters. Nevertheless, the goal of SC is to divide the data points (representing single cells) into several groups such that single cells in the same group are similar while single cells in different groups are dissimilar to each other [11]. Therefore, dissimilarities between single cells should not be neglected. In this study, we construct an appropriate matrix taking into account similarities as well as dissimilarities among cells and propose an improved spectral clustering method for cell clustering, called ISC. In our ISC algorithm, we incorporate the dissimilarity matrix to emphasize the dissimilarities between the natural groupings, and a parameter is provided to trade off the similarity matrix and dissimilarity matrix.

To evaluate the performance of the proposed method, we first apply it to breast cancer and analyze 549 cells from 11 patients and compare the result with traditional SC. Then our method is used to test other real experimental datasets with high-confidence cell labels. Our result shows that considering both similarity and dissimilarity improves the clustering performance. Furthermore, the clustering results reflect that the ISC achieves higher accuracy and satisfied robustness in classification of single cells.

## 2    Materials and Methods

### 2.1    Data Source

In this study, we first focus on our analysis on breast cancer cells. Original single-cell mRNA expression profiles were downloaded from the NCBI GEO database under the accession code GSE75688 [3]. The dataset contains eleven primary breast cancer (BC) patients including two lymph node metastases. Tumor cells were obtained by isolating tumor tissue, so they were gotten along with neighboring stromal and immune cells. 549 single-cell mRNA expression profiles were acquired.

In addition, we also directly obtained other sets of processed single cell gene expression datasets from previously published papers [14]. Buettner et al. generated sing-cell RNA-seq data from mouse embryonic stem (ES) cells with the Fluidigm C1 protocol. Then, they measured the transcriptional profile of 182 ES cells that had been staged for cell-cycle phase (G1, S and G2M), the cell-cycle stage of each cell was validated using the gold-standard Honechst staining [4]. Ting et al. achieved single-cell RNA-seq of mouse pancreatic circulating tumor (PCT) cells isolated without positives selection bias, along with mouse embryonic fibroblasts, white blood cells, an established genotype-matched cancer cell line and primary pancreatic tumors [15]. Treutlein et al. sequenced transcriptomes of 80 individual live cells of the developing mouse lung epithelium (LE) late in sacculation. Using known marker genes, they associated cells with five putative cell types [16]. Deng et al. performed RNA-Sequencing on individual cells isolated from mouse preimplantation embryos (PE) at different stages, different cell developmental stages composed of zygote, early 2-cell, mid 2-cell, late 2-cell, 4-cell, 8-cell and 16-cell [17].

### 2.2    Existing Spectral Clustering

$D = \{d_1, d_2, \cdots, d_n\}$ is a given set of data points, each data point $d_i$ is a $p$ dimensional column vector, $A = (a_{ij}) \in R^{n \times n}$ is a symmetric similarity matrix, where $a_{ij} \geq 0$ denotes a measure of the similarity between data points $d_i$ and $d_j$, a higher value of $a_{ij}$

indicates data points $d_i$ and $d_j$ are more similar. In traditional SC, we are trying to define a $k$ dimensional column feature vector $x_i$ for each data point $i$, where $k$ could be much less than $p$. Intuitively, if two data points are more similar, their feature vectors should be closer to each other in the feature space. The problem can be mathematically formulated as the following optimization problem:

$$J_1 = \min_{x_i \in R^k, i=1,2,\cdots n} \frac{1}{2} \sum_{i,j=1}^{n} a_{ij} ||x_i - x_j||^2$$

$$s.t. \sum_{i=1}^{n} x_i x_i^T = I_k$$

(1)

Let $D$ be a diagonal matrix where the $l$-th diagonal element equals to the sum of all elements in the $l$-th row of the similarity matrix, then one can compute the Laplace matrix as $L = D - A$. It can be proved that the eigenvalues of Laplace matrix are nonnegative and that the $k$ eigenvectors corresponding to the first $k$ smallest nonzero eigenvalues consist of the $n \times k$ feature matrix, whose $i$-th row is the best $k$ dimensional feature vector of data point $i$. With these $k$ dimensional features of all data points, any feature-based clustering method can be applied to cluster data points.

### 2.3    Improved Spectral Clustering

We improve the traditional SC by considering the dissimilarities between data points. A symmetric dissimilarity matrix $B = (b_{ij}) \in R^{n \times n}$ is presented to describe the dissimilarities between data points, where $b_{ij} \leq 0$, the smaller this value is, the more dissimilar between data points $d_i$ and $d_j$ are. We are also attempting to get a $k$ dimensional column feature vector $y_i$ for each data point $i$, where $k$ could be much less than $p$. Likewise, if two data points are more dissimilar, their feature vectors should be more distant to each other in the feature space. This problem can also be transformed into the following optimization problem:

$$J_2 = \min_{y_i \in R^k, i=1,2,\cdots,n} \frac{1}{2} \sum_{i,j=1}^{n} b_{ij} ||y_i - y_j||^2$$

$$s.t. \sum_{i=1}^{n} y_i y_i^T = I_k$$

(2)

Taking both similar and dissimilar representation problems into account, we are managing to find a $k$ dimensional column feature vector $z_i$ for each data point $i$, where $k$ could be much less than $p$. If two data points are more similar, their feature vectors should be closer to each other while if two data points are more dissimilar, their feature vectors should be more distant to each other in the feature space. Therefore, optimization problem (1) and (2) are merged into the following problem:

$$J_3 = \min_{z_i \in R^k, i=1,2,\cdots,n} \sum_{i,j=1}^{n} \left(\frac{1-\alpha}{2} a_{ij} + \frac{\alpha}{2} b_{ij}\right) \|z_i - z_j\|^2$$

$$s.t. \sum_{i=1}^{n} z_i z_i^T = I_k$$

$$(3)$$

where $0 \le \alpha \le 1$ is a parameter in the balance of similarity and dissimilarity described by feature vectors. Clearly when $\alpha = 0$, problem (3) is optimization problem (1), while when $\alpha = 1$, problem (3) is optimization problem (2). Let $W = (1 - \alpha)A + \alpha B$ is a weighted symmetric adjacency matrix of data points, if $w_{ij} > 0$ this means that the data points $d_i$ and $d_j$ are similar, if $w_{ij} < 0$ this means that the data points $d_i$ and $d_j$ are dissimilar, if $w_{ij} = 0$ this means that data points $d_i$ and $d_j$ are not related. Let $D' = diag(d'_{11}, d'_{22}, \cdots, d'_{nn})$ is a diagonal matrix with $d'_{ii} = \sum_{j=1}^{n} w_{ij}$, a generalized Laplace matrix $L'$ is defined as $L' = D' - W$. The problem (3) could be transformed into the following problem:

$$J_4 = \min_{z_i \in R^k, i=1,2,\cdots,n} tr(ZL'Z^T)$$

$$s.t. \; ZL'Z^T = I_k$$

$$(4)$$

where $Z = [z_1, z_2, \cdots, z_n] \in R^{k \times n}$ and $tr$ denotes the trace of a matrix. It can be proved that the first $k$ eigenvectors corresponding to the $k$ smallest eigenvalues of $L'$ is the solution to the problem (4). Specifically, each row of $Z$ is an eigenvector of matrix $L'$. Then we can use any feature-based clustering method on the first $k$ eigenvectors to perform cluster analysis.

## 2.4   Implementation of the ISC for Clustering Single Cells

After data preprocessing, the application of the proposed ISC for clustering single cells is to construct a suitable adjacency matrix for clustering. The detailed steps, illustrated in Fig. 1, are described as follows.

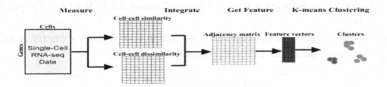

**Fig. 1.** The ISC framework for clustering of scRNA-seq data.

**Step 1: measuring pairwise similarity and dissimilarity.**
Pearson correlation coefficient (PCC) is a widely used measure of the linear correlation between two variables, we use it to calculate the similarity/dissimilarity among cells. The PCC of a pair of cells ($i$ and $j$) is defined as:

$$pcc(i,j) = \frac{1}{m-1}\sum_{t=1}^{m}(\frac{i_t - mean(i)}{std(i)})(\frac{j_t - mean(j)}{std(j)}) \tag{5}$$

where $m$ is the number of genes, $i_t$ is the expression level of gene $t$ in cell $i$, $mean(i)$ represents the mean gene expression level in cell $i$ and $std(i)$ represents the standard deviation of gene expression level in cell $i$.

Similarity $s(i,j)$ and dissimilarity $d(i,j)$ between cell $i$ and cell $j$ are defined as follows:

$$s(i,j) = \begin{cases} pcc(i,j) & \text{if } pcc(i,j) > 0 \\ 0 & \text{otherwise} \end{cases}, \quad d(i,j) = \begin{cases} pcc(i,j) & \text{if } pcc(i,j) < 0 \\ 0 & \text{otherwise} \end{cases} \tag{6}$$

The PCC values range from $-1$ to $1$, if the PCC between cell $i$ and cell $j$ is positive, which means the gene expression values of cell $i$ and cell $j$ tend to be simultaneously large, or small, that is to say, cell $i$ and cell $j$ have resembled expression patterns, the larger the PCC is, the stronger the similarity is. Likewise, if the PCC between cell $i$ and cell $j$ is negative, which represents the gene expression values of cell $i$ and cell $j$ tend to be opposite sides, in other words, there is a large dissimilarity between the expression patterns of cell $i$ and cell $j$, the smaller the PCC is, the greater the dissimilarity is.

**Step 2: integrating similarity matrix and dissimilarity matrix.**
For each cell $i$, the similarities between cell $i$ and all the other cells are sorted in ascending order, and the dissimilarities between cell $i$ and all the other cells are sorted in descending order. The similarity matrix $A = (a_{ij}) \in R^{n \times n}$ is constructed as follows: for cell $i$ and cell $j$, if cell $i$ is among the top $h$ similar cells of cell $j$, or cell $j$ is among the top $h$ similar cells of cell $i$, then $a_{ij} = a_{ji} = s(i,j) = s(j,i)$; otherwise, $a_{ij} = a_{ji} = 0$. Analogously, the dissimilarity matrix $B = (b_{ij}) \in R^{n \times n}$ is designed as follows: for cell $i$ and cell $j$, $b_{ij} = b_{ji} = d(i,j) = d(j,i)$ if $d(i,j)$ is in the top $q$ of the sorted dissimilarity list of cell $i$ or $d(j,i)$ is in the top $q$ of the sorted dissimilarity list of cell $j$; otherwise, $b_{ij} = b_{ji} = 0$.

The adjacency matrix $W$ is generated by integrating similarity matrix $A$ and dissimilarity matrix $B$ using the following equation:

$$W = (1 - \alpha)A + \alpha B \tag{7}$$

where $\alpha$ is selected from the set $\{0, 0.1, 0.2, 0.3, 0.4, 0.5, 0.6, 0.7, 0.8, 0.9, 1\}$, $\alpha$ is used to adjust the proportion of similarity and dissimilarity in the adjacency matrix.

**Step 3: calculating feature vectors for K-means clustering.**
The eigenvectors of the generalized Laplace matrix $L' = D' - W$ are calculated, where $W$ is the adjacency matrix defined in step 2 and $D'$ is a diagonal matrix that contains the row-sums of $W$ on the diagonal. The eigenvectors are then sorted in ascending order by their corresponding eigenvalues. The first $k$ eigen-vectors are considered as feature vectors of all cells. Then K-means clustering is performed on these feature vectors by using MATLAB's kmeans function.

## 2.5    Evaluation Metrics

In this study, four measures are used to compare the performance of ISC and competing SC methods, including Purity, Rand Index (RI), Adjusted Rand Index (ARI) and Normalized Mutual Information (NMI). Let $V = V_1 \cup V_2 \cup \cdots V_{C_V}$ $(V_i \cap V_j = \emptyset, i \neq j)$ be the set of our computed clusters with $C_V$ clusters and $U = U_1 \cup U_1 \cup \cdots \cup U_{C_U}$ $(U_i \cap U_j = \emptyset, i \neq j)$ be the set of genuine clusters with $C_U$ clusters. Purity tries to match clusters that have a maximum overlap. Each identified cluster $V_i(i = 1, 2, \cdots C_V)$ is assigned to the one which is most frequent in the cluster $U_j(j = 1, 2, \cdots C_U)$, and then the accuracy of this assignment is computed by counting the number of correctly assigned data points divided by the total number of data points (N):

$$Purity(V, U) = \frac{1}{N} \sum_{i=1}^{C_V} \max_j |V_i \cap U_j| \tag{8}$$

where $|V_i \cap U_j|$ is the number of elements in the intersection of two clusters $V_i$ and $U_j$.

RI measures the fraction of correctly classified pairs of data points to all pairs of data points. Mathematically, it is defined as follows:

$$RI(U, V) = \frac{2(a+b)}{N(N-1)} \tag{9}$$

where $a$ is the number of pairs that are in the same cluster under $V$ and $U$, $b$ is the number of pairs that are in different clusters under $V$ and $U$.

ARI remarks the normalized difference of the RI and its expected value under the null hypothesis that independent clusterings obey a generalized hypergeometric distribution [18]. It is defined as follows:

$$ARI(V, U) = \frac{2N(N-1)(a+b) - 4Q}{N^2(N-1)^2 - 4Q} \tag{10}$$
$$(Q = (a+c)(a+d) + (b+c)(b+d))$$

where $c$ is the number of pairs that are in the same cluster under $V$ but in different ones under $U$, $d$ is the number of pairs that are in different clusters under $V$ but in the same under $U$.

NMI provides a sound normalized indication of the shared information between a pair of clusterings as follows [19]:

$$NMI(V, U) = \frac{\sum_{i=1}^{C_V} \sum_{j=1}^{C_U} |V_i \cap U_j| \log \frac{N|V_i \cap U_j|}{|V_i| \times |U_j|}}{\max(-\sum_{i=1}^{C_V} |V_i| \log \frac{|V_i|}{N}, -\sum_{j=1}^{C_U} |U_j| \log \frac{|U_j|}{N})} \quad (11)$$

where the numerator is the mutual information between $V$ and $U$, and the denominator represents the entropy of the clusterings $V$ and $U$.

## 3   Results

### 3.1   Results on Breast Cancer Data

In this section, we apply our proposed ISC on primary breast cancer dataset to demonstrate its clustering performance. To eliminate noises or missing data that are contained in the dataset, a data preprocessing procedure is carried out first.

After applying the filtering criteria, 34 single cells with low sequencing quality were removed. Among the remaining 515 single cells, it was verified that there were 317 tumor cells, 175 tumor-associated immune cells and 23 non-carcinoma stromal cells, which together are considered as the gold standard. In order to enhance the reliability of results, lowly expressed genes were excluded before analysis. Finally, 11986 genes passed the quality control and the normalization of each gene were performed.

The parameter $\alpha$ is adjusted to balance the impact of similarity and dissimilarity on clustering results. A higher value of $\alpha$ would put a greater emphasis on the dissimilarity between clusters, while a lower value of $\alpha$ would more focus on similarity inside a cluster. Especially, when $\alpha = 0$, the ISC is the traditional SC. Figure 2 shows the performance of ISC with the change of parameter $\alpha$, where $h$ and $q$ are equal to 28. It is evident from Fig. 2 that as the parameter $\alpha$ increases, Purity, RI, ARI and NMI values all increase tremendously at the beginning and thereafter maintain steady and then decrease a little. Purity, RI, ARI and NMI all reach their maximum values when $\alpha$ is equal to 0.5. We can see that the ISC ($\alpha \neq 0$) performs better than the traditional SC ($\alpha = 0$). This illustrates that taking both the similarity within the cluster and the dissimilarity between clusters into account can increase the improvement over the traditional SC.

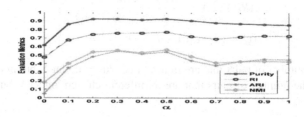

**Fig. 2.** The performance of ISC with the variation of parameter $\alpha$ when the values of $h$ and $q$ are fixed to 28.

Parameters $h$ and $q$ in step 2 of ISC respectively denote the width of similar neighborhoods and dissimilar neighborhoods. To studying the influence of these parameters on the results of ISC. In this study, we consider $h \in \{10, 12, 14, \cdots, 30\}$, and when the value of $h$ is fixed, $q$ is set equal to $h$, or equal to $h/2$. Thus, a total of 22 different combinations can be utilized for construction the adjacency matrix. When $\alpha = 0$ (only consider similarity) and $\alpha = 0.5$ (similarity and dissimilarity are equally important), the performance of ISC with different parameter combinations of $h$ and $q$ is shown in Table 1. We see that when $\alpha = 0.5$, the ISC does a better job of clustering cells by many combinations of $h$ and $q$ settings. All these suggest that the robustness of the ISC with respect to the changes of parameters $h$ and $q$.

**Table 1.** The performance of ISC with different parameter combinations of $h$ and $q$ when $\alpha = 0$ or 0.5

| $h$ | $q$ | $\alpha$ | Purity | RI | ARI | NMI |
|---|---|---|---|---|---|---|
| 10 | 0 | 0 | 0.7612 | 0.5902 | 0.1787 | 0.2617 |
| 10 | 5 | 0.5 | 0.8447 | 0.7514 | 0.5020 | 0.4069 |
| 10 | 10 | 0.5 | 0.9068 | 0.7239 | 0.4462 | 0.4881 |
| 12 | 0 | 0 | 0.7903 | 0.6127 | 0.2239 | 0.3074 |
| 12 | 6 | 0.5 | 0.7398 | 0.5861 | 0.1705 | 0.2513 |
| 12 | 12 | 0.5 | 0.9049 | 0.7221 | 0.4425 | 0.4837 |
| 14 | 0 | 0 | 0.7398 | 0.5713 | 0.1409 | 0.2236 |
| 14 | 7 | 0.5 | 0.7398 | 0.5881 | 0.1746 | 0.2565 |
| 14 | 14 | 0.5 | 0.9049 | 0.7232 | 0.4448 | 0.4852 |
| 16 | 0 | 0 | 0.7748 | 0.6138 | 0.2264 | 0.2993 |
| 16 | 8 | 0.5 | 0.7417 | 0.5891 | 0.1765 | 0.2580 |
| 16 | 16 | 0.5 | 0.9049 | 0.7224 | 0.4431 | 0.4867 |
| 18 | 0 | 0 | 0.7573 | 0.5873 | 0.1728 | 0.2581 |
| 18 | 9 | 0.5 | 0.7417 | 0.5950 | 0.1883 | 0.2670 |
| 18 | 18 | 0.5 | 0.9087 | 0.7275 | 0.4534 | 0.4967 |
| 20 | 0 | 0 | 0.7767 | 0.5986 | 0.1958 | 0.3045 |
| 20 | 10 | 0.5 | 0.9126 | 0.7299 | 0.4580 | 0.5048 |
| 20 | 20 | 0.5 | 0.9087 | 0.7278 | 0.4540 | 0.4970 |
| 22 | 0 | 0 | 0.7786 | 0.5996 | 0.1978 | 0.3119 |
| 22 | 11 | 0.5 | 0.8660 | 0.7813 | 0.5618 | 0.4460 |
| 22 | 22 | 0.5 | 0.9068 | 0.7255 | 0.4493 | 0.4921 |
| 24 | 0 | 0 | 0.6854 | 0.5283 | 0.0564 | 0.2601 |
| 24 | 12 | 0.5 | 0.9107 | 0.7271 | 0.4525 | 0.4991 |
| 24 | 24 | 0.5 | 0.9087 | 0.7272 | 0.4527 | 0.4991 |
| 26 | 0 | 0 | 0.6155 | 0.4731 | −0.0518 | 0.1874 |
| 26 | 13 | 0.5 | 0.9165 | 0.7338 | 0.4659 | 0.5146 |
| 26 | 26 | 0.5 | 0.9107 | 0.7299 | 0.4580 | 0.5017 |
| 28 | 0 | 0 | 0.6155 | 0.4778 | −0.0418 | 0.1491 |

(*continued*)

**Table 1.** (*continued*)

| h | q | α | Purity | RI | ARI | NMI |
|---|---|---|--------|------|--------|--------|
| 28 | 14 | 0.5 | 0.8816 | 0.8011 | 0.6017 | 0.4817 |
| **28** | **28** | **0.5** | **0.9243** | 0.7701 | 0.5393 | **0.5651** |
| 30 | 0 | 0 | 0.6155 | 0.4871 | −0.0247 | 0.2047 |
| **30** | **15** | **0.5** | 0.8990 | **0.8257** | **0.6510** | 0.5205 |
| 30 | 30 | 0.5 | 0.9165 | 0.7641 | 0.5272 | 0.5428 |

When $h = 12$, the traditional SC works best, the Purity, RI, ARI and NMI values are 0.7903, 0.6127, 0.2239 and 0.3074, respectively. When $h \geq 20$ ISC shows stable performance. When $h = 28$, $q = 28$ and $\alpha = 0.5$, ISC gains the best clustering results in terms of Purity and NMI, which are 0.9243 and 0.5651, respectively. When $h = 30$, $q = 15$ and $\alpha = 0.5$, ISC performs best in terms of RI and ARI, which are 0.8257 and 0.6510, respectively. Although, ISC provides the superior performance, the ARI value is still less than 0.7 and the NMI value is still less than 0.6. That might be because, among three types of cells from tumor tissue, tumor cells have distinct chromosomal expression patterns while immune cells and stromal cells have no apparent copy number variations patterns [3]. The latter two types of cells are easily confused with each other in feature-based clusterings.

Furthermore, to investigate whether the performance of the ISC significantly improved compared to the traditional SC, we use the non-parametric one-tailed Wilcoxon rank sum test. We use the P-value of the test as a measure of the extent of effect difference between evaluation metrics in the traditional SC and ISC. Through calculation and comparison, we find that evaluation metrics in the ISC exhibit significantly higher values than those in the traditional SC.

### 3.2 Results on Other Single-Cell RNA Sequence Data

To demonstrate the ISC performance compared with traditional SC, we apply the proposed method to other four single-cell RNA sequence datasets. These datasets are generated by different techniques in a variety of cell types in human or mouse. Some of cells are from different stages, some of them are from different conditions and others are from different lines [4, 15–17]. In the original papers, cell-labels are available and can be considered as gold standards. We directly download gene expression levels that have been pre-processed in previous study [14]. ES dataset consists of 182 cells at different cell-cycle phase and 8989 genes. PCT dataset consists of 114 cells coming from case samples, control samples and other cell lines. There are 14405 genes which passed the quality control. LE dataset consists of 80 individual live cells and 9352 genes which expressed in more than two cells and with non-zero variances. PE dataset consists of 135 individual cells isolated from mouse embryos at different preimplantation stages and 12548 genes for downstream analysis. we use the true cluster number to obtain the clustering results.

ISC with different parameterizations are applied to the filtered genes. Table 2 shows the best performance of traditional SC ($\alpha = 0$) and ISC ($\alpha \neq 0$). As shown in

Table 2, the performance of ISC was better than traditional SC in all the four measures. Further, we can observe that different datasets achieve their best performance with different parameter values. Although these parameters are required to be manually adjusted by the user, ISC outperforms traditional SC. By ISC, Purity and RI can be reached over 0.88, ISC increases ARI by at least 10% points except for ES dataset, the values of NMI also are increased drastically. Although ARI and NMI are increasing in PE dataset, they are still lower than 0.8, perhaps this is because only a small number of genes have expression changes between the 8-cells and 16-cells [20].

**Table 2.** The best performance of SC and ISC on other four datasets.

| Datasets | $h$ | $q$ | $\alpha$ | Purity | RI | ARI | NMI |
|----------|-----|-----|----------|--------|--------|--------|--------|
| ES  | 30 | 0  | 0   | 0.9506 | 0.9348 | 0.8533 | 0.8294 |
|     | 28 | 28 | 0.3 | 0.9506 | 0.9367 | 0.8577 | 0.8523 |
| PCT | 14 | 0  | 0   | 0.8947 | 0.905  | 0.7364 | 0.8619 |
|     | 16 | 16 | 0.6 | 0.9298 | 0.9441 | 0.8461 | 0.9016 |
| LE  | 16 | 0  | 0   | 0.9    | 0.9437 | 0.871  | 0.8247 |
|     | 24 | 24 | 0.7 | 0.9625 | 0.9854 | 0.9669 | 0.9308 |
| PE  | 14 | 0  | 0   | 0.8222 | 0.8498 | 0.533  | 0.7125 |
|     | 28 | 28 | 0.4 | 0.8815 | 0.8836 | 0.6393 | 0.7833 |

# 4 Conclusion

Advances in next-generation sequencing technologies have generated large volumes of scRNA-seq data, how to make full use of these data is very important. One of the most powerful utilization of single-cell data is to cluster cells on the basis of gene expression patterns. The performance of clustering may have a substantial impact on the outcome of downstream analysis. A lot of clustering algorithms for characterizing cell types have been proposed over the past few years.

Due to the high dimensionality of the single-cell data, the difference among the distances between cells become small, it is not reliable to identify groups of cells by using these high-dimensional data directly. Effective dimensionality reduction could make more accurate in cells classification. For example, SC reduces dimension based on the smallest eigenvalues of Laplace matrix and their corresponding eigenvectors. In SC, the Laplace matrix is deduced from the adjacency matrix. However, the conventional method for constructing the adjacency matrix highlights the similarities between cells and ignores the dissimilarities between cells. The dissimilarities between cells may reflect an expression pattern distinction between different cell types and are meaningful for clustering cells.

In this study we have proposed an improved method (ISC) for classification of single cells, by combining similarities and dissimilarities between cells. Furthermore, we apply this method to five publicly available scRNA-seq datasets including cells from different tissues, or from different stages, or from different cell lines. The results show that the performance of ISC is better than that of the traditional SC in terms of several metrics. The precision in distinguishing cell types based on the integration of similarities and dissimilarities is greater than only based on similarities. The performance of the proposed method with respect to the changes of parameters also shows the robustness of the improved clustering method.

Although ISC makes considerable improvement on identifying cell types, there remains room for improving the ability to detect cell types. Several problems are still open, which include how to establish appropriate measurements to measure similarities and dissimilarities, how many similar cells and dissimilar cells are to choose for constructing similarity matrix and dissimilarity matrix and how to balance similarity matrix and dissimilarity matrix to construct adjacency matrix. In future work, we will focus on the reasonable integration of ISC and other computational analysis methods, which may lead to promising results in identifying cell types.

**Acknowledgments.** This work was supported in part by Natural Science and Engineering Research Council of Canada (NSERC), China Scholarship Council (CSC), the National Natural Science Foundation of China under Grant No. 61571052 and by the Science Foundation of Wuhan Institute of Technology under Grant No. K201746.

# References

1. Liang, J., Cai, W., Sun, Z.: Single-cell sequencing technologies: current and future. J. Genet. Genom. **41**(10), 513–528 (2014)
2. Shekhar, K., Lapan, S.W., Whitney, I.E., Tran, N.M., Macosko, E.Z., Kowalczyk, M., et al.: Comprehensive classification of retinal bipolar neurons by single-cell transcriptomics. Cell **166**(5), 1308–23.e30 (2016)
3. Chung, W., Eum, H.H., Lee, H.O., Lee, K.M., Lee, H.B., Kim, K.T., et al.: Single-cell RNA-seq enables comprehensive tumour and immune cell profiling in primary breast cancer. Nat. Commun. **8**, 15081 (2017)
4. Buettner, F., Natarajan, K.N., Casale, F.P., Proserpio, V., Scialdone, A., Theis, F.J., et al.: Computational analysis of cell-to-cell heterogeneity in single-cell RNA-sequencing data reveals hidden subpopulations of cells. Nat. Biotechnol. **33**(2), 155–160 (2015)
5. Xu, C., Su, Z.: Identification of cell types from single-cell transcriptomes using a novel clustering method. Bioinformatics **31**(12), 1974–1980 (2015)
6. Tian, L.-P., Luo, P., Wang, H., Zheng, H., Wu, F.-X.: CASNMF: a converged algorithm for symmetrical nonnegative matrix factorization. Neurocomputing **275**, 2031–2040 (2018)
7. Li, L.X., Wu, L., Zhang, H.S., Wu, F.X.: A fast algorithm for nonnegative matrix factorization and its convergence. IEEE Trans. Neural Netw. Learn. Syst. **25**(10), 1855–1863 (2014)
8. Shao, C., Höfer, T.: Robust classification of single-cell transcriptome data by nonnegative matrix factorization. Bioinformatics **33**(2), 235–242 (2017)
9. Kiselev, V.Y., Kirschner, K., Schaub, M.T., Andrews, T., Yiu, A., Chandra, T., et al.: SC3: consensus clustering of single-cell RNA-seq data. Nat. Methods **14**(5), 483 (2017)
10. Kiselev, V.Y., Andrews, T.S., Hemberg, M.: Challenges in unsupervised clustering of single-cell RNA-seq data. Nat. Rev. Genet. **1** (2019)
11. Von Luxburg, U.: A tutorial on spectral clustering. Stat. Comput. **17**(4), 395–416 (2007)
12. Lu, C., Yan, S., Lin, Z.: Convex sparse spectral clustering: single-view to multi-view. IEEE Trans. Image Process. **25**(6), 2833–2843 (2016)

13. Wang, B., Zhu, J., Pierson, E., Ramazzotti, D., Batzoglou, S.: Visualization and analysis of single-cell RNA-seq data by kernel-based similarity learning. Nat. Methods **14**(4), 414–416 (2017)
14. Park, S., Zhao, H.: Spectral clustering based on learning similarity matrix. Bioinformatics **34**(12), 2069–2076 (2018)
15. Ting, D.T., Wittner, B.S., Ligorio, M., Jordan, N.V., Shah, A.M., Miyamoto, D.T., et al.: Single-cell RNA sequencing identifies extracellular matrix gene expression by pancreatic circulating tumor cells. Cell Rep. **8**(6), 1905–1918 (2014)
16. Treutlein, B., Brownfield, D.G., Wu, A.R., Neff, N.F., Mantalas, G.L., Espinoza, F.H., et al.: Reconstructing lineage hierarchies of the distal lung epithelium using single-cell RNA-seq. Nature **509**(7500), 371–375 (2014)
17. Deng, Q., Ramsköld, D., Reinius, B., Sandberg, R.: Single-cell RNA-seq reveals dynamic, random monoallelic gene expression in mammalian cells. Science **343**(6167), 193–196 (2014)
18. Wu, F.X., Zhang, W.J., Kusalik, A.J.: Dynamic model-based clustering for time-course gene expression data. J. Bioinform. Comput. Biol. **3**(4), 821–836 (2005)
19. Strehl, A., Ghosh, J.: Cluster ensembles—a knowledge reuse framework for combining multiple partitions. J. Mach. Learn. Res. **3**(11), 583–617 (2002)
20. Hamatani, T., Carter, M.G., Sharov, A.A., Ko, M.S.: Dynamics of global gene expression changes during mouse preimplantation development. Dev. Cell **6**(1), 117–131 (2004)

# A Novel Weight Learning Approach Based on Density for Accurate Prediction of Atherosclerosis

Jiang Xie[1(✉)], Ruiying Wu[1], Haitao Wang[1], Yanyan Kong[2], Haozhe Li[3], and Wu Zhang[1(✉)]

[1] School of Computer Engineering and Science, Shanghai University, Shanghai 200444, China
{jiangx,wzhang}@shu.edu.cn
[2] PET Center, Huashan Hospital, Fudan University, Shanghai 200235, China
[3] Christian Wuhan Britain-China School, Wuhan 430022, China

**Abstract.** Cardiovascular diseases (CVD) are the leading cause of death in the world. Based on density-based spatial clustering of applications with noise algorithm (DBSCAN), we proposed a weight learning approach to utilize the density information of the patient data. The proposed approach divided the sample points of dataset into three types with different weight of density, so that machine learning models achieved better performance in early diagnosis of CVD. Cross-validation on UCI dataset shown that the traditional machine learning models after weight learning can improve accuracy more than 10%.

**Keywords:** Weight learning · Machine learning · DBSCAN · Classification · Atherosclerosis

## 1 Introduction

Cardiovascular diseases (CVD) are the number one cause of death in global population [1–3]. In 2015, 17.7 million deaths were caused by CVD worldwide, representing 31% of all global deaths, among which 7.4 million deaths were due to coronary heart disease and 6.7 million deaths were due to stroke disease [4]. Atherosclerosis, a slow and progressive process of narrowing the artery and interrupting the flow of blood from the heart or to the brain, is responsible for a large proportion of CVD [5].

Up to now, various data mining approaches have been made for the risk prediction of CVD based on medical data. A three-step approach based on clustering supervised classification and the frequent item-sets search were adopted to predict if a patient might develop atherosclerosis according to the correlation between his or her habits and social environment [6]. A predictive risk assessment of atherosclerosis (PRAA), was used to predict the risk of atherosclerosis on the STULONG dataset [7]. The author proposed a PSO algorithm with a boosting approach to extract rules, by which the

---

J. Xie, R. Wu and H. Wang—These authors contributed equally to this work.

© Springer Nature Switzerland AG 2019
D.-S. Huang et al. (Eds.): ICIC 2019, LNCS 11644, pp. 190–200, 2019.
https://doi.org/10.1007/978-3-030-26969-2_18

patients with coronary artery disease can be recognized [8]. The method combined Iterative Principal Component Analysis (IPCA) and Multiclass SVM for classifying normal, risk and pathologic community in real time has performed with an overall accuracy of about 98.97% [9]. A ridge expectation maximization imputation (REMI) technique estimated the missing values in the atherosclerosis databases [10].

The above machine learning methods treated the sample points in the dataset as a uniform distribution, though they often exhibit an uneven density distribution, especially in medical datasets. Specifically, some sample points were more efficient to describe the overall distribution because they were distributed in a higher density. The machine learning model would be more effective if the density information of dataset were put into consideration.

For dataset with uneven distribution, people tend to consider density information when doing cluster research. Density-based clustering algorithms like density-based spatial clustering of applications with noise algorithm (DBSCAN) is the pioneer of density-based clustering, which can find arbitrarily shaped clusters and also handle noise or outliers correctly [11]. DBSCAN is capable of determining the number of clusters automatically with two parameters as input. While other clustering algorithms require priori knowledge of the data to cluster in one form or another, their performance often depends heavily on user-specified parameters [12].

In this paper, a novel weight learning approach was proposed based on density information to assist machine learning models to the accurate prediction of atherosclerosis. First, all sample points were divided into core points, boundary points, and noise points by DBSCAN. Next, a dataset containing density was built, in which above points were assigned different weights by weight learning algorithm. Finally, a new machine learning model was constructed combined the UCI features and the weight feature. The experiments demonstrated that the prediction performance can be improved after DBSCAN weight learning, and superior to other methods in terms of interpretability.

The rest of this paper is organized as follows: A brief overview on related work is presented in Sect. 2. The proposed method involving a missing value imputation, novel weight learning algorithm with DBSCAN, and feature engineering is discussed in Sect. 3. The introduction of dataset is in Sect. 4. Experiments are reported in Sect. 5. The paper ends by some conclusions in Sect. 6.

## 2   Related Work

DBSCAN is one of the most commonly used density-based clustering algorithms. Based on a density-based notion of clusters, DBSCAN is able to generate clusters of arbitrary shapes efficiently [13]. It can gather points that are closely packed together and marking as outliers' points that lie alone in low-density regions. DBSCAN only requires two input parameters: $\varepsilon$ (eps) and MinPts, where $\varepsilon$ (eps) describes the neighborhood distance threshold of a sample and MinPts is the minimum number of points required in the eps-neighborhood of a sample core points [14, 15].

Here are some definitions required for DBSCAN algorithm:

- ε-neighborhood: For a point $x_j \in D$, the ε-neighborhood denotes the set of points whose distance from $x_j$ is less than or equal to ε, that is

$$N(x_j) = \{x_i \in D | distance(x_i, x_j) \leq \varepsilon\} \tag{1}$$

- Directly density reachable: A point $x_j$ is directly density-reachable from the point $x_i$ if $x_j \in N(x_i)$.
- Density reachable: A point $x_i$ is density-reachable from the point $x_j$ if there is a chain of points $p_1, p_2, \ldots, p_T$, and $p_1 = x_i, p_T = x_j$, such that $x_i$ is directly density-reachable from the point $x_j$.
- Density connected: A point $x_i$ is a density connected to point $x_j$ if there is an intermediate point $x_k$ such that both $x_i$ and $x_j$ are density-reachable from the point $x_k$.

From the view of a DBSCAN method, every point in the dataset will divide into core point, border point and noise point [12].

- Core point: A point with threshold density greater than or equal to MinPts.
- Border point: A point with threshold density less than MinPts.
- Noise point: A point $p$ is a noise point if the threshold density $p$ is less than MinPts and all points in the ε - neighborhood of $p$ are border points.

Based on the above definitions, the DBSCAN algorithm can be abstracted into the following steps:

1. Find the points in the ε - neighborhood of every point and identify the core points with more than MinPts neighbors.
2. Find the connected components of core points on the neighbor graph, ignoring all non-core points.
3. Assign each non-core point to a nearby cluster if the cluster is an ε (eps) neighbor. Otherwise, assign it to noise.

The process is halted when all the points are either assigned to some cluster or marked noise.

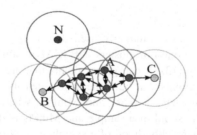

**Fig. 1.** Illustration of the DNSCAN cluster model [13] (Color figure online)

Figure 1 illustrates the concepts of DBSCAN. The MinPts parameter is 4, and the ε radius is indicated by the circles. Point N is a noise point; points B and C are border points; point A and other red points are core points [13].

## 3   Methods

In this paper, a novel weight learning approach was proposed based on density information to assist machine learning models to the accurate prediction of atherosclerosis. The specific flow of the approach is shown in Fig. 2:

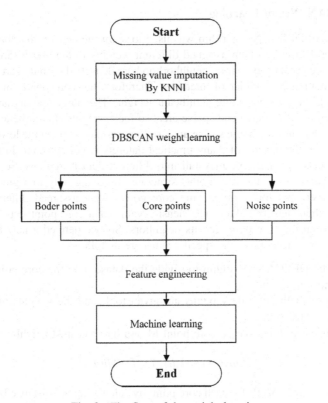

**Fig. 2.** The flow of the weight learning

### 3.1   Missing Value Imputation

Missing values is one of the inevitable problems that generally occurs in medical datasets. Ignoring all the missing cases is not recommend to deal with missing values. Other well-known and less computation methods for imputation of missing values are mean substitution and median substitution [16]. However, they all have a drawback that the relation structure among the data was ignored. The implicit knowledge in the missing attributes is also valuable in improving the prediction model. Sound data quality is the foundation for data mining, while the missing values impinge on the decision-making process strongly [17–19]. Thus, it is essential to find a proper way to impute the missing attributes.

K-nearest neighbor's imputation (KNNI), as the extension of K-nearest neighbor algorithm, has been successfully used in missing values field [20]. The main idea of KNNI is to calculate the distance between the missing points and complete points firstly, then selecting the $k$ nearest data points and impute the missing attributes through these neighbors [21]. In this paper, we used the KNNI algorithm to impute the missing values.

## 3.2   DBSCAN Weight Learning

In this research, DBSCAN algorithm was used to learn the density distribution information of the dataset and then assigned different weights to the above three types of points. The core points and their $\varepsilon$ - neighborhood are densely distributed areas, so their feature information is worthier of attention. Therefore, the core points in the model training should be assigned more significant weight. The border points are the transition between the dense area and sparse area, so their weights should be at a medium level. The noise points are in the sparse area, so they should be given the lowest weight.

Furthermore, the density of many practical datasets is not constant. In these datasets, data objects of same type may fall into different density regions. So, it is inappropriate to characterize the core points structures by using only one global density weight. To overcome this problem, the core points were assigned different weight according to their density values. The density value of a core point depends on the distance between the core point and its neighbors. So, we defined a new measure to assign weight to core points. The specific steps are as follows:

1. Perform the DBSCAN algorithm to divide the dataset into the core points, border points, and noise points.
2. The weights of the boundary points are given to 1, and the weights of the noise points are given to 0.
3. Get the distance $d_i$ between a core point M and its $i^{th}$ nearest neighbor:

$$dist(M) = \{d_i\}, 1 \leq i \leq MinPts. \tag{2}$$

4. Obtain the metrics $M(P)$ for each core point by calculating the distance between the core point and its neighbors; the $M(P)$ is defined as follows:

$$M(P) = \frac{1}{k} \sum_{k=1}^{MinPts} d_i \tag{3}$$

5. Divide the M(p) values of all core points into equal-sized buckets based on sample quantiles by quantile-based discretization. As a result, four interval values including (minValue, Q1], (Q1, Q2], (Q2, Q3], (Q3, maxValue) were obtained. After trying different weight-giving strategies such as exponential weights and multiple-level weights, no significant difference was found in the experimental results. Therefore,

weights assigned to the core points was according to their values corresponding to the sample points M(p). The weight allocation strategy is as follows:

$$WeightAssign \begin{cases} NoisePoint: 0 \\ BorderPoint: 1 \\ CorePoint: \begin{cases} M(p) \leq Q1: 2 \\ Q1 < M(p) \leq Q2: 4 \\ Q2 < M(p) \leq Q3: 6 \\ Q3 < M(p): 8 \end{cases} \end{cases} \tag{4}$$

After DBSCAN weight learning, weight is assigned to every sample, and input original machine learning algorithm as a new feature.

### 3.3   Features Engineering

Feature engineering is fundamental to the application of machine learning, and it is both challenging and expensive. The quality and quantity of features will have a significant influence on whether the model is good or not [22]. Generally speaking, feature engineering consists of feature selection and feature generation. Feature selection is one of the essential preprocessing techniques used in machine learning; it is capable of removing irrelevant data, increasing the learning accuracy, and reducing data dimension [23]. Feature generation attempts to create additional relevant features from the existing raw features in the data and to increase the predictive power of the learning algorithm [24].

The feature selection in this paper is employing Random Forests (RF) algorithm to remove unnecessary features and select the valuable features from different subsets. According to the importance score calculated by RF, the features that rank in the top 12 of 14 were kept. Because the importance score of the last two features were the lowest and the accuracy of model using 14 features is similar to that of model using 12 features.

The feature generation in this paper was implemented by discretizing the continuous medical data into categorical. Effective data discretization can remarkably improve model ability on prediction and make systems more robust [25]. Sturges' formula was a typical discretization method, which was derived from a binomial distribution and implicitly assumes an approximately normal distribution [26]. In this paper, Sturges' formula was used to determine the appropriate number of bins ($k$), in which $n$ represents the sample size.

$$k = \lceil log_2 n \rceil + 1 \tag{5}$$

### 3.4   Model Training and Metrics

The dataset after DBSCAN weight learning and feature engineering processing contained density distribution information and valuable features that can effectively increase predictive accuracy. In order to verify the validity of weight learning, various well-known data mining algorithms such as SVM, K-nearest neighbor (KNN), random forest (RF), gradient boosting decision tree (GBDT), and neural networks (NN) were built on the dataset. All the machine learning algorithms above were implemented by scikit-learn package [27].

After model training, different evaluation metrics including classification accuracy, F-measure, sensitivity, and specificity were used to verified their performances, among which the classification accuracy is the most common evaluation criterion used in the data mining field [28]. To calculate these measures, a well-known matrix called a confusion matrix (contingency table) was formed [29] (Fig. 3):

| | | True condition | |
|---|---|---|---|
| Total population | | Condition positive | Condition negative |
| **Predicted condition** | Predicted condition positive | **True positive,** Power | **False positive,** Type I error |
| | Predicted condition negative | **False negative,** Type II error | **True negative** |

**Fig. 3.**  confusion matrix

True Positive (TP), True Negative (TN), False Negative (FN), False Positive (FP)

$$accuracy = \frac{TP + TN}{TP + FP + TN + FN} \tag{6}$$

$$F - measure = \frac{2 * TP}{2 * TP + FP + FN} \tag{7}$$

$$precision = \frac{TP}{TP + FP} \tag{8}$$

$$sensitivity(recall) = \frac{TP}{TP + FN} \tag{9}$$

## 4   Dataset

In this work, the UCI heart disease dataset was used to evaluate the performance of the proposed weight learning approach. The UCI heart disease dataset is from Data Mining Repository of University of California, Irvine. The dataset includes 920 instances collected from Cleveland (303 samples), Hungarian (294 samples), VA Long Beach (200 samples) and Switzerland (123 samples). This dataset consists of 509 instances with coronary artery disease and 411 instances without coronary artery disease [30].

The dataset contains 76 attributes, but all published experiments refer to using 14 attributes of them. These attributes are listed as follows:

- Age: age in years.
- Sex: 0 indicates female, and 1 indicates male.
- CP: chest pain type, takes values equal to 1, 2, 3 or 4 indicating typical angina, atypical angina, nonanginal pain and asymptomatic, respectively.
- Trestbps: resting systolic blood pressure (in mmHg on admission to the hospital).
- chol: serum cholesterol in mg/dl.
- fbs: (fasting blood sugar > 120 mg/dl), takes 1 or 0 as yes or no, respectively.
- restecg: resting electrocardiographic results, takes values equal to 0, 1 or 2 indicating: normal, ST-T wave abnormality (T wave inversions and/or ST elevation or depression of > 0.05 mV) and probable or definite left ventricular hypertrophy by Estes' criteria, respectively.
- Thalach: maximum heart rate achieved.
- Exang: exercise-induced angina takes 1 or 0 as yes or no, respectively.
- Oldpeak: ST depression induced by exercise relative to rest (based on the ST segment on the electrocardiogram).
- Slope: the slope of the peak exercise ST segment, takes values equal to 1, 2 or 3 indicating upsloping, flat and down-sloping, respectively.
- Ca: value 0–3 indicating the number of major vessels colored by fluoroscopy.
- Thal: takes values equal to 3, 6 or 7 indicating normal, fixed defect and reversible defect, respectively.
- Num: diagnosis of heart disease. Value 0 is for nonexistence and values 1 is for disease existence.

## 5  Experiments

K-fold cross-validation technique was used to improve the reliability of results in the experiments. The cross-validation process was repeated $k$ times, with each of the $k$ subsamples used exactly once as the validation data. The number of folds was set to be 10 for the best approximation of error.

Table 1 shows the evaluation criteria obtained by the DBSCAN weight learning and original machine learning algorithms for UCI dataset.

**Table 1.** The effects of weight learning on algorithms

| Method | Accuracy | F-measure | Precision | Sensitivity |
|---|---|---|---|---|
| SVM | 81.51 | 83.55 | 82.58 | 84.67 |
| SVM weight learning | **93.04** | **92.03** | **93.47** | **90.69** |
| KNN | 80.32 | 82.18 | 82.92 | 81.72 |
| KNN weight learning | **92.72** | **91.61** | **93.65** | **89.71** |
| RF | 81.09 | 83.28 | 81.83 | 84.88 |
| RF weight learning | **94.46** | **93.68** | **95.11** | **92.40** |
| GBDT | 80.98 | 83.19 | 81.70 | 84.87 |
| GBDT weight learning | **94.68** | **93.95** | **94.87** | **93.14** |
| NN | 74.43 | 77.01 | 76.65 | 77.58 |
| NN weight learning | **91.19** | **89.99** | **90.32** | **89.96** |

As shown in Table 1, all list machine learning algorithms have improved the prediction accuracy after weight learning by 11.53% to 16.76%. It means that the weight learning could effectively utilize the implicit density distribution information in the dataset to get valuable feature information.

There are other researches created classification algorithms on UCI dataset. The accuracy, sensitivity, and specificity between weight learning method and some other methods were compared. The results are shown in Table 2.

**Table 2.** Comparison the proposed algorithm to other techniques

| Method | Accuracy | Precision | Sensitivity |
|---|---|---|---|
| RST [31] | 85.2 | – | – |
| Decision tree [31] | 85.6 | – | – |
| ANN [32] | 85.53 | – | – |
| NN-Alizadeh [33] | 85.43 | 90.2 | 73.5 |
| En-PSO2 [8] | 85.76 | 90.02 | 82.31 |
| BF-SVM [34] | 87.46 | – | – |
| ELM [10] | 89.86 | – | – |
| SVM weight learning | **93.04** | **93.47** | **90.69** |
| KNN weight learning | **92.72** | **93.65** | **89.71** |
| RF Weight learning | **94.46** | **95.11** | **92.40** |
| GBDT weight learning | **94.68** | **94.87** | **93.14** |
| NN weight learning | **91.19** | **90.32** | **89.96** |

As Table 2 illustrates, the accuracy of the approach in this paper is the highest among all algorithms listed in the table. The best accuracy of the weight learning method (GBDT Weight Learning) is 4.82% higher than the best accuracy of the other method (ELM). It means that the weight learning algorithm performs better on binary classification.

## 6   Conclusions

A novel weight learning approach based on density for accurate prediction of atherosclerosis was proposed in this paper. The main contribution of the research is providing a way to produce optimal features to cover more instances.

The missing value of the experimental dataset was imputed using the k-nearest neighbor's imputation approach. The proposed approach divided the dataset into three types of samples based on the different weight of density, so that a new weight feature was generated to assist machine learning models in learning the distribution of input variables. The feature generation in this paper was implemented by discretizing the continuous medical data into categorical. Overall, the accuracy of machine learning models after weight learning was improved more than 10%.

The proposed approach may also be applied to other disease samples. The future work will focus on the proposed weight learning algorithm on more challenging and comprehensive medical datasets for higher prediction accurate on cardiovascular diseases.

**Acknowledgements.** This research was supported by the National Key R&D Program of China [No. 2017YFB0701501], the National Natural Science Foundation of China [No. 61873156] and the Project of NSFS [No. 17ZR1409900].

# References

1. Dimmeler, S., Zeiher, A.M.: Circulating microRNAs: novel biomarkers for cardiovascular diseases? Eur. Heart J. **90**(8), 865–875 (2012)
2. Eeg-Olofsson, K., Cederholm, J., et al.: New aspects of HbA1c as a risk factor for cardiovascular diseases in type 2 diabetes: an observational study from the Swedish National Diabetes Register (NDR). J. Intern. Med. **268**(5), 471–482 (2010)
3. Nordestgaard, B.G., Varbo, A.: Triglycerides and cardiovascular disease. Lancet **384**(9943), 626–635 (2014)
4. Members, W.G., Mozaffarian, D., et al.: Heart disease and stroke statistics-2016 update: a report from the american heart association. Circulation **133**(4), e38 (2016)
5. Eleni, R., Adam, T., et al.: Blood pressure and incidence of twelve cardiovascular diseases: lifetime risks, healthy life-years lost, and age-specific associations in 1.25 million people. Lancet **383**(9932), 1899–1911 (2014)
6. Couturier, O., Delalin, H., et al.: A three-step approach for stulong database analysis: characterization of patients groups. In: Proceeding of the ECML/PKDD (2004)
7. Rao, V.S., Kumar, M.N.: Novel approaches for predicting risk factors of atherosclerosis. IEEE J. Biomed. Health Inform. **17**(1), 183–189 (2013)
8. Hedeshi, N.G., Abadeh, M.S.: Coronary artery disease detection using a fuzzy-boosting PSO Approach. Comput. Intell. Neurosci. **2014**, 1–12 (2014)
9. Kumar, P.R., Priya, M.: Classification of atherosclerotic and non-atherosclerotic individuals using multiclass state vector machine. Technol. Health Care **22**(4), 583–595 (2014)
10. Nikan, S., Gwadry-Sridhar, F., et al.: Machine learning application to predict the risk of coronary artery atherosclerosis. In: International Conference on Computational Science and Computational Intelligence, pp. 34–39 (2017)
11. Ester, M., Kriegel, H.P., et al.: A density-based algorithm for discovering clusters a density-based algorithm for discovering clusters in large spatial databases with noise. In: International Conference on Knowledge Discovery and Data Mining, pp. 226–231 (1996)
12. Kumar, K.M., Reddy, A.R.M.: A fast DBSCAN clustering algorithm by accelerating neighbor searching using Groups method. Pattern Recogn. **58**(3), 39–48 (2016)
13. Schubert, E., Sander, J., et al.: DBSCAN revisited, revisited: why and how you should (still) use DBSCAN. ACM Trans. Database Syst. **42**(3), 1–21 (2017)
14. Debnath, M., Tripathi, P.K., et al.: K-DBSCAN: identifying spatial clusters with differing density levels. In: International Workshop on Data Mining with Industrial Applications, pp. 51–60 (2016)
15. Dudik, J.M., Kurosu, A., et al.: A comparative analysis of DBSCAN, K-means, and quadratic variation algorithms for automatic identification of swallows from swallowing accelerometry signals. Comput. Biol. Med. **59**(8), 10–18 (2015)
16. Hron, K., Templ, M., et al.: Imputation of missing values for compositional data using classical and robust methods. Comput. Stat. Data Anal. **54**(12), 3095–3107 (2011)
17. Kang, P.: Locally linear reconstruction based missing value imputation for supervised learning. Neurocomputing **118**(11), 65–78 (2013)
18. Souto, D., Marcilio, C.P., et al.: Impact of missing data imputation methods on gene expression clustering and classification. BMC Bioinform. **16**(1), 1–9 (2015)

19. Zhu, X., Zhang, S., et al.: Missing value estimation for mixed-attribute data sets. IEEE Trans. Knowl. Data Eng. **23**(1), 110–121 (2010)
20. Eskelson, B.N.I., Temesgen, H., et al.: The roles of nearest neighbor methods in imputing missing data in forest inventory and monitoring databases. Scand. J. For. Res. **24**(3), 235–246 (2009)
21. Jiang, X., Haitao, W., et al.: A novel hybrid subset-learning method for predicting risk factors of atherosclerosis. In: IEEE International Conference on Bioinformatics and Biomedicine, pp. 2124–2131 (2017)
22. Cai, D., Chiyuan, Z., et al.: Unsupervised feature selection for multi-cluster data. In: ACM SIGKDD International Conference on Knowledge Discovery and Data Mining (2010)
23. Pal, M., Foody, G.M.: Feature selection for classification of hyperspectral data by SVM. IEEE Trans. Geosci. Remote Sens. **48**(5), 2297–2307 (2010)
24. Chandrashekar, G., Sahin, F.: A survey on feature selection methods. Comput. Electr. Eng. **40**(1), 16–28 (2014)
25. Klein, J.P., Wu, J.T.: Discretizing a continuous covariate in survival studies. Handb. Stat. **23**(03), 27–42 (2003)
26. Scott, D.W.: Sturges' rule. Wiley Interdisc. Rev. Comput. Stat. **1**(3), 303–306 (2010)
27. Scikit-learn Machine Learning in Python. https://scikit-learn.org/stable/
28. Vehtari, A., Gelman, A., et al.: Practical Bayesian model evaluation using leave-one-out cross-validation and WAIC. Stat. Comput. **27**(5), 1413–1432 (2017)
29. Confusion matrix. https://en.wikipedia.org/wiki/Confusion_matrix
30. Dua, D., Graff, C.: UCI Machine Learning Repository [http://archive.ics.uci.edu/ml]. University of California, School of Information and Computer Science (2017)
31. Setiawan, N.A., Venkatachalam, P.A., et al.: Rule selection for coronary artery disease diagnosis based on rough set. Int. J. Recent Trends Eng. **2**(5), 198–202 (2009)
32. Palaniappan, S., Awang, R.: Intelligent heart disease prediction system using data mining techniques. In: IEEE/ACS International Conference on Computer Systems and Applications (2008)
33. Alizadehsani, R., Habibi, J., et al.: A data mining approach for diagnosis of coronary artery disease. Comput. Methods Program. Biomed. **111**(1), 52–61 (2013)
34. Rajeswari, K., Vaithiyanathan, V., et al.: Feature selection for classification in medical data mining. Int. J. Emerg. Trends Technol. Comput. Sci. **2**(2), 492–497 (2013)

# An Effective Approach of Measuring Disease Similarities Based on the DNN Regression Model

Shuhui Su[1], Xiaoxiao(X.X.) Zhang[2], Lei Zhang[3], and Jian Liu[1(✉)]

[1] School of Computer Science and Technology,
Harbin Institute of Technology, Harbin 150001, China
jianliu@hit.edu.cn
[2] Changsha Medical University, Changsha 410219, China
[3] Zhejiang University of Science and Technology, Hangzhou 310023, China

**Abstract.** A large number of biomedical ontologies are created to provide controlled vocabularies for sharing biomedical knowledge in different biomedical domains. Quantitative measurement of disease associations based on biomedical ontologies could provide supports for discovering similar diseases caused by similar molecular process, which is beneficial to improve the corresponding medical diagnosis and treatment. Therefore, we deal with the effective measure of disease similarities in this paper. In particular, we propose a novel regression model based on the deep neural network (DNN) to improve the evaluation of similarities among diseases. We firstly extract the feature vectors of disease pairs, and then train the DNN-based regression model that learns from the information of training set to simulate the complex non-linear relationship among disease pairs. Finally, a comprehensive experimental evaluation is carried out to show the advantages as a solution for measuring disease similarities in terms of the receiver operating characteristic curve (ROC) and the precision-recall curve (PRC).

**Keywords:** Biomedical ontologies · Disease similarities · Regression model · Deep neural network

## 1 Introduction

The appearance of massive biomedical data brings significant opportunities for the life science research and disease diagnoses. Rich knowledge obtained from massive biomedical data brings great challenges, since many biologists usually express their knowledge and build the knowledge bases with their own terms leading to the difficulty in understanding and use. In order to create controlled vocabularies for biomedical knowledge share, a great amount of biomedical ontologies such as Disease Ontology (DO) [13, 35] and Human Phenotypic Ontology (HPO) [15] have been created. Biomedical ontologies reduce the complexity of biomedical concepts and makes contribution to the understanding of human diseases with controllable terms [6, 16, 26, 28, 36]. Nowadays, these biomedical ontologies have been applied in various practical applications [17, 32, 39, 41]. For example, clinical diagnosis and exon

© Springer Nature Switzerland AG 2019
D.-S. Huang et al. (Eds.): ICIC 2019, LNCS 11644, pp. 201–212, 2019.
https://doi.org/10.1007/978-3-030-26969-2_19

sequencing research could be effectively assisted by using HPO-based analysis tools [9, 19, 33, 38].

Using biomedical ontologies for measuring the associations among diseases has attracted a great attention. Quantitative measurement of the associations among diseases could help biomedical researchers understand human diseases more deeply since similar diseases may be caused by the similar molecular process. In other words, the similarity of diseases could represent their common attributes (e.g., phenotypes, related genes and drugs, etc.), which may enlarge the understanding of a disease's causes and improve its diagnoses and treatment plans. For instance, *SIGMAR1* is one of the common related genes between "Amnestic disorder" and "Alzheimer's disease", which infers that they might involve the same molecular process. The closer two diseases are and their more common information they share, the greater their similarity is and vice versa. Therefore, an effective quantitative method to measure the similarities of diseases could help the biomedical researchers gain more useful information of diseases with a close association from a mass of biomedical data, and carry out further experiments for analysis. This could effectively reduce the cost of experiments and increases the efficiency in identifying potential pathogenicity mechanisms and treatment plans.

Currently, DO is widely used to calculate the degree of correlation between diseases. For example, in order to improve the coverage of disease gene annotation, researchers could construct an etiology chain knowledge base to annotate human genes by using DO [30]. DO provides controlled vocabularies of diseases and medical data through external links, and it also provides precise, non-reproducible terms with high disease coverage. Previous attempts including the methods based on information content (IC) [1, 20, 21, 34, 37], the methods based on directed acyclic graph (DAG) of ontology [14, 42] and the methods based on biological process [4, 12, 26, 29], have been designed to measure the similarity among diseases by using DO. In particular, based on IC, Resnik's [34] measures the disease similarity with IC of the most informative common ancestor (MICA). Lin et al. [21] increase the efficiency of the Resnik's with the ratio of the MICA's IC and the sum of two DO terms' IC. Schlicher et al. [1] improve the Lin's with Bernoulli probability distribution by decreasing the impact of shallow annotations. However, IC-based methods rely on the semantic information and ignore the structural information of two terms in DAG, so it is difficult to represent the sematic differences between two disease pairs with the same MICA. For the DAG-based methods, Lee et al. [14] measure the disease similarity with the reciprocal of the shortest distance of two disease terms in DAG. Zhang et al. [40] integrate the shortest distance and the depth of the least common ancestor (LCA) with the exponential function. For the approaches based on biological process, BOG [27] computes the disease similarity by overlapping of related gene sets of two diseases. Besides, PSB [4] improve BOG's efficiency with the gene similarity network. However, these previous approaches have limitations in automatic recognition of potential similar diseases because they mainly depend on the knowledge rules manually constructed by researchers' experience. With the continuous updates of biomedical knowledge and explosive growth of biomedical data, it is of great significance for practical applications to provide a rapid and accurate recognition of potential similar diseases.

To improve the recognition efficiency, deep learning methods such as deep neural network (DNN) have attracted a lot of attention in bioinformatics community [2, 5, 7, 8, 11]. For example, medical data is used to classify or detect disease such as Alzheimer disease and pediatric diseases based on the DNN model [2, 8]. In these applications, DNN obtains high-quality classification and detection results. In fact, DNN is a powerful mathematical tool that can measure the non-linear relationship of various data [31]. By increasing the number of neurons in the hidden layer, the original function can be approximated [11]. Moreover, DNN can be applied not only to classification problems, but also to problems of regression prediction.

Therefore, in order to make full use of the advantages of deep learning models in mining potential association information among diseases and improve the quality of disease similarity measure, we propose a novel DNN-based regression model for measuring similarities among diseases in this paper. In particular, we extract the feature vectors of disease pairs and encode the disease pairs in training set by one-hot method. Furthermore, we apply the principal component analysis (PCA) to reduce the dimension of vectors in the training set. On this basis, we train the DNN-based regression model that excludes the interference of human and learns from the information of training set to simulate the complex non-linear relationship among disease pairs. Finally, comprehensive experiments are carried out to evaluate the performance of the proposed similarity measure approach, and experimental results shows our advantages in terms of ROC (receiver operating characteristic curve) and PRC (precision-recall curve), compared with previous measurement methods.

## 2    Methods

In this section, we introduce the details of our proposed method, including the data preprocessing, the model building and the cross-validation.

### 2.1    Data Preprocessing

There are two types of disease pair sets as our dataset. On one hand, the manually checked disease pairs with high similarity from database SIDD [3] is extracted into the dataset. The disease pairs from SIDD are marked as similar after validation. On the other hand, we extracted the disease pairs from the judgement of medical residents on relatedness of CUI pairs [10, 18, 23, 25, 40] into our dataset. The disease pairs with high mean score are marked as similar and those with low mean score are marked as not similar. Besides, we randomly 100 disease pairs as negative samples and put them into the dataset and divide the dataset in 8:2 ratio with the large data set as the training set and the small one as the test set.

The vectorization of diseases relies on the association between diseases and genes. We firstly integrate the association data extracted from HPO, DO and other biomedical databases. As shown in the Fig. 1, the information of diseases is extracted from DO, including DOID, name, synonyms and associated external database ID and then load in the database (depicted as DG). Disease-gene mapping relations come from mapping files of HPO, SIDD and Dancer database. For the mapping files of HPO, we associate disease's DOID with gene's ID (the ID in EntreZ) by ID in corresponding external databases such as ID in OMIM and Orphanet. For mapping files in Dancer, through the completely matching name and synonyms of diseases, we identify the disease and extracted the associations between disease's DOID and gene's ID. Finally, the associations whose format is "DOID: GeneID" from HPO, Dancer and SIDD are integrated and loaded in the database DG [22, 24]. According to the statistics, there are 11191 disease terms in DO and 4462 of them have disease-gene associations.

According to the extracted diseases' DOID and disease-gene associations (the format is "DOID: GeneID"), as shown in the Fig. 2, each disease is quantized in the form of One-Hot encoding, that is, if a disease is associated with a gene, the value of the gene's corresponding position in the disease vector should be 1, otherwise the value is 0 (for example, the disease with the DOID of 1059 is encoded as "(0, 1, 0, 0...)"). After encoding all diseases, the characteristic matrix $M$ of diseases (the size of matrix M is m $\times$ n, m is the number of diseases, n is the number of related genes) is obtained. Each row of the matrix $M$ represents a vector of a disease, and each column represents the associated condition of a related gene. Considering each disease pair in the training set, the One-Hot coded vector of the corresponding disease is extracted from the matrix $M$, and the two vectors are expanded and spliced to obtain a new vector (the size is 1 $\times$ 2n). After all the disease pairs are treated, all the vectors of disease pairs are reduced in dimension with PCA as the input of the model.

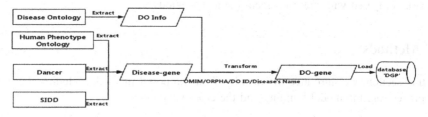

**Fig. 1.** Integrate association between disease and gene

## 2.2   Model Building and Cross-Validation

We design the regression model based on the neural networks. In order to assess the performance of the DNN, we compare the four types of DNN regression model as we allocated the number of hidden layers 1, 2, 3 and 4 as shown in the Fig. 3. Besides, each hidden layer is allocated different number of hidden nodes since the hidden layers and nodes influence the efficiency and the performance of DNN.

As shown in the Fig. 4, we address the structure of DNN-based regression model and the configuration of 4 regression models illustrating the number of hidden layers, hidden nodes, output nodes. The input layer is the vector matrix obtained after data preprocessing. The activation function of hidden layers is ReLu. In order to alleviate the over-fitting problem, the hidden layer is set dropout to 0.5 (The dotted arrow in Fig. 4 indicates that the weight is ignored in a round of update). In other words, half of the feature selectors are ignored in each training batch in the process of model update, so that the model has stronger generalization ability and is not easy to rely on some local features too much. In this paper, Adam algorithm is selected as the updating rule in the process of model training. As shown in Fig. 5, the training data is sparse, and the same update speed may not be suitable for all parameter updates. For example, some parameters need to be adjusted to a large extent, and some parameters only need to be fine-tuned. There is only one neuron node in the output layer, and the activation function of this layer is sigmoid function, whose output represents the probability of similarity of disease pairs, that is, the similarity of disease pairs.

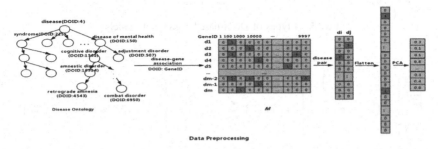

**Fig. 2.** Data vectorization

In order to evaluate the performance of regression model, we make a cross validation (CV). Especially, we apply the 5-fold cross to evaluate the proposed models. The training set is divided into 5 parts, 4 of those for training four regression model and one of them for testing the models. We repeat the procedures for testing each part of the training set. During the testing and training process, we determine the loss function is mean squared error (MSE) and R-square is the rule to judge the performance of the models. MSE and R-square are respectively described as

$$\text{MSE} = \frac{1}{n}\sum(y - \hat{y})^2 \tag{1}$$

$$\text{R} - \text{square} = 1 - \frac{\sum(y - \hat{y})^2}{\sum(y - \bar{y})^2} \tag{2}$$

Where $y$, $\hat{y}$ and $\bar{y}$ are the real value, the predicted value and the average value of all $y$, respectively, and n is the quantity of data in the training batch. R-square is a statistic that measures goodness of fit. The larger the value of R-square, the better the model explains the change in $y$. The range of R-square is clearly 0 to 1, and we will adopt the model with the highest R-square as our regression problem.

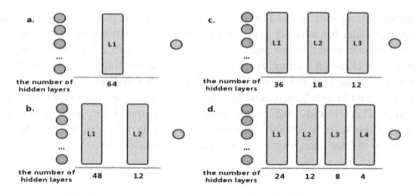

**Fig. 3.** 4 types of DNN-based regression models

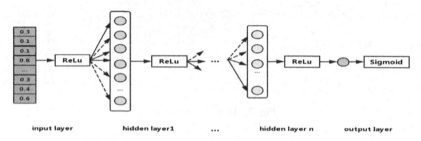

**Fig. 4.** The structure of DNN-based regression model

# 3    Results and Discussion

In our experiments, we compare the performance of 4 types of DNN-based regression models. After selecting the regression model with highest R-square, we compare the selected DNN-based regression model with previous disease similarity measurement methods, including Resnik's, Zhang's, BOG and SemFunSim. The experiments are run in the Windows 10 64-bit system with 2.50 GHz Intel Core i7 and 8 GB RAM.

## 3.1    Compare the Performance of the Four DNN-Based Model

We evaluate the four types of DNN-based regression models as shown in the Fig. 3 and represent the performance in the Fig. 5. It shows the average R-square by using 5-fold cross validation according to the four different DNN-based regression models configurations. the average R-square of 1, 2, 3 and 4 hidden layers models are 0.84, 0.91, 0.93 and 0.74 respectively. The regression model with 3 hidden layers obtains the best performance and is selected to compare with previous disease similarity measurement methods.

Obviously, increasing the number of hidden layers raises the average R-square from the three models with 1 to 3 hidden layers, which reveals that deepening the neural network enhances the ability of fitting data of a model. However, we observe the degradation of the average R-square of the model with 4 hidden layers, so we infer that the performance of the regression model with 4 hidden layers is saturated. The over-fitting problems occur when we add too many hidden layers because of less amount of training data and more complex neural network.

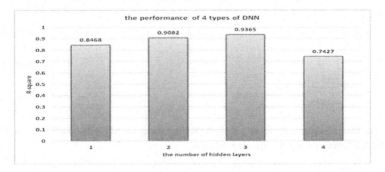

**Fig. 5.** The average of R-square of 4 types of DNN-based regression model by using 5-fold cross validation

## 3.2 Compare DNN-Based Regression Model with Previous Methods

We choose the DNN-based regression model with 3 hidden layers to compare with previous disease similarity measurement methods, including Resnik's, Zhang's, BOG, SemFunSim. We train the DNN-based regression model by using training set and test the model with testing set. As shown in the Fig. 6, we use 800 epochs and 60-size batch to train the model and describe the change of the loss, the R-square, the testing loss and the testing R-square during the different epochs. After training, their values are 0.0021, 0.9546, 0.0917 and 0.9171. The closer the R-square is to 1, the better the model fit the data. The training R-square is close to 1, which illustrates that the model fit the training data well. Moreover, the testing R-square is 0.91 that is close to 1, which reveals that the predicted values fit the real values well in the test set, though the testing R-square is lower than training one.

We compare the performance of 5 disease similarity measurement methods, including Resnik's, Zhang's, BOG, SemFunSim and DNN-based regression model. We calculate the similarity of disease pairs in the test set by using Resnik's method, Zhang's, BOG, SemFunSim and DNN-based regression model and the performance comparison are described by ROC and PRC.

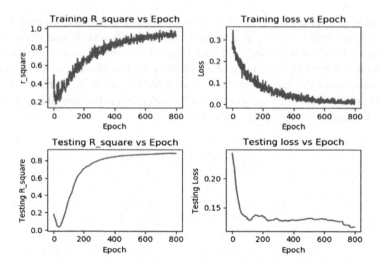

**Fig. 6.** The loss, R-square, the testing loss, the testing R-square of DNN-based regression model with 3 hidden layers during the training process

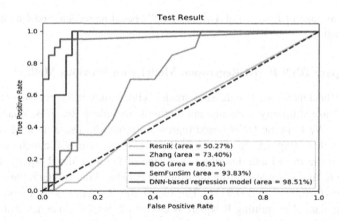

**Fig. 7.** Compare the DNN-based regression model with previous disease similarity measurement methods by the ROC and the AUC of the ROC

ROC is a curve drawn by false positive rate (FPR) as $X$ axis and true positive rate (TPR) as $Y$ axis according to a series of different thresholds. Generally, the closer to the upper left corner the ROC is, the better method is. For more directly, the area under curve (AUC) of the ROC was also given. The greater AUC is, the better the performance is. As shown in the Fig. 7, the AUCs of DNN-based regression model, Sem-FunSim, BOG, Zhang and Resnik are 98.51%, 93.83%, 86.91%, 73.40% and 50.27%, respectively and the DNN-based regression model outperforms the other methods in the terms of ROC and the AUC of ROC.

PR is a curve drawn by recall as $X$ axis and precision as $Y$ axis through different thresholds. Generally, the closer to the right corner the PR is, the better performance of the methods. As shown in the Fig. 8, the PRC of the DNN-based regression model almost wrap round those of Resnik's method, Zhang's, BOG and SemFunSim, which illustrates that the DNN-based regression model outperforms the other four methods in terms of the PRC. For presenting the performance of different methods more directly, the AUC of the PRC is given in the Fig. 8 and their values are 27.72%, 49.97%, 59.04%, 73.77% and 97.07%. Compared with previous four methods, DNN-based regression model obtains the higher precision and gets the higher recall as well, which reveals that the DNN-based regression model performs best among the 5 methods for similarity evaluation when the dataset of disease pairs is highly imbalanced and the positive samples, the similar disease pairs, are very rare.

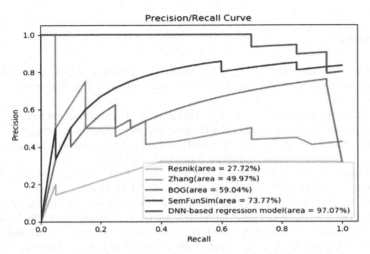

**Fig. 8.** Compare the DNN-based regression model with previous disease similarity measurement methods by the PRC and the AUC of PRC

# 4  Conclusion

The emergence of massive biomedical data offers tremendous opportunities for modern medical diagnosis and treatment. To provide the biomedical community with consistent and reusable descriptions of biomedical knowledge, vast amount of biological ontologies standardizing biomedical terms and advancing the medical knowledge shares are created. Currently, exploring the complex association among diseases in biological ontologies such as Disease Ontology has attracted a widespread attention. Therefore, in this paper, we propose a regression model based on the deep neural network to improve the disease similarity measure approach by using the Disease Ontology. We extract the feature vectors of the training set and train the DNN-based regression model. The experiments are carried out to evaluate the performance of different types of DNN-based regression model and select the model with the best performance. Furthermore, we compare the performance of the DNN-based regression model and previous methods, and the experimental results represent that the proposed DNN-based regression model outperforms previous methods in terms of ROC and PRC.

**Acknowledgements.** The authors thank the anonymous referees for their valuable comments and suggestions. The work was partially supported by the National Key R&D Program of China (2018YFC1603800, 2018YFC1603802, 2017YFC1200200 and 2017YFC1200205), National Natural Science Foundation of China (61602130 and 61872115), China Postdoctoral Science Foundation funded project (2015M581449 and 2016T90294), Heilongjiang Postdoctoral Fund (LBH-Z14089), Natural Science Foundation of Heilongjiang Province of China (QC2015067), and Fundamental Research Funds for the Central Universities (HIT.NSRIF.2017036).

# References

1. Schlicker, A., Domingues, F.S., Rahnenführer, J., Lengauer, T.: A new measure for functional similarity of gene products based on gene ontology. BMC Bioinformatics **7**, 302 (2006)
2. Anand, A., Haque, M.A., Alex, J.S.R., Venkatesan, N.: Evaluation of machine learning and deep learning algorithms combined with dimentionality reduction techniques for classification of Parkinson's disease. In: 2018 IEEE International Symposium on Signal Processing and Information Technology (ISSPIT). IEEE, Louisville (2018)
3. Bandyopadhyay, S., Mallick, K.: A new path based hybrid measure for gene ontology similarity. IEEE/ACM Trans. Comput. Biol. Bioinf. **11**(1), 116–127 (2014)
4. Cheng, L., Li, J., Ju, P., et al.: SemFunSim: a new method for measuring disease similarity by integrating semantic and gene functional association. PLoS One **9**(6), e99415 (2014)
5. Maji, D., Santara, A., Ghosh, S., Sheet, D., et al.: Deep neural network and random forest hybrid architecture for learning to detect retinal vessels in fundus images. In: 2015 37th Annual International Conference of the IEEE Engineering in Medicine and Biology Society (EMBC). IEEE, Milan (2015)
6. Denny, P., Feuermann, M., Hill, D.P., et al.: Exploring autophagy with gene ontology. Autophagy **14**(3), 1–18 (2018)
7. Kim, D., Kim, K.: Detection of early stage Alzheimer's disease using EEG relative power with deep neural network. In: 2018 40th Annual International Conference of the IEEE Engineering in Medicine and Biology Society (EMBC). IEEE, Honolulu (2018)
8. Lyu, G.: A review of Alzheimer's disease classification using neuropsychological data and machine learning. In: 2018 11th International Congress on Image and Signal Processing, BioMedical Engineering and Informatics. IEEE, Beijing (2018)
9. Groza, T., Köhler, S., et al.: The human phenotype ontology: semantic unification of common and rare disease. Am. J. Hum. Genet. **97**(1), 111–124 (2015)
10. Harrow, I., et al.: Matching disease and phenotype ontologies in the ontology alignment evaluation initiative. J. Biomed. Semant. **8**(1), 55 (2017)
11. Neagae, I., Faur, D., Vaduva, C., Datcu, M.: Exploratory visual analysis of multispectral EO images based on DNN. In: IGARSS 2018 - 2018 IEEE International Geoscience and Remote Sensing Symposium. IEEE, Valencia (2018)
12. Jeong, J.C., Chen, X.: A new semantic functional similarity over gene ontology. IEEE/ACM Trans. Comput. Biol. Bioinform. **12**(2), 322–334 (2015)

13. Kibbe, W.A., Arze, C., Felix, V., et al.: Disease ontology 2015 update: an expanded and updated database of human diseases for linking biomedical knowledge through disease data. Nucleic Acids Res. **43**(Database issue), D1071–D1078 (2015)

14. Kim, M.H., Lee, Y.J., Lee, J.H.: Information retrieval based on conceptual distance in is - a hierarchies. J. Doc. **49**(2), 188–207 (1993)

15. Köhler, S., Doelken, S.C., Mungall, C.J., et al.: The human phenotype ontology project: linking molecular biology and disease through phenotype data. Nucleic Acids Res. **42** (Database issue), 966–974 (2014)

16. Köhler, S., Schulz, M.H., Krawitz, P., et al.: Clinical diagnostics in human genetics with semantic similarity searches in ontologies. Am. J. Hum. Genet. **85**(4), 457–464 (2009)

17. Kozaki, K., Yamagata, Y., Mizoguchi, R., et al.: Disease compass - a navigation system for disease knowledge based on ontology and linked data techniques. J. Biomed. Semant. **8**(1), 22 (2017)

18. Lee, I., Blom, U.M., Wang, P.I., et al.: Prioritizing candidate disease genes by network-based boosting of genome-wide association data. Genome Res. **21**(7), 1109 (2011)

19. Lee, I., Lehner, B., Crombie, C., et al.: A single gene network accurately predicts phenotypic effects of gene perturbation in Caenorhabditis elegans. Nat. Genet. **40**(2), 181–188 (2008)

20. Li, B., Wang, J.Z., Feltus, F.A., et al.: Effectively integrating information content and structural relationship to improve the GO-based similarity measure between proteins, pp. 166–172 (2010)

21. Lin, D.: An information-theoretic definition of similarity. In: International Conference on Machine Learning, pp. 296–304 (1998)

22. Liu, J., Ma, Z.M., Feng, X.: Answering prdered tree pattern queries over fuzzy XML data. Knowl. Inf. Syst. **43**, 473 (2015)

23. Liu, J., Yan, D.: Answering approximate queries over XML data. IEEE Trans. Fuzzy Syst. **24**(2), 288–305 (2016)

24. Liu, J., Zhang, X., Zhang, L.: Tree pattern matching in heterogeneous fuzzy XML databases. Knowl. Based Syst. **122**, 119–130 (2017)

25. Liu, J., Zhang, X.: Efficient keyword search in fuzzy XML. Fuzzy Sets Syst. **317**, 68–87 (2017)

26. Lovering, R.C., Roncaglia, P., Howe, D.G., et al.: Improving interpretation of cardiac phenotypes and enhancing discovery with expanded knowledge in the gene ontology. Circ. Genom. Precis. Med. **11**(2), e001813 (2018)

27. Mathur, S., Dinakarpandian, D.: Finding disease similarity based on implicit semantic similarity. J. Biomed. Inform. **45**(2), 363–371 (2012)

28. Meehan, T.F., Vasilevsky, N.A., Mungall, C.J., et al.: Ontology based molecular signatures for immune cell types via gene expression analysis. BMC Bioinform. **14**(1), 263 (2013)

29. Ni, P., Wang, J., Zhong, P., et al.: Constructing disease similarity networks based on disease module theory. IEEE/ACM Trans. Comput. Biol. Bioinform. **PP**(99), 1 (2018)

30. Osborne, J.D., Flatow, J., Holko, M., et al.: Annotating the human genome with disease ontology. BMC Genom. **10**(S1), S6 (2009)

31. Forouzannezhad, P., Abbaspour, A., Li, C., Cabrerizo, M., et al.: A deep neural network approach for early diagnosis of mild cognitive impairment using multiple features. In: 2018 17th IEEE International Conference on Machine Learning and Applications (ICMLA). IEEE, Orlando (2018)

32. Patel, S., Roncaglia, P., Lovering, R.C.: Using gene ontology to describe the role of the neurexin-neuroligin-SHANK complex in human, mouse and rat and its relevance to autism. BMC Bioinform. **16**(1), 186 (2015)

33. Peng, J., Xue, H., Hui, W., et al.: An online tool for measuring and visualizing phenotype similarities using HPO. BMC Genom. **19**(Suppl 6), 185–193 (2018)

34. Resnik, P.: Using information content to evaluate semantic similarity in a taxonomy, pp. 448–453 (1995)
35. Schriml, L.M., et al.: Disease ontology: a backbone for disease semantic integration. Nucleic Acids Res. **40**(Database issue), 940–946 (2012)
36. Pakhomov, S.V.S., Finley, G., McEwan, R., Wang, Y., Melton, G.B.: Corpus domain effects on distributional semantic modeling of medical terms. Bioinformatics **32**(23), 3635–3644 (2016)
37. Wang, J.Z., Du, Z., Payattakool, R., et al.: A new method to measure the semantic similarity of GO terms. Bioinformatics **23**(10), 1274–1281 (2007)
38. Westbury, S.K., Turro, E., Greene, D., et al.: Human phenotype ontology annotation and cluster analysis to unravel genetic defects in 707 cases with unexplained bleeding and platelet disorders. Genome Med. **7**(1), 36 (2015)
39. Hu, Y., Zhou, M., Shi, H., et al.: Measuring disease similarity and predicting disease-related ncRNAs by a novel method. BMC Med. Genom. **10**(Suppl 5), 71 (2017)
40. Zhang, S., Shang, X., Wang, M., et al.: A new measure based on gene ontology for semantic similarity of genes. In: Wase International Conference on Information Engineering, pp. 85–88. IEEE Computer Society (2010)
41. Zhao, Y., Halang, W.: Rough concept lattice based ontology similarity measure. In: International Conference on Scalable Information Systems. Infoscale 2006, Hong Kong, p. 15. DBLP (2006)

# Herb Pair Danggui-Baishao: Pharmacological Mechanisms Underlying Primary Dysmenorrhea by Network Pharmacology Approach

Li-Ting Li[1], Hai-Yan Qiu[4], Mi-Mi Liu[1], and Yong-Ming Cai[1,2,3(✉)]

[1] Guangdong Pharmaceutical University, Guangzhou 510000,
Guangdong, China
ymbruce@qq.com
[2] Guangdong Provincial TCM Precision Medicine Big Data Engineering
Technology Research Center, Guangzhou 510000, Guangdong, China
[3] Guangdong University College of Precision Medicine Big Data Engineering
Research Center Based on Cloud Computing, Guangzhou 510000,
Guangdong, China
[4] Guangdong Huangcun Sports Training Center (Rehabilitation Center),
Guangzhou 510000, Guangdong, China

**Abstract.** Epidemiological studies have shown that primary dysmenorrhea is the most common disease in gynecology, and its incidence rate is between 20% and 90%, which is a common cause of affecting women's normal work and quality of life. Its high incidence rate, wide spread and economic losses and social harm have caused widespread concern worldwide. The present work adopted a network pharmacology-based approach to provide new insights into the active compounds and therapeutic targets of Danggui-Baishao herb pair for the treatment of primary dysmenorrhea. Fifteen active compounds of the herb pair possessing favorable pharmacokinetic profiles and biological activities were selected, interacting 17 dysmenorrhea-related targets to provide potential synergistic therapeutic actions. Systematic analysis of the constructed networks revealed that these targets such as ABCB1, ESR1, PGR, AKR1C3, PTGS2, CYP2C8, PTGS1 were mainly involved in Steroid hormone biosynthesis, Arachidonic acid metabolism, serotonergic synapse and Ovarian steroidogenesis through steroid metabolic process, steroid hormone mediated signaling pathway, cyclooxygenase pathway, response to estradiol and response to oxidative stress.

**Keywords:** Primary dysmenorrhea · Danggui-Baishao herb pair ·
Network pharmacology · Steroid hormone biosynthesis ·
Arachidonic acid metabolism

## 1 Introduction

Primary dysmenorrhea (PD), refers to dysmenorrhea caused by non-pelvic organic lesions, which is caused by endocrine dysfunction, symptoms of lower abdominal and lumbar spastic pain, severe headache, nausea, vomiting, bloating, backache and leg

D.-S. Huang et al. (Eds.): ICIC 2019, LNCS 11644, pp. 213–225, 2019.
https://doi.org/10.1007/978-3-030-26969-2_20

pain. Epidemiological studies have shown that PD is the most common disease in gynecology, and its incidence rate is between 20% and 90%, which is a common cause of affecting women's normal work and quality of life. Its high incidence rate, wide spread and economic losses and social harm have caused widespread concern worldwide [1, 2]. The pathogenesis of PD is complex, and the main pathogenesis is the increase of prostaglandin and leukotriene. Pathological changes includes intense contraction, ischemia and hypoxia of uterine smooth muscle and uterine wall spiral artery [3]. At present, most of the drugs used in the treatment of dysmenorrhea are nonsteroidal anti-inflammatory drugs (NSAIDS), associated with many adverse effects including headaches, drowsiness and indigestion [4].

In recent years, more and more attention has been paid to the search for drugs for gynecological diseases from traditional Chinese medicine (TCM) due to adverse reactions or unsatisfactory efficacy of chemical drugs [5, 6]. TCM has been widely used in China to prevent, diagnose, and treat diseases over 2000 years, and have been used for treating PD effectively without obvious side effects [7, 8]. Danggui-Baishao is a common clinical herb pair. Danggui is warm, nourishing blood while Baishao can nourish blood and astringe Yin. Both of the two herbs have the effect of regulating liver and relieving pain. Siwu Decoction, the most commonly used blood-tonifying prescription in clinic, Xiaoyao San, Tiaogan San and Danggui Shaoyao San, the most commonly used liver-soothing and spleen-invigorating prescription, are all composed of the core of the compatibility of the herb pair. Previous literature excavation showed that formulas based on Danggui-Baishao herb pair (DBHP) have obvious therapeutic effect on PD [9, 10]. However, because of the complexity of TCM formulas there are still no reports on the mechanism of action and compatibility rationality of that herb pair on multi-component, multi-target and multi-channel treatment of PD.

As a unique traditional medicine in China, TCM is the traditional theory of Chinese medicine, which is the material carrier for treating diseases that has the characteristics of multi-target, multi-link and multi-effect. The emergence of network pharmacology (NP) makes it possible to study the potential components and gene targets of TCM [11]. Based on high-throughput omics data analysis, virtual computing and network database retrieval, NP is the frontier research topic of TCM, which integrates systems biology and multi-directional pharmacology [12]. NP clarifies the synergistic effects and potential mechanisms between components-component networks, component-target networks, and targets-disease-pathways network at the molecular level through active molecular screening, target prediction, network construction and analysis, understanding the interaction between ingredients, genes, proteins and diseases [13]. Its systematic research method and the TCM holistic theory have the same goal, especially suitable for clarifying the relationship between TCM, complex drugs and complex diseases. Compared with traditional pharmacological methods, which are more time-consuming, less productive, and less comprehensive, NP combines systems biology and multi-directional pharmacology thinking, integrating biological networks and drug action networks, beyond single targets [14]. At present, NP has been applied in many TCM related research fileds successfully, such as predicting the target of TCM, predicting and identifying the active components of TCM, clarifying the mechanism of action of TCM, scientifically explaining the compatibility of TCM prescriptions, in-depth studying of TCM syndromes, and discovering new indications of TCM. For example, Li predicted the chemical constituents of Gegen-Qilian Decoction and clustered them with FDA-approved targets for diabetes drugs, and concluded 19 major active components, which were verified in cell experiments [15].

Therefore, in the present study, the pharmacological mechanisms of DBHP in treating PD were investigated by the network analysis and bioinformatics analysis with network pharmacology model based on chemical, pharmacokinetic and pharmacological data.

## 2 Materials and Methods

### 2.1 Acquisition of Chemical Compounds and Screening of Active Compounds

In the present study, we firstly searched for all the chemical composition of DBHP using TCMSP. Then, we screened active compounds for further target fishing. Drugs in human body need to be absorbed, distributed, metabolized and excreted (ADME) to reach the target organs to play its therapeutic role. Oral bioavailability (OB) and drug similarity (DL) are the key parameters for Chinese herbal medicine components to participate in the ADME process [16, 17]. OB is not only one of the most important pharmacokinetic parameters in the process of ADME, but also a key indicator for determining bioactive molecules and utilization. Oral drugs with high OB value can reach the human blood circulation more quickly after oral administration. DL denotes the similarity between compounds and known drug molecules. In this study, the screening conditions were set as OB $\geq$ 30% and DL $\geq$ 0.18 with reference to the values calculated by TCMSP.

### 2.2 Target Fishing and Mapping

Searching for the targets of candidate drugs solely by the experimental approaches is overspending, labor-intensive, and time-consuming. In the present work, an integrated in silico approach was introduced to identify the target proteins for the active ingredients of DBHP. We Used three databases to complement each other to obtain active ingredient targets, including TCMSP, DrugBank and SwissTargetPrediction. TTD, DrugBank and DisGeNET database were used to collect depression-related genes with the keyword "dysmenorrhea". Finally, screening common genes by mapping depression-related genes with formula target genes.

### 2.3 Compound-Target Network and Protein-Protein Interaction Network Construction

The selected compounds and target data are imported into Microsoft Excel to generate the "Compounds-Targets" relational pair of data tables, and then the "ID-Type" attribute table was created, in which the target is marked by its gene name. Finally, data and attribute tables were imported into Cytoscape 3.7.0 software to construct the "compound-target" network model.

To systematically illustrate the role of target proteins, we uploaded the potential target information obtained in item "2.3" to the STRING online software, and defined the research species as humans to obtain the PPI network map. The node1, node2 and Combined score information in data files exported from STRING database were imported into Cytoscape software for visual analysis, then the results of network topology analysis were obtained.

### 2.4    Compound-Target Molecular Docking Analysis

SystemsDock WebSite is a web server that uses high-precision docking simulations and molecular pathway maps based on network pharmacology prediction and analysis, with Dock-IN marker docking function to evaluate potential binding activity of protein-ligand. The PDB-ID of key target protein crystal structure was obtained from PDB database, and the 3D chemical structure of active ingredient was downloaded from PubChem. The data were imported into Systems Dock Web Site software for virtual molecular docking analysis of active ingredient and target protein.

### 2.5    GO Function and KEGG Pathway Enrichment Analysis

In order to further study the synergistic effect of DBHP on PD target groups, gene-based (GO) bioaccumulation analysis of target groups was carried out using DAVID platform, and the antidepressant biological processes of drug targets were interpreted. Then the metabolic pathway enrichment analysis of Kyoto Gene and Genome Encyclopedia (KEGG) was carried out. It is worth that DAVID Platform List and Background shoud set up "Homo Sapiens" (human) for operation. And the Bubble Chart was plotted using the OmicShare tools, a free online platform for data analysis, which can use the formula to calculate the enrichment factor (the number of genes belonging to the pathway in the target gene concentration/the number of all genes in the pathway in the background gene concentration).

## 3    Results

### 3.1    Active Ingredients in Danggui-Baishao Herb Pair

A total of 210 ingredients of the herb pair were found and 15 active ingredients were screened out according to ADME screening. Table 1 shows the related information of 15 active ingredients, the degree value represents the number of targets connected to the ingredients. As shown, paeoniflorgenone (BS-2, 87.59%) have the highest OB value, Lactiflorin (BS-4, 0.80) have the highest DL value, while kaempferol (BS-12) have 14 targets, which is the most.

**Table 1.** The active ingredients of "Danggui-Baishao" herb pair

| NO. | Name | OB (%) | DL | Degree |
|---|---|---|---|---|
| DG-1 | beta-sitosterol | 36.91 | 0.75 | 9 |
| DG-2 | Stigmasterol | 43.83 | 0.76 | 10 |
| BS-1 | 11alpha, 12alpha-epoxy-3beta-23-dihydroxy-30-norolean-20-en-28, 12beta-olide | 64.77 | 0.38 | 0 |
| BS-2 | paeoniflorgenone | 87.59 | 0.37 | 2 |
| BS-3 | (3S, 5R, 8R, 9R, 10S, 14S)-3, 17-dihydroxy-4, 4, 8, 10, 14-pentamethyl-2, 3, 5, 6, 7, 9-hexahydro-1H-cyclopenta[a] phenanthrene-15, 16-dione | 43.56 | 0.53 | 6 |
| BS-4 | Lactiflorin | 49.12 | 0.80 | 0 |
| BS-5 | paeoniflorin | 53.87 | 0.79 | 0 |
| BS-6 | paeoniflorin_qt | 68.18 | 0.40 | 0 |
| BS-7 | albiflorin_qt | 66.64 | 0.33 | 0 |
| BS-8 | benzoyl paeoniflorin | 31.27 | 0.75 | 2 |
| BS-9 | Mairin | 55.38 | 0.78 | 1 |
| BS-10 | beta-sitosterol | 36.91 | 0.75 | 9 |
| BS-11 | sitosterol | 36.91 | 0.75 | 8 |
| BS-12 | kaempferol | 41.88 | 0.24 | 14 |
| BS-13 | (+)-catechin | 54.83 | 0.24 | 3 |

Note: BS-: the active ingredient from Baishao; DG-: the active ingredient from Danggui; the Degree value represents the number of targets connected to the ingredient.

## 3.2 Target Proteins of Danggui-Baishao Herb Pair Against Primary Dysmenorrhea

We identified 43 PD-related targets, 219 targets of Danggui-Baishao herb pair. As shown in Fig. 1, there were 17 candidate targets connect compound and disease.

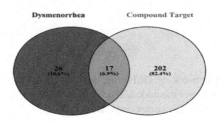

**Fig. 1.** Venns'diagram of compound-disease of DBHP

## 3.3 Compound-Target Network

Figure 2 shows the compound-target network of DBHP against PD. In the network analysis, in addition to the overall network feature analysis, a series of metrics that can quantitatively describe the internal structure of the network, such as degree, betweenness centrality and closeness centrality, are also proposed [18]. They have been applied to biological networks to identify drug targets [19], discover new indications for drugs [20], new mechanisms for the analysis of drug effects, and discovery of active compounds [21]. Therefore, we performed a detailed analysis of the degree and number of nodes in the compound-target network to determine the main active compounds and potential target proteins in DBHP (Table 2). The compound-target network embodies 27 nodes (10 active compounds and 17 potential targets) and 64 compound-target interactions. The degree value represents the number of nodes in the network that are connected to the node. According to the topology of the network, nodes with larger Degree values have a key role in the network.

As can be seen from Fig. 2, the components in DBHP can act on different or the same target of PD, fully reflecting the characteristics of multi-component and multi-target anti-PD of Danggui-Baishao herb pair. As shown, ESR1 (Estrogen receptor), ABCB1 (Multidrug resistance protein 1) and PGR (Progesterone receptor) all have 7 corresponding active compounds, but ABCB1 has the least AverageShortestPathLength, which means it was a more important node. Meanwhile the value of Betweenness centrality and ClosenessCentrality of ABCB1 are the highest in the three targets. All of these indicate ABCB1 may the most important target protein of DBHP against PD. The analysis of C-T network indicated that the active ingredients contained in the DBHP may act on a whole biological network system rather than a single target, which explains to some extent the complexity of multi-component and multi-target of TCM.

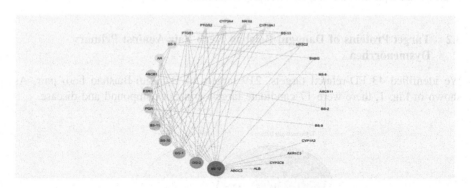

**Fig. 2.** Compound-target network of DBHP against PD. Note: The circular nodes represent compounds and targets, the larger the node and the darker the color, the greater the degree; Each edge represents interaction between compounds and targets.

**Table 2.** Network topology analysis of targets from "Danggui-Baishao" herb pair against PD

| UniProt ID | Gene name | AverageShortestPathLength | BetweennessCentrality | ClosenessCentrality | Degree |
|---|---|---|---|---|---|
| P03372 | ESR1 | 1.92307692 | 0.06663806 | 0.52 | 7 |
| P08183 | ABCB1 | 1.84615385 | 0.23285218 | 0.5416667 | 7 |
| P06401 | PGR | 1.92307692 | 0.11178259 | 0.52 | 7 |
| P10275 | AR | 2 | 0.03485951 | 0.5 | 6 |
| P11511 | CYP19A1 | 2.46153846 | 0.0120298 | 0.40625 | 5 |
| O75469 | NR1I2 | 2.07692308 | 0.01900602 | 0.4814815 | 5 |
| P08684 | CYP3A4 | 2.07692308 | 0.01900602 | 0.4814815 | 5 |
| P35354 | PTGS2 | 2.07692308 | 0.0314329 | 0.4814815 | 5 |
| P23219 | PTGS1 | 2.07692308 | 0.0314329 | 0.4814815 | 5 |
| P08235 | NR3C2 | 2.61538462 | 0.00436419 | 0.3823529 | 3 |
| P04278 | SHBG | 2.30769231 | 0.00736505 | 0.4333333 | 2 |
| O95342 | ABCB11 | 3.61538462 | 0.00153846 | 0.2765957 | 2 |
| Q92887 | ABCC2 | 2.53846154 | 0 | 0.3939394 | 1 |
| P02768 | ALB | 2.53846154 | 0 | 0.3939393 | 1 |
| P10632 | CYP2C8 | 2.53846154 | 0 | 0.3939394 | 1 |
| P42330 | AKR1C3 | 2.53846154 | 0 | 0.3939394 | 1 |
| P05177 | CYP1A2 | 2.53846154 | 0 | 0.3939394 | 1 |

## 3.4    Protein-Protein Interaction Network

Proteins rarely perform their assigned functions in an individual way. Therefore, the study and analysis of protein interactions and their networks are indispensable for understanding molecular functions, cellular tissues, and biological processes in life activities. The construction and analysis of protein-protein interaction network will help to systematically study the molecular mechanism of disease generation and development, and discover new therapeutic targets. To elucidate the role of target proteins at the system level, we used String database and Cytoscape software to construct and analyze PPI network (Fig. 3). The nodes in Figures represent proteins, while the edges represent inter-protein associations involving a total of 17 nodes and 67 edges.

ALB (Serum albumin, degree = 13), CYP3A4 (Cytochrome P450 3A4, degree = 12) and ESR1 (Estrogen receptor, degree = 10) interact strongly with other target proteins and has a large value, playing a key role in the PPI network.

## 3.5    Compound-Target Molecular Docking

The smaller the binding free energy, the stronger the ability of the ligand to bind to the receptor, so the larger the Score, the closer the combination. The docking score is greater than 4.25 and has a certain binding activity; higher than 5.0 indicates that the molecule has a good binding activity to the target; higher than 7.0 indicates that the binding model between the molecule and the target has strong stability and activity [22]. Table 3 indicates compound and have a good binding activity and provides a theoretical reference for further elucidation of the pharmacodynamics of DBHP.

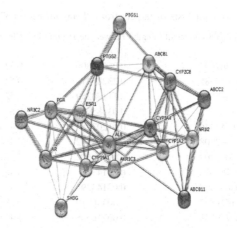

**Fig. 3.** PPI network of DBHP

**Table 3.** Docking score of active ingredients with its targetes of "Danggui-Baishao" herb pair

| Compound | Target | PDB ID | Docking score |
|----------|--------|--------|---------------|
| BS-10 DG-1 | ESR1 | 3CBP | 7.829 |
| DG-2 | | | 7.534 |
| BS-11 | | | 7.492 |
| BS-3 | | | 6.766 |
| BS-12 | | | 6.313 |

### 3.6   GO Function and KEGG Pathway Enrichment

GO (Gene Ontology) is a bioinformatics analysis model that defines the genes used to describe gene function and the relationships between these concepts, including Cellular Component (CC), Molecular Function (MF), Biological process (BP) three parts. The 17 targets were analyzed by David for GO enrichment analysis, according to the PValue, the meaningful items were determined, and the top 20 of them were selected for the key analysis (Table 4). There were 4 targets involved in steroid metabolic process; 4 targets involved in steroid hormone mediated signaling pathway; 3 targets involved in cyclooxygenase pathway; 3 targets involved in response to estradiol and 3 targets involved in response to oxidative stress. It suggests that the occurrence of human depression involves abnormal biological and molecular processes, and Danggui-Baishao herb pair can achieve antidepressant effect by regulating the above biological processes.

The biological pathway performs specific biological functions through the inter-action between different target proteins it constitutes, and is the physiological basis for understanding the clinical manifestations of the disease. Therefore, the action of drugs is not only related to target proteins, but also affected by the biological pathways of target proteins, which is especially true of multi-target TCM. By introducing the target gene of DBHP against PD into the DAVID database, using its KEGG pathway analysis function, the pathway was screened according to PValue (Fig. 4). The target of the main active components is distributed in different metabolic pathways. The coordination and co-regulation of these targets are the possible mechanisms of antidepressant action. A total of 4 targets were enriched in Steroid hormone biosynthesis, including AKR1C3, CYP3A4, CYP1A2, CYP19A1; AKR1C3, PTGS2, CYP2C8, PTGS1 were enriched in Arachidonic acid metabolism; AKR1C3, PTGS2, CYP19A1 were enriched in Ovarian steroidogenesis; and PTGS2, CYP2C8, PTGS1 were enriched in Serotonergic synapse.

**Table 4.** TOP20 of GO Erichment analysis of "Danggui-Baishao" herb pair

| GO ID | Term | Count | Genes | P Value |
|---|---|---|---|---|
| GO0042738 | Exogenous drug catabolic process | 4 | CYP3A4, NR1I2, CYP2C8, CYP1A2 | 1.3E-7 |
| GO0006805 | Xenobiotic metabolic process | 5 | CYP3A4, NR1I2, CYP2C8, PTGS1, CYP1A2 | 5.7E-7 |
| GO0008202 | Steroid metabolic process | 4 | AKR1C3, CYP3A4, NR1I2, CYP2C8 | 7.0E-6 |
| GO0055114 | Oxidation-reduction process | 7 | AKR1C3, CYP3A4, PTGS2, CYP2C8, PTGS1, CYP1A2, CYP19A1 | 7.1E-6 |
| GO0006367 | Transcription initiation from RNA polymerase II promoter | 5 | PGR, AR, NR1I2, NR3C2, ESR1 | 8.2E-6 |
| GO0043401 | Steroid hormone mediated signaling pathway | 4 | PGR, NR1I2, NR3C2, ESR1 | 1.6E-5 |
| GO0019371 | Cyclooxygenase pathway | 3 | AKR1C3, PTGS2, PTGS1 | 3.3E-5 |
| GO0070989 | Oxidative demethylation | 3 | CYP3A4, CYP2C8, CYP1A2 | 4.9E-5 |
| GO0008209 | Androgen metabolic process | 3 | CYP3A4, ESR1, CYP19A1 | 8.9E-5 |
| GO0006810 | Transport | 5 | AR, ABCB11, ALB, ABCB1, ABCC2 | 2.1E-4 |
| GO0017144 | Drug metabolic process | 3 | CYP3A4, CYP2C8, CYP1A2 | 2.6E-4 |
| GO0098869 | Cellular oxidant detoxification | 3 | PTGS2, ALB, PTGS1 | 1.7E-3 |
| GO0032355 | Response to estradiol | 3 | PTGS2, ESR1, CYP1A2 | 2.9E-3 |
| GO0006979 | Response to oxidative stress | 3 | PTGS2, PTGS1, ABCC2 | 4.2E-3 |
| GO0006706 | Steroid catabolic process | 2 | CYP3A4, CYP1A2 | 4.5E-3 |
| GO0042737 | Drug catabolic process | 2 | CYP3A4, CYP1A2 | 5.3E-3 |
| GO0046483 | Heterocycle metabolic process | 2 | CYP3A4, CYP1A2 | 5.3E-3 |
| GO0016098 | Monoterpenoid metabolic process | 2 | CYP3A4, CYP1A2 | 5.3E-3 |
| GO0002933 | Lipid hydroxylation | 2 | CYP3A4, CYP2C8 | 5.3E-3 |
| GO0042908 | Xenobiotic transport | 2 | NR1I2, ABCB1 | 6.2E-3 |

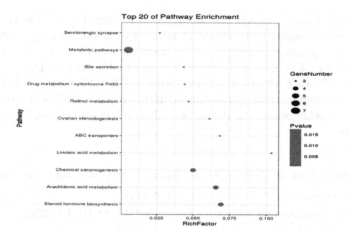

**Fig. 4.** TOP20 of KEGG Pathway Enrichment

# 4    Discussion

Prescription is the main form of TCM, and its complex functional system restricts the revelation of scientific connotation of prescription treatment of diseases. The herb pair is a relatively fixed combination of two medicinal properties commonly used in TCM, as well as the basic form of compatibility of TCM. It is an important basis for the study of prescription compatibility to study the structural characteristics, compatibility effects and material basis of the combination of herb pair [23]. The main pathogenesis of primary dysmenorrhea is related to prostaglandins, oxytocin, vasopressin, endorphin, endothelin, norepinephrine, calcium and magnesium ions and oxygen free radicals [24]. In the present study, the mechanisms of Danggui-Baishao herb pair in treating primary dysmenorrhea were investigated by the network analysis and bioinformatics analysis with network pharmacology model based on chemical, pharmacokinetic and pharmacological data. We used to select active compounds, predict targets, construct networks, systems docking, analyze GO enrichment and KEGG pathway enrichment, illuminating the molecular synergy of DBHP on PD, revealing the characteristics of its multi-target and multi-channel action.

In the present study, the potential antidepressant active ingredients of DBHP were obtained by ADME screening and using the TCMSP database. The established active ingredient-target network (Fig. 2) shows: BS-12 (degree = 12), DG-2 (degree = 10), DG-1 (degree = 9) and BS-10 (degree = 9) have a large number of anti-PD targets. This provides a new material basis for further study of the anti-PD effect of DBHP. Target studies indicate that ABCB1, ESR1, PGR, AKR1C3, PTGS2, CYP2C8, PTGS1may be the target genes of DBHP in the treatment of PD. Progesterone (PG), a natural progesterone secreted by the corpus luteum of the ovary, can promote the conversion of estradiol into inactive estrone, thereby inhibiting the action of estradiol, reducing the production of prostaglandins, and reducing the uterine smooth muscle relaxation and contraction activity, thereby relieving dysmenorrhea. It is worth mentioning that beta-sitosterol (DG-1, BS-10) and stigmasterol (DG-2) can act on ESR1.

Both β-sitosterol and stigmasterol are plant sterols and are natural estrogen receptor agonists that can be used as an alternative to estrogen receptor agonists [25]. It has been reported that the increase of estrogen content can indirectly promote the synthesis and release of PGF2α, leading to uterine contraction and dysmenorrhea caused by insufficient blood supply [26]. It is speculated that the active molecule may compete with the estrogen in the body for the corresponding receptor, producing estrogen-like effects, but does not increase the synthesis of PGF2α, thereby reducing the abnormal contraction of the uterus and exerting treatment for primary dysmenorrhea. The results of protein interaction network showed that there was a relationship between DBHP target proteins, which was a complex interaction network rather than acting alone.

The enrichment analysis of GO biological process indicated that the target gene of DBHP involved in steroid metabolic process, steroid hormone mediated signaling pathway, cyclooxygenase pathway, response to estradiol and response to oxidative stress. The results of KEGG metabolic pathway analysis showed that: DBHP can play an effect on PD role through Steroid hormone biosynthesis, Arachidonic acid metabolism, serotonergic synapse and Ovarian steroidogenesis. Arachidonic acid is a direct precursor of prostaglandin E2 (PGE2), which can cause pain by increasing prostaglandins. It is speculated that the active ingredient of DBHP can ease the dysmenorrhea by reducing the arachidonic acid content through the arachidonic acid metabolic pathway.

In summary, the network pharmacology approach constructed in this work have potential implications toward understanding the pharmacological mechanism of action and active substances of TCM, which may propel the new ways toward exploring new drug therapies for complex diseases. Meanwhile, due to the above findings mainly relied on theoretical analysis, more experiments are anticipated to support these findings as well as potential clinical applications.

# References

1. Ju, H., Jones, M., Mishra, G.: The prevalence and risk factors of dysmenorrhea. Epidemiol. Rev. **36**(1), 104–113 (2014). https://doi.org/10.1093/epirev/mxt009
2. Su, S., Duan, J., Wang, P., et al.: Metabolomic study of biochemical changes in the plasma and urine of primary dysmenorrhea patients using UPLC–MS coupled with a pattern recognition approach. J. Proteome Res. **12**, 852–865 (2013). https://doi.org/10.1021/pr300935x
3. Dawood, M.Y.: Primary dysmenorrhea: advances in pathogenesis and management. Obstet. Gynecol. **108**, 428–441 (2006). https://doi.org/10.1097/01.AOG.0000230214.26638.0c
4. Marjoribanks, J., Ayeleke, R.O.L., Farquhar, C., Proctor, M.: Nonsteroidal anti-inflammatory drugs for dysmenorrhoea. Cochrane Database Syst. Rev.**7**, article CD001751 (2015)
5. Taylor, D.K., Leppert, P.C.: Treatment for uterine fibroids: searching for effective drug therapies. Drug Dis. Today: Therap. Strat. **9**(1), e41–e49 (2012). https://doi.org/10.1016/j.ddstr.2012.06.001
6. Islam, M.S., Akhtar, M.M., Ciavattini, A.: Use of dietary phytochemicals to target inflammation, fibrosis, proliferation, and angiogenesis in uterine tissues: promising options for prevention and treatment of uterine fibroids? Mol. Nutr. Food Res. **58**(8), 1667–1684 (2014). https://doi.org/10.1002/mnfr.201400134

7. Lee, H., Choi, T.-Y.: Herbal medicine (Shaofu Zhuyu decoction) for treating primary dysmenorrhea: a systematic review of randomized clinical trials. Maturitas **86**, 64–73 (2016). https://doi.org/10.1016/j.maturitas.2016.01.012

8. Daily, J.W., Zhang, X.: Efficacy of ginger for alleviating the symptoms of primary dysmenorrhea: a systematic review and meta-analysis of randomized clinical trials. Pain Med. **16**(12), 2243–2255 (2015). https://doi.org/10.1111/pme.12853

9. Lee, M.S., Lee, H.W., Jun, J.H.: Herbal medicine (Danggui Shaoyao San) for treating primary dysmenorrhea: a systematic review and meta-analysis of randomized controlled trials. Maturitas **85**, 19–26 (2016). https://doi.org/10.1016/j.maturitas.2015.11.013

10. Wang, X.: Clinical study on the treatment of primary dysmenorrhea with Danggui peony powder. Cardiovasc. Dis. J. Integr. tradit. Chin. Western Med. **5**(35), 180 (2017). https://doi.org/10.16282/j.cnki.cn11-9336/r.2017.35.128

11. Zhang, Y.Q., Mao, X., Guo, Q.Y., et al.: Network pharmacology-based approaches capture essence of Chinese herbal medicines. Chin. Herb. Med. **8**(2), 107–116 (2016)

12. Hopkins, A.L.: Network pharmacology. Nat. Biotechnol. **25**(10), 1110 (2007). https://doi.org/10.1038/nbt1007-1110

13. Xu, T.F., Li, S.Z., Sun, F.Y., et al.: Systematically characterize the absorbed effective substances of Wutou decoction and their metabolic pathways in rat plasma using UHPLC-Q-TOF-MS combined with a target network pharmacological analysis. J. Pharm. Biomed. Anal. **141**, 95 (2017). https://doi.org/10.1016/j.jpba.2017.04.012

14. Xue, X.C., Hu, J.H.: Research methods and applications in network pharmacology. J. Pharm. Pract. **5**, 401–405 (2015). https://doi.org/10.3969/j.issn.1006-0111.2015.05.005

15. Li, H., Zhao, L., Zhang, B.: A network pharmacology approach to determine active compounds and action mechanisms of ge-gen-qin-lian decoction for treatment of type 2 diabetes. Evid.-Based Compl. Alt. Med. **2014**, 495840 (2014). https://doi.org/10.1155/2014/495840

16. Xu, H.Y., Liu, Z.M., Fu, Y., et al.: Exploiture and application of an internet-based computation platform for integrative pharmacology of traditional Chinese chronic obstructive pulmonary disease. Sci. Rep. **5**(8), 15290–15296 (2015). https://doi.org/10.19540/j.cnki.cjcmm.2017.0141

17. Shen, X., Zhao, Z., Wang, H., et al.: Elucidation of the anti-inflammatory mechanisms of Bupleuri and Scutellariae Radix using system pharmacological analyses. Mediat. Inflamm. (2017). https://doi.org/10.1155/2017/3709874

18. Zhang, M.L., Deng, J.Y., Fang, C.V., et al.: Molecular network analysis and applications. In: Knowledge-Based Bioinformatics. Wiley, Hoboken (2010). https://doi.org/10.1002/9780470669716.ch11

19. Hwang, W.C., Zhang, A., Ramanathan, M.: Identification of information flow-modulating drug targets: a novel bridging paradigm for drug discovery. Clin. Pharmacol. Ther. **84**(5), 563–572 (2008). https://doi.org/10.1038/clpt.2008.129

20. Wu, Z.K., Wang, Y., Chen, L.N.: Network-based drug repositioning. Mol. Bios. **9**(6), 1268–1281 (2013). https://doi.org/10.1039/C3MB25382A

21. Iorio, F., Bosotti, R., Scacheri, E., et al.: Discovery of drug mode of action and drug repositioning from transcriptional responses. Proc. Natl. Acad. Sci. U.S.A. **107**(33), 14621–14626 (2010). https://doi.org/10.4161/auto.6.8.13551

22. Hsin, K.Y., Ghosh, S., Kitano, H.: Combining machine learning systems and multiple docking simulation packages to improve docking prediction reliability for network pharmacology. PLoS One **8**(12), e83922 (2013). https://doi.org/10.1371/journal.pone.0083922

23. Jinao, D., Shulan, S., Yuping, D.: Modern understanding of compatibility of traditional Chinese medicines. J. Nanjing Univ. Tradit. Chin. Med. **25**(5), 330–333 (2009). https://doi.org/10.3969/j.issn.1000-5005.2009.05.003

24. Mrugacz, G., Grygoruk, C., Sieczyński, P., et al.: Etiopathogenesis of dysmenorrhea. Med. Wieku Rozwoj **17**(1), 85–89 (2013)

25. Sriraman, S., Ramanujam, G.M., Ramasamy, M., et al.: Identification of beta—sitosterol and stigmasterol in Bambusa bambos (L.) Voss leaf extract using HPLC and its estrogenic effect in vitro. J. Pharm. Biomed. Anal. **115**, 55–61 (2015). https://doi.org/10.1016/j.jpba.2015.06.024

26. Pu, B.C., Jiang, G.Y., Fang, L.: Research on pain related factors and associations of primary dysmenorrhea. Chin. Arch. Tradit. Chin. Med. **32**(6), 1368–1370 (2014). https://doi.org/10.13193/j.issn.1673-7717.2014.06.037

# End-to-End Learning Based Compound Activity Prediction Using Binding Pocket Information

Toshitaka Tanebe and Takashi Ishida[✉]

School of Computing, Tokyo Institute of Technology, Tokyo, Japan
ishida@c.titech.ac.jp

**Abstract.** Docking simulation is often performed for the activity prediction instead of ligand-based methods based on machine learning approaches, when we have no known drug information about a target protein. Because it calculates the binding energy by virtually docking a drug candidate compound with a binding pocket of a target protein, and does not require any other experimental information. However, the conformation search of a compound and evaluation of binding energy in a docking simulation are computationally heavy tasks, and thus it requires huge computation resources. Therefore, a machine learning-based method to predict the activity of a drug candidate compound against a novel target protein is highly required. Recently, Tsubaki et al. proposed an end-to-end learning method to predict the activity of compounds for novel target proteins. However, the prediction accuracy was insufficient because they used only amino acid sequence information for introducing protein information to a network. In this research, we proposed an end-to-end learning based compound activity prediction using binding pocket information of a target protein, which is more directly important to the activity. The proposed method predicts the activity by end-to-end learning using graph neural network for both compound structure and protein binding pocket structure. As a result of experiments on MUV dataset, the proposed method showed higher accuracy than existing method using only amino acid sequence information. In addition, proposed method achieved equivalent accuracy to docking simulation using AutoDock Vina with much shorter computing time.

**Keywords:** Drug discovery · Cheminformatics · Virtual screening · Docking simulation · Graph convolution · Deep neural network

## 1 Introduction

The cost of drug discovery has been increasing year by year. At present, it takes approximately ten years and US$2.6 billion to develop a new drug [1]. As one of the solutions to this problem, computer-aided efficiency methods now gain increased attention. At the initial stage of drug discovery, compound screening is often performed to select drug candidate compounds from a compound library containing millions to tens of millions of compounds. At this time, it is expected that the cost for drug discovery will be reduced by conducting virtual compound screening that predicts the

© Springer Nature Switzerland AG 2019
D.-S. Huang et al. (Eds.): ICIC 2019, LNCS 11644, pp. 226–234, 2019.
https://doi.org/10.1007/978-3-030-26969-2_21

activity of drug candidate compounds by a computer instead of biochemical experiments.

Compound virtual screening can be roughly divided into two types: ligand-based methods that use known drug information and supervised machine learning technique, and structure-based methods that use conformational information of a drug target protein. In the structure-based method, docking simulation is generally performed. It predicts a compound can bind with a target protein by virtually docking a drug candidate compound (ligand) with a binding site (pocket) on a target protein surface. Unlike ligand-based methods that require known drug information for a target protein, docking simulation is a useful even when known drug information for a target protein does not exist or is insufficient.

Many docking simulation software, such as AutoDock Vina [2], Glide [3], and eHiTS [4], have been developed and are now widely used. However, docking simulation is necessary to perform a conformational search by rotating and translating a compound in a protein pocket, and to evaluate the binding energy of protein-compound interaction using a complicated score function. Thus, these tasks are very computationally expensive (About 0.2 to 2.4 min for Glide to evaluate one compound using single CPU core [3], and several seconds for eHits at the fastest [4]). Although docking simulation of one compound is relatively fast, compound virtual screening requires searches from a large number of compound libraries. Therefore, even if a docking simulation can evaluate one compound in 10 s, screening for compound library consisting of 10 million compounds requires approximately 1200 CPU days.

In order to tackle the problem, several researchers proposed a machine learning-based virtual screening methods to predict the activity of a compound for a novel target protein. Such researches used protein information as well as compound structure information as the input of machine learning. These days, Tsubaki et al. proposed an end-to-end learning prediction method for the problem and achieved better accuracy than previous methods [6]. End-to-end learning that combines feature design and task learning has become popular in fields such as natural language processing. By directly learning the relationship between input and output, it is possible to obtain a better representation of input data than manually designed, and in fact, in machine translation etc., the accuracy is higher than that of the conventional method [5]. They used a graph neural network to a compound and a one-dimensional convolutional neural network to a protein amino acid sequence, and succeeded to avoid manual encoding of input vectors. However, their method seems to have room for improvement in prediction accuracy. They used amino acid information of a protein but pocket structure is considered to contain more useful information on ligand binding. In addition, when using amino acid sequence, it is assumed that there is sequence homology between the target protein and the protein in the training data set. However, using pocket structure, it may possible to predict the activity even for a new target protein without any sequence homology.

Here, we proposed a method to predict the activity of a compound by end-to-end learning using pocket structure information of a protein. The proposed method uses graph neural networks for both compound and protein pocket structures. As a result of benckmark on a data set for virtual screening performance evaluation, it is shown that the proposed method can predict with high accuracy compared to the existing method using amino acid sequence information as the input.

## 2  Methods

The proposed method uses pocket structure information instead of amino acid sequence information used in the previous method as the feature of a protein, and applies a graph neural network for not only a compound but also a protein. The generation of compound features is the same as the previous method by Tsubaki et al. [6], but protein feature generation is highly different. The proposed method consists of the following three parts.

  (i)  Graph generation from a compound structure and a protein pocket structure
 (ii)  Representation learning by graph neural network
(iii)  Activity prediction by a classifier

### 2.1  Graph Generation from a Compound Structure and Protein Pocket Structure

A compound structure was firstly converted from SMILES format string into a graph structure consisting of vertices (atom types) and edges (chemical bonds) by RDKit [7]. If a SMILES-formatted file contains dots (represents non-concatenation), it is excluded from the data set because the graph generation is not possible.

Protein pocket structure information was extracted as the information (residue type, coordinates) of ligand contact residues identified by LPC software [8]. And the information was converted into a graph structure with the following vertices and edges. The types of edge were set based on the research by Ito et al. [9]. By roughly grouping the distance between $C\alpha$ atoms, we expect that it can cope with the structural change of the protein pocket (Fig. 1).

**Vertices**: amino acid residue type (20 types)

**Edges**: distance type between $C\alpha$ atoms of amino acid residues (5 types: I–V)

$$\text{I} : 1.0-4.8\,\text{Å} \quad \text{II} : 4.8-7.0\,\text{Å} \quad \text{III} : 7.0-9.2\,\text{Å}$$
$$\text{IV} : 9.2-11.4\,\text{Å} \quad \text{V} : 11.4-13.6\,\text{Å}$$

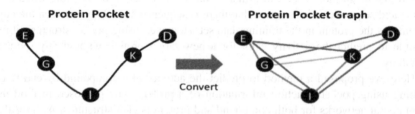

**Fig. 1.**  Conversion of protein pocket to graph structure.

## 2.2  Representation Learning by Graph Neural Network

The compound and protein pocket graphs generated by the procedure described above are each converted to real-valued vectors by using a graph neural network. The procedure of the graph neural network consists of the following three parts both compound graphs and protein pocket graphs.

(1) **Embedding**

   Define $r$-radius vertex and $r$-radius edge, which introduced the concept of $r$-radius subgraphs [10] to vertex and edge, respectively. r-radius subgraphs are subgraphs consisting of all neighboring vertices and edges within a radius $r$ ($r$ is number of hops) from a certain vertex. And, each $r$-radius vertex and $r$-radius edge are randomly assigned a vector.

(2) **Transition**

   Repeat the following operation (i) and (ii) for each $r$-radius vertex and each $r$-radius edge.

   (i)  Add adjacent $r$-radius vertex and $r$-radius edge vector

   (ii) The vector generated by the operation (i) is input to the non-linear function and updated

(3) **Averaging**

   Average all the $r$-radius vertex vectors generated by the operation (2) and output one real-valued vector (Fig. 2).

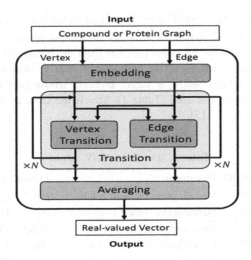

**Fig. 2.** Representation learning by graph neural network

## 2.3  Activity Prediction by Classifier

The activity of a compound is predicted using the $d$-dimensional compound vector $y_{molecule}$ and the protein vector $y_{protein}$ obtained by graph neural network.

First, we simply concatenate $y_{molecule}$ and $y_{protein}$ as follows: $[y_{molecule}; y_{protein}]$

Next, the input $z \in \mathbb{R}^2$ to the softmax layer is obtained by the following equation:

$$z = W_{output} [y_{molecule}; y_{protein}] + b_{output} \tag{1}$$

*Here*, $W_{output} \in \mathbb{R}^{2 \times 2d}, b_{output} \in \mathbb{R}^2$

Finally, we input $z = [y_0, y_1]$ into the softmax layer and perform binary classification on whether the ligand is active or not.

# 3    Experiments

## 3.1    Dataset

**Train Dataset.** We used a DUD-E (a Database of Useful Decoys: Enhanced) dataset [11] as the training dataset. DUD-E is a dataset for performance evaluation of the structure-based screening method created by Mysinger et al. 102 target proteins were selected in consideration of diversity, and active compounds and decoy compounds were prepared for each target. In total, there are 22,886 active compounds and more than 1 million decoy compounds. In this study, as in the previous method by Tsubaki et al., the one down-sampled at 1:1 ratio of active to decoy was used as a training dataset.

**Test Dataset.** We used MUV (Maximum Unbiased Validation) dataset [12] as our test dataset because the three-dimensional structure of a target protein is required. Rohrer et al. obtained assay data for 17 target proteins from the bioactivity data contained in PubChem [14], and assigned 30 active compounds and 15,000 decoy compounds to each target protein. In this study, 9 target proteins whose protein-ligand complex structure has been solved was used (Table 1). This is the same as the data set used in the study by Ragoza et al. [15].

**Table 1.** Details of MUV dataset

| MUV ID | PDB ID | Description | Ligand | Assay type |
|--------|--------|-------------|--------|------------|
| 600 | 1yow | Steroidogenic factor 1: inhibitors | P0E | Cell |
| 692 | 1yow | Steroidogenic factor 1: agonists | P0E | Cell |
| 859 | 5cxv | Muscarinic receptor M1 | 0HK | Cell |
| 852 | 4xe4 | Factor XIIa | NAG | Biochemical |
| 548 | 3poo | Protein kinase A | S69 | Biochemical |
| 832 | 1au8 | Cathepsin G | 0H8 | Biochemical |
| 689 | 2y6o | Ephrin receptor A4 | 1N1 | Biochemical |
| 846 | 5exm | Factor XIa | 5ST | Biochemical |
| 466 | 3v2y | Sphingosine 1-phosphate receptor | ML5 | Cell |

## 3.2   Evaluation Index

We use AUROC (Area Under Receiver Operating Characteristic) as an evaluation index. AUROC is an index using the area under ROC curve, and is an evaluation index mainly used for binary classification problems. ROC is a curve with true positive rate (TPR) on the vertical axis and false positive rate (FPR) on the horizontal axis. TPR is the proportion of positives correctly identified as positive in the dataset, and FPR is the proportion of negatives incorrectly identified as positive in the dataset. TPR and FPR can be calculated by the following formulas.

$$TPR = \frac{\#TP}{\#TP + \#FN} \tag{2}$$

$$FPR = \frac{\#FP}{\#FP + \#TN} \tag{3}$$

## 3.3   Hyperparameters

We used almost same neural network structure and hyperparameters used in previous method by Tsubaki et al. Thus, the hyperparameter values are same as previous research except for the parameters related to protein features (Protein pocket $r$-radius subgraphs). The value is determined by using the train dataset (30 proteins for training and 72 proteins for validation). Details of hyperparameters used in proposed method are shown in Table 2.

**Table 2.**   Details of the hyperparameters of the proposed method

| Hyperparameters | Value |
| --- | --- |
| Dimensions of feature vector | 10 |
| Compound r-radius subgraphs | 2 |
| Layers of compound graph neural network | 3 |
| Protein pocket r-radius subgraphs | 1 |
| Layers of protein pocket graph neural network | 3 |
| Learning rate | 0.001 |
| Learning rate decay | 0.5 |
| Decay interval | 10 |
| Optimization function | Adam |
| Epoch size | 100 |
| Batch size | 1 |

## 3.4   Docking Simulation by AutoDock Vina

We performed experiments on target proteins for which protein-ligand complex structures were obtained. Therefore, assuming that the central coordinates of the ligand is the central coordinates of the pocket, a $24\,\text{Å} \times 24\,\text{Å} \times 24\,\text{Å}$ cube centered on that coordinates is defined as the search range of AutoDock Vina. Also, we set *num_modes* to 100, *energy_range* to 3, *exhaustiveness* to 8.

## 4  Results and Discussion

### 4.1  Prediction Accuracy Evaluation

For checking the improvement of the prediction accuracy of proposed method, we performed performance evaluation on the MUV dataset. We compared the prediction accuracy of proposed method with those of the method by Tsubaki et al. and AutoDock Vina, which is one of the most major docking simulation software. Figure 3 shows the results by a boxplot. Proposed method achieved better accuracy than both Tsubaki et al. and Autodock Vina. Actually, the improvement of proposed method from Tsubaki et al.'s method is not large. However, as shown in Table 3, proposed method shows better accuracy for almost all targets (8/9 targets). In addition, from the results in Table 3, the proposed method shows the best accuracy with 6 target proteins out of 9 among 3 methods.

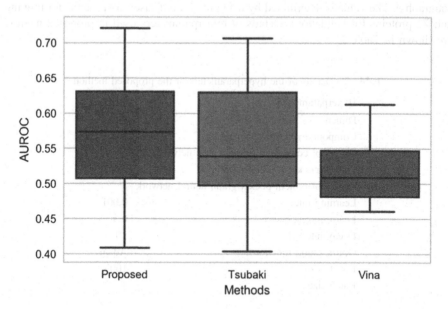

**Fig. 3.** Boxplot of prediction accuracy for MUV dataset

**Table 3.** Prediction results for MUV dataset

| MUV ID | Proposed | Tsubaki et al. | AutoDock Vina |
|--------|----------|----------------|---------------|
| 600 | 0.574 | 0.539 | 0.555 |
| 692 | 0.542 | 0.531 | 0.470 |
| 859 | 0.508 | 0.498 | 0.509 |
| 852 | 0.647 | 0.643 | 0.482 |
| 548 | 0.721 | 0.707 | 0.482 |
| 832 | 0.612 | 0.599 | 0.535 |
| 689 | 0.467 | 0.481 | 0.547 |
| 846 | 0.631 | 0.630 | 0.461 |
| 466 | 0.409 | 0.404 | 0.613 |
| Average | 0.568 | 0.559 | 0.517 |

## 4.2  Evaluation of Computing Time

The proposed method achieved to improve prediction accuracy. However, it is not so useful if it requires huge computing resources. Thus, we also evaluated the computing time. The prediction time was measured using supercomputer TSUBAME3.0 f-node. The details are shown in Table 4. The results of prediction time per a compound are shown in Table 5. The proposed method took longer time to prediction by Tsubaki et al. because the pocket graph used in the proposed method is more complicated than the 1-dimensional convolution neural network used in previous research. However, compared with AutoDock Vina, the prediction time is much faster (more than 1000-fold acceleration), and we think the performance is still sufficient for practical usages.

**Table 4.** Details of TSUBAME3.0 f-node

| CPU | Intel Xeon E5-2680 v4 2.4 GHz × 2 |
|-----|-----|
| Number of CPU core | 28 core |
| Memory | 240 GB |
| GPU | NVIDIA TESLA P100 for NVlink-Optimized Servers × 4 |

**Table 5.** Results of prediction time per compound

| Proposed | Tsubaki et al. | AutoDock Vina |
|----------|----------------|---------------|
| 0.011 [sec] | 0.0034 [sec] | 14.37 [sec] |

## 5  Conclusion and Future Work

In this research, we proposed a new end-to-end learning method for activity prediction using protein pocket structure information. The proposed method showed higher prediction accuracy than previous method. In addition, compared with docking simulation, the proposed method showed higher accuracy with much shorter computing time.

Currently, we use only protein pocket structure information as a protein feature but combination of amino acid information and protein pocket structure information may improve the prediction accuracy.

**Acknowledgement.** This work was supported by JSPS KAKENHI Grant Number 18K11524. The numerical calculations were carried out on the TSUBAME3.0 supercomputer at Tokyo Institute of Technology. (Part of) This work is conducted as research activities of AIST - Tokyo Tech Real World Big-Data Computation Open Innovation Laboratory (RWBC-OIL).

# References

1. Mullard, A.: New drugs cost US$2.6 billion to develop. Nat. Rev. Drug Discov. **13**(12), 877 (2014)
2. Trott, O., Olson, A.J.: AutoDock Vina: improving the speed and accuracy of docking with a new scoring function, efficient optimization, and multithreading. J. Comput. Chem. **31**(2), 455–461 (2010)
3. Friesner, R.A., et al.: Glide: a new approach for rapid, accurate docking and scoring. J. Med. Chem. **47**(7), 1739–1749 (2004)
4. Zsoldos, Z., Reid, D., Simon, A., Sadjad, S.B., Johnson, A.P.: eHiTS: a new fast, exhaustive flexible ligand docking system. J. Mol. Graph. Modell. **26**(1), 198–212 (2007)
5. Nakazawa, T.: New paradigm for machine translation: how the neural machine translation works. J. Inf. Process. Manage. **60**(5), 299–306 (2017)
6. Tsubaki, M., Tomii, K., Sese, J.: Compound–protein interaction prediction with end-to-end learning of neural networks for graphs and sequences. Bioinformatics **35**(2), 309–318 (2019)
7. Landrum, G.: RDKit: open-source cheminformatics
8. Sobolev, V., Sorokine, A., Prilusky, J., Abola, E.E., Edelman, M.: Automated analysis of interatomic contacts in proteins. Bioinformatics (Oxford, England) **15**(4), 327–332 (1999)
9. Ito, J.-I., Tabei, Y., Shimizu, K., et al.: PDB-scale analysis of known and putative ligand-binding sites with structural sketches. Proteins Struct. Funct. Bioinform. **80**(3), 747–763 (2012)
10. Costa, F., De Grave, K. (n.d.). Fast Neighborhood Sub-graph Pairwise Distance Kernel (2010)
11. Mysinger, M.M., Carchia, M., Irwin, J.J., Shoichet, B.K.: Directory of useful decoys, enhanced (DUD-E): better ligands and decoys for better benchmarking. J. Med. Chem. **55**(14), 6582–6594 (2012)
12. Rohrer, S.G., Baumann, K.: Maximum unbiased validation (MUV) data sets for virtual screening based on PubChem bioactivity data. J. Chem. Inf. Model. **49**(2), 169–184 (2009)
13. Liu, H., Sun, J., Guan, J., Zheng, J., Zhou, S.: Improv- ing compound–protein interaction prediction by building up highly credible negative samples. Bioinformatics **31**(12), i221–i229 (2015)
14. Wang, Y., et al.: PubChem's BioAssay database. Nucleic Acids Res. **40**(Database issue), D400–D412 (2012)
15. Ragoza, M., Hochuli, J., Idrobo, E., Sunseri, J., Koes, D.R.: Protein-ligand scoring with convolutional neural networks. J. Chem. Inf. Model. **57**(4), 942–957 (2017)

# A Novel Approach for Predicting LncRNA-Disease Associations by Structural Perturbation Method

Zhi Cao, Jun-Feng Zhang[⊠], Shu-Lin Wang, and Yue Liu

College of Computer Science and Electronics Engineering, Hunan University,
Changsha 410082, Hunan, China
18756576842@163.com

**Abstract.** Numerous experiments have demonstrated that long non-coding RNA (lncRNA) plays an important role in various systems of the human body. The prediction of lncRNA-disease associations is conducive to the diagnosis and prevention of complex diseases. However, the number of known disease-associated lncRNAs is very small. Therefore, predicting the associations between lncRNAs and diseases by computational models has become an urgent need. In this paper, we propose a model called SPMLDA (Structure Perturbation Method LncRNA-Disease Association), which establishes a bi-layer network by integrating disease similarity, lncRNA similarity and known lncRNA-disease associations. Then, we completed lncRNA-disease association matrix based on the structure perturbation model. SPMLDA obtained AUCs of 0.8823 ± 0.0034 based on LncRNADisease dataset and 0.8721 ± 0.0021 based on Lnc2Cancer dataset in the leave-one-out cross validation, which is higher than the state-of-the-art prediction methods. Finally, case studies of two complex human diseases further confirmed the superior performance of our model. SPMLDA could be an important resource with potential values for biomedical researches.

**Keywords:** LncRNA similarity · Disease similarity · Structural perturbation · Bi-layer network · Gaussian kernel similarity

## 1 Introduction

Among the non-coding RNAs, long non-coding RNA is the biggest category of non-coding RNA, and it is a new type of transcripts with a length of more than 200nt. LncRNAs have attracted more and more attention in various biological and medical researches, because it can regulate different gene expression and transcription [1]. LncRNAs play an important role in many physiological activities at various stages of biological life, including cell proliferation, cell differentiation, cell growth, metabolism and apoptosis [2]. In particular, a large number of lncRNAs are involved in complex human diseases such as cardiovascular diseases, neurological disorders and various types of cancer. Predicting the associations between lncRNAs and diseases is helpful for people to understand the pathogenesis of disease so as to prevent and diagnose diseases.

© Springer Nature Switzerland AG 2019
D.-S. Huang et al. (Eds.): ICIC 2019, LNCS 11644, pp. 235–246, 2019.
https://doi.org/10.1007/978-3-030-26969-2_22

So far, various lncRNA-related biological databases are available free of charge [3, 4]. However, experimentally verified disease-associated lncRNAs are rare. Clinical trials to explore the associations between the two require a lot of manpower and material resources. Therefore, using algorithms to predict their associations has become a hot topic. In recent years, several prediction methods have been proposed. It is roughly divided into the following three categories. In the first category, machine learning-based models were developed to predict potential lncRNA-disease associations based on known lncRNA-disease associations. For example, Chen and Yan proposed the computational method of LRLSLDA to identify potential disease-associated lncRNAs [5]. LRLSLDA is a new semi-supervised learning method in the Laplacian regularized least squares framework. In addition, this method did not require negative samples and could produce reliable results based on lncRNA expression profiles and known lncRNA-disease associations. The second category is based on random walks to predict their associations by integrating known lncRNA-disease associations. For example, Ganegoda proposed a kernel-based heterogeneous network model called KRWRH that combined lncRNA specific information, disease phenotype information and experimentally validated lncRNA-disease associations [6]. KRWRH used a Gaussian interaction profile kernel measure to calculate disease similarity and lncRNA similarity, and used a restart random walk method to predict lncRNA-disease associations. The third category is to predict the potential lncRNA-disease associations based on known disease-related genes. Li et al. proposed a simple computational method to predict the novel associations between lncRNAs and vascular diseases based on the genomic location of vascular disease-related genes and candidate lncRNAs [7].

In this article, we develop the lncRNA-disease association prediction method SPMLDA (Structure Perturbation Method LncRNA-Disease Association). We constructed bi-layer network by integrating the known lncRNA-disease association network, the lncRNA similarity network and the disease similarity network. Based on this bi-layer network, structure perturbation model was implemented. Finally, the lncRNA-disease similarity network was obtained and transformed into the lncRNA-disease association matrix. Our method obtained an AUC of 0.8823 and 0.8721 in the leave-one-out cross validation framework based on two databases, significantly improving previous classical methods. Furthermore, case studies of liver cancer and ovarian cancer were implemented and 90% of top 10 predicted lncRNAs have been confirmed by recent biological experiments. SPMLDA may be an important resource with the potential value of medical research.

## 2   Methods

The SPMLDA method will be introduced in details which mainly consists the following three steps shown in Fig. 1. Firstly, we calculate disease-disease similarity and lncRNA-lncRNA similarity. Secondly, bi-layer network was constructed by integrating the known lncRNA-disease association network, the lncRNA Gaussian kernel similarity network and the disease similarity network. Finally, we get the prediction association matrix by structure perturbation model.

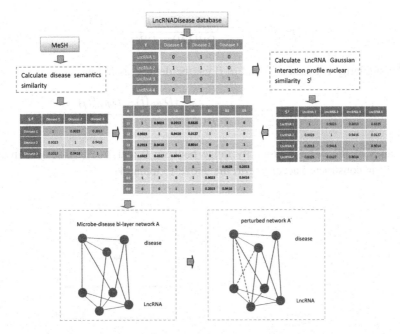

**Fig. 1.** Overall workflows of SPMLDA for discovering potential LncRNA-disease associations.

## 2.1 Disease Similarity

To more accurately represent the similarities between diseases, we calculated disease semantic similarities. We use directed acyclic graphs (*DAGs*) to represent diseases, including all annotation terms related to diseases. The disease DAGs was obtained from the database MeSH. Figure 2 shows DAG of "Liver Neoplasms (*LN*)" and "Stomach Neoplasms (*SN*)". $DAG(LN) = (T_{LN}, E_{LN})$ is a *DAG* map of liver neoplasms, where $T_{LN}$ represents a disease set and $E_{LN}$ represents a set of links. The semantic contribution of a node $t$ is calculated:

$$D_{LN}(t) = \begin{cases} 1 & if\ t = LN \\ max\{\beta * D_{LN}(t')|t' \in children\ of\ t\} & otherwise \end{cases} \quad (1)$$

where $\beta$ is a semantic contribution factor ($\beta = 0.5$) [8]. The overall semantic value of disease *LN*, is defined as

$$DV(LN) = \sum_{t \in T_{LN}} D_{LN}(t) \quad (2)$$

If two diseases share most of the same *DAG* map, they may be more similar. Then the semantic similarity of Liver Neoplasms and Stomach Neoplasms is defined as

$$SS(LN, SN) = \frac{\sum_{t \in T_{LN} \cap T_{SN}} (D_{LN}(t) + D_{SN}(t))}{DV(LN) + DV(SN)} \tag{3}$$

The similarity between $d(i)$ and $d(j)$ can be defined as follow

$$SS(d(i), d(j)) = \frac{\sum_{t \in T_{d(i)} \cap T_{d(j)}} (D_{d(i)}(t) + D_{d(j)}(t))}{DV(d(i)) + DV(d(j))} \tag{4}$$

The values of the disease similarity range between 0 and 1. The disease similarity matrix $S^d$ is constructed according to the above formula (4).

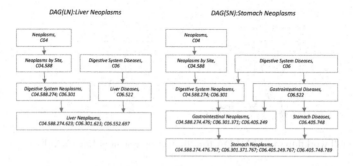

**Fig. 2.** The DAG of the diseases Liver Neoplasms and Stomach Neoplasms.

## 2.2 LncRNA Similarity

Gaussian interaction profile kernel similarity between lncRNA $l(i)$ and $l(j)$ can be defined as follow:

$$KL(l(i), l(j)) = exp\left(-\gamma_l \|l(i) - l(j)\|^2\right) \tag{5}$$

where $\gamma_l$ is responsible for controlling the kernel bandwidth. The update rule is as follows:

$$\gamma_l = \gamma_l' \bigg/ \left(\frac{1}{n_l} \sum_{j=1}^{n_l} \|l(i)\|^2\right) \tag{6}$$

here, $n_l$ is the number of lncRNAs. For calculation convenience, $\gamma_l'$ is set to 1. The lncRNA similarity matrix $S^l$ is constructed according to the above formula (5).

## 2.3 Construction of Bi-Layer Network

The original lncRNA-disease association matrix $Y$ (where $Y_{ij} = 1$ if lncRNA $m_i$ has a known association with disease $d_j$, otherwise $Y_{ij} = 0$) represents the lncRNA-disease network. The matrix $S^l$ represents the lncRNA similarity network and $S^d$ represents the disease similarity network. By integrating these three networks, we can get a bi-layer network $G$. Therefore, a bi-layer network of lncRNA-disease associations can be represented as an adjacency matrix $A_{N \times N}$ ($N$ is the total number of lncRNAs and diseases in the network).

$$A = \begin{bmatrix} S^l & Y \\ Y^T & S^d \end{bmatrix}$$ (7)

### 2.4    Structural Perturbation Method

A bi-layer network is represented as $G(N, E, W)$, where $N$ is the set of all nodes in the network, $E$ is the set of edges, and $W$ is the set of weights. We randomly select a fraction of the links and the nodes to constitute the perturbation set $\Delta E$, while the rest of the links is $E^R (E^R = E - \Delta E)$. $\Delta A$ and $A^R$ represent their corresponding adjacency matrices; obviously, $A = A^R + \Delta A$.

$A^R$ is a real symmetric matrix and it is diagonalizable. $A^R$ can be represented as:

$$P^{-1} A^R P = \Lambda.$$ (8)

The columns of the $P$ are the eigenvectors of the $A^R$, and the values on the main diagonal of $\Lambda$ are the eigenvalues of the $A^R$. Then, Eq. (8) can be transformed into:

$$A^R = \sum_{k=1}^{N} \lambda_k x_k x_k^T$$ (9)

where $\lambda_k$ and $x_k$ are the eigenvalue and the corresponding orthogonal normalized eigenvector for $A^R$, respectively.

We consider the set $\Delta E$ as a perturbation to the network $A^R$ and construct the perturbed matrix by first-order approximation. First-order approximation allows the eigenvalue to change but keep the eigenvector constant. After perturbation, the eigenvalue $\lambda_k$ is corrected to be $\lambda_k + \Delta \lambda_k$ and its corresponding eigenvector is corrected to be $x_k + \Delta x_k$. Left-multiplying the eigenfunction:

$$(A^R + \Delta A)(x_k + \Delta x_k) = (\lambda_k + \Delta \lambda_k)(x_k + \Delta x_k)$$ (10)

By $x_k^T$ and neglecting second-order terms $x_k^T \Delta A \Delta x_k$ and $\Delta \lambda_k x_k^T \Delta x_k$, we obtain:

$$x_k^T A^R (x_k + \Delta x_k) + x_k^T \Delta A x_k = \lambda_k x_k^T (x_k + \Delta x_k) + \Delta \lambda_k x_k^T x_k$$ (11)

Since $x_k^T A^R = \lambda_k x_k^T$, we obtain:

$$\Delta \lambda_k \approx \frac{x_k^T \Delta A x_k}{x_k^T x_k}$$ (12)

This formula is reminiscent of the expectation value of the first-order perturbation Hamiltonian in quantum mechanics. Perturbation theory comprises mathematical methods for finding an approximate solution to a problem. Using the perturbed eigenvalue while keeping eigenvector unchanged, the perturbed matrix can be obtained:

$$A' = \sum_{k=1}^{N} (\lambda_k + \Delta\lambda_k)x_k x_k^T \qquad (13)$$

which can be considered as the linear approximation of the given matrix $A$ [9].

The final prediction matrix $A'$ is obtained by averaging $t$ independent selections of perturbation set. Under this framework, the entries for $A'$ can be considered as score between a pair of nodes of the given matrix $A$.

## 3   Results

### 3.1   Datasets

We retrieve known lncRNA-disease associations from LncRNADisease and Lnc2Cancer database [10, 11]. To evaluate the performance of our proposed SPMLDA, we used three datasets. The first dataset is downloaded from LncRNADisease established in 2017, which contains 2947 experimentally validated lncRNA-disease associations. The second dataset is retrieved from Lnc2Cancer established in 2018, which contains 4989 lncRNA-disease associations. The third dataset is obtained from LncRNADisease established in 2019, which contains 10564 lncRNA-disease associations. The statistical data of the final datasets are shown in Table 1. The first and second data sets are used to measure method performance, and the third data set is used for case study verification.

**Table 1.** Details of final datasets.

| Datasets | No. of diseases | No. of LncRNAs | No. of interactions |
|---|---|---|---|
| LncRNADisease2017 | 328 | 881 | 2947 |
| Lnc2Cancer2018 | 165 | 1614 | 4989 |
| LncRNADisease2019 | 798 | 6085 | 10564 |

### 3.2   Performance Evaluation

To fully evaluate the performance of the model, we use leave-one-out cross validation (LOOCV) on SPMLDA. The LOOCV is based on known lncRNA-disease associations. Each time the training model is used, each pair of known associations is taken as the test sample in turn, and the remaining known associations are used as training samples. In order to reduce the bias caused by randomness, we performed LOOCV for 100 times. In the end, the result of LOOCV is $0.8823 \pm 0.0034$ based on LncRNA-Disease2017 dataset and $0.8721 \pm 0.0021$ based on Lnc2Cancer2018 dataset.

In order to intuitively evaluate the performance of the model, the receiver characteristic curve (ROC) was introduced, which is a common method for evaluating binary classifications. The true positive rate (TPR, sensitivity) and the false positive rate (FPR, 1-specificity) are two important indicators. Here, the ordinate of the ROC represents sensitivity and the abscissa represents 1-specificity. Sensitivity represents the percentage of the test samples that rank higher than the given threshold, while specificity represents the opposite. Calculated as follows:

$$Sensitivity = \frac{TP}{TP + FN} \tag{14}$$

$$Specificity = \frac{TN}{TN + FP} \tag{15}$$

where $TP$ means true positives, $FP$ refers to false positives, $TN$ is true negatives, and $FN$ represents false negatives. The area under ROC (AUC) is also used to measure performance. In general, AUC = 0.5 indicates random performance, and AUC = 1 indicates optimal performance. As shown in Fig. 3, we draw two ROC curve for our model.

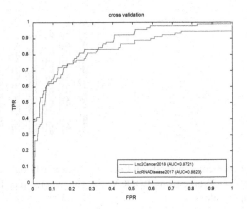

**Fig. 3.** The ROC curves for SPMLDA.

### 3.3 Compared with Other Methods

We compare SPMLDA with other two state-of-the-art computational methods (SIMCLDA, LRLSLDA) in terms of AUC and recall on two datasets. Lu et al. proposed a method (named SIMCLDA) for predicting potential lncRNA–disease associations based on inductive matrix completion. They computed Gaussian interaction profile kernel of lncRNAs from known lncRNA–disease interactions and functional similarity of diseases based on disease–gene and gene–gene ontology associations. Then, They extracted primary feature vectors from Gaussian interaction profile kernel of lncRNAs and functional similarity of diseases by principal component analysis, respectively [12]. Chen et al. proposed the assumption that similar diseases tend to be associated with functionally similar lncRNAs. They further developed the method of Laplacian Regularized Least Squares for LncRNA–Disease Association (LRLSLDA), which used Laplacian regularized least squares method, a semi-supervised learning method, to identify the possible associations between lncRNAs and diseases by incorporating the expression profiles of lncRNA [13].

On LncRNADisease2017 dataset, we can see that the AUC obtained by SPMLDA is $0.8823 \pm 0.0034$, which is significantly higher than others (SIMCLDA 0.8581, LRLSLDA 0.6815), suggesting that our method shows a great improvement in accuracy compared with these prediction methods (Fig. 4a). In addition, The average recalls across all the tested diseases at different top k values is shown in Fig. 4b. If a predicted association is ranked higher than the specified ranking threshold, it will be regarded as the correct retrieved association. SPMLDA outperforms other methods by predicting more associations. On Lnc2Cancer2018 dataset, SPMLDA performed better with the AUC of $0.8721 \pm 0.0021$ than others (SIMCLDA 0.8623, LRLSLDA 0.6754), whose results are shown in Fig. 5a. As shown in Fig. 5b, SPMLDA can get a higher recall across all the tested diseases at different top k values. Overall, our method is more effective than other methods.

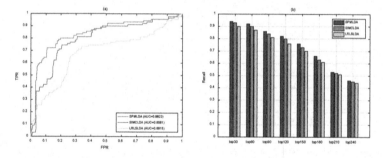

**Fig. 4.** Comparison of predicting methods on LncRNADisease2017 dataset. (a) Performance of all methods in terms of ROC curve using LOOCV. (b) The average recalls across all the tested diseases at different top k values.

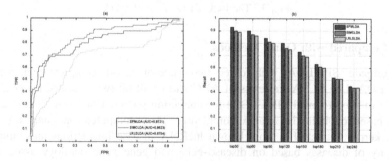

**Fig. 5.** Comparison of predicting methods on Lnc2Cancer2018 dataset. (a) Performance of all methods in terms of ROC curve using LOOCV. (b) The average recalls across all the tested diseases at different top k values.

### 3.4    Case Studies

We considered all known lncRNA-disease associations (in LncRNADisease2017) and train them as a complete set. The lncRNAs corresponding to each disease in the training results were ranked, and the top 10 lncRNAs related to liver cancer and ovarian cancer were shown in Fig. 6. These associations are verified by the training set.

After removing the known associations of the training set from results, take the top 10 to analyze the two complex human cancers. That is to say, the first ten lncRNAs we refer to were unknown before the prediction. We check the predicted associations by referring to available associations in LncRNADisease2019. According to LncRNA-Disease2019, the top 9 lncRNAs of the two diseases were separately verified.

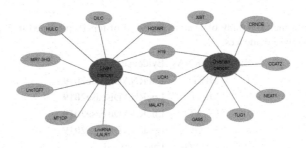

**Fig. 6.** Network view of the top 10 predicted lncRNAs for liver cancer and ovarian cancer based on LncRNADisease2017 dataset.

### 3.4.1    Liver Cancer

Liver cancer, also known as hepatic cancer and primary hepatic cancer, is cancer that starts in the liver. Human insulin-like growth factor II gene (IGF2) is overexpressed, and its imprinting is disrupted in many tumors, including liver cancer [14]. In liver CSCs, lncBRM associates with BRM to initiate the BRG1/BRM switch and the BRG1-embedded BAF complex triggers activation of YAP2 signalling [15]. The relationships between other lncRNAs and diseases listed in Table 2 have been confirmed by recent studies.

**Table 2.** Effect in the case study of liver cancer, 9 of top-10 predicted lncRNAs have been supported by literature evidences.

| Rank | LncRNAs | PMID |
|------|---------|------|
| 1 | IGF2-AS | LncRNADisease**2019** |
| 2 | LncBRM | LncRNADisease**2019** |
| 3 | LncAPC | LncRNADisease**2019** |
| 4 | MEG3 | LncRNADisease**2019** |
| 5 | TFDP1 | LncRNADisease**2019** |
| 6 | CAMTA1 | LncRNADisease**2019** |
| 7 | LINC00210 | LncRNADisease**2019** |
| 8 | NNT-AS1 | Unconfirmed |
| 9 | YAP1 | LncRNADisease**2019** |
| 10 | Uc.338 | LncRNADisease**2019** |

### 3.4.2   Ovarian Cancer

Ovarian cancer is a cancer that forms in or on an ovary. It results in abnormal cells that have the ability to invade or spread to other parts of the body. MEG3 protected ATG3 mRNA from degradation following treatment with actinomycin D. Xiu's results suggest that the lncRNA MEG3 acts as a tumor suppressor in EOC by regulating ATG3 activity and inducing autophagy [16]. Mitra et al. reported an integrated analysis of over 700 ovarian cancer molecular profiles, including genomic data sets, from four patient cohorts identifying lncRNA DNM3OS, and MIAT overexpression and their reproducible gene regulation in ovarian cancer EMT [17]. Similarly, the relationships between other lncRNAs and diseases listed in Table 3 have been confirmed by recent studies.

**Table 3.** Effect in the case study of ovarian cancer, 9 of top-10 predicted lncRNAs have been supported by literature evidences.

| Rank | LncRNAs | Evidence |
|------|---------|----------|
| 1 | MEG3 | LncRNADisease**2019** |
| 2 | MIAT | LncRNADisease**2019** |
| 3 | CCAT1 | LncRNADisease**2019** |
| 4 | HAGLR | LncRNADisease**2019** |
| 5 | LncAPC | Unconfirmed |
| 6 | SRA1 | LncRNADisease**2019** |
| 7 | LUCAT1 | LncRNADisease**2019** |
| 8 | NBAT1 | LncRNADisease**2019** |
| 9 | ZFAS1 | LncRNADisease**2019** |
| 10 | SNHG20 | LncRNADisease**2019** |

## 4   Conclusion

More and more evidence shows that it is very important to develop a powerful method to infer the potential associations between lncRNAs and diseases. In this paper, we propose a new calculation method, SPMLDA. First, we calculated the disease semantic similarity and formed the disease similarity matrix. Next, we calculated the Gaussian kernel similarity of lncRNA and formed the lncRNA similarity matrix. Then, the two matrices and the original adjacency matrix were transformed into bi-layer networks, and the perturbation model was implemented on the bi-layer network. Finally, a lncRNA-disease association matrix was obtained. Each value in the matrix represents the association of a disease with a lncRNA. The result of LOOCV is higher than the two state-of-the-art methods, which fully proves the feasibility of our model. Finally, case studies of two cancers confirmed that 90% of the top 10 lncRNAs were validated in the latest dataset. In addition, the method of this paper can be extended to other association prediction problems, such as drug-disease association prediction and microbe-disease association prediction.

There are three main reasons why SPMLDA has achieved good results. First, the reliable LncRNADisease dataset provides an officially validated lncRNA-disease association. Next, the use of Gaussian interaction profile kernel similarity to mature measure lncRNA similarity similarity. Last but not least, the use of structural perturbation model makes forecast results are satisfactory.

However, the performance of our model is still restricted by some limitations. First, disease-related information collected from the database is very sparse. However, we believe that with the in-depth research in this area, the database will be further improved to support more in-depth exploration. Second, the Gaussian interaction profile kernel similarity is extremely dependent on the known lncRNA-disease associations, and it is difficult to avoid bias brought by such an inference. The solution to this problem is to integrate databases from different sources. Finally, we believe that through further research and improvement in the future, our model performance will be further improved.

**Acknowledgement.** This work was supported by the grants of the National Science Foundation of China (Grant Nos. 61672011 and 61472467) and the National Key R&D Program of China (2017YFC1311003).

# References

1. Ponting, C.P., Oliver, P.L., Reik, W.: Evolution and functions of long noncoding RNAs. Cell **136**(4), 629–641 (2009)
2. Harries, L.W.: Long non-coding RNAs and human disease. Biochem. Soc. Trans. **40**(4), 902–906 (2012)
3. Amaral, P.P., Clark, M.B., Gascoigne, D.K., Dinger, M.E., Mattick, J.S.: lncRNAdb: a reference database for long noncoding RNAs. Nucleic Acids Res. **39**(Database issue), D146–D151 (2011)
4. Bu, D., et al.: NONCODE v30: integrative annotation of long noncoding RNAs. Nucleic Acids Res. **40**(Database issue), D210–D215 (2012)
5. Chen, X., Yan, G.Y.: Novel human lncRNA-disease association inference based on lncRNA expression profiles. Bioinformatics **29**(20), 2617–2624 (2013)
6. Ganegoda, G.U., Li, M., Wang, W., Feng, Q.: Heterogeneous network model to infer human disease-long intergenic non-coding RNA associations. IEEE Trans. Nanobiosci. **14**(2), 175–183 (2015)
7. Li, J., et al.: A bioinformatics method for predicting long noncoding RNAs associated with vascular disease. Sci. China Life Sci. **57**(8), 852–857 (2014)
8. Wang, D., Wang, J., Lu, M., Song, F., Cui, Q.: Inferring the human microRNA functional similarity and functional network based on microRNA-associated diseases. Bioinformatics **26**(13), 1644–1650 (2010)
9. Lu, L.Y., Pan, L.M., Zhou, T., Zhang, Y.C., Stanley, H.E.: Toward link predictability of complex networks. Proc. Natl. Acad. Sci. U.S.A. **112**(8), 2325–2330 (2015)
10. Chen, G., et al.: LncRNADisease: a database for long-non-coding RNA-associated diseases. Nucleic Acids Res. **41**(Database issue), D983–D986 (2013)
11. Gao, Y., et al.: Lnc2Cancer v2.0: updated database of experimentally supported long noncoding RNAs in human cancers. Nucleic Acids Res. **47**(D1), D1028–D1033 (2019)

12. Lu, C.Q., et al.: Prediction of lncRNA-disease associations based on inductive matrix completion. Bioinformatics **34**(19), 3357–3364 (2018)
13. Chen, X., Yan, G.-Y.: Novel human lncRNA–disease association inference based on lncRNA expression profiles. Bioinformatics **29**, 2617–2624 (2013). btt426
14. Vu, T.H., Chuyen, N.V., Li, T., Hoffman, A.R.: Loss of imprinting of IGF2 sense and antisense transcripts in Wilms' tumor. Cancer Res. **63**(8), 1900–1905 (2003)
15. Zhu, P., et al.: LncBRM initiates YAP1 signalling activation to drive self-renewal of liver cancer stem cells. Nat. Commun. **7**, 13608 (2016)
16. Xiu, Y.L., Sun, K.X., Chen, X.: Upregulation of the lncRNA Meg3 induces autophagy to inhibit tumorigenesis and progression of epithelial ovarian carcinoma by regulating activity of ATG3. Oncotarget **8**(19), 31714–31725 (2017)
17. Mitra, R., et al.: Decoding critical long non-coding RNA in ovarian cancer epithelial-to-mesenchymal transition. Nat. Commun. **8**, 1604 (2017)

# Improved Inductive Matrix Completion Method for Predicting MicroRNA-Disease Associations

Xin Ding[1], Jun-Feng Xia[2], Yu-Tian Wang[3], Jing Wang[1,4],
and Chun-Hou Zheng[1(✉)]

[1] College of Computer Science and Technology, Anhui University, Hefei, China
zhengch99@126.com
[2] Institute of Physical Science and Information Technology, Anhui University,
Hefei, China
[3] School of Software Engineering, Qufu Normal University, Qufu, China
[4] School of Computer and Information Engineering, Fuyang Normal University,
Fuyang, China

**Abstract.** Nowadays, plenty of evidence indicates that microRNAs (miRNAs) can result in various human complex diseases and may be as new biological markers to diagnose specific diseases. The reality is that biological experimental corroboration of disease-related miRNAs is time consuming and laborious. Therefore, the calculation methods for recognizing the potential relationship between the miRNA and the disease have become an increasingly significant hot topic in the world. In this paper, we exploited an improved calculation method based on inductive matrix completion to predict disease related miRNAs (IIMCMP). Firstly, we construct miRNA-disease adjacency matrix by adopting verified miRNA-disease associations. In addition, the proposed approach uses the three matrices, including the disease-miRNA association matrix, the integrated disease similarity matrix and the miRNA functional similarity matrix. Secondly, considering new diseases or new miRNAs, it is necessary to pre-process the adjacency matrix of biologically validated associations between diseases and miRNAs, so we calculate the interaction profile between miRNAs and diseases to update adjacency matrix of miRNA-disease association. Finally, inductive matrix completion algorithm is adopted to predict the probability score on the heterogeneous network between miRNAs and diseases. As a result, IIMCMP gained the AUC of 0.9016 which adopted new interaction likelihood profiles in the leave-one-out cross validation. In addition, two case studies and leave-one-out cross validation demonstrated that IIMCMP can achieve predominant and reliable performance assessment.

**Keywords:** MiRNA similarity · Disease similarity · Association · Improved inductive matrix completion

© Springer Nature Switzerland AG 2019
D.-S. Huang et al. (Eds.): ICIC 2019, LNCS 11644, pp. 247–255, 2019.
https://doi.org/10.1007/978-3-030-26969-2_23

# 1 Introduction

In biological, miRNAs are an important short non-coding RNAs consisting of approximately 21–24 endogenous noncoding RNAs. It usually inhibits gene expression at the post-transcriptional level of target mRNAs [1]. Biological processes such as immune response, cell differentiation, proliferation, and apoptosis, signal transduction, embryo development are all affected by miRNA regulation [2]. Chen et al. [3] who made full use of Gaussian interaction profile based on miRNA-disease associations proposed Within and Between Score for MiRNA-Disease Association prediction (WBSMDA). The advantage of WBSMDA is that it could detect those who do not have any knowledge or information of miRNAs and new diseases. Besides, lots of models based on machine learning have also appeared. Xu et al. [4] constructed a model based on MiRNA–Target Dysregulated Network (MTDN) to rank miRNAs in a specific phenotype disease. But due to lack of the type of negative sample training, the validity of this method was also affected. A Heterogeneous Graph Inference model for MiRNA-Disease Association prediction (HGIMDA) [5] achieved better performance.

In this paper, we introduce the model of Improved Inductive Matrix Completion Method for Predicting unknown microRNA-disease associations (IIMCMP). Concretely, the disease semantic similarity and the functional similarity of miRNA, along with the improved known disease-miRNA association, are regarded as inputs to an inductive matrix completion method to rank all disease-miRNA associations pairs. The predictive results show that the IIMCMP model based on K-nearest neighbor interaction profiles of each disease and each miRNA achieves an AUC of 0.9016 in the leave-one-out cross validation (LOOCV). In addition, analysis for two kinds of important diseases cases is implemented, the top 30 candidates for Esophageal neoplasm and the top 50 candidates for breast neoplasms by IIMCMP are extracted.

# 2 Materials and Methods

## 2.1 Human MiRNA–Disease Associations

In published predictive methods, a certain number of miRNA-disease associations pairs have been found by the biological test. In this paper, we get 5430 verified known human miRNA-disease associations dataset involved in human's 495 miRNAs and 383 diseases from the HMDD V2.0 database by screening redundant data. In our following predictive process, including the leave-one-out cross validation (LOOCV), this dataset is regarded as the human gold standard data set. As in previous predictive method, an adjacency matrix $Y$ is composed of two numbers 0 and 1. The element $Y_{ij}$ is 1, which means miRNA j is related to disease i. The remaining 0 is unknown link, which need to be detected.

## 2.2 MiRNA Functional Similarity

Through careful studying the existing predictive models, we find that the functional similarity scores of miRNAs developed by Wang et al. [6] are more accurately. we download them from http://www.cuilab.cn/files/images/cuilab/misim.zip. Therefore, we formulate the matrix FMS to denote the miRNA functional similarity. The element $FMS_{ij}$ of matrix FMS means the score of functional similarity between the miRNA $m_i$ and $m_j$.

## 2.3    Disease Semantic Similarity Measurement

According to the literature of Chen et al. [7], each disease can be represented by a Directed Acyclic Graph (DAG). The associations between different diseases were represented by the bridge between the DAGs. For a special disease $d_i$, $DAG(d_i) = DAG(d_i, T(d_i), E(d_i,))$ is calculated, in which $T(d_i)$ denotes the node set and $E(d_i)$ is the edge set. $T(d_i)$ is composed of node D itself and its ancestor nodes, $E(d_i)$ is comprised of the direct edges from parent nodes to child nodes. Each disease according to its own and ancestors finally obtain the semantic score, and each disease semantic computation formula is as follows:

$$\begin{cases} D1_D(d) = 1 & if\ d = D \\ D1_D(d) = \max\{\Delta * D1_D(d')|d' \in children\ of\ d\} & if\ d \neq D \end{cases} \quad (1)$$

where $\Delta$ denotes the semantic contribution decay factor.

Finally, the following formula describes the semantic score of disease D:

$$DV1(D) = \sum\nolimits_{d \in T(D)} D1_D(d) \quad (2)$$

where the disease itself and the ancestor nodes are expressed in T(D). According to our analysis, we conclude that the larger the DAG shared, the more similar they are. So, the sematic similarity between $d_i$ and $d_j$ is defined as follows:

$$SS1(d_i, d_j) = \frac{\sum_{t \in T\left(d_i \cap d_j\right)}\left(D1_{d_i}(t) + D1_{d_j}(t)\right)}{DV1(d_i) + DV1(d_j)} \quad (3)$$

## 2.4    Integrated Disease Semantic Similarity

We think that regarding directly the disease semantic similarity measurement of SS1 as the similarity of the diseases is not enough. There is such a situation that $d_i$ and $d_j$ were on the same layer of DAG(d) and the numbers of diseases DAGs are different, so we consider that the semantic value contribution of $d_i$ and $d_j$ should have a greater difference. According to the above description, disease semantic similarity measurement of SS2 [8] can be constructed. The following equation describes semantic contribution of disease t to D in DAGs.

$$D2_D(t) = -log\left[\frac{the\ number\ of\ DAGs\ including\ t}{the\ number\ of\ diseases}\right] \quad (4)$$

We recount the semantic similarity between $d_i$ and $d_j$ based on the following equation:

$$SS2(d_i, d_j) = \frac{\sum_{t \in T(d_i \cap d_j)} \left( D2_{d_i}(t) + D2_{d_j}(t) \right)}{DV2(d_i) + DV2(d_j)} \qquad (5)$$

Where $DV2(D) = \sum_{d \in T(D)} D2_D(d)$. Finally, we get the disease semantic similarity matrix represented by SS2. In the end, we obtain the result of the average of the SS1 and SS2 as the final disease semantic similarity matrix. Here, the similarity for disease pairs denoted by $S_d(d(i), d(j))$ was as follows:

$$S_d(d(i), d(j)) = \begin{cases} \frac{SS1(d_i, d_j) + SS2(d_i, d_j)}{2}, & d_i \text{ and } d_j \text{ have semantic similarity} \\ 0, & \text{oherwise} \end{cases} \qquad (6)$$

### 2.5   Improved MiRNA-Disease Associations

An adjacent matrix $Y \in R^{m \times n}$ represented the all miRNA-disease associations pairs, which has m miRNAs rows and n diseases columns. If miRNA $m_i$ and disease $d_j$ was linked through the biological experiment, we set $Y_{ij} = 1$ and 0 otherwise. In the adjacent matrix Y, 0 represents the horizontal direction disease j and vertical direction miRNA i may be relevant or irrelevant. There are too many 0 in the adjacent matrix Y that may affect performance. Therefore, replacing 0 in matrix Y by the continuous value between 0 and 1 should be helpful. We adopt an innovative technology named WKNKN (weighed K nearest known neighbors) [9] to implement this strategy. Before the introduction of WKNKN steps, there are two definitions should be denoted. We adopt the vector $Y(m_i)$ to denote the interaction profiles of $m_i$, which is the $i^{th}$ row vector of matrix Y. Similarly, we used the vector $Y(d_j)$ to denote the interaction $d_j$, which is the $j^{th}$ column vector of matrix Y. For 0 of Y matrix, their value is likely to change. The three steps of WKNKN are as follows. Step 1: Change in the direction of the abscissa Y matrix: For the interaction likelihood profile of $m_i$, we adopt K known miRNAs nearest to it in the functional similarity matrix with their interaction profiles to get its values. Step 2: Change in the direction of the ordinate Y matrix: For the interaction likelihood profile of $d_j$, we utilize K known diseases nearest to it in the semantic similarity matrix with their interaction profiles to obtain its values. Step 3: If the corresponding value is 0 in Y matrix, an average of the sum of two interaction likelihood profiles values is taken to replace 0.

## 3   IIMCMP

To solve some limitations, we develop an improved calculative method based on inductive matrix completion to predict disease related miRNAs (IIMCMP) in this paper. We propose IIMCMP model to predict disease-related miRNAs which is carried out by three inputs. One is miRNA functional similarity matrix, the other one is disease

integrated similarity matrix, the third one is improved known miRNA-disease associations by WKNKN. Figure 1 demonstrate the process of IIMCMP.

For our method, data integration and similarity calculation are the first step. Matrix
$Y \in R^{m \times n}$ means the adjacent matrix of the human miRNA-disease association. $S_d \in R^{n \times n}$ denotes the integrated disease similarity matrix. $S_m \in R^{m \times m}$ represents the
miRNA similarity matrix. We received a total of 5430 miRNA-disease association
pairs between 383 diseases and 495 miRNAs through the biological field test. Then, let
$S_m \in R^{m \times m}$ delegates the training feature matrix of 495 miRNAs, where the $i^{th}$ row of
adjacent matrix $S_m$ is the miRNA feature vector $m_i$, and let $S_d \in R^{n \times n}$ demonstrates the
training feature matrix of 383 diseases, where the $j^{th}$ row of adjacent matrix $S_d$ is the
disease feature vector $d_j$. Then we use the observed entries in matrix Y to uncover
completely a low-rank matrix Z. Which can be formed as $\mathbf{Z} = WH^T$, where $W \in R^{n \times r}$
and $H \in R^{m \times r}$, and r with small value. The matrix score of Z is calculated by the entry
score $(d_j, m_i)$ which represents the predicted score between $d_j$ and $m_i$. The factors
$W$ and $H$ are obtained as solutions of the following optimization problem:

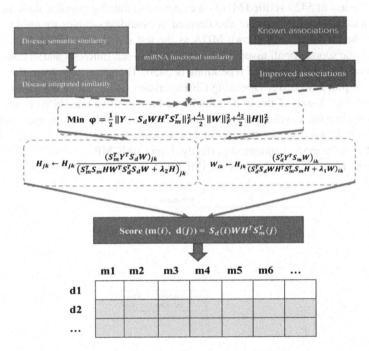

**Fig. 1.** Flowchart of IIMCMP prediction model to uncover the potential miRNA-disease
association.

$$\text{Min } \phi = \tfrac{1}{2}\left\|Y - S_d WH^T S_m^T\right\|_F^2 + \tfrac{\lambda_1}{2}\|W\|_F^2 + \tfrac{\lambda_2}{2}\|H\|_F^2 \tag{7}$$
$$\text{Such that, } W \geq 0, H \geq 0.$$

Where the regularization parameter $\lambda_1, \lambda_2$ trade off accrued losses on known entities
and the trace-norm constraint. To prevent over-fitting problem, we add $\tfrac{\lambda_1}{2}\|W\|_F^2$ and

$\frac{\lambda_2}{2}\|H\|_F^2$. In addition, we used a method developed by Jain et al. (Jain and Dhillon 2013) [10] to handle the minimum problem. The algorithm steps in detail of the method are given in Fig. 1. In the end, we can obtain the predicted scores of all miRNA-disease associations pairs through the following equation:

$$\text{Score}\,(\mathbf{d}(i),\,\mathbf{m}(j)) \;=\; S_d(i)\boldsymbol{W}\boldsymbol{H}^T S_m^T(j) \tag{8}$$

# 4   Result

## 4.1   Performance Evaluation

In this paper, we implement LOOCV to demonstrate the predictive accuracy of IIMCMP based on HMDD v2.0 and compared it with the method IMCMDA (See Fig. 2). In LOOCV, one of the 5430 miRNA-disease associations (MDAs) is regarded as test sample; the rest of 5429 verified MDAs are deemed as training samples; those unlabeled 184155 miRNA-disease pairs are also deemed as candidate samples we need to rank its scores. When we take each known MDA as the test sample in turn, by implementing IIMCMP, the scores for all association matrix Y between miRNAs and diseases would be returned. So, the number of repetitions is 5430. Therefore, we obtain the result of LOOCV to plot A Receiver Operating Characteristics (ROC) curve by setting different thresholds. The X-axis of the ROC namely the false positive rate (FPR) is also named specificity while the Y-axis namely the true positive rate (TPR) is also named sensitivity. If AUC = 0.5, we consider the model has a random performance, while AUC = 1 demonstrates the model implement perfectly. Using IIMCMP, we obtained the AUC of 0.9016 (see Fig. 2), while the AUC of IMCMDA we implemented is 0.8194.

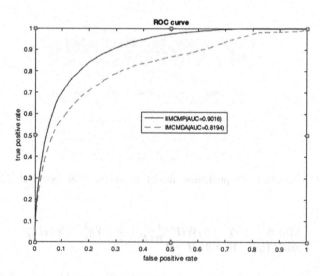

**Fig. 2.** The comparison result between IIMCMP and other previous model of IMCMDA in terms of LOOCV.

Furthermore, Precision-Recall (PR) curve is plotted according to the result of LOOCV in Fig. 3. As showed in the PR curve, obviously, our method is superior to IMCMDA. In the experiment, we use the blue curve to represent our method, and the yellow curve to denote IMCMDA.

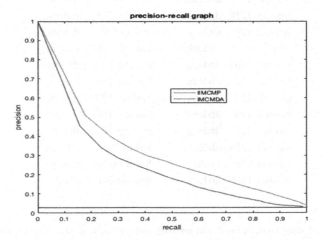

**Fig. 3.** Performance comparison between IIMCMP and IMCMDA in terms of PR curve.

## 4.2 Case Studies

Breast neoplasms are composed of many types of breast neoplasm cells, but breast neoplasms tumor cells with the ability to form new tumors are only a minority [11]. Here, IIMCMP is implemented to obtain the total ranking of the unidentified miRNA-disease pairs (See Table 1).

**Table 1.** The top 50 miRNA candidates for breast neoplasm predicted by IIMCMP in HMDD v2.0.

| miRNA | Evidence | miRNA | Evidence |
|---|---|---|---|
| has-mir-21 | dbdemc | has-mir-29b | dbdemc |
| has-mir-17 | dbdemc | has-mir-16 | dbdemc |
| has-mir-155 | dbdemc | has-let-7i | dbdemc |
| has-mir-20a | dbdemc | has-mir-210 | dbdemc |
| has-mir-19b | dbdemc | has-mir-9 | dbdemc |
| has-let-7a | miR2Disease | has-mir-106b | dbdemc |
| has-let-7e | dbdemc | has-mir-1 | dbdemc |
| has-mir-18a | dbdemc | has-mir-181b | dbdemc |
| has-mir-145e | dbdemc | has-mir-143 | dbdemc |
| has-mir-19a | dbdemc | has-mir-200b | dbdemc |
| has-let-7b | miR2Disease | has-mir-141 | dbdemc |

*(continued)*

254 X. Ding et al.

**Table 1.** (*continued*)

| miRNA | Evidence | miRNA | Evidence |
|---|---|---|---|
| has-mir-223 | dbdemc | has-mir-214 | dbdemc |
| has-let-7d | dbdemc | has-mir-127 | dbdemc |
| has-let-7c | dbdemc | has-mir-222 | dbdemc |
| has-mir-125b | dbdemc | has-mir-205 | dbdemc |
| has-mir-126 | dbdemc | has-mir-30c | dbdemc |
| has-let-7f | dbdemc | has-mir-92b | dbdemc |
| has-mir-146a | dbdemc | has-mir-24 | dbdemc |
| has-mir-34a | dbdemc | has-mir-29a | dbdemc |
| has-mir-132 | dbdemc | has-mir-150 | dbdemc |
| has-mir-199a | dbdemc | has-mir-106a | dbdemc |
| has-mir-221 | dbdemc | has-mir-101 | dbdemc |
| has-mir-125a | dbdemc | has-let-7g | dbdemc |
| has-mir-92a | hmdd | has-mir-15a | dbdemc |
| has-mir-146b | dbdemc | has-mir-34b | hmdd |

Esophageal neoplasm is fatal and terrible cancer but few people pay attention to the disease [12]. The predictive result of the top 30 esophageal neoplasm-associated miRNAs predicted by IIMCMP can be found listed (See Table 2).

**Table 2.** The selected top 30 predicted miRNAs linked with Esophageal neoplasm by sorting the association scores calculated by IIMCMP.

| miRNA | Evidence | miRNA | Evidence |
|---|---|---|---|
| has-mir-145 | dbdemc | has-let-7f | unconfirmed |
| has-mir-17 | dbdemc | has-let-7i | dbdemc |
| has-mir-18a | bdemc | has-mir-125a | dbdemc |
| has-mir-143 | dbdemc | has-mir-146a | dbdemc |
| has-let-7e | dbdemc | has-mir-141 | miR2Disease |
| has-mir-19a | dbdemc | has-mir-200b | dbdemc |
| has-let-7b | dbdemc | has-mir-92a | dbdemc |
| has-mir-34a | dbdemc | has-mir-221 | dbdemc |
| has-mir-223 | dbdemc | has-mir-210 | dbdemc |
| has-let-7d | dbdemc | has-mir-106a | dbdemc |
| has-mir-126 | dbdemc | has-mir-101 | dbdemc |
| has-let-7c | dbdemc | has-mir-29b | dbdemc |
| has-mir-132 | dbdemc | has-mir-214 | dbdemc |
| has-mir-125b | dbdemc | has-mir-16 | dbdemc |
| has-mir-199a | dbdemc | has-mir-146b | dbdemc |

# 5  Conclusion and Future Work

In this paper, to uncover the potential pairs between miRNAs and diseases, we demonstrate an improved computational method based on inductive matrix completion method. The prediction AUC of IIMCMP is 0.9016 based on LOOCV. Furthermore, we carry out two case studies on breast neoplasm and Esophageal neoplasm to evaluate the predictive performance of the IIMCMP. We think detecting the new miRNA records related disease will help to better understanding disease. Dynamic prediction of miRNA and disease relationship records are our next step plan. As the future work, other sources of information such as family clusters and miRNAs sequences should be added to our method to further improve predictive performance.

**Acknowledgement.** This work was supported by grants from the National Natural Science Foundation of China (No. 61873001), the Key Project of Anhui Provincial Education Department (No. KJ2017ZD01), and the Natural Science Foundation of Anhui Province (No. 1808085QF209).

# References

1. Ribeiro, A.O., Schoof, C.R., Izzotti, A., Pereira, L.V., Vasques, L.R.: MicroRNAs: modulators of cell identity, and their applications in tissue engineering. MicroRNA **3**(1), 45–53 (2014)
2. Lee, R.C., Feinbaum, R.L., Ambros, V.: The C. elegans heterochronic gene lin-4 encodes small RNAs with antisense complementarity to lin-14. Cell **75**(5), 843–854 (1993)
3. Chen, X., Yan, C.C., Zhang, X.: WBSMDA: within and between score for MiRNA-disease association prediction. Sci Rep. **6**(1), 21106 (2016)
4. Xu, J., et al.: Prioritizing candidate disease MiRNAs by topological features in the MiRNA target dysregulated network: case study of prostate cancer. Mol. Cancer Ther. **10**(10), 1857–1866 (2011)
5. Chen, X., Yan, C.C., Zhang, X., You, Z.H., Huang, Y.A., Yan, G.Y.: HGIMDA: heterogeneous graph inference for miRNA-disease association prediction. Oncotarget **7**(40), 65257–65269 (2016)
6. Wang, D., Wang, J.A.: Inferring the human microRNA functional similarity and functional network based on microRNA-associated diseases. Bioinformatics **26**(13), 1644–1650 (2010)
7. Chen, X., You, Z.H., Yan, G.Y., Gong, D.W.: IRWRLDA: improved random walk with restart for lncRNA-disease association prediction. Oncotarget. **7**(36), 57919–57931 (2016)
8. Chen, X., Liu, M.X., Yan, G.Y.: RWRMDA: predicting novel human microRNA-disease associations. Mol. BioSyst. **8**(10), 2792–2798 (2012)
9. Ezzat, A., Zhao, P., Wu, M., Li, X., Kwoh, C.K.: Drug-target interaction prediction with graph regularized matrix factorization. IEEE/ACM Trans. Comput. Biol. Bioinform. **11**(99), 646–656 (2017)
10. Jain, P., Dhillon, I.S.: Provable inductive matrix completion. Computer Science (2013)
11. Linehan, W.M., Grubb, R.L., Coleman, J.A., Zbar, B., Walther, M.C.M.: The genetic basis of cancer of kidney cancer: implications for gene-specific clinical management. BJU Int. **95**(s2), 2–7 (2005)
12. Siegel, R., Naishadham, D., Jemal, A.: Cancer statistics, 2012. CA Cancer J. Clin. **62**(1), 10–29 (2012)

# A Link and Weight-Based Ensemble Clustering for Patient Stratification

Yuan-Yuan Zhang[1], Chao Yang[1],
Jing Wang[1,2], and Chun-Hou Zheng[1(✉)]

[1] College of Computer Science and Technology, Anhui University, Hefei, China
zhengch99@126.com
[2] School of Computer and Information Engineering, Fuyang Normal University,
Fuyang, China

**Abstract.** Owing to its capability to combine multiple base clustering into a single robust consensus clustering, the ensemble clustering technique has attracted increasing attention over recent years. Although many successful clustering methods have been proposed, there is still room for improvement in the existing approaches. In this paper, we propose a novel ensemble clustering approach called a link and weight-based ensemble clustering (LWEC). We first generate a large number of similarity-indicators based on a scaled exponential similarity kernel. Then based on the similarity-indicators, an ensemble of diversified base clusterings is constructed. Further, we reckon how difficult it is to cluster an object by constructing the co-association matrix of the base clustering. And we regard related information as weights of objects. Experimental results on 35 high-dimensional cancer gene expression benchmark datasets and TCGA datasets demonstrate the efficiency and superiority of our approach.

**Keywords:** Ensemble clustering · Consensus clustering ·
Diversified similarity-parameter pairs generation · Weighted objects

## 1 Introduction

Clustering, an unsupervised machine learning analysis, which partition data into meaningful groups, has been increasing used for disease subtyping or patient stratification [1–3]. However, there is no single clustering algorithm that can handle all types of cluster shapes and structures presented in data. Each clustering algorithm has its own merits as well as its demerits [4–6]. We can often find that different clustering method or even the same method with different parameters, may bring about different patterns when applied to a given data set. Therefore, it is quite hard for user to determine which method would be the suitable one for a given clustering problem without the validation of ground truth.

Inspired by these works [2, 10, 11], in this paper, we propose a novel ensemble clustering approach based on link and weighted-objects (LWEC) for patient stratification. Our key idea is to translate the zero entries in the co-association matrix with the distance derived from the weighted objects, we implemented experiments on 35 high-dimensional benchmark datasets and 6 cancer types with three molecular data types

D.-S. Huang et al. (Eds.): ICIC 2019, LNCS 11644, pp. 256–264, 2019.
https://doi.org/10.1007/978-3-030-26969-2_24

from The Cancer Genome Atlas (TCGA). Extensive experimental results demonstrated the efficiency and robustness of our method.

## 2  Method

The overall process of our method is illustrated in Fig. 1. It includes following major steps: First, we exploit the random subspaces and the similarity-indicators which are introduced in [11] to create an ensemble of diversified base clusters. Based on weighted-object ensemble clustering method [10], we then reckon how hard it is to cluster an object by constructing the co-association matrix that summarizes the basic partitions, and we regard the corresponding information as weights related to objects. Next, in the light of the link-based cluster ensemble (LCE) method [2], we transform the zeros entries in the co-association matrix with the distance derived from the weighted objects. Finally, we employ a graph-based partitioning method to achieve the final clustering.

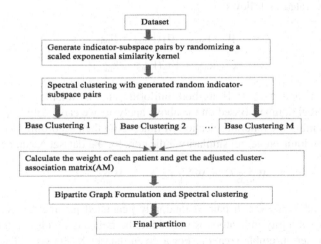

**Fig. 1.** Flow diagram of the proposed LWEC method

### 2.1  Creating a Cluster Ensemble

In ensemble clustering, the diversity among base clusterings plays a crucial role during the consensus process. Similarly to the multi-diversified ensemble clustering (MDEC) approach [11], in this paper, we combine considerable varied similarity-indicators with the random subspaces to generate the diversified base clusterings.

Consider $X = \{x_1, \ldots, x_N\}$ be a set of N samples and $\prod = \{\pi^1, \ldots, \pi^M\}$ be a cluster ensemble with M basic partitions. The similarity between $x_i$ and $x_j$ can be defined as $W(i, j)$. So the SES kernel function for samples $x_i$ and $x_j$ can be denoted as:

$$W^X(i,j) = \exp\left(-\frac{d(x_i, x_j)}{\mu \varepsilon_{ij}}\right) \tag{1}$$

Where $\mu$ is a hyperparameter, $\varepsilon_{ij}$ is used for eliminating the scaling problem, and $d(x_i, x_j)$ represents the Euclidean distance between $x_i$ and $x_j$. Then, $\varepsilon_{ij}$ can be computed as:

$$\varepsilon_{ij} = \frac{\text{mean}(d(x_i, N_{ki})) + \text{mean}(d(x_j, N_{kj})) + d(x_i, x_j)}{3} \tag{2}$$

where $\text{mean}(d(x_i, N_{ki}))$ is the average distance between $x_i$ and its k nearest neighbors. We can note that there are two free parameters in the SES kernel function, i.e., the number of nearest neighbors k and the hyperparameter $\mu$. In this way, we not only consider the neighborhood structure between data samples, but also realize the generation of multiple similarity-indicators through perturbing the two parameters. As the method proposed in [11], the two parameters ($\mu \in [\mu_{min}, \mu_{min}$ and $k \in [k_{min}, k_{max}])$ are randomly generated as follows:

$$\mu = \mu_{min} + \sigma_1(\mu_{max} - \mu_{min}) \tag{3}$$

$$k = k_{min} + \lfloor\sigma_2(k_{max} - k_{min})\rfloor \tag{4}$$

here $\sigma_1 \in [0, 1]$ and $\sigma_2 \in [0, 1]$ are both uniform random variables, and $\lfloor\rfloor$ represents the floor of a real number. Based on the above analysis, we can create M pairs of u and k by implementing the random selection M times. And then according to the generated $\mu$ and k, M randomized kernel similarity indicators for the dataset X can be denoted as:

$$W^X_{\mu_1,k_1}(\cdot,\cdot), W^X_{\mu_2,k_2}(\cdot,\cdot), \cdots, W^X_{\mu_M,k_M}(\cdot,\cdot), \tag{5}$$

Here $\mu_i$ and $k_i$ are the-th pair of randomly generated parameters. Next, with the generation of similarity-indicators, we combine the indicators with random subspaces for jointly cluster ensemble creation. For a given dataset X, the set of features can be denoted as $F = \{f_1, \cdots, f_D\}$, where $f_i$ represents the i-th feature. Thus yield M random indicator-subspace pairs as follows:

$$W^{X_1}_{\mu_1,k_1}(\cdot,\cdot), W^{X_2}_{\mu_2,k_2}(\cdot,\cdot), \cdots, W^{X_M}_{\mu_M,k_M}(\cdot,\cdot), \tag{6}$$

we first map $x_i$ and $x_j$ to the subspace related to the newly formed $X_m$, Therefore, based on M indicator-subspace pairs, we can get M similarity matrices as follow:

$$S = \left\{S^{(1)}, S^{(2)}, \cdots, S^{(M)}\right\}, \tag{7}$$

Here on the basic of the m-th indicator-subspace pair $W^{X_m}_{\mu_m,k_m}(\cdot,\cdot)$, the m-th similarity matrix (i.e., $S^{(m)}$) is derived, which can be denoted as:

$$S^{(m)} = \left\{ S_{ij}^{(m)} \right\}_{N*N} \tag{8}$$

In the light of the M varied similarity matrices in S, we can create M ensemble basic clusters as follows:

$$\Pi = \left\{ \pi^1, \ldots, \pi^M \right\} \tag{9}$$

In this way, we can obtain a binary matrix with $K_m$ columns from each basic clustering, where $K_m$ is the total number of clusters for each basic clustering, and there is only one element 1 in each row of the matrix, the rest are 0. At last we connect all the binary matrices together to form a large binary matrix. Take an ensemble of two base clusters $\Pi = \{\pi_1, \pi_2\}$ and six samples $(x_1, \cdots, x_6)$ as an example, as shown in Fig. 2a, the corresponding binary matrix is illustrated in Fig. 2b.

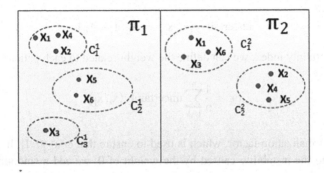

(a) Cluster ensemble

|     | $C_1^1$ | $C_2^1$ | $C_3^1$ | $C_1^2$ | $C_2^2$ |
|-----|---------|---------|---------|---------|---------|
| $x_1$ | 1 | 0 | 0 | 1 | 0 |
| $x_2$ | 1 | 0 | 0 | 0 | 1 |
| $x_3$ | 0 | 0 | 1 | 1 | 0 |
| $x_4$ | 1 | 0 | 0 | 0 | 1 |
| $x_5$ | 0 | 1 | 0 | 0 | 1 |
| $x_6$ | 0 | 1 | 0 | 1 | 0 |

(b) Binary matrix

**Fig. 2.** An example of (a) cluster ensemble of six samples $\{x_1 \cdots x_6\}$ that contains two base clustering ($\pi_1 = \{C_1^1, C_2^1, C_3^1\}$ and $\pi_2 = \{C_1^2, C_2^2\}$ (b) the corresponding binary matrix

## 2.2    Assigning Weights to Patients

After obtained the ensemble clustering $\Pi$, we then assign weights to patients. Similar to Weight-Object Ensemble Clustering (WOEC) algorithm [12]. And during the evaluation process, we assign larger weights to patients that are hard to be clustered, while those easy-to-cluster patients will be set smaller weights. The details are as follows: Suppose A is the $n \times n$ co-association matrix constructed from the basic clustering results, it can be computed as:

$$A_{ij} = \frac{H_{ij}}{R} \tag{11}$$

here $H_{ij}$ represents the number of times that samples $x_i$ and $x_j$ are grouped together in the same cluster, and $R = |M|$ is the size of ensemble. It's apparent that $A_{ij} \in [0, 1]$, and for any i from 1 to n, we set $A_{ii}$ to 1 by default. Therefore, the uncertainty of clustering two samples $x_i$ and $x_j$ can be computed as follows

$$\text{uncertainty}(x_i, x_j) = A_{ij}(1 - A_{ij}) \tag{12}$$

with the uncertainty index, we can define the weight related to each patients as follows:

$$\omega_i' = \frac{4}{n} \sum_{j=1}^{n} \text{uncertainty}(x_i, x_j) \tag{13}$$

here $\frac{4}{n}$ is a normalization factor, which is used to ensure that $\omega_i' \in [0, 1]$. In addition, in order to avoid the instability caused by the weight of 0, we add a smoothing term $\varepsilon$:

$$\omega_i = \frac{\omega_i' + \varepsilon}{1 + \varepsilon} \tag{14}$$

## 2.3    Generating an Adjusted Cluster-Association Matrix (AM)

Accordingly, the edge $R_{xy} \in R$ that connects clusters $C_x, C_y \in V$ can be calculated as:

$$R_{xy} = \frac{\sum_{x_k \in C_x \cap C_y} \omega_k}{\sum_{x_k \in C_x \cup C_y} \omega_k} \tag{15}$$

Therefore, the total RCT measure of clusters $C_x, C_y \in V$ related to all triples $(1 \dots z)$ can be estimated as:

$$RCT_{xy} = \sum_{p=1}^{z} RCT_{xy}^p \tag{16}$$

where $RCT_{xy}^p$ can be calculated as follows:

$$RCT_{xy}^p = \min\left(R_{xp}, R_{yp}\right) \tag{17}$$

Here $R_{xp}$, $R_{yp}$ are correlation between clusters $C_x$ and $C_p$, and $C_y$ and $C_p$, respectively. Next, the similarity between clusters $C_x$ and $C_y$ can be computed as:

$$\text{sim}\left(C_x, C_y\right) = \frac{RCT_{xy}}{RCT_{max}} * C \tag{18}$$

After obtaining the relations between clusters within the same partition, we can transform the original binary matrix into a new adjusted matrix (AM), the value of the element in the AM matrix $AM(x_i, cl) \in [0, 1]$, where $x_i \in X$ and $cl \in \left\{C_1^m, \cdots C_{k_m}^m\right\}$ is defined as follows:

$$AM(x_i, cl) = \begin{cases} 1 & \text{if } cl = C_*^m(x_i) \\ \text{sim}\left(cl, C_*^m(x_i)\right) & \text{otherwise} \end{cases} \tag{19}$$

Having obtained the adjusted cluster-association matrix AM, we then transform the matrix into a weighted bipartite graph, Given this graph G, we divide it into K disjoint subsets by performing the spectral clustering algorithm [15].

## 3 Results

### 3.1 Datasets and Experimental Setting

In this work, we employ 35 widely used cancer gene expression benchmark data sets to evaluate the performance of LWEC.

Besides the 35 benchmark datasets, 6 cancer types with three molecular data types from The Cancer Genome Atlas (TCGA) with survival information (available at the TCGA website https://cancergenome.nih.gov) are used for practical application evaluation.

In our experiments, the two kernel parameters $\mu$ and $k$ are randomly generated in the ranges of $[0.2, 0.8]$ and $\left[\sqrt{N}, 4\sqrt{N}\right]$, respectively. And to generate the basic partition, we set the ensemble size M to 100 and the sampling ratio $\theta$ to 0.6. In addition, the number of clusters in each basic partition is randomly selected from 2 to $\sqrt{N}$.

### 3.2 Evaluation Indicators

Since the real labels of benchmark datasets are available, we apply two widely used evaluation measures (i.e., normalized mutual information NMI [10] and adjusted rand index ARI) to objectively evaluate the quality of different clustering method. Therefore, the NMI score between $\pi'$ and $\pi^G$ can be calculated as follows:

$$\text{NMI}(\pi', \pi^G) = \frac{\sum_{i=1}^{n'} \sum_{j=1}^{n^G} n_{ij} \log \frac{n_{ij} n}{n'_i n_j^G}}{\sqrt{\sum_{i=1}^{n'} n'_i \log \frac{n'_i}{n} \sum_{j=1}^{n^G} n_j^G \log \frac{n_j^G}{n}}} \qquad (20)$$

The rand index (RI) is an initial form of the ARI. The ARI measures the similarity between two clusters by considering the number of common samples in a statistical way, which is computed as follows:

$$\text{ARI}(\pi', \pi^G) = \frac{2(N_{00}N_{11} - N_{01}N_{10})}{(N_{00} + N_{01})(N_{01} + N_{11}) + (N_{00} + N_{10})(N_{10} + N_{11})} \qquad (21)$$

We can note that better clusters leads to greater values of NMI and ARI. And there may be some negative values in the final ARI results, which means that partition result is even worse than random label allocation.

### 3.3    Results on Benchmark Cancer Gene Expression Data and Six Cancer Type from TCGA

We then evaluated LWEC and other clustering methods, namely, Agglomerative Hierarchical Clustering with Complete-Linkage (CL), Spectral Clustering (SC) [25], Consensus Clustering (CC) [10], the Link-based Cluster Ensemble (LCE) [2], the Entropy-based Consensus Clustering (ECC) [7] on 35 widely used benchmark cancer gene expression datasets. In order to achieve a fair comparison, we run each of the proposed algorithm for 50 times and then consider the average as the final result. Figure 3 shows the clustering performance of different methods measured by ARI and NMI respectively.

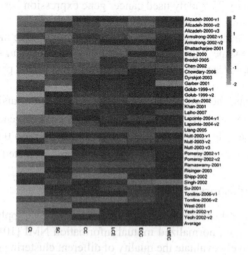

**Fig. 3.** The performance of different clustering algorithm measured by the normalized mutual information (NMI)

To test LWEA on real-world data sets, we applied it to six cancer type, with three molecular data types (mRNA expression, DNA methylation and miRNA expression) from TCGA project. Here we took advantage of survival analysis to evaluate the performance of different clustering algorithm in terms of $[-\log]\_10P$ with P the log-rank test P value. For each molecular data type, we compared the clustering performance of LWEC against other three approaches. We found that LWEC was superior to the other methods in identifying subtypes with significantly different survival status (see Fig. 4).

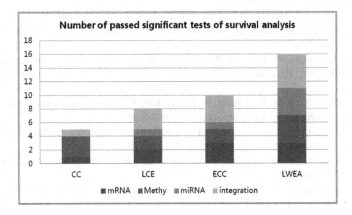

**Fig. 4.** The number of times that each clustering methods passes the significant tests of survival analysis

## 4   Conclusion

In this paper, we have proposed a novel ensemble clustering method (LWEC) based on the link of cluster and weighted-objects. We generate a large number of diversified indicators to solve high-dimensional data problems by randomizing a scaled exponential similarity kernel. Moreover, we also consider the weights of patients during the clustering process, which makes the clustering results more comprehensive. To sum up, the LWEC method provides a new way to integrate the results of basic partitions into the overall clustering process. We hope that LWEC can discover more meaningful cancer subtypes to promote the development.

**Acknowledgments.** This work was supported by grants from the National Natural Science Foundation of China (No. 61873001), the Key Project of Anhui Provincial Education Department (No. KJ2017ZD01), and the Natural Science Foundation of Anhui Province (No. 1808085QF209).

# References

1. Wang, T.: CA-Tree: a hierarchical structure for efficient and scalable coassociation-based cluster ensembles. IEEE Trans. Syst. Man Cybern. B Cybern. **41**(3), 686–698 (2011)
2. Iam-On, N., Boongoen, T., Garrett, S.: LCE: a link-based cluster ensemble method for improved gene expression data analysis. Bioinformatics **26**(12), 1513–1519 (2010)
3. Jain, A.K., Murty, M.N., Flynn, P.J.: Data clustering: a review. ACM Comput. Surv. **31**(3), 264–323 (1999)
4. Deluche, E., Onesti, E., Andre, F.: Precision medicine for metastatic breast cancer. Nat. Rev. Clin. Oncol. **12**(12), e2 (2015)
5. Lapointe, J., Li, C., Higgins, J.P., Van, d.R.M., Bair, E., Montgomery, K.: Gene expression profiling identifies clinically relevant subtypes of prostate cancer. Proc. Nat. Acad. Sci. **101**(3), 811–816 (2004)
6. Nguyen, T., Tagett, R., Diaz, D., Draghici, S.: A novel approach for data integration and disease subtyping. Genome Res. **27**, 2025–2039 (2017). https://doi.org/10.1101/gr.215129.116
7. Liu, H., Zhao, R., Fang, H., Cheng, F., Fu, Y., Liu, Y.: Entropy-based consensus clustering for patient stratification. Bioinformatics **33**, 2691–2698 (2017)
8. Huang, D., Lai, J., Wang, C.: Robust ensemble clustering using probability trajectories. IEEE Trans. Knowl. Data Eng. **28**(5), 1312–1326 (2016)
9. Liu, H., Ming, S., Sheng, L., Yun, F.: Infinite ensemble clustering. Data Min. Knowl. Disc. **32**(1), 1–32 (2017)
10. Ren, Y., Domeniconi, C., Zhang, G., Yu, G.: Weighted-object ensemble clustering. Knowl. Inf. Syst. **51**(2), 1–29 (2013)
11. Huang, D., Wang, C.D., Lai, J.H., Kwoh, C.K.: From subspaces to metrics and beyond: toward multi-diversified ensemble clustering of high-dimensional data (2017)
12. Huang, D., Wang, C., Lai, J.: Locally weighted ensemble clustering. IEEE Trans. Cybern. **48**(5), 1460–1473 (2016)

# HGMDA: HyperGraph for Predicting MiRNA-Disease Association

Qing-Wen Wu, Yu-Tian Wang, Zhen Gao, Ming-Wen Zhang,
Jian-Cheng Ni$^{(\boxtimes)}$, and Chun-Hou Zheng$^{(\boxtimes)}$

School of Software, Qufu Normal University, Qufu, China
nijch@163.com, zhengch99@126.com

**Abstract.** The prediction of potential associations between disease and microRNAs is of core importance for understanding disease etiology and pathogenesis. Many researchers have proposed different computational methods to predict potential associations between microRNAs and diseases. Considering the limitations in previous methods, we developed HyperGraph for MiRNA-Disease Association (HGMDA) to uncover the relationship between diseases and microRNAs. Firstly, the miRNA functional similarity, the disease semantic similarity, and known miRNA–disease associations were used to form an informative feature vector. Then the vector for known associated pairs obtained from the HMDD v2.0 database was used to construct hypergraph. Finally, inductive hypergraph learning was used for predicting miRNA-disease associations. Experimental results show that the proposed method is effective for miRNA-disease association predication.

**Keywords:** MiRNA · Disease · MiRNA-Disease Association · HyperGraph

## 1 Introduction

MicroRNAs (miRNAs) are a class of small ($\sim$22 nt) non-coding regulatory RNAs [1]. More and more evidences have confirmed that miRNAs are closely related to many kinds of human diseases. Therefore, identifying disease-related miRNAs is important and beneficial to treat, diagnose, and prevent human complex diseases.

Researchers have already developed some computational methods in predicting miRNA-disease associations. Xu et al. [2] constructed a miRNA-target gene dysregulated network and then applied support vector machine (SVM) classifier to distinguish positive disease miRNAs from negative ones based on the topological properties. However, it fails to predict related miRNAs of new diseases that have no known association. To overcome this limitation, Chen et al. [3] proposed a semi-supervised method named RLSMDA. Compared with previous methods, RLSMDA could identify related miRNAs for diseases without any known associated miRNAs. Xiao et al. [4] proposed the method based on non-negative matrix factorization from the similarity and association perspective of miRNAs and diseases. Jiang et al. [5] proposed the novel improved collaborative filtering-based miRNA-disease association prediction (ICFMDA) approach. ICFMDA defined significance SIG between pairs of diseases or miRNAs to model the preference on the choices of other entities.

© Springer Nature Switzerland AG 2019
D.-S. Huang et al. (Eds.): ICIC 2019, LNCS 11644, pp. 265–271, 2019.
https://doi.org/10.1007/978-3-030-26969-2_25

Although existing methods have made great contributions to uncover disease-related miRNAs, most methods are difficult to extract the deep feature representation and mine internal relationships from the multiple kinds of data. In this article, we propose a novel prediction method based on HyperGraph for MiRNA-Disease Association and refer to it as HGMDA. Using three different measures (graph theoretical measures, statistical measures and matrix factorization measures), we constructed informative feature vector for all the miRNA-disease pairs for hypergraph to predict potential miRNA-disease associations. To demonstrate the effectiveness of our proposed method, we apply global LOOCV and fivefold cross-validation to measure the prediction performance. We compare our method with three previous classical methods and the results indicate that our method could achieve comparable performance. Moreover, the results of case studies further verify the reliability and robustness of HGMDA. Together, all the results demonstrate that HGMDA can serve as an effective tool for discovering miRNA-disease associations.

## 2    Materials and Methods

### 2.1    Human MiRNA-Disease Associations

The information of 5,430 known human miRNA-disease associations between 495 miRNAs and 383 diseases was obtained from the HMDDv2.0 [6]. We used an adjacency matrix $A$ to describe the obtained miRNA-disease associations. Concretely, the element $A(m(i), d(j))$ is equal to 1 if miRNA $m(i)$ is verified to be associated with disease $d(j)$, and 0 otherwise.

### 2.2    MiRNA Functional Similarity

Here we directly downloaded the miRNA functional similarity scores from http://www.cuilab.cn/files/images/cuilab/misim.zip. Then, an adjacency matrix $SM$ was built to represent the similarity of miRNAs, where the element $SM(m(i), m(j))$ denotes the functional similarity score between miRNA $m(i)$ and miRNA $m(j)$.

### 2.3    Disease Semantic Similarity

The association between different diseases can be represented by a directed acyclic graph (DAG) according to the disease classification system in the Mesh (http://www.ncbi.nlm.nih.gov) database, in which nodes represent disease and links represent the association of two diseases. For a given disease $D$, DAG = $(D, T(D), E(D))$, where $T(D)$ represents its ancestor nodes and itself, and $E(D)$ represents all direct edges connecting the parent nodes to child nodes. The contribution values of disease $t$ to the semantic value of disease $d(i)$ can be calculated as follows:

$$D_{d(i)}(t) = -log\left(\frac{the\ number\ of\ DAGs\ including\ t}{the\ number\ of\ diseases}\right) \tag{1}$$

$$DV(d(i)) = \sum_{t \in D(d(i))} D_{d(i)}(t) \tag{2}$$

where $D(d(i))$ was the node set in $DAG(d(i))$ including node $d(i)$ itself. Therefore, the semantic similarity between disease $d(i)$ and $d(j)$ could be defined as follows:

$$SD(d(i), d(j)) = \frac{\sum_{t \in D(d(i)) \cap D(d(j))} \left( D_{d(i)}(t) + D_{d(j)}(t) \right)}{DV(d(i)) + DV(d(j))} \tag{3}$$

## 2.4  HGMDA

The HGMDA model is implemented by integrating matrix $A$, $SM$ and $SD$. Specifically, the implementation involves three parts.

First, feature engineering part. There are three types of vectors constructed during feature engineering. Type 1 features summarizes $A$, $SM$ and $SD$ from a statistical perspective. Type 2 features describes $SM$ and $SD$ using graph theories. Type 3 features focuses on each miRNA-disease pair in the association matrix $A$. Then a composite feature vector $X$ for each miRNA-disease pair is produced by concatenating these three feature types and used to construct hypergraph.

Second, hypergraph construction part. Given the feature factor $X = [x_1, \ldots, x_i, \ldots, x_n]^T \in \mathbb{R}^{n \times c}$, the corresponding labels matrix $Y = [y_1, \ldots, y_i, \ldots, y_l] \in \mathbb{R}^{n \times l}$, where each row $y_i$ labels the positive samples for the $i$-th category as 1, and all other samples as 0 accordingly. A hypergraph $G = (V, E, W)$ is constructed to formulate the relationship among these miRNA-disease pairs. In $V$, each vertex denotes miRNA-disease pair in $X$, and there are $n$ vertices in total. In $E$, the hyperedges are generated based on the pair-wise distances. In this work, we group all the miRNA-disease pairs into clusters by using the $K$-means algorithm, the miRNA-disease pairs in a cluster are connected by the corresponding edge, we set $k$ equal to 45. Finally, there are a total of $N_e$ hyperedges, the number of which is equal to the $k$ value. For $W$, the weight vector for $N_e$ hyperedges, each edge $e$ is given a weight $w(e)$. All the hyperedges are initialized with an equal weight, e.g., $w(e) = 1/n_e$.

The hypergraph $G$ can be denoted by a $|V| \times |E|$ incidence matrix $H$, in which each entry is defined by

$$h(v, e) = \begin{cases} 1 & \text{if } v \in e \\ 0 & \text{if } v \notin e \end{cases} \tag{4}$$

For a vertex $v \in V$, its degree is defined by $d(v) = \sum_{e \in E} w(e) h(v, e)$. For an edge $e \in E$, its degree is defined by $\delta(e) = \sum_{v \in V} h(v, e)$. Accordingly, two diagonal matrices $Dv$ and $De$ can be generated with every entry along the diagonal corresponds to the vertex degree and hyperedge degree, respectively.

Third, inductive hypergraph learning part. In this part, we learn the projection matrix $P$, which map the original feature $X$ to the label matrix $S = XP$. According to Zhang et al. [7] introduction, the cost function $F$ for learning the projection matrix $P$ can be formulated as:

$$F = \{\Omega(P) + \lambda R_{emp}(P) + \mu\Phi(P)\} \tag{5}$$

Specifically, hypergraph Laplacian regularizer $\Omega(P)$ is calculated as

$$\Omega(P) = \frac{1}{2}\sum_{k=1}^{l}\sum_{e\in E}\sum_{u,v\in V}\frac{W(e)H(u,e)H(v,e)}{\delta(e)}\left(\frac{(XP)(u,k)}{\sqrt{d(u)}} - \frac{(XP)(v,k)}{\sqrt{d(v)}}\right)^2 \tag{6}$$
$$= tr(P^T X^T \Delta XP)$$

where $\Delta = I - D_v^{-(1/2)}HWD_e^{-1}H^TD_v^{-(1/2)}$ is a positive semi-definite matrix, and it is usually called the hypergraph Laplacian.

The empirical loss term on $P$ is defined as

$$R_{emp}(P) = ||XP - Y||^2 \tag{7}$$

$\Phi(P)$ is a $l_{2,1}$ norm regularizer to avoid over-fitting for $P$, which is defined as:

$$\Phi(P) = ||P||_{2,1} \tag{8}$$

Consequently, Eq. (5) can be reformed as:

$$\arg\min_P \left\{tr(P^T X^T \Delta XP) + \lambda||XP - Y||^2 + \mu||P||_{2,1}\right\} \tag{9}$$

To solve the optimization task in Eq. (9), we derive $F$ to $P$, the close-form solution can be written as:

$$P = \lambda(X^T \Delta X + \lambda X^T X + \mu R)^{-1}X^T Y \tag{10}$$

where $R$ is a diagonal matrix and the $i$-th diagonal element is defined as

$$R_{i,i} = 1/(2||P(i,:)||_2) \quad i = 1, 2, \ldots\ldots, c \tag{11}$$

We initialize $R$ as an identity matrix here empirically and update $P$ by Eq. (10), then we fix $P$ and update $R$ by Eq. (11). The procedure is repeated until $R$ and $P$ are stable. Based on the learned $P$, the relevance score of the unknown miRNA-disease pair $x^{unk}$ can be obtained by

$$S\left(x^{unk}\right) = x^{unk}P \tag{12}$$

# 3   Experiments and Results

## 3.1   Performance Evaluation

We used global LOOCV and fivefold cross-validation to evaluate the performance of HGMDA. Specifically, global LOOCV selected a known miRNA-disease association in turn as a test sample, and the rest of the associations were considered as training samples. All untested association are used as candidate samples. The ranking was calculated by comparing the score of testing sample with the scores of candidate samples. The testing samples which obtained ranks higher than the given threshold were considered as successful predictions. To compare HGMDA's performance with previous models, we selected three classical computational methods: ICFMDA [5], RLSMDA [3], and SACMDA [8] to compete with HGMDA in cross-validation. Figure 1 shows the global LOOCV ROC curves for HGMDA and other methods. HGMDA, ICFMDA, RLSMDA and SACMDA obtained AUCs of 0.9073, 0.9067, 0.8426 and 0.8770, respectively.

As for fivefold cross-validation, all the known miRNA–disease associations were randomly divided into five subsets, four of which were used as known information to predict potential candidates, while the omitted subset was added to the testing dataset as positive samples. In order to make the validation more accurate, we repeated fivefold cross-validation procedure 100 times. The average AUC values of the four methods (HGMDA, ICFMDA, RLSMDA, SACMDA) are 0.9156 (+/−0.0010), 0.9045 (+/−0.0008), 0.8569 (+/−0.0020) and 0.8767 (+/−0.0011), respectively (see Fig. 1).

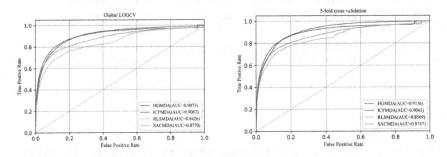

**Fig. 1.** Performance comparisons between HGMDA and other prediction models.

## 3.2   Case Studies

Case studies are conducted to further verify the capability of HGMDA to predict miRNA-disease associations. We implemented two different kinds of case studies in two different diseases. Case study for colon neoplasms was conducted as follows: based on HMDDv2.0 database, we regarded the miRNAs without known association with colon neoplasms as candidate miRNAs; after HGMDA finished the prediction work, we prioritized the candidate miRNAs according to the predicted scores by HGMDA. As a consequence, 10 out of top 10 predicted colon neoplasms related miRNAs were confirmed by dbDEMC [9] and miR2Disease [10] (See Table 1).

We took hepatocellular carcinoma as the second kind of case study, in which we removed all the known hepatocellular carcinoma related miRNAs to simulate a new disease as the input of HGMDA. Then, we verified the predicted potential hepatocellular carcinoma related miRNAs generated by the model. Finally, all 10 miRNAs were experimentally confirmed by HMDD v2.0, dbDEMC and miR2Disease (See Table 2).

**Table 1.** The top 10 predicted miRNAs associated with Colon Neoplasms

| miRNA | Evidence | miRNA | Evidence |
|-------|----------|-------|----------|
| hsa-mir-155 | miR2Disease; dbDEMC | hsa-mir-15a | dbDEMC |
| hsa-mir-146a | dbDEMC | hsa-mir-125b | dbDEMC |
| hsa-mir-20a | miR2Disease; dbDEMC | hsa-mir-29a | miR2Disease; dbDEMC |
| hsa-mir-221 | miR2Disease; dbDEMC | hsa-mir-16 | dbDEMC |
| hsa-mir-34a | miR2Disease; dbDEMC | hsa-mir-29b | miR2Disease; dbDEMC |

**Table 2.** The top 10 predicted miRNAs associated with Hepatocellular Carcinoma

| miRNA | Evidence | miRNA | Evidence |
|-------|----------|-------|----------|
| hsa-mir-155 | HMDD; miR2disease; dbDEMC | hsa-mir-126 | HMDD; miR2disease; dbDEMC |
| hsa-mir-146a | HMDD; miR2disease; dbDEMC | hsa-mir-34a | HMDD; miR2disease; dbDEMC |
| hsa-mir-17 | HMDD; miR2disease | hsa-mir-29a | HMDD; dbDEMC |
| hsa-mir-20a | HMDD; miR2disease; dbDEMC | hsa-mir-16 | HMDD; miR2disease; dbDEMC |
| hsa-mir-221 | HMDD; miR2disease; dbDEMC | hsa-mir-15a | HMDD;miR2disease; dbDEMC |

# 4    Discussion

In this study, we proposed a novel computational model to predict the underlying miRNA-disease associations based on hypergraph. The better performance of our model can be analyzed from two aspects. First, comprehensive statistical features and graph theoretic features were constructed from the integrated similarity matrices for miRNAs and diseases. Second, the model was based upon hypergraph, which can completely represent the complex relationships among miRNA-disease pairs. Because a hypergraph is a graph in which an edge can connect more than two vertices.

This method still has some limitations. HGMDA required its training data to have both positive and negative samples. In addition, there is an interesting issue that how to build an appropriate hypergraph structure from the miRNA functional similarity matrix and the disease semantic similarity matrix. It is a possible direction to further incorporate with hypergraph learning in the method. Therefore, we believe that our model would perform even better in future research.

**Acknowledgement.** This work was supported by grants from the National Natural Science Foundation of China (No. 61873001).

# References

1. Ambros, V.: microRNAs: tiny regulators with great potential. Cell **107**, 823–826 (2001)
2. Xu, J., et al.: Prioritizing candidate disease miRNAs by topological features in the miRNA target-dysregulated network: case study of prostate cancer. Mol. Cancer Ther. **10**, 1857–1866 (2011)
3. Chen, X., Yan, G.Y.: Semi-supervised learning for potential human microRNA-disease associations inference. Sci. Rep. **4**, 5501 (2014)
4. Xiao, Q., Luo, J., Liang, C., Cai, J., Ding, P.: A graph regularized non-negative matrix factorization method for identifying microRNA-disease associations. Bioinformatics **34**(2), 239–248 (2018)
5. Jiang, Y., Liu, B., et al.: Predict MiRNA-disease association with collaborative filtering. Neuroinformatics **16**, 363–372 (2018)
6. Li, Y., Qiu, C., et al.: HMDD v2.0: a database for experimentally supported human microRNA and disease associations. Nucleic Acids Res. **42**, D1070–D1074 (2014)
7. Zhang, Z., et al.: Inductive multi-hypergraph learning and its application on view-based 3D object classification. IEEE Trans. Image Process. **27**(12), 5957–5968 (2018)
8. Shao, B.Y., Liu, B.T., Yan, C.G.: SACMDA: MiRNA-disease association prediction with short acyclic connections in heterogeneous graph. Neuroinformatics **16**, 373–382 (2018)
9. Yang, Z., Ren, F., Liu, C., et al.: dbDEMC: a database of differentially expressed miRNAs in human cancers. BMC Genom. **11**(4 Suppl), S5 (2010)
10. Jiang, Q., et al.: miR2Disease: a manually curated database for microRNA deregulation in human disease. Nucleic Acids Res. **37**(Database issue), D98–D104 (2009)

# Discovering Driver Mutation Profiles in Cancer with a Local Centrality Score

Ying Hui[1], Pi-Jing Wei[1], Jun-Feng Xia[2], Hong-Bo Wang[3],
Jing Wang[1,4], and Chun-Hou Zheng[1,5(✉)]

[1] College of Computer Science and Technology, Anhui University,
Hefei 230601, Anhui, China
zhengch99@126.com
[2] Institute of Physical Science and Information Technology, Anhui University,
Hefei 230601, Anhui, China
[3] School of Computer Science and Technology, Hangzhou Dianzi University,
Hangzhou 310018, Zhejiang, China
[4] School of Computer and Information Engineering, Fuyang Normal University,
Fuyang 236037, Anhui, China
[5] Co-Innovation Center for Information Supply & Assurance Technology,
Anhui University, Hefei 230601, Anhui, China

**Abstract.** Although there are huge volumes of genomic data, how to integrate and analyze cancer omics data and identify driver genes is still a challenging task. Many published approaches have made great achievements in distinguishing driver genes from passenger genes, but the identification accuracy of driver genes needs to be improved. In this paper, we adopt a semi-local centrality measure to assess the impact of gene mutations on the changes in gene expression patterns. We consider mutated gene as source node and differentially expressed genes as target nodes in the transcriptional network. Firstly, we get differentially expression genes in the cohort by comparing tumor sample expression profiles with normal sample. Secondly, we construct a local network for each mutation gene using DEGs and mutation genes according to protein-protein interaction (PPI) network. Thirdly, we calculate each mutation genes' local centrality in the constructed network. Finally, we rank and select the driver genes from mutation genes according to its local centrality. We apply our method on five cancer datasets to identify influential genes in local network. Experimental results show that a stronger enrichment for true positive driver genes can be obtained.

**Keywords:** Driver mutation · Cancer · Centrality · Transcriptional network

## 1 Introduction

With the rapid development of genome technology, a huge amount of data has been produced [1, 2]. But how to integrative analyze cancer omics data in terms of somatic mutation, transcriptomic changes and epigenetic alterations is still an urgent task [3, 4]. One of the most challenging task is to identify and distinguish driver mutations that contribute to cancer initiation and progression, from numerous passenger mutations

© Springer Nature Switzerland AG 2019
D.-S. Huang et al. (Eds.): ICIC 2019, LNCS 11644, pp. 272–282, 2019.
https://doi.org/10.1007/978-3-030-26969-2_26

that have accumulated in cells but without carcinogenic consequences [5, 6]. From a clinical perspective, it is also crucial to identify cancer drivers which would hold significant value for defining personalized therapeutic target [7, 8]. Therefore, plentiful approaches have emerged to distinguish driver mutation genes from passenger mutation genes. Early approaches for driver analysis typically rely on the frequency of aberration of a given gene in a population of tumors (for example, [9–11]). However, those frequency-based methods failed to identify rare driver mutations which exhibit low population frequencies [12]. In order to discover novel driver mutations more effectively, researchers proposed many new methods enthusiastically. Recent methodologies for identifying driver genes can be mainly divided into two categories: the machine learning based methods [13–15] and the network based methods [16–22]. For example, the machine learning method, CHASM, adopts random forest which use alterations trained from known cancer-causing somatic missense mutations to classify driver mutations [13, 19]. However, the incomplete databases used by this model results in restrictions on application. As for network based method, they have become one of the most promising methods to understand cancer drivers due to their power to elucidate molecular mechanisms of disease development at the network level, e.g., DriverNet [16], MEMo [23] and Dendrix [24]. Although these methods have been successfully applied to the identification of cancer driver genes, they are not well suited for distinguishing rare driver genes. Moreover, the recognition accuracy of these methods needs to be improved.

The aforementioned methods analyze the constructed network generally based on statistical concepts. And there are also other measures using graph theory concepts which link network topological features to biologically informative properties. Different analysis of node centrality has been proposed to identify functionally-critical network components such as degree centrality, betweenness centrality and closeness centrality [25–27]. Those genes exhibiting high degree or high betweenness centrality scores have been proposed as candidate targets in gene-gene interaction network. But there are some disadvantages of above centrality measurement. For example, degree centrality may ignore some genes which have low degree but may have much higher influence than high degree genes [28]. As for betweenness and closeness centrality, they are well-known global metrics which can give better results due to the very high computational complexity [29–32]. But they are not easy to manage large-scale networks.

In this work, we adopt a semi-local centrality measure to assess the impact of single nucleotide variations (SNVs), copy number variations (CNVs) on the changes in gene expression patterns. Studies have shown that driver mutations are expected to alter gene expression of their cognate proteins [16]. Our method ranks potential driver genes based on their influence on the differentially expressed genes (DEGs) in the constructed local network. In this model, we consider mutated genes (e.g. SNVs; CNVs) as source node and DEGs as target nodes in the network. Because the distance between any source node and the target node is usually very short and thus implying high influence of the source node on the target [33]. Our method integrates the mutation and gene expression data to a gene-gene interaction network. And our local centrality measure only considers the nearest and the next nearest nodes from source node. In particular, we aim to identify influential mutated genes according to its local centrality in the interaction network. The main steps of our methods include: (1) get DEGs in the cohort by comparing tumor sample expression profiles with normal sample. (2) construct a local network for each mutation gene using DEGs and mutation genes according to PPI network. (3) calculate each mutation genes' local centrality in the constructed network. (4) rank and select the driver genes from mutation genes according to its local centrality. Our local centrality method considers both the nearest and the next nearest neighbors. So even a low degree gene also can be selected as candidate genes as it utilizes more information. And comparing with global centrality measures (e.t. betweenness centrality and closeness centrality), our local centrality measure has much lower computational complexity. Therefore, our method is a tradeoff between the above centrality methods and statistical network measures. To test the performance of our model, we analyze five datasets including Glioblastoma (GBM), Ovarian cancer (OVARIAN), Melanoma cancer (MELANOMA), Bladder cancer (BLCA) and Prostate cancer (PRAD) obtained from The Cancer Genome Atlas (TCGA). The results show that our model reveals notable improvements over other existing representative methods in terms of precision for discovering driver genes.

## 2 Materials and Methods

### 2.1 Datasets and Reference Network

We apply this model to 1435 tumor samples in TCGA data portal including 328 GBM samples, 316 Ovarian samples, 379 BLCA samples, 252 PRAD samples and 160 MELANOMA samples. The datasets consist of gene expression data and coding region mutation data for five cancer types which are obtained from [5]. In this paper, the coding region mutation data include SNV and CNV. And we integrate them to get the mutation matrix. The reference gene network we used in this study is HPRD (http://www.hprd.org). To help evaluating the quality of our results, we obtained a list of 616 cancer genes from the well-studied cancer gene database, Cancer Gene Census (CGC) and the version is (09/26/2016) [34].

### 2.2 Methods Overview

In this paper, we adopt a semi-local centrality model to assess the impact of gene mutations on the changes in gene expression patterns. We assume that the transcriptomic changes are due to mutation. Our method includes two main steps: (i) Identification of DEGs (target node) in the cohort. And construction of the local network for each mutated gene (source node). (ii) Identification of driver genes according to its local centrality which can assess the influential of nodes in the local network. An overview of the method workflow is presented in Fig. 1.

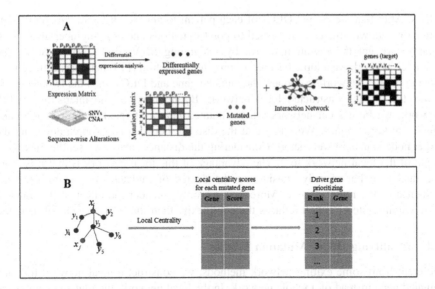

**Fig. 1.** A schematic of our method framework. (A) The data we used including gene expression, somatic SNVs and CNVs and PPI network. First, we use expression data to get the DEGs in the cohort and integrate SNVs, CNVs to get the mutation matrix. Second, DEGs and mutated genes are linked together by using PPI network to get the influence matrix. (B) Using this model, we calculate each mutated gene' local centrality score as its influential in the network. Take Fig. 1B as example, first, mutated gene $x_i$ is treated as target node and we construct the local network for $x_i$ within 2 steps. And we use a local centrality measure to calculate the influence score for this mutated gene. Finally, we rank the genes according to the score and select top-50 genes as the candidate-driver genes.

A general framework to assess the impact of mutations is to measure the ability on the changes in transition gene expression status from normal to tumor in gene network. Mutated genes are considered as source nodes and expression genes as target nodes. Moreover, any source node is usually very close to the target node which imply high influence of the source node on the target [33]. Based on the above information, in our model we construct a local network for each mutated gene. And we calculate each mutated gene's local centrality in the local network as the influential score. Finally, we rank mutation genes based on its influential score and select top-50 genes as the candidate-driver genes. A detailed introduction to our model is given in the next Section.

## 2.3 Identification of DEGs and Construction of Local Network

Gene expression data and coding region mutation data we used in this paper are obtained from reference [5] including single nucleotide variations (SNVs), copy number variations (CNVs) and expression data. To indicate the DEGs for each patient, we first calculate the log2 fold-change of gene expression between the paired tumor and normal samples. And then whose absolute value is greater than 1 are regarded as

DEGs. After that we collect DEGs of each patient to get the DEGs in the cohort. The mutation data we utilized is restricted to point mutations and copy-number alterations. That is, we get the mutation matrix by integrating SNVs and CNVs matrix. The reference network we adopted is protein-protein interaction (PPI) network. If there is an edge in the PPI network to connect the mutated gene and DEGs, then the two genes are connected in the constructed local network. By referring to the interactors in PPI network, we built local network where mutated gene is treated as source node and DEGs are target nodes. We know that the distance between any source node and the target node is usually very short. Considering the distance factor, we set the maximum depth of the local network to 2. The advantage of this local network is that it is more accurate than that of only considering the degree of mutated gene to measure the influence of the mutated gene. Moreover, we only consider the effect of the mutated gene within 2 steps, which reduces the complexity than that of using global network.

### 2.4    Identifying Driver Mutation Profiles

Different from some exiting network methods, we construct a local network for each mutated gene instead of a whole network. In the local network, the mutated gene is the source node and DEGs is the target nodes. We calculate the source node's influence in the constructed local network by using the local centrality measure $C_L(v)$ [28]. Which will give each mutated gene an influence score. The definition of local centrality $C_L(v)$ of node $v$ in local network is given below.

$$Q(u) = \sum_{\omega \in \Gamma(u)} N(\omega) \tag{1}$$

$$C_L(v) = \sum_{u \in \Gamma(v)} Q(u) \tag{2}$$

Where $\Gamma(u)$ and $\Gamma(v)$ is the set of the nearest neighbors of node $u$ and $v$. And $N(\omega)$ is the number of the nearest and the next nearest neighbors of node $\omega$. In this model, node $v$ denotes the mutated gene and we also call it as the source node. Others (e.g. $u$ and $\omega$) are DEGs, which are known as target nodes. Our main idea is to calculate the local centrality of the mutated gene as the influence score in local network. The higher the score, the greater the effect of the mutated gene on DEGs in the local network. There is a situation where a gene is both mutated and differentially expressed. These genes may be more important. So, we enhance the effect of both mutated and differentially expressed genes in the constructed local network. Take Fig. 1B as example, $y_3$ is a gene that both differentially expressed and mutated. So, when this gene is differentially expressed, it just acts as the target node. But when it is mutated, it acts as source node and the corresponding source nodes (e.t. $x_j$) when it is as target node act as target nodes of this gene. Thus, the local centrality score of those genes are increased. Using this model, we can get the local centrality score for each mutation gene. And then according to the scores, we rank mutation genes to identifying influential genes. We assume that the higher the ranking, the more likely it is to be a driver gene.

# 3 Results

## 3.1 The Comparison with Exited Methods

Most of the existed network methods for identifying driver genes are based on a global network. On the one hand, this global network will increase the computational complexity. On the other hand, the accuracy of this method needs to be improved. Our method applies a novel scheme: we firstly identify a local network for each mutated gene, where the mutated gene as the source node and DEGs as the target nodes; then, we use local centrality measure to calculate the influential of mutated genes in local network; finally, we rank mutated genes according to the local centrality score to decide which one is the driver gene. We select the top-50 ranked mutations as the candidate driver genes.

To evaluate the performance of our method, we compare it with two personalized-sample method (SCS [5] and OncoImpact [17]), an aggregate network approach (DriverNet) and a frequency-based method (Frequency [35]), and examine the proportion of candidate-driver genes found in CGC. As shown in Fig. 2, a stronger enrichment for true positive driver genes is obtained in our method's predictions. And in PRAD, OVARIAN and GBM datasets, the advantage of our method in predicting true positive driver genes is particularly obvious, almost 20% higher than other methods. The precision results show that our method is generally more accurate.

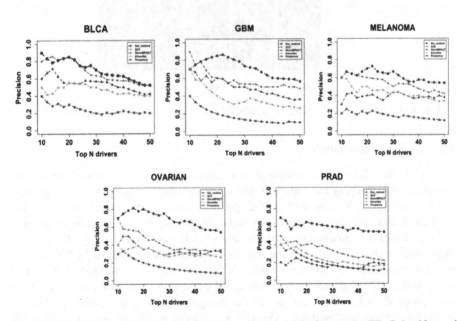

**Fig. 2.** The performance of five methods (Our method, SCS, OncoIMPACT, DriverNet and Frequency) on BLCA, GBM, MELANOMA, OVARIAN, PRAD datasets according to the cancer census genes set.

Furthermore, our method can not only identify frequently altered driver genes but also rare driver genes which are mutated in only a small number of patients. For example, in BLCA, TP53 (22.7%) is the most frequently mutated driver genes followed by ESR1 (0.26%), SRC (1.1%), EP300 (6.6%) which are altered in a very small number of patients but ranked high in our method. And the same is true for the other four datasets. Our method mainly ranks genes by their influence in network and is not sensitive to frequency.

In addition, we use the Drug-genes Interaction Database (DGIdb) online tool to analyze whether our candidate driver genes are clinically relevant genes or not [36]. The result is shown in Fig. 3. There are over 50% candidate driver genes are actionable targets in all five cancer datasets. And almost 20% other driver genes are druggable. This suggests that our candidate driver genes are clinically relevant.

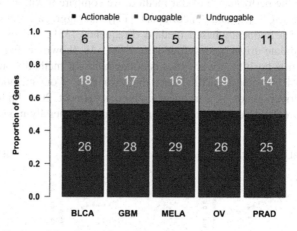

**Fig. 3.** Distribution of five datasets' candidate-driver genes in druggable genes datasets.

## 3.2   Enrichment Analysis

To test whether the top-50 candidate-driver genes for the five investigated cancers are collaboratively working for particular biological functions or pathways, we performe Gene Ontology (GO) term and KEGG pathway enrichment analysis by using the OmicShare tools (www.omicshare.com/tools). In this paper, five datasets are studied, but the results of enrichment analysis are not all listed here except BLCA. Go enrichment analysis reveals that the top-50 candidate-driver genes of BLCA are significantly enriched in 33 GO terms which is shown in Fig. 4A. The most enriched GO terms in the biological process are "biological regulation", "cellular process", "metabolic process" and so on. And in cellular component are "cell", "cell part" and in molecular function is "binding". The KEGG pathway results are shown in Fig. 4B. From this figure we can see that the most significantly enriched pathways in Human Diseases are Cancers and Infection diseases. According to the pathway enrichment table, we select the top 20 most significant pathways and display them in Fig. 4C.

Among the top 20 pathways, the most significant pathway is Pathway in cancer. It means that our candidate-driver genes in BLCA are significantly related to cancer.

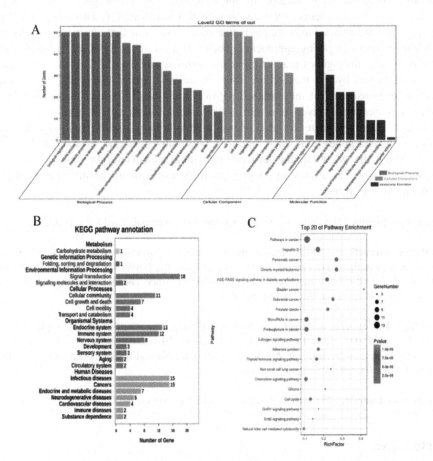

**Fig. 4.** GO term and KEGG pathway enrichment analysis on BLCA candidate-driver genes. (A) GO term enrichment analysis result of candidate-driver genes in BLCA. (B) KEGG pathway annotation result of driver-candidate genes in BLCA. The ordinate is the A level and the B level annotation of KEGG, the black typeface is the A level annotation name, and the color font is the B level. (C) Top 20 of pathway enrichment result of the candidate-driver genes.

## 3.3  Discussion and Conclusions

In this work, we propose a local centrality measure to assess the impact of gene mutations on the changes in gene expression patterns. We consider mutated gene as source node and DEGs as target nodes in the network and assume that the closer the source node is to the target node, the more influence it has on it [33]. So in the proposed model we construct a local network for each mutated gene to measure its influential by considering both the nearest and the next nearest neighbors. According to its local centrality scores, we rank genes and select top-50 genes as the candidate-drivers. We apply our method to five cancer datasets (BLCA, GBM, OVARIAN, MELANOMA, and PRAD). Through evaluation of the benchmarking driver genes, our method achieves a higher prediction accuracy than SCS, OncoIMPACT, DriverNet and Frequency. And the detailed average prediction result also show that our method performs better than other competing approaches. Then, we do some analysis on our candidate-drivers. Firstly, drug-genes interaction analysis shows that over 50% of the candidate-drivers identified by our method are clinically relevant genes. Secondly, enrichment analysis demonstrate that the enriched pathways are mostly related to cancer which means our candidate-driver gene are collaboratively working for particular cancer pathways. However, there are also some limitations in our work. E.g., our method prioritizes driver genes in a cohort and not a personalized-sample. So the drawback of our method is that we ignore the particular individual information. In future, we will try to propose a new framework to integrate more information and identify personalized driver mutation profiles.

**Acknowledgments.** The authors acknowledge the paper materials used for experiments.

**Author Contributions.** YH carried out the experiments, analyses presented in this work and wrote the manuscript. PJW carried out the data analysis. JX, HBW, JW and CHZ helped with project design, edited the manuscript and provided guidance and feedback throughout. All authors read and approved the final manuscript.

**Funding.** This research was funded by the National Natural Science Foundation of China (Nos. 61873001, 61672037, 61602142, 61861146002, and 61520106006), the Key Project of Anhui Provincial Education Department (No. KJ2017ZD01), the Natural Science Foundation of Anhui Province (1808085QF209).

**Conflicts of Interest.** The authors declare no conflict of interest.

# References

1. Zhang, J., Zhang, S.: The discovery of mutated driver pathways in cancer: models and algorithms. IEEE/ACM Trans. Comput. Biol. Bioinform. **PP**(99), 1 (2018)
2. Wang, D., et al.: An NMF-L2,1-norm constraint method for characteristic gene selection. PLoS ONE **11**(7), e0158494 (2016)
3. Campbell, I.M., et al.: Somatic mosaicism: implications for disease and transmission genetics. Trends Genet. **31**(7), 382–392 (2015)
4. Dees, N.D., et al.: MuSiC: identifying mutational significance in cancer genomes. Genome Res. **22**(8), 1589–1598 (2012)
5. Guo, W.F., et al.: Discovering personalized driver mutation profiles of single samples in cancer by network control strategy. Bioinformatics **34**(11), 1893 (2018)
6. Haber, D.A., Settleman, J.: Cancer: drivers and passengers. Nature **446**(7132), 145–146 (2007)
7. Chin, L., Andersen, J.N., Futreal, P.A.: Cancer genomics: from discovery science to personalized medicine. Nat. Med. **17**(3), 297–303 (2011)

8. Schilsky, R.L.: Personalized medicine in oncology: the future is now. Nat. Rev. Drug Discov. **9**(5), 363–366 (2010)
9. Chris, G., et al.: Statistical analysis of pathogenicity of somatic mutations in cancer. Genetics **173**(4), 2187 (2006)
10. Gad, G., et al.: Comment on "The consensus coding sequences of human breast and colorectal cancers". Science **317**(5844), 1500 (2007)
11. Ahrim, Y., Richard, S.: Identifying cancer driver genes in tumor genome sequencing studies. Bioinformatics **27**(2), 175–181 (2011)
12. Lawrence, M.S., et al.: Mutational heterogeneity in cancer and the search for new cancer-associated genes. Nature **499**(7457), 214 (2013)
13. Carter, H., et al.: Cancer-specific high-throughput annotation of somatic mutations: computational prediction of driver missense mutations. Cancer Res. **69**(16), 6660–6667 (2009)
14. Kumar, R.D., Swamidass, S.J., Bose, R.: Unsupervised detection of cancer driver mutations with parsimony-guided learning. Nat. Genet. **48**(10), 1288 (2016)
15. Mao, Y., et al.: CanDrA: cancer-specific driver missense mutation annotation with optimized features. PLoS ONE **8**(10), e77945 (2013)
16. Bashashati, A., et al.: DriverNet: uncovering the impact of somatic driver mutations on transcriptional networks in cancer. Genome Biol. **13**(12), R124 (2012)
17. Bertrand, D., et al.: Patient-specific driver gene prediction and risk assessment through integrated network analysis of cancer omics profiles. Nucleic Acids Res. **43**(7), e44 (2015)
18. Greenman, C., et al.: Patterns of somatic mutation in human cancer genomes. Eur. J. Cancer Suppl. **6**(9), 153–158 (2007)
19. Hou, J.P., Ma, J.: DawnRank: discovering personalized driver genes in cancer. Genome Med. **6**(7), 56 (2014)
20. Kang, H., et al.: Inferring sequential order of somatic mutations during tumorgenesis based on markov chain model. IEEE/ACM Trans. Comput. Biol. Bioinform. **12**(5), 1094–1103 (2015)
21. Suo, C., et al.: Integration of somatic mutation, expression and functional data reveals potential driver genes predictive of breast cancer survival. Bioinformatics **31**(16), 2607–2613 (2015)
22. Zhang, S.Y., et al.: m6A-Driver: identifying context-specific mRNA m6A methylation-driven gene interaction networks. PLoS Comput. Biol. **12**(12), e1005287 (2016)
23. Ciriello, G., et al.: Mutual exclusivity analysis identifies oncogenic network modules. Genome Res. **22**(2), 398 (2012)
24. Vandin, F., Upfal, E., Raphael, B.J.: De novo discovery of mutated driver pathways in cancer. Genome Res. **22**(2), 375–385 (2012)
25. Azuaje, F.J., et al.: Information encoded in a network of inflammation proteins predicts clinical outcome after myocardial infarction. BMC Med. Genomics **4**(1), 59 (2011)
26. Dewey, F.E., et al.: Gene coexpression network topology of cardiac development, hypertrophy, and failure. Circ. Cardiovasc. Genet. **4**(1), 26 (2011)
27. Azuaje, F., et al.: Analysis of a gene co-expression network establishes robust association; between Col5a2 and ischemic heart disease. BMC Med. Genomics **6**(1), 13 (2013)
28. Chen, D., et al.: Identifying influential nodes in complex networks. Phys. Stat. Mech. Appl. **391**(4), 1777–1787 (2012)
29. Lü, L., et al.: Leaders in social networks, the delicious case. PLoS ONE **6**(6), e21202 (2011)
30. Brin, S., Page, L.: The anatomy of a large-scale hypertextual Web search engine. Comput. Netw. **56**(18), 3825–3833 (2012)
31. Radicchi, F., et al.: Diffusion of scientific credits and the ranking of scientists. Phys. Rev. E Stat. Nonlinear Soft Matter Phys. **80**(2), 056103 (2009)

32. Lee, S.H., et al.: Googling Social Interactions: Web Search Engine Based Social Network Construction. PLoS ONE **5**(7), e11233 (2010)
33. Shrestha, R., et al.: HIT'nDRIVE: patient-specific multi-driver gene prioritization for precision oncology. Genome Res. **27**(9), 1573 (2017)
34. Futreal, P.A., et al.: A census of human cancer genes. Nat. Rev. Cancer **4**(3), 177 (2004)
35. Wei, X., et al.: Exome sequencing identifies GRIN2A as frequently mutated in melanoma. Nat. Genet. **43**(5), 442–446 (2011)
36. Cotto, K.C., et al.: DGIdb 3.0: a redesign and expansion of the drug–gene interaction database. Nucleic Acids Res. **46**, D1068–D1073 (2017)

# LRMDA: Using Logistic Regression and Random Walk with Restart for MiRNA-Disease Association Prediction

Zhengwei Li[1,2,3,4]($\boxtimes$), Ru Nie[1]($\boxtimes$), Zhuhong You[5], Yan Zhao[6],
Xin Ge[1], and Yang Wang[2]

[1] School of Computer Science and Technology, China University of Mining
and Technology, Xuzhou 221116, China
{zwli, nr}@cumt.edu.cn
[2] KUNPAND Communications (Kunshan) Co., Ltd., Suzhou 215300, China
[3] Institute of Machine Learning and Systems Biology, College of Electronics
and Information Engineering, Tongji University, Shanghai 201804, China
[4] Mine Digitization Engineering Research Center of Ministry of Education,
China University of Mining and Technology, Xuzhou 221116, China
[5] Xinjiang Technical Institute of Physics and Chemistry,
Chinese Academy of Science, Urumqi 830011, China
[6] School of Information and Electrical Engineering, China University of Mining
and Technology, Xuzhou 21116, China

**Abstract.** MicroRNAs (MiRNAs) have received much attention in recent years because growing evidences indicate that they play critical roles in tumor initiation and progression. Predicting underlying disease-related miRNAs from existing huge amount of biological data is a hot topic in biomedical research. Herein, we presented a novel computational model of logistic regression and random walk with restart algorithm for miRNA-disease association prediction (LRMDA) through integrating multi-source data. The model employs random walk with restart to fuse the association distribution between miRNAs and diseases and obtains highly discriminative feature from those heterogeneous data. To evaluate the performance of LRMDA, we performed 5-fold cross validation to compare it with several state-of-the-art models. As a result, our model achieves mean AUC of $0.9230 \pm 0.0059$. Besides, we carried out case study for predicting potential miRNAs related to Esophageal Neoplasms (EN). The achieved results indicate that 90% out of the top 50 prioritized miRNAs for EN are confirmed by biological experiments and further demonstrates the feasibility of our method. Therefore, LRMDA could potentially aid future research efforts for miRNA-disease association identification.

**Keywords:** MiRNA-disease association prediction · Multi-source data · Random walk with restart · Logistic regression

© Springer Nature Switzerland AG 2019
D.-S. Huang et al. (Eds.): ICIC 2019, LNCS 11644, pp. 283–293, 2019.
https://doi.org/10.1007/978-3-030-26969-2_27

# 1  Introduction

MicroRNAs (MiRNAs) are a class of small noncoding RNAs that play important roles in gene regulation by targeting mRNAs for cleavage or translational repression at the post-transcriptional level [1–3]. Since the first two miRNAs (Caenorhabditis elegans lin-4 and let-7) were discovered in the past decades, a large number of miRNAs (for example, more than 1900 human miRNAs according to miRBase22 [4]) have been identified continually in eukaryotic organisms ranging from nematodes to mammals [5, 6]. Extensive research suggests that miRNAs have great effects in almost all cellular processes, such as regulation of cell proliferation, differentiation, angiogenesis, migration, and apoptosis [7–9]. Therefore, it is not surprising that the mutation or mis-expression of miRNAs may be closely involved in various human cancers as tumor suppressors and oncogenes. Calin et al. was first to notice that miR-15a and miR-16-1 are deleted in more than 55% of B-cell chronic lymphocytic leukemia (B-CLL) patients and recognized that miRNAs were related to cancer formation [10]. Another example is a down-regulated expression of miRNA-181d, probably through reverse regulation on NKAIN2 (Na+/K+ transporting ATPase interacting 2), which could suppress pancreatic cancer development [11]. Accordingly, identifying the associations between miRNAs and diseases could effectively advance our understanding of disease pathogenesis at the molecular level and contribute to the discovery of biomarker for disease diagnosis and treatment [12–14].

However, biological experiments for identifying miRNA-disease associations are extremely expensive and time consuming. As more and more experiment-supported data of miRNA-disease associations are collected, developing computational methods for identifying miRNA-disease associations has become an economical and practical way. In recent years, many computational models for potential miRNA-disease association prediction have been proposed based on the assumption that miRNAs with similar function are likely to be related to phenotypically similar diseases, or vice versa [15–17]. Jiang et al. presented a hypergeometric distribution-based network model by combing two different feature vectors for inferring potential disease-related miRNAs based on support vector machine (SVM) [18]. Because the model mainly utilizes the neighbor information of each miRNA, the predictive performance is very limited. Shi et al. constructed a novel framework to identify miRNA-disease associations by the aid of the functional connections between miRNA targets and disease genes and achieved satisfactory results in identifying known cancer-related miRNAs for nine human cancers [19]. By integrating various datasets, Chen et al. developed a model named WBSMDA to rank miRNA-disease associations by within and between score [20]. Although WBSMDA has achieved good results than previous methods, there is still room for improvement based on cross validation. Further, Chen et al. also proposed another model EGBMMDA for predicting miRNA-disease association by constructing highly discriminative feature vector to train a regression tree within a gradient boosting framework [21].

In this study, we presented an efficient computational model based on logistic regression and random walk with restart for predicting miRNA-disease associations (LRMDA). To capture highly discriminative feature vector, we first constructed a heterogeneous network by integrating not only the disease semantic similarity, miRNA functional similarity, experimentally validated miRNA-disease associations, but also Gaussian kernel profile similarity of miRNAs and diseases respectively. Furthermore, we applied a random walk with restart (RWR) algorithm on the network to mine potential global distribution between miRNAs and diseases as much as possible. Then, we finally used a binary logistic regression classifier to identify whether a miRNA is associated with a given disease. In the 5-fold cross validation, our model achieved mean AUC of $0.9230 \pm 0.0059$, which was superior to four other classical miRNA-disease association predictive models (MCMDA [22], EPMDA [23], EGBMMDA [21], and DRMDA [14]). In the case studies of the Esophageal Neoplasms, 45 out of top 50 prioritized miRNAs for the disease were confirmed by recent experimental discoveries. All the above results indicate that LRMDA is an effective and stable model for predicting miRNA-disease associations.

## 2   Materials and Methods

### 2.1   Golden Standard Data Sets

The human disease-miRNA associations used in this work were downloaded from HMDD 2.0 database [24], including 5430 experimentally verified associations, also called known associations, between 383 diseases and 495 miRNAs. Herein, we introduced an adjacency matrix $A \in R^{n_d \times n_m}$ to express the associations, where $n_d$ and $n_m$ are the number of the corresponding diseases and miRNAs, respectively. If there is a relationship between disease $d_i$ and miRNA $m_j$, the value of element $A_{ij}$ will be set to 1, otherwise 0. Wang et al. pioneered a method to calculate the miRNA functional similarity which is available at http://www.cuilab.cn/files/images/cuilab/misim.zip [25]. We constructed matrix $M$ to express those functional similarity scores, where the element $M_{ij}$ denotes the similarity of miRNA pair $(m_i, m_j)$. Similarly, the disease semantic similarity scores were calculated in the light of the method adopted in the literature [26, 27]. The matrix $D$ was constructed to represent their semantic similarity, where the element $D_{ij}$ denotes the similarity between disease pair $(m_i, m_j)$. Furthermore, according to [20], we also calculated Gaussian kernel similarity for miRNAs and diseases respectively. For example, the Gaussian kernel similarity between disease $d_i$ and $d_j$ can be calculated as follows:

$$GD_{ij} = \exp(-\beta_d \|G(d_i) - G(d_j)\|^2) \tag{1}$$

$$\beta_d = \frac{1}{(\frac{1}{n_d} \sum_{k=1}^{n_d} \|G(d_k)\|^2)} \tag{2}$$

where the binary vector $G(d_i)$ is to denote whether $d_i$ is associated with each disease or not, and $\beta_d$ is the kernel bandwidth.

## 2.2    Positive and Negative Samples

In our study, 5430 high-quality experimentally verified associations between 495 diseases and 383 miRNAs, coming from miR2Disease [12] and HMDD [13] databases, were used as the positive samples. Due to lack of experimental validation negative miRNA-disease associations, we randomly select 5430 unknown miRNA-disease associations as negative samples from all possible ones between those miRNAs and diseases (excluding the 5430 known associations). To improve the credibility of the results, we made 100 times such stochastic selection operations. In the field of bioinformatics, similar selection strategy for setting up negative samples were frequently used [28–31].

## 2.3    Feature Construction

As a simulation of a random walker's transition from its current location to neighbors in a network, random walk with restart (RWR) algorithm [32] adopts a global perspective to fully mine the hidden information in network and obtains the ranked list for potential biological elements (including protein, drug, gene, etc.). At first, we employed the RWR to get the ranked scores between a given miRNA $m_i$ and other miRNAs based on the $M$ matrix. Next we chose top-k miRNAs as the "friends" of miRNA $m_i$ according to the association scores in decent order. Then we adopted $FO_{ij} = [FO_{i,1}, FO_{i,2}]$ as its sub-feature vector, where $FO_{i,1}$ and $FO_{i,2}$ denote the accumulated association values of neighbor miRNAs with label 1 and 0 for a given disease $d_j$. But in practice, it is unreasonable to set to 0 to all labels for all unconfirmed miRNAs for a given disease. Herein, we adopted the method [33] to utilize miRNA family and cluster information to calculate the prior probability for unknown miRNAs. If miRNA $m_i$ belongs to some kind of miRNA family or cluster, then we calculated the posterior probability

$$p_i = \frac{N_i}{T_i} \tag{3}$$

where $N_i$ is the number of $d_j$-related miRNAs and $T_i$ is the number of all disease miRNAs in the corresponding families or clusters. Then we generated a random probability following standard uniform distribution for miRNA $m_i$. If it is less than $p_i$, then assign 1 as the prior label for miRNA $m_i$, otherwise 0. Based on the above steps, we could obtain the corresponding estimation label for each unknown miRNA for the given disease $d_j$ and subsequently obtained the new sub-feature vector $FE_{ij} = [FE_{i,1}, FE_{i,2}]$. Similarly, on the basis of the Gaussian kernel similarity for miRNAs and diseases, we could generate another two sub-feature rectors for disease $d_j$, namely $GO_{ij} = [GO_{i,1}, GO_{i,2}]$ and $GE_{ij} = [GE_{i,1}, GE_{i,2}]$. Therefore, the final feature vector for miRNA $m_i$ and disease $d_j$ could be expressed by

$$F_{ij} = [FO_{i,1}, FO_{i,2}, FE_{i,1}, FE_{i,2}, GO_{i,1}, GO_{i,2}, GE_{i,1}, GE_{i,2}] \tag{4}$$

where the label of $F_{ij}$ will be 1 if the association between miRNA $m_i$ and disease $d_j$ is experimentally validated, otherwise 0. By repeating these steps for all the miRNAs and diseases, we could obtain the whole feature matrix for further classification.

## 2.4    Logistic Regression

As a widely used binary classification algorithm, logistic regression (LR) technique is a multivariable analysis that concentrates on all the available predictive factors in a problem at the same time. At first, LR classifier is constructed with a modification of linear regression.

$$z = w_0 + w_1 x_1 + w_2 x_2 + \cdots + w_n x_n + b \tag{5}$$

where $W = [w_0, w_1, \cdots, w_n]^T$ and $b$ are weight vector and bias, respectively. We multiplied each variable in feature vector by a weight and then add them up. To scale the value of $z$ into a number between 0 and 1, we put $z$ into the sigmoid transforming function:

$$\sigma(z) = \frac{\exp(z)}{1 + \exp(z)} \tag{6}$$

Then, the formula of logit transformation can be calculated by:

$$g(z) = \ln(\frac{\sigma(z)}{1 - \sigma(z)}) \tag{7}$$

In our proposed model, the maximum likelihood estimation and the gradient ascent algorithm were adopted to calculate the corresponding optimum values of $W$ and $b$.

## 2.5    Evaluation Criteria

To evaluate the performance of our predictive methods, four criteria, including the accuracy (Acc), sensitivity (Sen), precision (Pre), and Matthews's correlation coefficient (MCC), were employed in this work, which can be calculated as follows:

$$\text{Acc} = \frac{TP + TN}{TP + FP + TN + FN} \tag{8}$$

$$\text{Pre} = \frac{TP}{TP + FP} \tag{9}$$

$$\text{Sen} = \frac{TP}{TP + FN} \tag{10}$$

$$\text{MCC} = \frac{(TP \times TN) - (FP \times FN)}{\sqrt{(TP + FN) \times (TN + FP) \times (TP + FP) \times (TN + FN)}} \tag{11}$$

where *TP* (true positive) represents the number of the known miRNA-disease associations predicted correctly while *FP* (false positive) denotes the number of the unknown miRNA-disease associations falsely predicted to be the known miRNA-disease associations. Similarly, *TN* (true negative) stands for the number of the unknown miRNA-disease associations predicted correctly, and *FN* (false negative) indicates the number of the known miRNA-disease associations falsely predicted to be the unknown miRNA-disease associations. Receiver-operating characteristics (ROC) curve is a standard technique for summarizing classifier performance, which can be drawn by plotting the true positive rate (TPR) against the false positive rate (FPR) according to different thresholds. Better performance is reflected in curves with a stronger bend toward the upper-left corner of ROC curve. AUC (Area Under Curve) is defined as the area enclosed by the coordinate axis under the ROC curve, which is commonly used to measure the quality of a classification model. Typically, the AUC value is between 0.5 and 1.0, with a larger AUC representing better performance.

## 3 Experimental Results

### 3.1 Performance Evaluation

In order to decrease data dependence and avoid over-fitting for predictive model, we adopted 5-fold cross-validation based on the data set we just created. At first, the whole data set was roughly separated into five equal subsets, four of which were randomly chosen for training, and the rest for test. To verify the effectiveness of the proposed method, we repeated such random selection for 100 times. Table 1 shows the predictive results of LRMDA based on 5-fold cross-validation. In addition, the ROC curve (See Fig. 1) is used to visually evaluate the performance of LRMDA and the AUC was calculated correspondingly. As a result, LRMDA obtained the mean AUC of $0.9230 \pm 0.0059$, which is obviously higher than several other existing computational models (see Table 2).

**Table 1.** Performance of the proposed method through 5-fold cross validation

| Test set | Accuracy (%) | Precision (%) | Recall (%) | MCC (%) |
|---|---|---|---|---|
| 1 | 88.67 | 88.84 | 88.67 | 88.66 |
| 2 | 89.22 | 89.32 | 89.23 | 89.22 |
| 3 | 89.73 | 89.94 | 89.73 | 89.72 |
| 4 | 88.66 | 88.94 | 88.67 | 88.65 |
| 5 | 87.47 | 87.62 | 87.48 | 87.46 |
| Average | **88.75 ± 0.84** | **89.35 ± 0.85** | **89.13 ± 0.83** | **89.08 ± 0.84** |

### 3.2 Case Study on Esophageal Neoplasms

Besides, to further validate the effectiveness of our method, we carried it out on esophageal neoplasms (EN). Here, all experimentally verified miRNA-disease

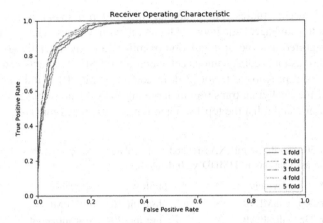

**Fig. 1.** ROC curves of LRMDA. As a result, LRMDA achieved mean AUC of 0.9230 ± 0.0059 in 5-fold cross validation, slightly outperforming several previous computational models in predictive accuracy.

**Table 2.** Performance comparisons between LRMDA and four other existing computational models (MCMDA, EPMDA, EGBMMDA, and DRMDA) for predicting miRNA-disease association in terms of mean AUC based on 5-fold cross validations

| Method | Mean AUC |
|---|---|
| MCMDA | 0.8767 ± 0.0011 |
| EPMDA | 0.8917 ± 0.0004 |
| EGBMMDA | 0.9048 ± 0.0012 |
| DRMDA | 0.9156 ± 0.0006 |
| LRMDA | 0.9230 ± 0.0059 |

associations in the HMDD v.2.0 [24] were set aside for training. The top 50 miRNAs with the highest predicted values were chosen as the most potential miRNAs associated with this disease, and then we confirmed them by means of the two other well-known databases, namely dbDEMC [34] and miR2Disease [35].

EN is the seventh most frequent malignant cancers (more than 572,000 new cases) and ranks sixth in mortality worldwide in 2018 (over 509,000 deaths). About 70% of cases occur in men, and there is a 2-fold to 3-fold difference in incidence and mortality rates between the sexes overall [36]. Early detection and treatment of EN could greatly improve the survival rate. MiRNA expression profiles are important diagnostic and prognostic markers for EN. For example, 7 serum miRNAs (miR-10a, miR-22, miR-100, miR-148b, miR-223, miR-133a, and miR-127-3p) clearly distinguish stage I/II EN patients from controls [37]. Therefore, we took EN as a case study for LRMDA and prioritized the potential miRNAs related to this disease. As a result, 9 out of the top 10 and 45 out of the top 50 miRNAs potentially associated with the disease were confirmed by experimental findings collected in dbDEMC and/or miR2Disease (see Table 3). In fact, four of the rest five unvalidated miRNAs in Table 3 were also confirmed by recent studies. For example, miR-499 rs3746444 is associated with susceptibility to esophageal squamous cell carcinoma in Chinese Han population [38]. Chen et al. reported that the up-regulation of miR-92a is significantly correlated with poor survival of esophageal squamous cell carcinoma (ESCC) patients and can be used as an independent prognostic factor [39]. In addition, miR-204 could inhibit invasion and epithelial-mesenchymal transition by targeting FOXM1 in esophageal cancer [40].

The column 1 and 3 list the top 1–25 and top 26–50 related miRNAs, respectively.

**Table 3.** Top 50 prioritized miRNAs related to Esophageal Neoplasms based on the known miRNA-disease associations in HMDD v2.0 database

| miRNA | Evidence | miRNA | Evidence |
|---|---|---|---|
| hsa-mir-206 | dbMEMC | hsa-mir-16 | dbMEMC |
| hsa-mir-23a | dbMEMC | hsa-mir-204 | **unconfirmed** |
| hsa-mir-150 | dbMEMC | hsa-mir-210 | dbMEMC |
| hsa-mir-142 | dbMEMC | hsa-mir-24 | dbMEMC |
| hsa-mir-107 | dbMEMC, miR2Disease | hsa-mir-100 | dbMEMC |
| hsa-mir-15b | dbMEMC | hsa-mir-375 | dbMEMC, miR2Disease |
| hsa-mir-31 | dbMEMC | hsa-mir-200c | dbMEMC |
| hsa-mir-103a | **unconfirmed** | hsa-mir-196a | dbMEMC, miR2Disease |
| hsa-mir-133b | dbMEMC | hsa-mir-221 | dbMEMC |
| hsa-mir-143 | dbMEMC | hsa-mir-203 | dbMEMC, miR2Disease |
| hsa-mir-182 | dbMEMC | hsa-mir-148a | dbMEMC |
| hsa-mir-124 | dbMEMC | hsa-mir-106a | dbMEMC |
| hsa-mir-15a | dbMEMC | hsa-mir-125b | dbMEMC |
| hsa-mir-181a | dbMEMC | hsa-mir-27a | dbMEMC |
| hsa-mir-183 | dbMEMC | hsa-mir-92a | **unconfirmed** |
| hsa-mir-122 | **unconfirmed** | hsa-mir-146b | dbMEMC |
| hsa-mir-195 | dbMEMC | hsa-mir-155 | dbMEMC |
| hsa-mir-26a | dbMEMC | hsa-mir-205 | dbMEMC, miR2Disease |
| hsa-mir-7 | dbMEMC | hsa-mir-499a | **unconfirmed** |
| hsa-mir-146a | dbMEMC | hsa-mir-29a | dbMEMC |
| hsa-mir-21 | dbMEMC, miR2Disease | hsa-mir-141 | dbMEMC |
| hsa-mir-222 | dbMEMC | hsa-mir-223 | dbMEMC, miR2Disease |
| hsa-mir-30a | dbMEMC | hsa-mir-19b | dbMEMC |
| hsa-mir-29c | dbMEMC | hsa-mir-18a | dbMEMC |
| hsa-mir-181b | dbMEMC | hsa-let-7 g | dbMEMC |

## 4    Conclusion

To identify potential miRNA-disease associations, we proposed a novel computational model named LRMDA. The model integrated multi-source heterogeneous data of experimentally verified miRNA-disease associations, miRNA functional similarity, disease semantic similarity and Gaussian kernel profile similarity of miRNAs and diseases to construct an integrated network and then exploit the random walk with restart algorithm to extract highly discriminative global feature for each miRNA-disease pair. Then, we employed binary logistic regression classifier to identify the potential miRNA-disease associations. In the 5-fold cross validation, LRMDA demonstrated promising predictive performance than the four other classical approaches. Moreover, for the case study of Esophageal Neoplasms, 45 out of the top 50 prioritized miRNAs have been experimentally validated according to the miR2Diseaes and dbDEMC databases. The fusion of multisource data and plenty of known miRNA-disease associations provided by HMDD 2.0 guarantee the effectiveness of our model. With more and more experimentally confirmed associations to be collected in the future, predicting potential disease-related miRNAs by computational methods will become more feasible and reliable.

**Acknowledgements.** This work was supported in part by the National Natural Science Foundation of China (Grant No. 61873270, 61732012), the Jiangsu Postdoctoral Innovation Plan (Grant No. 1701031C), and the Jiangsu Shuangchuang Talents Program.

# References

1. Victor, A.: The functions of animal microRNAs. Nature **431**(7006), 350–355 (2004)
2. Chen, X., Xie, D., Wang, L., Zhao, Q., You, Z.-H., Liu, H.: BNPMDA: Bipartite Network Projection for MiRNA–Disease Association prediction. Bioinformatics **34**, 3178–3186 (2018)
3. Wang, L., et al.: LMTRDA: using logistic model tree to predict miRNA-disease associations by fusing multi-source information of sequences and similarities. PLoS Comput. Biol. **15**(3), e1006865 (2019)
4. Kozomara, A., Birgaoanu, M., Griffiths-Jones, S.: miRBase: from microRNA sequences to function. Nucleic Acids Res. **47**(D1), D155–D162 (2018)
5. Lee, R.C., Feinbaum, R.L., Ambros, V.: The C. elegans heterochronic gene lin-4 encodes small RNAs with antisense complementarity to lin-14. Cell **75**(5), 843 (1993)
6. Kozomara, A., Griffiths-Jones, S.: miRBase: annotating high confidence microRNAs using deep sequencing data. Nucleic Acids Res. **42**(Database issue), D68–D73 (2014)
7. Cheng, A.M., Byrom, M.W., Jeffrey, S., Ford, L.P.: Antisense inhibition of human miRNAs and indications for an involvement of miRNA in cell growth and apoptosis. Nucleic Acids Res. **33**(4), 1290–1297 (2005)
8. Miska, E.A.: How microRNAs control cell division, differentiation and death. Curr. Opin. Gen. Dev. **15**(5), 563–568 (2005)
9. Karp, X., Ambros, V.: Encountering MicroRNAs in cell fate signaling. Science **310**(5752), 1288–1289 (2005)
10. Acunzo, M., Croce, C.M.: Downregulation of miR-15a and miR-16-1 at 13q14 in chronic lymphocytic leukemia. Clin. Chem. **62**(4), 655–656 (2016)
11. Zhang, G., et al.: Downregulation of microRNA-181d had suppressive effect on pancreatic cancer development through inverse regulation of KNAIN2. Tumour Biol. J. Int. Soc. Oncodevelopmental Biol. Med. **39**(4), 1010428317698364 (2017)
12. You, Z.-H., et al.: PRMDA: personalized recommendation-based MiRNA-disease association prediction. Oncotarget **8**(49), 85568 (2017)
13. Chen, X., et al.: A novel computational model based on super-disease and miRNA for potential miRNA–disease association prediction. Mol. BioSyst. **13**(6), 1202–1212 (2017)

14. Chen, X., Gong, Y., Zhang, D.H., You, Z.H., Li, Z.W.: DRMDA: deep representations-based miRNA-disease association prediction. J. Cell Mol. Med. **22**(1), 472–485 (2018)
15. Bandyopadhyay, S., Mitra, R., Maulik, U., Zhang, M.Q.: Development of the human cancer microRNA network. Silence **1**(1), 6 (2010)
16. You, Z.H., et al.: PBMDA: a novel and effective path-based computational model for miRNA-disease association prediction. PLoS Comput. Biol. **13**(3), e1005455 (2017)
17. Chen, X., Huang, L.: LRSSLMDA: Laplacian Regularized Sparse Subspace Learning for MiRNA-Disease Association prediction. PLoS Comput. Biol. **13**(12), e1005912 (2017)
18. Jiang, Q., et al.: Prioritization of disease microRNAs through a human phenome-microRNAome network. BMC Syst. Biol. **4**(s1), S2 (2010)
19. Shi, H., et al.: Walking the interactome to identify human miRNA-disease associations through the functional link between miRNA targets and disease genes. BMC Syst. Biol. **7**, 101 (2013). https://doi.org/10.1186/1752-0509-7-101
20. Chen, X., et al.: WBSMDA: Within and Between Score for MiRNA-Disease Association prediction. Sci. Rep. **6**(1), 21106 (2016)
21. Chen, X., Huang, L., Xie, D., Zhao, Q.: EGBMMDA: Extreme Gradient Boosting Machine for MiRNA-Disease Association prediction. Cell Death Dis. **9**(1), 3 (2018)
22. Li, J.-Q., Rong, Z.-H., Chen, X., Yan, G.-Y., You, Z.-H.: MCMDA: matrix completion for MiRNA-disease association prediction. Oncotarget **8**(13), 21187–21199 (2017)
23. Huang, Y.-A., et al.: EPMDA: an expression-profile based computational model for microRNA-disease association prediction. Oncotarget **8**(50), 87033 (2017)
24. Li, Y., et al.: HMDD v2.0: a database for experimentally supported human microRNA and disease associations. Nucleic Acids Res. **42**(Database issue), D1070–D1074 (2014)
25. Wang, D., Wang, J., Lu, M., Song, F., Cui, Q.: Inferring the human microRNA functional similarity and functional network based on microRNA-associated diseases. Bioinformatics **26**(13), 1644–1650 (2010)
26. Xuan, P., et al.: Prediction of microRNAs associated with human diseases based on weighted k most similar neighbors. PLoS ONE **8**(8), e70204 (2013)
27. Chen, X., Zhang, D.-H., You, Z.-H.: A heterogeneous label propagation approach to explore the potential associations between miRNA and disease. J. Transl. Med. **16**(1), 348 (2018)
28. Jiang, Q., Wang, G., Jin, S., Li, Y., Wang, Y.: Predicting human microRNA-disease associations based on support vector machine. Int. J. Data Min. Bioinform. **8**(3), 282–293 (2013)
29. Li, Z.W., et al.: Accurate prediction of protein-protein interactions by integrating potential evolutionary information embedded in PSSM profile and discriminative vector machine classifier. Oncotarget **8**(14), 23638 (2017)
30. An, J.Y., et al.: Identification of self-interacting proteins by exploring evolutionary information embedded in PSI-BLAST-constructed position specific scoring matrix. Onco-target **7**(50), 82440–82449 (2016)
31. Li, Z., et al.: In silico prediction of drug-target interaction networks based on drug chemical structure and protein sequences. Sci. Rep. **7**(1), 11174 (2017)
32. Köhler, S., Bauer, S., Horn, D., Robinson, P.N.: Walking the interactome for prioritization of candidate disease genes. Am. J. Hum. Genet. **82**(4), 949–958 (2008)
33. Wei, L., Wu, S., Zhang, J., Xu, Y.: Random walk based global feature for disease gene identification. In: Tan, T., Li, X., Chen, X., Zhou, J., Yang, J., Cheng, H. (eds.) CCPR 2016. CCIS, vol. 663, pp. 464–473. Springer, Singapore (2016). https://doi.org/10.1007/978-981-10-3005-5_38
34. Yang, Z., et al.: dbDEMC: a database of differentially expressed miRNAs in human cancers. BMC Genom. **11**(Suppl 4), S5 (2010)

35. Jiang, Q., et al.: miR2Disease: a manually curated database for microRNA deregulation in human disease. Nucleic Acids Res. **37**(Database), D98–D104 (2009)

36. Bray, F., Ferlay, J., Soerjomataram, I., Siegel, R.L., Torre, L.A., Jemal, A.: Global cancer statistics 2018: GLOBOCAN estimates of incidence and mortality worldwide for 36 cancers in 185 countries. CA Cancer J. Clin. **68**(6), 394–424 (2018)

37. Zhang, C., et al.: Expression profile of MicroRNAs in serum: a fingerprint for esophageal squamous cell carcinoma. Clin. Chem. **56**(12), 1871–1879 (2010)

38. Shen, F., et al.: Genetic variants in miR-196a2 and miR-499 are associated with susceptibility to esophageal squamous cell carcinoma in Chinese Han population. Tumor Biol. **37**(4), 4777–4784 (2016)

39. Zhao-Li, C., et al.: microRNA-92a promotes lymph node metastasis of human esophageal squamous cell carcinoma via E-cadherin. J. Biol. Chem. **286**(12), 10725–10734 (2011)

40. Sun, Y., Yu, X., Bai, Q.: miR-204 inhibits invasion and epithelial-mesenchymal transition by targeting FOXM1 in esophageal cancer. Int. J. Clin. Exp. Pathol. **8**(10), 12775–12783 (2015)

# Distinguishing Driver Missense Mutations from Benign Polymorphisms in Breast Cancer

Xiyu Zhang, Ruoqing Xu, Yannan Bin, and Zhenyu Yue[✉]

Institutes of Physical Science and Information Technology,
School of Life Sciences, Anhui University, Hefei 230601, Anhui, China
zhenyuyue@126.com

**Abstract.** In genetics, a missense mutation is a point mutation in which a single nucleotide transition or transversion results in a different amino acid substitution. Most human cancers, including breast cancer, contain different mutations and majority of these are missense mutations locating in different position in DNA. There exist many bioinformatics algorithms used for predicting and prioritizing disease variants, but very little work has been specifically done for identifying driver mutations in breast cancer. Here, we present a driver missense mutation prediction method which specializes in breast cancer. This predictor was established based on a data set extracted from the DoCM and dbCPM databases, which contains 365 pathogenic and 408 benign mutations for breast cancer, respectively. We applied various approaches to obtain the features of missense mutations. In order to optimize the model, we conducted a feature selection. Then, the classifier was constructed by using a machine learning algorithm called Naïve Bayes. Ultimately, our method clearly outperformed other previous predictors when comparing them on a test set.

**Keywords:** Breast cancer · Machine learning · Naïve Bayes · Missense mutations · Cancer driver

## 1 Introduction

Worldwide, breast cancer is the leading type of cancer in women, accounting for 25% of all cases. In 2019, it is estimated that there will be 27.12 million new cases and 42260 deaths due to this malignant disease [1]. Breast cancer is more common in developed countries and is over 100 times more common in women than in men. It has become a major public health problem in the current society. As a malignant tumor, breast cancer is associated with multiple mutations [2]. These mutations are comprised of a small fraction of diver mutations that can induce the occurrence of cancer, while a major of mutations are regarded as passengers that have no direct effect on cancer [3, 4]. Therefore, identifying driver mutations has become a hot trend in breast cancer prevention and treatment. So far, there are several strategies have been applied to distinguish driver mutations [5]. Recently, scientists discovered a number of driver

---

X. Zhang and R. Xu—These authors contributed to this work equally.

© Springer Nature Switzerland AG 2019
D.-S. Huang et al. (Eds.): ICIC 2019, LNCS 11644, pp. 294–302, 2019.
https://doi.org/10.1007/978-3-030-26969-2_28

mutations in the genes highly associated with breast cancer, such as *BRCA1*, *BRCA2* and *PALB2* which is newly discovered [6]. Nowadays, in the era of big data and high-throughput sequencing, it is necessary to apply statistical and computational methods to solve these problems.

Lately, numerous missense mutation prediction methods were developed, more "cancer-focused" tools were designed to distinguish somatic missense driver mutations from passengers, such as CHASM [7], CanDrA [8], fathmm [9], and transFIC [10]. In this article, we introduced a new algorithm based on existing tools to predict the effect of missense mutations in breast cancer. Firstly, we extracted a set of mutations from DoCM [11], HGMD [12] and dbCPM [13], then arranged them into training and test sets. From other "cancer-focused" tools, we summarized some features that contributed to the prediction models. We utilized these features to construct our own method. Furthermore, we compared the prediction performance between our model and previous tools. As a result, our model obtained a higher accuracy and more balanced sensitivity and specificity in identifying driver missense mutations for breast cancer, which showed to make a significant improvement upon existing methods.

## 2  Materials and Methods

### 2.1  Data Collection

The mutation data were collected from the Database of Curated Mutations (DoCM), the Human Gene Mutation Database (HGMD) and the database of Cancer Passenger Mutations (dbCPM). We extracted mutations using the following three criteria: (1) breast cancer related, (2) missense mutations, and (3) no redundancy, i.e. a mutation only appears once. Consequently, we regarded the dataset collected from DoCM (365 pathogenic mutations) as the positive samples of the training set, while the mutations from HGMD (471 pathogenic mutations) were used for testing. We collected a total of 489 passenger mutations from the dbCPM [13] database. Then we used the data from dbCPM version 1.0 as the negative samples in training set and the newly added data in version 1.1 as negative samples in test set. The number of mutations for each dataset are shown in Table 1.

**Table 1.** Summary of mutation datasets used in this study.

| Dataset | Sample | Mutation | Source |
|---|---|---|---|
| Training set | Positive | 365 | DoCM |
| | Negative | 408 | dbCPM v1.0 |
| Test set | Positive | 471 | HGMD |
| | Negative | 81 | dbCPM v1.1 |

*Note:* dbCPM v1.1, the data in dbCPM version 1.1 and not in version 1.0.

## 2.2   Feature Selection

We collected various levels of features that contribute to the models from the existing tools for cancer missense mutation prediction. Then we selected the appropriate feature combination for driver mutation prediction in breast cancer after quantified them. All the features considered are shown in Table 2.

**Table 2.**  Summary of the features used in this study.

| Level | Feature | Type | Number |
|---|---|---|---|
| DNA | Conservation | Numeric | 2 |
|  | Functional region annotation | Discrete | 4 |
| Protein | AAFactor [14] | Numeric | 5 |
|  | Pfam domain | Discrete | 1 |
|  | Secondary structure | Numeric | 5 |

*Note:* DNA conservation was qualified by two ways, Phylop and PhastCons; Functional region annotation consists of DNase, Histone, RNA-binding proteins (RBPs) and transcription factor binding sites (TFBs); AAFactor includes Factor I, Factor II, Factor III, Factor IV, Factor V; Pfam domain was measured based on Ensembl; Secondary structures include P(H), P(E), P(C), P(D) and ASA.

The following processing has been done on the collected features. If a feature is distributed between 0 and 1, it is not normalized. Otherwise, we normalized it to between 0 and 1 (minuses the minimum and is divided by the range). To further improve the representation ability of features, we employed a two-step feature selection strategy to generate an optimal feature set containing only six most discriminative features. The first step is to sort all 17 features according to the importance of the feature, and the second is to search for the best feature subset from the ranked features. This two-step strategy will be described in detail in the following.

For the first step, we adopted all the 17 features using Weka [15] (Version 3.8.3). In this study, we utilized package "LibSVM" (Version 1.0.10) [16, 17] to rank these features. We generated a ranked list with respect to their ability of classification (see Table 3). In this list, a higher rank indicates that it is better for the classification of breast cancer mutations, which can be further considered the feature as more informative than others.

For the second step, we used the sequential forward search (SFS) [18] to select the optimal feature subset. For SFS, features were added one-by-one from the higher rank to the lower rank each time, and were used to construct the Naïve Bayes-based prediction model on the 10-fold cross validation. Finally, the feature subset, with which the prediction model would achieve the best performance, was recognized as the optimal set.

Through these two-step strategy, we have selected a subset of features contained six, which are phylop (Computation of p-values for conservation or acceleration, either lineage-specific or across all branches), P(D) (Probability for disordered), phastCons

(Conservation scoring and identification of conserved elements), pfam (pfam domains affected due to mutation), P(E) (Probability for Beta-strand), DNase (DNase I hypersensitive sites), and P(C) (Probability for coil of protein secondary structure).

### 2.3    Model Construction

We adopted the selected optimal feature subset to construct a predicting model using an effective and widely used classification algorithm. Specifically, in this process, package "Naïve Bayes" [19, 20] in Weka (Version 3.8.3) was used to generate the final model. In order to get better generalization performance and improve the credibility of the model, we adopted 10-fold cross validation in the construction of the model.

### 2.4    Performance Measurement

For measuring the proposed prediction model, we used four metrics commonly used in binary classification tasks, including Sensitivity (SE), Specificity (SP), Mathew's correlation coefficient (MCC), and the area under ROC curve (AUC) [21–23]. They were calculated as follows:

$$\begin{cases} SEN = \frac{TP}{TP+FN} \times 100\% \\ SPE = \frac{TN}{TN+FP} \times 100\% \\ MCC = \frac{TP \times TN - FP \times FN}{\sqrt{(TP+FN)(TP+FP)(TN+FP)(TN+FN)}} \times 100\% \end{cases}$$

where the TP, TN, FP and FN represent the numbers of true positives, true negatives, false positives and false negatives, respectively. The two metrics, SEN and SPE, measure the prediction ability of a predictor for the positives and negatives, respectively, while the MCC is used to evaluate the overall performance of a predictor.

In addition, we calculated the area under ROC curve (AUC) to evaluate the predictive performance. A receiver operating characteristic curve, or ROC curve, is a graphical plot that illustrates the diagnostic ability of a binary classifier system as its discrimination threshold is varied. The value of AUC ranges from 0.5 to 1. When the AUC score of a predictor is close to 1, the predictor is considered as a perfect predictor; when the AUC score is 0.5, it corresponds to a random predictor.

## 3    Results and Discussion

### 3.1    Analysis of the Features

In order to construct a better model, we chose the features that contribute to the model as much as possible. Here, we interpreted the encoded features from a statistical perspective using the Wilcoxon signed-rank test, which is a non-parametric statistical hypothesis test. The statistical results of these features and the corresponding potential biological significance are summarized as follows.

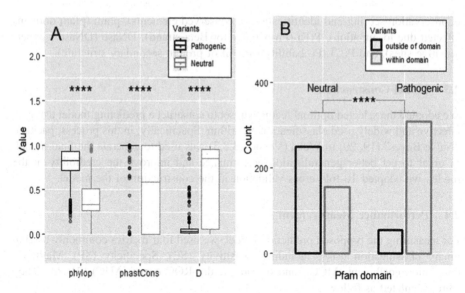

**Fig. 1.** Graphical interpretation of four statistically significant features. ****indicates that the difference is extremely significant, p value $\leq$ 0.0001.

We concluded that among the 17 encoded features, there are ten with statistically significant differences, i.e. pfam, phylop, phastCons, Factor III, DNase, ASA, P(C), P(E), P(H), P(D), when the threshold of p value was set to 0.01. As depicted in Fig. 1, the smallest p values [24] are obtained from pfam, phylop, phastCons, P(D).

As can be seen in Fig. 1 A, among the three features, phylop, phastCons and P(D), the corresponding distribution of values between pathogenic and neutral mutations are significantly different. From 1 B, we can clearly see the pathogenic missense mutations tend to more likely occur in pfam domains. We could draw a conclusion that a mutation happened at a highly conservative DNA site or locate in an important position of protein structure [25], may be more likely to be a driver mutation for breast cancer.

### 3.2    Model Construction

In this study, we attempted to use a variety of machine learning algorithms to build prediction models, such as Random Forest, J48, Naïve Bayes, and so on. Based on the performance metrics of these predictors, we decided to adopt Naïve Bayes algorithm [26] as the final prediction model. Our experiment results showed that the models built using Bayesian have better classification performance than any other algorithms used here on the training set using 10-fold cross validation.

### 3.3    Comparison of the Optimal Feature Subset and Individual Features

As mentioned earlier in the article, our optimal features subset consists of six individual features: pfam, phylop, phastCons, DNase, P(E), P(D). AUC and the ranks of individual features on the training set are list in Table 3. To evaluate the effects of the

**Table 3.** Performance of individual features on the training set.

| Feature | AUC | Rank |
|---------|-----|------|
| phylop | 0.829 | 1 |
| P(D) | 0.787 | 2 |
| phastCons | 0.750 | 3 |
| pfam | 0.732 | 4 |
| P(E) | 0.645 | 5 |
| DNase | 0.634 | 6 |
| P(C) | 0.619 | 7 |
| ASA | 0.605 | 8 |
| P(H) | 0.598 | 9 |
| Factor I | 0.526 | 10 |
| Factor V | 0.526 | 11 |
| histone | 0.503 | 12 |
| Factor III | 0.501 | 13 |
| TFBs | 0.500 | 14 |
| RBPs | 0.500 | 15 |
| Factor II | 0.500 | 16 |
| Factor IV | 0.492 | 17 |

feature selection, we compared the predictive performance between our feature subset and the six individual features. For the purpose of making a fair comparison, we performed 10-fold cross validation tests using the training dataset. The results are shown in Table 4.

We could draw conclusions clearly from Table 4 that the feature set achieved the best predictive performance globally in terms of AUC and MCC. Specially, our features achieved an AUC of 0.921 and an MCC of 0.673, which were 3.1–29.5% and 1.5–39.5% higher than that of the other features, respectively. In the term of SEN, P(E)

**Table 4.** The comparison of the optimal feature subset and individual features.

| Feature | SEN | SPE | AUC | MCC |
|---------|-----|-----|-----|-----|
| pfam | 0.615 | 0.849 | 0.694 | 0.474 |
| phylop | 0.833 | 0.824 | 0.890 | 0.658 |
| phastCons | 0.531 | **0.964** | 0.890 | 0.658 |
| DNase | 0.497 | 0.769 | 0.608 | 0.278 |
| P(E) | **0.852** | 0.427 | 0.626 | 0.312 |
| P(D) | 0.634 | 0.934 | 0.808 | 0.590 |
| Optimal feature subset | 0.748 | 0.921 | **0.921** | **0.673** |

*Note*: Sen, sensitivity; Spe, specificity; MCC, Matthews's correlation coefficient; AUC, area under ROC curve. The maximum values in each evaluation metric are marked in bold.

achieved the highest score of 0.852, which is slightly higher than our feature subset and phastCons reached the highest score of 0.964 in the respects of SPE. However, we could conclude that the individual feature failed to achieve the best balance between SEN and SPE. But our feature subset was much better in the way of balance. This may suggest that the established model which is constructed by the selected features subset could identify true driver mutations in breast cancer more possibly, which means that it may get more true positives and at the meanwhile produce fewer false positives.

### 3.4    Comparison of Performance Between Our Method and Other Tools

For comparison, we utilized four representative and widely used predictors. According to Bailey et al. [27], CHASM [7], CanDrA [8], FATHMM [9] and transFIC [10] are cancer-focused algorithms which could make prediction for driver missense mutations. TransFIC was applied to PolyPhen2 predictions due to fewer missing values. When evaluating the performance of them, we also used area under ROC curve (AUC) and other performance metrics which can be derived from a confusion matrix. The results on the test set for our method and other tools were shown in Table 5.

**Table 5.** Performance evaluation on the test set.

| Method | Sen | Spe | MCC | AUC |
|---|---|---|---|---|
| CanDra | 0.907 | 0.059 | −0.030 | 0.483 |
| CHASM | **0.876** | 0.059 | −0.078 | 0.467 |
| Fathmm | 0.610 | 0.784 | 0.273 | 0.697 |
| transFIC | 0.403 | **0.878** | 0.249 | 0.640 |
| Our method | 0.826 | 0.679 | **0.414** | **0.824** |

*Note*: Sen, sensitivity; Spe, specificity; MCC, Matthews correlation coefficient; AUC, area under ROC curve. The maximum values in each evaluation metric are marked in bold.

Our method reached 0.824 in the area under the ROC curve (AUC). Compared with other four predictors, our model obtained the highest AUC and more balanced sensitivity and specificity for identifying driver missense mutations in breast cancer, which was shown to make a significant improvement upon existing methods.

## 4    Conclusion

Breast cancer is becoming more and more common as a malignant tumor, and the cure rate of early breast cancer is 60%–70%. Consequently, early screening is the key to improve clinical treatment. In this article, we presented an accurate predictor based on the machine learning method, which achieved the effective discrimination performance between driver and passenger missense mutations, specifically for breast cancer. Moreover, we find that pathogenic mutations of breast cancer tend to locate in highly conservative DNA sites and protein domains. The proposed prediction method might contribute to identify potentially disease-causing mutations from a computational perspective and we hope that it will be helpful for high-throughput screening from massive breast cancer mutation data.

**Acknowledgements.** This work was supported by the National Natural Science Foundation of China (61672037, 21601001, and 11835014), the Anhui Provincial Out-standing Young Talent Support Plan (gxyqZD2017005), the Young Wanjiang Scholar Program of Anhui Province, the Recruitment Program for Leading Talent Team of Anhui Province (2019-16), the China Post-doctoral Science Foundation Grant (2018M630699) and the Anhui Provincial Postdoctoral Science Foundation Grant (2017B325).

# References

1. Siegel, R.L., Miller, K.D., Jemal, A.: Cancer statistics, 2019. CA Cancer J. Clin. **69**(1), 7–34 (2019)
2. Early Breast Cancer Trialists' Collaborative Group: Favourable and unfavourable effects on long-term survival of radiotherapy for early breast cancer: an overview of the randomised trials. Lancet **355**(9217), 1757–1770 (2000)
3. Pierce, L.J., Phillips, K.A., Griffith, K.A., et al.: Local therapy in BRCA1 and BRCA2 mutation carriers with operable breast cancer: comparison of breast conservation and mastectomy. Breast Cancer Res. Treat. **121**(2), 389–398 (2010)
4. Early Breast Cancer Trialists' Collaborative Group: Effects of radiotherapy and of differences in the extent of surgery for early breast cancer on local recurrence and 15-year survival: an overview of the randomised trials. Lancet **366**(9503), 2087–2106 (2005)
5. Haber, D.A., Settleman, J.: Cancer: drivers and passengers. Nature **446**(7132), 145–146 (2007)
6. Hart, S.N., Hoskin, T., Shimelis, H., et al.: Comprehensive annotation of BRCA1 and BRCA2 missense variants by functionally validated sequence-based computational prediction models. Genet. Med. **21**(1), 71–80 (2019)
7. Carter, H., Chen, S., Isik, L., et al.: Cancer-specific high throughput annotation of somatic mutations: computational prediction of driver missense mutations. Cancer Res. **69**, 6660–6667 (2009)
8. Mao, Y., Chen, H., Liang, H., et al.: CanDrA: cancer-specific driver missense mutation annotation with optimized features. PLoS ONE **8**, e77945 (2013)
9. Shihab, H.A., Gough, J., Cooper, D.N., et al.: Predicting the functional, molecular, and phenotypic consequences of amino acid substitutions using hidden Markov models. Hum. Mutat. **34**, 57–65 (2013)
10. Gonzalez-Perez, A., Deu-Pons, J., Lopez-Bigas, N.: Improving the prediction of the functional impact of cancer mutations by baseline tolerance transformation. Genome Med. **4**, 89 (2012)
11. Ainscough, B.J., Griffith, M., Coffman, A.C., et al.: DoCM: a database of curated mutations in cancer. Nat. Methods **13**(10), 806–807 (2016)
12. Stenson, P.D., Mort, M., Ball, E.V., et al.: The Human Gene Mutation Database: towards a comprehensive repository of inherited mutation data for medical research, genetic diagnosis and next-generation sequencing studies. Hum. Genet. **136**(6), 665–677 (2017)
13. Yue, Z., Zhao, L., Xia, J.: dbCPM: a manually curated database for exploring the cancer passenger mutations. Briefings Bioinform. bby105 (2018)

14. Atchley, W.R., Zhao, J., Fernandes, A.D., et al.: Solving the protein sequence metric problem. Proc. Natl. Acad. Sci. **102**(18), 6395–6400 (2005)

15. Holmes, G., Donkin, A., Witten, I.H.: WEKA: a machine learning workbench. (Working paper 94/09). 1994, Hamilton, New Zealand: University of Waikato, Department of Computer Science

16. Sarojini, B., Ramaraj, N., Nickolas, S.: Enhancing the performance of LibSVM classifier by Kernel f-score feature selection. In: Ranka, S., Aluru, S., Buyya, R., Chung, Y.-C., Dua, S., Grama, A., Gupta, S.K.S., Kumar, R., Phoha, V.V. (eds.) IC3 2009. CCIS, vol. 40, pp. 533–543. Springer, Heidelberg (2009). https://doi.org/10.1007/978-3-642-03547-0_51

17. Chang, C.C., Lin, C.J.: LIBSVM: a library for support vector machines. ACM Trans. Intell. Syst. Technol. (TIST) **2**(3), 27 (2011)

18. Pudil, P., Novovičová, J., Kittler, J.: Floating search methods in feature selection. Pattern Recogn. Lett. **15**(11), 1119–1125 (1994)

19. Lewis, D.D.: Naive (Bayes) at forty: The independence assumption in information retrieval. In: Nédellec, C., Rouveirol, C. (eds.) ECML 1998. LNCS, vol. 1398, pp. 4–15. Springer, Heidelberg (1998). https://doi.org/10.1007/BFb0026666

20. Rish, I.: An empirical study of the Naïve Bayes classifier. In: IJCAI 2001 Workshop on Empirical Methods in Artificial Intelligence, vol. 3, no. 22, 41–46 (2001)

21. Purves, R.D.: Optimum numerical integration methods for estimation of area-under-the-curve (AUC) and area-under-the-moment-curve (AUMC). J. Pharmacokinet. Biopharm. **20**(3), 211–226 (1992)

22. Chen, Z., Liu, X., Li, F., et al.: Large-scale comparative assessment of computational predictors for lysine post-translational modification sites. Briefings Bioinform. bby089 (2018)

23. Huang, J., Ling, C.X.: Using AUC and accuracy in evaluating learning algorithms. IEEE Trans. Knowl. Data Eng. **17**(3), 299–310 (2005)

24. Westfall, P.H., Young, S.S.: Resampling-Based Multiple Testing: Examples and Methods for P-Value Adjustment. Wiley, New York (1993)

25. Yang, Y., Chen, B., Tan, G., et al.: Structure-based prediction of the effects of a missense variant on protein stability. Amino Acids **44**(3), 847–855 (2013)

26. Patil, T.R., Sherekar, S.S.: Performance analysis of Naïve Bayes and J48 classification algorithm for data classification. Int. J. Comput. Sci. Appl. **6**(2), 256–261 (2013)

27. Bailey, M.H., Tokheim, C., Porta-Pardo, E., et al.: Comprehensive characterization of cancer driver genes and mutations. Cell **173**(2), 371–385.e18 (2018)

# A Novel Method to Predict Protein Regions Driving Cancer Through Integration of Multi-omics Data

Xinguo Lu[1], Xinyu Wang[1], Ping Liu[2(✉)], Zhenghao Zhu[1], and Li Ding[1]

[1] College of Computer Science and Electronic Engineering, Hunan University, Changsha, China
[2] Hunan Want Want Hospital, Changsha, China
lp-simple123@126.com

**Abstract.** Identifying cancer drivers is critical to advancing cancer research and personalized medicine. Most methods for identifying cancer drivers focus on the entire genes or a single mutation site. But not all mutations in a gene have the same effect, the consequences of which usually depend on the position in the protein and amino acid change. The intermediate level of analysis between individual locations and the entire gene may give us better statistics and better resolution than the former. Here, we developed prDriver, a Bayesian hierarchical modeling method that identifies regions of proteins with high functional impact scores and significant effects on gene expression levels. Our study highlights the importance of integrating multi-omics data in predicting cancer driver and provides a statistically rigorous solution for cancer target discovery and development.

**Keywords:** Cancer · Driver region · Bayesian hierarchical model

## 1 Introduction

The formation of cancer is due to a functional acquired genomic aberrations. In recent years, with the emergence of a large number of genomics research methods, many large-scale cancer research open source projects have also been proposed, such as The Cancer Genome Atlas (TCGA), the International Cancer Genome Consortium (ICGC) and the Therapeutically Applicable Research to Generate Effective Treatments (TARGET). These large cancer research programs bring a wealth of high-throughput data to the exploration of cancer pathogenesis [1]. Genomic alterations in tumors can be divided into two categories: those that play a leading role in tumorigenesis (driver) and those that have no direct impact on the overall progression of tumor development (passenger). Previous studies have shown that only a small percentage of genomic alterations are the cancer driver, and the rest are cancer passenger conferring no selective advantage [2]. It is easy to define "driver" in physiological roles, but systematically identifying driver from large-scale cancer genome data remains a huge challenge.

© Springer Nature Switzerland AG 2019
D.-S. Huang et al. (Eds.): ICIC 2019, LNCS 11644, pp. 303–312, 2019.
https://doi.org/10.1007/978-3-030-26969-2_29

A number of computational methods and tools have been developed to distinguish between cancer drivers and passengers, using multi-omics data. Such methods rely on the hypothesis that the mutation rate of cancer driver should be significantly higher than the background mutation rate, such as MutSigCV [3], or those that accumulate highly destructive mutations, such as OncodriveFM [4]. OncodriveCLUST [5] focuses on the internal distribution of protein mutations and measures the bias of genes towards mutation clustering with respect to a background model. These methods only consider local sequence features and ignore the gene regulatory effects associated with mutations. Some studies have begun to use gene expression data to systematically predict cancer drivers. For example, xSeq [6], a statistical model, systematically studies the effects of somatic mutations on gene expression levels, according to known pathway networks. This method biases the unexplored networks. Most importantly, these methods do not take into account the functional impact scores generated by SIFT [7] and PROVEAN [8], which are recognized as useful for identifying functional mutations.

Here, we present prDriver, a novel method that identifies protein regions that have high functional impact scores and significant effects on gene expression levels in tumor samples. We hypothesize that different regions of the same protein play different roles in the process of carcinogenesis. Because missense mutations on different protein regions may affect or destroy different functions. We use a Bayesian hierarchical model that combines somatic mutation data, gene expression data and structural information to predict cancer driver and provide a deeper insights into the mechanisms of cancer drivers.

## 2 Materials and Methods

### 2.1 Datasets

The experimental data is from the TCGA project. Data (https://www.synapse.org/#!Synapse:syn1461151) includes somatic mutations and gene expression data for Breast invasive carcinoma (BRCA). We only care about missense mutations with a frequency greater than two, with a total of 750 mutations in 454 patients. We choose to use the annotated Pfam domains as protein candidate regions. The eDriver [9] provides the start and end positions of the pfam domain on the amino acid sequence. There are a total of 36,626 Pfam domains available for us to use. We use the PROVEAN tool to calculate the functional impact scores for mutations, an online tool that provides us with SIFT scores and PROVEAN scores.

### 2.2 Method Overview

In recent years, researchers have gradually found that the formation of cancer may be related to the joint action of multiple somatic mutations [10]. The driving region is an intermediate level between the driving gene and the driving mutation, which provides us with better statistical performance and better resolution [9]. We are based on the hypothesis that different regions of the same protein have different effects on carcinogenesis. If the hypothesis is established, the influence or disruption of the missense mutation on the protein region will have different effects on the expression level of the gene. Therefore, we believe that the region under positive selection will significantly affect the expression level of genes and have a high functional impact score.

We use the Bayesian hierarchical model developed for eQTL data [11, 12] as a framework for combining somatic mutation data, gene expression data, and protein structure data to identify regions of the protein that best predict gene expression. The functional impact potential is used as a priori parameter to adjust the weight coefficient. Finally, the correlation matrix between the protein region and the gene is obtained. We use the number of genes associated with the region as its score, and select the top ranked region as the candidate driver.

## 2.3 Functional Impact Potentials for Regions

Functional impact scores represent evolutionary and structural features of single nucleotide variants (SNVs). Many tools have been developed to characterize the functional effects of SNVs based on physicochemical properties or evolutionary conservation of amino acid sequences affected by SNVs, such as SIFT, MutationAssessor and PROVEAN [13]. SIFT predicts the effect of amino acid substitutions on protein function by aligning homologous protein sequences [7]. We use the functional impact scores as the features of the SNV. Based on the features of the SNV, we define functional impact potential (*FIP*), which indicate the probability that each SNV $n$ affects gene $g$ expression levels.

$$FIP_{g,n} = sigmoid\left(\sum_k \beta_k f_{n,k}\right) \tag{1}$$

$$sigmoid(t) = \frac{1}{1 + exp(-t)} \tag{2}$$

Where $f_{n,k}$ represents the $k^{th}$ feature of SNV $n$. $\beta_k$ is a priori parameter that reflects the relative contribution of each feature. The value of the parameter $\beta$ can be automatically estimated from the data. The sigmoid function is used to prevent the functional impact potential from increasing unboundedly.

We define the functional impact of the region as a collection of functional effects of the SNV in the region and use the sigmoid function. We define the regional functional impact potential (*RFIP*) as follows:

$$RFIP_{g,r} = sigmoid\left(\sum_n \left(2 \times sigmoid\left(\sum_k \beta_k f_{n,k}\right) - 1\right)\right) \tag{3}$$

Where $n$ is the SNV in the calculated region $r$. Therefore, regions containing a large number of SNVs with high functional impact potential may have a higher functional impact potential. This method makes the region containing more SNVs tend to have a high functional impact potential, which is reasonable because it is more polymorphic.

## 2.4    Prediction of the Cancer Driver

We establish a linear model where the expression level of gene $g$ is a linear combination of potential regions $r$. Considering the gene expression levels and potential regions in $N$ samples, the linear relationship is as follows:

$$y_g = \sum_{m=1}^{M} w_{g,m} r_m + \varepsilon_g \tag{4}$$

Where $y_g = [y_{g_1}, y_{g_2}, y_{g_3}, \ldots, y_{g_N}]$ represents the expression level of gene $g$ in $N$ samples, $r_m = [r_{m_1}, r_{m_2}, r_{m_3}, \ldots, r_{m_N}]$ represents the mutation states of the potential region in the same $N$ samples. In sample $n$, if a missense mutation occurs on the region $r_m$, $r_{m_n}$ is 1; otherwise, it is 0. $w_{g,m}$ is a weight coefficient indicating the correlation between the gene expression level and the mutation states of the potential region. $\varepsilon_g$ represents a zero mean Gaussian noise.

Our goal is to evaluate weight coefficients to identify regions that affect gene expression levels. More specifically, maximize the values of the following formulas:

$$P(w, y | r) = P(y | r, w) P(w) \tag{5}$$

$$P(y | r, w) \sim N\left(\sum_i w_i r_i, \sigma^2\right) \tag{6}$$

Where $P(w)$ represents the prior probability distribution of the weight coefficient $w$. We use the functional impact potential as a priori parameter of the weight coefficient $w$ to establish the following model:

$$P(w) \propto exp(-|RFIP \times w|) \tag{7}$$

Considering that only a small part of the variation is cancer driver, we hope that the smaller weight coefficients shrink to zero, and the larger weight coefficients have a stronger bias towards non-zero [14]. Therefore, an L1 penalty is added to the prior score to make it an L1 prior.

In order to avoid singularity problems and degeneracy problems, we introduce an L2 regularization term [15]. We estimate the functional impact coefficient $\beta$ and the weight coefficient $w$ by solving the following optimization problems.

$$min_\beta \left\{ \left\| y_g - \sum_m w_{g,m} r_m \right\|_2 + \lambda_{1,g} \sum_m |RFIP_{g,m} \cdot w_{g,m}| + \lambda_{2,g} \sum_m \sum_n \sum_k \beta_{m,n,k}^2 \right\} \tag{8}$$

Since the optimization problem is a convex optimization problem, the local minimum is the global minimum. We use an iterative coordinate descent algorithm to solve this optimization problem. This method is a non-gradient optimization method that searches in the direction of one coordinate in each iteration and uses a different coordinate to achieve a local minimum of the objective function. Iteration is between two steps [16]. In the one step, the current $\beta$ is given to optimize $w$; the other step is given the current $w$ to optimize $\beta$. We calculate the regularization parameters $\lambda_{1,g}, \lambda_{2,g}$ through 10 fold cross validation procedure.

By solving the above optimization problem, we obtained an association matrix between protein regions and gene expression levels. The association matrix represents the correlation between each pair of regions and genes, that is, the extent to which regional variation affects the expression level of the gene. Here, we calculate the number of genes associated with the region by a non-zero weight coefficient and use it as a score. The greater the score, the more likely the region is to become a cancer driver. So, we give a threshold, and the region score greater than the threshold is considered to be the driver.

# 3 Result

## 3.1 prDriver Identifies Known Cancer Driver Regions

We validated the validity of prDriver using the BRCA dataset from the TCGA database. We need to filter out mutations that are less than two in frequency. The mutation has an accidental occurrence only once. 750 missense mutations from 454 patients were used. In order to reduce the amount of data for gene expression, we selected 3,030 of these genes, which are known to play important functional roles, such as signaling genes [17]. For candidate protein regions, we used the annotated Pfam domain and map missense mutations to the Pfam domain based on location. The potential functional impact scores were calculated using SIFT scores and PROVEAN scores, and missing values were filled with average values.

**Table 1.** Candidate driver regions.

| Rank | Gene | Region (Pfam) | Brief description |
|------|------|---------------|-------------------|
| 1 | AKT1 | PH | Involved in intracellular signaling or as constituents of the cytoskeleton |
| 2 | TP53 | P53 | Conserving stability by preventing genome mutation |
| 3 | BTNL8 | V-set | Found in tyrosine-protein kinase receptors; and in the programmed cell death protein 1 |
| 4 | ERBB2 | Pkinase | Moving a phosphate group onto proteins, in a process called phosphorylation |
| 5 | PIK3CA | PI3Ka | Suggested to be involved in substrate presentation |
| 6 | OR11H1 | 7tm_1 | Transducing extracellular signals through interaction with guanine nucleotide-binding (G) proteins |
| 7 | OR11H12 | 7tm_1 | Transducing extracellular signals through interaction with guanine nucleotide-binding (G) proteins |
| 8 | MAP2K4 | Pkinase_Tyr | Transfering a phosphate group from ATP to a protein in a cell |
| 9 | FRG1B | FRG1 | Function is still not clear |
| 10 | GUCY2C | Pkinase | Moving a phosphate group onto proteins, in a process called phosphorylation |
| 11 | KRAS | Arf | Functioning as regulators of vesicular traffic and actin remodelling |

*(continued)*

**Table 1.** (*continued*)

| Rank | Gene | Region (Pfam) | Brief description |
|------|------|---------------|-------------------|
| 12 | POTEE | Actin | Essential for such important cellular functions as the mobility and contraction of cells during cell division |
| 13 | AL445665.1 | FTHFS | Formate–tetrahydrofolate ligase. Function is still not clear |
| 14 | USP32 | UCH | Ubiquitin carboxyl-terminal hydrolase. Function is still not clear |
| 15 | PIK3CA | PI3K_C2 | Involved in targeting proteins to cell membranes |

When prDriver was applied to the above data, we predicted 25 protein regions located on 25 genes (see Table 1, only the first 15 regions are listed). 7 of the 25 genes we predicted include in the Cancer Gene Census (CGC) [18], such as the AKT1 and TP53 genes with high scores in our experiments. We predicted that the P53 DNA-binding domain on the TP53 gene is cancer driver. Studies have shown that

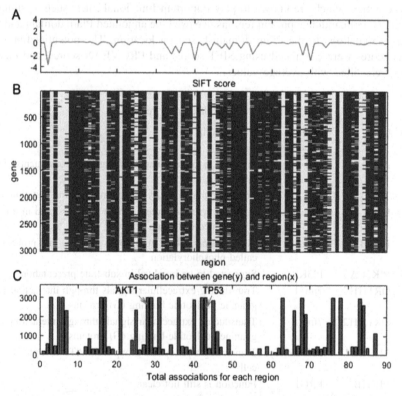

**Fig. 1.** The output of prDriver on BRCA. (A) SIFT scores for 89 protein regions. (B) An association matrix between protein regions and genes, with yellow dots indicating non-zero correlation coefficients. (C) Score of the protein region, and regions located on the known genes AKT1 and TP53 were labeled. (Color figure online)

suppression of p53 in human breast cancer cells is shown to lead to increased CXCR5 chemokine receptor gene expression and activated cell migration in response to chemokine CXCL13 [19].

We show the correlation matrix between protein regions and genes, as well as the score for each region (see Fig. 1). The predicted protein regions have higher scores than other candidate protein regions, that is, these protein regions can significantly affect gene expression levels and have a high functional impact.

### 3.2 Compare with State-of-the-Art Approach

We compare prDriver with three other methods, MutSigCV [3], OncodriveCLUST [5] and eDriver [9]. A list of driver genes identified by the three known methods on the BRCA dataset from the TCGA database was obtained by the database DriverDB [20]. MutSigCV identified 158 driver genes. OncodriveCLUST and eDriver identified 26 and 273 driver genes, respectively. As shown in the Venn diagram [21], prDriver identified 25 driver genes, 14 of which were determined by at least one other method, making them more convincing as driver genes (see Fig. 2). At the same time, it can be seen that prDriver is a good complement to other methods.

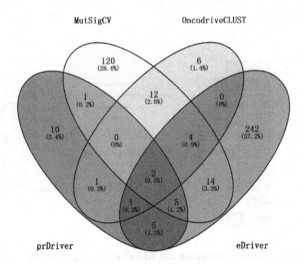

**Fig. 2.** Venn diagram showing the overlap between the four different methods in their predictions.

We filtered out 36 BRCA known driver genes in the CGC list. As shown in the Table 2, the proportion of known driver genes among the driver genes identified by the four methods is shown. In this ratio, the prDriver method is 24%, the largest of the four methods. In other words, the candidate driver genes recognized by prDriver are more reliable. The known driver gene accounted for only 4.76% in the eDriver method, which recognized 36.11% of the BRCA known driver genes, but the number of candidate driver genes was too large, resulting in high false positives. The number of candidate driver genes identified by OncodriveCLUST is similar to prDriver, but it only recognizes two known genes. So, prDriver is better than other methods in terms of accuracy.

**Table 2.** Comparison of the accuracy of the four methods.

| Method | Driver gene | CGC (BRCA) | Proportion |
|---|---|---|---|
| MutSigCV | 158 | 15 | 9.49% |
| OncodriveCLUST | 26 | 2 | 7.69% |
| eDriver | 273 | 13 | 4.76% |
| prDriver | 25 | 6 | 24% |

### 3.3 Enrichment Analysis

To demonstrate the biological specificity of the driver genes predicted by prDriver, we used GO annotation analysis and KEGG pathway annotation analysis through the R language ClusterProfiler package.

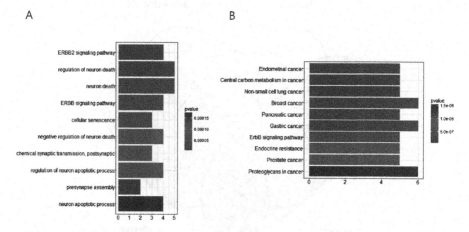

**Fig. 3.** GO biological function annotation and KEGG pathway annotation enrichment analysis. (A) Top 10 GO biological function annotation bar chart. The length of the bar chart represents the number of driver gene annotations to this GO term, and the color of the bar chart indicates the degree of enrichment of the gene on the GO term. (B) Top 10 GO biological function annotation bar chart. (Color figure online)

There are three main branches in GO annotation analysis: molecular function, biological process and cellular composition. We mainly consider gene annotation related to biological processes. As shown in the Fig. 3A, the genes predicted by prDriver on the BRCA data set are mainly enriched in important biological processes such as signaling pathways, gene regulation, cell migration, and receptor binding.

These processes are highly correlated with the formation of cancer. The KEGG database system analyzes the metabolic pathways of gene products in cells and the function of these gene products. Figure 3B shows the KEGG pathway annotation enrichment results for candidate genes identified by prDriver on the BRCA data set. As can be seen from the figure, candidate genes are significantly enriched in the cancer pathway, particularly including BRCA cancer.

# 4 Discussion

We proposed prDriver to distinguish between cancer drivers and passengers. Using the Bayesian hierarchical model as the statistical framework, combining somatic mutation data, gene expression data and protein structure data, the method predicts the protein regions with high functional impact scores and significant effects on gene expression levels. The score is calculated based on the obtained correlation matrix between the region and the gene. We predict that regions with scores above a given threshold are cancer driver. We applied the prDriver to the BRCA dataset with the annotated Pfam domain as the candidate region, using the SIFT score and the PROVEAN score as the functional impact score. Through the analysis of the results, our method can identify known driver regions, and the prediction accuracy is higher than MutSigCV, eDriver and OncodriveCLUST. Then, we performed GO biological process analysis and KEGG pathway annotation analysis on candidate driver genes. The results indicate that they are closely related to gene regulation processes, cancer formation and the like.

**Acknowledgments.** This work was supported by National Natural Science Foundation of China (Grant No. 61502159) and Natural Science Foundation of Hunan Province, China (Grant No. 2018JJ2053).

# References

1. Lu, X., Lu, J., Liao, B., Li, X., Qian, X., Li, K.: Driver pattern identification over the gene co-expression of drug response in ovarian cancer by integrating high throughput genomics data. Sci. Rep. **7**(1), 16188 (2017)
2. Stratton, M.R., Campbell, P.J., Futreal, P.A.: The cancer genome. Nature **458**(7239), 719 (2009)
3. Lawrence, M.S., et al.: Mutational heterogeneity in cancer and the search for new cancer-associated genes. Nature **499**(7457), 214 (2013)
4. Gonzalez-Perez, A., Lopez-Bigas, N.: Functional impact bias reveals cancer drivers. Nucleic Acids Res. **40**(21), e169–e169 (2012)
5. Tamborero, D., Gonzalez-Perez, A., Lopez-Bigas, N.: OncodriveCLUST: exploiting the positional clustering of somatic mutations to identify cancer genes. Bioinformatics **29**(18), 2238–2244 (2013)
6. Ding, J., et al.: Systematic analysis of somatic mutations impacting gene expression in 12 tumour types. Nat. Commun. **6**, 8554 (2015)
7. Ng, P.C., Henikoff, S.: SIFT: predicting amino acid changes that affect protein function. Nucleic Acids Res. **31**(13), 3812–3814 (2003)

8. Choi, Y., Chan, A.P.: PROVEAN web server: a tool to predict the functional effect of amino acid substitutions and indels. Bioinformatics **31**(16), 2745–2747 (2015)

9. Porta-Pardo, E., Godzik, A.: e-Driver: a novel method to identify protein regions driving cancer. Bioinformatics **30**(21), 3109–3114 (2014)

10. Lu, X., Qian, X., Li, X., Miao, Q., Peng, S.: DMCM: a data-adaptive mutation clustering Method to identify cancer-related mutation clusters. Bioinformatics **35**(3), 389–397 (2018)

11. Lee, S.I., et al.: Learning a prior on regulatory potential from eQTL data. PLoS Genet. **5**(1), e1000358 (2009)

12. Wang, Z., et al.: Cancer driver mutation prediction through Bayesian integration of multi-omic data. PLoS ONE **13**(5), e0196939 (2018)

13. Cheng, F., Zhao, J., Zhao, Z.: Advances in computational approaches for prioritizing driver mutations and significantly mutated genes in cancer genomes. Briefings Bioinform. **17**(4), 642–656 (2015)

14. Tibshirani, R.: Regression shrinkage and selection via the lasso. J. Royal Stat. Soc. Ser. B (Methodol.) **58**(1), 267–288 (1996)

15. Nie, F., Huang, H., Cai, X., Ding, C.H.: Efficient and robust feature selection via joint $\ell2$, 1-norms minimization. In: Advances in neural information processing systems, pp. 1813–1821 (2010)

16. Wright, S.J.: Coordinate descent algorithms. Math. Program. **151**(1), 3–34 (2015)

17. Logsdon, B.A., Gentles, A.J., Miller, C.P., Blau, C.A., Becker, P.S., Lee, S.I.: Sparse expression bases in cancer reveal tumor drivers. Nucleic Acids Res. **43**(3), 1332–1344 (2015)

18. Sondka, Z., Bamford, S., Cole, C.G., Ward, S.A., Dunham, I., Forbes, S.A.: The COSMIC cancer gene census: describing genetic dysfunction across all human cancers. Nat. Rev. Cancer **18**, 696–705 (2018)

19. Mitkin, N.A., et al.: p53-dependent expression of CXCR1 chemokine receptor in MCF-7 breast cancer cells. Sci. Rep. **5**, 9330 (2015)

20. Cheng, W.C., et al.: DriverDB: an exome sequencing database for cancer driver gene identification. Nucleic Acids Res. **42**(D1), D1048–D1054 (2014)

21. Venn's diagrams. http://bioinfogp.cnb.csic.es/tools/venny/index.html. Accessed 25 Mar 2019

# *In Silico* Identification of Anticancer Peptides with Stacking Heterogeneous Ensemble Learning Model and Sequence Information

Hai-Cheng Yi[1,2], Zhu-Hong You[1,2(✉)], Yan-Bin Wang[1],
Zhan-Heng Chen[1,2], Zhen-Hao Guo[1,2], and Hui-Juan Zhu[3]

[1] The Xinjiang Technical Institute of Physics and Chemistry,
Chinese Academy of Sciences, Urumqi 830011, China
zhuhongyou@ms.xjb.ac.cn
[2] University of Chinese Academy of Sciences, Beijing 100049, China
[3] School of Computer Science and Communication Engineering,
Jiangsu University, Zhen Jiang 212013, China

**Abstract.** Cancer is a well-known dreadful killer of human being's health, which has led to countless deaths and misery. Traditional treatment can also affect the normal cells while killing cancer cells. Meanwhile, physical or chemical techniques are costly and inefficient. Fortunately, anticancer peptides are a promising treatment, with specifically targeted, low production cost and other advantages. In order to effectively identify the anticancer peptides, we proposed a stacking heterogeneous ensemble learning model, ACP-SE, for predicting anticancer peptides. More specifically, to fully exploit protein sequence information, we developed an efficient feature representation approach by integrating binary profile feature and conjoint triad feature. Then we use a stacking ensemble strategy to combine the three heterogeneous classifiers and get the final prediction results. It was demonstrated that the proposed ACP-SE remarkably outperformed other comparison methods.

**Keywords:** Anticancer peptides · Machine learning · Stacking ensemble · Binary Profile Feature · Conjoint Triad Feature

## 1 Introduction

Cancer is one of the most terrible health killers faced by mankind, which accounting for millions of deaths around the world one year [1, 2]. Traditional therapy has strong side effects [3, 4]. With the discovery of anticancer peptides (ACPs), a kind of short peptides generally with a length of 10 to 50 amino acids, a novel alternative therapy to cure cancer has been provided, which open promising perspective for the cancer treatment and has various attractive advantages [5, 6]. It's high specificity, high tumor penetration, ease of synthesis and modification, and low production cost [7]. ACPs can selectively kill cancer cells with no effect on normal cells, because they could interact with the anionic cell membrane components of only cancer cells [4, 8]. In the recent years, ACPs-based strategies have been used to against many types of cancer successfully [9–13]. In considering the importance of ACPs to humankind's health, it's necessary and urgent to

© Springer Nature Switzerland AG 2019
D.-S. Huang et al. (Eds.): ICIC 2019, LNCS 11644, pp. 313–323, 2019.
https://doi.org/10.1007/978-3-030-26969-2_30

develop highly efficient identification and prediction method. Some valuable studies have been made in the identification and prediction of ACP [14–20].

In this study, we proposed a novel stacking ensemble machine learning model integrated SVM [21, 22], Random Forest (RF) [23, 24] and Naïve Bayes (NB) [25] model to predict anticancer peptides, named ACP-SE. The efficient features extracted from peptides sequences are fed as input into base predictors, including kernel SVM predictor, Random Forest predictor and Naïve Bayes predictor, as level 1. And then, Logistic Regression (LR) automatically learn how to achieve better performance [26]. The experimental results proved that our method is suitable for anticancer prediction task with notably prediction performance.

## 2   Materials and Methodology

### 2.1   Construction of Datasets

We constructed a novel benchmark dataset in this work for ACPs identification, as previous studies suggested, the new dataset comprised of both positive data set and negative data set, while positive samples are experimentally validated ACPs and antimicrobial peptides without anticancer function are collected as negative samples. To avoid the bias of dataset, the well-known tool CD-HIT [29] was further used to remove those peptides with similarity more than 90%. As a result, we finally obtained a dataset containing 740 samples, of which 376 were positive samples and 364 were negative samples [34]. The positive anticancer peptides samples can be represented as $P^+$, and the negative non-anticancer peptides can be represented as $P^-$. So, the whole dataset can be represented as $P$.

$$P = P^+ \cup P^- \tag{1}$$

Moreover, there is no overlap between positive and negative dataset.

$$\emptyset = P^+ \cap P^- \tag{2}$$

### 2.2   Representation of the Peptide's Sequences

A peptide sequence can be represented as:

$$P = p_1 p_2 p_3 p, \ldots p_l \tag{3}$$

Where $p_1$ represents the $1_{st}$ residue in the peptide P, $p_2$ denotes the $2_{nd}$ residue in the peptide $P$, and so on. $l$ represents the length of $P$.

*Binary Profile Features (BPF).* As mentioned above, there are 20 different amino acids in the standard amino acid alphabet [A, C, D, E, F, G, H, I, K, L, M, N, P, Q, R, S, T, V, W, Y]. Each amino acid type is encoded with the following feature vector composed of 0/1. Subsequently, for a given peptide sequence $P$, its N-terminus with the length of $k$ amino acids was encoded as the following feature vector:

$$BPF(k) = [f(p_1), f(p_2), \ldots f(p_k)]$$    (4)

where $k$ represents the length of the N-terminus of the peptide $P$ [20]. Thus, the dimension of is $20 \times k$. In this study, we set $k$ equal 7 to get better prediction performance, so a given peptide sequence was encoded to a $1 \times 140$ feature vector.

*Conjoint Triad Feature (CTF).* We also encoded peptides sequence by using the $k$-mer sparse matrix previously proposed in [27, 28]. Conjoint triad (3-mer) of peptides is composed by 3 amino acids [30]. The 20 amino acids were reduced into 7 groups [14, 31, 32]. Suppose an above-mentioned peptide sequence length is $L$, there would be $7^k$ different possible $k$-mer and $L - k + 1$ step appearing in the RNA sequence.

In this study, the value of $k$ is set to 3 to process peptide sequence. The $k$-mer sparse matrix $M$ can be defined as follows.

$$M = (a_{ij})_{7^k} \times (L - k + 1)$$    (5)

$$a_{ij} = \begin{cases} 1, \text{if } m_j m_{j+1} m_{j+2} = k - mer(i) \\ 0, \quad else \end{cases}$$    (6)

The Conjoint triad sparse matrix $M$ is a low-rank matrix, which almost retained all the raw information, including sequence frequency, position and order hidden information. Then, singular value decomposition (SVD) [33] is used to reducing one two-dimensional matrix M into a $1 \times 343$ feature vector.

## 2.3 Base Classifiers

In this study, three different type of classifiers was used as base classifiers, including SVM classifier, Random Forest classifier and Naïve Bayes classifier. Each classifier has its own advantages and disadvantages.

**Naive Bayes.** Method is a supervised learning algorithm based on Bayes theorem, which assumes that each pair of features is independent of each other. Given a class $y$ and a related feature vector from $x_1$ to $x_n$, the Bias theorem illustrates the following relations:

$$P(y|x_1 x_2 \ldots x_n) = \frac{p(y)P(x_1 x_2 \ldots x_n|y)}{p(x_1 x_2 \ldots x_n)}$$    (7)

$$P(x_i|y) = \frac{1}{\sqrt{2\pi\sigma_y^2}} exp\left(-\frac{(x^2 - \mu_y)^2}{2\sigma_y^2}\right)$$    (8)

Where parameter $\sigma_y$ and $\mu_y$ was estimated by using maximum likelihood method.

**Random Forest.** Algorithm also is a popular model in many research fields, which is an important ensemble learning method based on Bagging strategy. It can be used for classification, regression and other problems. Unlike the original paper [23], the implementation of the Scikit-learn is to take the average of each classifier's prediction probability, rather than to allow each classifier to vote on the category.

**SVM.** Has been widely used in the bioinformatics field, which is a powerful model for binary classification and regression estimation task. In binary classification problem, for a given train feature vector $x$ and label vector $y$:

$$x_i \in R^p, i = 1, 2, \ldots n \tag{9}$$

$$y = \{-1, 1\}^n \tag{10}$$

the support vector classification model can solve the problem:

$$min \frac{1}{2} w^T w + C \sum_{i=1}^{n} \zeta_i \tag{11}$$

While the decision function can be defined as:

$$sgn\left(\sum_{i=1}^{n} y_i x_i K(x_i, x) + \rho\right) \tag{12}$$

We employed Radial Basis Function (RBF) kernel in this work, which can be defined as:

$$f(x) = e^{-\gamma \|x - x'\|^2} \tag{13}$$

### 2.4 Stacking Ensemble Strategy

Moreover, we employed stacking ensemble strategy to integrate three base classifiers, the base classifiers output their prediction probability for each fold, which will further be fed as input into Logistic Regression (LR) to automatically carry out how to integrate base classifiers. The logistic regression is got from Scikit-learn [24], and can be defined as follow.

$$P_w(\pm 1|p) = \frac{1}{1 + e^{-w^T p(\pm 1|p)}} \tag{14}$$

where the $p$ is the probability outputs of the individual classifiers and $w$ is the weight for each single classifier.

### 2.5 Performance Evaluation Criteria

In this study, we purposed a novel stacking ensemble model ACP-SE, using efficiency feature to identify and predict new anticancer peptides. We used five-fold cross-validation to evaluate the performance of ACP-SE. We follow the widely used evaluation criteria, including accuracy (Acc), true positive rate (TPR), true negative rate (TNR), positive predictive value (PPV) and Matthews Correlation Coefficient (MCC) defined as:

$$Acc = \frac{TN + TP}{TN + TP + FN + FP} \tag{15}$$

$$TPR = \frac{TP}{TP + FN} \tag{16}$$

$$TNR = \frac{TN}{TN + FP} \tag{17}$$

$$PPV = \frac{TP}{TP + FP} \tag{18}$$

$$MCC = \frac{TP \times TN - FP \times FN}{\sqrt{(TP + FP)(TP + FN)(TN + FP)(TN + FN)}} \tag{19}$$

Where $TN$ indicates the true negative number, $TP$ denotes the true positive number, $FN$ represents the false negative number and $FP$ stands for the false positive number. Certainly, the Receiver Operating Characteristic (ROC) curve and the area under the ROC curve (AUC) are also adopted to evaluate the performance.

## 3   Results and Discussion

### 3.1   Evaluation of ACP-SE's Capability to Predict Anticancer Peptides

First, we executed our model ACP-SE on ACP740 dataset to evaluate its ability of predicting anticancer peptides. The five-fold cross validation details are offered in the Table 1.

**Table 1.** The fivefold cross-validation details on ACP740 dataset.

| Fold set | Acc (%) | TPR (%) | TNR (%) | PPV (%) | MCC (%) |
|---|---|---|---|---|---|
| 1 | 83.78 | 87.14 | 80.26 | 87.5 | 67.84 |
| 2 | 81.76 | 86.36 | 76.00 | 87.67 | 64.04 |
| 3 | 86.49 | 85.71 | 88.00 | 84.93 | 72.98 |
| 4 | 87.84 | 91.30 | 84.00 | 91.78 | 75.95 |
| 5 | 83.11 | 82.89 | 84.00 | 82.19 | 66.21 |
| Average | 84.59 ± 2.50 | 86.68 ± 3.04 | 82.45 ± 4.53 | 86.82 ± 3.56 | 69.40 ± 4.93 |

The average accuracy of five-fold cross-validation is 84.59% with standard deviation 2.50%, the average TPR is 86.68% with standard deviation 3.04%, the average TNR is 82.45% with standard deviation 4.53%, the mean PPV is 86.82% with standard deviation 3.56% and the MCC is 69.40% with standard deviation 4.93%. ACP-SE showed an outstanding capability to identify anticancer peptides, performed an AUC 0.907 as shown in Fig. 1, has achieved the best performance on ACP740 dataset among all methods.

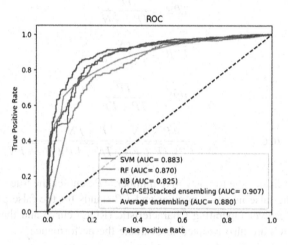

**Fig. 1.** Comparison of different ensemble strategies, individual model and ACP-SE.

## 4 Conclusion

In this study, we proposed a stacking heterogeneous ensemble learning model, ACP-SE, for predicting anticancer peptides. More specifically, we developed an efficient feature representation approach by integrating binary profile feature and conjoint triad sparse matrix to fully exploit peptides sequence information. Then we use a stacking ensemble strategy to combine the three heterogeneous classifiers including SVM, Random Forest and Naïve Bayes model and obtain the final prediction results. Meanwhile, experimental results show that the proposed method can achieve outstanding performance than other existing methods. It is anticipated that ACP-SE will become a very useful high-throughput and cost-effective tool, being widely used in anticancer peptides prediction as well as cancer treatment research. In recent years, machine learning and bioinformatics have made many inspiring advances [34–79], and we will conduct further research based on these advances.

**Acknowledgments.** This work is supported by the National Science Foundation of China, under Grants 61572506, in part by the NSFC Excellent Young Scholars Program, under Grants 61722212, in part by the Pioneer Hundred Talents Program of Chinese Academy of Sciences.

**Author Contributions.** Hai-Cheng Yi and Zhu-Hong You conceived the algorithm, carried out analyses, prepared the data sets, carried out experiments, and wrote the manuscript; Yan-Bin Wang, Zhan-Heng Chen, Zhen-Hao Guo and Hui-Juan Zhu designed, performed and analyzed experiments and wrote the manuscript; All authors read and approved the final manuscript.

**Conflicts of interest.** The authors declare no conflict of interest.

# References

1. Ferlay, J., Shin, H.R., Bray, F., Forman, D., Mathers, C., Parkin, D.M.: Estimates of worldwide burden of cancer in 2008: GLOBOCAN 2008. Int. J. Cancer **127**(12), 2893–2917 (2010)
2. Siegel, R.L., Miller, K.D.: Cancer statistics, 2018. CA Cancer J. Clin. **68**, 284–296 (2018)
3. Holohan, C., Schaeybroeck, S.V., Longley, D.B., Johnston, P.G.: Cancer drug resistance: an evolving paradigm. Nat. Rev. Cancer **13**(10), 714–726 (2013)
4. Hoskin, D.W., Ramamoorthy, A.: Studies on anticancer activities of antimicrobial peptides. BBA Biomembr. **1778**(2), 357–375 (2008)
5. Gaspar, D., Veiga, A.S., Castanho, M.A.R.B.: From antimicrobial to anticancer peptides. Rev. Front. Microbiol. **4**(4), 294 (2013)
6. Huang, Y., Feng, Q., Yan, Q., Hao, X., Chen, Y.: Alpha-helical cationic anticancer peptides: a promising candidate for novel anticancer drugs. Mini. Rev. Med. Chem. **15**(1), 73–81 (2015)
7. Otvos, L.: Peptide-based drug design: here and now. Methods Mol. Biol. **494**(8), 1 (2008)
8. Mader, J.S., Hoskin, D.W.: Cationic antimicrobial peptides as novel cytotoxic agents for cancer treatment. Expert Opin. Investig. Drugs **15**(8), 933–946 (2006)
9. Hariharan, S., et al.: Assessment of the biological and pharmacological effects of the alpha nu beta3 and alpha nu beta5 integrin receptor antagonist, cilengitide (EMD 121974), in patients with advanced solid tumors. Ann. Oncol. **18**(8), 1400–1407 (2007)
10. Gregorc, V., et al.: Phase I study of NGR-hTNF, a selective vascular targeting agent, in combination with cisplatin in refractory solid tumours. Clin. Cancer Res. **17**, 1964–1972 (2011)
11. Barras, D., Widmann, C.: Promises of apoptosis-inducing peptides in cancer therapeutics. Curr. Pharm. Biotechnol. **12**(8), 1153–1165 (2011)
12. Boohaker, R.J., Lee, M.W., Vishnubhotla, P., Perez, J.M., Khaled, A.R.: The use of therapeutic peptides to target and to kill cancer cells. Curr. Med. Chem. **19**(22), 3794–3804 (2012)
13. Thundimadathil, J.: Cancer treatment using peptides: current therapies and future prospects. J. Amino Acids **2012**(1), 967347 (2012)
14. Tyagi, A., Kapoor, P., Kumar, R., Chaudhary, K., Gautam, A., Raghava, G.P.: In silico models for designing and discovering novel anticancer peptides. Sci. Rep. **3**(10), 2984 (2013)
15. Hajisharifi, Z., Piryaiee, M., Mohammad, B.M., Behbahani, M., Mohabatkar, H.: Predicting anticancer peptides with Chou's pseudo amino acid composition and investigating their mutagenicity via Ames test. J. Theor. Biol. **341**, 34–40 (2014)

16. Chou, K.C.: Using amphiphilic pseudo amino acid composition to predict enzyme subfamily classes. **21**, 10–19 (2005)
17. Shen, H.B., Chou, K.C.: Using ensemble classifier to identify membrane protein types. Amino Acids **32**(4), 483–488 (2007)
18. Vijayakumar, S., Lakshmi, P.T.V.: ACPP: a web server for prediction and design of anti-cancer peptides. Int. J. Pept. Res. Ther. **21**(1), 99–106 (2015)
19. Wei, C., Hui, D., Feng, P., Hao, L., Chou, K.C.: iACP: a sequence-based tool for identifying anticancer peptides. Oncotarget **7**(13), 16895–16909 (2016)
20. Wei, L., Zhou, C., Chen, H., Song, J., Su, R.: ACPred-FL: a sequence-based predictor based on effective feature representation to improve the prediction of anti-cancer peptides. Bioinformatics **34**, 4007–4016 (2018)
21. Vapnik, V.N.: Statistical learning theory. Encycl. Sci. Learn. **41**(4), 3185 (1998)
22. Chang, C.-C., Lin, C.-J.: LIBSVM: a library for support vector machines. ACM Trans. Intell. Syst. Technol. **2**(3), 1–27 (2011)
23. Breiman, L.: Random forest. Mach. Learn. **45**, 5–32 (2001)
24. Pedregosa, F., et al.: Scikit-learn: machine learning in Python. J. Mach. Learn. Res. **12**(10), 2825–2830 (2013)
25. Zhang, H.: The optimality of Naive Bayes. In: The International Flairs Conference (2004)
26. Schmidt, M., Roux, N.L., Bach, F.: Minimizing finite sums with the stochastic average gradient. Math. Program. **162**(5), 1–30 (2016)
27. You, Z.H., Zhou, M., Luo, X., Li, S.: Highly efficient framework for predicting interactions between proteins. IEEE Trans. Cybern. **PP**(99), 1–13 (2016)
28. Yi, H.-C., You, Z.-H., Huang, D.-S., Li, X., Jiang, T.-H., Li, L.-P.: A deep learning framework for robust and accurate prediction of ncRNA-protein interactions using evolutionary information. Mol. Ther. Nucleic Acids **11**, 337–344 (2018)
29. Li, W., Godzik, A.: Cd-hit: a fast program for clustering and comparing large sets of protein or nucleotide sequences. Bioinformatics **22**(13), 1658 (2006)
30. Muppirala, U.K., Honavar, V.G., Dobbs, D.: Predicting RNA-protein interactions using only sequence information. BMC Bioinformatics **12**(1), 489 (2011)
31. Suresh, V., Liu, L., Adjeroh, D., Zhou, X.: RPI-Pred: predicting ncRNA-protein interaction using sequence and structural information. Nucleic Acids Res. **43**(3), 1370–1379 (2015)
32. Pan, X., Fan, Y.X., Yan, J., Shen, H.B.: IPMiner: hidden ncRNA-protein interaction sequential pattern mining with stacked autoencoder for accurate computational prediction. BMC Genom. **17**(1), 582 (2016)
33. Kolda, T.G., O'Leary, D.P.: A semidiscrete matrix decomposition for latent semantic indexing in information retrieval. Proc. ACM Trans. Inf. Syst. **16**(4), 322–346 (1996)
34. Yi, H.-C., et al.: ACP-DL: a deep learning long short-term memory model to predict anticancer peptides using high efficiency feature representation. Mol. Ther. Nucleic Acids **17**, 1–9 (2019)
35. Wang, L., et al.: MTRDA: using logistic model tree to predict miRNA-disease associations by fusing multi-source information of se-quences and similarities. PLoS Comput. Biol. **15**(3), e1006865 (2019)
36. Chen, Z.-H., You, Z.-H., Li, L.-P., Wang, Y.-B., Wong, L., Yi, H.-C.: Prediction of self-interacting proteins from protein sequence information based on random projection model and fast fourier transform. Int. J. Mol. Sci. **20**(4), 930 (2019)
37. Chen, Z.-H., Li, L.-P., He, Z., Zhou, J.-R., Li, Y., Wong, L.: An improved deep forest model for predicting self-interacting proteins from protein sequence using wavelet transformation. Front. Genet. **10**, 90 (2019)

38. Zhu, H.-J., You, Z.-H., Zhu, Z.-X., Shi, W.-L., Chen, X., Cheng, L.: DroidDet: effective and robust detection of android malware using static analysis along with rotation forest model. Neuro-Comput. **272**, 638–646 (2018)

39. You, Z.-H., Huang, W., Zhang, S., Huang, Y.-A., Yu, C.-Q., Li, L.-P.: An efficient ensemble learning approach for predicting protein-protein interactions by integrating protein primary sequence and evolutionary information. IEEE/ACM Trans. Comput. Biol. Bioinform. **16**, 809–817 (2018)

40. Wang, Y.-B., You, Z.-H., Li, X., Jiang, T.-H., Cheng, L., Chen, Z.-H.: Prediction of protein self-interactions using stacked long short-term memory from protein sequences information. BMC Syst. Biol. **12**(8), 129 (2018)

41. Wang, Y., et al.: Predicting protein interactions using a deep learning method-stacked sparse autoencoder combined with a probabilistic classification vector machine. Complexity **2018**, 12 (2018)

42. Wang, L.: Using two-dimensional principal component analysis and rotation forest for prediction of protein-protein interactions. Sci. Rep. **8**(1), 12874 (2018)

43. Wang, L., et al.: An improved efficient rotation forest algorithm to predict the interactions among proteins. Soft. Comput. **22**(10), 3373–3381 (2018)

44. Wang, L., You, Z.-H., Huang, D.-S., Zhou, F.: Combining high speed ELM learning with a deep convolutional neural network feature encoding for predicting protein-RNA interactions. IEEE/ACM Trans. Comput. Biol. Bioinform. **PP**, 1 (2018)

45. Wang, L., et al.: A computational-based method for predicting drug–target interactions by using stacked autoencoder deep neural network. J. Comput. Biol. **25**(3), 361–373 (2018)

46. Song, X.-Y., Chen, Z.-H., Sun, X.-Y., You, Z.-H., Li, L.-P., Zhao, Y.: An ensemble classifier with random projection for predicting protein-protein interactions using sequence and evolutionary information. Appl. Sci. **8**(1), 89 (2018)

47. Qu, J., et al.: In silico prediction of small molecule-miRNA associations based on HeteSim algorithm. Molecular Therapy-Nucleic Acids **14**, 274–286 (2019)

48. Qu, J., Chen, X., Sun, Y.Z., Li, J.Q., Ming, Z.: Inferring potential small molecule–miRNA association based on triple layer heterogeneous network. J. Cheminformatics **10**(1), 30 (2018)

49. Luo, X., et al.: Incorporation of efficient second-order solvers into latent factor models for accurate prediction of missing QoS data. IEEE Trans. Cybern. **48**(4), 1216–1228 (2018)

50. Li, L.-P., Wang, Y.-B., You, Z.-H., Li, Y., An, J.-Y.: PCLPred: a bioinformatics method for predicting protein-protein interactions by combining relevance vector machine model with low-rank matrix approximation. Int. J. Mol. Sci. **19**(4), 1029 (2018)

51. Huang, Y.-A., You, Z.-H., Chen, X.: A systematic prediction of drug-target interactions using molecular fingerprints and protein sequences. Curr. Protein Pept. Sci. **19**(5), 468–478 (2018)

52. Chen, X., Zhang, D.-H., You, Z.-H.: A heterogeneous label propagation approach to explore the potential associations between miRNA and disease. J. Transl. Med. **16**(1), 348 (2018)

53. Chen, X., Xie, D., Wang, L., Zhao, Q., You, Z.-H., Liu, H.: BNPMDA: bipartite network projection for MiRNA–disease association prediction. Bioinformatics **1**, 9 (2018)

54. Chen, X., Wang, C.-C., Yin, J., You, Z.-H.: Novel human miRNA-disease association inference based on random forest. Mol. Ther.-Nucl. Acids **13**, 568–579 (2018)

55. Chen, X., Gong, Y., Zhang, D.H., You, Z.H., Li, Z.W.: DRMDA: deep representations-based miRNA–disease association prediction. J. Cell Mol. Med. **22**(1), 472–485 (2018)

56. Zhu, L., Deng, S.-P., You, Z.-H., Huang, D.-S.: Identifying spurious interactions in the protein-protein interaction networks using local similarity preserving embedding. IEEE/ACM Trans. Comput. Biol. Bioinform. (TCBB) **14**(2), 345–352 (2017)

57. Zhu, H.-J., Jiang, T.-H., Ma, B., You, Z.-H., Shi, W.-L., Cheng, L.: HEMD: a highly efficient random forest-based malware detection framework for Android. Neural Comput. Appl. **30**(11), 3353–3361 (2018)
58. Zhang, S., Zhu, Y., You, Z., Wu, X.: Fusion of superpixel, expectation maximization and PHOG for recognizing cucumber diseases. Comput. Electron. Agric. **140**, 338–347 (2017)
59. Zhang, S., Zhang, C., Zhu, Y.: You, Z,: Discriminant WSRC for large-scale plant species recognition. Comput. Intell. Neurosci. **2017**, 10 (2017)
60. Zhang, S., You, Z., Wu, X.: Plant disease leaf image segmentation based on superpixel clustering and EM algorithm. Neural Comput. Appl. 1–8 (2017)
61. Zhang, S., Wu, X., You, Z., Zhang, L.: Leaf image based cucumber disease recognition using sparse representation classification. Comput. Electron. Agricul. **134**, 135–141 (2017)
62. Zhang, S., Wu, X., You, Z.: Jaccard distance based weighted sparse representation for coarse-to-fine plant species recognition. PLoS ONE **12**(6), e0178317 (2017)
63. You, Z.-H., Zhou, M., Luo, X., Li, S.: Highly efficient framework for predicting interactions between proteins. IEEE Trans. Cybern. **47**(3), 731–743 (2017)
64. You, Z.-H., et al.: PRMDA: personalized recommendation-based MiRNA-disease association prediction. Oncotarget **8**(49), 85568 (2017)
65. You, Z.-H., et al.: PBMDA: a novel and effective path-based computational model for miRNA-disease association prediction. PLoS Comput. Biol. **13**(3), e1005455 (2017)
66. You, Z.H., Li, X., Chan, K.C.: An improved sequence-based prediction protocol for protein-protein interactions using amino acids substitution matrix and rotation forest ensemble classifiers. Neurocomputing **228**, 277–282 (2017)
67. Wen, Y.-T., et al.: Prediction of protein-protein inter-actions by label propagation with protein evolutionary and chemical information derived from heterogeneous network. J. Theor. Biol. **430**, 9–20 (2017)
68. Wang, Y.-B., You, Z.-H., Li, L.-P., Huang, Y.-A., Yi, H.-C.: Detection of interactions between proteins by using legendre moments descriptor to extract discriminatory information embedded in PSSM. Molecules **22**(8), 1366 (2017)
69. Wang, Y.B., et al.: Predicting protein-protein interactions from protein sequences by a stacked sparse autoencoder deep neural network. Mol. BioSyst. **13**(7), 1336–1344 (2017)
70. Wang, Y., You, Z., Li, X., Chen, X., Jiang, T., Zhang, J.: PCVMZM: using the probabilistic classification vector machines model combined with a zernike moments descriptor to predict protein-protein interactions from protein sequences. Int. J. Mol. Sci. **18**(5), 1029 (2017)
71. Wang, L., et al.: Advancing the prediction accuracy of protein-protein interactions by utilizing evolutionary information from position-specific scoring matrix and ensemble classifier. J. Theor. Biol. **418**, 105–110 (2017)
72. Wang, L., et al.: Computational methods for the prediction of drug-target interactions from drug fingerprints and protein sequences by stacked auto-encoder deep neural network. In: Cai, Z., Daescu, O., Li, M. (eds.) ISBRA 2017. LNCS, vol. 10330, pp. 46–58. Springer, Cham (2017). https://doi.org/10.1007/978-3-319-59575-7_5
73. Li, S., Zhou, M., Luo, X., You, Z.-H.: Distributed winner-take-all in dynamic networks. IEEE Trans. Autom. Control **62**(2), 577–589 (2017)
74. Li, J.-Q., You, Z.-H., Li, X., Ming, Z., Chen, X.: PSPEL: in silico prediction of self-interacting proteins from amino acids sequences using ensemble learning. IEEE/ACM Trans. Comput. Biol. Bioinform. (TCBB) **14**(5), 1165–1172 (2017)
75. Chen, X., Xie, D., Zhao, Q., You, Z.-H.: MicroRNAs and complex diseases: from experimental results to computational models. Briefings Bioinform. **20**, 515–539 (2017)
76. Luo, X., Zhou, M., Li, S., You, Z., Xia, Y., Zhu, Q.: A nonnegative latent factor model for large-scale sparse matrices in recommender systems via alternating direction method. IEEE Trans. Neural Networks Learn. Syst. **27**(3), 579–592 (2016)

77. Luo, X., et al.: An incremental-and-static-combined scheme for matrix-factorization-based collaborative filtering. IEEE Trans. Autom. Sci. Eng. **13**(1), 333–343 (2016)
78. Li, S., You, Z.H., Guo, H., Luo, X., Zhao, Z.Q.: Inverse-free extreme learning machine with optimal information updating. IEEE Trans. Cybern. **46**(5), 1229 (2016)
79. Ji, Z., Wang, B., Deng, S., You, Z.: Predicting dynamic deformation of retaining structure by LSSVR-based time series method. Neurocomputing **137**, 165–172 (2014)

# Effective Analysis of Hot Spots in Hub Protein Interfaces Based on Random Forest

Xiaoli Lin[1,2](✉) and Fengli Zhou[2]

[1] Hubei Key Laboratory of Intelligent Information Processing
and Real-Time Industrial System, School of Computer Science and Technology,
Wuhan University of Science and Technology, Wuhan 430065, China
aneya@163.com
[2] Information and Engineering Department of City College,
Wuhan University of Science and Technology, Wuhan 430083, China
thinkview@163.com

**Abstract.** In the protein-protein interactions, hub proteins are the key factor to maintain stability of protein-protein interactions and exert the protein biological function. Conventional methods mainly focus on the topological structure and gene expression of hub proteins, while we mainly discuss the hot spots of hub protein interfaces. In order to evaluate the performance of the classification models, the importance of feature variables is analyzed by using the average precision descent curve and the average Gini coefficient descent curve. In addition, the margin box-plot is used to measure the certainty of the classification models. The experimental results show that the error rate of random forest method is lower, and our classification model has higher reliability.

**Keywords:** Hub protein · Hot spots · Random forest · Classification · PPIs

## 1 Introduction

Proteins interact with each other to realize life activities [1], and researching protein-protein interactions and protein structure through computational methods is critical to understand protein function [2, 3]. As a special protein, a hub protein has many interacting partners and plays an important role in the interaction network [4, 5]. Hub proteins help to explain the molecular mechanism of exerting biological function, to understand the micro process of life activity and provide theoretical guidance of the drug design based on protein structure. In recent years, hub proteins have been studied in many aspects. Many researchers have devoted to exploring the relationship between lethality and topological centrality. It has been found that hub proteins are more likely to be lethal proteins than other proteins in protein-protein interaction networks. Yu [6] et al. introduced the concept of marginal essentiality by large-scale phenotypic experiments, and found that proteins with high marginal importance tended to be hub proteins. He [7] et al. also studied the central-lethality rule, and found the relationship between the structural and the functional of hub nodes in PPIN. These indicate that hub proteins are more important than non-hub proteins in stabilizing the whole network structure. Researches on the hub protein can reveal the biological significance of

D.-S. Huang et al. (Eds.): ICIC 2019, LNCS 11644, pp. 324–332, 2019.
https://doi.org/10.1007/978-3-030-26969-2_31

network structure. Yu [8] et al. studied the relationship between hub protein and bottleneck protein. The bottleneck protein is defined as a highly intermediate protein, which has surprising functions and is a dynamic and key binding protein. Raman [9] et al. considered protein networks of 20 different organisms to assess the relationship between centrality and lethality. They found that the important nodes in the network have significantly higher intermediate centrality, which exceed the average value of the network. McCormack [10] et al. indicated that except for high connectivity, hub proteins are often characterized by other network properties, in which centrality is the most basic. Han [11] et al. studied how hub proteins contribute to the genetic robustness of interactions and how they dynamically regulate networks in time and in space. Then, the concepts of DateHub and PartyHub are proposed according to the characteristics of hub protein and its different roles in different time and different space in the protein-protein interaction network.

Some very crucial residues are known as hot spots which contribute the majority of the binding-free-energies [12, 13]. Hot spots are tightly packed together to form the special areas that named hot regions. Hot regions are importance areas where the receptors bind to the ligands with high-affinity. Also, hot regions are the particular functional areas that promote the stability of protein-protein interactions. The hot spots in different hub protein interfaces can bind to different protein partners. Therefore, it is very important for understanding protein functionalities to study the hot spots of hub protein interfaces [14]. The methods in existing literature mainly focus on the topological structure and gene expression of hub protein, while we mainly discuss the hot spots of hub protein interfaces.

Although more and more structures and attributes of protein are found, a large amount of information is unavailable and redundant, resulting in the extreme difficulty of identifying hub protein interfaces by using traditional methods and techniques. The development of high-quality predictive models and analysis algorithms is an essential task. In this paper, we use random forest to classify hot spot residues and non-hot spot residues in hub proteins.

# 2    Method

## 2.1    Dataset

The datasets of hub protein used in this paper are derived from Ekman [15] protein interaction network. In this network, the proteins are annotated as PartyHub proteins, DateHub proteins and NonHub proteins according to the gene OLNs. In order to analyze the hot spots of hub protein interfaces, the three-dimensional structures of hub proteins should be used. Therefore, OLNs information of genes and ID information of protein databases need to be referenced. The interfaces are obtained from the Tuncbag's datasets [16]. Solvent accessible surface area (ASA) and other attributes of the interface in hub protein dataset were obtained from HotPoint server [17]. Finally, 2906 residues were obtained in the DateHub dataset, including 1056 hot spot residues and 1850 non-hot spot residues. In addition, 3013 residues were obtained in the PartyHub dataset, including 1033 hot spot residues and 1980 non-hot spot residues. The proportion of hot spots and non-hot spots is shown in Table 1. Because there are fewer published results about hot spots of hub protein, we firstly measured our methods using common datasets [18–20].

**Table 1.** Proportion of hot spots and non-hot spots in DateHub and PartyHub

| Dataset | Number of hot spot | Number of non-hot spot | Total | Proportion |
|---------|---------|---------|---------|---------|
| DateHub | 1056 | 1850 | 2906 | 0.57 |
| PartyHub | 1033 | 1980 | 3013 | 0.52 |

## 2.2   Algorithm

Random forest algorithm is a flexible and easy-to-use ensemble learning method for both classification and regression. Even without parametric optimization, it can get good results in most cases. It is also one of the most commonly used algorithms because it is very easy to operate by constructing a multitude of decision trees at training time and outputting the class [21, 22]. Generally, the more trees there are, the better the performance and the more stable the prediction will be, but the calculation speed will also slow down. Therefore, it is better to choose hundreds of trees in practice.

For most of the data, the classification effect of random forest is better. It can deal with high-dimensional features, and is not easy to produce over-fitting, because random forest uses the voting mechanism of multiple decision trees to resist the overfitting of decision trees. The training speed is relatively fast, especially for large data. When deciding on a category, it can assess the importance of variables. In addition, it can process both discrete data and continuous data, and data sets need not be standardized. Therefore, this paper chooses the random forest for predicting hot spot residues and non-spot residues in PPIs.

Firstly, we adopt the sampling with replacement from the original data set to construct the sub-data set, which has the same amount of data as the original data set. Elements in different sub-datasets can be duplicated, and elements in the same sub-dataset can also be duplicated. Secondly, the sub-decision tree is constructed by using the sub-data set. The data is put into each sub-decision tree, and each sub-decision tree outputs a result. Finally, the output of random forest can be obtained by voting on the judgment result of sub-decision tree.

Similar to random selection of data sets, each splitting process of sub-trees in random forests does not use all the features, while randomly select certain features from all the features, and then select the optimal features in the randomly selected features. In this way, decision trees in random forests can be different from each other and improve the diversity of the system, thus improving the classification performance.

The basic principle of predicting hot spot residues using random forest algorithm is shown in Fig. 1. There are two stopping criterions of splitting node. One of the criterions is that the training samples belong to the same class, and the other is that the node does not have other features for the splitting test.

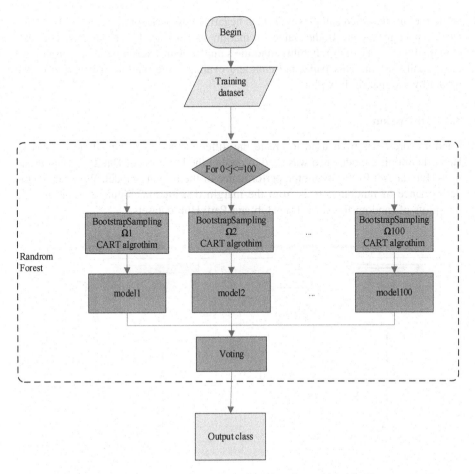

**Fig. 1.** The basic principle of random forest algorithm.

# 3 Experiments

## 3.1 Evaluation

In order to measure the certainty of the classification model in hub protein datasets, this paper uses the marginal distribution to evaluate the random forest classification model. The evaluation value is calculated according to the correct number of classified samples and the maximum misclassified samples. The marginal distribution of the classification model is defined as:

$$\text{margin}(x_i) = \text{support}_c(x_i) - \max support_j(x_i) \quad j \neq c \tag{1}$$

where, $c$ represents the correct classification. The margin of sample $x_i$ is equal to that the number of samples correctly classified minus the number of samples misclassified into category $j$. Therefore, the correctly classified samples will establish positive edges, while the misclassified samples will form negative edges. The return value of formula (1) is an edge vector. If the value of margin is close to 1, the samples correctly classified have very high reliability, and the samples with uncertain classification have only small margin. This paper presents and evaluates the margin distribution of the model by the margin box-plot.

## 3.2    Discussion

Random forest has been used to create the different models using different datasets. Here, the training model also was used to predict the hot spots of DateHub dataset and PartyHub dataset for verifying the performance of the training model. We evaluate the performance of random forests from the margin distribution. Figure 2 is the margin box-plots of random forest on DateHub and PartyHub datasets.

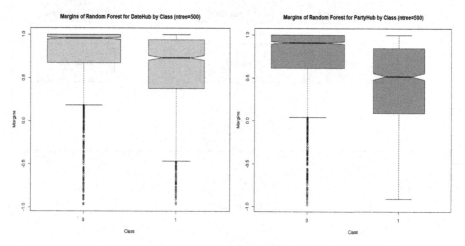

**Fig. 2.**  Margin box-plots of random forest on DateHub and PartyHub.

It can be seen from the two margin box-plots that the grooves of the positive class and negative class can overlap, which indicates that there is no significant difference in their median values. The width of negative boxplot is wider than that of positive boxplot, which indicates that the number of non-hot spot samples is more than that of hot spot samples. In addition, it can be seen that the distribution of hot spots is more uniform and wider than that of non-hot spots. The difference between the margin box-plots is that the distribution of hot spots in PartyHub is wider and more uniform than in DateHub. And there are no outliers among hot spots in PartyHub, but there are more outliers among non-hot spots. The outliers of hot spots are fewer than those of non-hot spots in DateHub. In both margin box-plots, the maximum margin values of positive class and negative class are close to 1. The margin box-plots show that the classification results of the models have higher reliability.

Table 2 lists the classification performance of DateHub dataset and PartyHub dataset with different parameters based on the random forest. From the results, we can see that hot spots and non-hot spots can be misjudged, but OOB (Out of Bag) decreases with the increase of the number of decision trees. The OOB of PartyHub dataset is higher than that of DateHub dataset. The running time of the algorithm will also increase with the increase of the number of decision trees. When the number of decision trees is set to 500 and 1000, the change is not obvious.

**Table 2.** Classification performance of two datasets with different parameters

| Ntree | Dataset | Misjudged rate | | OOB |
|---|---|---|---|---|
| | | Hot spot | Non-hot spot | |
| 50 | **DateHub** | 0.176 | 0.070 | 10.87% |
| | **PartyHub** | 0.254 | 0.075 | 13.51% |
| 100 | **DateHub** | 0.145 | 0.070 | 9.7% |
| | **PartyHub** | 0.252 | 0.069 | 13.07% |
| 300 | **DateHub** | 0.141 | 0.064 | 9.22% |
| | **PartyHub** | 0.233 | 0.067 | 12.47% |
| 500 | **DateHub** | 0.142 | 0.063 | 9.19% |
| | **PartyHub** | 0.220 | 0.073 | 12.27% |
| 1000 | **DateHub** | 0.134 | 0.062 | 8.81% |
| | **PartyHub** | 0.212 | 0.074 | 12.03% |

Then, we evaluated the importance of the characteristics of random forests in the training process. There are two ways to measure the importance of feature variables. The first measure is the average precise reduction, which is calculated from the permutation of OOB. For each tree, the classification error rate of OOB is recorded, and the same operation is performed after replacing the variables of each prediction model. Then the difference between two trees is averaged and normalized by standard deviation of the difference. The second measure is the reduction of average impurity, which is to split variables and calculate the average value of all trees. At the same time, the impurity of classification nodes is measured by Gini coefficient.

Figures 3 and 4 show the importance curves of characteristic variables of random forests with DateHub and PartyHub datasets. Mean Decrease Accuracy is the decline curve of average accuracy and Mean Decrease Gini is the decline curve of average Gini coefficient. As can be seen from two figures, *BtRASA* plays a very important role in both datasets.

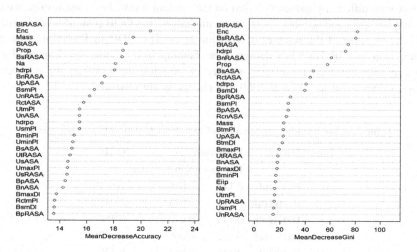

**Fig. 3.** The importance curves of characteristic variables on DateHub dataset.

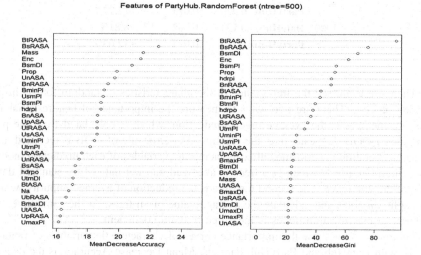

**Fig. 4.** The importance curves of characteristic variables on PartyHub dataset.

## 4    Conclusions

In this paper, the random forest method was applied to predict the hot spots on hub protein binding. The training model was applied to DateHub dataset and PartyHub datasets. The experimental results show that the random forest method is feasible and effective for predicting hot spot residues on hub protein interfaces. The classification performance of the random forest was also analyzed, and the importance of characteristic variables in the classification process was analyzed by using the average precision descent curve and the average Gini coefficient descent curve. In addition, the certainty of the classification model was measured by the margin box-plots, and the error rate was measured by OOB. These illustrate that the proposed method is effective in detecting the hot spots on the hub protein interfaces. The following work is to further analyze the distribution and formation of hot spots in the different interfaces of hub proteins. Because PartyHub proteins are more conservative than DateHub proteins, the study of conservation is also helpful to analyze the distribution mechanism of hot spots in DateHub protein interfaces and PartyHub protein interfaces.

**Acknowledgment.** The authors thank the members of Machine Learning and Artificial Intelligence Laboratory, School of Computer Science and Technology, Wuhan University of Science and Technology, for their helpful discussion within seminars. This work was supported in part by Hubei Province Natural Science Foundation of China (No. 2018CFB526), by National Natural Science Foundation of China (No. 61502356).

# References

1. Keskin, O., Tuncbag, N., et al.: Predicting protein-protein interactions from the molecular to the proteome level. Chem. Rev. **116**(8), 4884–4909 (2016)
2. Scott, D.E., Bayly, A.R., et al.: Small molecules, big targets: drug discovery faces the protein-protein interaction challenge. Nat. Rev. Drug Discovery **15**(8), 533–550 (2016)
3. Huang, D.S., Zhang, L., et al.: Prediction of protein-protein interactions based on protein-protein correlation using least squares regression. Curr. Protein Pept. Sci. **15**(6), 553–560 (2014)
4. Vandereyken, K., Leene, J.V., et al.: Hub protein controversy: taking a closer look at plant stress response hub. Front. Plant Sci. **9**, 694 (2018)
5. Mirzarezaee, M., Araabi, B.N., et al.: Features analysis for identification of date and party hubs in protein interaction network of Saccharomyces Cerevisiae. BMC Syst. Biol. **4**, 172 (2010)
6. Yu, H., Greenbaum, D., Xin, L.H., et al.: Genomic analysis of essentiality within protein networks. Trends Genet. **20**(6), 227–231 (2004)
7. He, X., Zhang, J.: Why do hubs tend to be essential in protein networks? PLoS Genet. **2**(6), 88–96 (2006)
8. Yu, H., Kim, P.M., Sprecher, E., et al.: The importance of bottlenecks in protein networks: correlation with gene essentiality and expression dynamics. PLoS Comput. Biol. **3**(4), e59 (2007)
9. Raman, K., Damaraju, N., Joshi, G.K.: The organisational structure of protein networks: revisiting the centrality-lethality hypothesis. Syst. Synth. Biol. **8**, 3–81 (2014)
10. McCormack, M.E., Lopez, J.A., Crocker, T.H., et al.: Making the right connections: network biology and plant immune system dynamics. Curr. Plant Biol. **5**, 2–12 (2016)
11. Han, J.D., Bertin, N., Hao, T., et al.: Evidence for dynamically organized modularity in the yeast protein-protein interaction network. Nature **430**, 88–93 (2004)
12. Yamamoto, N.: Hot spot of structural ambivalence in prion protein revealed by secondary structure principal component analysis. J. Phys. Chem. **118**(33), 9826–9833 (2014)
13. Lin, X.L., Zhang, X.L.: Prediction of hot regions in PPIs based on improved local community structure detecting. IEEE/ACM Trans. Comput. Biol. Bioinform. **99**, 1–1(2018)

14. Engin, G., Attila, G., Ozlem, K.: Analysis of hot region organization in hub protein. Ann. Biomed. Eng. **38**(6), 2068–2078 (2010)
15. Ekman, D., Light, S., Bjorklund, A.K., et al.: What properties characterize the hub proteins of the protein-protein interaction network of Saccharomyces cerevisiae? Genome Biol. **7**(6), R45 (2006)
16. Tuncbag, N.A., Gursoy, E., et al.: Architectures and functional coverage of protein-protein interfaces. J. Mol. Biol. **381**(3), 785–802 (2008)
17. Tuncbag, N., Keskin, O., Gursoy, A.: HotPoint: hot spot prediction server for protein interfaces. Nucleic Acids Res. **38**, w402–w406 (2010)
18. Mustapha, I.B., Saeed, F.: Bioactive molecule prediction using extreme gradient boosting. Molecules **21**(8), 983 (2016)
19. Thorn, K.S., Bogan, A.A.: ASEdb: a data base of alanine mutations and their effects on the free energy of binding in protein interactions. Bioinformatics **17**(3), 284–285 (2001)
20. Moal, I.H., Fernández-Recio, J.: SKEMPI: a structural kinetic and energetic database of mutant protein interactions and its use in empirical models. Bioinformatics **28**(20), 2600–2607 (2012)
21. Ho, T.K.: The random subspace method for constructing decision forests. IEEE Trans. Pattern Anal. Mach. Intell. **20**(8), 832–844 (1998)
22. Breiman, L.: Random forests. Mach. Learn. **45**(1), 5–32 (2001)

# Prediction of Human LncRNAs Based on Integrated Information Entropy Features

Junyi Li[1,2(✉)], Huinian Li[1], Li Zhang[1], Qingzhe Xu[1], Yuan Ping[1],
Xiaozhu Jing[1], Wei Jiang[1], Bo Liu[2], and Yadong Wang[1,2(✉)]

[1] School of Computer Science and Technology, Harbin Institute of Technology
(Shenzhen), Shenzhen 518055, Guangdong, China
{lijunyi,ydwang}@hit.edu.cn
[2] Center for Bioinformatics, School of Computer Science and Technology,
Harbin Institute of Technology, Harbin 150001, Heilongjiang, China

**Abstract.** The prediction of long non-coding RNA (lncRNA) has attracted great attention from researchers, as more and more evidence indicate that various complex human diseases are closely related to lncRNAs. In the era of bio-big data, in addition to the prediction of lncRNAs by biological experimental methods, many computational methods based on machine learning have been proposed to make better use of the sequence resources of existing long non-coding RNAs. We use a lncRNA prediction method by integrating information-entropy-based features and machine learning algorithms. We calculate generalized topological entropy and generate 6 novel features for lncRNA sequences. By employing these 6 features and other features such as open reading frame (ORF), we apply Supporting Vector Machine (SVM), XGBoost and Random Forest (RF) algorithms to distinguish human lncRNAs. We compare our method with the one which has more Kmer features and results show that our method has higher Area Under the Curve (AUC) up to 99.7905%. We develop an accurate and efficient method which has novel information entropy features to analyze and classify lncRNAs. Our method is also extendable for research on other functional elements in DNA sequences.

**Keywords:** Long non-coding RNA · Information entropy ·
Generalized topological entropy · Machine learning

## 1 Background

According to the central dogma of molecular biology, genetic information is stored in protein-coding genes [1]. Therefore, non-coding RNAs have been considered to be transcriptional noises for a long time. In the past decade, this traditional view has been challenged [2]. There is increasing evidence shows that non-coding RNAs play a key role in a variety of basic and important biological processes [3]. Moreover, the proportion of non-protein coding sequences increases with the complexity of the organism [4]. Non-coding RNAs can be further divided into short non-coding RNAs and long

---

J. Li and H. Li—Equally contributed to this work.

© Springer Nature Switzerland AG 2019
D.-S. Huang et al. (Eds.): ICIC 2019, LNCS 11644, pp. 333–343, 2019.
https://doi.org/10.1007/978-3-030-26969-2_32

non-coding RNAs (lncRNAs) based on whether the length of the transcript exceeds more than 200 nucleotides (nt).

Recently, long non-coding RNA has attracted great attention from researchers, as more and more research results indicate that mutations and dysregulation of these long non-coding RNAs are associated with the development of various complex human diseases such as cancers, Alzheimer's disease and cardiovascular diseases [5]. Therefore, accurate prediction of lncRNAs is very important in lncRNA studies [6–8].

Various lncRNA prediction methods have been proposed by using experimental techniques and biological data. For example, the discovery of two well-known lncRNAs, H19 and X-inactive specific transcripts can be traced back to traditional genetic mapping in the early 1990s [9]. Guttman et al. developed a functional genomics approach that assigns putative functions to each large intervening lncRNA [10]. Cabili et al. proposed a comprehensive approach to construct large non-coding RNA catalogs of human intercropping, including more than 8000 large intervening lengths in 24 different human cell types and tissues based on chromatin markers [11].

However, the method of biological experiment is costly, time-consuming and laborious, which is not conducive to large-scale application. In the era of bio-big data, in order to make better use of the existing sequence resources of lncRNA, many computational methods based on machine learning have been proposed by researchers.

In 2013, CPAT was implemented by Wang et al., which is a potential evaluation tool for protein coding and includes the feature of Open Reading Frame (ORF) [12]. In molecular biology, ORF starts from the start codon and is a base sequence in the DNA sequence that encodes a protein potential and is interrupted by a stop codon. The classification model of CPAT is a Supporting Vector Machine (SVM) basis function kernel with standard radial. In 2014, PLEK was implemented by Li et al., which used a Kmer scheme and a sliding window to analyze transcripts [13]. The classification model of PLEK is an SVM with a radial kernel function.

In 2015, LncRNA-ID was implemented by Achawanantakun et al. [14]. LncRNA-ID can be classified according to ORF, ribosome interaction and protein conservation. The use of random forests improves the classification model of LncRNA-ID, which helps LncRNA-ID to efficiently process unbalanced training data.

In 2017, Schneider et al. proposed a SVM-based method for the prediction of lncRNAs [15]. It uses the kmer protocol and features derived from the ORF to analyze the transcript. These features are divided into two groups. The first set derives from the four characteristics of the ORF: the first ORF length; the relative length of the first ORF; the longest ORF length; the longest ORF relative length. The second group is based on the kmer feature extraction scheme, where k = 2, 3, 4, a total of 336 nucleotide patterns of different frequencies: 16 dinucleotide pattern frequencies; 64 trinucleotide mode frequencies; and 256 four nucleotide mode frequencies. The first ORF relative length and the frequency of the nucleotide pattern selected by PCA were used as features for these two sets of features.

In our study, we use a lncRNA prediction method by integrating information-entropy-based features and machine learning algorithms. We calculate generalized topological entropy and generate 6 novel features for lncRNA sequences. By employing these 6 features and other features such as ORF, we apply SVM, XGBoost and Random Forest algorithms to distinguish human lncRNAs. We compare our method with the one which has more kmer-based features and results show that our method has higher Area Under the Curve (AUC) score up to 99.7905%. Our accurate and efficient method which has novel information entropy features and is extendable for research on other functional elements in DNA sequences.

## 2    Materials and Methods

### 2.1    Data Sets

We use the dataset from the Ensembl [16] database for model training: human (Homo sapiens) assemblies GRCh37 (release-75) and GRCh38 (release-91). These categorical FASTA files of transcripts contain lncRNAs and protein-encoding transcripts (PCTs) (shown in Table 1). In this project, we consider lncRNAs as positive samples and PCTs as negative samples.

**Table 1.** Categorical original FASTA files of transcripts.

| Transcripts types | GRCh37 ncRNAs | GRCh37 PCTs | GRCh38 ncRNAs | GRCh38 PCTs |
|---|---|---|---|---|
| Number | 34917 | 104763 | 37297 | 104817 |

### 2.2    Data Processing with CD-HIT

CD-HIT is a widely used program for clustering biological sequences to reduce sequence redundancy and improve the performance of other sequence analyses. CD-HIT was originally developed to clustering protein sequences to create a reduced reference database [17, 18] and then extended to support clustering nucleotide sequences and compare two data sets [19].

Currently, the CD-HIT package has many programs: cd-hit, cd-hit-2d, cd-hit-est, cd-hit-est-2d, cd-hit-para and so on. In this project, we use cd-hit-est to clustering nucleic acid sequences. The purpose is to perform de-redundancy operations on nucleic acid sequences to ensure the accuracy of the model for machine learning training.

Data processing is briefly described as shown in Fig. 1.

Firstly, we remove all sequences shorter than 200 nt from the original files (as shown in Table 1). Secondly, we use the "cd-hit-est" program in the CD-HIT package to perform deduplication operations. Thirdly, balance the data sets by down-sampling method. In the fourth step, feature extraction is performed to obtain training data. Table 2 shows the changes in the number of nucleic acid sequences in the FASTA file after data processing.

### 2.3    Novel Features Extracted from Modified Topological Entropy and Modified Generalized Topological Entropy

Koslicki proposed topological entropy and defined it as follows [20]:

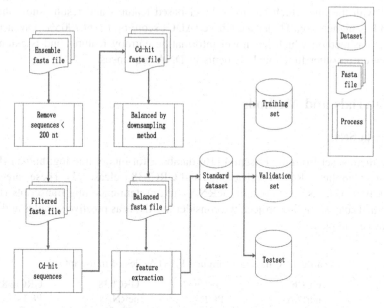

**Fig. 1.** Data processing flow chart

**Table 2.** Categorical FASTA files of transcripts after data processing.

| Transcripts types | GRCh37 ncRNAs | GRCh37 PCTs | GRCh38 ncRNAs | GRCh38 PCTs |
|---|---|---|---|---|
| After removing short | 24513 | 94830 | 28628 | 94527 |
| After deduplication | 21965 | 41134 | 24863 | 41200 |
| After data balancing | 21965 | 21965 | 24863 | 24863 |

$$H_{top(w)} = \frac{\log_4\left(p_{w_1^{4^n+n-1}}(n)\right)}{n} \tag{1}$$

The length of a finite sequence is $\omega$ and the length of a sub sequence is n, where $4^n + n - 1 \leq |\omega| \leq 4^{n+1} + (n+1) - 1$ where $p_{w_1^{4^n+n-1}}(n)$ is the number of sub sequences of length n within first $4^{n_\omega} + n_\omega - 1$ bp of $\omega$. In our project, we choose $n = 3, 4, 5$ to calculate three novel features.

Our previous work shows generalized topological entropy is a complete form of topological entropy [21] and it is defined as:

$$H_{n_\omega}^{(k)}(\omega) = \frac{1}{k} \sum_{i=n_\omega-k+1}^{n_\omega} \frac{\log_4(p_\omega(i))}{i}$$ (2)

In Eq. 2, $n_\omega$ fulfills $4^{n_\omega} + n_\omega - 1 \leq |\omega| \leq 4^{n_\omega+1} + (n_\omega + 1) - 1$ and $k \leq n$. And $p_\omega(i)$ is the number of different sub sequences within $\omega$.

We modify both topological entropy and generalized topological entropy in order to highlight the characteristics of repetition subsequences. In our calculation we remove subsequences with lower appearance frequencies. That means, this sub sequence will not be included in the entropy calculation if the frequency of a subsequence is smaller than $4^{n_\omega}/\omega$. From Eq. (2), $k = 3, 4, 5$ was chosen and 3 novel features based on modified generalized topological entropy are calculated.

## 2.4 Combination of Information Theoretic Features

It is very difficult to perform lncRNA prediction based only on the 6 previously-extracted features. The best approach is to combine them with other commonly used informational theory features and ORF-related features of lncRNAs to obtain better performance classifiers. Common features based on information theory and entropy have been proposed in computational biology and bioinformatics to analyze and measure structural properties in the transcripts. Different complexity calculations reveal different aspects of transcript specificity. In our project, we also employ useful theoretical information features used by Henkel et al. [22]. All features used in this paper are 35 dimensional including: 1 sequence length feature, 4 ORF features [15], 4 Shannon entropy (SE) features [23], 3 topological entropy (TE) features [20], 3 generalized topology Entropy (GTE) features [21], 17 mutual information (MI) features [24] and 3 Kullback-Leibler divergence (KLD) features [25]. In order to better illustrate the superiority of our research, the Kmer feature was chosen as a comparative test. In the comparative experiment, there are a total of 84 nucleotide patterns with different frequencies when k is 1, 2 and 3. They are 4 single nucleotide pattern frequency, 16 dinucleotide pattern frequencies and 64 trinucleotide pattern frequencies. Calculation of all features are listed in the additional files.

## 2.5 Support Vector Machine, Random Forest and XGBoost Algorithms for Classification Procedure

SVM [26], Random Forest [27] and XGBoost [28] are widely used machine learning algorithms, which we use to identify lncRNAs and PCTs. The SVM algorithm is a supervised learning model related to the relevant learning algorithm, which can analyze data, identify patterns, and use for classification and regression analysis. The RF algorithm is an integrated learning method for classification tasks. It constructs a large number of decision trees when training data, and outputs the classes of each tree. The XGBoost algorithm predicts output variables based on various rules organized in a tree structure. Moreover, XGBoost's learning method does not require linear features or linear interaction between features. It is a gradient enhancement which can accelerate tree construction and propose a new tree search distributed algorithm. In our work, all samples are described by the 35 features. And the entire process used in our study is shown in Fig. 2.

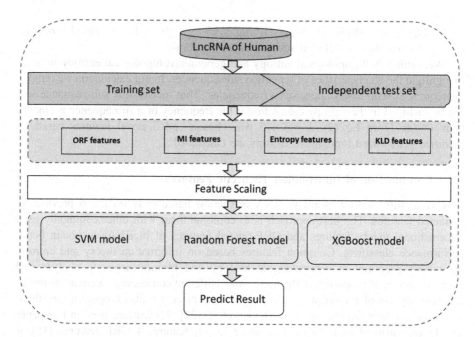

**Fig. 2.** The flowchart of human LncRNA prediction based on combination of information entropy and ORF features.

## 3   Results and Discussions

### 3.1   Feature Selection by SVM

The algorithms of RF and XGboost implemented in Python already have built-in functions for automatically selecting parameters. In order to better train a good machine learning model, we do not select features in advance and use 35 features as input to train the classifier. However, SVM does not have the function of automatically selecting the number of features. We select features in order to improve training speed and efficiency. The feature selection results are shown in Fig. 3.

Figure 3 shows that the first four importance features are: length, fourth of generalized topological entropy, and longest ORF Relative length (lp), the length of the longest ORF (ll). And the two versions of human data have a certain degree of consistency in the selection of features. In the Kmer comparison experiment we designed, we use the same method for feature selection and the selected feature importance are listed in the additional files.

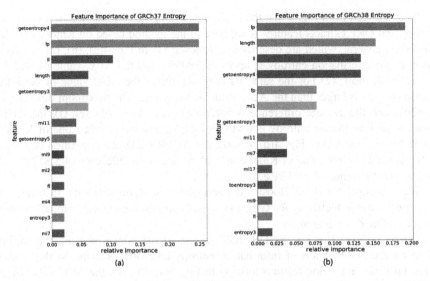

**Fig. 3.** (a) Feature importance of human GRCh37 data based on information entropy and ORF; (b) Feature importance of human GRCh38 data based on information entropy and ORF

## 3.2 Machine Learning Model Training Results Comparison

We apply SVM, XGBoost and Random Forest algorithms with 35 features to distinguish human lncRNAs for GRCh37 version and compare with the ones with Kmer features.

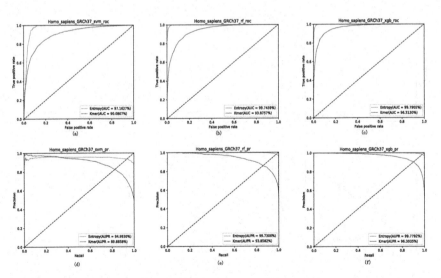

**Fig. 4.** Experimental results based on GRCh37 version of human species: (a) ROC curve of svm algorithm; (b) ROC curve of random forest algorithm; (c) ROC curve of xgboost algorithm; (d) PR curve of svm algorithm (e) PR curve of random forest algorithm; (f) PR curve of xgboost algorithm

It can be seen from Fig. 4 that the method based on the combination of information entropy and ORF extracts features is superior to the method based on Kmer extraction features in general, which are described as follows: (1) In Fig. 4(a), (b), (c), the AUC value of the information entropy is up to 99.7905%, and the AUC value of Kmer is 96.3130% at most; (2) For the same training algorithm, the AUC value of the information entropy is larger than the AUC value of Kmer one. The maximum difference is 7.0820% and the average difference is 5.4766%; (3) In Fig. 4(d), (e), (f), the AUPR value of the information entropy is up to 99.7792%, and the AUPR value of Kmer is 96.3035% at most; (4) In Fig. 4(d), (e), (f), the AUPR value of information entropy is larger than the AUPR value of Kmer one, with a maximum difference of 5.8724% and an average difference of 4.8184%.

We also apply SVM, XGBoost and Random Forest algorithms with 35 features to distinguish human lncRNAs for GRCh38 version and do the similar comparison with the ones with Kmer features.

Figure 5 shows that for the GRCh38 version of the human species, the method based on the combination of information entropy and ORF is better to the method based on Kmer extraction features too: (1) In Fig. 5(a), (b), (c), the AUC value of the information entropy is 99.7887% at the maximum, and the AUC value of Kmer is 97.3003% at the maximum; (2) In Fig. 5(a), (b), (c), the AUC value of the information entropy method is larger than the AUC value of Kmer one as the maximum difference is 6.6198% and the average difference is 4.6982%; (3) In Fig. 5(d), (e), (f), the AUPR value of the information entropy is up to 99.7606%, and the AUPR value of Kmer is 97.3299% at most; (4) In the Fig. 5(d), (e), (f), the AUPR value of information entropy is larger than the AUPR value of Kmer one, with a maximum difference of 4.8293% and an average difference of 3.8553%.

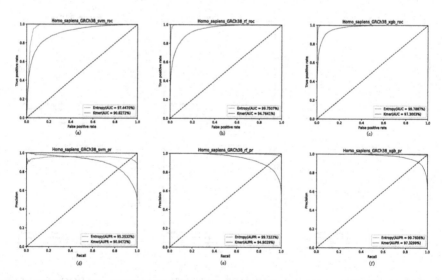

**Fig. 5.** Experimental results based on GRCh38 version of human species: (a) ROC curve of svm algorithm; (b) ROC curve of random forest algorithm; (c) ROC curve of xgboost algorithm; (d) PR curve of svm algorithm (e) PR curve of random forest algorithm; (f) PR curve of xgboost algorithm

To further investigate our integrated features and method, we apply current existing methods PLEK and CPAT for comparison. The results are shown in Fig. 6. It demonstrates that Xgboost for integrated features has the best AUC and PR values. In Fig. 6(a) and (b), the AUC value of PLEK is 1.0562% greater than that of Kmer_RF, while it is 1.1904% less than that of CPAT. The PR value of PLEK is 1.3487% greater than that of CPAT. In Fig. 6(c) and (d), the AUC value of PLEK is 1.0155% greater than that of Kmer_RF, while it is 0.5216% less than that of CPAT. The PR value of PLEK is 2.27% greater than that of CPAT. It is worth noting that the running time of PLEK on these 35 features is 9 days and the other methods are much less time-consuming.

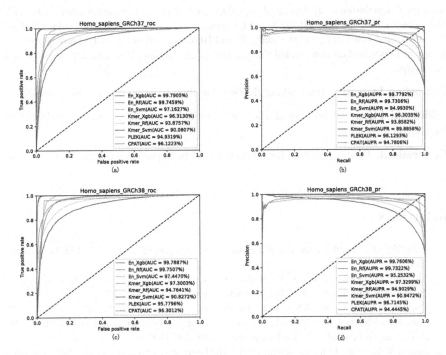

**Fig. 6.** (a) ROC curve of GRCh37; (b) PR curve of GRCh37; (c) ROC curve of GRCh38; (d) PR curve of GRCh38

# 4    Conclusions

In this paper, an effective lncRNA predictor is proposed. In order to obtain more accurate and realistic prediction results, we use the CD_HIT tool to perform de-redundancy operations on nucleic acid sequences. Characteristics features are extracted from the nucleic acid sequence itself, and the topological entropy and generalized topological entropy are regarded as new information theoretical features. We combine 35 features to train the classifier. Feature selection and classifier training are performed using SVM, random forest and XGBoost machine learning methods. Compared with the Kmer control experiment, we use 49 fewer features and speed up the training process. One advantage of our approach is that we only use features that are calculated directly from the sequence itself. Our method not only achieves good performance in lncRNA prediction, but also is extendable for research on other functional elements in DNA sequences.

**Acknowledgements.** This work was supported by the grants from the National "863" Key Basic Research Development Program (2014AA021505), Key research and development plan of the Ministry of science and technology (2017YFSF120182) and the startup grant of Harbin Institute of Technology (Shenzhen).

**Authors' Contributions.** JL and HL designed the study, performed bioinformatics analysis and drafted the manuscript. HL, YP, QZX and LZ participated in entropy relevance studies. XJ and WJ participated in the design of the study. JL and HL conceived of the study, participated in its design and coordination and drafted the manuscript. All authors read and approved the final manuscript.

**Competing Interests.** The authors declare that they have no competing interests.

**Additional Files.** All additional files are available at: https://github.com/lihuinian/lncRNAIdentification.

**Declarations.** Publication cost for this article has been paid by corresponding authors.

# References

1. Yanofsky, C.: Establishing the triplet nature of the genetic code. Cell **128**(5), 815–818 (2007)
2. Mohanty, V., Gokmen-Polar, Y., Badve, S., Janga, S.C.: Role of lncRNAs in health and disease-size and shape matter. Brief Funct. Genomics **14**(2), 115–129 (2015)
3. Esteller, M.: Non-coding RNAs in human disease. Nat. Rev. Genet. **12**(12), 861–874 (2011)
4. Taft, R.J., Pheasant, M., Mattick, J.S.: The relationship between non-protein-coding DNA and eukaryotic complexity. BioEssays **29**(3), 288–299 (2007)
5. Gupta, R.A., et al.: Long non-coding RNA HOTAIR reprograms chromatin state to promote cancer metastasis. Nature **464**(7291), 1071-U1148 (2010)
6. Ferre, F., Colantoni, A., Helmer-Citterich, M.: Revealing protein-lncRNA interaction. Brief. Bioinform. **17**(1), 106–116 (2016)
7. Li, J.W., et al.: LncTar: a tool for predicting the RNA targets of long noncoding RNAs. Brief. Bioinform. **16**(5), 806–812 (2015)
8. Yotsukura, S., Duverle, D., Hancock, T., Natsume-Kitatani, Y., Mamitsuka, H.: Computational recognition for long non-coding RNA (lncRNA): software and databases. Brief Bioinform. **18**(1), 9–27 (2017)
9. Brown, C.J., et al.: The human Xist Gene - analysis of a 17 Kb inactive X-specific Rna that contains conserved repeats and is highly localized within the nucleus. Cell **71**(3), 527–542 (1992)
10. Guttman, M., et al.: Chromatin signature reveals over a thousand highly conserved large non-coding RNAs in mammals. Nature **458**(7235), 223–227 (2009)

11. Cabili, M.N., et al.: Integrative annotation of human large intergenic noncoding RNAs reveals global properties and specific subclasses. Genes Dev. **25**(18), 1915–1927 (2011)

12. Wang, L., Park, H.J., Dasari, S., Wang, S.Q., Kocher, J.P., Li, W.: CPAT: coding-potential assessment tool using an alignment-free logistic regression model. Nucleic Acids Res. **41**(6), e74 (2013)

13. Li, A.M., Zhang, J.Y., Zhou, Z.Y.: PLEK: a tool for predicting long non-coding RNAs and messenger RNAs based on an improved k-mer scheme. BMC Bioinform. **15**, 311 (2014)

14. Achawanantakun, R., Chen, J., Sun, Y.N., Zhang, Y.: LncRNA-ID: Long non-coding RNA IDentification using balanced random forests. Bioinformatics **31**(24), 3897–3905 (2015)

15. Schneider, H.W., Raiol, T., Brigido, M.M., Walter, M.E.M.T., Stadler, P.F.: A support vector machine based method to distinguish long non-coding RNAs from protein coding transcripts. BMC Genomics **18**, 804 (2017)

16. Zerbino, D.R., et al.: Ensembl 2018. Nucleic Acids Res. **46**(D1), D754–D761 (2018)

17. Li, W.Z., Jaroszewski, L., Godzik, A.: Clustering of highly homologous sequences to reduce the size of large protein databases. Bioinformatics **17**(3), 282–283 (2001)

18. Li, W.Z., Jaroszewski, L., Godzik, A.: Tolerating some redundancy significantly speeds up clustering of large protein databases. Bioinformatics **18**(1), 77–82 (2002)

19. Li, W.Z., Godzik, A.: Cd-hit: a fast program for clustering and comparing large sets of protein or nucleotide sequences. Bioinformatics **22**(13), 1658–1659 (2006)

20. Koslicki, D.: Topological entropy of DNA sequences. Bioinformatics **27**(8), 1061–1067 (2011)

21. Jin, S.L., et al.: A generalized topological entropy for analyzing the complexity of DNA sequences. PloS One, **9**(2), e88519 (2014)

22. Nigatu, D., Sobetzko, P., Yousef, M., Henkel, W.: Sequence-based information-theoretic features for gene essentiality prediction. BMC Bioinform. **18**, 473 (2017)

23. Shannon, C.E.: The mathematical theory of communication (Reprinted). M D Comput. **14** (4), 306–317 (1997)

24. Church, K.W., Hanks, P.: Word association norms, mutual information, and lexicography. Comput. Linguist. **16**(1), 22–29 (1990)

25. Kullback, S., Leibler, R.A.: On information and sufficiency. Ann. Math. Stat. **22**(1), 79–86 (1951)

26. Platt, J.: Sequential minimal optimization: A fast algorithm for training support vector machines (1998)

27. Ho, T.K.: The random subspace method for constructing decision forests. IEEE Trans. Pattern Anal. Mach. Intell. **20**(8), 832–844 (1998)

28. Chen, T., Guestrin, C.: XGBoost: a scalable tree boosting system. In: Proceedings of the 22nd ACM SIGKDD International Conference on Knowledge Discovery and Data Mining, pp. 785–794. ACM, San Francisco (2016)

# A Gated Recurrent Unit Model for Drug Repositioning by Combining Comprehensive Similarity Measures and Gaussian Interaction Profile Kernel

Tao Wang[1], Hai-Cheng Yi[2,3(✉)], Zhu-Hong You[2], Li-Ping Li[2],
Yan-Bin Wang[2], Lun Hu[4], and Leon Wong[2]

[1] Department of Network Security, Xijing University, Xi'an 710123, China
[2] Xinjiang Technical Institute of Physics and Chemistry,
Chinese Academy of Sciences, Urumqi 830011, China
[3] University of Chinese Academy of Sciences, Beijing 100049, China
yihaicheng17@mails.ucas.ac.cn
[4] Department of Computing, The Hong Kong Polytechnic University,
Kowloon, Hong Kong

**Abstract.** Drug repositioning can find new uses for existing drugs and accelerate the processing of new drugs research and developments. It is noteworthy that the number of successful drug repositioning stories is increasing rapidly. Various computational methods have been presented to predict novel drug-disease associations for drug repositioning based on similarity measures among drugs and diseases or heterogeneous networks. However, there are some known associations between drugs and diseases that previous studies not utilized. In this work, we proposed a GRU model to predict potential drug-disease interactions by using comprehensive similarity. 10-fold cross-validation and common evaluation indicators are used to evaluate the performance of our model. Our model outperformed existing methods. The experimental results proved our model is a useful tool for drug repositioning and biochemical medicine research.

**Keywords:** Drug repositioning · Gated recurrent units · Fingerprint · Similarity measures

## 1 Introduction

Although the impressive advances have been witnessed in life sciences and technology and genomics over the past years. To bring a new drug to patients still takes $\sim 15$ years and 800 million to billion dollars [1–3]. Traditional drug research and development (R&D) process requires testing for side efforts and safety through cellular model systems, extensive animal model and clinical trial experimental validation. The average cost of new drug discovery has significantly increased and more than 90% of drug

---

The authors wish it to be known that, in their opinion, the first two authors should be regarded as joint First Authors.

D.-S. Huang et al. (Eds.): ICIC 2019, LNCS 11644, pp. 344–353, 2019.
https://doi.org/10.1007/978-3-030-26969-2_33

candidates fail during development, which caused pharmaceutical R&D tremendously expensive, time costing and high risky [3, 4]. This further directly led to a small quantity and high price of new drugs on the market. Drug repositioning or drug repurposing, identifying new clinical indications for those approved drugs has been used as an important strategy to maximize the potential usage of the existing drugs and increase the number of new drugs [5–7].

In recent years, the establishment of online public databases on pharmacochemical properties, drug molecules chemical structure, drug-drug interactions, disease-disease interactions, related genomic sequences and side efforts has promoted the study of drug-disease interactions and drug repositioning [8–26]. Such as KEGG [9], OMIM [10], CMap [11], DrugBank [12], STITCH [13] and ChEMBL [14]. The goal of drug repositioning is to find potential indications for existing approved drugs and apply the new identified drug candidates to the clinical treatment for other disease than originally targeted disease.

In this study, we present a novel deep learning model for the potential **Drug-Disease Interactions Pred**iction, named **DDIPred**, which applied gated recurrent neural network for the prediction of the new use of existing drugs using comprehensive similarity measures and gaussian interaction profile kernel. In order to verify the performance of our model, we evaluate DDIPred on two gold standard datasets, common evaluation criterions and 10-fold cross-validation are also adopted. The experimental results demonstrate that our model outperforms state-of-the-art methods and has the superior capability to discover potential new use of drugs.

## 2   Materials and Methodology

### 2.1   Construction of Datasets

To evaluate the performance of our model, we selected two widely used benchmark datasets including Fdataset and Cdataset. The gold standard dataset Fdataset is obtained from Gottlieb *et al.*'s work [18], which is made up of multiple data sources. More concretely, for this dataset, there are 1933 known associations between drugs and diseases and 593 drugs from DrugBank [27] and 313 diseases registered in OMIM [10] (the Online Mendelian Inheritance in Man). We also carried out another benchmark dataset Cdataset at the same time, this dataset is firstly presented in Luo *et al.*'s paper [5]. The details of these two datasets are shown in Table 1.

**Table 1.** The details of the two benchmark drug-disease associations datasets.

| Dataset | Number of drugs | Number of diseases | Interaction pairs |
|---------|-----------------|--------------------|--------------------|
| Fdataset | 593 | 313 | 1933 |
| Cdataset | 663 | 409 | 2532 |

## 2.2 Similarity Measures

Follow the description above, the drugs similarity is calculated based on the chemical structure information, which comes from drug-related properties [5]. More concretely, the similarity between two drugs is calculated by the Chemical Development Kit [28] of their 2D chemical fingerprints, which use the Simplified Molecular Input Line Entry Specification (SMILES) [29] of all drugs. The similarity is adjusted using the Logistic regression function which has been used to modify the diseases-genes associations similarity by [30]. The function can be defined as follow:

$$L(x) = \frac{1}{1 + e^{(ax+b)}} \tag{1}$$

Where $x$ represents the similarity value, $a$ and $b$ are adjusting parameters. And then, the drugs are clustered based on known drug-disease associations by using a graph clustering method, ClusterONE [31], which has been employed to detect valuable modules for drug repositioning [5, 22, 32]. The cohesiveness of a cluster M could be defined by ClusterONE as follows:

$$f(M) = \frac{C_{in}(M)}{(C_{in}(M) + C_{bound}(M) + P(M))} \tag{2}$$

Where $C_{in}(M)$ indicates the total weight of edges within a set of vertices $M$, $C_{bound}(M)$ stands for the total weight of edges connecting this set to the remaining of group, and $P(M)$ is the penalty term [5].

## 2.3 Gaussian Interaction Profile Kernel

For diseases, we adopted gaussian interaction profile kernel [33] to obtain the representation of disease-disease associations [34]. Based on the assumption that the diseases with a similar interaction pattern with drugs are likely to show similar interaction behavior with new drugs [33]. Similar assumptions can also be applied to drugs. Suppose $(D_i, D_j)$ indicates two different drugs, while $(d_i, d_j)$ represents two different diseases. Their gaussian interaction profile kernel similarity can calculation as follows:

$$KG_{disease}(d_i, d_j) = \exp\left(-\alpha_d \|y_{d_i} - y_{d_j}\|^2\right) \tag{3}$$

$$\alpha_d = \frac{\alpha_d'}{\left(\frac{1}{nd}\sum_{i=1}^{nd} \|y_{d_i}\|^2\right)} \tag{4}$$

Here, the $nd$ stands for the number of the diseases. And the parameter $\alpha_d$ controls the kernel bandwidth. For simplicity, the $\alpha_d'$ is set to 0.5.

## 2.4    Implementation of Gated Recurrent Neural Network

In order to overcome several known defects of standard Recurrent Neural network (RNN) model, a series of improved models has been proposed in deep learning field. Among them, the Long short-term memory (LSTM) [35, 36] and other similar variant models have the best performance and are widely used in many fields [37]. The main reason for their effectiveness is the pull-in of gated mechanisms. The Gated Recurrent Units (GRU) was proposed by Cho *et al.* [38], which have only resetting gate and updating gate and all memory contents are fully open to each timestep. We follow the similar calculation process in [39].

## 2.5    Performance Evaluation Metrics

In order to comprehensively evaluate the performance of our model, we follow the widely used evaluation indicators and strategies. The 10-fold cross-validation was applied to evaluate the performance of DDIPred. We follow the extensive used evaluation criteria, which can be defined as:

$$\text{Acc} = \frac{TN + TP}{TN + TP + FN + FP} \tag{5}$$

$$\text{TPR} = \frac{TP}{TP + FN} \tag{6}$$

$$\text{TNR} = \frac{TN}{TN + FP} \tag{7}$$

$$\text{PPV} = \frac{TP}{TP + FP} \tag{8}$$

$$\text{MCC} = \frac{TP \times TN - FP \times FN}{\sqrt{(TP + FP)(TP + FN)(TN + FP)(TN + FN)}} \tag{9}$$

Where *TN* stands for the true negative number, *TP* represents the true positive number, *FN* denotes the false negative number and *FP* indicates the false positive number. Certainly, the Receiver Operating Characteristic (ROC) curve and the area under the ROC curve (AUC) are also adopted to evaluate the performance.

## 3    Results and Discussion

### 3.1    Drug-Disease Interactions Prediction Capability Evaluation of DDIPred

First, we did a comparison between our model and widely used machine learning model SVM, which is often used as a baseline model and usually has great performance in various fields. The feature input, 10-fold cross validation set, evaluation metrics and other experimental conditions are exactly same between DDIPred and

SVM model. The parameters of SVM are determined by grid search. The results are shown in Table 2. Our model has significantly improved all indicators.

**Table 2.** 10-fold cross-validation performance of DDIPred and SVM on gold standard datasets.

| Datasets | Methods | Acc (%) | TPR (%) | TNR (%) | PPV (%) | MCC (%) |
|---|---|---|---|---|---|---|
| Cdataset | SVM | 72.57 | 70.99 | 76.41 | 68.70 | 45.25 |
| | DDIPred | 81.48 | 80.59 | 83.01 | 80.03 | 63.06 |
| Fdataset | SVM | 70.15 | 69.06 | 73.00 | 67.34 | 40.36 |
| | DDIPred | 77.83 | 77.13 | 79.22 | 76.57 | 55.80 |

## 3.2   Comparison with Other State-of-the-Art Methods

Furthermore, we compared our model with other state-of-the-art methods on same datasets under same experimental conditions. The comparison results are reported at Table 3. We compared the AUC of our model and previous studies including DrugNet [24] and HGBI [23]. Considering the difference of experimental evaluation indicators in different research, we only compared the AUC value reported in every study, which can best reflect the performance of model. As shown in Table 3, our model performs best on both datasets.

**Table 3.** Comparison of the AUC of previous studies and DDIPred.

| Predictors | Cdataset | Fdataset |
|---|---|---|
| DrugNet | 0.804 | 0.778 |
| HGBI | 0.858 | 0.829 |
| DDIPred | **0.871** | **0.838** |

# 4   Conclusion

In this work, we proposed a novel deep learning model DDIPred using comprehensive similarity measure and gaussian interaction profile kernel and gated recurrent neural networks to predict potential drug-disease associations, which could find new indications of existing drugs and can accelerate the process of drug research and development. The experimental results proved that our feature representation techniques can provide discriminative feature and our deep learning model is up to computational drug repositioning task. In recent years, machine learning and bioinformatics have made many inspiring advances [40–85], and we will conduct further research based on these advances.

**Acknowledgments.** This work is supported by the National Science Foundation of China, under Grants 61572506, in part by the NSFC Excellent Young Scholars Program, under Grants 61722212, in part by the Pioneer Hundred Talents Program of Chinese Academy of Sciences.

**Author Contributions.** Hai-Cheng Yi, Zhu-Hong You conceived the algorithm, carried out analyses, prepared the datasets, carried out experiments, and wrote the manuscript; Li-Ping Li, Yan-Bin Wang, Lun Hu and Leon Wong designed, performed and analyzed experiments and wrote the manuscript; All authors read and approved the final manuscript.

**Conflicts of Interest.** The authors declare no conflict of interest.

# References

1. Ashburn, T.T., Thor, K.B.: Drug repositioning: identifying and developing new uses for existing drugs. Nat. Rev. Drug Discovery **3**, 673 (2004)
2. Booth, B., Zemmel, R.: Prospects for productivity. Nat. Rev. Drug Discovery **3**, 451 (2004)
3. Dudley, J.T., Deshpande, T., Butte, A.J.: Exploiting drug–disease relationships for computational drug repositioning. Brief. Bioinform. **12**(4), 303–311 (2011)
4. Nagaraj, A.B., et al.: Using a novel computational drug-repositioning approach (DrugPredict) to rapidly identify potent drug candidates for cancer treatment. Oncogene **37**(3), 403–414 (2018)
5. Luo, H., et al.: Drug repositioning based on comprehensive similarity measures and bi-random walk algorithm. Bioinformatics **32**(17), 2664 (2016)
6. Luo, H., Li, M., Wang, S., Liu, Q., Li, Y., Wang, J.: Computational drug repositioning using low-rank matrix approximation and randomized algorithms. Bioinformatics **34**(11), 1904–1912 (2018)
7. Tartaglia, L.A.: Complementary new approaches enable repositioning of failed drug candidates. Expert Opin. Investig. Drugs **15**(11), 1295–1298 (2006)
8. Chen, X., et al.: NRDTD: a database for clinically or experimentally supported non-coding RNAs and drug targets associations. Database **2017** (2017)
9. Kanehisa, M., Goto, S., Furumichi, M., Tanabe, M., Hirakawa, M.: KEGG for representation and analysis of molecular networks involving diseases and drugs. Nucleic Acids Res. **38** (suppl_1), D355–D360 (2009)
10. Hamosh, A., Scott, A.F., Amberger, J., Bocchini, C., Valle, D., Mckusick, V.A.: Online mendelian inheritance in man (OMIM), a knowledgebase of human genes and genetic disorders. Nucleic Acids Res. **33**(1), 514–517 (2005)
11. Lamb, J., et al.: The connectivity map: using gene-expression signatures to connect small molecules, genes, and disease. Science **313**(5795), 1929–1935 (2006)
12. Knox, C., et al.: DrugBank 3.0: a comprehensive resource for 'Omics' research on drugs. Nucleic Acids Res. **39**, 1035 (2011). (Database issue)
13. Kuhn, M., et al.: STITCH 4: integration of protein-chemical interactions with user data. Nucleic Acids Res. **42**, 401–407 (2014). (Database issue)
14. Gaulton, A., et al.: ChEMBL: a large-scale bioactivity database for drug discovery. Nucleic Acids Res. **40**, 1100–1107 (2012)
15. Meng, F.-R., You, Z.-H., Chen, X., Zhou, Y., An, J.-Y.: Prediction of drug–target interaction networks from the integration of protein sequences and drug chemical structures. Molecules **22**(7), 1119 (2017)
16. Luo, H, et al.: DRAR-CPI a server for identifying drug repositioning potential and adverse drug reactions via the chemical–protein interactome. Nucleic Acids Res. **39**(suppl_2), W492–W498 (2011)
17. Chiang, A.P., Butte, A.J.: Systematic evaluation of drug–disease relationships to identify leads for novel drug uses. Clin. Pharmacol. Ther. **86**(5), 507–510 (2009)

18. Gottlieb, A., Stein, G.Y., Ruppin, E., Sharan, R.: PREDICT: a method for inferring novel drug indications with application to personalized medicine. Mol. Syst. Biol. **7**(1), 496 (2011)
19. Francesco, N., et al.: Drug repositioning: a machine-learning approach through data integration. J. Cheminform. **5**(1), 30 (2013)
20. Iorio, F., et al.: Discovery of drug mode of action and drug repositioning from transcriptional responses. Proc. Natl. Acad. Sci. **107**(33), 14621–14626 (2010)
21. Cheng, F., et al.: Prediction of drug-target interactions and drug repositioning via network-based inference. PLoS Comput. Biol. **8**(5), e1002503 (2012)
22. Wu, C., Gudivada, R.C., Aronow, B.J., Jegga, A.G.: Computational drug repositioning through heterogeneous network clustering. BMC Syst. Biol. **7**(5), 1–9 (2013)
23. Wang, W., Yang, S., Zhang, X., Li, J.: Drug repositioning by integrating target information through a heterogeneous network model. Bioinformatics **30**(20), 2923–2930 (2014)
24. Martínez, V., Navarro, C., Cano, C., Fajardo, W., Blanco, A.: DrugNet: network-based drug–disease prioritization by integrating heterogeneous data. Artif. Intell. Med. **63**(1), 41–49 (2015)
25. Yi, H.-C., You, Z.-H., Huang, D.-S., Li, X., Jiang, T.-H., Li, L.-P.: A deep learning framework for robust and accurate prediction of ncRNA-protein interactions using evolutionary information. Mol. Ther. - Nucleic Acids **11**, 337–344 (2018)
26. You, Z.-H., Zhan, Z.-H., Li, L.-P., Zhou, Y., Yi, H.-C.: Accurate prediction of ncRNA-protein interactions from the integration of sequence and evolutionary information. Front. Genet. **9**, 458 (2018)
27. Wishart, D.S., et al.: DrugBank: a knowledgebase for drugs, drug actions and drug targets. Nucleic Acids Res. **36**, 901–906 (2008). (Database issue)
28. Steinbeck, C., Han, Y., Kuhn, S., Horlacher, O., Luttmann, E., Willighagen, E.: The Chemistry Development Kit (CDK): an open-source java library for chemo-and bioinformatics. J. Chem. Inf. Comput. Sci. **43**(2), 493–500 (2003)
29. Weininger, D.: SMILES, a chemical language and information system. 1. Introduction to methodology and encoding rules. J. Chem. Inf. Comput. Sci. **28**(1), 31–36 (1988)
30. Vanunu, O., Magger, O., Ruppin, E., Shlomi, T., Sharan, R.: Associating genes and protein complexes with disease via network propagation. PLoS Comput. Biol. **6**(1), e1000641 (2010)
31. Nepusz, T., Yu, H., Paccanaro, A.: Detecting overlapping protein complexes in protein-protein interaction networks. Nat. Methods **9**(5), 471 (2012)
32. Yu, L., Huang, J., Ma, Z., Zhang, J., Zou, Y., Gao, L.: Inferring drug-disease associations based on known protein complexes. BMC Med. Genomics **8**(2), S2 (2015)
33. van Laarhoven, T., Nabuurs, S.B., Marchiori, E.: Gaussian interaction profile kernels for predicting drug–target interaction. Bioinformatics **27**(21), 3036–3043 (2011)
34. Chen, X., et al.: A novel computational model based on super-disease and miRNA for potential miRNA–disease association prediction. Mol. BioSyst. **13**(6), 1202–1212 (2017)
35. Hochreiter, S., Schmidhuber, J.: Long short-term memory. Neural Comput. **9**(8), 1735–1780 (1997)
36. Gers, F.A., Schmidhuber, J., Cummins, F.: Learning to forget: continual prediction with LSTM (1999)
37. Shen, Z., Bao, W., Huang, D.-S.: Recurrent neural network for predicting transcription factor binding sites. Scientific Rep. **8**(1), 15270 (2018)
38. Cho, K., Van Merriënboer, B., Bahdanau, D., Bengio, Y.: On the properties of neural machine translation: Encoder-decoder approaches. arXiv preprint. arXiv:14091259 (2014)
39. Chung J, Gulcehre C, Cho K, Bengio Y: Empirical evaluation of gated recurrent neural networks on sequence modeling. arXiv preprint. arXiv:14123555 (2014)

40. Yi, H.-C., et al.: ACP-DL: a deep learning long short-term memory model to predict anticancer peptides using high efficiency feature representation. Mol. Ther.- Nucleic Acids (2019)
41. Wang, L., et al.: MTRDA: using logistic model tree to predict miRNA-disease associations by fusing multi-source information of sequences and similarities. PLoS Comput. Biol. **15**(3), e1006865 (2019)
42. Chen, Z.-H., You, Z.-H., Li, L.-P., Wang, Y.-B., Wong, L., Yi, H.-C.: Prediction of self-interacting proteins from protein sequence information based on random projection model and fast fourier transform. Int. J. Mol. Sci. **20**(4), 930 (2019)
43. Chen, Z.-H., Li, L.-P., He, Z., Zhou, J.-R., Li, Y., Wong, L.: An improved deep forest model for predicting self-interacting proteins from protein sequence using wavelet transformation. Front. Genet. **10** (2019)
44. Zhu, H.-J., You, Z.-H., Zhu, Z.-X., Shi, W.-L., Chen, X., Cheng, L.: DroidDet: effective and robust detection of android malware using static analysis along with rotation forest model. Neurocomputing **272**, 638–646 (2018)
45. You, Z.-H., Huang, W., Zhang, S., Huang, Y.-A., Yu, C.-Q., Li, L.-P.: An efficient ensemble learning approach for predicting protein-protein interactions by integrating protein primary sequence and evolutionary information. IEEE/ACM Trans. Comput. Biol. Bioinform. **16**(3), 809–817 (2018)
46. Wang, Y.-B., You, Z.-H., Li, X., Jiang, T.-H., Cheng, L., Chen, Z.-H.: Prediction of protein self-interactions using stacked long short-term memory from protein sequences information. BMC Syst. Biol. **12**(8), 129 (2018)
47. Wang, Y., et al.: Predicting protein interactions using a deep learning method-stacked sparse autoencoder combined with a probabilistic classification vector machine. Complexity **2018** (2018)
48. Wang, L., et al.: Using two-dimensional principal component analysis and rotation forest for prediction of protein-protein interactions. Scientific Rep. **8**(1), 12874 (2018)
49. Wang, L., et al.: An improved efficient rotation forest algorithm to predict the interactions among proteins. Soft. Comput. **22**(10), 3373–3381 (2018)
50. Wang, L., You, Z.-H., Huang, D.-S., Zhou, F.: Combining high speed ELM learning with a deep convolutional neural network feature encoding for predicting protein-RNA Interactions. IEEE/ACM Trans. Comput. Biol. Bioinform. (2018)
51. Wang, L., et al.: A computational-based method for predicting drug–target interactions by using stacked autoencoder deep neural network. J. Comput. Biol. **25**(3), 361–373 (2018)
52. Song, X.-Y., Chen, Z.-H., Sun, X.-Y., You, Z.-H., Li, L.-P., Zhao, Y.: An ensemble classifier with random projection for predicting protein-protein interactions using sequence and evolutionary information. Appl. Sci. **8**(1), 89 (2018)
53. Qu, J., et al.: In silico prediction of small molecule-miRNA associations based on HeteSim algorithm. Mol. Ther.-Nucleic Acids (2018)
54. Qu, J., Chen, X., Sun, Y.Z., Li, J.Q., Ming, Z.: Inferring potential small molecule–miRNA association based on triple layer heterogeneous network. J. Cheminform. **10**(1), 30 (2018)
55. Luo, X., Zhou, M., Li, S., Xia, Y., You, Z.-H., Zhu, Q., Leung, H.: Incorporation of efficient second-order solvers into latent factor models for accurate prediction of missing QoS data. IEEE Trans. Cybern. **48**(4), 1216–1228 (2018)
56. Li, L.-P., Wang, Y.-B., You, Z.-H., Li, Y., An, J.-Y.: PCLPred: a bioinformatics method for predicting protein-protein interactions by combining relevance vector machine model with low-rank matrix approximation. Int. J. Mol. Sci. **19**(4), 1029 (2018)
57. Huang, Y.-A., You, Z.-H., Chen, X.: A systematic prediction of drug-target interactions using molecular fingerprints and protein sequences. Curr. Protein Pept. Sci. **19**(5), 468–478 (2018)

58. Chen, X., Zhang, D.-H., You, Z.-H.: A heterogeneous label propagation approach to explore the potential associations between miRNA and disease. J. Transl. Med. **16**(1), 348 (2018)
59. Chen, X., Xie, D., Wang, L., Zhao, Q., You, Z.-H., Liu, H.: BNPMDA: bipartite network projection for miRNA–disease association prediction. Bioinformatics **1**, 9 (2018)
60. Chen, X., Wang, C.-C., Yin, J., You, Z.-H.: Novel human miRNA-disease association inference based on random forest. Mol. Ther.-Nucleic Acids **13**, 568–579 (2018)
61. Chen, X., Gong, Y., Zhang, D.H., You, Z.H., Li, Z.W.: DRMDA: deep representations-based miRNA–disease association prediction. J. Cell Mol. Med. **22**(1), 472–485 (2018)
62. Zhu, L., Deng, S.-P., You, Z.-H., Huang, D.-S.: Identifying spurious interactions in the protein-protein interaction networks using local similarity preserving embedding. IEEE/ACM Trans. Comput. Biol. Bioinform. **14**(2), 345–352 (2017)
63. Zhu, H.-J., Jiang, T.-H., Ma, B., You, Z.-H., Shi, W.-L., Cheng, L.: HEMD: a highly efficient random forest-based malware detection framework for Android. Neural Comput. Appl. **30**(11), 1–9 (2017)
64. Zhang, S., Zhu, Y., You, Z., Wu, X.: Fusion of superpixel, expectation maximization and PHOG for recognizing cucumber diseases. Comput. Electron. Agric. **140**, 338–347 (2017)
65. Zhang, S., Zhang, C., Zhu, Y., You, Z.: Discriminant WSRC for large-scale plant species recognition. Computational intelligence and neuroscience, **2017**, (2017)
66. Zhang, S., You, Z., Wu, X.: Plant disease leaf image segmentation based on superpixel clustering and EM algorithm. Neural Comput. Appl. **31**, 1225–1232 (2019)
67. Zhang, S., Wu, X., You, Z., Zhang, L.: Leaf image based cucumber disease recognition using sparse representation classification. Comput. Electron. Agric. **134**, 135–141 (2017)
68. Zhang, S., Wu, X., You, Z.: Jaccard distance based weighted sparse representation for coarse-to-fine plant species recognition. PLoS ONE **12**(6), e0178317 (2017)
69. You, Z.-H., Zhou, M., Luo, X., Li, S.: Highly efficient framework for predicting interactions between proteins. IEEE Trans. Cybern. **47**(3), 731–743 (2017)
70. You, Z.-H., et al.: PRMDA: personalized recommendation-based MiRNA-disease association prediction. Oncotarget **8**(49), 85568 (2017)
71. You, Z.-H., et al.: PBMDA: a novel and effective path-based computational model for miRNA-disease association prediction. PLoS Comput. Biol. **13**(3), e1005455 (2017)
72. You, Z.H., Li, X., Chan, K.C.: An improved sequence-based prediction protocol for protein-protein interactions using amino acids substitution matrix and rotation forest ensemble classifiers. Neurocomputing **228**, 277–282 (2017)
73. Wen, Y.-T., Lei, H.-J., You, Z.-H., Lei, B.-Y., Chen, X., Li, L.-P.: Prediction of protein-protein interactions by label propagation with protein evolutionary and chemical information derived from heterogeneous network. J. Theor. Biol. **430**, 9–20 (2017)
74. Wang, Y.-B., You, Z.-H., Li, L.-P., Huang, Y.-A., Yi, H.-C.: Detection of interactions between proteins by using legendre moments descriptor to extract discriminatory information embedded in pssm. Molecules **22**(8), 1366 (2017)
75. Wang, Y.B., et al.: Predicting protein-protein interactions from protein sequences by a stacked sparse autoencoder deep neural network. Mol. BioSyst. **13**(7), 1336–1344 (2017)
76. Wang, Y., You, Z., Li, X., Chen, X., Jiang, T., Zhang, J.: PCVMZM: using the probabilistic classification vector machines model combined with a zernike moments descriptor to predict protein-protein interactions from protein sequences. Int. J. Mol. Sci. **18**(5), 1029 (2017)
77. Wang, L., et al.: Advancing the prediction accuracy of protein-protein interactions by utilizing evolutionary information from position-specific scoring matrix and ensemble classifier. J. Theor. Biol. **418**, 105–110 (2017)

78. Wang, L., et al.: Computational methods for the prediction of drug-target interactions from drug fingerprints and protein sequences by stacked auto-encoder deep neural network. In: Cai, Z., Daescu, O., Li, M. (eds.) ISBRA 2017. LNCS, vol. 10330, pp. 46–58. Springer, Cham (2017). https://doi.org/10.1007/978-3-319-59575-7_5
79. Li, S., Zhou, M., Luo, X., You, Z.-H.: Distributed winner-take-all in dynamic networks. IEEE Trans. Automat. Contr. **62**(2), 577–589 (2017)
80. Li, J.-Q., You, Z.-H., Li, X., Ming, Z., Chen, X.: PSPEL: in silico prediction of self-interacting proteins from amino acids sequences using ensemble learning. IEEE/ACM Trans. Comput. Biol. Bioinform. **14**(5), 1165–1172 (2017)
81. Chen, X., Xie, D., Zhao, Q., You, Z.-H.: MicroRNAs and complex diseases: from experimental results to computational models. Brief. Bioinform. **20**(2), 515–539 (2017)
82. Luo, X., Zhou, M., Li, S., You, Z., Xia, Y., Zhu, Q.: A nonnegative latent factor model for large-scale sparse matrices in recommender systems via alternating direction method. IEEE Trans. Neural Netw. Learn. Syst. **27**(3), 579–592 (2016)
83. Luo, X., et al.: An incremental-and-static-combined scheme for matrix-factorization-based collaborative filtering. IEEE Trans. Autom. Sci. Eng. **13**(1), 333–343 (2016)
84. Li, S., You, Z.H., Guo, H., Luo, X., Zhao, Z.Q.: Inverse-free extreme learning machine with optimal information updating. IEEE Trans. Cybern. **46**(5), 1229 (2016)
85. Ji, Z., Wang, B., Deng, S., You, Z.: Predicting dynamic deformation of retaining structure by LSSVR-based time series method. Neurocomputing **137**, 165–172 (2014)

# A Novel Approach to Predicting MiRNA-Disease Associations

Guo Mao and Shu-Lin Wang[✉]

College of Computer Science and Electronics Engineering, Hunan University,
Changsha 410082, Hunan, China
smartforesting@gmail.com

**Abstract.** MiRNA is a kind of RNA that cannot be translated into protein and is a kind of non-coding RNA molecules. In the current research, miRNA is a hot topic: many variants and dysregulations of miRNA are closely related to human diseases and they participate in the occurrence of various diseases. Predicting and confirming disease-related miRNAs helps to understand the pathogenesis of disease at miRNA molecular level. Matrix completion methods are often used to predict potential associations between miRNAs and diseases. In the methods, we want to fill in the missing entries of the matrix based on the small portion of entries observed. Thus, we developed a new approach for the prediction of miRNA–disease associations by using inductive matrix completion (named MGAPG) of an accelerated proximal gradient algorithm in this work. To evaluate the performance of MGAPG, a comparison between the MGAPG and other algorithms reveals the reliable performance of MGAPG which performed well in a 5-fold cross-validation. The AUC values of MGAPG are higher than some well-known methods, indicating its outstanding performance. Moreover, in the case of colon cancer associated with miRNA-diseases known in the DBDEMC, HMDD and Mir2disease databases, 90% of the top 20 predictive miRNA were confirmed by the experimental report. These results fully demonstrate the reliability of MGAPG predictive ability.

**Keywords:** Inductive matrix completion · MicroRNA · Disease · MicroRNA-disease association

## 1 Introduction

MicroRNAs (miRNAs) play critical roles by regulating gene expression at the post-transcriptional level. And a great deal of evidence proved that miRNA mutation is related to a variety of human diseases through many factors, such as schizophrenia [9]. Thus, more potential miRNA-disease associations that have been discovered will effectively contribute to the diagnosis and treatment of disease and facilitate the exploration of disease pathogenesis. A miRNA-disease association database has been constructed, for instance, Li et al. [8] and Jiang et al. [7] constructed two comprehensive databases, namely, Human miRNA associated Disease Database (abbreviated as HMDD) and miR2Disease. Yang et al. [15]. proposed a database (dbDEMC) for the differential expression of miRNAs in human cancer to explore the expression of

© Springer Nature Switzerland AG 2019
D.-S. Huang et al. (Eds.): ICIC 2019, LNCS 11644, pp. 354–365, 2019.
https://doi.org/10.1007/978-3-030-26969-2_34

abnormal miRNAs in different cancer conditions. However, These databases can only verify a small number of miRNA-disease associations. The biological experimental method that identifies disease-related miRNA is expensive and time-consuming. Therefore, efficient computational methods have received much attention in predicting disease-associated miRNAs. Among the computational methods proposed, many measures that the similarity computation for miRNAs and diseases to accomplish association prediction. Zou et al. [17] reviewed the main similarity computation methods. These methods can be approximately divided into four categories: machine learning-based methods, neighborhood-based methods, random walk-based methods, and path-based methods.

Machine-learning-based models are beneficial to improve the prediction performance. Chen et al. [4] proposed a semi-supervised method to integrate known miRNA-disease associations, disease-disease similarity datasets and miRNA-miRNA functional similarity networks based on regularized least squares(abbreviated as RLSMDA). A Naive Bayes model using genomic data is adopted by Jiang et al. to predict disease-related miRNAs. Chen et al. [2] utilized a restricted Boltzmann machine (RBM) for multiple types of miRNA-disease association prediction (RBMMMDA); however, this method required a lot of parameters. To improve predicted performance, random walk-based methods were adopted by some researchers. For example, Chen et al. [3] proposed the RWRMDA model of adopting random walks on the miRNA functional similarity network. Gu et al. [6] develop the Network Consistency Projection for Human miRNA-Disease Associations (NCPMDA) method to reveal the potential associations between miRNAs and diseases, NCPMDA does not require negative sample and it is a non-parametric universal network-based method. Xuan et al. [14] proposed the HDMP model that evaluated the k most functionally similar neighbors by integrating disease terms and phenotype similarity. PBMDA [16] for predicting miRNA–disease association is an effective path-based model. This model uses the depth-first search algorithm by integrating the miRNA functional similarity, known human miRNA–disease associations, disease semantic similarity, and Gaussian interaction profile kernel similarity for miRNAs and diseases. CPTL [11], a Collective Prediction based on Transduction Learning, to systematically prioritize miRNAs related to the disease. However, their performance for sparse known associations validated by biology experiment is still not satisfactory.

In this study, we formulate miRNA-disease association prediction as a recommendation system problem. We propose the use of an inductive matrix completion (IMC) method [13] for predicting miRNA-disease associations (named MGAPG) by using informative feature vectors corresponding to the top singular vectors of the miRNA and disease feature matrices. The well-known Netflix problem is a good example: in the area of recommender systems, users submit ratings on a subset of entries in a database, and the vendor provides recommendations based on the user's preferences. Recently, recommendation system methods have been applied to association prediction in a variety of bioinformatics problems. Luo et al. [10] propose a drug repositioning recommendation system (DRRS) to predict novel drug indications by integrating related data sources and validated information on drugs and diseases. MGAPG uses principle components analysis (PCA) [1] to extract informative feature vectors, based on disease semantic similarity and miRNA functional similarity matric.

Then, MGAPG completes the association matrix using the primary feature vectors from the constructed feature matrices. MGAPG outperforms the other methods in 5-fold cross-validation experiments. Moreover, case studies show that MGAPG is capable of inferring potential miRNAs for renal cancer, gastric cancer, and prostate cancer. In summary, combining effective features extracted the disease semantic similarity and miRNA functional similarity into fast IMC computation leads to MGAPG performance enhancement.

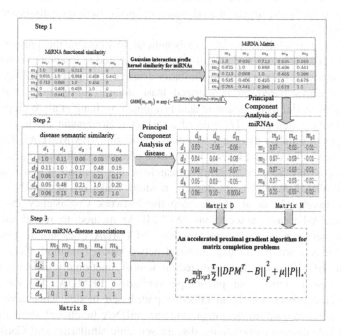

**Fig. 1.** Scheme of MGAPG. Step 1: MGAPG computes the Gaussian interaction profile kernel similarity of miRNAs functional similarity. Step 2: MGAPG extracts primary feature vectors from miRNA (MS) and disease (DS) by PCA, respectively. Step 3: MGAPG calculates the interaction profile for a new miRNA with the interaction profiles of its neighbors. Step 4: MGAPG completes the association matrix with An accelerated proximal gradient(APG) algorithm for matrix completion problems using primary feature vectors and constructed interaction profiles.

# 2   Materials and Methods

## 2.1   Method Overview

MGAPG mainly consists of four steps, as shown in Fig. 1. First, MGAPG computes the Gaussian interaction profile kernel similarity of miRNAs functional similarity. Second, MGAPG extracts primary feature vectors from miRNA(MS) and disease(DS) by PCA, respectively. Three, MGAPG calculates the interaction profile for a new miRNA with the interaction profiles of its neighbors. Last, MGAPG completes the association matrix with An accelerated proximal gradient(APG) algorithm for matrix completion problems using primary feature vectors and constructed interaction profiles. The total algorithm flow of MGAPG is shown in Algorithm 1.

---

**Algorithm 1:** MGAPG Algorithm

---

**Input:** known miRNA-disease associations matrix B, disease semantic similarity matrix D, and miRNA functional similarity matrix MS.

**Output:** Predicted association matrix PM

1: Calculate the Gaussian interaction profile kernel (namely matrix M) of miRNAs based on miRNA functional similarity matrix MS.
2: Extract the primary feature vectors from $D \in \mathcal{R}^{d \times d}$ and $M \in \mathcal{R}^{m \times m}$ by using PCA of employing singular value decomposition (SVD)
3: Construct new interaction profile for a new miRNA
4: Recover a (low-rank) data matrix of incomplete miRNA-disease associations matrix B using an accelerated proximal gradient algorithm (APG), namely, the matrix completion problem can be cast the following minimization problem:

$$\min_{P \in \mathcal{R}^{d l \times d p}} \frac{\tau}{2} ||DPM^T - B||_F^2 + \mu ||P||_*$$

---

## 2.2    Dataset and Preprocessing

**Disease Semantic Similarity Model.** We downloaded DAGs of the disease from the MESH database (http://www.ncbi.nlm.nih.gov/), which afforded a strict system for disease classification. To represent the semantic similarity of the disease, we use the matrix DD representation, where DD(i, j) represents the semantic similarity between disease i and disease j.

**MiRNA Functional Similarity.** Wang et al. [12] proposed a miRNAs similarity measure and constructed a miRNAs similarity matrix MISIM containing 577 miRNAs. And this research starts from the related research, from the http://www.cuilab.cn/fles/images/cuilab/misim.zip Downloads relevant miRNA functional similarity metrics.

**Human miRNA–Disease Associations.** The human miRNA-disease-association dataset used in this study was derived from the HMDDv2.0 database [8], which included a total of 6441 associations between 577 miRNAs and 336 human diseases and lied an important data fundamental for further miRNA-related computational research.

## 2.3    Gaussian Interaction Profile Kernel Similarity of MiRNAs

Based on the opinion that miRNA functional similarity tends to the disease similarity, and vice versa. The binary vector IP ($m_i$) indicates whether the miRNA $m_i$ interacts with each disease or not and the interaction profile of miRNA $m_i$. Subsequently, the Gaussian kernel similarity is computed the kernel for the two miRNA $m_i$ and $m_j$ based on their interaction profiles, which are defined as follows:

$$GMM(m_i, m_j) = \exp\left(-\frac{\sum_{i=1}^{n}\|IP(m_i)\|^2 \times \left(\|IP(m_i) - IP(m_j)\|^2\right)}{n}\right),$$ (1)

where the interaction profile of miRNA $m_i(IP(m_i))$ is the ith row vector of the functional similarity matrix miRNAs, n is the total number of rows of the miRNA matrix.

## 2.4 Extracting Primary Feature Vectors

Since most of the data sources for the formation of miRNAs and disease feature matrices MS and DS are high-dimensional, so we apply principal component Analysis (PCA) to each of these data sources to build a robust and useful miRNA and disease feature matrix and use singular value decomposition (SVD) to perform PCA. Suppose the SVD of D is given by:

$$D = U\Sigma V^T, \Sigma = Diag(\sigma),$$ (2)

where U and V are respectively a $\times$ b and c $\times$ b matrices with orthogonal columns, $\sigma \in \mathcal{R}^b$ is the vector of positive singular values arranged in descending order $\sigma_1 \geq \sigma_2 \geq \ldots \geq \sigma_b > 0$, with $q \leq \min\{a, c\}$. M and D respectively represent the singular eigenvectors corresponding to the singular values of the former $dp$ and $dl$ of MS and DS, and it contains most important information.

## 2.5 MGIMC Method

**Introduction of the Matrix Completion.** Matrix completion is the task of filling in the missing entries of a partially observed matrix. A wide range of datasets are naturally organization matrix form. Thus matrix completion often seeks to find the lowest rank matrix, or if the rank of the completed matrix is known, a matrix of rank that matches the known entries. The matrix completion problem is in general NP-hard, but there are tractable algorithms that achieve exact reconstruction with high probability. In statistical learning point of view, the matrix completion problem is an application of matrix regularization which is a generalization of vector regularization. The matrix completion problem can be cast as the following minimization problem:

$$\min_{X \in \mathcal{R}^{m \times n}} \{rank(X) : X_{ij} = M_{ij}, (i,j) \in \Omega\},$$ (3)

where $M$ is the unknown matrix with p available sampled entries and $\Omega$ is a set of pairs of indices of cardinality p. Recently, others proved that a random low-rank matrix can be recovered exactly with high probability from a rather small random sample a rather small random sample of its entries, and it can be done by solving a convex relaxation problem:

$$\min_{X \in \Re^{m \times n}} \left\{ \|X\|_*: X_{ij} = M_{ij}, \ (i,j) \in \Omega \right\}, \tag{4}$$

where $\|.\|_*$ is the nuclear norm defined as the sum of the singular values.

**An Accelerated Proximal Gradient Algorithm for Matrix Completion Problems.**
We complete B based on the low-rank assumption in [13], the column vectors in B lie in the subspace spanned by the column vectors in D, and the row vectors in B lie in the subspace spanned by the column vectors in $M$. We want to recover all entries in the set $\Omega$ of the observed item. Then, the problem can be defined as:

$$\min_{P \in \mathbb{R}^{dl \times dp}} \lambda \|P\|_* + \frac{1}{2} \|DPM^T - B\|_F^2, \tag{5}$$

where $P$ is the objective matrix to complete $B$. Among them $M \in R^{m \times dp}$ and $D \in R^{d \times dl}$, where dp and dl separately are the dimensions of miRNAs and the latent characteristic space of the disease, and $\lambda$ is the regularization parameter.

In this paper, we consider a more general unconstrained nonsmooth convex minimization problem of the form:

$$\min_{P \in \mathbb{R}^{dl \times dp}} F(P) := f(P) + G(P), \tag{6}$$

the problem (4) is a special case of (5) with $f(P) = \frac{1}{2}\|DPM^T - B\|_F^2$ and $G(P) = \lambda\|P\|_*$ with $P \in \mathbb{R}^{dl \times dp}$.

For any $Y \in \mathbb{R}^{dl \times dp}$, consider the following quadratic approximation of $F(\cdot)$ at $Y$:

$$\begin{aligned} Q(P, Y) &:= f(Y) + \ <\nabla f(Y), P - Y> \ + \frac{1}{2} \| P - Y \|_F^2 + G(P) \\ &= \frac{1}{2} \| P - Y + \nabla f(Y) \|_F^2 + G(P) + f(Y) - \frac{1}{2} \| \nabla f(Y) \|_F^2, \end{aligned} \tag{7}$$

where the term $<\nabla f(Y), P - Y>$ is actually the nuclear norm $\|P - Y + \nabla f(Y)\|_*$, $\nabla f(Y)$ is the first-order reciprocal of $F(P)$ at $Y$, $H = Y - \nabla f(Y)$. Since (6) is a strongly convex function of $P$, Formula (7) has a unique minimizer which we denote by

$$S(H) := \operatorname{argmin}\{Q(P, Y)|P \in \mathbb{R}^{dl \times dp}\} \tag{8}$$

The detailed process of the algorithm can be found in the Algorithm 2.

---

**Algorithm 2** : APG algorithm

---

For a given $\tau = 1$, initialize threshold $\varepsilon$, choose $P^0 \in \mathcal{R}^{dl \times dp}$, $t^0 = 1$. For $k = 0,1,2...$, generate $P^{k+1}$ from $P^k$ according to the following iteration:

Step 1. Set $Y^k = P^k + \frac{t^{k-1}-1}{t^k}(p^k - p^{k-1})$.

Step 2.  Set $G^k = Y^k - (s^k)^{-1}D^T(DPM^T - B)M$ , where $s^k = L_f$ . Compute $S_{sk}(G^k)$ from the SVD of $G^k$.

Step 3. Set $P^{k+1} = S_{sk}(G^k)$.

Step 4. Compute $t^{k+1} = \left(\sqrt{(t^k)^4 + 4(t^k)^2} - (t^k)^2\right)/2$

Step 5.  If $|f(P^{k+1}) - f(P^k)| \geq \varepsilon$, then skip to Step 1.

Step 6.  Predicted association matrix $PM = DPM^T$

---

# 3  Results

## 3.1  Performance on the Matrix Completion

The predictive performance of the MGAPG model is evaluated by implementing a 5-fold cross-validation in the experimentally validated disease-related miRNA. For a specific disease d, miRNAs with known correlations for the disease d are randomly divided into five subsets, four of which are used for training and the last subset for testing.

Table 1 shows ROC curves, PRE and REC for both diseases. AUC values for both diseases performed well. The REC is rising as k increases, but the PRE decreases. This suggests that the highest ranked correlation is more likely to be a potential correlation.

## 3.2  Comparison with Other Existing Methods

We compare the MGAPG with four widely applied miRNA-disease prediction algorithms: (i) RWRMDA [3]; (ii) HDMP [14]; (iii) NCPMDA [6]; (iv) CPTL [11]. Table 2 shows the predictive performance of different diseases measured by AUC. The highest value per line is shown in bold. For all six tested diseases, MGAPG is superior to almost all other algorithms, including RWRMDA, HDMP, NCPMDA, and CPTL. The average AUC values of RWRMDA, HDMP, NCPMDA, CPTL, and MGAPG were 80.1%, 81.3%, 79.7%, 71.9% and 86.1%, respectively. The average AUC of SPM was higher than the 6%, 4.8%, 6.4% and 14.2% of the other four methods, respectively. The highest AUC value for Colorectal neoplasm was 0.943. We compare MGAPG with NCPMDA in AUC, see Fig. 2. The AUC value of NCPMDA corresponding to all diseases is 0.757 in a 5-fold cross-validation, whereas the AUC value of MGAPG is 0.869, which improves by 11.2%. Thus, MGAPG shows very good results.

**Table 1.** ROC curves, precision, and recall for MGAPG by using 5-fold cross-validation.

| Disease Name | ROC | PRE | REC |
|---|---|---|---|
| Breast neo-plasm | | | |
| Colo-rectal neo-plasms | | | |

**Table 2.** The AUC values for five methods by using 5-fold cross-validation

| Disease name | AUC | | | | |
|---|---|---|---|---|---|
| | RWRMDA | HDMP | NCPMDA | CPTL | **MGAPG** |
| Breast neoplasm | 0.785 | 0.801 | 0.675 | 0.732 | **0.920** |
| Hepatocellular carcinoma | 0.749 | 0.759 | 0.761 | 0.685 | **0.823** |
| Colorectal neoplasm | 0.793 | 0.802 | 0.812 | 0.715 | **0.943** |
| Lung neoplasm | 0.827 | 0.835 | 0.754 | 0.753 | **0.843** |
| Stomach neoplasm | 0.779 | 0.787 | **0.895** | 0.675 | 0.798 |
| Pancreatic neoplasm | 0.871 | 0.895 | 0.798 | 0.756 | **0.901** |

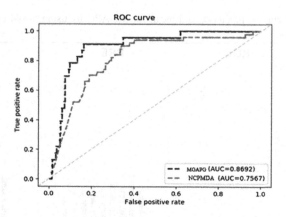

**Fig. 2.** The ROC across all tested disease.

### 3.3 Case Study: Colon Cancer

In order to demonstrate the MGIMC model, we have removed the known-proven miRNAs-disease associated with predictive disease for the predictive power of diseases that do not have any known associated miRNAs. This operation ensures that we use only the similarity of the disease information and other known miRNAs disease associations to predict disease-related miRNAs. We use colon cancer as a case study.

Colorectal cancer (CRC), also known as bowel cancer and colon cancer(CN), is cancer that develops from the colon or rectum (part of the large intestine). On a global scale, colorectal cancer is the third most common type, accounting for about 10% of all cases. In 2012, 1.4 million new cases and 694,000 deaths were reported in the disease. It is more common in developed countries, where more than 65% of cases are found. Women are less common than men. With the development of high-throughput sequencing technology, more and more CN related miRNAs have been identified. For example, by downgrading the IRS-1 protein in colon cancer cell lines, miR-145 was found to act as a potent inhibitor. It was also found miR-106a, which is always lost in CN tissue, target mRNA at E2F1 and inhibit tumor cell growth. A case study of potential miRNA predictions for the implementation of SPYSMDA in CN, based on DBDEMC and mir2disease evidence (see Table 3). Among them, hsa-mir-155 can up-regulate the expression level of tight junction protein-1 (claudin-1) in CRC cells, thus inducing the expression of ZEB-1 and inhibiting the expression of E-cadherin, resulting in EMT, leading to lymph nodes or distant metastasis of tumor cells.

**Table 3.** Case study on colon neoplasms.

| Top 1–10 | | Top 10–20 | |
|---|---|---|---|
| miRNA | Evidence | miRNA | Evidence |
| hsa-mir-21 | dbDEMC;miR2Disease | hsa-mir-145 | dbDEMC |
| hsa-mir-155 | dbDEMC; miR2Disease | hsa-mir-142 | unconfirmed |
| hsa-mir-146a | dbDEMC | hsa-mir-106a | dbDEMC; miR2Disease |
| hsa-mir-451a | unconfirmed | hsa-let-7d | dbDEMC |

(*continued*)

**Table 3.** (*continued*)

| Top 1–10 | | Top 10–20 | |
|---|---|---|---|
| miRNA | Evidence | miRNA | Evidence |
| hsa-mir-125b | dbDEMC | hsa-mir-223 | dbDEMC; miR2Disease |
| hsa-mir-221 | dbDEMC; miR2Disease | hsa-mir-19a | dbDEMC; miR2Disease |
| hsa-mir-34a | dbDEMC; miR2Disease | hsa-mir-199a | unconfirmed |
| hsa-mir-129 | HMDD,miR2Disease | hsa-let-7g | dbDEMC; miR2Disease |
| hsa-mir-27a | HMDD,miR2Disease | hsa-mir-125b | dbDEMC |
| hsa-let-7d | HMDD,miR2Disease | hsa-mir-29a | dbDEMC; miR2Disease |

According to the above analysis, the MGIMC model has obtained convincing results in comprehensive prediction, disease-specific miRNAs prediction. The MGIMC model can be used as a useful bioinformatics resource for biomedical research, and can successfully predict disease-related miRNAs and help to find therapeutic targets for miRNAs.

## 4 Discussion and Conclusion

More and more studies have shown that miRNA plays an important role in various biological processes. Therefore, it is of great significance to identify the relationship between miRNA and disease. Although experimental methods can determine the relationship between them, these methods are often time-consuming and only apply to small-scale datasets. In this study, we have proposed a computational method MGAPG to predict miRNA-disease associations from known data using IMC, based on the assumption that functionally similar miRNAs tend to interact with phenotypically similar diseases. We compared MGAPG with RWRMDA, HDMP, NCPMDA, and CPTL, The results demonstrate that MGAPG performs better in AUC values. In addition, we have conducted a case study of colon cancer and found that 88% of respondents through the literature excavation can determine the candidate miRNA of each kind of disease of interest.

Our study has two major contributions in predicting miRNA-disease associations. First of all, we can find more precise primary feature vector from MS and DS to improve the accuracy. Second, we predict the miRNA-Disease Association by using the matrix completion of an accelerated proximal gradient algorithm. MGIMC may be a useful tool to study the relationship between miRNA-diseases. The basic idea for completing miRNA-disease association prediction using an induction matrix is to find a low-rank matrix that can integrate prior knowledge about miRNA and disease to complete the miRNA-disease association matrix. By decomposing the matrix into a low-rank matrix, matrix decomposition provides a framework for dimension reduction and matrix completion, which can also be applied to miRNA-disease correlation prediction.

Nevertheless, the inductive matrix completion method described in this paper for miRNA-disease association prediction is flexible and can incorporate feature vectors from multiple sources. Moreover, the nuclear norm regularization adopted in (5) ensures convex optimization. Computational methods have been developed to detect interactions between miRNA and related proteins [5]. Accurate predictions of the relationship between miRNA-diseases through MGAPG can help reduce false positives in miRNA-protein interactions and identify disease-related miRNA-protein interactions. Thus, there are several possible extensions to the miRNA method provided here: The framework itself can be extended to solve the sparsity problems inherent in the real-World Association matrix and will integrate multiple data sources into our approach to improve the accuracy of predictions.

**Acknowledgement.** This work was supported by the grants of the National Science Foundation of China (Grant Nos. 61472467 and 61672011) and the National Key R&D Program of China (2017YFC1311003). And i would like to thank Yingting Jiang for providing help in grammar and spelling.

# References

1. Bryant, F.B., Yarnold, P.R.: Principal-components analysis and exploratory and confirmatory factor analysis. In: Reading and Understanding Multivariate Statistics, pp. 99–136 (1995)
2. Chen, X., Clarence Yan, C., Zhang, X., et al.: RBMMMDA: predicting multiple types of disease-microRNA associations. Sci. Rep. **5**, 13877 (2015)
3. Chen, X., Liu, M.X., Yan, G.Y.: RWRMDA: predicting novel human microRNA-disease associations. Mol. BioSyst. **8**, 2792–2798 (2012)
4. Chen, X., Yan, G.-Y.: Semi-supervised learning for potential human microRNA-disease associations inference. Sci. Rep. **4**, 5501 (2014)
5. Ge, M., Li, A., Wang, M.: A bipartite network-based method for prediction of long noncoding RNA-protein interactions. Genomics Proteomics Bioinformatics **14**, 62–71 (2016)
6. Gu, C., Liao, B., Li, X., et al.: Network consistency projection for human miRNA-Disease associations inference. Sci. Rep. **6**, 36054 (2016)
7. Jiang, Q., Wang, Y., Hao, Y., et al.: miR2Disease: a manually curated database for microRNA deregulation in human disease. Nucleic Acids Res. **37**, D98–104 (2009)
8. Li, Y., Qiu, C., Tu, J., et al.: HMDD v2.0: a database for experimentally supported human microRNA and disease associations. Nucleic Acids Res. **42**, 1070–1074 (2014)
9. Lu, M., Zhang, Q., Deng, M., et al.: An analysis of human microRNA and disease associations. PLoS ONE **3**, e3420 (2008)
10. Luo, H., Li, M., Wang, S., et al.: Computational drug repositioning using low-rank matrix approximation and randomized algorithms. Bioinformatics **34**, 1904–1912 (2018)
11. Luo, J., Ding, P., Liang, C., et al.: Collective prediction of disease-associated miRNAs based on transduction learning. IEEE/ACM Trans. Comput. Biol. Bioinform. **14**, 1468–1475 (2017)
12. Wang, D., Wang, J., Lu, M., et al.: Inferring the human microRNA functional similarity and functional network based on microRNA-associated diseases. Bioinformatics **26**, 1644–1650 (2010)
13. Xu, M., Jin, R., Zhi, Z.H.: Speedup matrix completion with side information: application to multi-label learning. In: Jordan, M.I., LeCun, Y., Solla, S.A. (eds.) Advances in Neural Information Processing System, pp. 2301–2309. MIT Press (2013)

14. Xuan, P., Han, K., Guom, M. et al.: Correction: prediction of microRNAs associated with human diseases based on weighted k most similar neighbors. PLoS One **8**, (2013)
15. Yang, Z., Wu, L., Wang, A., et al.: dbDEMC 2.0: updated database of differentially expressed miRNAs in human cancers. Nucleic Acids Res. **45**, D812–D818 (2017)
16. You, Z.H., Huang, Z.A., Zhu, Z., et al.: PBMDA: a novel and effective path-based computational model for miRNA-disease association prediction. PLoS Comput. Biol. **13**, e1005455 (2017)
17. Zou, Q., Li, J., Song, L., et al.: Similarity computation strategies in the microRNA-disease network: a survey. Brief. Funct. Genomics **15**, 55–64 (2016)

# Hierarchical Attention Network for Predicting DNA-Protein Binding Sites

Wenbo Yu[1](✉), Chang-An Yuan[2], Xiao Qin[2], Zhi-Kai Huang[3],
and Li Shang[4]

[1] Institute of Machine Learning and Systems Biology, School of Electronics
and Information Engineering, Tongji University, Shanghai, China
`yu_wen_bo@outlook.com`
[2] Science Computing and Intelligent Information Processing of GuangXi Higher
Education Key Laboratory, Nanning Normal University, Nanning,
Guangxi, China
[3] College of Mechanical and Electrical Engineering,
Nanchang Institute of Technology, Nanchang 330099, Jiangxi, China
[4] Department of Communication Technology,
College of Electronic Information Engineering,
Suzhou Vocational University, Suzhou 215104, Jiangsu, China

**Abstract.** Discovering DNA-protein binding sites, also known as motif discovery, is the foundation for further analyses of transcription factors (TFs). Deep learning algorithms such as convolutional neural networks (CNN) and recurrent neural networks (RNN) are introduced to motif discovery task and have achieved state-of–art performance. However, these methods still have limitations such as neglecting the context information in large-scale sequencing data. Thus, inspired by the similarity between DNA sequence and human language, in this paper we propose a hierarchical attention network for predicting DNA-protein binding sites which is based on a natural language processing method for document classification. The proposed method is tested on real ChIP-seq datasets and the experimental results show a considerable improvement compared with two well-tested deep learning-based sequence model, DeepBind and Deepsea.

**Keywords:** NLP · DNA · Transcription factor · Binding specificity

## 1 Introduction

Transcription factors (TFs) are proteins that directly interpret the genome, performing the first step in decoding the DNA sequence [1–3]. It recognizes specific DNA sequences to control chromatin and transcription, forming a complex system that guides expression of the genome [4, 5]. Such specific DNA sequences are called transcription factors binding sites (TFBSs) and play important roles in vital movement. It has proved that transcription factors binding sites are also associated with various human diseases and phenotypes. Mutation of TFs and TFBSs are often highly deleterious and may lead to diseases [6]. Thus, by discovering TFBSs, also called motif discovering, is helpful for further study in gene expression and therapy to diseases

© Springer Nature Switzerland AG 2019
D.-S. Huang et al. (Eds.): ICIC 2019, LNCS 11644, pp. 366–373, 2019.
https://doi.org/10.1007/978-3-030-26969-2_35

caused by gene mutation. Identification of DNA-binding motif provides a gateway to further analyses [7].

Data for study in transcription factors binding is pretty much nowadays due to the fast development of high-throughput sequencing technology which make obtaining DNA data a much easier work [8]. Traditionally, transcription factors binding sites are displayed by sequence logo which based on weight matrices (PWMs) [9, 10]. The PWM is computed by multiplying the scores for each base of a sequence and means a predicted relative affinity of the TF to that sequence. However, such model has an obvious defect that it is not able to process large-scale data from high-throughput sequencing technology [4, 11–13].

Thanks to the rapid development of deep learning in recent year, new computational methods such as convolutional neural network (CNN) and recurrent neural network (RNN) have been applied to many fields such as natural language processing and computer graphic and have made great achievement [14]. The recent applications of deep learning methods to processing biological data have also shown impressive improvement on performance [15–18]. Among these applications, DeepBind is one of the earliest attempts to apply deep learning to the motif discovery task and has proved to be an effective model. DeepBind's success mainly based on the appropriate use of CNN, which is a variant of multilayer artificial neural network specialized in computer vision. By converting 1-D genomic sequence with four nucleotides {A, C, G, T} into 2-D image-like data through one-hot encoding method, CNN can be adapted to modeling DNA sequence protein-binding which is analogous to a two-class image classification problem [16]. Another impressive application of deep learning to bioinformatics field is Deepsea [19, 20]. Deepsea is an algorithmic framework for predicting the chromatin effects of sequence alterations with single nucleotide sensitivity that directly learns a regulatory sequence code from large-scale chromatin-profiling data. It also achieved a great performance by using CNN and related deep learning algorithm.

However, their performance results require much improvement, and in this study, we aim to enhance the solution to this issue with an innovative approach. Our idea is to transform the enhancer sequences into vectors using word embedding and then proceed to classify them with method of natural language processing (NLP). DNA sequence and human language are both sequence data which express certain meanings by the combination of a number of certain elements, which in human language is words or alphabet and in DNA is the four nucleic acids bases A, C, G, T. This idea has indeed been used in past experiments, where researches attempt to apply existing natural language processing algorithms to the study of biological sequences [21]. It was first presented by E. Asgari et al. and applied successfully in many latter biological applications [22]. Moreover, the word feature namely k-mer has also been applied in RNA sequence description and protein structure. In this paper, we proposed a hierarchical attention networks for motif discovery task [23], which based on a computational method for document classification proposed by Zichao Yang et al.

## 2 Proposed Approach

We propose a hierarchical attention network (HAN) for discovering DNA-protein binding sites. The model we used is based on the hierarchical attention network for document classification proposed by Zichao Yang et al. which includes two levels of attention mechanisms. The attention mechanism was used in two hierarchical level, word and sentence. By letting the model pay different attention on different part of sentence or document, the application of attention mechanism could generate a more reasonable representation of sentence or document. We believe the attention method is also suitable for bioinformatics data such as DNA sequence because different part of the DNA sequence may have different amount of influence to the expression of the whole sequence. DNA sequence also has the hierarchical structure like word and sentence in natural language, thus the hierarchy mechanism in the model could improve the performance of processing DNA data [23].

In order to transform DNA sequence into a form that could be read by our model based on natural language processing method while not losing information in the DNA sequence, we use k-mer and word embedding when processing the sequence data. The DNA sequence is first sliced by k-mer and word vectors is generated by word embedding according to the sliced result. The vectors are then fed into the hierarchical attention network for training. We also tried several different k-mer methods with different k lengths and gram lengths to figure out the best k length and gram length.

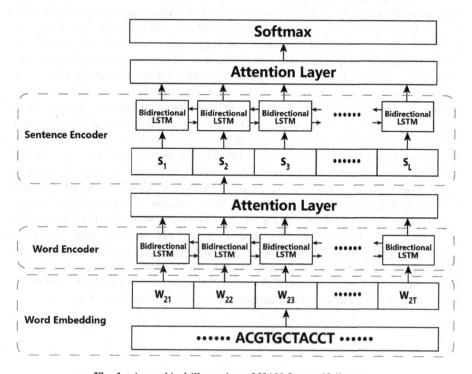

**Fig. 1.** A graphical illustration of HAN for motif discovery

Figure 1 shows the structure of our proposed model. The result of word embed-dding is fed into a hierarchical attention network. The network use a bidirectional LSTM based recurrent neural network as word encoder which encodes the word embedding into vectors containing information of context and the word itself. The followed attention layer extract words that are important to the meaning of the sentence and aggregate the representation of those informative words to form a sentence vector because not all words contribute equally to the representation of the sentence meaning. How attention mechanism work is specifically showed below.

$$u_{it} = \tanh(W_w h_{it} + b_w) \tag{1}$$

$$a_{it} = \frac{\exp(u_{it}^T u_w)}{\sum_t \exp(u_{it}^T u_w)} \tag{2}$$

$$s_i = \sum_t \alpha_{it} h_{it} \tag{3}$$

Where $h_{it}$ is the output of word encoder which is generated according to the input word embedding. The word annotation hit is then fed into a one-layer MLP to get $u_{it}$ as a hidden representation of hit. After that, we measure the importance of the word as the similarity of $u_{it}$ with a word level context vector $u_w$ and get a normalized importance weight $\alpha$ it through a softmax function. Finally, the sentence vector $s_i$ is computed as a weighted sum of the word annotations based on the weights. The context vector $u_w$ can be seen as a high-level representation of a fixed query "what is the informative word" over the words like that used in memory networks. The word context vector $u_w$ is randomly initialized and jointly learned during the training process.

The structure of sentence encoder and the following attention layer is as same as those for word except the sentence encoder take the sentence vector as input and finally output a vector representing the whole documentation. The documentation vector is fed into a softmax layer to get the final classification.

## 3  Experiments and Results

### 3.1  Data Sets

We collected 50 public ChIP-seq datasets from ENCODE to evaluate the performance of our proposed method. For each ChIP-seq dataset, 1000–5000 top ranking peaks were chosen as the foreground (positive) set in which each sequence consists of 200 bps. On the other hand, the way of generating background (negative) sequences is also crucial. It is widely recognized that the background sequences have to be selected to match the statistical properties of the foreground set, otherwise the elicited motifs could be biased [24, 25]. To satisfy such requirements, equal numbers of background sequences were generated by matching the length, GC content and repeat fraction of the foreground set following. According to this guideline, we selected the upstream and downstream DNA sequences of DNA-protein binding sites as negative sequences. Moreover, as mentioned before, the data sets are processed by several different k-mer methods with various k length and gram length to evaluate how value of k and gram influence the performance of the model [26, 27].

## 3.2    Evaluation Metrics

Area under the curve (AUC), a widely used evaluation metric in both machine learning and motif discovery, is one of the two evaluation metrics used in this paper [2, 28]. It is equal to the area under receiver operating characteristic curve (ROC curve), which is a graphical plot that illustrates the performance of a binary classifier, and indicates the probability that a classifier will rank a randomly chosen positive instance higher than a randomly chosen negative one [29, 30].

Another evaluation metrics applied in this paper is average precision (AP), which is also a commonly used metric to measure the ability of a proposed classifier [31]. Average precision is a measure that combines recall and precision for ranked retrieval results. For one information need, the average precision is the mean of the precision scores after each relevant document is retrieved [32].

## 3.3    Results

We compared the performance of HAN for predicting DNA-Protein binding sites with those of DeepBind and Deepsea. The AUCs and APs of proposed method HAN and competing methods DeepBind and Deepsea are calculated as the metrics of their performances. Each method is tested on the same 50 different datasets. The result of comparison is illustrated in Figs. 2 and 3.

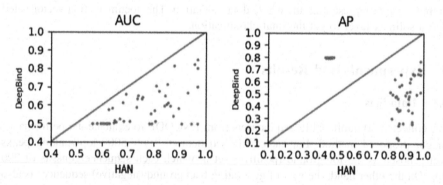

**Fig. 2.** The AUC and AP of our proposed HAN method across 50 experiments in the motif discovery task compared with DeepBind

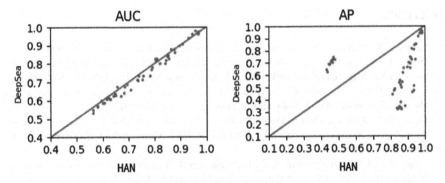

**Fig. 3.** The AUC and AP of our proposed HAN method across 50 experiments in the motif discovery task compared with Deepsea

From Figs. 2 and 3, which displays the AUC and AP comparisons between HAN and DeepBind, and between HAN and Deepsea, we observe that the AUCs and APs of MSC are much higher than those of DeepBind and Deepsea on most datasets, which means HAN could achieve a better performance than these two algorithms in most cases. In addition, APs of HAN are within a narrow range while these of DeepBind or Deepsea distribute in a quite larger area, which means the accuracy of our proposed HAN method are more stable than the other two compared methods across the various datasets.

The experiment results prove that with the help of hierarchical network structure and attention mechanism, our computation model extract critical information from DNA sequence more efficiently and process the data with a more reasonable method.

## 4    Conclusions and Future Work

In this paper, we propose a hierarchical attention network for predicting DNA-protein binding sites inspired by the model in natural language process area and has shown an impressive improvement on performance compared with DeepBind and DeepSea through a series of experiments. Our proposed HAN method obtains a higher accuracy than the two competing methods and shows a stable performance on various datasets. The results of experiments prove that the application of algorithms for NLP such as hierarchical attention network to motif discovery field are practicable and effective.

Although our proposed HAN method has achieved a relatively better performance, it still have some drawbacks such as that it shows poor performance on some datasets despite it perform well on most datasets, which means HAN for predicting DNA-protein binding sites still need to be improved.

**Acknowledgements.** This work was supported by the grants of the National Science Foundation of China, Nos. 61861146002, 61520106006, 61772370, 61873270, 61702371, 61672382, 61672203, 61572447, 61772357, and 61732012, China Post-doctoral Science Foundation Grant, No. 2017M611619, and supported by "BAGUI Scholar" Program and the Scientific & Technological Base and Talent Special Program, GuiKe AD18126015 of the Guangxi Zhuang Autonomous Region of China.

# References

1. Lambert, S.A., et al.: The human transcription factors. Cell **172**, 650–665 (2018)
2. Huang, D.-S., Du, J.-X.: A constructive hybrid structure optimization methodology for radial basis probabilistic neural networks. IEEE Trans. Neural Netw. **19**, 2099–2115 (2008)
3. Bao, W., Huang, Z., Yuan, C.-A., Huang, D.-S.: Pupylation sites prediction with ensemble classification model. Int. J. Data Min. Bioinform. **18**, 91–104 (2017)
4. Deng, S.-P., Zhu, L., Huang, D.-S.: Predicting hub genes associated with cervical cancer through gene co-expression networks. IEEE/ACM Trans. Comput. Biol. Bioinform. (TCBB) **13**, 27–35 (2016)
5. Vaquerizas, J.M., Kummerfeld, S.K., Teichmann, S.A., Luscombe, N.M.J.N.R.G.: A census of human transcription factors: function, expression and evolution. Nat. Rev. Genet. **10**, 252 (2009)
6. Huang, D.-S., Zhang, L., Han, K., Deng, S., Yang, K., Zhang, H.: Prediction of protein-protein interactions based on protein-protein correlation using least squares regression. Curr. Protein Pept. Sci. **15**, 553–560 (2014)
7. Elnitski, L., Jin, V.X., Farnham, P.J., Jones, S.J.J.G.R.: Locating mammalian transcription factor binding sites: a survey of computational and experimental techniques. Genome Res. **16**, 1455–1464 (2006)
8. Berger, M.F., Philippakis, A.A., Qureshi, A.M., He, F.S., Estep III, P.W., Bulyk, M.L.J.N.B.: Compact, universal DNA microarrays to comprehensively determine transcription-factor binding site specificities. Nat. Biotechnol. **24**, 1429 (2006)
9. Stormo, G.D.J.B.: DNA binding sites: representation and discovery. Bioinformatics **16**, 16–23 (2000)
10. Weirauch, M.T., et al.: Evaluation of methods for modeling transcription factor sequence specificity. Nat. Biotechnol. **31**, 126 (2013)
11. Furey, T.S.J.N.R.G.: ChIP–seq and beyond: new and improved methodologies to detect and characterize protein–DNA interactions. Nat. Rev. Genet. **13**, 840 (2012)
12. Yu, H.-J., Huang, D.-S.: Normalized feature vectors: a novel alignment-free sequence comparison method based on the numbers of adjacent amino acids. IEEE/ACM Trans. Comput. Biol. Bioinform. (TCBB) **10**, 457–467 (2013)
13. Zhu, L., Deng, S.-P., Huang, D.-S.: A two-stage geometric method for pruning unreliable links in protein-protein networks. IEEE Trans. Nanobiosci. **14**, 528–534 (2015)
14. Bao, W., Jiang, Z., Huang, D.-S.: Novel human microbe-disease association prediction using network consistency projection. BMC Bioinform. **18**, 543 (2017)
15. Liu, B., Li, K., Huang, D.-S., Chou, K.-C.: iEnhancer-EL: identifying enhancers and their strength with ensemble learning approach. Bioinformatics **34**(22), 3835–3842 (2018)
16. Alipanahi, B., Delong, A., Weirauch, M.T., Frey, B.J.J.N.B.: Predicting the sequence specificities of DNA-and RNA-binding proteins by deep learning. Nat. Biotechnol. **33**, 831 (2015)
17. Shen, Z., Zhang, Y.-H., Han, K., Nandi, A.K., Honig, B., Huang, D.-S.: miRNA-disease association prediction with collaborative matrix factorization. Complexity **2017**, 9 (2017)
18. Zhu, L., Guo, W.-L., Deng, S.-P., Huang, D.-S.: ChIP-PIT: enhancing the analysis of ChIP-Seq data using convex-relaxed pair-wise interaction tensor decomposition. IEEE/ACM Trans. Comput. Biol. Bioinform. **13**, 55–63 (2016)
19. Zhou, J., Troyanskaya, O.G.J.N.M.: Predicting effects of noncoding variants with deep learning–based sequence model. Nat. Methods **12**, 931 (2015)
20. Huang, D.-S., Jiang, W.: A general CPL-AdS methodology for fixing dynamic parameters in dual environments. IEEE Trans. Syst. Man Cybern. B (Cybern.) **42**, 1489–1500 (2012)

21. Le, N.Q.K., Yapp, E.K.Y., Ho, Q.-T., Nagasundaram, N., Ou, Y.-Y., Yeh, H.-Y.J.A.B.: iEnhancer-5Step: Identifying enhancers using hidden information of DNA sequences via Chou's 5-step rule and word embedding. Anal. Biochem. **571**, 53–61 (2019)
22. Asgari, E., Mofrad, M.R.J.P.O.: Continuous distributed representation of biological sequences for deep proteomics and genomics. PloS One **10**, e0141287 (2015)
23. Yang, Z., Yang, D., Dyer, C., He, X., Smola, A., Hovy, E.: Hierarchical attention networks for document classification. In: Proceedings of the 2016 Conference of the North American Chapter of the Association for Computational Linguistics: Human Language Technologies, pp. 1480–1489 (2016)
24. Fletez-Brant, C., Lee, D., McCallion, A.S., Beer, M.A.J.N.A.R.: kmer-SVM: a web server for identifying predictive regulatory sequence features in genomic data sets. Nucleic Acids Res. **41**, W544–W556 (2013)
25. Orenstein, Y., Shamir, R.J.N.A.R.: A comparative analysis of transcription factor binding models learned from PBM, HT-SELEX and ChIP data. Nucleic Acids Res. **42**, e63–e63 (2014)
26. Lee, D., et al.: A method to predict the impact of regulatory variants from DNA sequence. Nat. Genet. **47**, 955 (2015)
27. Yao, Z., MacQuarrie, K.L., Fong, A.P., Tapscott, S.J., Ruzzo, W.L., Gentleman, R.C.J.B.: Discriminative motif analysis of high-throughput dataset. Bioinformatics **30**, 775–783 (2013)
28. Zeng, H., Edwards, M.D., Liu, G., Gifford, D.K.J.B.: Convolutional neural network architectures for predicting DNA–protein binding. Bioinformatics **32**, i121–i127 (2016)
29. Fawcett, T.J.P.R.L.: An introduction to ROC analysis. Pattern Recogn. Lett. **27**, 861–874 (2006)
30. Zhu, L., Zhang, H.-B., Huang, D.-S.: Direct AUC optimization of regulatory motifs. Bioinformatics **33**, i243–i251 (2017)
31. Aslam, J.A., Yilmaz, E., Pavlu, V.: A geometric interpretation of r-precision and its correlation with average precision. In: Proceedings of the 28th Annual International ACM SIGIR Conference on Research and Development in Information Retrieval, pp. 573–574. ACM
32. Davis, J., Goadrich, M.: The relationship between precision-recall and ROC curves. In: Proceedings of the 23rd International Conference on Machine Learning, pp. 233–240. ACM

# Motif Discovery via Convolutional Networks with K-mer Embedding

Dailun Wang[1(⊠)], Qinhu Zhang[2], Chang-An Yuan[2],
Xiao Qin[2], Zhi-Kai Huang[3], and Li Shang[4]

[1] Institute of Machine Learning and Systems Biology,
School of Electronics and Information Engineering,
Tongji University, Shanghai, China
dalen_wang@163.com
[2] Science Computing and Intelligent Information Processing
of GuangXi Higher Education Key Laboratory,
Nanning Normal University, Nanning, Guangxi, China
[3] College of Mechanical and Electrical Engineering,
Nanchang Institute of Technology, Nanchang 330099, Jiangxi, China
[4] Department of Communication Technology, College of Electronic Information
Engineering, Suzhou Vocational University, Suzhou 215104, Jiangsu, China

**Abstract.** With the rapid development of deep learning, some discriminative motif discovery methods based on deep neural network are gradually becoming the mainstream, which also bringing huge improvement of prediction accuracy. In this paper, we propose a convolutional neural network based architecture (eCNN), combining embedding layer with GloVe. Firstly, eCNN divides each single sequence of ChIP-seq datasets into multiple subsequences called k-mers by a sliding window, and then encoding k-mers into a relatively low dimension vectors by GloVe, and finally scores each vector using multiple convolutional networks. The experiment shows that our architecture can get good results on the task of motif discovery.

**Keywords:** Convolutional neural network · CNN · K-mer embedding · eCNN · GloVe · Motif discovery

## 1 Introduction

Discover transcription factor binding site (TFBS), also known as motif discovery, is crucial for further understanding of the transcriptional regulation mechanism in gene expression. In the past decade, with the development of high-throughput sequencing technology, a variety of experimental methods for extracting binding regions have been proposed. In particular, ChIP-seq [1] binds chromatin immunoprecipitation and high-throughput sequencing, greatly increasing the amount of available data and facilitating the study of protein-DNA binding in vivo. On the other hand, protein binding microarray (PBM) [2] can measure the binding of transcription factors to all possible DNA sequence variants of a given length k in vitro, providing a good information source for the direct establishment of TFBS model. However, due to the low resolution

© Springer Nature Switzerland AG 2019
D.-S. Huang et al. (Eds.): ICIC 2019, LNCS 11644, pp. 374–382, 2019.
https://doi.org/10.1007/978-3-030-26969-2_36

of high-throughput sequencing experiments [3], sequencing reads obtained directly from chip-seq or PBM cannot accurately represent TFBS. Therefore, a series of computational methods for accurate modeling of protein-DNA binding are proposed, which can be roughly divided into unsupervised algorithm and supervised algorithm. Given a set of DNA sequences bound by a specific transcription factor (TF), unsupervised algorithms usually attempt to simulate TF binding preferences by using a position weight matrix (PWM) [4, 5] or a consistent sequence [6, 7]. In contrast, the supervised algorithm collects a large number of positive and negative sequences and differentiates TF binding preferences by using classification or regression models. Not surprisingly, the supervised algorithm proved to be more accurate in predicting TF-DNA binding [8].

However, these traditional unsupervised or supervised methods usually have great limitations, such as poor ability to process large-scale sequencing data, poor generalization ability and long-time consumption. In recent years, deep learning such as DeepBind [9] and DeepSea [10] has been successfully applied to the modeling of TF binding preference, overcoming the above limitations of traditional methods. Although the success of these deep learning approaches, they generally follow the fully supervised learning paradigm and ignore the weak supervised information inherent in DNA sequences.

In this paper, we firstly focus on a method named GloVe which has been commonly used in natural language processing. In addition, we conduct this method on embedding layer of our architecture and take another popular method named Word2Vec for a contrast. Then, we score all the vectors using multiple convolutional networks and finally discuss the influence of network architecture on experimental results.

## 2  Materials and Methods

In this section, we introduce our architecture of deep neural network named deepbind and discuss the benefit of k-mer embedding detailly, which has an ability to encode k-mers into a relatively low dimension vector whatever you want by unsupervised learning. Additionally, convolutional neural network, which we applied to detect motifs sequence, plays an important role in achieving superior performance in classification tasks.

### 2.1  Dataset

In order to access the performance of our method on real world datasets, we chose 50 ChIP-seq datasets from one group (HAIB_TFBS) [11] of ENCODE. The length of each single sequence in 50 ChIP-seq datasets are the same, which is 200 bps. Each ChIP-seq dataset has about 70 k–100 k samples while the ratio of positive samples to negative samples is about 1:2 to 1:3.

## 2.2   K-mer Embedding

This section explains why and how we embed k-mers into a low-dimensional vector space. As mentioned above [12, 13], traditional kmer-based methods such as soft-margin SVM only calculate the vector of k-mer frequencies while ignoring the co-occurrence relationship of k-mers. The k-mer feature [14] is now commonly used in natural language processing and retrieval. Moreover, global statistical information is contained in the co-occurrence matrix and the information can be used to construct better feature representations. As is known to all, two famous models are widely used today to apply embedding layer, GloVe (Global Vector) [15, 16] and Word2Vec [17]. Here, we apply the GloVe for our k-mer embedding process while Word2Vec for a contrast in the experiment. The advantage of GloVe compared to other methods is that it combines the advantages of both global matrix factorization and local context window methods.

Learning embedding representations relies on the statistics of k-mer occurrences [18]. Therefore, we denote the matrix of kmer-kmer co-occurrence [19] counts by X, whose entry $X_{ij}$ represents the number of times that k-mer $j$ occurs in the context window of k-mer $i$. $i, j \in [1, V]$ are two k-mer indexes, where $V = 4^k$ is the vocabulary size. According to the GloVe model, the cost function can be minimized as following:

$$J = \sum_{\substack{i, j = 1 \\ X_{ij} \neq 0}}^{V} f(X_{ij})(w_i^T \tilde{w}_j + b_i + \tilde{b}_j - log X_{ij})^2 \tag{1}$$

Where $w \in \mathbb{R}^d$ represents k-mer vectors, $\tilde{w} \in \mathbb{R}^d$ represents separate context vectors, and $b, \tilde{b} \in \mathbb{R}$ are biases. Therefore, the weighting function f can be described as,

$$f(x) = \begin{cases} (x/x_{max})^\alpha & if\ x < x_{max} \\ 1 & otherwise \end{cases} \tag{2}$$

where $x_{max}$ is a cutoff value and $\alpha$ controls the fractional power scaling which is commonly set to 0.75.

Given these vectors, we can fulfill the embedding stage of feature learning $g_{embed}$: $C^L \rightarrow \mathbb{R}^{d \times L}$ by embedding every k-mer into the vector space $\mathbb{R}^d$:

$$h_{embed}(x) = [w_{x1}, w_{x2}, \ldots, w_{xL}] \tag{3}$$

Where $x = [x_1, x_2, \ldots, x_L] \in C^L$. According to the output $d \times L$ matrix, we continue with the convolutional layers and max-pooling layers to extract the spatial features.

## 2.3   Network Architecture

Given a single DNA sequence with length of $L_0$, we first split it into k-mers via the sliding window [20]. A common approach is to extract the subsequences of length k with stride s, resulting in a k-mer sequence with length $L = \lfloor (L_0 - k)/s \rfloor + 1$, wherein each single k-mer subsequence can be represented as an output $d \times L$ matrix. Suppose that we have N sequences in one ChIP-seq dataset, each with a binary label representing whether there is a protein-DNA binding site [21]. Then, we use a convolutional network-based architecture to detect sequence motifs as shown in Fig. 1.

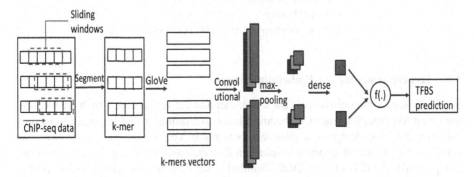

**Fig. 1.** A graphical illustration of eCNN

Finally, during the supervised training stage, we deal with binary classification as a logistic regression by sigmoid function [22].

## 3  Results and Discussion

In order to verify our eCNN, we conducted a series of classification experiments using the ChIP-seq datasets we have mentioned above [23, 24]. First, in Sect. 3.1, we evaluate and compare the performance of GloVe and Word2Vec for our embedding layer. Then, in Sect. 3.2, we discuss the choice of k of k-mer and the vector size d. In addition, in Sect. 3.3, we discuss the effects of different depth of convolutional networks on the experimental result.

### 3.1  Comparison of Embedding Layer

In this section, we tried two different methods GloVe and Word2Vec to conduct our embedding layer. As we mentioned above [25], our goal is encoding k-mers to relatively low-dimension vectors to facilitate the following computation. As we know, both these two methods have such ability to encode words information to any size of vector you want. Therefore, we applied these two methods on five different ChIP-seq datasets from one group (HAIB_TFBS) of ENCODE. Here, we reported the classification performance measured on ROCAUC and PRAUC on five datasets in Table 1. Table 1 lists the average PR AUC on all three folds on two methods, from which we easily can find that GloVe has competitive performance to Word2Vec under PR AUC metric. According to this metric, GloVe performs only slightly better on average than Word2Vec on 6 different ChIP-seq datasets but with more computation resources consumed.

**Table 1.** Performance of Two methods on 6 ChIP-seq datasets on PR AUC metric

| Datasets | Methods | |
|---|---|---|
| | GloVe | Word2Vec |
| BatfPcr1xPkRep1 | 0.86158 | 0.86112 |
| Bcl3V0416PkRep1 | 0.83397 | 0.83126 |
| Bcl11aPcr1xPkRep1 | 0.79456 | 0.78419 |
| Bclaf101388Rep1 | 0.82651 | 0.82196 |
| Ebfsc137065Rep1 | 0.87496 | 0.87445 |

## 3.2 Hyperparameters of Embedding Layer

In this section, we mainly discuss the k of k-mer and the vector size, which are the most important hyperparameters of embedding layer. According to the relationship of one-hot encoding, we designed a relatively reasonable vector size to suit k. Here, we conduct two groups of hyperparameters on five different ChIP-seq datasets from one group (HAIB_TFBS) of ENCODE. Similarly, we reported the classification performance measured on PRAUC [26] on five datasets in Table 2. We can find that lager k slightly contributes our experimental result but with huge increaseament of computation resources.

**Table 2.** Two choice of k and d on 6 ChIP-seq datasets on PR AUC metric

| Datasets | K and d values | |
|---|---|---|
| | 4 and 50 | 5 and 100 |
| BatfPcr1xPkRep1 | 0.85898 | 0.86158 |
| Bcl3V0416PkRep1 | 0.83225 | 0.83397 |
| Bcl11aPcr1xPkRep1 | 0.78342 | 0.79456 |
| Bclaf101388Rep1 | 0.81895 | 0.82651 |
| Ebfsc137065Rep1 | 0.86490 | 0.87496 |

## 3.3 Depth of Convolutional Network

In this section, we changed the depth of convolutional network and show how it influences experimental result. In eCNN, we apply three of the triple-layer structure which is composed by 1D-convolutional layer, max-pooling layer with dropout. Here, we only applied one depth to make a contrast with eCNN on 50 different ChIP-seq datasets from one group (HAIB_TFBS) of ENCODE to show the influence of the depth on the whole architecture. From Fig. 2(a) and (b), we find that the performance of three-layer eCNN has little improvement compared to the one-layer one.

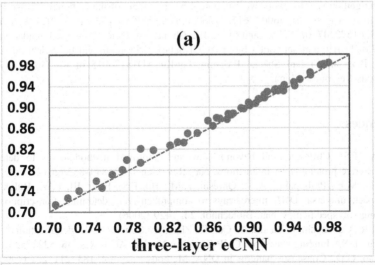

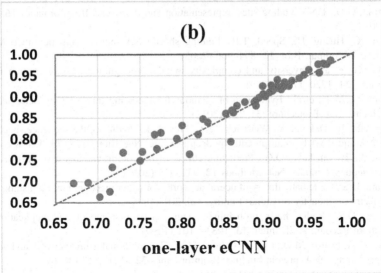

**Fig. 2.** ROC AUC metric comparison of three-layer eCNN and one-layer eCNN on 50 ChIP-seq datasets

## 4 Conclusion

This paper proposes a new architecture named eCNN that combines embedding layer and convolutional network. We mainly discuss the model selection of embedding layer and the choice of some important hyperparameters. Similarly, the depth of convolutional network also influences the performance of our method. The experiment on ChIP-seq datasets also shows that the new architecture can get good results on the task of motif discovery.

**Acknowledgements.** This work was supported by the grants of the National Science Foundation of China, Nos. 61861146002, 61520106006, 61772370, 61873270, 61702371, 61672382, 61672203, 61572447, 61772357, and 61732012, China Post-doctoral Science Foundation Grant, No. 2017M611619, and supported by "BAGUI Scholar" Program and the Scientific & Technological Base and Talent Special Program, GuiKe AD18126015 of the Guangxi Zhuang Autonomous Region of China.

# References

1. Furey, T.S.: ChIP–seq and beyond: new and improved methodologies to detect and characterize protein–DNA interactions. Nat. Rev. Genet. **13**, 840–852 (2012)
2. Berger, M.F., Philippakis, A.A., Qureshi, A.M., He, F.S., Estep III, P.W., Bulyk, M.L.: Compact, universal DNA microarrays to comprehensively determine transcription-factor binding site specificities. Nat. Biotechnol. **24**, 1429 (2006)
3. Jothi, R., Cuddapah, S., Barski, A., Cui, K., Zhao, K.: Genome-wide identification of in vivo protein–DNA binding sites from ChIP-Seq data. Nucleic Acids Res. **36**, 5221–5231 (2008)
4. Stormo, G.D.: Consensus patterns in DNA. Methods Enzymol. **183**, 211–221 (1990)
5. Stormo, G.D.: DNA binding sites: representation and discovery. Bioinformatics **16**, 16–23 (2000)
6. Zhao, X., Huang, H., Speed, T.P.: Finding short DNA motifs using permuted Markov models. J. Comput. Biol. **12**, 894–906 (2005)
7. Badis, G., et al.: Diversity and complexity in DNA recognition by transcription factors. Science **324**, 1720–1723 (2009)
8. Weirauch, M.T., et al.: Evaluation of methods for modeling transcription factor sequence specificity. Nat. Biotechnol. **31**, 126 (2013)
9. Alipanahi, B., Delong, A., Weirauch, M.T., Frey, B.J.: Predicting the sequence specificities of DNA-and RNA-binding proteins by deep learning. Nat. Biotechnol. **33**, 831–838 (2015)
10. Zhou, J., Troyanskaya, O.G.: Predicting effects of noncoding variants with deep learning-based sequence model. Nat. Methods **12**, 931–934 (2015)
11. Huang, D.S.: Systematic theory of neural networks for pattern recognition. Publishing House of Electronic Industry of China, Beijing, vol. 201 (1996)
12. Huang, D.S.: Radial basis probabilistic neural networks: model and application. Int. J. Pattern Recogn. Artif. Intell. **13**, 1083–1101 (1999)
13. Zeng, H., Edwards, M.D., Liu, G., Gifford, D.K.: Convolutional neural network architectures for predicting DNA–protein binding. Bioinformatics **32**, i121–i127 (2016)
14. Quang, D., Xie, X.: DanQ: a hybrid convolutional and recurrent deep neural network for quantifying the function of DNA sequences. Nucleic Acids Res. **44**, e107–e107 (2016)
15. Kelley, D.R., Snoek, J., Rinn, J.L.: Basset: learning the regulatory code of the accessible genome with deep convolutional neural networks. Genome Res. **26**, 990–999 (2016)
16. Hassanzadeh, H.R., Wang, M.D.: DeeperBind: Enhancing prediction of sequence specificities of DNA binding proteins. In: IEEE International Conference on Bioinformatics and Biomedicine, pp. 178–183 (2017)
17. Dietterich, T.G., Lathrop, R.H., Lozano-Pérez, T.: Solving the multiple instance problem with axis-parallel rectangles. Artif. Intell. **89**, 31–71 (1997)
18. Amores, J.: Multiple instance classification: review, taxonomy and comparative study. Artif. Intell. **201**, 81–105 (2013)

19. Wu, J., Yu, Y., Huang, C., Yu, K.: Deep multiple instance learning for image classification and auto-annotation. In: Proceedings of the IEEE Conference on Computer Vision and Pattern Recognition, pp. 3460–3469 (2015)
20. Van de Sande, K.E., Uijlings, J.R., Gevers, T., Smeulders, A.W.: Segmentation as selective search for object recognition. In: 2011 IEEE International Conference on Computer Vision (ICCV), pp. 1879–1886 (2011)
21. Zitnick, C.L., Dollár, P.: Edge boxes: locating object proposals from edges. In: Fleet, D., Pajdla, T., Schiele, B., Tuytelaars, T. (eds.) ECCV 2014. LNCS, vol. 8693, pp. 391–405. Springer, Cham (2014). https://doi.org/10.1007/978-3-319-10602-1_26
22. Gao, Z., Ruan, J.: Computational modeling of in vivo and in vitro protein-DNA interactions by multiple instance learning. Bioinformatics 33(14), 2097–2105 (2017)
23. Annala, M., Laurila, K., Lähdesmäki, H., Nykter, M.: A linear model for transcription factor binding affinity prediction in protein binding microarrays. PLoS ONE 6, e20059 (2011)
24. Maron, O., Ratan, A.L.: Multiple-instance learning for natural scene classification. In: Fifteenth International Conference on Machine Learning, pp. 341–349 (1998)
25. Park, Y., Kellis, M.: Deep learning for regulatory genomics. Nature Biotechnol. 33, 825–826 (2015)
26. Glorot, X., Bordes, A., Bengio, Y.: Deep sparse rectifier neural networks. In: Proceedings of the Fourteenth International Conference on Artificial Intelligence and Statistics, pp. 315–323 (2011)
27. Shen, Z., Bao, W.-Z., Huang, D.S.: Recurrent neural network for predicting transcription factor binding sites. Sci. Rep. 8, 15270 (2018)
28. Zhang, H., Zhu, L., Huang, D.S.: DiscMLA: an efficient discriminative motif learning algorithm over high-throughput datasets. IEEE/ACM Trans. Comput. Biol. Bioinform. 15 (6), 1810–1820 (2018)
29. Guo, W.-L., Huang, D.S.: An efficient method to transcription factor binding sites imputation via simultaneous completion of multiple matrices with positional consistency. Mol. BioSyst. 13(9), 1827–1837 (2017). https://doi.org/10.1039/c7mb00155j
30. Shen, Z., Zhang, Y.-H., Han, K., Nandi, A.K., Honig, B., Huang, D.S.: miRNA-disease association prediction with collaborative matrix factorization. Complexity 2017(2017), 1–9 2017
31. Yuan, L., Yuan, C.-A., Huang, D.S.: FAACOSE: a fast adaptive ant colony optimization algorithm for detecting SNP epistasis. Complexity 2017(2017), 1–10 (2017)
32. Yuan, L., et al.: Nonconvex penalty based low-rank representation and sparse regression for eQTL mapping. IEEE/ACM Trans. Comput. Biol. Bioinform. 14(5), 1154–1164 (2017)
33. Deng, S.-P., Cao, S., Huang, D.S., Wang, Y.-P.: Identifying stages of kidney renal cell carcinoma by combining gene expression and DNA methylation data. IEEE/ACM Trans. Comput. Biol. Bioinform. 14(5), 1147–1153 (2017)
34. Jiang, W., Huang, D.S., Li, S.: Random-walk based solution to triple level stochastic point location problem. IEEE Trans. Cybern. 46(6), 1438–1451 (2016)
35. Deng, S.-P., Zhu, L., Huang, D.S.: Predicting hub genes associated with cervical cancer through gene co-expression networks. IEEE/ACM Trans. Comput. Biol. Bioinform. 13(1), 27–35 (2016)
36. Deng, S.-P., Huang, D.S.: An integrated strategy for functional analysis of microbial communities based on gene ontology and 16S rRNA gene. Int. J. Data Min. Bioinform. (IJDMB) 13(1), 63–74 (2015)
37. Deng, S.-P., Zhu, L., Huang, D.S.: Mining the bladder cancer-associated genes by an integrated strategy for the construction and analysis of differential co-expression networks. BMC Genomics 16(Suppl 3), S4 (2015)

38. Deng, S.-P., Huang, D.S.: SFAPS: an R package for structure/function analysis of protein sequences based on informational spectrum method. Methods **69**(3), 207–212 (2014)
39. Huang, D.S., Zhang, L., Han, K., Deng, S., Yang, K., Zhang, H.: Prediction of protein-protein interactions based on protein-protein correlation using least squares regression. Curr. Protein Pept. Sci. **15**(6), 553–560 (2014)
40. Huang, D.S., Yu, H.-J.: Normalized feature vectors: a novel alignment-free sequence comparison method based on the numbers of adjacent amino acids. IEEE/ACM Trans. Comput. Biol. Bioinform. **10**(2), 457–467 (2013)

# Whole-Genome Shotgun Sequence
# of *Natronobacterium gregoryi* SP2

Lixu Jiang[1], Hao Xu[1], Zhixi Yun[1], Jiayi Yin[1], Juanjuan Kang[1],
Bifang He[1,2], and Jian Huang[1(✉)]

[1] Center for Informational Biology,
University of Electronic Science and Technology of China,
No. 2006, Xiyuan Ave, West Hi-Tech Zone, Chengdu 611731, China
hj@uestc.edu.cn
[2] School of Medicine, Guizhou University, Guiyang 550025, China

**Abstract.** The archaeon *Natronobacterium gregoryi* SP2 is both an extreme halophile and alkaliphile. Its Argonaute protein functions as a potential tool for genome editing is controversial in recent years. Here, the genome of *N. gregoryi* SP2 purchased from China General Microbiology Culture Collection Center under the accession number 1.1967 is presented with Illumina sequencing and the draft genome sequences have been deposited at GenBank. The genome statistics of three submissions for *N.gregoryi* SP2 are summarized. Clusters of Orthologous Groups, Gene ontology and Kyoto Encyclopedia of Genes and Genomes pathway classification analyses of all genes in *N.gregoryi* SP2 were performed. We also identified and annotated three proteins in the genome as type II secretion system protein E, SirA family protein and uncharacterized membrane protein YeiH, which were only annotated as hypothetical proteins previously.

**Keywords:** *N.gregoryi* SP2 · Illumina sequencing · GO · COG · KEGG ·
Hypothetical protein · Genome map · Phylogenetic tree

## 1 Introduction

The *Natronobacterium gregoryi (N.gregoryi)* SP2 (NCIMB2189/ATCC43098), a member of the Natronobacterium genus [1], was isolated from Lake Magadi, Kenya. *N.gregoryi* SP2 is an obligate halophile and alkaliphile with the cell size of $0.5 - 0.7 \times 10 - 15$ μm, thriving at an optimum saline concentration of 20%, temperature of 37 °C and pH of 9.5 [2]. The strain SP2 is Gram-stain-negative, catalase-positive, oxidase-positive, red pigmented and aerobic archaeon, which has a sequence similarity value of 97.3% with the strain B23 of *Natronobacterium texcoconense* sp. nov. based on phylogenetic analysis of 16S rRNA gene sequences [2]. In this study, *N.gregoryi* SP2 grew in 20% NaCl for almost 5 days with the provided culture medium for genome sequencing.

The strain has been in the spotlight since a report on *Natronobacterium gregoryi* Argonaute (NgAgo)-mediated genome editing published by Chunyu Han group in 2016 [3]. The NgAgo/gDNA system may offer an attractive alternative for genome

© Springer Nature Switzerland AG 2019
D.-S. Huang et al. (Eds.): ICIC 2019, LNCS 11644, pp. 383–393, 2019.
https://doi.org/10.1007/978-3-030-26969-2_37

manipulation. Soon, some research groups questioned the genome editing ability of NgAgo because they found that it did not perform successfully in eukaryotic cells [4–7]. The controversy over the genome editing functionality of NgAgo continued more than a year [8–11] and the paper of Han group was finally retracted. However, a very recent study revealed that NgAgo enhances homologous sequence-guided gene editing in bacteria [12]. Another recent report shows that NgAgo is a novel DNA endonuclease defined by a characteristic repA domain. NgAgo cleaves DNA through both a conserved catalytic tetrad in PIWI and a novel repA domain. The result provides insight into poorly characterized NgAgo for development of subsequent gene-editing tool [13].

Here, we submitted the whole sequences of 149 contigs to GenBank and the raw data to Sequence Read Archive (SRA) after whole genome sequencing. Three hypothetical proteins from the '.gbff' file provided by NCBI FTP Site (ftp://ftp.ncbi.nlm.nih.gov/genomes/all/GCA/002/855/455/GCA_002855455.1_ASM285545v1) were annotated. COG, GO and KEGG pathway classification analysis were employed in *N.gregoryi* SP2. What's more, we compared the genome statistics of three submissions from different research groups, and draw the phylogenetic tree based on 16S rRNA sequences and the genome circular maps of the three genome submissions.

## 2 Whole-Genome Shotgun Sequence

### Genome Statistics

Here, we present the whole-genome sequence of *N.gregoryi* SP2. Archaeal colonies were identified by amplification of 16S rRNA of the colonies and alignment with 16S rRNA of reference genome. The 16S rRNA gene sequences of *N.gregoryi* SP2 were amplified using primers archaeal 344F (5′-ACGGGGYGCAGCAGGCGCGA-3′) and archaeal 915R (5′-GTGCTCCCCCGCCAATTCCT-3′) and sequenced from the 344F end (Shanghai Majorbio Bio-pharm Technology Co., Ltd). DNA for sequencing was extracted from pure colonies with the Wizard® Genomic DNA Purification Kit (Promega, Madison, WI) following the manufacturer's recommended protocol. A paired-end library with an average insert size 440 bp of genome fragments (Covaris M220) was prepared by TruSeq™ DNA Sample Prep Kit (Illumina, Inc.). PCR was performed with Hiseq PE Cluster Kit v4 cBot (Illumina, Inc.). Genome of the strain SP2 was sequenced by the Illumina HiSeq 4000 platform using 150 × 2 paired-end reads with HiSeq 3000/4000 SBS Kits according to the protocol provided by the producer. Using readfq.v5, the raw reads were trimmed to obtain high-quality sequences. The trimming included removing reads with adapter contamination, reads with a certain proportion of low-quality bases (default setting) and reads with a certain proportion of Ns (default setting) [14]. Quality control was then performed by using FastQC v0.11.7 [15]. Finally a total of 8,252,309 paired-end reads were produced. Assembly was performed with SOAPdenovo v.2.04 [16] with a genome coverage value of 640.0x. The draft genome sequence of *N.gregoryi* SP2 consists of 3,695,310 bp with a G + C content of 62.4%. The assembly had 112 scaffolds, N50 scaffold size of 77,395 bp and the maximum scaffold size of 309,005 bp. The number of contigs is 149, the size of contig N50 is 68,510 bp and the largest contig has 172,207 bp. These contigs were then annotated using the GeneMarkS + v4.4 suite [17] implemented in the NCBI Prokaryotic Genome Annotation Pipeline (PGAP) [18] with default settings. We identified 3,568 coding genes, 51 RNAs, 3 rRNA operons, 46 tRNAs, 2 ncRNAs and 200 pseudogenes. In addition, 21 simple sequence repeats (SSRs), 21 interspersed repeats and 118 tandem repeats were also predicted.

**Table 1.** Genome statistics of three submissions for *N.gregoryi* SP2.

| Features | [a]JGI | [b]UC | [c]UESTC |
|---|---|---|---|
| GenBank accession no. | CP003377.1 | AOIC00000000 | PKKI00000000 |
| BioProject accession no. | PRJNA60135 | PRJNA174939 | PRJNA423232 |
| Assembly level | Complete genome | Contig | Contig |
| Genome coverage | 30x | 200x | 640x |
| Genome size (bp) | 3,788,356 | 3,694,030 | 3,695,310 |
| GC% | 62.2 | 62.3 | 62.4 |
| CDS (total) | 3,710 | 3,712 | 3,768 |
| CDS (coding) | 3,588 | 3,524 | 3,568 |
| No. of genes | 3,770 | 3,766 | 3,819 |
| No. of pseudo genes | 122 | 188 | 200 |
| No. of RNAs | 60 | 54 | 51 |
| No. of ncRNAs | 2 | 2 | 2 |
| No. of tRNAs | 49 | 47 | 46 |
| No. of rRNAs (5S, 16S, 23S) | 3,3,3 | 3,1,1 | 1,1,1 |
| Annotation pipeline | PGAP 4.1 | PGAP 4.1 | PGAP 4.4 |

[a]DOE Joint Genome Institute, [b]University of California, [c]University of Electronic Science and Technology of China

At the time of writing, there are three whole-genome sequences of strain SP2 in NCBI (Table 1). The sequencing platforms of genomes from JGI, UC and UESTC are 454/Illumina, Illumina GAIIX/HiSeq and Illumina HiSeq 4000 respectively. The features of three genome statistics are summarized in Table 1. The submission of Gen-Bank accession number CP003377.1 is the only complete genome and it was updated from GenBank accession number of FORO00000000 which assembly level was contig. The genome assembly level of the other two submissions is contig. The submission from UESTC has the highest genome coverage of 640.0x described as Table 1. UC's and UESTC's genomes have 128 and 149 contigs respectively and their contig N50 values are 48,343 and 68,510 respectively. Although the three genomes were assembled in different level, their genome features are basically the same except for some minimal differences. Comparative genome analysis shows that the gene organizations of the three genomes for *N.gregoryi* SP2 are similar.

## COG, GO, KEGG Classification Analysis
Based on a BLAST search (E-value $\leq 10^{-5}$) against the string database (v9.05), there were 4020 ORFs hits were searched against the Clusters of Orthologous Groups of

proteins (COG) database to predict and classify their possible functions based on the conserved domain alignment. In total, 1,332 genes were successfully annotated and grouped into 21 COG functional categories, including "Amino acid transport and metabolism", "Translation, ribosomal structure and biogenesis", "Energy production and conversion", and "Inorganic ion transport and metabolism", in which the cluster of 'Amino acid transport and metabolism' occupied the largest number (164; 12.3%), followed by 'Translation, ribosomal structure and biogenesis' (134; 10.1%) (Fig. 1). The columns represent the number of genes in each subcategory.

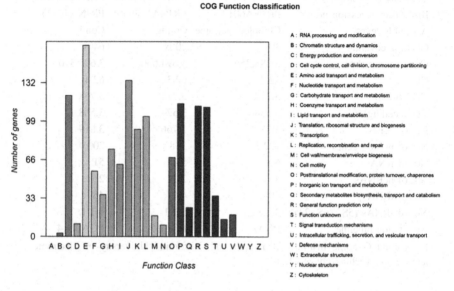

**Fig. 1.** Histogram presentation of COG classification of all unigenes. A total of 1332 genes were successfully annotated and grouped into 21 COG functional categories.

All genes were also subject to Gene ontology (GO) classification analysis to predict their potential biological functions. There are 1294 genes successfully annotated. These genes belong to three categories: cellular component, molecular function and biological process (Fig. 2). High percentages of genes related to 'single-organism process', 'cellular process', 'metabolic process', 'binding' and 'catalytic activity' were observed to be represented in the biological process and molecular function category. The cellular component category consists of 473 genes annotated with Gene Ontology by Blast2GO against the NCBI-nr protein database with an E-value threshold of 1E-5. The molecular function category is comprised of 1150 genes and the biological process category contains 1195 genes. Cell or cell part, binding, catalytic activity, cellular process, single-organism process and metabolic process are the majority of the categories from each GO cluster. These genes related to cellular structure, molecular interaction and metabolism were mostly involved in the life cycle of this strain. In contrast, among the subcategories with the fewest members were 'immune system process', 'reproduction', 'multi-organism process' of the biological processes ontology and 'virion part', 'virion' of the cellular components.

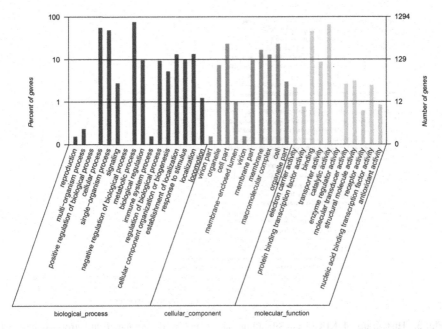

**Fig. 2.** GO classification of all unigenes. A total of 1294 genes were annotated and grouped into three GO categories: biological process, cellular component and molecular function. The x-axis represents the Gene ontology. The left y-axis indicates the percentages of a specific category of genes in that category. The right y-axis indicates the number of genes in a category.

Kyoto Encyclopedia of Genes and Genomes (KEGG) pathway analysis was performed for classification. In the KEGG classification, genes were annotated and grouped into 170 KEGG Pathways. Among the 170 KEGG pathways, 29 pathways in which the genes are more than 20 were shown in Fig. 3. The metabolic pathway (ko01100) consisted of a large number of 476 genes which was the largest group. Besides, KEGG pathways involved in the "Biosynthesis of secondary metabolites" (ko01110), "Microbial metabolism in diverse environments" (ko01120),

"Biosynthesis of amino acids" (ko01230) and "Carbon metabolism" (ko01200) were also considerably enriched. Pathway-based analysis helps to further understand the biological pathways of all genes.

**KEGG Pathway**

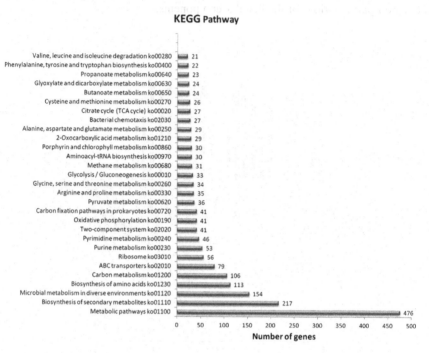

**Fig. 3.** Histogram of KEGG classification. The genes were annotated and grouped into 170 KEGG pathways. The bar chart shows 29 KEGG pathways which include more than 20 genes.

## Annotation of the Hypothetical Proteins

The genome of *N.gregoryi* SP2 submitted by us was annotated with Prokaryotic Genome Annotation Pipeline 4.4 and 3,568 proteins were found. Among them, 1091 proteins were annotated as hypothetical proteins. As the structural and functional domains of proteins are conserved in evolution, we can compare the hypothetical proteins with known protein sequences or conserved domains and infer their families and potential functions. We blasted the 1091 proteins against the non-redundant protein sequences of Halobacteria and got 8 hits with E-value < 1e-20 and identity > 90%. There are no specific annotations for these 8 hypothetical proteins in reference genome annotation file of *N.gregoryi* SP2.

CD-Search [19] was further used to search conserved domains or functional units within the 8 protein sequences against the CDD v3.17 database. As shown in Table 2 and Fig. 4, we identified 3 conserved domains, i.e. VirB11, TusA, and UPF0126 in 3 hypothetical proteins within the e-value threshold of 0.01. The 3 hypothetical proteins could be annotated as type II secretion system protein, SirA family protein, and uncharacterized membrane protein YeiH of *N.gregoryi* SP2 respectively as they have

**Table 2.** The annotations of the 3 hypothetical proteins

| Hypothetical proteins | | BLASTP | | | CD-Search | | |
|---|---|---|---|---|---|---|---|
| Contig | Location | Accession | E-value | | Accession | Domain | E-value |
| 7 | 14169..15842 | gb\|EMA38323.1 | 0 | | COG0630 | VirB11 | 7.2e-118 |
| 14 | 27563..27808 | ref\|WP_006672247.1 | 3e-45 | | COG0425 | TusA | 2.86e-26 |
| 20 | 527..1141 | gb\|SDR16177.1 | 1e-125 | | pfam03458 | UPF0126 | 1.73e-13 |

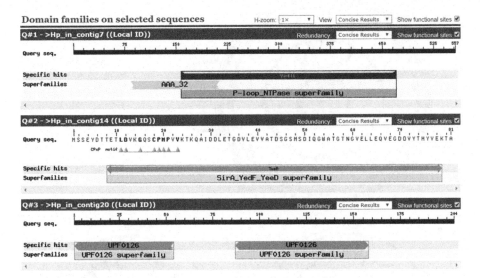

**Fig. 4.** Conserved domains found in hypothetical proteins. All search results type is "specific hits" in the concise display. The track of "Query seq" represents the length and sequence of corresponding query protein. The small triangles under sequence represent conserved motifs, such as catalytic and binding sites. The specific hits of CD-Search represent a high confidence level for the inferred function of the hypothetical proteins indicated by its E-values.

relevant conserved domain VirB11, which is similar to that of *Halobiforma lacisalsi*; TusA, which is similar to that of *Halobiforma nitratireducens;* and UPF0126, which is similar to *Natronobacterium texcoconense*.

# 3   Three Submissions of *N. gregoryi* SP2 at NCBI

As described previously, there are now three whole genome sequencing results of *N.gregoryi* SP2 at NCBI, which are submitted by JGI, UC, and UESTC respectively. As shown in Fig. 5, the corresponding 3 genome circular maps were drawn using DNAPlotter (http://www.sanger.ac.uk/science/tools/dnaplotter). The genome size of the JGI submission is 3,788,356 bp (Fig. 5A); the UC submission has 128 contigs with 3,694,030 bp in total (Fig. 5B); and the UESTC submission has 149 contigs with 3,695,310 bp in total (Fig. 5C). The complete genome of JGI submission contains

3,770 genes in total, including 3,588 proteins, 122 pseudo genes, 2 ncRNAs, 49 tRNAs, and 9 rRNAs. The G + C content of JGI submission is 62.2% and its coding regions cover 82.4% of the complete genome. The genome submitted by UC has a G + C content of 62.3%, and 3,766 genes is annotated, including 3,524 proteins, 188 pseudo genes, 2 ncRNAs, 47 tRNAs, and 5 rRNAs. Our submission shows a G + C

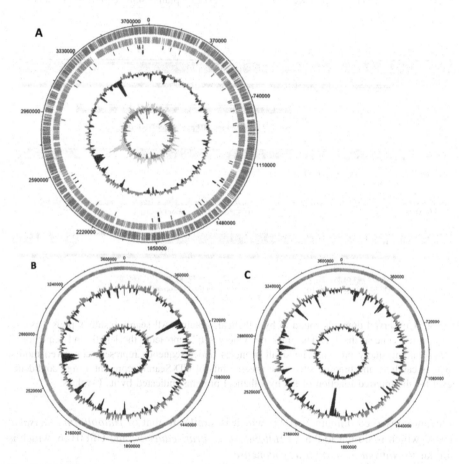

**Fig. 5.** Circular genome maps of three submission of *N. gregoryi* SP2 at NCBI. (A) Complete genome submitted by JGI. The outer scale indicates the coordinates in base pairs. The open reading frames (ORF) are shown on the first two rings; first ring (red) is forward ORF and second ring (bright blue) is reverse ORF. The third and fourth circle shows the tRNA (blue) and rRNA genes (black). The next circle shows the GC content values. Purple and green colors indicate negative and positive sign, respectively. The inner-most circle shows GC skew, gray indicating negative values and yellow for positive values. (B) The UC submission. The first ring is composed of 128 contigs. The next circle displays the GC percentage plot (deep yellow above average, purple below average). The inner circle displays the GC skew. (C) The UESTC submission. The legend of this map is the same to Fig. 5B, except that 149 contigs makes the first ring. (Color figure online)

content of 62.4%, and 3,819 genes is annotated, including 3,568 proteins, 200 pseudo genes, 2 ncRNAs, 46 tRNAs, and 3 rRNAs. Although three submissions are a little bit different from gene numbers, they share a distinct similarity of G + C content.

A phylogenetic tree based on 16S rRNA sequences of Halobacteria was constructed by MEGA X [20] (Fig. 6). The evolutionary history was inferred using the Neighbor-Joining method. The percentage of replicate trees in which the associated taxa clustered together in the bootstrap test are shown next to the branches. The tree is drawn to scale, with branch lengths in the same units as those of the evolutionary distances used to infer the phylogenetic tree. The evolutionary distances were computed using the Maximum Composite Likelihood method [21] and are in the units of the number of base substitutions per site. This analysis involved 27 nucleotide sequences of 16S rRNA. The tree showed a branch of *N.gregoryi* SP2 with other strains investigating phylogenetic incongruence using tree–tree distances. The optimal tree with the sum of branch length = 0.41783639 is shown. The percentage of replicate trees in which the

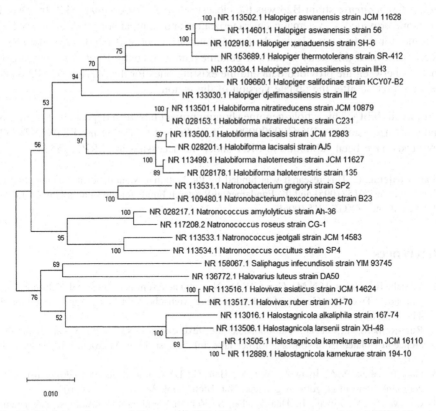

**Fig. 6.** Phylogenetic tree of *N.gregoryi* SP2. The tree was constructed by MEGA X using neighbor-joining as statistical method based on 16S rRNA gene sequences. The parameters were set: 'Maximum Composite Likelihood method' in substitution model and '1000 bootstrap replications' in phylogeny test. The scale length is 0.010.

associated taxa clustered together in the bootstrap test (1000 replicates) are shown next to the branches. The tree revealed that the closest relative of *N.gregoryi* SP2 was the *Natronobacterium texcoconense* strain B23 with the similarity of 97.6%.

## 4 Conclusion

In this study, we sequenced the whole genome of *N.gregoryi* SP2. We also compared the other two submissions of *N.gregoryi* SP2 genome from two different groups with our genome data and found that they have similar genome statistical features. The COG, GO, KEGG pathway classifications were performed based on the whole genome sequence of *N.gregoryi* SP2. We also identified and annotated three proteins in the genome as type II secretion system protein E, SirA family protein and uncharacterized membrane protein YeiH using BLASTP program and CD Search. The 3 genes were only annotated as hypothetical proteins automatically through PGAP workflow in all the submissions. In addition, our evolution analysis indicated that *Natronobacterium texcoconense* strain B23 was the closest relative of *N.gregoryi* SP2. In a word, the work provided a systematic characterization and comparison of *N.gregoryi* SP2 genomes for the first time, laying the foundation for further research of *N.gregoryi* SP2. The limitation of this study is that our genome data is at contig level. The comparison analysis between the provided data and the existing ones for the *N.gregoryi* SP2 could be more precise after the update version of the genome.

**Data availability.** This whole-genome shotgun project has been deposited at GenBank under the accession no. PKKI00000000 and BioProject accession no. PRJNA423232. The reads have been submitted to SRA under the accession no. SRP127538.

**Acknowledgment.** This research was funded by the National Natural Science Foundation of China [grant no. 61571095] and the Fundamental Research Funds for the Central Universities of China [grant no. ZYGX2015Z006].

## References

1. Tindall, B.J., Ross, H.N.M., Grant, W.D.: Natronobacterium gen. nov. and Natronococcus gen. nov., Two New Genera of Haloalkaliphilic Archaebacteria. Syst. Appl. Microbiol. **5**, 41–57 (1984)
2. Ruiz-Romero, E., et al.: Natronobacterium texcoconense sp. nov., a haloalkaliphilic archaeon isolated from soil of a former lake. Int. J. Syst. Evol. Microbiol. **63**, 4163–4166 (2013)
3. Gao, F., Shen, X.Z., Jiang, F., Wu, Y., Han, C.: DNA-guided genome editing using the Natronobacterium gregoryi Argonaute. Nat. Biotechnol. **34**, 768–773 (2016)
4. Cai, M., Si, Y., Zhang, J., Tian, Z., Du, S.: Zebrafish embryonic slow muscle is a rapid system for genetic analysis of sarcomere organization by CRISPR/Cas9, but not NgAgo. Mar. Biotechnol. **20**, 168–181 (2018)

5. Javidi-Parsijani, P., Niu, G., Davis, M., Lu, P., Atala, A., Lu, B.: No evidence of genome editing activity from Natronobacterium gregoryi Argonaute (NgAgo) in human cells. PLoS ONE **12**, e0177444 (2017)

6. Khin, N.C., Lowe, J.L., Jensen, L.M., Burgio, G.: No evidence for genome editing in mouse zygotes and HEK293T human cell line using the DNA-guided Natronobacterium gregoryi Argonaute (NgAgo). PLoS ONE **12**, e0178768 (2017)

7. Qi, J., et al.: NgAgo-based fabp11a gene knockdown causes eye developmental defects in zebrafish. Cell Res. **26**, 1349–1352 (2016)

8. Blow, N.P.D.: To edit or not: the Ngago story. Biotechniques **61**, 172–174 (2016)

9. Burgess, S., et al.: Erratum to: questions about NgAgo. Protein Cell **8**, 77 (2017)

10. Wei, Q., et al.: An NgAgo tool for genome editing: did CRISPR/Cas9 just find a competitor? Genes Dis. **3**, 169–170 (2016)

11. Zhang, X.: NgAgo: a hope or a hype? Protein Cell **7**, 849 (2016)

12. Fu, L., et al.: The prokaryotic Argonaute proteins enhance homology sequence-directed recombination in bacteria. Nucleic Acids Res. **47**, 3568–3579 (2019)

13. https://www.biorxiv.org/content/10.1101/597237v1

14. Tong, C., Wu, Z., Zhao, X., Xue, H.: Arginine catabolic mobile elements in livestock-associated methicillin-resistant staphylococcal isolates from bovine mastitic milk in China. Front. Microbiol. **9**, 1031 (2018)

15. http://www.bioinformatics.babraham.ac.uk/projects/fastqc/

16. Wang, B., et al.: SOAPdenovo2: an empirically improved memory-efficient short-read de novo assembler. GigaScience **1**, 18 (2012)

17. Besemer, J., Lomsadze, A., Borodovsky, M.: GeneMarkS: a self-training method for prediction of gene starts in microbial genomes implications for finding sequence motifs in regulatory regions. Nucleic Acids Res. **29**, 2607–2618 (2001)

18. Tatusova, T., et al.: NCBI prokaryotic genome annotation pipeline. Nucleic Acids Res. **44**, 6614–6624 (2016)

19. Marchler-Bauer, A., Bryant, S.H.: CD-Search: protein domain annotations on the fly. Nucleic Acids Res. **32**, W327–W331 (2004)

20. Kumar, S., Stecher, G., Li, M., Knyaz, C., Tamura, K.: MEGA X: molecular evolutionary genetics analysis across computing platforms. Mol. Biol. Evol. **35**, 1547–1549 (2018)

21. Tamura, K., Nei, M., Kumar, S.: Prospects for inferring very large phylogenies by using the neighbor-joining method. Proc. Nat. Acad. Sci. U.S.A. **101**, 11030–11035 (2004)

# The Detection of Gene Modules with Overlapping Characteristic via Integrating Multi-omics Data in Six Cancers

Xinguo Lu[1], Qiumai Miao[1], Ping Liu[2(✉)], Li Ding[1], Zhenghao Zhu[1], Min Liu[3], and Shulin Wang[1]

[1] College of Computer Science and Electronic Engineering, Hunan University, Changsha 410082, China
[2] Hunan Want Want Hospital, Changsha 410006, China
lp-simple123@126.com
[3] State Key Laboratory of Radiation Medicine and Protection, School for Radiological and Interdisciplinary Sciences (RAD-X) and Collaborative Innovation Center of Radiation Medicine of Jiangsu Higher Education Institutions, Soochow University, Suzhou 215123, China

**Abstract.** A large amount of applications on high throughput data are applied for cancer diagnosis, clinical treatment and prognosis prediction. Module network inference is an established and effective method to identify the biomarks for specific cancer and uncover the oncogenesis mechanism. Exploiting the overlapping characteristic between modules, rather than detecting disjoint modules, may broaden our understanding of the molecular dysfunction that govern tumor initiation, progression and maintenance. To this end, we propose a novel framework to identify gene modules with overlapping characteristic by integrating gene expression data and protein-protein interaction data. In our framework, social community and overlapping characteristic are introduced to construct disjoint and overlapping modules which can represent the relationship between diverse modules. Applying this framework on six cancer datasets from The Cancer Genome Atlas, we obtain functional gene modules in each cancer, which are more significantly enriched in the known pathways than identified by other state-of-the-art methods. Meanwhile, those identified modules can significantly distinguish the survival prognostic of patients by Kaplan-Meier analysis, which is critical for cancer therapy. Furthermore, identified driver genes in the network can be considered as biomarkers which can distinguish the tumor and normal samples. Uncovering the overlapping feature in gene modules will help elucidate the relationship between modules and could have important therapeutic implication and a more comprehensive interpretation on carcinogenesis.

**Keywords:** Module network · Gene expression · Protein-protein interaction · Overlapping characteristic

D.-S. Huang et al. (Eds.): ICIC 2019, LNCS 11644, pp. 394–405, 2019.
https://doi.org/10.1007/978-3-030-26969-2_38

# 1  Introduction

The landscape driver mutation identification in cancer is extraordinarily complex, making it difficult to distinguish alterations from passenger mutation [1]. Although an increasing number of disease biomarks have been identified through integrating high-throughput omics data, their reproducibility and overlap are poor. Most of these methods are module-based, and discovered modules are disjoint. This poor reproducibility is possibly due to the fact that individual biomarks are often detected without considering the overlapping characteristic in modules and their communication. For example, overlapping characteristic for gene TP53 and GATA3 are illustrated in different function pathways in breast cancer [2] in which TP53, BRCA2 and GATA3 response to gamma radiation while gene TP53, PIK3CA and GATA3 regulate the process of neuron apoptotic. Thus, overlapping genes can play different roles in different modules. One powerful method for systematically presenting the overlapping characteristic in modules is to introduce the concept of overlapping community detection [3], which is effectively used in social networks [4, 5], and biological networks [6] and other complex networks. By capturing the role of overlapping genes in different gene modules, it is possible to identify additional biomarks that confer an effective complement to understand cancer mechanism.

Recent advances in next-generation sequencing (NGS) [7] technologies have generated massive profiles for better characterizing the molecular signatures of human cancers. Some public databases have been established and have provided multi-omics data consisting of genomic, epigenomic, transcriptomic, and proteomic data [8]. Large cross-institutional projects, such as the Cancer Genome Atlas (TCGA) [9], a collaboration between the National Cancer Institute (NCI) and National Human Genome Research Institute (NHGRI), has generated comprehensive, multi-dimensional maps of the key genomic changes in 33 types of cancer including glioblastoma, lung, ovarian, breast, endometrial, kidney, colorectal cancer and so on. Due to the high dimension and high heterogeneity of these data, researchers address a great challenge to obtain the mutant biomarkers that promotes the growth of tumorigenesis proliferation from these massive high-throughput data [10, 11]. In particular, recent methods have attempted to solve the fact that some co-regulation modules and their components are identified from biological networks through data mining and integration of high-throughput omics data, which will have better performance. Several analytical approaches have been developed to detect driver genes associated with oncogenesis. One of these widely applied algorithms, CONEXIC (COpy Number and Expression In Cancer) was developed to integrate copy number change and gene expression data to identify potential driver genes located in regions that are amplified or deleted in tumors [12]. Additional approaches have been developed to capture the biomarks for regulating the detected modules. These approaches generally rely on integrating multi-Omics information (e.g. gene expression data, copy number variation data (CNV), methylation profile), and generating functional modules by some clustering methods (e.g. spectral clustering, k-means), and computing the consensus regulator-to-module score through the regulation program [12–14].

Although many functional independent modules are identified and in some cases module-inference approach is effective, they fail to discover dependencies between modules. Besides, in the real biological networks, gene modules may exhibit overlapping features, i.e. some genes may contribute to multiple groups and the nodes in the network may have multiple functions. Analogously, there are significant overlapping communities in many real-world networks [15–17]. For example, a person belongs to more than one social group such as family group and friend group, which corresponds to genes in our study belonging to multiple modules [18]. Actually, overlapping community detection algorithms have also been applied in many aspects including bioinformatics [6]. In addition, protein-protein interaction (PPI) data is not considered in most module-based analyses [19]. We here extend upon these analyses by performing a systematic cancer analysis of overlapping characteristic in modules and integrating PPI data that makes an effective complement to gene properties.

Here, we propose an approach for the systematic and comprehensive analysis of integrating gene expression data and PPI data. Our aim is to reveal driver genes that control key biological process in a cancer study. In our approach, the "ganesh" application [20] is used to generate interactive networks by clustering of gene expression data in each cancer to obtain gene-to-gene relationships. Functional overlapping modules are detected from the network by overlapping community detection method. Driver genes are obtained by calculating the scores of candidate regulators assigned into functional overlapping modules [21, 22].

From an analysis of 2101 tumor/normal samples of six cancer types profiled by the TCGA consortium and PPI data from STRING database [23], we identified 745 driver genes with high regulation score. These genes can accurately classify tumor and normal samples in each cancer. In addition, functional modules detected by our method are more significantly enriched in many Gene Ontology (GO) terms [24] with lower p-value.

## 2    Materials and Methods

Complex gene network inference has become one of the main problems in the biological function discovery and comprehensive carcinogenesis interpretation. Overlapping community detection in social networks is proposed to identify a set of clusters that are not necessarily disjoint. In our work, there is a similar network structure between social network and gene network as some genes take part in different functional processes resulting in gene intersection for these distinct clusters. Thus, overlapping communities are used to construct modules in complex gene network. And we propose a framework to identify gene modules with functional overlapping characteristic by integrating gene expression data and PPI data. The framework consists of three parts: network construction, modules detection and driver genes identification. The detailed method description is demonstrated as follows (Fig. 1).

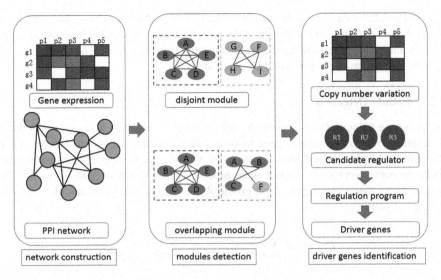

**Fig. 1.** Schematic of the framework.

## 2.1  Network Construction

A weighted, undirected network of genes in each cancer is built by integrating gene expression and PPI data. For the purpose, in an expression matrix with N genes and M conditions, the cluster assignment of each gene and condition is conducted in 5 times with "ganesh" software (default parameters), which yields 5 partitions. In order to generate the correlation coefficient of gene pairs, we calculate the average number of times a pair of genes appearing in a module. For the k-th "ganesh" run, we specify a corresponding matrix $C^{(k)}$, which captures the assignment and the detail of cluster. $C^{(k)}$ is a $N \times S_k$ matrix where N is the number of genes and $S_k$ is the number of clusters in the k-th run. In matrix $C^{(k)}$, $C_{iS}^{K} = 1$, if gene i belongs to cluster S, and $C_{iS}^{K} = 0$, otherwise. Next, an $N \times N$ co-clustering matrix $A^{(k)}$ for the k-th run is defined as:

$$A^{(k)} = C^{(k)} \left(C^{(k)}\right)^{T} \tag{1}$$

Where $A^{(k)} = 1$, if gene i and gene j belong to the same cluster, and $A^{(k)} = 0$, otherwise. Then, the co-occurrence frequency matrix $A_m$ is obtained from the mean of $A^{(k)}$ over all K runs. We convert adjacency matrix $A_m$ to an interaction matrix I. In order to complete the network information, PPI data from the STRING database [23] is added to I matrix for obtaining the correlation between genes. The correction gene network matrix $I_{ij}^{*}$ is calculated as:

$$I_{ij}^{*} = \begin{cases} P_{ij} & \text{if } I_{ij} < P_{ij} \\ I_{ij} & \text{otherwise} \end{cases} \tag{2}$$

Where $P_{ij}$ is the weight of gene pair from the PPI (STRING) database. The final gene network G is constructed by filtering gene pairs with threshold 0.4, where the value of the element of $I^*$ is the correlation coefficient of the gene pair. For each cancer type, the specific gene network is generated by these steps.

## 2.2   Module Detection

Given a gene network G $(v, \xi, \omega)$ with genes $v$, edges $\xi$, and weight $\omega$, our method can be applied to detect functional gene modules. The algorithm builds on the concept of the cohesiveness score and uses a greedy growth process to discover groups that are likely to correspond to functional gene modules in each cancer gene network. The cohesiveness assess the degree for a group of genes to form a gene module, and it is defined as follows.

$$CS(M) = \sum_{i \in M} \frac{\omega_i^{in}(M)}{\omega_i^{in}(M) + \omega_i^{out}(M) + p|M|} \tag{3}$$

Where $\omega_i^{in}(M)$ is the total weight of edges contained in cluster M, and $\omega_i^{out}(M)$ denotes the total weight of edges which interact the cluster with the remaining network. In order to consider the undiscovered interactions in the gene network, we set up penalty term p| M|. The default setting of p|M| is 0.2. Our method identifies cancer-related mutation clusters for a given data set by 3 steps.

Step1: Grow groups with high cohesiveness.

Initially, the genes with the highest degree are selected as the first seed, and grows a cohesive group using growth greedy procedure. Whenever the growth process is finished, the algorithm selects the next seed by considering all the genes that have not been included in any of the gene modules found so far and taking the one with the largest amount of connections again. The whole procedure ends when there are no genes remaining to consider.

Step2: Merge groups with high overlapping score.

Highly overlapping pairs of locally optimal cohesive groups are merged. Merging groups is based on the value of overlap score, which measures overlapping intensity. A value close to zero indicates that the groups have a few common genes whereas a value close to one corresponds to almost identical groups. The calculation of overlap score $\omega$ about two gene sets S1 and S2 is given as follows:

$$\omega(S1, S2) = \frac{|S1 \cap S2|^2}{|S1||S2|} \tag{4}$$

Given a set of cohesive groups, we first calculate the overlap scores for each pair of groups and construct an overlap graph in which each vertex represents a cohesive group, and two groups are connected by an edge if they overlap substantially. Groups connect to each other are then merged into a gene module candidates. If a group has no connection with other groups, it will be promoted a gene module candidate without any additional merging.

Step3: Remove insufficient modules.

Gene module candidates which contain less than 10 genes are removed to ensure that modules have sufficient genes for further statistical analysis. The density δ of module M is defined as follows:

$$\delta(M) = \frac{n(n-1)}{2} \tag{5}$$

Where n is the number of genes in the candidate module. We discard candidate modules whose density is below a given threshold 0.5.

## 2.3    The Identification of Driver Genes

The candidate regulators represent potential regulator to their corresponding gene modules. In the cancer module network, we identify a list of amplified and deleted genes if their CNVs frequencies are high across the tumors connected to one or more modules of co-expressed genes. The mutation frequency of each gene is calculated as follows:

$$fre = \frac{\#Mutgenes}{N} \tag{6}$$

Where #Mutgenes represents the number of mutant genes across samples, N is the total number of samples. In our works, we employ a fuzzy decision tree method to calculate a probabilistic score for each candidate regulator by regulation program [21, 22], which reveal potential regulation for module activity. Then the set of regulators assigned to each cluster of co-expressed genes can be ranked according to their global probabilistic score and a cutoff level can be defined, leaving only very high-scoring regulators called driver genes.

## 2.4    Differentially Expressed Genes Selection

We apply our approach to cancer datasets obtained from the Cancer Genome Atlas (TCGA). We use the Illumina HiSeqv2 of 2101 tumor/normal samples from six cancer types (see Table 1). Due to the high heterogeneous data between each cancer, we use the Earth mover's distance [25] to compute the overall diversity between the distributions of a gene's expression in two classes, normal vs tumor, identifying differentially expressed genes (p-values < 0.001).

**Table 1.** Cancer types and number of samples for tumor and normal tissues from TCGA database

| Cancer type | TCGA ID data | No. tumor samples | No. normal samples |
|---|---|---|---|
| Bladder Urothelial Carcinoma | BLCA | 407 | 19 |
| Breast invasive carcinoma cancer | BRCA | 485 | 8 |
| Colon adenocarcinoma | COAD | 288 | 41 |
| Esophageal carcinoma | ESCA | 185 | 11 |
| Head and Neck squamous cell carcinoma | HNSC | 522 | 44 |
| Kidney Chromophobe | KICH | 66 | 25 |

## 2.5 Survival Analysis

We use the function coxph (R package survival) to implement the Cox proportional hazard model [26] to analyze the association of gene profile of each module with the patient survival. Only patients with fully characterized tumors and with at least 30 days of overall survival (OS) are included in the study.

We use the median cut to generate high- and low-expression patient groups.

$$\text{group} = \begin{cases} \text{high} & \textit{if } X_i < X_{ic} \\ \text{low} & \text{if } X_i > X_{ic} \end{cases} \tag{7}$$

Where $X_i$ is the average expression level of genes in the i-th patient, $X_{ic}$ is the average expression level of gene within the c-th gene module in the i-th patient. Patients are grouped into high- and low-expression groups based on the mean expression value of the module. The difference between these two groups of patients is obtained by using the Kaplan-Meier estimator and log-rank method.

## 3 Results

### 3.1 Survival Time Analysis in Functional Modules

Gene and modules are associated with patient survival time [27]. The Kaplan-Meier survival analysis is used to predict the survival time (see METHODS). For each module, we predict the patient survival time by using the gene expression and clinical data. We find that both the disjoint and overlapping modules divide the patients into two classes whose survival time is significant different (p-value < = 0.01). The top most significant survival associated modules in 6 cancers are shown in Fig. 2. In Bladder Urothelial Carcinoma, the detected module is significantly divide the patients into two groups (p = 0.0001). The module obtained from Breast invasive carcinoma cancer also reveal the significantly different survival time (p-value = 0.01). In Colon adenocarcinoma, the patients in the given module are divided into two groups with significantly different survival time (p-value = 0.005). In Esophageal carcinoma, we find the module could significantly classify patients according to the mean expression values of genes (p-value = 0.003). In Head and Neck squamous cell carcinoma, the identified functional module is significantly associated with the survival time of patients (p-value = 0.002). The most significant survival associated module is obtained from Kidney Chromophobe (p-value = 1.825E-08). The analysis results show that the modules obtained from six cancers could distinguish likely patient prognostic survival time according to their average expression values, with the statistical significance (p < = 0.01).

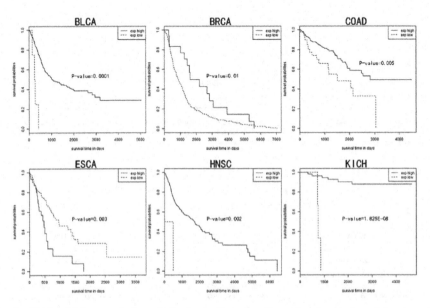

**Fig. 2.** Kaplan-Meier survival analysis for patients based on the detected modules in each cancer.

### 3.2 Patients Classification with Driver Genes

We identify 745 driver genes by fuzzy decision tree method in six cancers. To validate whether our cancer-specific driver genes obtained from integration analysis are also applicable to distinguish tumor versus normal samples, we classify the samples using SVM [28]. SVM is a linear maximum-margin model for classification [29]. The gene expression level of each driver gene across all tumor or normal samples from diverse cancers data is input to train SVM. The area under curve (AUC) is employed to measure the performance. We select 80% of the samples in each tumor or normal separately to train the classifier, and the rest of the samples are used to test the classification performance. The results on TCGA data are shown in Fig. 3. For each cancer type, ROC curves and AUC values are shown for the top five cancer-specific driver genes (with high-regulation ability) with the highest AUC performance. PEX19 and SCAMP3 are the best predictors in breast invasive carcinoma cancer (AUC = 0.969 and AUC = 0.954), ASXL1 and PLCG1 in colon adenocarcinoma (AUC = 0.929 and AUC = 0.903), SLC4A4 and GPR87 in esophageal carcinoma (AUC = 0.932 and AUC = 0.919), ATP6VOD2 and MED30 in head and neck squamous cell carcinoma (AUC = 0.958 and AUC = 0.957), MRAP2 and EDARADD in kidney chromophobe (AUC = 0.984 and AUC = 0.954). The best performance are obtained in bladder urothelial carcinoma dataset. The AUC value of the top five drivers of the cancer are higher than 0.9, and the AUC value of gene FGFR1 reaches 1 (Fig. 3).

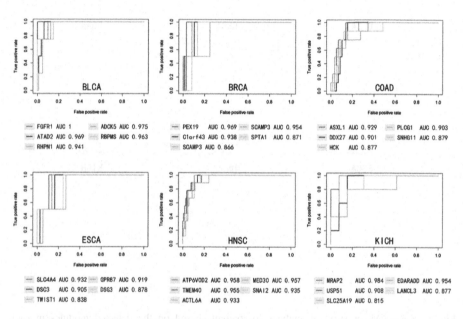

**Fig. 3.** ROC Curves and AUC values for the top ten driver cancer-specific genes.

## 3.3   Comparison with Other Tools

To compare the Gene Ontology (GO) categories between our method, Lemon-Tree and K-Means, we build a list of all common categories for a given p-value threshold scores in each cancer by using BINGO [30]. We select the highest score for each GO category and we count the number of GO categories having a higher score for our approach, Lemon-Tree and K-Means, and calculate the sum of scores for each GO category and each method. The results show in Fig. 4 indicate that clusters detected by our method have a higher number of GO categories with lower p-values than Lemon-Tree and K-Means, and that globally the p-values are lower for our method modules.

There are two possible reasons why our work is better than other methods. The first is that our method integrate protein-protein interactions in gene-to-gene relationships, which enhances the interactions with low-connectivity and high-function gene pairs.

The second reason is that our framework considers overlapping between modules. These genes with overlapping characteristic can produce different functions through the combination of different modules.

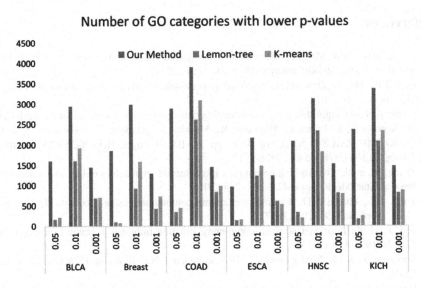

**Fig. 4.** Gene Ontology (GO) enrichment of the co-expressed gene clusters, shows by counting the number of GO categories having a lower p-value.

# 4 Conclusion

In this paper, we propose an overlapping community-based framework to identify gene modules. Our method efficiently handles with the heterogeneity transcriptomic and proteomic data and successfully clusters genes by two-steps. This method shows its validity in functional module detection. This method uses "ganesh" to generate inter-action network with gene expression data and PPI data, and overlapping community detection method is applied to build consensus gene module with overlapping characteristic in gene network. We apply the approach to six cancer datasets from TCGA. The comparisons with other proposed methods shows its good performance. From diverse cancer networks, we identify the consistent gene modules. These modules enrich many GO terms significantly and contain more GO categories with lower p-value than other methods. The survival analysis of the modules shows that the identified modules can distinguish different patients. The results not only show the efficiency of our proposed method, but also provide more useful information for cancers. Its ease of implementation and high efficiency may accelerate the study of complex biological networks.

**Acknowledgement.** The authors are grateful to Anagha who provided the initial co-clustering method with GaneSh Java package and Tamaaas who provided the module network construction method with cluster-one Java package. This work was supported by National Natural Science Foundation of China (Grant Nos. 61502159, 61472467 and 61672011), Natural Science Foundation of Hunan Province, China (Grant No. 2018JJ2053) and National Key R\&D Program of China (2017YFC1311003).

# References

1. Lu, X., et al.: DMCM: a data-adaptive mutation clustering method to identify cancer-related mutation clusters. Bioinformatics **35**(3), 389–397 (2019)
2. Hou, J.P., Ma, J.: DawnRank: discovering personalized driver genes in cancer. Genome Med. **6**(7), 1–16 (2014)
3. Gregory, S.: An algorithm to find overlapping community structure in networks. In: Kok, J. N., Koronacki, J., Lopez de Mantaras, R., Matwin, S., Mladenič, D., Skowron, A. (eds.) PKDD 2007. LNCS (LNAI), vol. 4702, pp. 91–102. Springer, Heidelberg (2007). https://doi.org/10.1007/978-3-540-74976-9_12
4. Zhang, X., et al.: Overlapping community identification approach in online social networks. Phys. A Stat. Mech. Appl. **421**, 233–248 (2015)
5. Yang, J.X., Zhang, X.D.: Finding overlapping communities using seed set. Phys. A Stat. Mech. Appl. **467**, 96–106 (2017)
6. Lee, J., et al.: Improved network community structure improves function prediction. Sci. Rep. **3**(2197), 1–9 (2013)
7. Shendure, J., Ji, H.: Next-generation DNA sequencing. Nat. Biotechnol. **26**(10), 1135–1145 (2008)
8. Hawkins, R.D., et al.: Next-generation genomics: an integrative approach. Nat. Rev. Genet. **11**(7), 476–486 (2010)
9. Tomczak, K., et al.: The cancer genome atlas (TCGA): an immeasurable source of knowledge. Contemp. Oncol. **19**(1A), 68–77 (2015)
10. Ulitsky, I., Shamir, R.: Identification of functional modules using network topology and high-throughput data. BMC Syst. Biol. **1**(8), 1–17 (2007)
11. Ching, T., et al.: Cox-nnet: an artificial neural network method for prognosis prediction of high-throughput omics data. PLoS Comput. Biol. **14**(4), e1006076 (2018)
12. Bonnet, E., et al.: Integrative multi-omics module network inference with lemon-tree. PLoS Comput. Biol. **11**(2), e1003983 (2015)
13. Lu, X., et al.: Driver pattern identification over the gene co-expression of drug response in ovarian cancer by integrating high throughput genomics data. Sci. Rep. **7**(1), 16188–16204 (2017)
14. Lu, X., et al.: The integrative method based on the module-network for identifying driver genes in cancer subtypes. Molecules **23**(2), 183–197 (2018)
15. Ding, Z., et al.: Overlapping community detection based on network decomposition. Sci. Rep. **6**(24115), 1–11 (2016)
16. Wang, Z. et al.: Discovering and profiling overlapping communities in location-based social networks. IEEE Trans. Syst. Man Cybern. Syst. **44**(4), 499–509 (2014)
17. Gui, Q., et al.: A New Method for Overlapping Community Detection Based on Complete Subgraph and Label Propagation, pp. 127–134 (2018)
18. Amelio, A., Pizzuti, C.: Community mining in signed networks: a multiobjective approach. In: 2013 IEEE/ACM International Conference on Advances in Social Networks Analysis and Mining (ASONAM), pp. 95–99 (2013)

19. Zhang, J., et al.: Identification of mutated core cancer modules by integrating somatic mutation, copy number variation, and gene expression data. BMC Syst. Biol. **7**(S4), 1–12 (2013)
20. Joshi, A., et al.: Analysis of a Gibbs sampler method for model-based clustering of gene expression data. Bioinformatics **24**(2), 176–183 (2008)
21. Segal, E., et al.: Module networks: identifying regulatory modules and their condition-specific regulators from gene expression data. Nat. Genet. **34**(4), 166–176 (2003)
22. Joshi, A., et al.: Module networks revisited: computational assessment and prioritization of model predictions. Bioinformatics **25**(4), 490–496 (2009)
23. Szklarczyk, D., et al.: The STRING database in 2017: quality-controlled protein-protein association networks, made broadly accessible. Nucleic Acids Res. **45**, 362–368 (2017). (Database issue)
24. Gene ontology consortium: the gene ontology (GO) database and informatics resource. Nucleic Acids Res. **32**, 258–261 (2004). (Database issue)
25. Nabavi, S., et al.: EMDomics: a robust and powerful method for the identification of genes differentially expressed between heterogeneous classes. Bioinformatics **32**(4), 533–541 (2016)
26. A Package for Survival Analysis in S. R package version 2.37.7. https://cran.r-project.org/src/contrib/survival_2.44-1.1.tar.gz. Accessed 01 Apr 2019
27. Heagerty, P.J., Lumley, T., Pepe, M.S.: Time-dependent ROC curves for censored survival data and a diagnostic marker. Biometrics **56**(2), 337–344 (2000)
28. Lu, X., Peng, X., Deng, Y., Feng, B., Liu, P., Liao, B.: A novel feature selection method based on correlation-based feature selection in cancer recognition. J. Comput. Theor. Nanosci. **11**(2), 427–433 (2014)
29. Burges, C.J.C.: A tutorial on support vector machines for pattern recognition. Data Min. Knowl. Discov. **2**(2), 121–167 (1998)
30. Maere, S., et al.: BiNGO: a cytoscape plugin to assess overrepresentation of gene ontology categories in biological networks. Bioinformatics **21**(16), 3448–3449 (2005)

# Combining High Speed ELM with a CNN Feature Encoding to Predict LncRNA-Disease Associations

Zhen-Hao Guo[1,2], Zhu-Hong You[1,2(✉)], Li-Ping Li[1],
Yan-Bin Wang[1], and Zhan-Heng Chen[1]

[1] Xinjiang Technical Institute of Physics and Chemistry,
Chinese Academy of Sciences, Urumqi 830011, China
zhuhongyou@ms.xjb.ac.cn, zhuhongyou@gmail.com
[2] University of Chinese Academy of Sciences, Beijing 100049, China

**Abstract.** Accumulated evidence indicates that lncRNAs are critical for many biological processes, especially diseases. Therefore, identifying potential lncRNA-disease associations is significant for disease prevention, diagnosis, treatment and understanding of cell life activities at the molecular level. Although novel technologies have generated considerable associations for various lncRNAs and diseases, it has inevitable drawbacks such as high cost, time consumption, and error rate. For this reason, integrating various biological databases to predict the potential association of lncRNA and disease is of great attraction. In this paper, we proposed the model called ECLDA to predict lncRNA-disease associations by combining CNN and highspeed ELM. Firstly, the feature vectors are constructed by integrating lncRNA functional similarity, disease semantic similarity and Gaussian interaction profile kernel similarity. Secondly, CNN is carried out to mine local and higher-level abstract features of the vectors. Finally, high speed ELM is used to identify the novel lncRNA disease associations. The ECLDA computational model achieved AUCs of 0.9014 in 5-fold cross validation. The results showed that ECLDA is expected to be a practical tool for biomedical research in the future.

**Keywords:** LncRNA · Disease · Association prediction · CNN · ELM

## 1 Introduction

In vivo, RNA is a product of DNA transcription and is divided into two categories: mRNA with protein-coding function (messenger RNA) and non-coding RNA (ncRNA) with no protein-coding level [1]. With the increase in RNA complex data obtained, the researchers found that although the human genome has a large number of transcripts, the proportion of mRNA in transcripts only accounts for 1.5%, the amount of ncRNAs in the transcriptome is over 98% in total, suggesting that ncRNAs may be the organism with abundant biological functions [2, 3]. In recent years, with the application and development of a new generation of high-throughput sequencing technology, there is a great number of research indicate that lncRNAs are associated with a large number of human diseases such as cancers [4], blood diseases [5] and neurodegeneration [6].

© Springer Nature Switzerland AG 2019
D.-S. Huang et al. (Eds.): ICIC 2019, LNCS 11644, pp. 406–417, 2019.
https://doi.org/10.1007/978-3-030-26969-2_39

Therefore, the identification of potential human lncRNA-disease associations not only helps to understand the molecular mechanisms of human disease, but also helps researchers for the diagnosis, treatment and prevention of human diseases [7, 8].

It is unrealistic to make experimental studies on a large scale to detect new associations between small molecule transcripts and translations in cells. Several computational models have been proposed to alleviate this situation. In the field of predicting protein-protein interactions (PPI), You *et al.* proposed some machine learning-based models to detect undiscovered interactions [9–39]. Chen et al. proposed some computational models to predict new miRNA-disease associations [40–48]. Several models for predicting ncRNA-protein interactions have been proposed in recent years [49, 50]. Forecasting drug-target interactions based on databases has become more and more popular in recent times. Methods of network and machine learning are constantly being proposed [51–59].

In this paper, we developed a novel prediction model named ECLDA to detect potential association. Firstly, we downloaded lncRNA-disease associations from LncRNADisease database [60] and Medical Subject Headings (MeSH) descriptors from U.S. National Library of Medicine. Then the representation vector of lncRNA and disease is constructed by combining various similarity matrices. In addition, after being extracted by CNN and reduced in dimension, the vector was placed in the classifier. ELM was used as the model to train and predict lncRNA-disease associations. As a result, ECLDA obtained AUCs of 0.9007 under 5-fold cross validation of known experimentally confirmed lncRNA-disease association in the LncRNADisease database (Fig. 1).

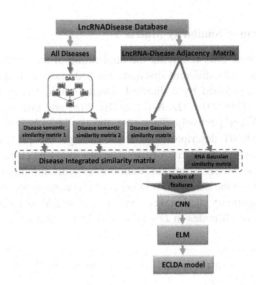

**Fig. 1.** Flowchart of the proposed model.

## 2  Materials

### 2.1  LncRNA-Disease Associations

Known lncRNA-disease associations were downloaded from the LncRNADisease database in August 2018. After removing duplicates due to different evidence, we got 881 different lncRNAs and 328 diseases form 1765 independent lncRNA–disease pairs. Then an adjacency matrix $A$ called association matrix with 328 (number of diseases) rows and 881 (number of lncRNAs) columns could be constructed to store information of lncRNA-disease associations. If the $i$-th disease and $j$-th lncRNA was experimentally associated, the element $A(d(i), l(j))$ was defined as 1, otherwise 0.

### 2.2  Disease MeSH Descriptors and Directed Acyclic Graph

Medical Subject Headings (MeSH) is a comprehensive controlled vocabulary for the purpose of indexing journal articles and books in the life sciences. The top-level categories in the MeSH descriptor including 16 types such as Anatomy [A], Organisms [B], Diseases [C], Chemicals Drugs [D] and so on. Disease-related descriptors were downloaded from National Library of Medicine. The descriptor consists of letters and numbers, with every three bits separated by '.'. By removing the last three bits of the descriptor, a new, shorter descriptor is obtained called the ancestor node. A Directed Acyclic Graph (DAG) consists of the disease itself and its ancestors.

## 3  Methods

### 3.1  Disease Semantic Similarity Matrix 1

As we have mentioned, after removing the duplication of associations due to different evidence, there were 328 different diseases. Each disease with one or more MeSH descriptors can be described by a directed acyclic graph. For instance, disease $D$'s DAG is denoted as $DAG(D) = (D, N(D), E(D))$. $N(d)$ and $E(d)$ are the point set and edge set of $D$'s DAG respectively. Therefore, the similarity between 2 diseases could be calculated by MeSH descriptor and DAG through the previous literature [61]. Specifically, the semantic similarity of two diseases could be calculated as follows. If either of them had no MeSH descriptor, then the semantic similarity of them would be defined as 0. Otherwise the disease $D$'s DAG could be constructed by Mesh descriptors and the semantic similarity of diseases could be calculated as follows. The semantic contribution values of all nodes in $D$'s DAG can be calculated as follows:

$$\begin{cases} D1_D(t) = 1 & \text{if } t = D \\ D1_D(t) = \max\{\Delta * D1_D(t') | t' \in \text{children of } t\} & \text{if } t \neq D \end{cases} \quad (1)$$

where $\Delta$ is the attenuation factor. Obviously in the DAG of disease $D$, compared to other nodes, d has the highest semantic contribution to itself and defines its value as 1. The semantic contribution of other nodes to disease $D$ is attenuated as the distance increases. Therefore, the total semantic contribution of all nodes is calculated as follows:

$$DV1(D) = \sum_{t \in N_A} D1_D(t) \tag{2}$$

Similar to the Jaccard similarity formula, the semantic similarity of any two diseases is calculated as follows:

$$S1(A, B) = \frac{\sum_{t \in N_A \cap N_B} (D1_A(t) + D1_B(t))}{DV1(A) + DV1(B)} \tag{3}$$

$t$ is the intersection of $A$'s DAG and $B$'s DAG.

## 3.2   Disease Semantic Similarity Matrix 2

The semantic similarity calculation matrix 1 only considers the distance factors from different nodes to disease $D$. Therefore, it does not highlight the importance of frequency information. The contribution of higher frequency diseases in DAG should be more important than the lower frequency of diseases. In conclusion, Drawing on the viewpoint of information theory, some scholars have proposed a new way to measure semantic similarity [62]. The new contribution of node $t$'s semantic value to disease $A$ can be calculated as follows:

$$D2_D(t) = -\log(\frac{\text{the number of DAGs including t}}{\text{the number of disease}}) \tag{4}$$

Then we can define the new semantic value of disease $D$, $DV2(D)$ as

$$DV2(D) = \sum_{t \in N_D} D2_D(t) \tag{5}$$

The new semantic similarity of disease $A$ and disease $B$ can be defined as follows:

$$S2(A, B) = \frac{\sum_{t \in N_A \cap N_B} (D2_A(t) + D2_B(t))}{DV2(A) + DV2(B)} \tag{6}$$

## 3.3   Disease Gaussian Interaction Profile Kernel Similarity Matrix

As we have mentioned that the adjacency matrix $A$ with 328 (number of diseases) rows and 881 (number of lncRNAs) columns could be constructed to store information of lncRNA-disease associations. Any disease can be represented by a row of the matrix $A$ called vector $d_A$. The value of vector's each dimension is a Boolean. Then Gaussian interaction profile kernel similarity between disease $A$ and disease $B$ can be defined as follows:

$$DG(A, B) = \exp(-\alpha_d \|d_A - d_B\|^2) \tag{7}$$

Parameter $\alpha_d$ can be implemented to tune the kernel bandwidth and hyperparameters $\alpha_d$ can be adjusted as follows:

$$\alpha_d = \alpha_d'(\frac{1}{nd} \sum_{A=1}^{nd} \|d_A\|^2) \tag{8}$$

Here for simplicity, we set $\alpha_d' = 0.5$. $nd$ is the number of the diseases, it equals to 328.

### 3.4 Disease Integrated Similarity Matrix

In order to make full use of the above three similarity matrices, the integrated disease similarity matrix $DS$ could be constructed. The element of $DS$ $(A, B)$ was defined as follows:

$$DS(A, B) = \begin{cases} \frac{S1(A,B) + S2(A,B)}{2} & \text{if A and B has semantic similarity} \\ DG(A, B) & \text{otherwise} \end{cases} \tag{9}$$

### 3.5 LncRNA Gaussian Interaction Profile Kernel Similarity Matrix

Similar to disease, each lncRNA can be represented as an 881-dimensional vector with 1 or 0 for each component. The Gaussian profile kernel similarity between lncRNA $C$ and lncRNA $D$ could be calculated similarly to the method of disease:

$$RG(C, D) = \exp(-\alpha_l \|l_C - l_D\|^2) \tag{10}$$

Parameter $\alpha_l$ can be implemented to tune the kernel bandwidth and hyperparameters $\alpha_l$ can be adjusted as follows:

$$\alpha_l = \alpha_l'(\frac{1}{nl} \sum_{C=1}^{nl} \|l_C\|^2) \tag{11}$$

Here for simplicity, we set $\alpha_l' = 0.5$, and $nl$ is the number of the lncRNAs, it equals to 881. Finally, $RG$ was regarded as the lncRNA similarity matrix $(RS)$. After constructing the similarity matrix with adjacency matric $A$, the representation vector of each lncRNA or disease will not change with cross-validation. The impact of this on the results will be discussed in a follow-up article.

## 3.6    Convolutional Neural Networks (CNN)

Deep learning is based on a set of algorithms on a multi-layer neural network that solves text, sound, and image processing problems. CNN is a special feedforward neural network, which was proposed by Lecun *et al.* inspired by biological experiments. Compared with other deep learning algorithms, CNN extracts local and global features through effective operations such as convolution and pooling, and then uses the fully connected layer for classification, recognition, prediction or decision making (Figs. 2 and 3).

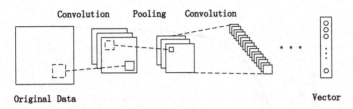

**Fig. 2.** The convolution layer is used to highlight data features with different filters. The definition of the convolution operation is shown in the figure below.

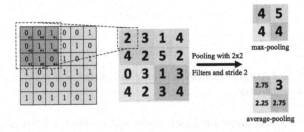

**Fig. 3.** Convolution and pooling process

## 3.7    Extreme Learning Machine (ELM)

The extreme learning machine is a single hidden layer feedforward neural network algorithm proposed by Huang [63]. Traditional artificial neural networks have one or more hidden layers, and the back-propagation algorithm determines the weights and thresholds between different layers. These iterative steps take a lot of computing resources and time. The ELM has only one hidden layer. Therefore, when EML is trained, it is required to set the number of hidden layer nodes. There are no other hyperparameters here. When training, only the number of hidden layer neuron nodes needs to be set, and other parameters are determined. Therefore, ELM is very fast and saves a lot of time when training.

# 4   Results

## 4.1   K-Fold Cross Validation

Cross-validation method is proposed for the over-fitting problem of the empirical risk minimization algorithm, which is very common in classification problems. In the k-fold cross-validation, the original data samples are randomly divided into k sub-sample sets of the same size. Each subset is sequentially carried out as a test set and the rest as a training set training classifier. Given that LOOCV will consume a lot of time, we did the experiments with five-fold cross validation. The result is shown below (Fig. 4):

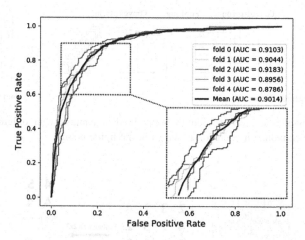

**Fig. 4.** The ROCs and AUCs of ECLDA in 5-fold cross validation on the current data set (3530 lncRNA-disease associations)

The ROCs in 5-fold cross validation performed on the LncRNA disease database were shown in Fig. 7. We obtained the results of average Acc., Sen., Spec., Prec., MCC and AUC of 82.97%, 84.93%, 81.02%, 81.82%, 66.06% and 90.14% when using the proposed method to predict lncRNA-disease associations. The high AUCs showed that LDARF combining Multiple similarities and rotation forest is feasible and effective to predict lncRNA disease associations. At the same time, smaller fluctuations of these standards implied that the proposed model was robust and stable (Table 1).

**Table 1.** 5-Fold Cross-Validation Results Performed by ECLDA on the current data set (3530 lncRNA-disease associations)

| Fold | Acc. (%) | Sen. (%) | Spec. (%) | Prec. (%) | MCC (%) | AUC (%) |
|---|---|---|---|---|---|---|
| 0 | 82.72 | 86.69 | 78.75 | 80.31 | 65.65 | 91.03 |
| 1 | 82.15 | 84.99 | 79.32 | 80.43 | 64.41 | 90.44 |
| 2 | 84.84 | 83.85 | 85.84 | 85.55 | 69.70 | 91.83 |
| 3 | 82.86 | 82.44 | 83.29 | 83.14 | 65.72 | 89.56 |
| 4 | 82.29 | 86.69 | 77.90 | 79.69 | 64.84 | 87.86 |
| **Average** | **82.97** | **84.93** | **81.02** | **81.82** | **66.06** | **90.14** |

## 4.2 Compared with Other Classifiers

The machine learning model has been used to solve problems such as classification and regression for a long time. In this section we compared ELM with several other common classifiers, and the results based on the five-fold cross-validation were as follows (Fig. 5):

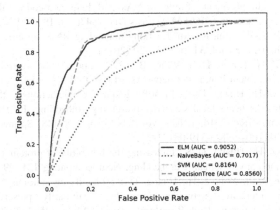

**Fig. 5.** Comparison with NaiveBayes, SVM and DecisionTree in 5-fold cross validation on the current data set (3530 lncRNA-disease associations)

## 5 Conclusion

In this paper, we proposed a computational model based on CNN and ELM to predict the association between lncRNA and disease. CNN is a very useful and efficient deep learning model to extract feature and ELM was chosen as the classifier because of its high speed. To evaluate the model, we made experiments in LOOCV and 5-fold cross validation under different data sets to compare with previous methods and other classifiers. It is obvious that the proposed new model has achieved remarkable results. However, there are still some limitations existed in the model. The negative samples were randomly selected. Randomly selecting unverified samples as negative samples makes the mixed noise affect the predictor's prediction results. Less essential features of the introduction of lncRNA, such as sequence information, fail to solve new user problems. In conclusion, the promising experimental results show that the CELDA will be an excellent addition to the biomedical research in the future.

## References

1. Djebali, S., et al.: Landscape of transcription in human cells. Nature **489**(7414), 101 (2012)
2. Bertone, P., et al.: Global identification of human transcribed sequences with genome tiling arrays. Science **306**(5705), 2242–2246 (2004)
3. ENCODE Project Consortium: Identification and analysis of functional elements in 1% of the human genome by the ENCODE pilot project. Nature **447**(7146), 799 (2007)

4. Schmitt, A.M., Chang, H.Y.: Long noncoding RNAs in cancer pathways. Cancer Cell **29**(4), 452–463 (2016)

5. Alvarez-Dominguez, J.R., Lodish, H.F.: Emerging mechanisms of long noncoding RNA function during normal and malignant hematopoiesis. Blood **130**(18), 1965–1975 (2017)

6. Johnson, R.: Long non-coding RNAs in Huntington's disease neurodegeneration. Neurobiol. Dis. **46**(2), 245–254 (2012)

7. Hrdlickova, B., de Almeida, R.C., Borek, Z., Withoff, S.: Genetic variation in the non-coding genome: Involvement of micro-RNAs and long non-coding RNAs in disease. Biochim. et Biophys. Acta (BBA)-Mol. Basis Dis. **1842**(10), 1910–1922 (2014)

8. Qiu, M.-T., Hu, J.-W., Yin, R., Xu, L.: Long noncoding RNA: an emerging paradigm of cancer research. Tumor Biol. **34**(2), 613–620 (2013)

9. Zhu, L., You, Z.-H., Huang, D.-S., Wang, B.: t-LSE: a novel robust geometric approach for modeling protein-protein interaction networks. PLoS ONE **8**(4), e58368 (2013)

10. Zhu, L., You, Z.-H., Huang, D.-S.: Identifying spurious interactions in the protein-protein interaction networks using local similarity preserving embedding. In: Basu, M., Pan, Y., Wang, J. (eds.) ISBRA 2014. LNCS, vol. 8492, pp. 138–148. Springer, Cham (2014). https://doi.org/10.1007/978-3-319-08171-7_13

11. Zhu, L., You, Z.-H., Huang, D.-S.: Increasing the reliability of protein–protein interaction networks via non-convex semantic embedding. Neurocomputing **121**, 99–107 (2013)

12. Zhu, L., Deng, S.-P., You, Z.-H., Huang, D.-S.: Identifying spurious interactions in the protein-protein interaction networks using local similarity preserving embedding. IEEE/ACM Trans. Comput. Biol. Bioinform. (TCBB) **14**(2), 345–352 (2017)

13. You, Z.-H., Zhou, M., Luo, X., Li, S.: Highly efficient framework for predicting interactions between proteins. IEEE Trans. Cybern. **47**(3), 731–743 (2017)

14. You, Z.-H., Yu, J.-Z., Zhu, L., Li, S., Wen, Z.-K.: A MapReduce based parallel SVM for large-scale predicting protein-protein interactions. Neurocomputing **145**, 37–43 (2014)

15. You, Z.-H., Ming, Z., Huang, H., Peng, X.: A novel method to predict protein-protein interactions based on the information of protein sequence. In: 2012 IEEE International Conference on Control System, Computing and Engineering (ICCSCE), pp. 210–215. IEEE (2012)

16. You, Z.-H., Li, S., Gao, X., Luo, X., Ji, Z.: Large-scale protein-protein interactions detection by integrating big biosensing data with computational model. BioMed. Res. Int. **2014** (2014)

17. You, Z.-H., Li, L., Ji, Z., Li, M., Guo, S.: Prediction of protein-protein interactions from amino acid sequences using extreme learning machine combined with auto covariance descriptor. In: 2013 IEEE Workshop on Memetic Computing (MC), pp. 80–85. IEEE (2013)

18. You, Z.-H., et al.: Detecting protein-protein interactions with a novel matrix-based protein sequence representation and support vector machines. Biomed. Res. Int. **2015**, 1 (2015)

19. You, Z.-H., Lei, Y.-K., Zhu, L., Xia, J., Wang, B.: Prediction of protein-protein interactions from amino acid sequences with ensemble extreme learning machines and principal component analysis. BMC Bioinform. **14**(Suppl 8), S10 (2013)

20. You, Z.-H., Lei, Y.-K., Gui, J., Huang, D.-S., Zhou, X.: Using manifold embedding for assessing and predicting protein interactions from high-throughput experimental data. Bioinformatics **26**(21), 2744–2751 (2010)

21. You, Z.-H., Huang, W., Zhang, S., Huang, Y.-A., Yu, C.-Q., Li, L.-P.: An efficient ensemble learning approach for predicting protein-protein interactions by integrating protein primary sequence and evolutionary information. IEEE/ACM Trans. Comput. Biol. Bioinform. **16**(3), 809–817 (2018)

22. You, Z.-H., Chan, K.C., Hu, P.: Predicting protein-protein interactions from primary protein sequences using a novel multi-scale local feature representation scheme and the random forest. PLoS ONE **10**(5), e0125811 (2015)

23. You, Z.H., Zhu, L., Zheng, C.H., Yu, H.J., Deng, S.P., Ji, Z.: Prediction of protein-protein interactions from amino acid sequences using a novel multi-scale continuous and discontinuous feature set. BMC Bioinform. **15**(S15), S9 (2014)
24. Wong, L., You, Z.-H., Ming, Z., Li, J., Chen, X., Huang, Y.-A.: Detection of interactions between proteins through rotation forest and local phase quantization descriptors. Int. J. Mol. Sci. **17**(1), 21 (2015)
25. Wen, Y.-T., Lei, H.-J., You, Z.-H., Lei, B.-Y., Chen, X., Li, L.-P.: Prediction of protein-protein interactions by label propagation with protein evolutionary and chemical information derived from heterogeneous network. J. Theor. Biol. **430**, 9–20 (2017)
26. Wang, Y.-B., You, Z.-H., Li, L.-P., Huang, Y.-A., Yi, H.-C.: Detection of interactions between proteins by using legendre moments descriptor to extract discriminatory information embedded in pssm. Molecules **22**(8), 1366 (2017)
27. Wang, Y.-B., You, Z.-H., Li, L.-P., Huang, D.-S., Zhou, F.-F., Yang, S.: Improving prediction of self-interacting proteins using stacked sparse auto-encoder with PSSM profiles. Int. J. Biol. Sci. **14**(8), 983–991 (2018)
28. Wang, Y.B., et al.: Predicting protein-protein interactions from protein sequences by a stacked sparse autoencoder deep neural network. Mol. BioSyst. **13**(7), 1336–1344 (2017)
29. Wang, Y., et al.: Predicting protein interactions using a deep learning method-stacked sparse autoencoder combined with a probabilistic classification vector machine. Complexity **2018** (2018)
30. Wang, L., et al.: Advancing the prediction accuracy of protein-protein interactions by utilizing evolutionary information from position-specific scoring matrix and ensemble classifier. J. Theor. Biol. **418**, 105–110 (2017)
31. Li, Z.-W., You, Z.-H., Chen, X., Gui, J., Nie, R.: Highly accurate prediction of protein-protein interactions via incorporating evolutionary information and physicochemical characteristics. Int. J. Mol. Sci. **17**(9), 1396 (2016)
32. Li, Z.W., et al.: Accurate prediction of protein-protein interactions by integrating potential evolutionary information embedded in PSSM profile and discriminative vector machine classifier. Oncotarget **8**(14), 23638 (2017)
33. Lei, Y.-K., You, Z.-H., Ji, Z., Zhu, L., Huang, D.-S.: Assessing and predicting protein interactions by combining manifold embedding with multiple information integration. BMC Bioinform. **13**(Suppl 7), S3 (2012)
34. Lei, Y.-K., You, Z.-H., Dong, T., Jiang, Y.-X., Yang, J.-A.: Increasing reliability of protein interactome by fast manifold embedding. Pattern Recogn. Lett. **34**(4), 372–379 (2013)
35. Huang, Y.-A., et al.: Construction of reliable protein–protein interaction networks using weighted sparse representation based classifier with pseudo substitution matrix representation features. Neurocomputing **218**, 131–138 (2016)
36. Huang, Y.-A., You, Z.-H., Chen, X., Yan, G.-Y.: Improved protein-protein interactions prediction via weighted sparse representation model combining continuous wavelet descriptor and PseAA composition. BMC Syst. Biol. **10**(4), 120 (2016)
37. Huang, Q., You, Z., Zhang, X., Zhou, Y.: Prediction of protein-protein interactions with clustered amino acids and weighted sparse representation. Int. J. Mol. Sci. **16**(5), 10855–10869 (2015)
38. Chen, Z.-H., You, Z.-H., Li, L.-P., Wang, Y.-B., Li, X.: RP-FIRF: prediction of self-interacting proteins using random projection classifier combining with finite impulse response filter. In: Huang, D.-S., Jo, K.-H., Zhang, X.-L. (eds.) ICIC 2018. LNCS, vol. 10955, pp. 232–240. Springer, Cham (2018). https://doi.org/10.1007/978-3-319-95933-7_29
39. An, J.Y., Meng, F.R., You, Z.H., Chen, X., Yan, G.Y., Hu, J.P.: Improving protein–protein interactions prediction accuracy using protein evolutionary information and relevance vector machine model. Protein Sci. **25**(10), 1825–1833 (2016)

40. You, Z.-H., et al.: PRMDA: personalized recommendation-based MiRNA-disease association prediction. Oncotarget **8**(49), 85568 (2017)
41. Wang, L., et al.: MTRDA: using logistic model tree to predict miRNA-disease associations by fusing multi-source information of sequences and similarities. PLoS Comput. Biol. **15**(3), e1006865 (2019)
42. Huang, Y.-A., You, Z.-H., Chen, X., Huang, Z.-A., Zhang, S., Yan, G.-Y.: Prediction of microbe–disease association from the integration of neighbor and graph with collaborative recommendation model. J. Transl. Med. **15**(1), 209 (2017)
43. Chen, X., Zhang, D.-H., You, Z.-H.: A heterogeneous label propagation approach to explore the potential associations between miRNA and disease. J. Transl. Med. **16**(1), 348 (2018)
44. Chen, X., Xie, D., Zhao, Q., You, Z.-H.: MicroRNAs and complex diseases: from experimental results to computational models. Briefings Bioinform. **20**(2), 515–539 (2017)
45. Chen, X., Wang, C.-C., Yin, J., You, Z.-H.: Novel human miRNA-disease association inference based on random forest. Mol. Ther.-Nucleic Acids **13**, 568–579 (2018)
46. Chen, X., Huang, Y.-A., Wang, X.-S., You, Z.-H., Chan, K.C.: FMLNCSIM: fuzzy measure-based lncRNA functional similarity calculation model. Oncotarget **7**(29), 45948 (2016)
47. Chen, X., You, Z.H., Yan, G.Y., Gong, D.W.: IRWRLDA: improved random walk with restart for lncRNA-disease association prediction. Oncotarget **7**(36), 57919–57931 (2016)
48. Huang, Y., Chen, X., You, Z., Huang, D., Chan, K.: ILNCSIM: improved lncRNA functional similarity calculation model. Oncotarget **7**(18), 25902 (2016)
49. Zhan, Z.-H., You, Z.-H., Zhou, Y., Li, L.-P., Li, Z.-W.: Efficient framework for predicting ncRNA-protein interactions based on sequence information by deep learning. In: Huang, D.-S., Jo, K.-H., Zhang, X.-L. (eds.) ICIC 2018. LNCS, vol. 10955, pp. 337–344. Springer, Cham (2018). https://doi.org/10.1007/978-3-319-95933-7_41
50. Yi, H.-C., You, Z.-H., Huang, D.-S., Li, X., Jiang, T.-H., Li, L.-P.: A deep learning framework for robust and accurate prediction of ncRNA-protein interactions using evolutionary information. Mol. Ther.-Nucleic Acids **11**, 337–344 (2018)
51. Chan, K.C., You, Z.-H.: Large-scale prediction of drug-target interactions from deep representations. In: 2016 International Joint Conference on Neural Networks (IJCNN), pp. 1236–1243. IEEE (2016)
52. Chen, X., et al.: NRDTD: a database for clinically or experimentally supported non-coding RNAs and drug targets associations. Database **2017** (2017)
53. Huang, Y.-A., You, Z.-H., Chen, X.: A systematic prediction of drug-target interactions using molecular fingerprints and protein sequences. Curr. Protein Pept. Sci. **19**(5), 468–478 (2018)
54. Li, Z., et al.: In silico prediction of drug-target interaction networks based on drug chemical structure and protein sequences. Sci. Rep. **7**(1), 11174 (2017)
55. Meng, F.-R., You, Z.-H., Chen, X., Zhou, Y., An, J.-Y.: Prediction of drug–target interaction networks from the integration of protein sequences and drug chemical structures. Molecules **22**(7), 1119 (2017)
56. Sun, X., Bao, J., You, Z., Chen, X., Cui, J.: Modeling of signaling crosstalk-mediated drug resistance and its implications on drug combination. Oncotarget **7**(39), 63995 (2016)
57. Wang, L., et al.: Computational methods for the prediction of drug-target interactions from drug fingerprints and protein sequences by stacked auto-encoder deep neural network. In: Cai, Z., Daescu, O., Li, M. (eds.) ISBRA 2017. LNCS, vol. 10330, pp. 46–58. Springer, Cham (2017). https://doi.org/10.1007/978-3-319-59575-7_5
58. Wang, L., et al.: A computational-based method for predicting drug–target interactions by using stacked autoencoder deep neural network. J. Comput. Biol. **25**(3), 361–373 (2018)

59. Wang, L., You, Z.-H., Chen, X., Yan, X., Liu, G., Zhang, W.: Rfdt: a rotation forest-based predictor for predicting drug-target interactions using drug structure and protein sequence information. Curr. Protein Pept. Sci. **19**(5), 445–454 (2018)
60. Chen, G., et al.: LncRNADisease: a database for long-non-coding RNA-associated diseases. Nucleic Acids Res. **41**(D1), D983–D986 (2012)
61. Wang, D., Wang, J., Lu, M., Song, F., Cui, Q.: Inferring the human microRNA functional similarity and functional network based on microRNA-associated diseases. Bioinformatics **26**(13), 1644–1650 (2010)
62. Chen, X., Yan, C.C., Luo, C., Ji, W., Zhang, Y., Dai, Q.: Constructing lncRNA functional similarity network based on lncRNA-disease associations and disease semantic similarity. Sci. Rep. **5**, 11338 (2015)
63. Huang, G.-B., Zhu, Q.-Y., Siew, C.-K.: Extreme learning machine: theory and applications. Neurocomputing **70**(1–3), 489–501 (2006)

# A Prediction Method of DNA-Binding Proteins Based on Evolutionary Information

Weizhong Lu[1,2], Zhengwei Song[1], Yijie Ding[1,2(✉)], Hongjie Wu[1,2], and Hongmei Huang[1]

[1] School of Electronic and Information Engineering,
Suzhou University of Science and Technology, Suzhou 215009, China
wuxi_dyj@163.com
[2] Suzhou Key Laboratory of Virtual Reality Intelligent Interaction
and Application Technology, Suzhou University of Science and Technology,
Suzhou 215009, China

**Abstract.** DNA is the carrier of genetic information in organisms, and DNA-binding protein is one type of unwinding enzymes, which plays a key role in various biological molecular functions. That has greatly promoted the research of various methods for identifying DNA-binding proteins. In recent years, researchers have developed a Machine Learning-based method to predict DNA-binding proteins quickly and accurately. Although the prediction accuracy of current methods is considerable, the performance of their prediction can be further improved. In this paper, a DNA-binding proteins prediction model based on PSSM (Position Specific Scoring Matrix) features and Random Forest classifier is proposed. The results of experiments show that the proposed method can achieve great prediction performance on PDB1075 and PDB186 datasets, whose accuracy is 82.14% and 79.0%, respectively. Experiments show that the method can be compared with other methods, and even surpass the previous methods on some datasets.

**Keywords:** DNA-binding proteins · Protein sequence · PSSM · Random Forest · Jackknife test

## 1 Introduction

The activities of life related to DNA are very important in organisms, including DNA damage detection, DNA replication, gene transcription regulation and so on. These activities all occur with the assistance of specific proteins, but they are also regulated by protein-DNA interaction, which is achieved by the specific or non-specific binding of proteins and DNA chains. The proteins that are bind to DNA as well as regulate DNA-related life activities are called DNA-binding proteins [1].

DNA-binding proteins, also known as helical unstable proteins, are one type of unwinding enzymes, playing a critical role in many biological molecular functions, which has greatly promoted the research of various methods for identifying DNA-binding proteins. The experimental methods used to identify DNA-binding proteins include recognition techniques based on physical, chemical and biological experiments

© Springer Nature Switzerland AG 2019
D.-S. Huang et al. (Eds.): ICIC 2019, LNCS 11644, pp. 418–429, 2019.
https://doi.org/10.1007/978-3-030-26969-2_40

that can accurately predict the structure and function of proteins, but they have some shortcomings, such as high actual cost, time-consuming and strict requirements of experimental environment. Therefore, it is unrealistic to use experimental methods to identify all DNA-binding proteins. Thus, scholars at home and abroad have proposed many computational-based methods to reduce time costs, which can be divided into two types: one is based on sequence information, and the other is based on structural information.

The method based on structure information usually has better performance, but the structure information of proteins is difficult to obtain, so the development of the method is limited to a certain extent.

DNA-binding proteins prediction methods based on sequence information can be divided into template library-based methods and machine learning-based methods. The method based on the template library first found homologous proteins from the template library by using the similarity of protein sequences, and then predicted whether the target protein is a DNA-binding protein according to the homologous proteins. The prediction method based on machine learning mainly solves the two problems of vectorized expression of proteins and selection of classifiers. With the increase of DNA-binding proteins with known structures, it is possible to predict DNA-binding proteins by using machine learning methods and structural information. Sequence information commonly used for DNA-binding proteins prediction can be generally divided into four categories, namely, combinatorial characteristics, physical and chemical properties, evolutionary characteristics, and structural and functional characteristics [29].

Although there are many methods for predicting DNA-binding proteins based on sequence information now, their performance of prediction can be further improved. Protein representation is a battle of the improvement of protein structure prediction performance based on sequence information. More research is still needed to effectively represent protein sequence information. To solve this problem, we propose a DNA-binding proteins prediction model based on PSSM features and RF classifier, which extracts three features from PSSM matrix, namely PsePSSM, PSSM-AB and PSSM-DWT. RF classifier is used to predict DNA-binding proteins, and then evaluate the benchmark datasets based on Jackknife test and independent test respectively. Experiments show that good prediction performance can be obtained by using the method proposed in this paper when predicting DNA-binding proteins.

## 2   Modeling of DNA-Binding Protein Structure Prediction Method

The process of researching DNA-binding proteins structure prediction method can be divided into three stages: model building, training and protein prediction. In the stage of model building, three features are first extracted from PSSM matrix, namely PsePSSM, PSSM-DWT and PSSM-AB, and then NMBAC features are extracted from six physicochemical properties. The prediction model is obtained by inputting these features into the RF classifier. In the stage of protein prediction, representation algorithm based on PSSM features is used to describe the protein sequence predicted, and then RF model is used to train DNA-binding protein prediction. Figure 1 shows the framework of the method. The evolutionary features selected in this paper are as follows.

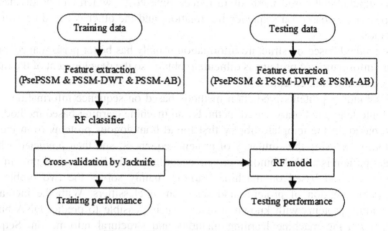

**Fig. 1.** The framework of our method

## 2.1 PSSM

PSSM is Position Specific Scoring Matrix, which stores the evolutionary information of protein sequence, generated by three iterations of a PSI-BLAST [13] search of the SWISS-PROT [14] protein knowledgebase (PSI-BLAST aims to find the best result through multiple iterations, using the first search result to form a location-specific score matrix that will be used for the second search, and the second search result will be used for the third search, and so on until the best search result is found. The search result is the best after three iterations, so it performs three PSI-BLAST iterative searches on its settings.), and its E-value threshold is set to 0.001. Supposing that the amino acids residue of a protein sequence is L, and the size of the PSSM for the protein is L×20 (L rows and 20 columns), The matrix's formula is as follows:

$$PSSM_{original} = \begin{bmatrix} P_{1,1} & \cdots & P_{1,20} \\ \vdots & \ddots & \vdots \\ P_{L,1} & \cdots & P_{L,20} \end{bmatrix}_{L \times 20} \quad (1)$$

where $PSSM_{original}(i,j)$ is as follows:

$$PSSM_{original}(i,j) = \sum_{k=1}^{20} \omega(i,k) \times D(k,j), i = 1,\ldots,L. j = 1,\ldots,20 \quad (2)$$

where $\omega(i,k)$ is the frequency of the k-th amino acid type at the position i, and $D(k,j)$ is the rate of mutation from the k-th amino acid to the j-th amino acid in a protein sequence in Dayhoff's substitution matrix (mutation matrix). The bigger values of mutation matrix make known more conserved positions intensely. If not, the result will be reverse.

## 2.2 PsePSSM

PsePSSM features were directed against membrane protein prediction, which were proposed by Chou and Shen [22] and were usually used to explore the evolutionary information as well as the sequence-order information embedded in PSSMs [20]. However, it is not suitable for representing the DNA-binding proteins sequences directly. For the convenience of later calculation and work, the original PSSM of proteins ought to be further normalized.

$$f_{i,j} = \frac{p_{i,j} - \frac{1}{20}\sum_{k=1}^{20} p_{i,k}}{\sqrt{\frac{1}{20}\sum_{l=1}^{20}\left(p_{i,l} - \frac{1}{20}\sum_{k=1}^{20} p_{i,k}\right)^2}}, i = 1, \ldots, L; j = 1, \ldots, 20 \qquad (3)$$

where $f_{i,j}$ is the normalized scores of PSSM having a zero mean over the 20 amino acids, $p_{i,j}$ is the original scores of PSSM. A positive score indicates that the homologous mutation occurs more frequently in the multiple alignment than expected by chance while a negative score is the reverse.

The normalization of PSSM can be described as follows:

$$P_{normalized} = \begin{bmatrix} f_{1,1} & \cdots & f_{1,20} \\ \vdots & \ddots & \vdots \\ f_{i,1} & \cdots & f_{i,20} \\ \vdots & \ddots & \vdots \\ f_{L,1} & \cdots & f_{L,20} \end{bmatrix}_{L \times 20} \qquad (4)$$

## 2.3 PSSM-DWT

The full name of DWT is Discrete Wavelet Transform. WT (Wavelet Transform) is defined as the projection of a signal $f(t)$ onto wavelet function. The formulation is as follows:

$$T(a, b) = \frac{1}{\sqrt{a}} \int_0^t f(t)\psi\left(\frac{t-b}{a}\right) d_t \qquad (5)$$

where a is a scale variable, b is a translation variable, $\psi((t-b)/a)$ is the analyzing wavelet function. T $(a, b)$ is the transform coefficients, found for both specific wavelet periods and specific locations on the signal. DWT can resolve the amino acid sequences into coefficients with different dilations, and then it will remove the noise component from the profiles. In order to perform DWT by supposing that the discrete signal $f(t)$ is $x[n]$, Nanni et al. [11] proposed an efficient algorithm. The formula is as follows:

$$y_{j,low}[n] = \sum_{k=1}^{N} x[k]g[2n - k] \tag{6}$$

$$y_{j,high}[n] = \sum_{k=1}^{N} x[k]h[2n - k] \tag{7}$$

where N is discrete signal's length, g is the low-pass filter while h is high-pass filter. $y_{j,low}[n]$ is the approximative coefficient of the signal, the approximative coefficient is low-frequency components, and $y_{j,high}[n]$ is the detailed coefficient, which is high-frequency components. To further improve the frequency resolution and the approximation coefficients that are decomposed with low-pass filter and high-pass filter, the decomposition are done repeatedly and then down-sample. More detailed characteristics of the signal can be observed with the increase of decomposition level j.

In this work, we use 4-level DWT, and then calculate the maximum values, minimum values, mean values and standard deviation values. Besides, the discrete signals (20 types of discrete signals constitute PSSM) of PSSM are analyzed with above 4 levels DWT. Figure 2 shows the architecture diagram of a 4-level DWT.

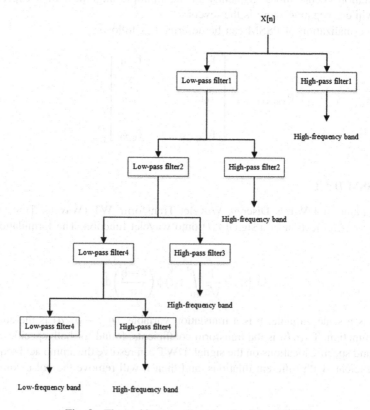

**Fig. 2.** The architecture diagram of a 4-level DWT

## 2.4   PSSM-AB

AB method is Average Blocks [30]. Because the number of amino acids in each protein is different, the eigenvectors will have different sizes when PSSM is directly transformed into eigenvectors. To solve the problem, the features are averaged over a local region in PSSMs, which is called as AB method. Each block has 5% protein sequence. Therefore, regardless of the length of a protein sequence, dividing a protein sequence into 20 blocks, and each block consists of 20 features derived from 20 columns in PSSMs. The mathematical formula is as follows:

$$AB(k) = \frac{20}{N} \sum_{p=1}^{\frac{N}{20}} Mt\left(p + (i-1) \times \frac{20}{N}, j\right)$$

$$i = 1, \ldots, 20; \; j = 1, \ldots, 20; \; k = j + 20 \times (i-1) \tag{8}$$

where $N/20$ is the size of the j-th block, occupying 5% of a protein sequence, and $Mt(p + (i-1) \times 20/N, j)$ refers to a $1 \times 20$ vector extracted from PSSM profile at the i-th position in the j-th block. For each protein sequence having 20 blocks, and each protein sequence of the two datasets is transformed into a 400-dimensional vector by AB feature extraction method.

## 3   Experiment Results

The experimental process is as follows:

(1)  establish a prediction model of DNA-binding proteins on PDB1075 and PDB186 datasets.
(2)  determine the evolutionary characteristics used in experimen: PsePSSM, PSSM-DWT and PSSM-AB, and the types of evaluation indicators used to evaluate the performance of the prediction model.
(3)  compare the performance of different feature combinations on PDB1075 dataset via Jackknife test (commonly known as LOOCV test), as well as compare the performance of these methods with several existing methods. Besides, the performance of methods on PDB186 dataset is compared with other existing methods to prove the validity of DNA-binding proteins prediction model and method.

### 3.1   Datasets

In the selection of data sets, two datasets widely used in the field of DNA-binding proteins are selected as the basic data for experiments, namely PDB1075 [10] and PDB186 [11] datasets. The protein sequences in these two datasets are derived from the international protein database PDB (protein data bank: https://www.resb.org/), and the PDB1075 dataset is constructed by Liu et al. including 525 DNA-binding proteins and 550 non-DNA-binding proteins. The PDB186 dataset was constructed by Lou et al. including 93 DNA-binding proteins and 93 non-DNA-binding proteins. In this paper, PDB1075 dataset is used as the training set and PDB186 dataset is the independent test dataset.

## 3.2    Measurements

Two commonly used evaluation methods, including Jackknife test and independent test, are used to evaluate the performance of the prediction model. In addition, Accuracy (ACC), Matthews Correlation Coefficient (MCC), Sensitivity (SN), Specificity (SP) are used as evaluation indicators. These evaluation indicators have been widely used in various studies of biological sequence classification. The mathematical definitions of these indicators are as follows.

$$
\begin{cases}
SN = \dfrac{TP}{TP+FN} \\
SP = \dfrac{TN}{TN+FP} \\
ACC = \dfrac{TP+TN}{TP+FP+TN+FN} \\
MCC = \dfrac{TP\times TN - FP\times FN}{\sqrt{(TP+FN)\times(TN+FP)\times(TP+FP)\times(TN+FN)}}
\end{cases}
\tag{9}
$$

Where TP denotes the number of positive samples correctly predicted, TN denotes the number of negative samples predicted correctly, FP denotes the number of positive samples predicted mistakenly and FN denotes the number of negative samples predicted mistakenly. SN represents the percentage of correctly predicted samples in positive samples, SP represents the percentage of correctly predicted samples in negative samples, and ACC represents the percentage of correctly predicted samples in all samples. MCC can be used to reflect the quality of the prediction model, and its range of values is [−1,1]. When MCC = −1, all samples are misclassified. When MCC = 0, the performance of prediction results is the same as that of random selection. When MCC = 1, all samples are correctly predicted.

## 3.3    Experimental Results and Analysis

**The Performance of Different Features on Training Dataset.** Three features, namely PSPSPSSM, PSSM-AB and PSSM-DWT, are extracted from PDB1075 dataset, testing the combined features on PDB1075 dataset via Jackknife test to obtain the best combination of performance and better prediction results. Table 1 shows the performance of different feature combinations via Jackknife test.

**Table 1.** The performance of different features on PDB1075 dataset via Jackknife test evaluation

| Feature | ACC(%) | MCC | SN(%) | SP(%) | AUC |
|---|---|---|---|---|---|
| PsePSSM | 77.12 | 0.5457 | 81.52 | 72.91 | 0.8574 |
| PSSM-AB | 77.21 | 0.5450 | 79.05 | 75.45 | 0.8483 |
| PSSM-DWT | 74.98 | 0.4993 | 74.67 | 75.27 | 0.8337 |
| PsePSSM+PSSM-AB | 81.67 | 0.6337 | 82.48 | 80.91 | 0.8974 |
| PsePSSM+PSSM-DWT | 81.77 | 0.6357 | 82.86 | 80.73 | 0.8935 |
| PSSM-AB+PSSM-DWT | 77.21 | 0.5447 | 78.48 | 76.00 | 0.8443 |
| PsePSSM+PSSM-AB+PSSM-DWT | **82.42** | **0.6489** | **83.81** | **81.09** | **0.8994** |

The indices of seven different combinations of features obtained via Jackknife test evaluation listed in Table 1 show that each feature is useful for predicting DNA-binding proteins, the effect of the three indices combined independently are approximate, but the performance is obviously improved when combined together, and the performance of combination including PsePSSM feature is great, so PsePSSM is an important feature. The performance of PSePSSM+PSSM-AB and PSPSSM+PSSM-DWT is similar and the performance of PSSM-AB+PSSM-DWT is slightly lower than that of the first two combinations, and the performance of PsePSSM+PSSM-AB +PSSM-DWT is the best, whose ACC (82.42%), MCC (0.6489), SN (83.81%), SP (81.09) and AUC (0.8994).

To evaluate the performance of the prediction model, we use AUROC feature curve to analyze the classification effect of the model. The AUROC feature curve is composed of ROC curve (Receiver Operating Characteristic Curve) and AUC (Area Under Curve). ROC curve (also known as Sensitivity curve) reflects the same sensitivity, which is the response to the same signal stimulus, but the results are obtained under several different criteria. The curve is generally above the line y = x, so its range is [0.5, 1]. If the curve is closer to the upper left corner, the performance of the classifier will be better; AUC is defined as the area enclosed by the coordinate axis under the ROC curve. The larger the AUC value is, the better the performance of the classifier obtains. Thus, the AUROC curves of seven different feature combinations obtained via Jackknife cross-validation on PDB1075 dataset are compared, shown in Fig. 3.

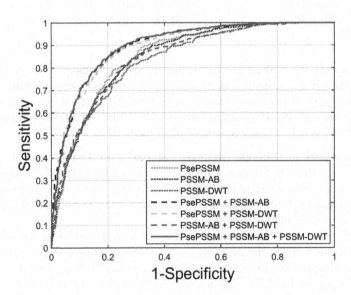

**Fig. 3.** The AUROC of seven different features combinations via Jackknife cross-validation on PDB1075 dataset

From the figure above, we can find that the ROC curve is biased to the longitudinal direction when the three features are combined, and the value of AUC is the largest as well as the best performance in the meantime.

**The Performance of Different Methods Compared on Training Dataset.** We compare the performance of the proposed method on PDB1075 dataset with other existing methods, including DNA-Prot [15], DNA binder [15], IDNA-Prot|dis [10], IDNA-Prot [16], PseDNA-Pro [17], IDNAPro-PseAAC [18] and Kmer1+ACC [19]. Table 2 shows the performance of the proposed method and other existing methods on PDB1075 dataset via Jackknife test evaluation.

**Table 2.** The performance of the method and other existing methods on PDB1075 dataset via jackknife test evaluation.

| Methods | ACC (%) | MCC | SN (%) | SP (%) |
|---|---|---|---|---|
| DNA-Prot | 72.55 | 0.44 | 82.67 | 59.76 |
| IDNA-Prot | 75.40 | 0.50 | **83.81** | 64.73 |
| IDNA-Prot|dis | 77.30 | 0.54 | 79.40 | 75.27 |
| PseDNA-Pro | 76.55 | 0.53 | 79.61 | 73.63 |
| DNA binder (dimension = 400) | 73.58 | 0.47 | 66.47 | 80.36 |
| DNA binder (dimension = 21) | 73.95 | 0.48 | 68.57 | 79.09 |
| IDNAPro-PseAAC | 76.56 | 0.53 | 75.62 | 77.45 |
| Kmer1+ACC | 75.23 | 0.50 | 76.76 | 73.76 |
| Our method | **82.14** | **0.64** | 82.86 | **81.45** |

By analyzing the evaluation index values of several methods listed in Table 2 on PDB1075 dataset, the evaluation index values of the method proposed in this paper are significantly higher than those of other existing methods, among which ACC (82.14%), MCC (0.64) and SP (81.45%), improved by 4.84%, 0.10 and 1.09% respectively, and SN is 83.81%, which is better than most other methods as well.

**The Performance of Different Methods Compared on Test Dataset.** In the independent test set, PDB1075 dataset is used as the training set and PDB186 dataset is test dataset, and the performance of this method is compared with several existing methods on PDB186 dataset. Table 3 shows the performance of the proposed method and other existing methods on PDB186 dataset.

Table 3 shows the evaluation index values of several methods tested on PDB186 dataset. We can find that the evaluation index values of the method are significantly higher than those of other existing methods, where ACC (79.0%), MCC (0.616) and SP (95.7%), improved by 2.1%, 0.078 and 12.9% respectively compared with other methods, having better performance than other existing methods.

**Table 3.** The performance of the method and other existing methods on PDB186 dataset.

| Methods | ACC (%) | MCC | SN (%) | SP (%) |
|---|---|---|---|---|
| IDNA-Prot\|dis | 72.0 | 0.445 | 79.5 | 64.5 |
| IDNA-Prot | 67.2 | 0.344 | 67.7 | 66.7 |
| DNA-Prot | 61.8 | 0.240 | 69.9 | 53.8 |
| DNAbinder | 60.8 | 0.216 | 57.0 | 64.5 |
| DNABIND | 67.7 | 0.355 | 66.7 | 68.8 |
| DNA-Threader | 59.7 | 0.279 | 23.7 | **95.7** |
| DBPPred | 76.9 | 0.538 | 79.6 | 74.2 |
| IDNAPro-PseAAC-EL | 71.5 | 0.442 | 82.8 | 60.2 |
| Kmer1+ACC | 71.0 | 0.431 | 82.8 | 59.1 |
| Our method | **79.0** | **0.616** | **95.7** | 62.4 |

## 4  Conclusion

In this paper, a DNA-binding proteins prediction model based on PSSM features and RF classifier is proposed. Three features are extracted from the model: PsePSSM, PSSM-AB and PSSM-DWT. Then combining the three evolutionary features tested on PDB1075 dataset and PDDB186 dataset, and analyzing and comparing the test results.

In Jackknife test, the evaluation index values of several different methods on PDB1075 dataset are listed. Experiments show that the performance of this method is better than other existing methods, obtaining ACC (82.14%), MCC (0.64), SP (81.45%), improved by 4.84%–9.59%, 0.10–0.20, 1.09–21.69% compared with other methods respectively.

In independent test, the evaluation index values obtained by several different methods on PDB186 dataset are given, in which ACC (79.0%), MCC (0.616), SN (95.7%), improved by 7.0–19.3%, 0.078–0.5%, 12.9–72.0% compared with other methods, respectively.

Experiments show that the method proposed in this paper has great performance on PDB1075 and PDB186 datasets, proving the rationality of feature extraction algorithm and the validity of DNA-binding proteins prediction model.

In the future work, we intend to consider combining other biological relevant features, such as amino acid compositions and protein secondary structures etc. for refining the feature representation and classification algorithm based on Machine Learning, which may can help us improve the performance of DNA-binding proteins prediction.

**Acknowledgements.** This paper is supported by the National Natural Science Foundation of China (61772357, 61502329, 61672371, and 61876217), Jiangsu Province 333 Talent Project, Top Talent Project (DZXX-010), Suzhou Foresight Research Project (SYG201704, SNG201610, and SZS201609).

# References

1. Luscombe, N.M., Austin, S.E., Thomton, J.M.: An overview of the structures of protein-DNA complexes. Genome Biol. **1**(1), 1–37 (2000)
2. Lou, W., Wang, X., Chen, F., et al.: Sequence based prediction of DNA-binding proteins based on hybrid feature selection using random forest and gaussian naïve bayes. PLoS ONE **9**(1), e86703 (2014)
3. Breiman, L.: Random forests. Mach. Learn. **45**(1), 5–32 (2001)
4. Stawiski, E.W., Gregoret, L.M., Mandel-Gutfreund, Y.: Annotating nucleic acid-binding function based on protein structure. J. Mol. Biol. **326**(4), 1–1079 (2003)
5. Shanahan, H.P., Garcia, M.A., Jones, S., et al.: Identifying DNA-binding proteins using structural motifs and the electrostatic potential. Nucleic Acids Res. **32**(16), 4732–4741 (2004)
6. Gao, M., Skolnick, J.: DBD-Hunter: a knowledge-based method for the prediction of DNA-protein interactions[J]. Nucleic Acids Res. **36**(12), 3978–3992 (2008)
7. Szilágyi, A., Skolnick, J.: Efficient prediction of nucleic acid binding function from low-resolution protein structures. J. Mol. Biol. **358**(3), 1–933 (2006)
8. Nimrod, G., Schushan, M., Szilagyi, A., et al.: iDBPs: a web server for the identification of DNA binding proteins. Bioinformatics **26**(5), 692–693 (2010)
9. Zhao, H., Yang, Y., Zhou, Y.: Structure-based prediction of DNA-binding proteins by structural alignment and a volume-fraction corrected DFIRE-based energy function. Bioinformatics **26**(15), 1857–1863 (2010)
10. Liu, B., Xu, J., Lan, X., et al.: iDNA-Prot|dis: identifying DNA-binding proteins by incorporating amino acid distance-pairs and reduced alphabet profile into the general pseudo amino acid composition. PLoS ONE **9**(9), e106691 (2014)
11. Nanni, L., Brahnam, S., Lumini, A.: Wavelet images and Chou's pseudo amino acid composition for protein classification. Amino Acids **43**(2), 657–665 (2012)
12. Schaffer, A.A.: Improving the accuracy of PSI-BLAST protein database searches with composition-based statistics and other refinements. Nucleic Acids Res. **29**(14), 2994–3005 (2001)
13. Boeckmann, B.: The SWISS-PROT protein knowledgebase and its supplement TrEMBL in 2003. Nucleic Acids Res. **31**(1), 365–370 (2003)
14. Kumar, K.K., Pugalenthi, G., Suganthan, P.N.: DNA-Prot: identification of DNA binding proteins from protein sequence information using random forest. J. Biomol. Struct. Dyn. **26**(6), 679–686 (2009)
15. Kumar, M., Gromiha, M.M., Raghava, G.P.: Identification of DNA-binding proteins using support vector machines and evolutionary profiles. BMC Bioinform. **8**(1), 463 (2007)
16. Wei-Zhong, L., Jian-An, F., Xuan, X., et al.: iDNA-Prot: identification of DNA binding proteins using random forest with grey model. PLoS One **6**(9), e24756 (2011)
17. Liu, B., Xu, J., Fan, S., et al.: PseDNA-Pro: DNA-binding protein identification by combining chou's PseAAC and physicochemical distance transformation. Mol. Inform. **34**(1), 8–17 (2015)
18. Liu, B., Wang, S., Wang, X.: DNA binding protein identification by combining pseudo amino acid composition and profile-based protein representation. Sci. Rep. **5**(4), 108–142 (2015)
19. Dong, Q., Wang, S., Wang, K., et al.: Identification of DNA-binding proteins by auto-cross covariance transformation. In: IEEE International Conference on Bioinformatics & Biomedicine. IEEE (2015)

20. Chou, K.C., Shen, H.B.: MemType-2L: a web server for predicting membrane proteins and their types by incorporating evolution information through Pse-PSSM. Biochem. Biophys. Res. Commun. **360**(2), 1–345 (2007)
21. Chiu, T.P., Rao, S., Mann, R.S., et al.: Genome-wide prediction of minor-groove electrostatic potential enables biophysical modeling of protein–DNA binding. Nucleic Acids Res. **45**(21), 12565–12576 (2017)
22. Liu, B.: Identification of DNA-binding proteins by incorporating evolutionary information into pseudo amino acid composition via the top-n-gram approach. J. Biomol. Struct. Dyn. **33**(8), 1720–1730 (2015)
23. Wu, J., Liu, H., Duan, X., et al.: Prediction of DNA-binding residues in proteins from amino acid sequences using a random forest model with a hybrid feature. Bioinformatics **25**(1), 30–35 (2009)
24. Xu, R., Zhou, J., Wang, H., et al.: Identifying DNA-binding proteins by combining support vector machine and PSSM distance transformation. BMC Syst. Biol. **9**(S1), S10 (2015)
25. Yang, R., Wu, H., Fu, Q., Ding, T., Chen, C.: Optimizing HP model using reinforcement learning. In: Huang, D.-S., Jo, K.-H., Zhang, X.-L. (eds.) ICIC 2018. LNCS, vol. 10955, pp. 383–388. Springer, Cham (2018). https://doi.org/10.1007/978-3-319-95933-7_46
26. Chen, C., Wu, H., Bian, K.: β-barrel transmembrane protein predicting using support vector machine. In: Huang, D.-S., Hussain, A., Han, K., Gromiha, M.M. (eds.) ICIC 2017. LNCS (LNAI), vol. 10363, pp. 360–368. Springer, Cham (2017). https://doi.org/10.1007/978-3-319-63315-2_31
27. Wu, H., Li, H., Jiang, M., et al.: Identify high-quality protein structural models by enhanced K-means. Biomed. Res. Int. **2017**(18), 1–9 (2017)
28. Huang, H.L., Lin, I.C., Liou, Y.F., et al.: Predicting and analyzing DNA-binding domains using a systematic approach to identifying a set of informative physicochemical and biochemical properties. BMC Bioinform. **12**(S1), S47 (2011)
29. Ji-Yong, A., Zhu-Hong, Y., Fan-Rong, M., et al.: RVMAB: using the relevance vector machine model combined with average blocks to predict the interactions of proteins from protein sequences. Int. J. Mol. Sci. **17**(5), 757 (2016)
30. Cong, S., Yijie, D., Jijun, T., et al.: Identification of DNA–protein binding sites through multi-scale local average blocks on sequence information. Molecules **22**(12), 2079 (2017)

# Research on RNA Secondary Structure Prediction Based on Decision Tree

Weizhong Lu[1,2], Yan Cao[1], Hongjie Wu[1,2(✉)],
Hongmei Huang[1], and Yijie Ding[1,2]

[1] School of Electronic and Information Engineering,
Suzhou University of Science and Technology, Suzhou 215009, China
Hongjie.wu@qq.com
[2] Jiangsu Province Key Laboratory of Intelligent Building Energy Efficiency,
Suzhou University of Science and Technology, Suzhou 215009, China

**Abstract.** The secondary structure of RNA is closely related to its role in biological function, and it is difficult to predict the secondary structure of RNA sequence containing pseudoknots due to its complex structure. In this paper, a DT (Decision Tree) based RNA secondary structure prediction model is proposed, and a training algorithm is constructed. By adjusting the size of the window, the secondary structure prediction of RNA sequences containing pseudoknots is realized. The comparison experiments are carried out on the authoritative dataset RNA STRAND with classical classification algorithms such as LR (Logistic Regression), RF (Random forests) and SVM (Support Vector Machine). The experimental results show that compared with SVM, LR and RF, DT has more advantages in classification accuracy and robustness, and its accuracy rate has increased by 8.68%.

**Keywords:** Decision tree · RNA secondary structure prediction · Pseudoknots · Base

## 1 Introduction

RNA (Ribonucleic Acid) has many different biological functions in biological cells, especially for viruses such as HIV and Ebola, RNA plays an important role as a carrier of genetic information. RNA sequence is a single strand composed of a large number of nucleotides. Predicting and obtaining the secondary structure or even the tertiary structure of RNA molecules is of great significance for understanding the biological function of RNA molecules and the decomposition of viruses. Pseudoknots are the most extensive units in RNA structure. They are very complex and stable RNA structures. The prediction of RNA secondary structure with pseudoknots has been proved to be a NP-complete problem [1].

At present, the algorithms for predicting RNA secondary structure mainly fall into the following two categories.

The first type is a prediction algorithm based on sequence comparison, which uses the conservation of RNA structure and uses homologous RNA sequences with high sequence similarity as alignment templates to predict RNA secondary structure.

© Springer Nature Switzerland AG 2019
D.-S. Huang et al. (Eds.): ICIC 2019, LNCS 11644, pp. 430–439, 2019.
https://doi.org/10.1007/978-3-030-26969-2_41

At present, the most effective algorithms are Needleman-Wunsch algorithm and Smith-Waterman algorithm [5]. The advantage of this method is that the prediction accuracy is high, but the homologous sequence of the predicted sequence is needed, so it is not suitable for prediction of a single RNA [6], and multi-sequence comparison has the problems of great consumption of computing resources and the like.

The second type is the prediction algorithm based on the minimum free energy model, which can predict just based on the given sequence information without prior knowledge, and has greater practicability. However, the method has the following problems: the prediction accuracy is affected by the energy model, the prediction accuracy is not as high as that of the sequence comparison method, the longer sequence cannot be predicted well, and the inconsistency between the real structure and the minimum free energy structure often occurs in the actual situation.

In addition, Rivas and Eddy proposed a predigesting thermodynamic model with pseudoknots, and proposed a polynomial time algorithm based on energy for the first time to solve RNA secondary structure of the predigesting model [2]. Its time complexity is $O(n^6)$ and space complexity is $O(n^4)$. The algebraic dynamic programming algorithm published by Jens and Robert is used to predict RNA secondary plane pseudoknots with simple structure. Its time complexity is $O(n^4)$ and its space complexity is $O(n^2)$ [3]. Based on the idea of iteratively forming stable stems, RenJihong and BaharakRastegari proposed a new algorithm HotKnots [4] to predict RNA secondary structure including pseudoknots.

According to the basic characteristics of RNA planar pseudoknot, this paper takes the result of vectorization of the primary sequence of RNA with pseudoknots is used as model input, and combines DT algorithm to model and predict the secondary structure of RNA sequence with pseudoknots. The algorithm has the advantages of fast learning speed, easy implementation and easy use.

## 2 Prediction of RNA Secondary Structure

### 2.1 Introduction of RNA Secondary Structure

RNA is a single-stranded molecule, which makes bases that can be paired with each other meet through its own folding back to form hydrogen bonds (A to U; G to C) and form double helix structures at the same time. These double helix structures are called "stems". Irregular loops formed by unpaired base regions are excluded from the double helix structure. A continuous sequence adjacent to each other is folded back due to the complementation of both ends, thus forming a hairpin-like structure, which is vividly called a hairpin ring with a stem and a loop.

In addition, it is necessary to specifically explain the pseudoknots. if the paired bases in an RNA sequence have paired base pairs <i, j> and <m, n> , and i < m < j < n (i, j, m, n represent the position of a base in an RNA sequence), then the RNA molecular chain has a pseudoknot [8].

The secondary structure of RNA can be shown in Fig. 1, and the schematic diagram of the secondary structure of RNA with pseudoknots is shown in Fig. 2. In which the P structure is a planar Pseudoknot and N-PP is a non-planar Pseudoknot.

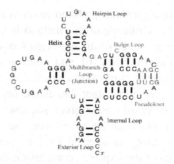

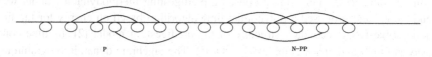

**Fig. 1.** Schematic diagram of RNA secondary structure

**Fig. 2.** A schematic diagram of a dummy junction

## 2.2 E-NSSEL Secondary Structure Label

In article [12], a visual labeling method is proposed: E-NSSEL labeling. This method is simple and easy to understand. Its encoding method is to remove some redundant marks and add false junction marks on NSSEL label. The E-NSSEL label is described as follows:

1. +Stem represents the base self-sequence in the stem;
2. +pseudoknots represent pseudoknot base sequence (or single node) near the 5' end;
3. -Stem represents the base self-sequence in the negative stem;
4. -pseudoknots represent pseudo-knot base sequences (or single nodes) near the 3' end;
5. Loop represents all base segments that are unpaired.

   With this coding method, the subsequence of each type of tag can be represented by 5 numerical sequences such as 1 to 5, i.e. the secondary structure of an RNA sequence can be uniquely represented by an E-NSSEL tag, whereas an E-NSSEL tag uniquely represents the secondary structure of an RNA sequence.

   For example, the E-NSSEL expression in Fig. 1 is as follows:

   51115111155111555555555555533353333511155555333511552222333335111544 453333353555.

## 2.3 Prediction of RNA Secondary Structure

In this paper, the E-NSSEL labeling method is used to predict the RNA secondary structure with pseudoknots. The RNA secondary structure prediction problem can be divided into the following two sub-problems:

(1) Can judge whether each base of an RNA molecular chain has a base matched with it;

(2) If the result of (1) is yes, the pairing type can be determined.

Therefore, the prediction of RNA secondary structure can be regarded as a five-class problem. The RNA base sequence $(x_1, \ldots, x_n)$ containing pseudoknots is regarded as the input of the problem, and then finds out whether each base in the RNA sequence has a pair base $Y_{ij}$. $Y_{ij} = 0$ indicates that the base is a single base; otherwise, it indicates that the i-th base of the sequence is matched with the j-th base. When predicting the secondary structure of RNA sequences with pseudoknots, in addition to predicting whether or not a base pair exists, it also requires a pseudoknot condition that can determine whether the pairing is satisfied. Therefore, it is more difficult to predict the secondary structure of data sets with false knots than data sets without pseudoknot.

## 2.4    Evaluation Index

Accuracy (acc) is a commonly used evaluation index in classification problems, sensitivity (sen) and specificity (ppv) are commonly used evaluation indexes in RNA secondary structure prediction, Matthews correlation coefficient mcc is an evaluation index combining sensitivity and specificity. This paper uses these four evaluation indexes to evaluate the model, and their calculation methods are as follows:

$$acc = \frac{TP + TN}{TP + TN + FP + FN} \tag{1}$$

$$sen = \frac{TP}{TP + FN} \tag{2}$$

$$ppv = \frac{TP}{TP + FP} \tag{3}$$

$$mcc = \frac{TP \times TN - FP \times FN}{\sqrt{(TP + FP)(TP + FN)(TN + FP)(TN + FN)}} \tag{4}$$

Where TP (true positives) indicates the number of correctly predicted paired bases; FN (false negatives) indicates the number of incorrectly predicted unpaired bases; FP (false positives) indicates the number of incorrectly predicted paired bases; TN (true negatives) indicates the number of correctly predicted unpaired bases. The higher the value of the four indicators, the better the prediction effect of the model.

## 3    Decision Tree Algorithm

J.R. Quinlan proposed ID3 decision tree algorithm in 1986, this algorithm use top-down greedy algorithm to construct a tree structure. Each internal node selects an optimal attribute to split, each branch corresponds to an attribute value, and each leaf node represents the category of samples along this path, thus recursively building trees until the termination condition is met. DT do not require prior knowledge and are easier to understand than neural networks. In this experiment, information entropy is used as a measure to construct a tree with the fastest entropy decrease, and the entropy at the leaf node is zero. At this time, the instances in each leaf node belong to one class.

## 3.1   Information Gain

Entropy of random variable x is defined as

$$H(X) = -\sum_{i=1}^{n} p_i log p_i \tag{5}$$

In Eq. (5), P(X) represents the probability of occurrence of event X. The greater the entropy, the greater the uncertainty of random variables.

The information gain indicates the degree to which the information uncertainty of class Y is reduced when the information of X is known. It is defined that the information gain g(D, A) of the feature a to the training set D is the difference between the entropy H(D) of the set D and the conditional entropy H(D|A) of D under the given condition of the feature A, i.e. (see in Table 1).

$$g(D, A) = H(D) - H(D|A) \tag{6}$$

**Table 1.** Algorithm of Information Gain

| Algorithm 1 : Algorithm of Information Gain |
| --- |
| 1   input : Training set D, feature A |
| 2   Calculate the entropy H(D) of data set D:<br><br>$H(D) = -\sum_{k=1}^{K} \frac{|C_k|}{|D|} \log_2 \frac{|C_k|}{|D|}$ |
| 3   Calculate the conditional entropy H(D\|A) of feature A on training data set D:<br><br>$H(D\|A) = \sum_{i=1}^{n} \frac{|D_i|}{|D|} \sum_{k=1}^{K} \frac{|D_i|}{|D_i|} \log_2 \frac{|D_{ik}|}{|D_i|}$ |
| 4   Calculate the information gain:<br><br>g(D,A)=H(D)-H(D\|A) |
| 5   output : Feature A's information gain g(D, A) for training data set D |

Among them, |D| is the sample size, with K categories, k = 1, 2, ..., K, $|C_k|$ is the number of samples belonging to the class. Let feature A have n different values, and according to feature A have n different values, D is divided into n subsets $D_1, D_2, \ldots, D_n$, $D_{ik}$ is a set belonging to class $C_k$ in subset $D_i$.

## 3.2 Decision Tree Algorithm

The DT decides which attribute to choose to build a tree according to the value of the maximum information gain each time. The description of the decision tree algorithm is shown in Table 2.

**Table 2.** DT algorithm

| Algorithm 2: DT algorithm |
| --- |

| | |
| --- | --- |
| 1 | Set D as the training sample set, Target_attribute as the predicted target Attribute of this tree, and Attribute as the list of attributes other than target attributes for learning |
| 2 | ID3(D, Target_attribute, Attribute) |
| 3 | Create the Root node of the tree |
| 4 | If both D are positive, then return label =+ single-node tree Root |
| | If both D are negative, then the single-node tree Root with label =- is returned |
| | If D is empty, return the root node tree Root, the maximum Target_attribute value in label=D |
| | Else start |
| | A<- Attribute with the largest information gain of classified D |
| | Root's decision attribute <-A |
| 5 | For each possible value v of a, let it be a subset of v that satisfies the property of a in D |
| 6 | If $D_v$ is empty, then add a leaf node to the new branch |
| 7 | The most common Target_attribute value in node label=D |
| 8 | Else, add A subtree ID3($D_v$, Target_attribute, Attribute-{A}) to the new branch. |
| 9 | End, and return Root |

# 4 RNA Sequence Modeling Based on Decision Tree

## 4.1 Feature Extraction and Analysis

Decision tree algorithm is a typical multi-classification problem. Similar to neural network, the primary sequence of RNA can be regarded as information related to the secondary structure, while the secondary structure is related to the long-range and short-range information of bases. Therefore, when we predict the pairing of a certain base, we need to input several neighboring bases around the base as the characteristics of the base to be predicted. In this article, we call such an input sample a window.

Table 3 shows the experimental results of sen, ppv, mcc and acc indexes of decision tree algorithm under different window values of data set RFA. In order to obtain better experimental results and reduce the unpredictable number of bases at the beginning and end, the window with the best index should be selected as the size of the experimental input window when the number of windows is the smallest. Many experiments show that when the window length is 15, the prediction effect is better. That is, when it is necessary to predict the pairing of a certain base, we will input a sample with a total of 15 bases of 7 adjacent bases around a certain base. The input of this window is represented by a 0-1 code of $15 \times 4b = 60$, which means that each bit in the window represents a base and a 4-bit orthogonal 0-1 code represents one of the base types (ACGU). The corresponding relationship between each base character and binary code is: A-1000; C-0100; G-0010; U-0001; others -0000.

**Table 3.** Index results under different window sizes

| Size | sen | ppv | mcc | acc |
|---|---|---|---|---|
| 3 | 0.197 | 0.409 | −0.03 | 0.524 |
| 5 | 0.505 | 0.523 | 0.168 | 0.596 |
| 7 | 0.718 | 0.699 | 0.469 | 0.737 |
| 9 | 0.760 | 0.722 | 0.523 | 0.763 |
| 11 | 0.778 | 0.740 | 0.555 | 0.779 |
| 13 | 0.770 | 0.723 | 0.542 | 0.773 |
| 15 | 0.779 | 0.744 | 0.565 | 0.784 |
| 17 | 0.777 | 0.737 | 0.561 | 0.782 |
| 19 | 0.804 | 0.734 | 0.582 | 0.792 |
| 21 | 1 | 0.863 | 0 | 0.863 |

The output of the window is one of the categories to which the base in the middle position belongs in its molecular secondary structure, i.e. E-NSSEL tag 5.

## 4.2   Decision Tree Training Algorithm

Algorithm input: the input of the algorithm is a four-dimensional vector obtained by transforming the i-th base of RNA primary sequence and a four-dimensional vector obtained by transforming the front and back 7 bases of the base, i.e. a $1 \times 60$ vector for each input sequence. The format is as follows.

```
[0  0  1  0  0  1  0  0  0  1  0  0  0  1  0
 0  0  0  1  0  0  0  1  0  0  0  0  1  0  0
 0  0  0  0  0  1  0  0  1  0  0  1  0  0  1
 0  0  0  0  1  0  0  0  0  0  1  0  0  1  0]
```

Algorithm output: the output Y of the model is a two-dimensional array, and Y[i, j] represents the predicted category of the i-th and j-th bases in the sequence. if it is 0, it means no pairing, otherwise, it means the i-th and j-th bases are paired.

Because of the large number of characteristic attributes of training data, decision tree algorithm is easy to fall into over-fitting. Therefore, this paper uses the method of limiting the maximum leaf node number to prevent the model from falling into the state of over-fitting. Through a large number of experiments and comparisons, it is found that when the maximum number of large leaf nodes is limited to 3000, a better prediction effect can be obtained and the model can be prevented from falling into an over-fitting state.

## 5  Experiment and Analysis

### 5.1  Dataset

The data set used in this paper is from the authoritative data set RNA STRAND, from which five subsets are selected: TMR (The tmRNA website), SPR (Sprinzl tRNA Database), SRP (Signal recognition particle database), RFA (The RNA family database) and ASE (RNase P Database). The five data sets in Table 4.

**Table 4.** Dataset

| Datasets | TMR | SPR | SRP | RFA | ASE |
| --- | --- | --- | --- | --- | --- |
| Total number | 721 | 622 | 383 | 313 | 454 |
| Pseudoknots | 713 | 0 | 0 | 29 | 416 |
| Average length | 361.1 | 77.3 | 224.7 | 118.9 | 332.6 |
| Max length | 463 | 93 | 533 | 553 | 486 |
| Min length | 102 | 54 | 66 | 40 | 189 |

Where Total number is the total number of sequences in the dataset, Pseudoknots is the number of sequences with false knots, Average length is the average length of the sequences, Max length is the maximum sequence length, and Min length is the length of the minimum sequence.

### 5.2  Analysis of Experimental Results

Table 5 shows the results of secondary structure prediction of RNA sequences with pseudoknots by four methods: LR, SVM, RF and DT.

According to the experimental results in Table 3, both decision tree and random forest algorithm can achieve good prediction effect in the training process, but from the perspective of prediction, decision tree method is obviously superior to other three classification algorithms. This result can be attributed to the decision tree being good at dealing with non-linear relations and being less susceptible to extreme data. Therefore, it can be concluded that when RNA secondary structure is predicted in samples with

**Table 5.** Comparison of experimental results

| Dataset | Method | sen | ppv | mcc | acc |
|---------|--------|-------|-------|--------|-------|
| SPR | LR | 0.621 | 0.635 | 0.357 | 0.683 |
|     | SVM | 0.786 | 0.839 | 0.657 | 0.829 |
|     | RF | 0.589 | 0.851 | 0.529 | 0.758 |
|     | DT | 0.832 | 0.776 | 0.628 | 0.814 |
| TMR | LR | 0.244 | 0.418 | 0.010 | 0.552 |
|     | SVM | 0.502 | 0.689 | 0.348 | 0.683 |
|     | RF | 0.268 | 0.747 | 0.253 | 0.618 |
|     | DT | 0.656 | 0.672 | 0.425 | 0.720 |
| SRP | LR | 0.526 | 0.444 | −0.045 | 0.474 |
|     | SVM | 0.682 | 0.566 | 0.167 | 0.581 |
|     | RF | 0.662 | 0.598 | 0.193 | 0.597 |
|     | DT | 0.775 | 0.633 | 0.301 | 0.650 |
| RFA | LR | 0.128 | 0.426 | −0.014 | 0.536 |
|     | SVM | 0.151 | 0.748 | 0.182 | 0.581 |
|     | RF | 0.329 | 0.847 | 0.360 | 0.660 |
|     | DT | 0.812 | 0.759 | 0.604 | 0.803 |
| ASE | LR | 0.512 | 0.450 | 0.044 | 0.524 |
|     | SVM | 0.707 | 0.657 | 0.355 | 0.677 |
|     | RF | 0.633 | 0.634 | 0.290 | 0.645 |
|     | DT | 0.804 | 0.693 | 0.457 | 0.725 |

false knots, the prediction accuracy of decision tree algorithm is higher than that of other three classifiers. Compared with traditional models based on minimum free energy, decision tree successfully solves the problem of time complexity, and the input method using sliding window is not affected by RNA sequence length.

## 6   Conclusion

In this paper, the decision tree algorithm is used to predict the secondary structure of RNA with pseudoknots. The experimental results show that:

(1) Using decision tree algorithm to predict RNA secondary structure is more accurate than SVM, random forest and logistic regression algorithm.
(2) The algorithm can obtain ideal prediction accuracy, and can effectively solve the computational complexity and long-chain molecule prediction problems existing in traditional algorithms.

The decision tree model is used to predict RNA secondary structure, which gives full play to the advantages of decision tree algorithm and combines the characteristics of RNA secondary structure to successfully realize RNA secondary structure prediction including false knots. Since the first 7 bases and the last 7 bases of RNA molecular chain cannot be predicted for the time being when processing the original data, the author will continue to conduct in-depth research to further solve this problem. The direction of further improvement in this paper is to use relevant technical means to balance the proportion of samples or to combine other algorithms such as depth learning algorithm to further improve the accuracy of data.

**Acknowledgements.** This paper is supported by the National Natural Science Foundation of China (61772357, 61502329, 61672371, and 61876217), Jiangsu Province 333 Talent Project, Top Talent Project (DZXX-010), Suzhou Foresight Research Project (SYG201704, SNG201610, and SZS201609).

# References

1. Rivas, E., Eddy, S.R.: A dynamic programming algorithm for RNA structure prediction including pseudoknots. J. Mol. Biol. **285**(5), 1–2068 (1999)
2. Rodland, A.E.: Pseudoknots in RNA secondary structures: representation, enumeration, and prevalence. J. Comput. Biol. **13**(6), 1197–1213 (2006)
3. Reeder, J., Giegerich, R.: Design, implementation and evaluation of a practical pseudoknot folding algorithm based on thermodynamics. BMC Bioinformatics **5**(1), 104 (2004)
4. Ieong, S., Kao, M.Y., Lam, T.W., et al.: Predicting RNA secondary structures with arbitrary pseudoknots by maximizing the number of stacking Pairs. J. Comput. Biol. **10**(6), 981–995 (2003)
5. Ren, J., Rastegari, B., Hoos, H.H.: HotKnots: heuristic prediction of RNA secondary structures including pseudoknots. Rna-a Publ. Rna Soc. **11**(10), 1494–1504 (2005)
6. Yu, N., Zhao, W., et al.: Evaluation of RNA secondary structure prediction for both base-pairing and topology. Biophysics **4**(3), 123–132 (2018). English edition
7. Liu, Y., Zhao, Q., Zhang, H., et al.: A New method to predict RNA secondary structure based on RNA folding simulation. IEEE/ACM Trans. Comput. Biol. Bioinf. **13**(5), 990–995 (2016)
8. Madera, M., Calmus, R., Thiltgen, G., et al.: Improving protein secondary structure prediction using a simple k-mer model. Bioinformatics **26**(5), 596–602 (2010)
9. Yonemoto, H., Asai, K., Hamada, M.: A semi-supervised learning approach for RNA secondary structure prediction. Comput. Biol. Chem. **57**, 72–79 (2015)
10. Liu, Z., Zhu, D.: New algorithm for predicting RNA secondary structure with pseudoknots. Mater. Sci. Inform. Technol. II **2**, 1796–1799 (2012)
11. Sprinzl, M., Horn, C., Brown, M., et al.: Compilation of tRNA sequences and sequences of tRNA genes. Nucleic Acids Res. **26**(1), 148–153 (1998)
12. Zhang, X., Deng, Z., Song, D.: Neural network approach to predict RNA secondary structures. J. Tsinghua Univ. **46**(10), 1793–1796 (2006)
13. Yang, R., Wu, H., Fu, Q., et al.: Optimizing HP model using reinforcement learning (2018)
14. Chen, C., Wu, H., Bian, K.: β-barrel transmembrane protein predicting using support vector machine. In: Huang, D.-S., Hussain, A., Han, K., Gromiha, M.M. (eds.) ICIC 2017. LNCS (LNAI), vol. 10363, pp. 360–368. Springer, Cham (2017). https://doi.org/10.1007/978-3-319-63315-2_31
15. Wu, H., Li, H., Jiang, M., et al.: Identify high-quality protein structural models by enhanced K-Means. Biomed. Res. Int. **2017**(18), 1–9 (2017)

# Improving Hot Region Prediction by Combining Gaussian Naive Bayes and DBSCAN

Jing Hu[1,2], Longwei Zhou[1,2(✉)], Xiaolong Zhang[1,2],
and Nansheng Chen[3]

[1] School of Computer Science and Technology,
Wuhan University of Science and Technology, Wuhan, China
895231916@qq.com
[2] Hubei Province Key Laboratory of Intelligent Information Processing
and Real-Time Industrial System, Wuhan 430065, Hubei, China
[3] Molecular Biology and Biochemistry, Simon Fraser University,
Burnaby, BC, Canada

**Abstract.** It's crucial to recognize hot regions from protein-protein interactions in the process of designing drug and protein. Every hot region consists of at least three hot spots, which contribute significantly to binding free energy in the protein-protein interaction. However, biological experiments take much time and effort; therefore, we need to train better predictive models to find hot regions based on machine learning methods.

We construct several models to predict hot spot residues with several machine learning methods. The result shows that these methods have close performance in the evaluation of F-measure, however the other evaluations of recall and precision have differences. Due to the highest recall, the result of gaussian naïve Bayes can predict more hot spot residues than other methods. Furthermore, according to the attribute of hot region packed tightly, we use cluster algorithm to exclude the residue which is mistaken as hot spot residue. The result demonstrates that the method combining gaussian naïve Bayes and DBSCAN can predict most of the standard hot regions correctly, and the evaluation of F-measure can reach at 0.809.

**Keywords:** Hot region · Protein-protein interactions · Hot spot · DBSCAN · Gaussian Naïve Bayes

## 1 Introduction

Protein-protein interactions (PPIs) is associated with the most fundamental cellular functions about signal transduction, cellular motion and hormone-receptor interactions [1]. The basis of two parts of proteins' combination is from the affinity of them, but some residues in the protein-protein interaction interface provide most of binding free energy, thus, this kind of residues are called as hot spots [2, 3]. Furthermore, hot spot residues are usually crowded tightly in a small region in protein-protein interaction interface called hot region [4]. It's essential to find hot regions for drug and protein design [5, 6].

© Springer Nature Switzerland AG 2019
D.-S. Huang et al. (Eds.): ICIC 2019, LNCS 11644, pp. 440–452, 2019.
https://doi.org/10.1007/978-3-030-26969-2_42

Hot regions are consisted of at least three favorable hot spot residues in protein-protein interaction interface. Xia identifies potential hot spot residues by constructing sequence, structure, and neighborhood features. Xia extracted features by maximum relevance and minimum redundancy and used support vector machine (SVM) to establish prediction method [7]. An empirical model was constructed to predict hot spot residues combining solvent accessibility and inter-residue potentials to improve the accuracy [8]. Huang proposed an improved ensemble method with SMOTE to deal with the imbalance data for a better performance based on structural features [9]. Hu presented a prediction method from sequence features only by a new ensemble learning method [10].

It's favorable to construct a better model to predict hot spot residues. Cukuroglu studied hot regions by analyzing hot spots and the rule of hot region formation, and then built a database called Hot Region [11]. Pons proposed a network-based method and used small-world residue networks to predict protein-binding areas [12]. An approach based on complex network and community detection was proposed to predict hot regions by Nan with a retrieving strategy which exploited false positive (FP) and false negative (FN) residues [13]. Lin used a new clustering algorithm named Local community structure detecting (LCSD) to detect and analyze the conformation of hot regions with an improved maximum relevance minimum redundancy algorithm to improve the feature selection of hot spot residues [14].

Hot spot residues are the basic of Hot regions, so we need to distinguish interface residues correctly as much as possible in order to find hot regions. Based on the features of structural, sequential features and the interaction of two amino acids, we construct several machine learning models to predict hot spot residues. Recently SKEMPI 2.0 was presented, it contained about two times alanine mutations data than SKEMPI 1.0 [15, 16]. Limited to the size of dataset that amino acids are mutated into alanine and the imbalance of the hot spot residues and nonhot spot residues, the accuracy of prediction of hot spot residues and hot regions is not very high. According to the result predicted by machine learning methods, we used DBSCAN algorithm to cluster and obtain hot regions [17]. The experiment shows that the classification algorithm with higher recall and the cluster algorithm improving precision can predict hot regions effectively.

## 2   Methods

In our method to predict hot regions, firstly, construct the standard hot regions from the SKEMPI 2.0 with hot spot residues; then we use DBSCAN algorithm to calculate the predicted hot regions from our predicted hot spot residues with gaussian naïve Bayes; eventually, we can fetch the prediction accuracy compared the predicted hot regions to the standard hot regions.

### 2.1   Binding Free Energy Changes

We collect those data having single amino acid mutated into alanine from SKEMPI 2.0 (https://life.bsc.es/pid/skempi2/) which contains affinities of wild type complexes and affinities of mutated complexes measured by biological experiment from scientific literatures. Because binding free energy changes of multiple mutated residues are not accumulated based on single mutated residue, we can't deduce more samples from multiple mutated samples. Some samples whose affinities are not measured will be removed, too. According to the binding affinity ($K_d$) measured by biological experiment like Isothermal titration calorimetry, Surface plasmon resonance etc. [18, 19], binding free energy is calculated from Eq. (1). R is the gas constant 8.314/4184 kcal/(K*mol) and T is the experimental temperature that is in the range of 273 K to 323 K. Based on the calculated $\Delta G_{mut}$ and $\Delta G_{wt}$, binding free energy changes can be calculated from Eq. (2).

$$\Delta G = - RTln(Kd) \tag{1}$$

$$\Delta\Delta G = \Delta G_{mut} - \Delta G_{wt} \tag{2}$$

## 2.2    Feature Selection

For this paper, all protein complexes about structural information are from Protein Data Bank [20]. We extracted features including solvent accessible surface area, relative accessible surface area, protrusion index, depth index and the count of binding sites with aimed amino acid from PSAIA [21], conservation score from ConSurf server [22], hydrophobic index, the attribute of amino acid side chain and the interaction number of two amino acids. The detailed features are showed in Appendix 1 (https://github.com/nsiakjdw/Paper-appendix.git). As the appendix 1 shows, the attribution of amino acid side chain is discrete variable, we encode the feature with one-hot. Finally, we obtain 83 features. As we know, not each feature is contributed to construct a better model, we acquire the optimal feature subset with maximum relevant minimum redundancy (mRMR) algorithm [23], the mutual information I(x,y) is labeled as:

$$I(\text{x,y}) = \iint P(x,y) \log \frac{P(x,y)}{p(x)p(y)} dxdy \tag{3}$$

the maximum correlation criterion and the minimum redundancy criterion are defined as:

$$\max D(F,c), D = \frac{1}{F} \sum\nolimits_{X_i \in F} I(X_i, c) \tag{4}$$

$$\min R(F), R = \frac{1}{F^2} \sum\nolimits_{Xi, Xj \in F} I(Xi, Xj) \tag{5}$$

According to the training set, a feature list is produced. To discover the highest F-score combination, we apply incremental feature selection and get a rank of all features in descending. Every time we combine the feature ranked the top with its next one to obtain the F-measure in machine learning models, then, select the set of features with the best F-measure as the result.

## 2.3    Prediction of Hot Regions

There are non-hot spot residues and hot spot residues in the dataset. However, we need to detect more hot spots as much as possible in the hot regions which only contains hot spot residues.

### 2.3.1  Naïve Bayes Classifier

We construct a gaussian naïve Bayes classifier given a set of training examples with class labels, and then use the model to distinguish non-hot spot residues and hot spot residues [24–26]. One example is a tuple of features $(x_1, x_2, \ldots, x_n)$ of one sample $(x_i)$ and the class label c of the sample, so X is all samples and C is the classification variable. In our experiment, we assume that there are two classes: c = 0 (non-hot spot residue) and c=1 (hot spot residue). According to Bayes Rule, the probability of one sample $E = (x_1, x_2, \ldots, x_n)$, being class c of sample E is:

$$p(c|E) = \frac{p(E|c)p(c)}{p(E)} \tag{6}$$

Sample E is classified as class 1 if and only if $f_b(E)$ is more than 1, otherwise, sample E will be classified as class 0:

$$f_b(E) = \frac{p(C = 1|E)}{p(C = 0|E)} \geq 1 \tag{7}$$

Where $f_b(E)$ is called a Bayesian classifier.

Assume that all features are independent given the value of the class variable; that is,

$$p(E|c) = p(x_1, x_2, \ldots, x_n|c) = \prod_{i=1}^{n} p(x_i|c) \tag{8}$$

The resulting classifier is then:

$$f_{nb}(E) = \frac{p(C = 1)}{p(C = 0)} \prod_{i=1}^{n} \frac{p(x_i|C = 1)}{p(x_i|C = 0)} \tag{9}$$

The function $f_{nb}(E)$ is called a naïve Bayesian classifier, or simply naïve Bayes (NB). When we use the gaussian distribution to calculate the $p(x_i|C)$, the classifier is gaussian naïve Bayes. Given probability distribution is under gaussian distribution, the function is:

$$g(x_i, \mu, \sigma) = \frac{1}{\sqrt{2\pi}\sigma} e^{-\frac{(x_i-\mu)}{2\sigma^2}} \tag{10}$$

### 2.3.2  Cluster

Clustering is a process to group data into multiple sub-groups or clusters so that objects within a cluster may have strong similarities. The distribution of hot spot residues in hot region is flowing structure. It's the reason that we select density-based spatial clustering of applications with noise (DBSCAN). DBSCAN algorithm is suitable to cluster this kind of data. There are two hyper parameters "Min" and "ε" to be measured in this algorithm. "Min" represents the density of residue measured by the number of residues of it, "ε" represents the radii of residue O as the center of a circle.

For a dataset D composed of residues, Given the parameters "Min" and "ε", the residue with more than or equal to "Min" will be regarded as core residue in its ε-neighborhood. After checking all residues, the core residues and their ε-neighborhood residues will make up the dense regions, which are the clusters we need.

The detailed process about clustering is: all residues should be defined as "unvisited" in the first stage. Then, we need to select randomly one residue p as the center of circle and calculate the number of residues in the neighborhood of it to distinguish whether the residue is core residue or not. If it's core residue, label the residue as "visited" and select the neighborhood residues as the next detected object. If existing core residues in them, the process will not continue until the cluster C can't extend. Then return to the beginning, select randomly one residue p in the remaining residues labeled as "unvisited" as the center of circle, repeat the process until all residues are "visited". Eventually, we will obtain several clusters and noises.

---

Algorithm Procedure

---

Calculate binding free energy changes in SKEMPI 2.0
Collect features of every amino acid
Obtain feature rank from MRMR algorithm
Split training set and construct classifier with gaussian naïve Bayes
Predict samples in testing set and obtain the predicted hot spot residues as the data for DBSCAN.
Input:
the coordinates of all amino acid D and radii parameter ε and density threshold MinPts.
Method:
Label all residues as unvisited.
Do:
    Select one unvisited object p and label it as visited.
    If p has at least MinPts ε-neighborhood residues:
        Create a new cluster C and add p into C, create another set N to save ε-neighborhood residues of p.
      For p′ in N:
        If p′ is unvisited:
          Label p′ as visited.
          If p′ has at least MinPts ε-neighborhood residues:
            Add MinPts ε-neighborhood residues of p′ into N.
          If p′ is not in any C:
            Add p′ in C.
      End for.
      Output C.
    Else:
      Label p as noise.
Until all residues are labeled as visited.

# 3  Experiment Results and Analysis

## 3.1  Dataset

The dataset we use in this paper are from the latest database SKEMPI 2.0 (Structural database of Kinetics and Energetics of Mutant Protein Interactions). The SKEMPI 2.0 database contains data on the changes in thermodynamic parameters and kinetic rate constants upon mutation, for protein-protein interactions for which a structure of the complex has been solved and is available in the Protein Data Bank (PDB). We select alanine mutation residues from SKEMPI 2.0, and then replace the repetitive items with the average value of their different binding free energy values. After that, we had 180 complexes, each complex is composed of a bunch of interface residues, these interface residues are the residues that the decrement of the accessible surface area are more than 1 Å during the process of forming the complex. The energy values binding for each residue were obtained by alanine mutation experiments.

In the different kinds of biological experiments about same complexes, we choose the average of $\Delta\Delta G$ as the final result. Most of the latest work about hot spots prediction adopted one strategy which discarded the interface residues in the range of 0.4 to 2, then the interfaces residue whose binding free energy changes were more than 2 kcal/mol were labeled as hot spots and those whose binding free energy changes were less than 0.4 kcal/mol were labeled as non-hot spots. Here, we choose 1.0 kcal/mol as the threshold to split hot spot residues and non-hot spot residues for using the whole dataset effectively. The definition of 2.0 kcal/mol will make dataset extremely imbalanced and the evaluation of recall precision decrease sharply.

Therefore, we assume that an interface residue will be defined as a hot spot if its corresponding binding free energy changes is higher than 1.0 kcal/mol, while an interface residue with binding free energy changes less than or equal to 1.0 kcal/mol is considered as a non-hot spot residue. Finally, we got 2326 interface residues in 180 complexes as our dataset, there are 813 hot spot residues and 1513 non-hot spot residues in the dataset. The detailed distribution of amino acids in the dataset is in Table 1. The result in Table 1 shows that amino acids with aromatic side chain are more possible to be hot spot residue and GLU, LYS, ARG and TYR are enriched in the hot spot residues. Otherwise, ARG, GLU, TYR and SER are easier to occur in the interface residues.

Over two fifth of the data in SKEMPI 2.0 is from SKEMPI 1.0 and a great quantity of researches have been done in SKEMPI 1.0. Then our work is to predict hot regions in the expanded part of SKEMPI 2.0. The complex whose number of experimental residues is less than 3 will be put in the training set to improve the stability of the model. The left part of data in the expanded part of SKEMPI 2.0 is regarded as the testing set, the remained data of SKEMPI 2.0 is labeled as the training set. The specific distribution of hot spots and non-hot spots in the training set and testing set is in Table 2.

**Table 1.** Distribution of data from SKEMPI 2.0

| Amino acid | Non-hot spots | Hot spots | All residues | Ratio of hot spots | Property of side chain |
|---|---|---|---|---|---|
| SER | 143 | 21 | 164 | 0.128 | Hydroxyl-containing |
| CYS | 6 | 1 | 7 | 0.143 | Sulfur-containing |
| GLN | 107 | 29 | 136 | 0.213 | Amid |
| THR | 114 | 31 | 144 | 0.214 | Hydroxyl-containing |
| PRO | 42 | 17 | 59 | 0.288 | Cyclic |
| ASN | 101 | 41 | 142 | 0.289 | Amid |
| GLY | 48 | 20 | 68 | 0.294 | Aliphatic |
| VAL | 70 | 30 | 100 | 0.3 | Aliphatic |
| GLU | 154 | 68 | 222 | 0.306 | Acid |
| HIS | 60 | 28 | 88 | 0.318 | Basic aromatic |
| MET | 23 | 11 | 34 | 0.326 | Sulfur-containing |
| LYS | 131 | 64 | 195 | 0.328 | Basic |
| ARG | 146 | 80 | 226 | 0.354 | Basic |
| ASP | 101 | 41 | 142 | 0.409 | Acid |
| LEU | 48 | 41 | 89 | 0.419 | Aliphatic |
| ILE | 72 | 52 | 124 | 0.461 | Aliphatic |
| PHE | 51 | 55 | 106 | 0.519 | Aromatic |
| TYR | 75 | 104 | 179 | 0.581 | Aromatic |
| TRP | 21 | 50 | 71 | 0.704 | Aromatic |
| All | 1513 | 813 | 2623 | 0.35 | None |

**Table 2.** Training set and testing set

| Dataset | Non-hot spots | Hot spots | All residues | Complexes |
|---|---|---|---|---|
| Training set | 864 | 390 | 1254 | 101 |
| Testing set | 649 | 423 | 1072 | 79 |

## 3.2  Evaluation

In order to evaluate the performance of the predictive models, three criteria are used for predicting both hot spots and hot regions:

$$Recall = TP/(TP + FN) \tag{11}$$

$$Precision = TP/(TP + FP) \tag{12}$$

$$F\text{-measure} = 2 * Recall * Precision(Recall + Precision) \tag{13}$$

When predicting hot spots in hot regions, the following notations are used:

True Positive (TP): The number of hot spots in predicted hot regions and also in standard hot regions;

False Negative (FN): The number of hot spots that are not in predicted hot regions but in standard hot regions;

False Positive (FP): The number of hot spots in predicted hot regions but not in standard hot regions;

Precision represents the accuracy of the hot spot prediction, and Recall represents the coverage of predicted hot spots in standard hot regions. With a good balance between Precision and Recall, the F-measure offers a better overall accuracy of hot spot prediction.

However, for prediction of hot regions, the above notations assume different meanings, as follows:

True Positive (TP): The number of hot regions in predicted hot regions and also in standard hot regions;

False Negative (FN): The number of hot regions that are not in predicted hot regions but in standard hot regions;

False Positive (FP): The number of hot regions in predicted hot regions but not in standard hot regions;

Similarly, Precision represents the accuracy of the hot region prediction, and Recall represents the coverage of predicted hot regions in standard hot regions. With a good balance between Precision and Recall, the F-measure offers a better overall accuracy in predicting hot regions than solely using either Precision or Recall.

## 3.3 Experimental Results

In this section, we need to predict hot spot residues correctly as much as possible. The feature rank calculated with mRMR algorithm is suppled in Appendix 1 (https://github.com/nsiakjdw/Paper-appendix.git) for selecting optimal feature set. In this paper, we adopted four widely used machine learning methods including Xgboost, random forest, support vector machine (SVM) and artificial neural network to compare with gaussian naïve Bayes [27–30]. Because of the automatic feature fusion of neural networks, mMMR algorithm was not used in artificial neural networks method. In the artificial neural networks, we construct networks containing five layers with activation function rectified linear units (ReLu) [31].

The result of hot spot residues prediction with different methods is presented in Table 3. Xgboost algorithm has highest accuracy and precision and GNB has highest recall and F-measure. The data demonstrates that GNB can predict correctly more hot spot residues and Xgboost can predict correctly more non-hot spot residues. However, non-hot spot residues are more than hot spot residues in the training set, which cause that accuracy of Xgboost is greater than that of GNB. Thus, more predicted hot spot residues are kept in the result so that we can obtain more hot regions in clustering process.

**Table 3.** Comparison results with different methods to predict hot spots

| Methods | Hot spot | | | |
|---------|----------|--------|-----------|-----------|
|         | Accuracy | Recall | Precision | F-measure |
| GNB     | 0.674    | **0.792** | 0.561   | **0.657** |
| SVM     | 0.696    | 0.6    | 0.618     | 0.609     |
| Xgboost | **0.726** | 0.596 | **0.672** | 0.632    |
| RF      | 0.693    | 0.596  | 0.615     | 0.605     |
| ANN     | 0.715    | 0.679  | 0.632     | 0.655     |

In the clustering section, we use DBSCAN algorithm to cluster the predicted hot spot residues. The results are showed in Table 4. The highest F-measure as the most important evaluation are selected and the parameters "Min" and "ε" are decided by grid search. According to the result of GNB and DBSCAN algorithm can outperform that of other methods greatly. From Fig. 1, GNB algorithm barely have more hot spot residues of single hot region than other methods in 47 standard hot regions. Because the composition of hot region needs enough hot spot residues, if the result predicted as hot spot residues is too less to form hot region, the evaluation of recall will be low. In addition, due to the lack of true positive residues, some hot regions are wrong to be clustered when we have to form true positive hot regions with relative unconstrained parameters.

**Table 4.** Comparison results with different methods to predict hot regions

| Methods | Hot region | | |
|---------|--------|-----------|-----------|
|         | Recall | Precision | F-measure |
| GNB     | **0.766** | **0.923** | **0.809** |
| SVM     | 0.617  | 0.725     | 0.667     |
| Xgboost | 0.574  | 0.6       | 0.587     |
| RF      | 0.617  | 0.644     | 0.63      |
| ANN     | 0.596  | 0.538     | 0.567     |

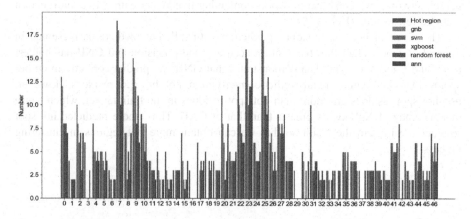

**Fig. 1.** Distribution of hot spots in hot regions. The x-axis corresponding 47 standard hot regions in appendix 2. The bar shows the number of hot spots in the 47 standard hot region and the number of true positive hot spots in the predicted hot regions.

## 3.4 Standard Hot Regions and Predicted Hot Regions

In this paper, we adopted the standard definition of hot regions from Keskin [4]. A hot region is defined as follows: every hot region contains at least three hot spot residues, and each hot spot residue is assumed to be within a hot region if it has at least two hot spot neighbors, and each hot spot residue is assumed to be a perfect sphere with a specific volume. The Cα-atoms of the hot spot residues are the centers of these spheres. The radii of the spheres are extracted from their sphere volumes which had been measured [32]. If the distance between the centers of two spheres (two Cα-atoms of two hot spots) is less than the sum of the radii of the two spheres plus a tolerance distance (2 Å), the two hot spot residues are flagged to be clustered and to form a network in the hot region. The prediction accuracy of each predictive models with different amino acid mutation will be compared to the standard hot regions. Finally, 47 standard hot regions are detected. The detailed result are in Appendix 2 (https://github.com/nsiakjdw/Paper-appendix.git).

In the process of setting parameters "Min" and "ε", hot region is consisted of at least three hot spot residues and the biggest distance between two contacted amino acid is 9.5 Å, so we set that Min is greater than 3 and ε is less than 9.5 Å. Based on the results of all methods except GNB clustered, DBSCAN can get a best F-measure when Min is 9.5 Å and ε is 3. However, the parameters Min and ε are 8.5 Å and 4, it's the reason that GNB regards more non-hot spot residues as hot spot residues, the cluster algorithm will perform well when improve the factors of forming hot regions for recognizing non-hot spot residues as noises.

## 3.5 Comparison of Prediction Results Visualization

We assume that hot region is predicted correctly only when 60% of hot spot residues in the standard hot regions occur in the predicted hot region. Hence, only the result of predicted with GNB are clustered to form a hot region regarded as true positive hot region, and the result of other machine learning methods not only the evaluation of recall will be lower and the increasement of mistaken hot regions will cause the evaluation of precision decreasing. In the Fig. 2, we showed the protein complex 3HQY with all hot spot residues and predicted hot spot residues with PyMol [33] software. In the protein complex 3QHY, there are 16 hot spot residues in the standard hot region. GNB algorithm can predict correctly 15 true positive hot spot residues in the standard hot region and only two non-hot spot residues come out in the predicted hot region. The results of other methods except ANN have higher accuracy in non-hot spot residues, but more hot spot residues do not come out in the predicted hot region. ANN can predict 14 hot spot residues and only regard one non-hot spot residue as hot spot residue.

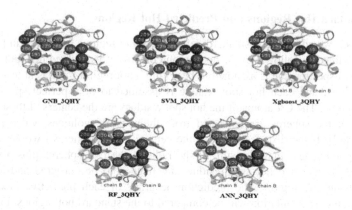

**Fig. 2.** Visual prediction results of 2NYY from different methods with PyMol. All residues are in the chain A. Red spheres are predicted correctly in the standard hot region, Blue spheres are not predicted in the standard hot region, yellow spheres are mistaken as hot spot residues (Color figure online)

## 4   Conclusion

In this paper, we collect alanine mutations from the latest presented SKEMPI 2.0 database. In this dataset, aromatic amino acids are more likely to be hot spot residues When the threshold of hot spot residue is 1 kcal/mol. Furthermore, 70.4% of TRP are hot spot residues. In the first stage, we use mRMR algorithm to calculate the importance of every feature based on the mutual information and RctmPI is the most important feature. In the next stage of predicting hot spot residues, the performance of all methods about F-measure are closed, but GNB has a best performance about recall so that hot regions can be made up of enough true positive hot spot residues. In the final stage, DBSCAN algorithm is selected to cluster the data for forming hot regions. DBSCAN can obtain a better performance with higher penalty for reducing mistaken residues.

**Acknowledgement.** This work is supported by the National Natural Science Foundation of China (No. 61702385), Science and Technology Research Plan of Hubei Ministry of Education (Q20171105) and China Scholarship Council. Conflict of Interest: none declared.

## References

1. Chothia, C., Janin, J.: Principles of protein-protein recognition. Nature **256**(5520), 705–708 (1975)
2. Clackson, T., Wells, J.: A hot spot of binding energy in a hormone-receptor interface. Science **267**(5196), 383–386 (1995)
3. Bogan, A.A., Thorn, K.S.: Anatomy of hot spots in protein interfaces. J. Mol. Biol. **280**(1), 1–9 (1998)

4. Keskin, O., Ma, B., Nussinov, R.: Hot regions in protein–protein interactions: the organization and contribution of structurally conserved hot spot residues. J. Mol. Biol. **345**(5), 1–1294 (2005)
5. Gul, S., Hadian, K.: Protein-protein interaction modulator drug discovery: past efforts and future opportunities using a rich source of low- and high-throughput screening assays. Expert Opin. Drug Discov. **9**(12), 1393–1404 (2014)
6. Cukuroglu, E., Engin, H.B., Gursoy, A., Keskin, O.: Hot spots in protein–protein interfaces: towards drug discovery. Prog. Biophys. Mol. Biol. **116**(2–3), 165–173 (2014)
7. Xia, J.F., Zhao, X.M., Song, J., Huang, D.S.: APIS: accurate prediction of hot spots in protein interfaces by combining protrusion index with solvent accessibility. BMC Bioinformatics **11**(1), 174 (2010)
8. Tuncbag, N., Gursoy, A., Keskin, O.: Identification of computational hot spots in protein interfaces: combining solvent accessibility and inter-residue potentials improves the accuracy. Bioinformatics **25**(12), 1513–1520 (2009)
9. Huang, Q., Zhang, X.: An improved ensemble learning method with SMOTE for protein interaction hot spots prediction. In: IEEE International Conference on Bioinformatics & Biomedicine. IEEE (2017)
10. Hu, S.S., Chen, P., Wang, B., Li, J.: Protein binding hot spots prediction from sequence only by a new ensemble learning method. Amino Acids **49**, 1–13 (2017)
11. Cukuroglu, E., Gursoy, A., Keskin, O.: Analysis of hot region organization in hub proteins. Ann. Biomed. Eng. **38**, 2068–2078 (2010)
12. Pons, C.: Prediction of protein-binding areas by small-world residue networks and application to docking. BMC Bioinformatics **12**, 378 (2011)
13. Nan, D., Zhang, X.: Prediction of hot regions in protein-protein interactions based on complex network and community detection. In: IEEE International Conference on Bioinformatics & Biomedicine. IEEE (2014)
14. Lin, X., Zhang, X.: Prediction of hot regions in PPIs based on improved local community structure detecting. IEEE/ACM Trans. Comput. Biol. Bioinform., 1 (2018)
15. Moal, I.H., Fernandez-Recio, J.: SKEMPI: a structural kinetic and energetic database of mutant protein interactions and its use in empirical models. Bioinformatics **28**(20), 2600–2607 (2012)
16. Jankauskaite, J., Jimenezgarcia, B., Dapkunas, J., Fernandezrecio, J., Moal, I.H.: SKEMPI 2.0: an updated benchmark of changes in protein–protein binding energy, kinetics and thermodynamics upon mutation. Bioinformatics **35**(3), 462–469 (2019)
17. Ester, M.: A Density-Based Algorithm for Discovering Clusters in Large Conference on Knowledge Discovery and Data Mining (KDD). AAAI Press (1996)
18. Pierce, M.M., Raman, C.S., Nall, B.T.: Isothermal titration calorimetry of protein-protein interactions. Methods **19**(2), 213–221 (1999)
19. Wang, Y., Shen, B.-J., Sebald, W.: A mixed-charge pair in human interleukin 4 dominates high-affinity interaction with the receptor $\alpha$ chain. Proc. Natl. Acad. Sci. **94**(5), 1657–1662 (1997)
20. Berman, H., Bourne, P., Westbrook, J., Zardecki, C.: The Protein Data Bank. Protein Structure. Springer International Publishing (2003)
21. Mihel, J., Šikić, M., Tomić, S., Jeren, B., Vlahoviček, K.: PSAIA – protein structure and interaction analyzer. BMC Struct. Biol. **8**(1), 21 (2008)
22. Ashkenazy, H., Abadi, S., Martz, E., Chay, O., Mayrose, I., Pupko, T., et al.: ConSurf 2016: an improved methodology to estimate and visualize evolutionary conservation in macromolecules. Nucleic Acids Res. **44**, 344–350 (2016)

23. Peng, H., Long, F., Ding, C.: Feature selection based on mutual information: criteria of max-dependency, max-relevance, and min-redundancy. IEEE Trans. Pattern Anal. Mach. Intell. **27**(8), 1226–1238 (2005)
24. Chan, T.F., Golub, G.H., LeVeque, R.J.: Updating formulae and a pairwise algorithm for computing sample variances. In: COMPSTAT 1982 5th Symposium held at Toulouse 1982. Physical-Verlag HD (1982)
25. Mitchell, T.: Machine Learning. McGraw-Hill, Maidenhead (1997)
26. Zhang, H.: The Optimality of Naive Bayes. In: The Florida AI Research Society, pp. 562–567 (2004)
27. Chen, T., Guestrin, C.: XGBoost: a scalable tree boosting system. In: KDD, pp. 785–794 (2016)
28. Cutler, A., Cutler, D.R., Stevens, J.R.: Random forests. Mach. Learn. **45**(1), 157–176 (2011)
29. Chang, C.C., Lin, C.J.: LIBSVM: a library for support vector machines. ACM Trans. Intell. Syst. Technol. **2**, 27:1–27:27 (2011)
30. Pao, Y.H.: Adaptive Pattern Recognition and Neural Networks, vol. 12, pp. 31–67. Reading Addison Wesley, Massachusetts (1989)
31. Hinton, G.E.: Rectified linear units improve restricted boltzmann machines vinod nair. In: International Conference on International Conference on Machine Learning. Omnipress (2010)
32. Miller, S., Lesk, A.M., Janin, J., Chothia, C.: The accessible surface area and stability of oligomeric proteins. Nature **328**(6133), 834–836 (1987)
33. Python Molecule. https://pymol.org/2/. Accessed 10 Mar 2019

# An Efficient LightGBM Model to Predict Protein Self-interacting Using Chebyshev Moments and Bi-gram

Zhao-Hui Zhan[1], Zhu-Hong You[2]([✉]), Yong Zhou[1],
Kai Zheng[1], and Zheng-Wei Li[1]

[1] School of Computer Science and Technology,
China University of Mining and Technology, Xuzhou 221116, China
[2] The Xinjiang Technical Institute of Physics and Chemistry,
Chinese Academy of Science, Urumqi 830011, China
zhuhongyou@ms.xjb.ac.cn

**Abstract.** Protein self-interactions (SIPs) play significant roles in most life activities. Although numerous computational methods have been developed to predict SIPs, there is still a need of efficient and accurate techniques to improve the performance of SIPs prediction. In this paper, we proposed a machine learning scheme named LGCM for accurate SIP predictions based on protein sequence information. More specifically, an novel feature descriptor employing bi-gram and Chebyshev moments algorithm was developed with the extraction of discriminative sequence information. Then, we fed the integrated protein features into LightGBM classifier as input to train automatic LGCM model. It was demonstrated by rigorous cross-validations that the proposed approach LGCM had a superior prediction performance than other previous methods for SIP predictions with the accuracy of 96.90% and 98.29% on yeast and human datasets, respectively. Experiment results anticipated the effectiveness and reliability of LGCM and played a definite guiding role in future bioinformatics research.

**Keywords:** Self-interacting protein · PSSM · Bi-gram ·
Chebyshev moments · LightGBM

## 1 Introduction

Protein self-interaction (SIP) is a particular type of protein-protein interaction (PPI) [1–3]. Recent studies have shown that PPI plays a basic but significant role which can conduct signal, regulate gene expression and activate enzyme in most biological processes of living organisms [4–10]. As an idiographic case of PPIs, SIPs tend to have a specific disposition in protein interaction networks [2, 11–15]. The traditional biological experiment approach for predicting SIP is complicated in obtaining materials which resulting in time-consuming and labor-intensive [16–24]. Therefore, a particular approach is required to be presented for predicting SIPs as an urgent priority [25].

In this study, we proposed a protein sequence-based computational method named LGCM for SIP predictions by employing an improved gradient boosting decision tree

© Springer Nature Switzerland AG 2019
D.-S. Huang et al. (Eds.): ICIC 2019, LNCS 11644, pp. 453–459, 2019.
https://doi.org/10.1007/978-3-030-26969-2_43

classifier. More specifically, protein original sequences were completed convert into position-specific scoring matrices (PSSMs) firstly into which the required significant information from protein sequences integrated. Afterwards, both bi-gram and Chebyshev moments algorithm were employed to extract feature vectors from PSSMs. Moreover, the integrated protein features were fed into the LightGBM classifier as a gradient boosting decision tree machine learning model to train SIPs prediction scheme. In order to evaluate the SIP predictive accuracy of proposed model LGCM, two extensive employed datasets *yeast* and *human* were trained and forecasted through a five-fold cross-validation which was capable of avoiding over-fitting [22, 23, 26–29].

## 2    Materials and Methods

### 2.1    Datasets

In this study, two extensive employed datasets *human* and *yeast* were adopted to assess the predictive performance of machine learning model LGCM [30–32]. The *human* GS dataset is consist with 1441 positive SIP samples and 15938 negative non-SIP samples. And the *yeast* GS dataset is consist of 710 positive SIP samples and 5511 negative non-SIP samples [33].

### 2.2    Protein Feature Extraction of Bi-gram Algorithm

In this study, original protein sequences were represented by PSSMs and the bi-gram features were extracted from resulting PSSMs correspondingly [34–36]. More specifically, let $B$ be a $20 \times 20$-dimension bi-gram occurrence matrix with the element $b_{m,n}$ interpreted as the occurrence probability of the transition from $m_{th}$ to $n_{th}$ amino acid [37]. Therefore, the frequency value of bi-gram algorithm can be calculated as following equations:

$$B = \left\{ b_{m,n} \; 1 \leq \mathrm{m} \leq 20, 1 \leq \mathrm{n} \leq 20 \right\} \tag{1}$$

$$b_{m,n} = \sum_{a=1}^{l-1} P_{a,m} P_{a+1,n} \; (a \leq m \leq 20, 1 \leq n \leq 20) \tag{2}$$

where $l$, $P_{a,m}$ and $P_{a+1,n}$ were elements in the corresponding PSSM [38, 39]. Then, a widely used dimension reducing approach named Principal component analysis (PCA) was employed to reduce the dimension of resulting feature vectors and preserve the variables of the original data as much as possible [40]. As a result, two low 150-dimension linearly independent feature vectors were calculated after PCA dimensionality reduction.

### 2.3    Protein Feature Extraction of Chebyshev Moments

The Chebyshev polynomials are solutions of the Chebyshev differential equation and can be expressed by explicit formulas [41, 42]:

$$T_s(x) = \frac{\left(x + \sqrt{x^2 - 1}\right)^s + \left(x - \sqrt{x^2 - 1}\right)^s}{2}$$

$$= \sum_{k=0}^{s/2} \binom{s}{2k} (x^2 - 1)^k x^{s-2k} \tag{3}$$

$$= \frac{s}{2} \sum_{t=0}^{s/2} (-1)^t \frac{(s - t - 1)!}{t!(s - 2t)!} (2x)^{s-2t}$$

and the last version cannot be used for $s = 0$. Moreover, the orthogonality relation is:

$$\int_{-1}^{1} T_s(x) T_t(x) \frac{1}{\sqrt{1 - x^2}} dx = \begin{cases} 0 & s \neq t \\ \pi & s = t = 0 \\ \pi/2 & s = t \neq 0 \end{cases} \tag{4}$$

where the interval of orthogonality $\Omega_1 = \langle -1, 1 \rangle$. As a result, the Chebyshev moments are expressed in the basic version as [43]:

$$\varphi_{st} = \iint_{\Omega} T_s(x) T_t(y) f(x, y) dx dy$$

$$= \sum_{p=0}^{t} \sum_{q=0}^{s} a_{tp} a_{sq} t_{pq} \ (s, t = 0, 1, 2, \ldots) \tag{5}$$

where the parameters are all mentioned above. In consequence, two Chebyshev moments feature vectors with 21-dimension were extracted by the above formulas on datasets *human* and *yeast*. In general, a total feature descriptor of 171-dimension for SIP conjoined bi-gram feature and Chebyshev moments feature was obtained.

## 2.4   LightGBM Classification Model

LightGBM optimizes the training speed and memory of the machine learning model substantially in many aspects [44, 45]. In the process of learning decision trees as the most time-consuming part, histogram-based algorithm is employed to find the best split points. On the other hand, when splitting sample data instances, LightGBM employed the leaf-wise decision tree growth strategy. In addition, to address the limitations existing in the boosting classifier at this stage, two techniques for data preprocessing called Gradient-based One-Side Sampling (GOSS) and Exclusive Feature Bundling (EFB) are employed in LightGBM.

## 3   Experimental Results

Under the five-fold cross-validation, five widely used evaluation criteria were calculated including accuracy, precision, sensitivity, specificity and MCC and the detail prediction numerical value on datasets *yeast* and *human* were shown in Tables 1 and 2 [46–52].

**Table 1.** 5-fold cross-validation results on dataset *Yeast*.

| Testing set | Acc. (%) | Prec. (%) | Sen. (%) | Spec. (%) | MCC (%) |
|---|---|---|---|---|---|
| 1 | 97.35 | 99.10 | 77.46 | 99.91 | 86.31 |
| 2 | 96.54 | 98.06 | 71.13 | 99.82 | 81.86 |
| 3 | 96.70 | 100.0 | 71.13 | 100.0 | 82.81 |
| 4 | 96.62 | 99.02 | 71.13 | 99.91 | 82.33 |
| 5 | 97.27 | 97.37 | 78.17 | 99.73 | 85.85 |
| **Average** | **96.90** | **98.71** | **73.80** | **99.87** | **83.83** |

**Table 2.** 5-fold cross-validation results on dataset *Human*.

| Testing set | Acc. (%) | Prec. (%) | Sen. (%) | Spec. (%) | MCC (%) |
|---|---|---|---|---|---|
| 1 | 98.33 | 100.0 | 79.93 | 100.0 | 88.60 |
| 2 | 98.25 | 100.0 | 78.82 | 100.0 | 87.94 |
| 3 | 98.36 | 99.57 | 80.56 | 99.97 | 88.76 |
| 4 | 98.13 | 99.56 | 77.78 | 99.97 | 87.10 |
| 5 | 98.36 | 99.57 | 80.56 | 99.97 | 88.76 |
| **Average** | **98.29** | **99.74** | **79.53** | **99.98** | **88.23** |

## 4   Conclusions

In this study, we proposed a tree-based machine learning model for predicting protein self-interacting using efficient features based on raw protein sequence information, named LGCM. More concretely, we integrated amino acid position information and location-based evolutionary information by combining Chebyshev moments and bi-gram feature representation approaches. Then an improved gradient boosting decision tree framework LightGBM classifier was implemented. Meanwhile, to evaluate the predictive performance of the proposed LGCM, we further compared our model with other state-of-the-art methods on two golden standard datasets *yeast* and *human*. It was confirmed by rigorous five-fold cross-validations that the proposed LGCM evidently outperformed other comparison methods, on benchmark *yeast* dataset, with an accuracy of 96.90% and an AUC of 0.9975, on benchmark dataset *human* dataset, with an accuracy of 98.29% and an AUC as high as 0.9987, respectively. Experimental results provided that our LGCM is a competitive model for predict self-interacting proteins and can promote related research.

## References

1. You, Z.-H., Li, X., Chan, K.C.: An improved sequence-based prediction protocol for protein-protein interactions using amino acids substitution matrix and rotation forest ensemble classifiers: Elsevier Science Publishers B. V. (2017)
2. An, J.-Y., You, Z.-H., Chen, X., Huang, D.-S., Yan, G., Wang, D.-F.: Robust and accurate prediction of protein self-interactions from amino acids sequence using evolutionary information. Mol. BioSyst. **12**(12), 3702 (2016)

3. Gao, Z.-G., Lei, W., Xia, S.-X., You, Z.-H., Xin, Y., Yong, Z.: Ens-PPI: a novel ensemble classifier for predicting the interactions of proteins using autocovariance transformation from PSSM. BioMed Res. Int. **2016**(4), 1–8 (2016)
4. Wang, Y.-B., You, Z.-H., Li, X., Jiang, T.-H., Cheng, L., Chen, Z.-H.: Prediction of protein self-interactions using stacked long short-term memory from protein sequences information. BMC Syst. Biol. **12**(8), 129 (2018)
5. Song, X.-Y., Chen, Z.-H., Sun, X.-Y., You, Z.-H., Li, L.-P., Zhao, Y.: An ensemble classifier with random projection for predicting protein-protein interactions using sequence and evolutionary information. Appl. Sci. **8**(1), 89 (2018)
6. Li, L.-P., Wang, Y.-B., You, Z.-H., Li, Y., An, J.-Y.: PCLPred: a bioinformatics method for predicting protein-protein interactions by combining relevance vector machine model with low-rank matrix approximation. Int. J. Mol. Sci. **19**(4), 1029 (2018)
7. You, Z.-H., Zhou, M., Luo, X., Li, S.: Highly efficient framework for predicting interactions between proteins. IEEE Trans. Cybern. **47**(3), 731–743 (2017)
8. Wen, Y.-T., Lei, H.-J., You, Z.-H., Lei, B.-Y., Chen, X., Li, L.-P.: Prediction of protein-protein interactions by label propagation with protein evolutionary and chemical information derived from heterogeneous network. J. Theor. Biol. **430**, 9–20 (2017)
9. Li, Z.-W., You, Z.-H., Chen, X., Gui, J., Nie, R.: Highly accurate prediction of protein-protein interactions via incorporating evolutionary information and physicochemical characteristics. Int. J. Mol. Sci. **17**(9), 1396 (2016)
10. An, J.-Y., You, Z.-H., Meng, F.-R., Xu, S.-J., Wang, Y.: RVMAB: using the relevance vector machine model combined with average blocks to predict the interactions of proteins from protein sequences. Int. J. Mol. Sci. **17**(5), 757 (2016)
11. Huang, Y.-A., You, Z.-H., Chen, X., Yan, G.-Y.: Improved protein-protein interactions prediction via weighted sparse representation model combining continuous wavelet descriptor and PseAA composition. BMC Syst. Biol. **10**(4), 120 (2016)
12. You, Z.-H., Chan, K.C., Hu, P.: Predicting protein-protein interactions from primary protein sequences using a novel multi-scale local feature representation scheme and the random forest. PLoS One **10**(5), e0125811 (2015)
13. Lei, Y.-K., You, Z.-H., Dong, T., Jiang, Y.-X., Yang, J.-A.: Increasing reliability of protein interactome by fast manifold embedding. Pattern Recogn. Lett. **34**(4), 372–379 (2013)
14. Xia, J.-F., Wu, M., You, Z.-H., Zhao, X.-M., Li, X.-L.: Prediction of β-hairpins in proteins using physicochemical properties and structure information. Protein Pept. Lett. **17**(9), 1123–1128 (2010)
15. Akiva, E., Itzhaki, Z., Margalit, H.: Built-in loops allow versatility in domain-domain interactions: lessons from self-interacting domains. Proc. Natl. Acad. Sci. U.S.A. **105**(36), 13292–13297 (2008)
16. You, Z.-H., Huang, W., Zhang, S., Huang, Y.-A., Yu, C.-Q., Li, L.-P.: An efficient ensemble learning approach for predicting protein-protein interactions by integrating protein primary sequence and evolutionary information. IEEE/ACM Trans. Comput. Biol. Bioinf. (2018)
17. Chen, Z.-H., You, Z.-H., Li, L.-P., Wang, Y.-B., Li, X.: RP-FIRF: prediction of self-interacting proteins using random projection classifier combining with finite impulse response filter. In: Huang, D.-S., Jo, K.-H., Zhang, X.-L. (eds.) ICIC 2018. LNCS, vol. 10955, pp. 232–240. Springer, Cham (2018). https://doi.org/10.1007/978-3-319-95933-7_29
18. Wang, Y.-B., You, Z.-H., Li, L.-P., Huang, Y.-A., Yi, H.-C.: Detection of interactions between proteins by using legendre moments descriptor to extract discriminatory information embedded in pssm. Molecules **22**(8), 1366 (2017)
19. Li, J.-Q., You, Z.-H., Li, X., Ming, Z., Chen, X.: PSPEL: In silico prediction of self-interacting proteins from amino acids sequences using ensemble learning. IEEE/ACM Trans. Comput. Biol. Bioinf. (TCBB) **14**(5), 1165–1172 (2017)

20. Bao, W., You, Z.-H., Huang, D.-S.: CIPPN: computational identification of protein pupylation sites by using neural network. Oncotarget **8**(65), 108867 (2017)
21. Koike, R., Kidera, A., Ota, M.: Alteration of oligomeric state and domain architecture is essential for functional transformation between transferase and hydrolase with the same scaffold. Protein Sci. **18**(10), 2060 (2009)
22. You, Z.-H., Li, L., Ji, Z., Li, M., Guo, S.: Prediction of protein-protein interactions from amino acid sequences using extreme learning machine combined with auto covariance descriptor. In: 2013 IEEE Workshop on Memetic Computing (MC), pp. 80–85. IEEE (2013)
23. Huang, Q., You, Z., Zhang, X., Zhou, Y.: Prediction of protein-protein interactions with clustered amino acids and weighted sparse representation. Int. J. Mol. Sci. **16**(5), 10855–10869 (2015)
24. Luo, X., Ming, Z., You, Z., Li, S., Xia, Y., Leung, H.: Improving network topology-based protein interactome mapping via collaborative filtering. Knowl.-Based Syst. **90**, 23–32 (2015)
25. Liu, X., Yang, S., Li, C., Zhang, Z., Song, J.: SPAR: a random forest-based predictor for self-interacting proteins with fine-grained domain information. Amino Acids **48**(7), 1655–1665 (2016)
26. You, Z.-H., Zhu, L., Zheng, C.-H., Yu, H.-J., Deng, S.-P., Ji, Z.: Prediction of protein-protein interactions from amino acid sequences using a novel multi-scale continuous and discontinuous feature set. BMC Bioinformatics **15**(S15), S9 (2014)
27. Zhu, L., You, Z.-H., Huang, D.-S.: Increasing the reliability of protein–protein interaction networks via non-convex semantic embedding. Neurocomputing **121**, 99–107 (2013)
28. Chen, X., Xie, D., Zhao, Q., You, Z.-H.: MicroRNAs and complex diseases: from experimental results to computational models. Briefings in bioinformatics (2017)
29. Luo, X., et al.: An incremental-and-static-combined scheme for matrix-factorization-based collaborative filtering. IEEE Trans. Autom. Sci. Eng. **13**(1), 333–343 (2016)
30. An, J.-Y., et al.: Identification of self-interacting proteins by exploring evolutionary information embedded in PSI-BLAST-constructed position specific scoring matrix. Oncotarget **7**(50), 82440–82449 (2016)
31. Wang, Y.-B., You, Z.-H., Li, L.-P., Huang, D.-S., Zhou, F.-F., Yang, S.: Improving prediction of self-interacting proteins using stacked sparse auto-encoder with PSSM profiles. Int. J. Biol. Sci. **14**(8), 983–991 (2018)
32. You, Z.-H., Ming, Z., Huang, H., Peng, X.: A novel method to predict protein-protein interactions based on the information of protein sequence. In: 2012 IEEE International Conference on Control System, Computing and Engineering (ICCSCE), pp. 210–215. IEEE (2012)
33. Consortium UP: UniProt: a hub for protein information. Nucleic Acids Research 43 (Database issue), p. D204 (2015)
34. Sharma, A., Lyons, J., Dehzangi, A., Paliwal, K.K.: A feature extraction technique using bi-gram probabilities of position specific scoring matrix for protein fold recognition. J. Theor. Biol. **320**(1), 41 (2013)
35. Paliwal, K.K., Sharma, A., Lyons, J., Dehzangi, A.: A tri-gram based feature extraction technique using linear probabilities of position specific scoring matrix for protein fold recognition. IEEE Trans. Nanobiosci. **320**(1), 41 (2013)
36. Chen, X., Yan, C.-C., Zhang, X., You, Z.-H.: Long non-coding RNAs and complex diseases: from experimental results to computational models. Brief. Bioinform. **18**(4), 558 (2016)
37. Zhan, Z.-H., You, Z.-H., Zhou, Y., Li, L.-P., Li, Z.-W.: Efficient framework for predicting ncRNA-Protein interactions based on sequence information by deep learning. In: International Conference on Intelligent Computing, pp. 337–344 (2018)

38. Wang, Y.-B., et al.: Predicting protein-protein interactions from protein sequences by a stacked sparse autoencoder deep neural network. Mol. BioSyst. **13**(7), 1336–1344 (2017)
39. Wang, Y., You, Z., Li, X., Chen, X., Jiang, T., Zhang, J.: PCVMZM: using the probabilistic classification vector machines model combined with a zernike moments descriptor to predict protein-protein interactions from protein sequences. Int. J. Mol. Sci. **18**(5), 1029 (2017)
40. Zhang, S.-L., Ye, F., Yuan, X.-G.: Using principal component analysis and support vector machine to predict protein structural class for low-similarity sequences via PSSM. J. Biomol. Struct. Dyn. **29**(6), 1138–1146 (2012)
41. Yap, P.T., Raveendran, P., Ong, S.H.: Chebyshev moments as a new set of moments for image reconstruction. **4**, 2856–2860 (2001)
42. Askey, R.: Chebyshev polynomials from approximation theory to algebra and number theory. Bull. London Math. Soc. **23**(1), 105–115 (1990)
43. Kotoulas, L., Andreadis, I.: Fast computation of Chebyshev moments. IEEE Trans. Circuits Syst. Video Technol. **16**(7), 884–888 (2006)
44. Ke, G., et al.: LightGBM: a highly efficient gradient boosting decision tree (2017)
45. Friedman, J.H.: Greedy function approximation: a gradient boosting machine. Ann. Stat. **29**(5), 1189–1232 (2001)
46. Wang, L., You, Z.-H., Huang, D.-S., Zhou, F.: Combining high speed ELM learning with a deep convolutional neural network feature encoding for predicting protein-RNA interactions. IEEE/ACM Trans. Comput. Biol. Bioinf. (2018)
47. Chen, X., et al.: WBSMDA: within and between score for MiRNA-Disease association prediction. Sci. Rep. **6**, 21106 (2016)
48. Huang, Y.-A., Chan, K.C., You, Z.-H.: Constructing prediction models from expression profiles for large scale lncRNA–miRNA interaction profiling. Bioinformatics **34**(5), 812–819 (2017)
49. Li, J.-Q., Rong, Z.-H., Chen, X., Yan, G.-Y., You, Z.-H.: MCMDA: matrix completion for MiRNA-Disease association prediction. Oncotarget **8**(13), 21187 (2017)
50. You, Z.-H., Ming, Z., Li, L., Huang, Q.-Y.: Research on signaling pathways reconstruction by integrating high content RNAi screening and functional gene network. In: Huang, D.-S., Jo, K.-H., Zhou, Y.-Q., Han, K. (eds.) ICIC 2013. LNCS (LNAI), vol. 7996, pp. 1–10. Springer, Heidelberg (2013). https://doi.org/10.1007/978-3-642-39482-9_1
51. Huang, Y.-A., You, Z.-H., Chen, X.: A systematic prediction of drug-target interactions using molecular fingerprints and protein sequences. Curr. Protein Pept. Sci. **19**(5), 468–478 (2018)
52. Zhu, H.-J., You, Z.-H., Zhu, Z.-X., Shi, W.-L., Chen, X., Cheng, L.: DroidDet: effective and robust detection of android malware using static analysis along with rotation forest model. Neurocomputing **272**, 638–646 (2018)

# Combining Evolutionary Information and Sparse Bayesian Probability Model to Accurately Predict Self-interacting Proteins

Yan-Bin Wang[1], Zhu-Hong You[1(✉)], Hai-cheng Yi[2],
Zhan-Heng Chen[1], Zhen-Hao Guo[1], and Kai Zheng[3]

[1] University of Chinese Academy of Sciences, Beijing 100049, China
zhuhongyou@ms.xjb.ac.cn
[2] Xinjiang Technical Institute of Physics and Chemistry,
Chinese Academy of Science, Urumqi 830011, China
[3] School of Computer Science and Technology,
China University of Mining and Technology, Xuzhou, China

**Abstract.** Self-interacting proteins (SIPs) play a crucial role in investigation of various biochemical developments. In this work, a novel computational method was proposed for accelerating SIPs validation only using protein sequence. Firstly, the protein sequence was represented as Position-Specific Weight Matrix (PSWM) containing protein evolutionary information. Then, we incorporated the Legendre Moment (LM) and Sparse Principal Component Analysis (SPCA) to extract essential and anti-noise evolutionary feature from the PSWM. Finally, we utilized robust Probabilistic Classification Vector Machine (PCVM) classifier to carry out prediction. In the cross-validated experiment, the proposed method exhibits high accuracy performance with 95.54% accuracy on *S.erevisiae* dataset, which is a significant improvement compared to several competing SIPs predictors. The empirical test reveal that the proposed method can efficiently extracts salient features from protein sequences and accurately predict potential SIPs.

**Keywords:** Self-interacting proteins · Protein-protein interactions · Probabilistic Classification Vector Machine · Legendre Moment

## 1 Introduction

Self-interaction protein (SIP) is a particular case of protein-protein interactions (PPIs). There's many evidence that SIP plays a critical role in vital movement [1, 2]. Detecting SIPs is thus becoming a focal point of attention. However, SIPs detection is encountering bottlenecks in the process of scale and high-precision. Firstly, high throughput method has limited capacity for large-scale detecting SIPs. Secondly, many computational methods [3–15] are built on datasets that do not contain SIP data, which makes them difficult to predict SIP. Thirdly, some methods specific to SIP prediction have

---

Y.-B. Wang and Z.-H. You—These authors contributed equally to this work.

D.-S. Huang et al. (Eds.): ICIC 2019, LNCS 11644, pp. 460–467, 2019.
https://doi.org/10.1007/978-3-030-26969-2_44

been proposed. For example, Liu *et al.* [16] constructed SLIPPER model for human SIP prediction. Song *et al.* [17] designed SPAR using random forest classifier. Chen *et al.* [18] developed RP-FFT by using fast random projection and Fourier transform. Soon afterwards, they proposed an improved method by utilizing deep forest [19]. In addition, there are other computational models [20–32], However, there is still much room for improvement in the performance of these methods.

In this paper, a novel computational method was proposed to accuracy predict SIP. More specifically, protein sequence is firstly converted into PSWM. Second, the effective LM is employed for abstracting salient and invariant feature from PSWM. And then, the SPCA took charge of the extraction of main features. Finally, the optimized features are fed into the PCVM model was employed to carry out prediction. The flow of our proposed scheme is shown in Fig. 1. The prediction results shows a significant improvement with 95.54% accuracy on *S.erevisiae* benchmark dataset, respectively.

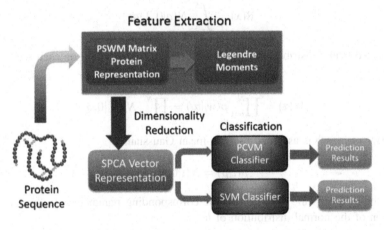

**Fig. 1.** The flow of our proposed method.

## 2 Methods

### 2.1 Protein Characterization

In this work, we developed an efficient protein feature extraction method based on Position Specific Weight Matrix (PSWM) [9, 23, 34–43], Legendre Moment (LM) [11, 44–57] and the Sparse Principal Component Analysis (SPCA) [9, 23, 34].

The PSWM is generated by employing Position-specific iterated BLAST (PSI-BLAST), which obtains protein evolutionary. An PSWM corresponding to a protein sequence is an $Z \times 20$ matrix. The number of rows of PSWM represent the size of the protein sequence and the columns denote 20 amino acids. The parameters in PSI-BLAST are set to default, except for three of iterations and the 0.001 e-value. Then, The LM is intended for extracting prominent features from PSWM. The LM has two parameters, $M$ and $N$, used for control extracted feature number. Here, we assigned 30 to $M$ and $N$, hence, we obtained 625 feature for each protein. Finally, to improve the quality of features and reduce computational consumption, SPCA was employed to choose the optimal feature space. As a results, each protein was characterized by 200-D vector.

## 2.2 Classification Algorithm

The Probabilistic Classification Vector Machine (PCVM) [33, 35] consists of linear addition of $N$ basis functions:

$$f(x; w) = \sum_{i=1}^{N} w_i \complement_{i,\theta}(x) + s \tag{1}$$

where $\complement_{i,\theta}(x)$ are basis functions, $w$ are the weight of the prediction model.

A function $\aleph(x)$ is used for mapping linear output into binary output,

$$\aleph(x) = \int_{-\infty}^{x} N(a|0, 1) dt \tag{2}$$

A truncated Gauss distribution is assigned as a prior distribution on each weight $w_i$ as follow:

$$p(W|\alpha) = \prod_{i=1}^{N} p(w_i|\alpha_i) = \prod_{i=1}^{N} N_t(w_i|0, \alpha_i^{-1}) \tag{3}$$

The prior of bias $b$ is assigned by zero-mean Gaussian:

$$p(t|\beta) = N(t|0, \beta^{-1})) \tag{4}$$

where the $\alpha_i$ means the precision of the corresponding parameter $w_i$, $\beta$ refers to the precision of the normal distribution of $h$.

## 3 Results and Discussion

### 3.1 Evaluation Measures

The proposed method was test on golden standard $S.erevisiae$ dataset [9], which contain 5,511 negative non-SIPs and 710 confident SIPs. To objectively evaluate our methods, follow criteria: Accuracy (Ac), Specificity (Sp), Precision (Pe), and Matthew`s correlation coefficient (Mcc), are calculated for assessing the proposed model. The formulas are as follows: In addition, the Receiver Operating Characteristic (ROC) [23, 34] curve is created and the area under the ROC curve (AUC) [48] also is computed for further assessing the method.

## 3.2    Results of Fivefold Cross Validation

The five-fold experiment results are shown in Table 1. From Table 1, we can see the average accuracy on *S.erevisiae* dataset is as high as 95.54% and the standard deviation only is 0.2. Specifically, the accuracy of the fivefold experiments is 95.34%, 95.58%, 95.66%, 95.26% and 95.90%, respectively. Moreover, our method also achieves high AUC values with average 0.9758.

**Table 1.** The results of the PCVM on the *S.erevisiae* dataset.

| Model | Testing set | Ac(%) | Sp(%) | Pe(%) | Mcc(%) | AUC |
|-------|-------------|-------|-------|-------|--------|-----|
| PCVM | 1 | 95.34 | 99.46 | 93.81 | 75.52 | 0.9777 |
| | 2 | 95.58 | 99.45 | 94.83 | 79.19 | 0.9833 |
| | 3 | 95.66 | 99.73 | 96.97 | 77.79 | 0.9750 |
| | 4 | 95.26 | 99.01 | 88.17 | 73.07 | 0.9620 |
| | 5 | 95.90 | 99.64 | 95.45 | 76.65 | 0.9808 |
| | Average | 95.54 ± 0.2 | 99.46 ± 0.2 | 93.84 ± 3.3 | 76.44 ± 2.3 | 0.9758 |

# 4    Conclusions

This paper describes a novel computational method for predicting SIPs using protein sequence. The main idea is use the PSI-BLAST to transform protein sequences into PSWMs, which contains multiple valuable protein information. Then combination of LM and SPCA is employed to exploit significant feature from the PSWM. Finally, we train PCVM classifier to accomplish SIPs prediction task. Test results on gold standard datasets shown that the proposed method could identify SIPs with high accuracy and has the widespread application value.

**Acknowledgments.** This work is supported in part by the National Science Foundation of China, under Grants 61373086, 11301517 and 61572506. The authors would like to thank all the editors and anonymous reviewers for their constructive advices.

**Author Contributions.** YBW and ZHY considered the algorithm, arranged the datasets, carried out the experiments, HCY and ZHC wrote the manuscript, designed. ZHG and KZ make analyses. All authors read and approved the final manuscript.

**Funding.** The publication costs for this article were funded by the corresponding author's institution.

**Availability of Data and Materials:** Source code of our models and training/testing datasets are available at: https://figshare.com/s/1ff62d10d3bcb94e2bba.

**Competing Interests:** The authors declare that they have no competing interests.

**Consent for Publication:** Not applicable.

**Ethics Approval and Consent to Participate:** Not applicable.

# References

1. Chen, Y., Dokholyan, N.V.: Natural selection against protein aggregation on self-interacting and essential proteins in yeast, fly, and worm. Mol. Biol. Evol. **25**(8), 1530–1533 (2008)
2. Gautier, A., Nakata, E., Lukinavicius, G., Tan, K.T., Johnsson, K.: Selective cross-linking of interacting proteins using self-labeling tags. J. Am. Chem. Soc. **131**(49), 17954–17962 (2009)
3. Franceschini, A., et al.: STRING v9.1: protein-protein interaction networks, with increased coverage and integration. Nucleic Acids Res. **41**, 808–815 (2012)
4. Rhodes, D.R., et al.: Probabilistic model of the human protein-protein interaction network. Nat. Biotechnol. **23**(8), 951–959 (2005)
5. You, Z.H., Yu, J.Z., Zhu, L., Li, S., Wen, Z.K.: A MapReduce based parallel SVM for large-scale predicting protein–protein interactions. Neurocomputing **145**(18), 37–43 (2014)
6. Zhu, L., You, Z.H., Huang, D.S.: Increasing the reliability of protein–protein interaction networks via non-convex semantic embedding. Neurocomputing **121**(18), 99–107 (2013)
7. You, Z., Li, S., Gao, X., Luo, X., Ji, Z.: Large-scale protein-protein interactions detection by integrating big biosensing data with computational model. Biomed. Res. Int. **2014**, 598129 (2014)
8. You, Z., et al.: Detecting protein-protein interactions with a novel matrix-based protein sequence representation and support vector machines. Biomed. Res. Int. **2015**, 867516 (2015)
9. Wang, Y.B., et al.: Predicting protein-protein interactions from protein sequences by a stacked sparse autoencoder deep neural network. Mol. BioSyst. **13**(7), 1336–1344 (2017)
10. Wang, L., et al.: Advancing the prediction accuracy of protein-protein interactions by utilizing evolutionary information from position-specific scoring matrix and ensemble classifier. J. Theor. Biol. **418**(4), 105–110 (2017)
11. Li, Z.W., You, Z.H., Chen, X., Gui, J., Nie, R.: Highly accurate prediction of protein-protein interactions via incorporating evolutionary information and physicochemical characteristics. Int. J. Mol. Sci. **17**(9), 1396 (2016)
12. Huang, Y.A., You, Z.H., Xing, C., Yan, G.Y.: Improved protein-protein interactions prediction via weighted sparse representation model combining continuous wavelet descriptor and PseAA composition. BMC Syst. Biol. **10**(Suppl. 4), 485–494 (2016)
13. Zhu, L., Deng, S.P., You, Z.H., Huang, D.S.: Identifying spurious interactions in the protein-protein interaction networks using local similarity preserving embedding. IEEE/ACM Trans. Comput. Biol. Bioinf. **14**(2), 345–352 (2017)
14. Huang, Y., You, Z., Li, J., Wong, L., Cai, S.: Predicting protein-protein interactions from amino acid sequences using SaE-ELM combined with continuous wavelet descriptor and PseAA composition. In: International Conference on Intelligent Computing, pp. 634–645 (2015)
15. Li, L., Wang, Y., You, Z., Li, Y., An, J.: PCLPred: a bioinformatics method for predicting protein-protein interactions by combining relevance vector machine model with low-rank matrix approximation. Int. J. Mol. Sci. **19**(4), 1029 (2018)
16. Liu, Z., et al.: Proteome-wide prediction of self-interacting proteins based on multiple properties. Mol. Cell. Proteomics **12**(6), 1689–1700 (2013)
17. Liu, X., Yang, S., Li, C., Zhang, Z., Song, J.: SPAR: a random forest-based predictor for self-interacting proteins with fine-grained domain information. Amino Acids **48**(7), 1655–1665 (2016)

18. Chen, Z., Li, L., He, Z., Zhou, J., Li, Y., Wong, L.: An improved deep forest model for predicting self-interacting proteins from protein sequence using wavelet transformation. Front. Genet. **10**, 90 (2019)
19. Chen, Z.-H., You, Z.-H., Li, L.-P., Wang, Y.-B., Wong, L., Yi, H.-C.: Prediction of self-interacting proteins from protein sequence information based on random projection model and fast fourier transform. Int. J. Mol. Sci. **20**(4), 930 (2019)
20. Li, J., You, Z., Li, X., Ming, Z., Chen, X.: PSPEL: in silico prediction of self-interacting proteins from amino acids sequences using ensemble learning. IEEE/ACM Trans. Comput. Biol. Bioinform. **14**(5), 1165–1172 (2017)
21. An, J.Y., et al.: Identification of self-interacting proteins by exploring evolutionary information embedded in PSI-BLAST-constructed position specific scoring matrix. Oncotarget **7**(50), 82440–82449 (2016)
22. An, J.Y., You, Z.H., Chen, X., Huang, D.S., Yan, G., Wang, D.F.: Robust and accurate prediction of protein self-interactions from amino acids sequence using evolutionary information. Mol. BioSyst. **12**(12), 3702 (2016)
23. Wang, Y., You, Z.: Improving prediction of self-interacting proteins using stacked sparse auto-encoder with PSSM profiles. Int. J. Biol. Sci. **14**(8), 983–991 (2018)
24. Wang, Y., You, Z., Li, X., Jiang, T., Cheng, L., Chen, Z.: Prediction of protein self-interactions using stacked long short-term memory from protein sequences information. BMC Syst. Biol. **12**(8), 129 (2018)
25. Chen, Z.-H., You, Z.-H., Li, L.-P., Wang, Y.-B., Li, X.: RP-FIRF: prediction of self-interacting proteins using random projection classifier combining with finite impulse response filter. In: Huang, D.-S., Jo, K.-H., Zhang, X.-L. (eds.) ICIC 2018. LNCS, vol. 10955, pp. 232–240. Springer, Cham (2018). https://doi.org/10.1007/978-3-319-95933-7_29
26. Huang, S.Y., Zou, X.: An iterative knowledge-based scoring function to predict protein-ligand interactions: I. Derivation of interaction potentials. J. Comput. Chem. **27**(15), 1866–1875 (2006)
27. You, Z.H., Lei, Y.K., Zhu, L., Xia, J., Wang, B.: Prediction of protein-protein interactions from amino acid sequences with ensemble extreme learning machines and principal component analysis. BMC Bioinform. **14**(S8), S10 (2013)
28. You, Z., Li, L., Ji, Z., Li, M., Guo, S.: Prediction of protein-protein interactions from amino acid sequences using extreme learning machine combined with auto covariance descriptor. In: Memetic Computing, pp. 80–85 (2013)
29. Wang, L., et al.: An ensemble approach for large-scale identification of protein-protein interactions using the alignments of multiple sequences. Oncotarget **8**(3), 5149–5159 (2017)
30. Yi, H.C., You, Z.H., Huang, D.S., Li, X., Jiang, T.H., Li, L.P.: A deep learning framework for robust and accurate prediction of ncRNA-Protein interactions using evolutionary information. Mol. Ther. Nucleic Acids **11**, 337–344 (2018)
31. Wen, Y., Lei, H., You, Z., Lei, B., Chen, X., Li, L.: Prediction of protein-protein interactions by label propagation with protein evolutionary and chemical information derived from heterogeneous network. J. Theor. Biol. **430**(10), 9–20 (2017)
32. An, J.Y., Zhang, L., Zhou, Y., Zhao, Y.J., Wang, D.F.: Computational methods using weighed-extreme learning machine to predict protein self-interactions with protein evolutionary information. J. Cheminform. **9**(1), 47 (2017)
33. Wang, Y., You, Z., Li, X., Chen, X., Jiang, T., Zhang, J.: PCVMZM: using the probabilistic classification vector machines model combined with a zernike moments descriptor to predict protein-protein interactions from protein sequences. Int. J. Mol. Sci. **18**(5), 1029 (2017)
34. Wang, Y.-B., You, Z.-H., Li, L.-P., Huang, Y.-A., Yi, H.-C.: Detection of interactions between proteins by using legendre moments descriptor to extract discriminatory information embedded in pssm. Molecules **22**(8), 1366 (2017)

35. Wang, Y., et al.: Predicting protein interactions using a deep learning method-stacked sparse autoencoder combined with a probabilistic classification vector machine. Complexity **2018**, 12 (2018)

36. Wang, Y., You, Z.-H., Yang, S., Li, X., Jiang, T.-H., Zhou, X.: A high efficient biological language model for predicting protein-protein interactions. Cells **8**(2), 122 (2019)

37. Zhu, L., You, Z.-H., Huang, D.-S.: Identifying spurious interactions in the protein-protein interaction networks using local similarity preserving embedding. In: Basu, M., Pan, Y., Wang, J. (eds.) ISBRA 2014. LNCS, vol. 8492, pp. 138–148. Springer, Cham (2014). https://doi.org/10.1007/978-3-319-08171-7_13

38. You, Z.-H., Zhou, M., Luo, X., Li, S.: Highly efficient framework for predicting interactions between proteins. IEEE Trans. Cybern. **47**(3), 731–743 (2017)

39. Luo, X., Ming, Z., You, Z., Li, S., Xia, Y., Leung, H.: Improving network topology-based protein interactome mapping via collaborative filtering. Knowl.-Based Syst. **90**, 23–32 (2015)

40. Li, Z.W., et al.: Accurate prediction of protein-protein interactions by integrating potential evolutionary information embedded in PSSM profile and discriminative vector machine classifier. Oncotarget **8**(14), 23638 (2017)

41. An, J.Y., You, Z.H., Meng, F.R., Xu, S.J., Wang, Y.: RVMAB: using the relevance vector machine model combined with average blocks to predict the interactions of proteins from protein sequences. Int. J. Mol. Sci. **17**(5), 757 (2016)

42. You, Z.-H., Lei, Y.-K., Gui, J., Huang, D.-S., Zhou, X.: Using manifold embedding for assessing and predicting protein interactions from high-throughput experimental data. Bioinformatics **26**(21), 2744–2751 (2010)

43. You, Z.-H., Huang, W., Zhang, S., Huang, Y.-A., Yu, C.-Q., Li, L.-P.: An efficient ensemble learning approach for predicting protein-protein interactions by integrating protein primary sequence and evolutionary information. IEEE/ACM Trans. Comput. Biol. Bioinform. (2018)

44. You, Z.H., Li, X., Chan, K.C.: An improved sequence-based prediction protocol for protein-protein interactions using amino acids substitution matrix and rotation forest ensemble classifiers: Elsevier Science Publishers B. V. (2017)

45. You, Z., Ming, Z., Niu, B., Deng, S., Zhu, Z.: A SVM-based system for predicting protein-protein interactions using a novel representation of protein sequences. In: Huang, D.-S., Bevilacqua, V., Figueroa, J.C., Premaratne, P. (eds.) ICIC 2013. LNCS, vol. 7995, pp. 629–637. Springer, Heidelberg (2013). https://doi.org/10.1007/978-3-642-39479-9_73

46. Wong, L., You, Z.-H., Ming, Z., Li, J., Chen, X., Huang, Y.-A.: Detection of interactions between proteins through rotation forest and local phase quantization descriptors. Int. J. Mol. Sci. **17**(1), 21 (2015)

47. Wen, Y.-T., Lei, H.-J., You, Z.-H., Lei, B.-Y., Chen, X., Li, L.-P.: Prediction of protein-protein interactions by label propagation with protein evolutionary and chemical information derived from heterogeneous network. J. Theor. Biol. **430**, 9–20 (2017)

48. Zhu, L., Deng, S.-P., You, Z.-H., Huang, D.-S.: Identifying spurious interactions in the protein-protein interaction networks using local similarity preserving embedding. IEEE/ACM Trans. Comput. Biol. Bioinform. (TCBB) **14**(2), 345–352 (2017)

49. An, J.Y., Meng, F.R., You, Z.H., Fang, Y.H., Zhao, Y.J., Zhang, M.: Using the relevance vector machine model combined with local phase quantization to predict protein-protein interactions from protein sequences. Biomed. Res. Int. **2016**(6868), 1–9 (2016)

50. An, J.Y., Meng, F.R., You, Z.H., Chen, X., Yan, G.Y., Hu, J.P.: Improving protein–protein interactions prediction accuracy using protein evolutionary information and relevance vector machine model. Protein Sci. **25**(10), 1825–1833 (2016)

51. Huang, Y.-A., You, Z.-H., Chen, X., Chan, K., Luo, X.: Sequence-based prediction of protein-protein interactions using weighted sparse representation model combined with global encoding. BMC Bioinformatics 17(1), 184 (2016)
52. Huang, Y.-A., You, Z.-H., Chen, X., Yan, G.-Y.: Improved protein-protein interactions prediction via weighted sparse representation model combining continuous wavelet descriptor and PseAA composition. BMC Syst. Biol. 10(4), 120 (2016)
53. Huang, Y.-A., et al.: Construction of reliable protein–protein interaction networks using weighted sparse representation based classifier with pseudo substitution matrix representation features. Neurocomputing 218, 131–138 (2016)
54. Lei, Y.-K., You, Z.-H., Dong, T., Jiang, Y.-X., Yang, J.-A.: Increasing reliability of protein interactome by fast manifold embedding. Pattern Recogn. Lett. 34(4), 372–379 (2013)
55. Lei, Y.-K., You, Z.-H., Ji, Z., Zhu, L., Huang, D.-S.: Assessing and predicting protein interactions by combining manifold embedding with multiple information integration. BMC Bioinformatics 13(Suppl. 7), S3 (2012)
56. Li, J., Shi, X., You, Z., Chen, Z., Lin, Q., Fang, M.: Using weighted extreme learning machine combined with scale-invariant feature transform to predict protein-protein interactions from protein evolutionary information. In: Huang, D.-S., Bevilacqua, V., Premaratne, P., Gupta, P. (eds.) ICIC 2018. LNCS, vol. 10954, pp. 527–532. Springer, Cham (2018). https://doi.org/10.1007/978-3-319-95930-6_49
57. Song, X.-Y., Chen, Z.-H., Sun, X.-Y., You, Z.-H., Li, L.-P., Zhao, Y.: An ensemble classifier with random projection for predicting protein-protein interactions using sequence and evolutionary information. Appl. Sci. 8(1), 89 (2018)
58. Yi, H.-C., et al.: ACP-DL: a deep learning long short-term memory model to predict anticancer peptides using high efficiency feature representation. Mol. Ther. Nucleic Acids 17, 1–9 (2019)

# Identification of DNA-Binding Proteins via Fuzzy Multiple Kernel Model and Sequence Information

Yijie Ding[1], Jijun Tang[2,3], and Fei Guo[2(✉)]

[1] School of Electronic and Information Engineering,
Suzhou University of Science and Technology, Suzhou,
People's Republic of China
[2] School of Computer Science and Technology,
College of Intelligence and Computing, Tianjin University, No. 135,
Yaguan Road, Tianjin Haihe Education Park, Tianjin,
People's Republic of China
fguo@tju.edu.cn
[3] Department of Computer Science and Engineering,
University of South Carolina, Columbia, USA

**Abstract.** DNA-binding proteins is the molecular basis for understanding the basic processes of life activities. Many diseases are associated with DNA binding proteins. The methods of detecting DNA-binding proteins are mainly realized by biochemical experiment, which is time consuming and extremely expensive. A lot of computational methods based on Machine Learning (ML) algorithm have been developed to detect DNA-binding proteins. In this study, we propose a novel DNA-binding proteins model via a Fuzzy Multiple Kernel Support Vector Machine. The multiple features of sequence and evolutionary are extracted and constructed as multiple kernels, respectively. Next, these corresponding kernels are integrated by Multiple Kernel Learning (MKL) algorithm. At last, Fuzzy Support Vector Machine (FSVM) is employed to build an effective DNA-binding protein predictor. Comparing with other outstanding methods, our proposed approach achieves good results. The accuracy of our model are 82.98% and 81.70% on PDB1075 (benchmark data set of DNA-binding proteins) and PDB186 (independent test set), respectively. Our approach is comparable to previous methods.

**Keywords:** DNA-binding protein prediction · Protein sequence ·
Multiple Kernel Learning · Fuzzy Support Vector Machine · Feature extraction

## 1 Introduction

DNA and protein are the two most important types of biological macromolecules that constitute organisms. The specific or non-specific recognition and interaction between the two macromolecules play an important role in the regulation of genome expression. What's more, DNA-binding proteins is also the molecular basis for understanding the basic processes of life activities. With the development of genomics research, scientists

© Springer Nature Switzerland AG 2019
D.-S. Huang et al. (Eds.): ICIC 2019, LNCS 11644, pp. 468–479, 2019.
https://doi.org/10.1007/978-3-030-26969-2_45

have discovered that DNA can not only encode proteins, but also bind to proteins, regulate gene activity, and affect gene expression by transcribed RNA. In addition, DNA can serve as a substrate for various chemical modifications and play a role in gene silencing. The occurrence and development of many diseases are associated with abnormalities in DNA binding proteins. The study of DNA-binding proteins has also received increasing attention, and it is important to detect which proteins can bind to DNA. However, Biochemical methods are time- and money-consuming. Lots of computational methods based on Machine Learning (ML) algorithm have been developed to detect DNA-binding proteins. ML-based methods can be divided into two kinds of methods: structure-based model and sequence-based model.

The structure-based models employ structure feature to achieve better performance of DNA-binding protein prediction. Bhardwaj et al. [1] used positive potential surface patches, overall charge and surface composition feature to train a Support Vector Machine (SVM) [2] classifier. Nimrod et al. [3] employed the average surface electrostatic potentials and dipole moments of the protein to train a predictor via Random Forest (RF). Ahmad et al. [4] used three protein features, including net charge, electric dipole moment and quadrupole moment tensors to construct a neural network model for DNA-binding proteins identification. The number of known protein structures is much smaller than the number of protein sequences. Therefore, the structure-based methods cannot be applied to all proteins.

The sequence of protein is easier to obtain for further feature extraction. Cai et al. [5] trained a SVM model via amino acid composition of the protein. Liu et al. [6] employed Pseudo Amino Acid Composition (PseAAC) to represent protein feature. Yu et al. [7] also used amino acid compositions and physicochemical properties of sequence to train SVM classifier. The Position Specific Scoring Matrix (PSSM) of protein sequence can describe the evolutionary conservation of its protein family. PSI-BLAST software [8] can calculate evolutionary conservation information. To identify DNA-binding proteins, Kumar et al. [9] used PSSM to develop a SVM classifier called DNAbinder. Liu et al. [10] proposed a model called iDNAPro-PseAAC, which includes the PseAAC and evolutionary information. Their results of prediction showed that PSSM can significantly improve the performance of detecting DNA-binding proteins. The Local-DPP [11] can capture local conservation information, together with the sequence-order information of PSSM to predict DNA-binding proteins. DBPPred [12] employed RF to select optimal feature subset and trained predictive model by Gaussian Naive Bayes.

Inspired by the previous works, we propose a novel DNA-binding proteins model via a Fuzzy Support Vector Machine (FSVM) based on Multiple Kernel Learning (MKL). The sequence and evolution features are extracted and constructed as multiple kernels, respectively. Then, these corresponding kernels are linear weighted by MKL algorithm. Fuzzy membership score of each training sample is calculated by integrated kernel. Finally, FSVM is employed to train a DNA-binding protein predictor. There are few sequence-based methods that can filter out or reduce the noise.

## 2 Methods

Detecting DNA-binding proteins can be regarded as a traditional binary classification problem. Figure 1 shows the framework of our method. Firstly, we extract sequence and evolution features by six feature extraction algorithms. Then, six types of kernels are constructed by Radial Basis Function (RBF) from above features. MKL algorithm calculates the weights of these kernels, which are linear combined into an integrated kernel. Next, membership scores are calculated for each training sample. Finally, the integrated kernel and membership scores are employed to train a FSVM model for identifying DNA-binding proteins.

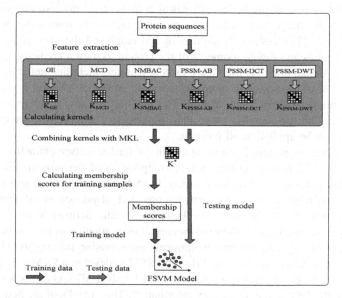

**Fig. 1.** The diagram of DNA-binding protein prediction.

### 2.1 Feature Extraction

One challenge of identifying DNA-binding proteins is to extract feature from proteins. Here, we employ six kinds of feature extraction methods to extract sequence information. These feature extraction algorithms include: Global Encoding (GE) [13], Multi-scale Continuous and Discontinuous descriptor (MCD) [14], Normalized Moreau-Broto Auto correlation (NMBAC) [15, 16], Position Specific Scoring Matrix based Average Blocks (PSSM-AB) [17], PSSM-based Discrete Cosine Transform (PSSM-DCT) [18] and PSSM-based Discrete Wavelet Transform (PSSM-DWT) [19]. Above related feature extraction algorithms have been described in the related literatures. After feature extraction, we use a Radial Basis Function (RBF) and six types of features to construct the corresponding kernel matrix, respectively.

The RBF kernel is defined as follows:

$$K_{ij} = K(\mathbf{x}_i, \mathbf{x}_j) = exp(-\gamma \parallel \mathbf{x}_i - \mathbf{x}_j \parallel^2), \ i, j = 1, 2, \ldots, N \tag{1}$$

where $\gamma$ is the Gaussian kernel bandwidth. $\mathbf{x}_i$ and $\mathbf{x}_j$ are the feature vector of sample $i$ and $j$, $N$ is the number of samples.

We can obtain a kernel set $\mathbf{K}$ as follows:

$$\mathbf{K} = \{\mathbf{K}_{GE}, \mathbf{K}_{MCD}, \mathbf{K}_{NMBAC}, \mathbf{K}_{PSSM\text{-}AB}, \mathbf{K}_{PSSM\text{-}DCT}, \mathbf{K}_{PSSM\text{-}DWT}\} \tag{2}$$

## 2.2 Multiple Kernel Learning

Different kernels can be integrated by MKL, which estimates the weight coefficient of each kernel (kernel set: $\mathbf{K} = \{\mathbf{K}_1, \mathbf{K}_2, \ldots, \mathbf{K}_H\}$). The $H$ is the number of basic kernels defined from feature space. The optimal kernel $\mathbf{K}^*$ is obtained as follows:

$$\mathbf{K}^* = \sum_{h=1}^{H} \beta_h \mathbf{K}_h, \quad \mathbf{K}_h \in \Re^{N \times N} \tag{3}$$

Kullback-Leibler divergence [20], also called relative entropy, is a measure of the similarity for two probability distributions. When the value of Kullback-Leibler divergence is 0, two distributions are identical (ideal case). Kullback-Leibler divergence can be defined as follows:

$$D_{KL}(P \| Q) = \sum_{x} P(x) log \frac{P(x)}{Q(x)} \tag{4}$$

Kullback-Leibler divergence has two main properties: asymmetry ($D_{KL}(P \| Q) \neq D_{KL}(Q \| P)$) and non-negative ($D_{KL}(P \| Q) > 0$). To solve the asymmetry problem, Jensen-Shannon (JS) divergence is proposed for more exact measure of similarity as follows:

$$JS(P \| Q) = \frac{1}{2} D_{KL}(P(x) \| \frac{P(x) + Q(x)}{2}) + \frac{1}{2} D_{KL}(Q(x) \| \frac{P(x) + Q(x)}{2}) \tag{5}$$

The range of JS divergence is 0 to 1. The same is 0, the opposite is 1. In addition, JS divergence is symmetrical ($JS(P \| Q) = JS(Q \| P)$).

The elements in the kernel matrix can be regarded as random variables, which are obeyed some probability distribution. In addition, the kernel matrices are symmetric matrices, we define vector $\mathbf{u} \in \mathbf{R}^{1 \times (N(N-1)/2)}$ as the upper triangular elements. Thus, $\mathbf{u}$ can be regarded as random variables of kernel matrix. We define $\mathbf{u}^p = \mathbf{u}/\|\mathbf{u}\|_1$ as the probability value of $\mathbf{u}$. The probability distribution of kernel matrix $\mathbf{K}_1, \mathbf{K}_2, \ldots, \mathbf{K}_H$ is $\mathbf{u}_1^p, \mathbf{u}_2^p, \ldots, \mathbf{u}_H^p$, respectively.

Inspired by kernel alignment [21, 22], we define objective optimization function as follows:

$$\min_{\beta, \mathbf{u}^{p*}} JS(\mathbf{u}^{p*} \| \mathbf{u}^p_{ideal})$$

$$\text{subject to } \mathbf{u}^{p*} = \sum_{h=1}^{H} \beta_h \mathbf{u}^p_h,$$

$$\beta_h \geq 0, \ h = 1, 2, \ldots, H,$$

$$\sum_{h=1}^{H} \beta_h = 1 \tag{6}$$

where $\mathbf{u}^p_{ideal}$ is the probability values of ideal kernel $\mathbf{K}_{ideal}$ (upper triangular elements), which is calculated by $\mathbf{K}_{ideal} = \mathbf{y}\mathbf{y}^T$ ($\mathbf{y} \in \mathbf{R}^{N \times 1}$ is the labels of training set).

There must be some commonality between multiple kernels, and these commonalities are often able to reflect the inner relationship of heterogeneous data. Generally, the effective kernel is more related to ideal kernel. However, the noise kernel is less dependent on ideal kernel. To promote any two effective kernel matrices (with high similarity) closer (both having higher weights), we minimize the following objective function as follows:

$$\sum_{i,j}^{H} (\beta_i - \beta_j)^2 W_{ij} = \sum_{i,j}^{H} (\beta_i^2 + \beta_j^2 - 2\beta_i\beta_j) W_{ij}$$

$$= \sum_{i}^{H} \beta_i^2 D_{ii} + \sum_{j}^{H} \beta_j^2 D_{jj} - 2 \sum_{i,j}^{H} \beta_i\beta_j W_{ij} \tag{7}$$

$$= 2\beta^T \mathbf{L}\beta$$

where $i, j = 1, \ldots, H$, $\mathbf{W} \in \mathbf{R}^{H \times H}$ is the similarity between kernel matrices. $\mathbf{L} \in \mathbf{R}^{H \times H}$ is graph Laplacian matrix, which is obtained by $\mathbf{L} = \mathbf{D} - \mathbf{W}$ ($D_{ii} = \sum_{j=1}^{H} W_{ij}$, $\mathbf{D} \in \mathbf{R}^{H \times H}$ is a diagonal matrix). $\mathbf{W}$ is calculated by Cosine similarity as follows:

$$W_{ij} = Cosine(\mathbf{K}_i, \mathbf{K}_j) = \frac{\langle \mathbf{K}_i, \mathbf{K}_j \rangle_F}{\|\mathbf{K}_i\|_F \|\mathbf{K}_j\|_F} \tag{8}$$

where $\langle \mathbf{K}_i, \mathbf{K}_j \rangle_F = Trace(\mathbf{K}_i^T \mathbf{K}_j)$ is the Frobenius inner product and $\| \mathbf{K}_i \|_F = \sqrt{\langle \mathbf{K}_i, \mathbf{K}_i \rangle_F}$ is Frobenius norm.

Then, the solution of Eq. 6 can be obtained by solving the problem as follows:

$$\min_{\beta, \mathbf{u}^{p*}} JS(\mathbf{u}^{p*} \| \mathbf{u}^p_{ideal}) + v_1 \beta^T \mathbf{L}\beta + v_2 \|\beta\|_2^2$$

$$\text{subject to } \mathbf{u}^{p*} = \sum_{h=1}^{H} \beta_h \mathbf{u}^p_h,$$

$$\beta_h \geq 0, \ h = 1, 2, \ldots, H,$$

$$\sum_{h=1}^{H} \beta_h = 1 \tag{9}$$

where $v_1$ is graph regularization term, $v_2$ is $L2$ norm regularization term. In addition, this two parameters are optimized by 5-fold cross validation using a grid search. Above method is called Jensen-Shannon Divergence-based Multiple Kernel Learning (JSD-MKL) algorithm.

## 2.3  Fuzzy Support Vector Machine

SVM is a classification and regression paradigm, which was developed by Vapnik [2]. It has been applied to many biological problems. In real-world problems, it is well known that some training points are noises or outliers. The standard SVM algorithm treating every training point equally may lead to overfitting. Different training points make different contributions to the final decision function. Lin and Wang [23] proposed an Fuzzy SVM (FSVM) model to filter outliers by giving different fuzzy membership values for each training point.

Given a set of labeled training samples $\{\mathbf{x}_i, y_i, s_i\}$, $i = 1, 2, \ldots, N$ with input data $\mathbf{x}_i \in R^d$, output labels $y_i \in \{+1, -1\}$ and $s_i \in [0, 1]$, which is a fuzzy membership. A fuzzy membership value represents the degree of $\mathbf{x}_i$ belonging to $y_i$. For FSVM, the goal is to find an optimal hyperplane $\omega^T \phi(\mathbf{x}) + b = 0$, which separates training points into two classes with the maximal margin. $\omega$, $b$ and $\phi$ is the normal vector of the hyperplane, bias and feature map, respectively. The optimization problem of finding hyperplane can be defined as follows:

$$
\begin{aligned}
min \ &\tfrac{1}{2}\|\omega\|^2 + C \sum_{i=1}^{N} s_i \xi_i \\
subject \ to \ &y_i(\omega^T \phi(\mathbf{x}) + b) \geq 1 - \xi_i, \\
&\xi_i \geq 0, i = 1, 2, \ldots, N
\end{aligned}
\tag{10}
$$

where $\xi_i$ is the slack variables for each training point. $\xi_i$ is a measure of error for classifying the point $\mathbf{x}_i$. $C$ is the regularization parameter for imposing a tradeoff between the generalization and training error. $s_i$ is fuzzy membership value for sample $i$. $s_i \xi_i$ can be regarded as a measure of error with different weights. Low $s_i$ reduces the effect of point $\mathbf{x}_i$, which is less important for hyperplane. In other words, $s_i C$ is the cost assigned for a misclassification, important data points have a higher cost, and less important samples have lower cost.

For FSVM, the dual problem is denoted as follows:

$$
\begin{aligned}
max \ &\sum_{i=1}^{N} \alpha_i - \tfrac{1}{2} \sum_{i=1}^{N} \sum_{j=1}^{N} \alpha_i \alpha_j \cdot y_i y_j \cdot K(\mathbf{x}_i, \mathbf{x}_j) \\
subject \ to \ &0 \leq \alpha_i \leq s_i C, \\
&\sum_{i=1}^{N} \alpha_i y_i = 0, i = 1, 2, \ldots, N
\end{aligned}
\tag{11}
$$

The classification decision function implemented by FSVM is represented as follows:

$$f(x) = sign[\sum_{i=1}^{N} y_i \alpha_i \cdot K(\mathbf{x}, \mathbf{x}_i) + b] \tag{12}$$

The kernel width parameter $\gamma$ and regularization parameter $C$ are optimized by 5-fold cross validation using a grid search.

## 2.4  Fuzzy Membership Score

Different training samples have different contributions (parameter $C$) to final decision model. In this study, we employ the optimal kernels $\mathbf{K}^*$ to calculate fuzzy membership. For the training point $t$, we define the membership scores as follows:

$$score_t = \frac{1}{N^2}(\sum_{y_t=y_i} K^*(\mathbf{x}_t, \mathbf{x}_i) - \sum_{y_t \neq y_i} K^*(\mathbf{x}_t, \mathbf{x}_i)) \tag{13}$$

The kernel $K^*(\mathbf{x}_t, \mathbf{x}_i)$ measures the similarity between point $\mathbf{x}_t$ and $\mathbf{x}_i$. The similarity represented by kernel is large for the same class, and it is small for patterns from different classes. In other words, samples with a larger score should make a greater contribution to prediction. Oppositely, the points with smaller score should be considered as outliers (noise samples). We map scores into fuzzy membership values (0–1), as follows:

$$s_t = \frac{1}{1 + exp(-score_t)}, \ t = 1, 2, \ldots, N \tag{14}$$

In addition, the overview of our method is shown in Fig. 1 and Algorithm 1.

---

**Algorithm 1**: Algorithm of our proposed method

**Input**: A training set with N protein sequences and labels;

**Output**: The prediction labels of testing set;

1: Extracting GE, MCD, NMBAC, PSSM-AB, PSSM-DCT and PSSM-DWT feature vector;

2: Employing RBF to construct kernels in Eq 2;

3: Using Eq 9 (JSD-MKL) to calculate kernel weights $\boldsymbol{\beta}$;

4: Calculating $\mathbf{K}^*$ by Eq 3;

5: Using Eq 14 and 13 to calculate membership score ($s_i, \ i = 1, 2, \ldots, N$) of training set;

6: Solving Eq 11 (Training FSVM);

7: Eq 12 is utilized to predict labels of testing set;

---

# 3  Results

In this section, we employ two benchmark data sets to evaluate our method. First, we analyze the performance of different kernels. Then, our method is compared with other methods under Jackknife test on PDB1075 data set. Finally, PDB186 data set as independent test set is used to test the robustness of method.

## 3.1  Data Sets

PDB1075 and PDB186 are two benchmark data sets, which have been used to test predictive model of DNA-binding proteins. The data set of PDB1075 was extracted from Protein Data Bank (PDB) [24]. The PDB1075 consists of 525 DNA-bind proteins and 550 non-DNA-binding proteins. Lou et al. [12] constructed PDB186 data set, which contains 93 DNA-bind proteins and 93 non-DNA-bind proteins.

## 3.2  Measurements

Accuracy (ACC), Specificity (SP), Sensitivity (SN) and Matthew's correlation coefficient (MCC) are employed to evaluate the performance of our method. They are calculated as follows:

$$ACC = \frac{TP + TN}{TP + FP + TN + FN}$$
$$SN = \frac{TP}{TP + FN}$$
$$Spec = \frac{TN}{TN + FP}$$
$$MCC = \frac{TP \times TN - FP \times FN}{\sqrt{(TP + FN) \times (TN + FP) \times (TP + FP) \times (TN + FN)}}$$

(15)

where $TP$, $TN$, $FN$ and $FP$ denotes the number of true positive, true negative, false negative and false positive, respectively. In addition, Area Under ROC curve (AUC) is also used to evaluate our model.

## 3.3  Performance of Analysis on PDB1075

Different kernels have different performance of prediction, we test (Jackknife test evaluation) these kernels on PDB1075 data set, as shown in Table 1. PSSM contains evolutionary information of the protein family. The performance of PSSM information is better than sequence except PSSM-DCT feature (MCC: 0.4233). PSSM-DWT (MCC: 0.4991) and PSSM-AB (MCC: 0.5416) obtain better performance. The performance of multiple kernel integration is generally better than single kernel. Mean weighted kernels (SVM) integrates above six kernel information via average weight and achieves better performance (MCC: 0.6130) than any single kernel. Comparing with mean weighted kernels (SVM), JSD-MKL (SVM) achieves higher value of MCC (0.6243). To filter outliers, FSVM can weight each training sample by fuzzy membership. Mean weighted kernels (FSVM) (MCC: 0.6556) and JSD-MKL (FSVM) (MCC: 0.6593) both employ multiple kernel information and fuzzy membership to improve performance. Moreover, JSD-MKL (FSVM) obtains best MCC of 0.6593. The performance also can be seen from the ROC curve in Fig. 2. JSD-MKL (FSVM) achieves the best AUC of 0.9099.

**Table 1.** The performance of different kernels (features) on PDB1075 data set (Jackknife test).

| Kernel type | Model | ACC | SN | Spec | MCC | AUC |
|---|---|---|---|---|---|---|
| $K_{GE}$ | SVM | 0.7060 | 0.7086 | 0.7036 | 0.4121 | 0.7861 |
| $K_{MCD}$ | SVM | 0.7163 | 0.7086 | 0.7236 | 0.4322 | 0.7830 |
| $K_{NMBAC}$ | SVM | 0.7442 | 0.7676 | 0.7218 | 0.4896 | 0.8108 |
| $K_{PSSM-AB}$ | SVM | 0.7702 | 0.7924 | 0.7491 | 0.5416 | 0.8375 |
| $K_{PSSM-DCT}$ | SVM | 0.7116 | 0.7143 | 0.7091 | 0.4233 | 0.7881 |
| $K_{PSSM-DWT}$ | SVM | 0.7498 | 0.7314 | 0.7673 | 0.4991 | 0.8340 |
| Mean weighted kernels | SVM | 0.8065 | 0.8095 | 0.8036 | 0.6130 | 0.9008 |
| JSD-MKL | SVM | 0.8121 | 0.8171 | 0.8073 | 0.6243 | 0.9015 |
| Mean weighted kernels | FSVM | 0.8279 | 0.8190 | **0.8364** | 0.6556 | 0.9084 |
| JSD-MKL | FSVM | **0.8298** | **0.8248** | 0.8345 | **0.6593** | **0.9099** |

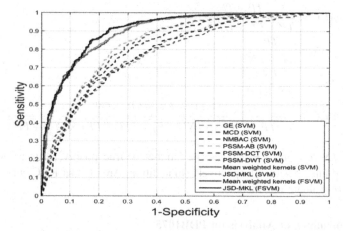

**Fig. 2.** The ROC curve of different kernels (features) on PDB1075 data set (Jackknife test).

Our method (FSVM with JSD-MKL) is also compared to existing methods on PDB1075 data set (Jackknife test). Table 2 shows the results of comparison. The existing methods include IDNA-Prot|dis [10], PseDNA-Pro [6], IDNA-Prot [25], DNA-Prot [26], DNAbinder [9], iDNAPro-PseAAC [27], Kmer1+ACC [28] and Local-DPP [11]. Local-DPP (MCC: 0.59), IDNA-Prot|dis (MCC: 0.54), PseDNA-Pro (MCC: 0.53) and iDNAPro-PseAAC (MCC: 0.53) achieve good performance. Our proposed model obtains best ACC (82.98%), Spec (83.45%) and MCC (0.66) on PDB1075 data set (Jackknife test).

**Table 2.** Compared with existing methods on PDB1075 data set (Jackknife test).

| Methods | ACC (%) | MCC | SN (%) | Spec (%) |
|---|---|---|---|---|
| IDNA-Prot\|dis | 77.30 | 0.54 | 79.40 | 75.27 |
| PseDNA-Pro | 76.55 | 0.53 | 79.61 | 73.63 |
| IDNA-Prot | 75.40 | 0.50 | 83.81 | 64.73 |
| DNA-Prot | 72.55 | 0.44 | 82.67 | 59.76 |
| DNAbinder | 73.95 | 0.48 | 68.57 | 79.09 |
| iDNAPro-PseAAC | 76.56 | 0.53 | 75.62 | 77.45 |
| Kmer1+ACC | 75.23 | 0.50 | 76.76 | 73.76 |
| Local-DPP | 79.10 | 0.59 | **84.80** | 73.60 |
| Our method (FSVM with JSD-MKL) | **82.98** | **0.66** | 82.48 | **83.45** |

Same as previous methods, the PSSM or sequence information of protein is also employed to construct predictive model. Compared with other sequence-based methods, the magnitude of the improvement (our method) is limited. In addition, the final parameter $\gamma$ is $2^{-2}$ (GE), $2^{-5}$ (MCD), $2^{-1}$ (NMBAC), $2^{-3}$ (PSSM-AB), $2^{-3}$ (PSSM-DCT), $2^{-4}$ (PSSM-DWT). The regularization parameter of $C$ is 1.5.

### 3.4   Independent Data Set of PDB186

To further test the performance of methods, our model is also compared to other methods on the independent date set of PDB186 (PDB1075 is training set), which are shown in Table 3.

**Table 3.** Compared with existing methods on PDB186 data set.

| Methods | ACC (%) | MCC | SN (%) | Spec (%) |
|---|---|---|---|---|
| IDNA-Prot\|dis | 72.0 | 0.445 | 79.5 | 64.5 |
| IDNA-Prot | 67.2 | 0.344 | 67.7 | 66.7 |
| DNA-Prot | 61.8 | 0.240 | 69.9 | 53.8 |
| DNAbinder | 60.8 | 0.216 | 57.0 | 64.5 |
| DBPPred | 76.9 | 0.538 | 79.6 | **74.2** |
| iDNAPro-PseAAC | 71.5 | 0.442 | 82.8 | 60.2 |
| Kmer1+ACC | 71.0 | 0.431 | 82.8 | 59.1 |
| Local-DPP | 79.0 | 0.625 | 92.5 | 65.6 |
| Our method (FSVM with JSD-MKL) | **81.7** | **0.657** | **94.6** | 68.8 |

The performance of Local-DPP (MCC: 0.625) is better than other baseline methods. The DBPPred obtains the best Spec of 74.2%. DBPPred (MCC: 0.538) employed random forest to select optimal feature subset and train predictive model by Gaussian Naive Bayes. By employing residue distance pair coupling information, iDNAPro-PseAAC (MCC: 0.442) improved the performance of PseAAC feature. Our method

(FSVM with JSD-MKL) achieves 81.7% of ACC, 0.657 of MCC and 94.6% of SN. Our model still performs better than most of existing methods. The data and program are available at https://figshare.com/s/01ceacb8d0cf9dbfbc41.

## 4   Conclusion and Discussion

The challenges of predicting DNA-binding proteins included: (1) fully describing the information of proteins (multiple sources of information); (2) noise samples. For the problem (1), we used JSD-MKL algorithm to combine different kernels. For the problem (2), a fuzzy-based model was constructed to handle the outliers. Above two technologies achieved better results on independent data sets (MCC: 0.657). In the future, we will use sparse representation method to filter the noise points to improve the performance of DNA-binding proteins predictor.

**Acknowledgments.** This work is supported by a grant from the National Science Foundation of China (NSFC 61772362).

## References

1. Bhardwaj, N., Langlois, R.E., Zhao, G., Lu, H.: Kernel-based machine learning protocol for predicting DNA-binding proteins. Nucleic Acids Res. **33**(20), 6486–6493 (2005)
2. Cortes, C., Vapnik, V.: Support-vector networks. Mach. Learn. **20**, 273–297 (1995)
3. Nimrod, G., Schushan, M., Szilágyi, A., Leslie, C.: iDBPs: a web server for the identification of DNA binding proteins. Bioinformatics **26**(5), 692–693 (2010)
4. Ahmad, S., Sarai, A.: Moment-based prediction of DNA-binding proteins. J. Mol. Biol. **341**(1), 65–71 (2004)
5. Cai, Y.D., Lin, S.L.: Support vector machines for predicting rRNA-, RNA-, and DNA-binding proteins from amino acid sequence. Biochim. Biophys. Acta **1648**(1), 127–133 (2003)
6. Liu, B., Xu, J., Fan, S., Xu, R., Zhou, J., Wang, X.: PseDNA-Pro: DNA-binding protein identification by combining chou's PseAAC and physicochemical distance transformation. Mol. Inform. **34**(1), 8–17 (2015)
7. Yu, X., Cao, J., Cai, Y., Shi, T., Li, Y.: Predicting rRNA-, RNA-, and DNA-binding proteins from primary structure with support vector machines. J. Theor. Biol. **240**(2), 175–184 (2006)
8. Lipman, D.J., et al.: Gapped BLAST and PSI-BLAST: a new generation of protein database search programs. Nucleic Acids Res. **25**(17), 3389–3402 (1997)
9. Kumar, M., Gromiha, M.M., Raghava, G.P.: Identification of DNA-binding proteins using support vector machines and evolutionary profiles. BMC Bioinformatics **8**, 463 (2007)
10. Liu, B., et al.: iDNA-prot|dis: identifying DNA-binding proteins by incorporating amino acid distance-pairs and reduced alphabet profile into the general pseudo amino acid composition. PLoS One **9**, e106691 (2014)
11. Wei, L., Tang, J., Quan, Z.: Local-DPP: an improved DNA-binding protein prediction method by exploring local evolutionary information. Inf. Sci. **384**, 135–144 (2016)
12. Lou, W., Wang, X., Chen, F., Chen, Y., Jiang, B., Zhang, H.: Sequence based prediction of DNA-binding proteins based on hybrid feature selection using random forest and Gaussian Naïve Bayes. PLoS One **9**, e86703 (2014)

13. Li, X., Liao, B., Shu, Y., Zeng, Q., Luo, J.: Protein functional class prediction using global encoding of amino acid sequence. J. Theor. Biol. **261**(2), 290–293 (2009)
14. You, Z.H., Zhu, L., Zheng, C.H., Yu, H.J., Deng, S.P., Ji, Z.: Prediction of protein-protein interactions from amino acid sequences using a novel multi-scale continuous and discontinuous feature set. BMC Bioinformatics **15**, S9 (2014)
15. Ding, Y.J., Tang, J.J., Guo, F.: Predicting protein-protein interactions via multivariate mutual information of protein sequences. BMC Bioinformatics **17**, 398 (2016)
16. Feng, Z.P., Zhang, C.T.: Prediction of membrane protein types based on the hydrophobic index of amino acids. J. Protein Chem. **19**(4), 269–275 (2000)
17. Jeong, J.C., Lin, X., Chen, X.W.: On position-specific scoring matrix for protein function prediction. IEEE/ACM Trans. Comput. Biol. Bioinf. **8**(2), 308–315 (2011)
18. Huang, Y.A., You, Z.H., Gao, X., Wong, L., Wang, L.: Using weighted sparse representation model combined with discrete cosine transformation to predict protein-protein interactions from protein sequence. Biomed. Res. Int. **19**, 902198 (2015)
19. Nanni, L., Brahnam, S., Lumini, A.: Wavelet images and Chou's pseudo amino acid composition for protein classification. Amino Acids **43**, 657–665 (2012)
20. Endres, D.M., Schindelin, J.E.: A new metric for probability distributions. IEEE Trans. Inf. Theory **49**(7), 1858–1860 (2003)
21. Cristianini, N., Kandola, J., Elisseeff, A.: On kernel-target alignment. Adv. Neural. Inf. Process. Syst. **179**(5), 367–373 (2001)
22. Cortes, C., Mohri, M., Rostamizadeh, A.: Algorithms for learning kernels based on centered alignment. J. Mach. Learn. Res. **13**(2), 795–828 (2012)
23. Lin, C.F., Wang, S.D.: Fuzzy support vector machines. IEEE Trans. Neural Networks **13**(2), 464–471 (2002)
24. Rose, P.W., Prlić, A., Bi, C., et al.: The RCSB Protein Data Bank: views of structural biology for basic and applied research and education. Nucleic Acids Res. **43**(Database issue), 345–356 (2015)
25. Lin, W., Fang, J., Xiao, X., Chou, K.: iDNA-Prot: identification of DNA binding proteins using random forest with grey model. PLoS ONE **6**, e24756 (2011)
26. Kumar, K.K., Pugalenthi, G., Suganthan, P.N.: DNA-Prot: identification of DNA binding proteins from protein sequence information using random forest. J. Biomol. Struct. Dyn. **26**(6), 679–686 (2009)
27. Liu, B., Wang, S., Wang, X.: DNA binding protein identification by combining pseudo amino acid composition and profile-based protein representation. Sci. Rep. **5**, 15479 (2015)
28. Xu, R., Zhou, J., Wang, H., He, Y., Wang, X., Liu, B.: Identifying DNA-binding proteins by combining support vector machine and PSSM distance transformation. BMC Syst. Biol. **9**, S10 (2015)

# Research on HP Model Optimization Method Based on Reinforcement Learning

Zhou Fengli$^{(\boxtimes)}$ and Lin Xiaoli

Faculty of Information Engineering,
City College Wuhan University of Science and Technology,
Wuhan 430083, China
thinkview@163.com, aneya@163.com

**Abstract.** Protein structure prediction is an important factor in the area of bioinformatics. Predicting the two-dimensional structure of proteins based on the hydrophobic polarity model (HP model) is a typical non-deterministic polynomial(NP)-hard problem. Currently, HP model optimization methods include the greedy algorithm, particle swarm optimization, genetic algorithm, ant colony algorithm and the Monte-Carlo simulation method. However, the robustness of these methods are not sufficient, and it is easy to fall into a local optimum. Therefore, a HP model optimization method, based on reinforcement learning was proposed. In the full state space, a reward function based on energy function was designed and a rigid overlap detection rule was introduced. By using the characteristics of the continuous Markov optimal decision and maximizing global cumulative return, the global evolutionary relationship in biological sequences was fully exploited, and effective and stable predictions were retrieved. Eight classical sequences from publications and Uniref50 were selected as experimental objects. The robustness, convergence and running time were compared with the greedy algorithm and particle swarm optimization algorithm, respectively. Both reinforcement method and swarm optimization method can find all the lowest energy structures for these eight sequences, while the greedy algorithm only detects 62.5%. Compared with particle swarm optimization, the running time of the reinforcement method is 63.9% lower than that of particle swarm optimization.

**Keywords:** Reinforcement learning · HP model · Structure prediction

## 1 Introduction

The biological functions of proteins are determined by their spatial folding structure. Understanding the folding process of proteins is one of the most challenging problems in the field of bioinformatics [1]. Now relevant researches have proposed many computable theoretical models at home and abroad, such as Hydrophobic Polarity (HP) model, AB amorphous lattice model and continuous model of Euclidean space [2]. HP model is a widely recognized protein folding model that its prediction results have important reference value for the prediction of a protein's three-dimensional structure [3]. The two major difficulties in both HP prediction and protein's three-dimensional structure prediction are exploration problem and evaluation problem.

© Springer Nature Switzerland AG 2019
D.-S. Huang et al. (Eds.): ICIC 2019, LNCS 11644, pp. 480–492, 2019.
https://doi.org/10.1007/978-3-030-26969-2_46

At present, the methods used to solve HP model optimization mainly include Monte-Carlo simulation method, ant colony algorithm (ACO), genetic algorithm (GA) and particle swarm optimization (PSO).

Reinforcement learning has been successfully applied in many aspects of the biological field, such as biological sequence comparison, genome sequencing, etc. [4]. The advantage of reinforcement learning is that the training process does not require external supervision, the interaction between Agent and environment is self-learning with experience, and the global optimal solution is explored according to the reward [5].

Therefore, this paper has proposed an HP model optimization method based on reinforcement learning. By using the characteristics of continuous Markov optimal decision and maximizing global cumulative return, a reward function based on energy function has been designed in the full state space, it has adopted Q-learning training algorithm and introduced a rigid overlap detection rule, the global evolutionary relationship in biological sequences has been fully exploited, and then effective and stable predictions will be retrieved.

## 2  HP Model Optimization Method Based on Reinforcement Learning

### 2.1  Reinforcement Learning

Reinforcement Learning (RL) is a kind of online learning method that is mapping from environmental state to action and obtains the maximum expected cumulative reward [6]. In addition to the Agent and the environment, a reinforcement learning system includes three basic elements of action, state value function and reward (reward function).

The Markov decision process (MDP) can be represented by a five-dimensional array (S, A, T, R, V), S is state set, A is action set, T is state transfer function, R is reward function obtained under state $s$ by action $a$, and V is objective function [7]. Markov property means that the agent in current state $s$ by current action $a$ reaches the next state $s'$ and the immediate reward is only related to current state $s$ and action $a$.

The Q-learning algorithm is a model-independent reinforcement learning algorithm based on action value function. It is also a classic TD control algorithm. Its basic form can be expressed as Eq. (1).

$$Q(s_t, a_t) \leftarrow Q(s_t, a_t) + \alpha\left[r_{t+1} + \gamma_a^{\max} Q(s_{t+1}, a_{t+1}) - Q(s_t, a_t)\right] \tag{1}$$

The algorithm will create a Q value table for discrete problems and the general initial value is set to 0. In the learning process, assuming that the agent is in state $s_t$ at time $t$, action $a_t$ will be adopted to migrate current state to the next state $s_{t+1}$ and the immediate reward $r_{t+1}$ will be obtained, then updating the Q value table by formula (1). Finally it will converge to the optimal action value function and obtain the optimal action strategy according to the optimal action value function through continuous iteration [8].

## 2.2   HP Model Optimization Framework Based on Reinforcement Learning

In recent years, some scholars had proposed some simplified models for protein folding problems, and the most typical model was the two-dimensional HP lattice model proposed by Dill et al. [9] According to the different hydrophilic and hydrophobic properties of amino acid, it can be divided into two categories, one is hydrophobic amino acid which is represented by black sphere and indicated by H, and the other one is hydrophilic amino acid which is represented by white sphere and represented by P, so any protein chain can be represented as a finite string of H or P. A legal protein space configuration must satisfy the following three constraints.

(1) The center of each ball in the sequence must be placed on the integer coordinates of the two-dimensional space.
(2) Any two adjacent balls placed on the chain must be adjacent to each other in two-dimensional space, and the distance between adjacent small balls is 1.
(3) Each integer grid point in two-dimensional space can only put one small ball at most, and any two small balls do not overlap.

It can be understood as a Markov decision process by using the reinforcement learning method to solve the HP two-dimensional sequence model optimization problem, which is solved by the Q-learning algorithm. The framework is shown in Fig. 1.

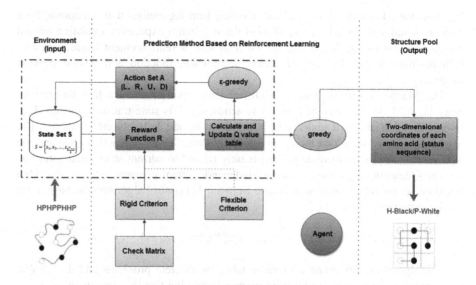

**Fig. 1.** HP model optimization method framework based on reinforcement learning

(1)  Environment (Sequence Input)

According to the hydrophilic and hydrophobic properties, amino acids are divided into H (hydrophobic) and P (hydrophilic), and the amino acid sequence can be converted into HP sequence. So it can be replaced by 0 or 1 in the operation, and the entire state set S composed by HP sequence is corresponding to the environment part of reinforcement learning.

(2) Action Set A

Action Set A consists of four actions that can be used to solve the problem, and they are corresponding to the four possible directions $L$ (left), $U$ (top), $R$ (right), $D$ (bottom) that are used to solve the encoding. Which means $A = \{a_1, a_2, a_3, a_4\}$, where $a_1 = L$, $a_2 = U$, $a_3 = R$, $a_4 = D$.

In the training step, the Agent uses $\varepsilon - greedy$ strategy to select action ($\varepsilon \in [0, 1]$), it indicates that the agent will explore with a probability value $\varepsilon$ and take other actions at random. The remaining probability value of removing "greedy" is $1 - \varepsilon$, that means taking the optimal action and constantly calculating the updated Q value [10].

(3) Structure Pool (Result Output)

After the training the Agent will adopt the greedy strategy to select the optimal action according to Q-valued function and obtain the optimal structure of the HP model.

## 2.3    State Set S of the HP Model

For a two-dimensional sequence of length n, its state space S consists of $\frac{4^{n-1}}{3}$ states. When the state of first amino acid is fixing, all possible states of each subsequent amino acid are the sets of previous amino acid's four states (Up, Down, Left, Right), which is shown in Eq. (2).

$$1 + 4^1 + 4^2 + \ldots + 4^{n-1} = \frac{1 * (1 - 4^n)}{1 - 4} = \frac{4^n - 1}{3} \tag{2}$$

Where $S = \left\{s_1, s_2, \ldots, s_{\frac{4^n-1}{3}}\right\}$ has been called full state space.

During the learning process, it is assumed that the initial state of Agent is $s_1$, the successor state $s_{i_k} \in S\left(i_k \in \left[1, \frac{4^{n-1}}{3}\right]\right)$ will be reached after accessing states $s_1, s_{i_1}, s_{i_2}, \ldots, s_{i_{k-1}}$. And the path from initial state to final state will represent the possible two-dimensional structure of the protein sequence.

At the same time, the state transition function of the HP model can be defined as $T : s \to s'$, which is $T(s, a) = s'$. That means taking action $a$ in state $s$ can reach the successor state $s'$, the concrete expression is shown in formula (3).

$$T\left(s_{\frac{4^k-1}{3}+i}, a_j\right) = s_{\frac{4^{k+1}-1}{3}+4(i-1)+j} \tag{3}$$

Where $k \in [0, n-1]$, $i \in [1, 4^k]$ and $j \in [1, 4]$.

At a given moment, the Agent can move to one of the four possible successor states by performing one of four possible actions under state $s \in S$, the process will repeat until it reaches the end state. It should be noted that taking different actions in different states will transfer to different subsequent states. Each $s' \in S$ can be accessed from state $s$, that means $s' \in U_{a \in A} T(s, a)$ is a successor state of $s$ [11].

## 2.4    Reward Function R Based on HP Model Energy Function

The definition of energy only considers the hydrophobic force. Each H pair that is not adjacent in the sequence and adjacent in the two-dimensional space can generate energy of $-1$, and the energy is 0 in other cases. The entire structure's energy can be calculated by the sum of all energy of the H pairs that are not adjacent in sequence and physically adjacent to each other in the legal configuration.

A formalized description for the legal conformational energy E of protein which chain length is n has been shown in Eq. (4).

$$E = \sum_{i,j=1, i<j-1}^{n} w_{i,j} \tag{4}$$

Where $i$ and $j$ represent the position of the amino acid in the sequence, if $i$ and $j$ are both H and $d(i,j) = 1$, $w_{i,j} = -1$, otherwise $w_{i,j} = 0$.

During the training process, Agent can easily place the amino acid in the crystal lattice where the amino acid was placed before, and this is not allowed in the actual situation. This article has solved the above overlapping problem by reward setting and divided the definition of reward function into two categories.

First category: Flexible Criterion.

(1)  When Agent is selecting the action, it allows the subsequent amino acid to be placed on the lattice position where the amino acid was placed before. But a given penalty (negative reward) will allow the Agent to judge and self-optimize until a maximum reward has been received. It means that the next state will place the amino acid in the invalid position and the reward is set to $-10$ before the termination state is reached.

(2)  The next state is that puts the amino acid in a blank position and sets the reward to 0 before reaching the termination state.

(3)  When the termination state is reached, the absolute value of the final structure' energy is used as a reward.

The reward setting is shown in Eq. (5).

$$R = \begin{cases} -10, i \in (1 \sim n - 1) \text{ is in an invalid location.} \\ 0, i \in (1 \sim n - 1) \text{ is in a blank position.} \\ |E|, i = n \end{cases} \tag{5}$$

Where $n$ is the number of amino acids and $i$ is the $i$-th amino acid.

Second category: Rigid Criterion.

Rigid criterion is different from the flexibility criterion. When the agent selects action to put the next amino acid that may cause the next amino acid to be placed on the lattice of the existing amino acid, and this action is called invalid action. So an action needs to be re-selected until a valid action is selected. Here a check matrix named *check* is introduced that is a two-dimensional matrix of $2n - 1$ rows and $2n - 1$ columns for the sequence of length $n$. And the lattice position which the amino acid has been placed is marked and called invalid position, so this position can no longer be placed and showed in Eq. (6).

$$check = \begin{cases} 1, (i,j) \text{ is in a repeating position.} \\ 0, (i,j) \text{ is in a blank position.} \end{cases} \quad (6)$$

Where $(i,j)$ represents the placement of current amino acid in the check matrix.

(1) The reward is set to 0 before the termination state is reached.
(2) When the termination state is reached, the absolute value of the final structure' energy is used as a reward.

The reward setting is shown in Eq. (7).

$$R = \begin{cases} 0, i \in (1 \sim n - 1) \\ |E|, i = n \end{cases} \quad (7)$$

## 2.5   HP Model Training Algorithm Based on Reinforcement Learning

In the whole state space, the algorithm for solving the 2D-HP protein's folding problem that has utilized Q-learning algorithm in reinforcement learning based on rigid criterion by using the reward function. Its description is as follows.

---

**HP Model Training Algorithm.**

---

$T = zeros\left(\frac{4^{n-1}-1}{3}, 4\right)$
$Q = zeros\left(\frac{4^{n-1}-1}{3}, 4\right)$
$A = [L, R, U, D]$
while ($Iter <=$ Iteration) {
    One-Episode:
    Initialize $s$
    while ($s \mathrel{!=}$ terminal) {
    $a \leftarrow Q(\varepsilon - \text{greedy})$
    $s' \leftarrow T(s, a)$
    If($Check[s'] == 1$)
        $a \leftarrow Q(\varepsilon - \text{greedy})$
        $s' \leftarrow T(s, a)$
    $r \leftarrow R(s, a)$
    $Q(s_t, a_t) \leftarrow (s_t, a_t) + \alpha[r_{t+1} + \gamma_a^{\max} Q(s_{t+1}, a_{t+1}) - Q(s_t, a_t)]$
    $s \leftarrow s'$ } }

Where $T$ is a function matrix of state transfer, $Q$ is state action value matrix, Iteration is number of iterations, *Iter* is iterative counting, One − Episode is episode training function, $s$ is current state, *Terminal* is terminal state, $s'$ is next state, *Check* is check matrix, and $R$ is reward function.

# 3 Experiment

## 3.1 Parameters Analysis

The step parameter and exploration probability affect the accuracy and convergence of value function. The optimal $\alpha$ and $\varepsilon$ need to be adjusted and determined by experiments in different cases. By taking sequence 1 which has been show in Table 2 as the experimental object, the average number of trainings that required to test the lowest energy structure under the different step parameter and exploration probability have been compared. The experiment has compared 5 rounds and the number of trainings per round is 300,000, the test frequency is 1000 per time and the data has been recorded in Table 1.

**Table 1.** Comparison of average number required for lowest energy under different parameters (unit: thousand)

| $\varepsilon$ | $\alpha$ | | | | | | |
|---|---|---|---|---|---|---|---|
| | 0.01 | 0.1 | 0.3 | 0.5 | 0.7 | 0.9 | 0.99 |
| 0.2 | – | 122 | 25 | 21 | 18 | 18 | 16 |
| 0.5 | 92 | 10 | 6 | 4 | 4 | 2 | 2 |
| 0.8 | 88 | 15 | 6 | 3 | 2 | 2 | 2 |

It can be seen from Table 1 that when $\alpha$ is the same, the required number is very close under $\varepsilon = 0.5$ and $\varepsilon = 0.8$. And the value function will fluctuate when it tends to converge under $\alpha = 0.99$, so the values of parameters is $\alpha = 0.9$ and $\varepsilon = 0.5$ by comprehensive considering.

## 3.2 Comparative Experiment Between Rigid Criterion and Flexible Criterion

According to the two different reward settings based on rigid criterion and flexible criterion, the sequence has been tested for training. In this paper, three proceedings' sequence and five classic Uniref50 dataset sequences have been selected as experimental subjects, and its known and test information are shown in Table 2.

**Table 2.** HP sequence set for testing

| Serial number | HP sequence | Length | Sequence source | Known energy | Flexibe energy | Rigid energy |
|---|---|---|---|---|---|---|
| 1 | HPHHHPHHPH | 10 | Literature [13] | −4 | −4 | −4 |
| 2 | HPPHPPPPHPPHP | 13 | Literature [12] | −4 | −4 | −4 |
| 3 | HHPHPPHPHPHHPH | 14 | Literature [13] | −6 | −6 | −6 |
| 4 | HHHPHHPPPHHPPH | 14 | E5RJN1* | −6 | −6 | −6 |
| 5 | HPHHHPHPPPPHHH | 14 | E5RJN1* | −5 | −5 | −5 |
| 6 | HHHHHPPPPHPHHP | 14 | G3V446* | −5 | −5 | −5 |
| 7 | HHPPPHPHHPHPPH | 14 | K7EL94* | −5 | −5 | −5 |
| 8 | PHHPPHHPPHHPHP | 14 | H0YAK6* | −5 | −5 | −5 |

This paper has tested eight sequences separately. In order to avoid contingency, each sequence has been tested for five rounds according to the rigid and flexible criteria. The iteration number per round is set to 300,000. The current energy has been tested once from every 1,000 trainings. The number of training iterations required to obtain the lowest energy that is also convergence state has been shown in Table 3.

It can be seen from Tables 2 and 3 that the rigid criterion can train to the convergence state and obtain the optimal structure with lowest energy more quickly for all sequences. But the advantage of rigid criterion is not obvious for shorter sequences. With the increasing of sequence's length, the iterative times required for flexible criterion multiply relative to rigid criterion, so rigid criterion has outstanding advantage. The reinforcement learning method based on rigid criterion can obtain the lowest energy by training 32,000, 16,000 and 9000 times respectively for sequence 5, sequence 6 and sequence 8, which is 79.5%, 89.6% and 83.6% less than 156,000, 154,000 and 55,000 times required for the reinforcement learning method based on flexible criterion. All sequences have been stably predicted for the lowest energy structure after 50,000 trainings under the reinforcement learning method based on rigid criterion.

**Table 3.** Iteration number for sequence to converge under different criterias (unit: thousand)

| Serial number | Criterion | 1 | 2 | 3 | 4 | 5 | Avg |
|---|---|---|---|---|---|---|---|
| 1 | Rigidity | 2 | 2 | 2 | 3 | 3 | 2 |
| | Flexibility | 6 | 5 | 4 | 4 | 6 | 5 |
| 2 | Rigidity | 44 | 22 | 34 | 15 | 45 | 32 |
| | Flexibility | 116 | 87 | 190 | 162 | 136 | 138 |
| 3 | Rigidity | 41 | 23 | 19 | 47 | 20 | 30 |
| | Flexibility | 229 | 210 | 92 | 212 | 82 | 165 |
| 4 | Rigidity | 32 | 27 | 26 | 19 | 20 | 25 |
| | Flexibility | 56 | 37 | 123 | 64 | 52 | 66 |

(*continued*)

**Table 3.** (*continued*)

| Serial number | Criterion | 1 | 2 | 3 | 4 | 5 | Avg |
|---|---|---|---|---|---|---|---|
| 5 | Rigidity | 31 | 28 | 46 | 35 | 22 | 32 |
|   | Flexibility | 155 | 94 | 220 | 149 | 161 | 156 |
| 6 | Rigidity | 14 | 26 | 9 | 16 | 15 | 16 |
|   | Flexibility | 214 | 54 | 289 | 77 | 134 | 154 |
| 7 | Rigidity | 18 | 18 | 18 | 30 | 18 | 20 |
|   | Flexibility | 27 | 57 | 144 | 55 | 37 | 64 |
| 8 | Rigidity | 9 | 9 | 10 | 4 | 13 | 9 |
|   | Flexibility | 49 | 61 | 54 | 45 | 65 | 55 |

According to the comprehensive comparison, the results that have been tested by reward function using rigid criterion are always better than flexible criterion and are consistent with expected results.

### 3.3    Comparative Experiment Between Reinforcement Learning and Greedy Algorithm

This paper has compared the reinforcement learning method based on rigid criterion with greedy algorithm and the experimental objects are the above eight sequences. In order to avoid contingency, each sequence has been trained for 5 rounds and the training iterations per round is set to 300,000. Taking one sample per 1000 iterations and a total of 300 samples will be taken. The times' number that can get the lowest energy in 300 samples has been recorded in Table 4.

It can be seen from Table 4 that the average time obtained the lowest energy by reinforcement learning method is much larger than that of greedy algorithm for a sequence with a longer length, and the greedy algorithm hardly trains the lowest energy for sequences 2, 3 and 4. The Agent can learn more times to select better actions in the sequence training and obtain a lower energy structure for reinforcement learning method. So it can be considered that the training effect of reinforcement learning method is stable and better. Greedy algorithm is not ideal for the training of other sequences except sequence 1. It can only be trained occasionally to get the lowest energy, and the accuracy of the lowest energy structure cannot be guaranteed.

**Table 4.** Number of times obtained the lowest energy in 300 samples

| Serial number | Method | 1 | 2 | 3 | 4 | 5 | Avg |
|---|---|---|---|---|---|---|---|
| 1 | Reinforcement | 60 | 63 | 72 | 65 | 66 | 65 |
|   | Greedy | 73 | 95 | 97 | 96 | 77 | 88 |
| 2 | Reinforcement | 10 | 4 | 17 | 10 | 8 | 10 |
|   | Greedy | 0 | 1 | 0 | 0 | 1 | 0 |

(*continued*)

**Table 4.** (*continued*)

| Serial number | Method | 1 | 2 | 3 | 4 | 5 | Avg |
|---|---|---|---|---|---|---|---|
| 3 | Reinforcement | 12 | 11 | 5 | 18 | 14 | 12 |
|   | Greedy | 1 | 0 | 0 | 0 | 1 | 0 |
| 4 | Reinforcement | 5 | 5 | 2 | 1 | 4 | 3 |
|   | Greedy | 0 | 0 | 1 | 0 | 1 | 0 |
| 5 | Reinforcement | 6 | 10 | 9 | 14 | 16 | 11 |
|   | Greedy | 1 | 3 | 4 | 1 | 4 | 3 |
| 6 | Reinforcement | 27 | 18 | 27 | 24 | 18 | 23 |
|   | Greedy | 3 | 9 | 6 | 6 | 5 | 6 |
| 7 | Reinforcement | 21 | 15 | 17 | 9 | 14 | 15 |
|   | Greedy | 8 | 10 | 5 | 6 | 8 | 7 |
| 8 | Reinforcement | 17 | 20 | 32 | 29 | 25 | 25 |
|   | Greedy | 5 | 17 | 11 | 9 | 6 | 10 |

## 3.4 Comparative Experiment Between Reinforcement Learning and Particle Swarm Optimization

Particle swarm optimization is a kind of evolutionary algorithm which starts from the random solution and finds the optimal solution through iteration. It evaluates the solution's quality through fitness and is a global random search algorithm essentially. The above eight sequences have still been used as experimental objects. The experiments have compared the time that required to find the optimal structure with the lowest energy by reinforcement learning method based on rigid criterion and particle swarm optimization algorithm. The operating environment of all experiments is the same, and the running time has been measured by program running time. The total measurements are five rounds in order to avoid contingency and the time has been recorded in Table 5.

Both the reinforcement learning method and the particle swarm optimization algorithm can find the lowest energy of all sequences. But the time needed for PSO algorithm to find the lowest energy of other sequences except sequence 1 is several times over that for RL method. For Sequences 3, 4 and 7, the time needed to obtain the lowest energy by RL method is much smaller than PSO algorithm. And RL method which used rigid criterion can obtain the lowest energy by training 8.3, 7.3 and 5.7 s respectively, while PSO algorithm needs 70.3, 58.1 and 31.1 s. So the time used in RL method is 88.2%, 88.1% and 81.7% shorter than PSO algorithm. And the efficiency of RL method is much higher than that of PSO algorithm, the running time is reduced to 63.9% on the average.

**Table 5.** Time comparison of RL and PSO algorithm (unit: second)

| Serial number | Method | 1 | 2 | 3 | 4 | 5 | Avg |
|---|---|---|---|---|---|---|---|
| 1 | Reinforcement | 0.4 | 0.4 | 0.4 | 0.5 | 0.5 | 0.4 |
|   | Particle swarm | 0.1 | 0.3 | 0.1 | 0.1 | 0.3 | 0.2 |
| 2 | Reinforcement | 11.9 | 6.0 | 7.8 | 4.6 | 12.2 | 8.5 |
|   | Particle swarm | 61.5 | 63.0 | 9.5 | 60.4 | 11.0 | 41.1 |
| 3 | Reinforcement | 11.7 | 6.6 | 6.0 | 12.7 | 4.7 | 8.3 |
|   | Particle swarm | 73.1 | 51.7 | 49.4 | 84.8 | 92.6 | 70.3 |
| 4 | Reinforcement | 11.0 | 7.6 | 7.5 | 5.0 | 5.2 | 7.3 |
|   | Particle swarm | 67.3 | 80.1 | 67.2 | 30.4 | 45.3 | 58.1 |
| 5 | Reinforcement | 7.4 | 6.7 | 12.4 | 10.1 | 6.1 | 8.5 |
|   | Particle swarm | 31.2 | 19.7 | 9.8 | 10.1 | 42.5 | 22.7 |
| 6 | Reinforcement | 3.8 | 7.1 | 2.4 | 5.1 | 4.4 | 4.6 |
|   | Particle swarm | 34.8 | 20.8 | 10.3 | 13.5 | 11.9 | 18.3 |
| 7 | Reinforcement | 4.7 | 4.6 | 5.4 | 8.6 | 5.3 | 5.7 |
|   | Particle swarm | 7.9 | 28.2 | 83.0 | 22.6 | 13.9 | 31.1 |
| 8 | Reinforcement | 2.5 | 2.6 | 2.8 | 1.4 | 3.3 | 2.5 |
|   | Particle swarm | 25.6 | 19.1 | 18.0 | 7.6 | 27.0 | 19.5 |

## 3.5  Case

Sequence 4 is a fragment of focal adhesion kinase (FAK) and is a cytoplasmic non-receptor protein tyrosine kinase. FAK is very important in cell signal transduction, it is the center of intracellular and extracellular signal transduction, and mediates multiple signaling pathways. Reinforcement learning method based on rigid criterion can predict the optimal structure which means the lowest global energy obtained by sequence 4 and the result is shown in Fig. 2.

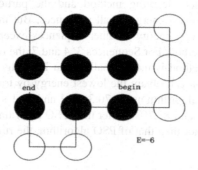

**Fig. 2.** Optimal structure of sequence 4 predicted by RL method (rigid criterion)

# 4   Conclusion and Outlook

This paper has proposed a model based on reinforcement learning to solve the prediction problem of HP model, which is also a basic problem in computational molecular biology. The state set and state transition space can be calculated according to HP sequence length and the reward functions can also be set in different situations. Agent have used $\varepsilon - greedy$ strategy to select the exploration action to get the reward and continuously updated the Q value. Finally, the optimal solution will be selected from the converged Q value table and the optimal HP structure will be obtained. Then eight sequences have been selected as experimental objects. The experimental results show that RL method can converge to optimal value function and obtain the optimal structure of HP model under the condition of rigid criterion. Compared with greedy algorithm, the proposed method has strong robustness and convergence. Compared with the particle swarm optimization algorithm, the method in this paper has taken less time and is more efficient. This paper is a new attempt to use RL method in the field of protein structure prediction. It will be a role model for using RL in protein three-dimensional structure prediction and other fields of biological information.

This method will have further optimization space. On the one hand, the method has good calculation performance and stable convergence for short-length sequences, but it also can be calculated for long-length sequences that need high memory requirements for computers; on the other hand, a large number of training events need to be considered in order to obtain accurate results that will bring slow convergence process and speed. The follow-up study will further enhance the forecasting effect from these two aspects.

**Acknowledgment.** This work was supported in part by Hubei Province Natural Science Foundation of China (No. 2018CFB526).

# References

1. Márquez-Chamorro, A.E., Asencio-Cortés, G., Santiesteban-Toca, C.E., et al.: Soft computing methods for the prediction of protein tertiary structures: a survey. Appl. Soft Comput. **35**(C), 398–410 (2015)
2. Qian, J., Xin, J., Lee, S.J., et al.: Protein secondary structure prediction: a survey of the state of the art. J. Mol. Graph. Model. **76**, 379–402 (2017)
3. Huang, Y.A., You, Z.H., Chen, X., et al.: Sequence-based prediction of protein-protein interactions using weighted sparse representation model combined with global encoding. BMC Bioinformatics **17**(1), 1–11 (2016)
4. Hwang, K.S., Jiang, W.C., Chen, Y.J.: Model learning and knowledge sharing for a multiagent system with Dyna-Q learning. IEEE Trans. Cybern. **45**(5), 964–976 (2015)
5. Liu, Q., Zhai, J.W., Zhang, Z.Z., et al.: Summary of deep reinforcement learning. J. Comput. **1**, 1–27 (2018)
6. Hu, L.B., Chen, J.P., Fu, Q.M., et al.: An adaptive learning adaptive control method for building energy saving. Comput. Eng. Appl. **53**(21), 239–246 (2017)
7. Liu, Z.B., Zeng, X.Q., Liu, H.Y., et al.: Two-layer heuristic reinforcement learning method based on BP neural network. Comput. Res. Dev. **52**(3), 579–587 (2015)

8.  Liu, X.W., Gao, C.M.: Combining behavior tree and Q-learning to optimize agent behavior decision in UT2004. Comput. Eng. Appl. **52**(3), 113–118 (2016)
9.  Zhang, L., Kong, L., Han, X., et al.: Structural class prediction of protein using novel feature extraction method from chaos game representation of predicted secondary structure. J. Theor. Biol. **400**, 1–10 (2016)
10. Yu, J., Liu, Q., Fu, Q.M., et al.: Bayesian Q learning method based on priority scanning Dyna structure. J. Commun. **11**, 129–139 (2013)
11. Fu, Q.M., Liu, Q., Wang, H., et al.: An out-of-strategy Q($\lambda$) algorithm based on linear function approximation. J. Comput. **37**(3), 677–686 (2014)
12. Chen, M.: Quasi-physical and quasi-human algorithm for solving protein folding problems. Huazhong University of Science and Technology (2007)
13. Lu, Q.G., Chen, D.F., Mao, L.M., et al.: Protein structure prediction based on simplified energy function and genetic algorithm. In: China Artificial Intelligence Annual Conference (2015)

# EnsembleKQC: An Unsupervised Ensemble Learning Method for Quality Control of Single Cell RNA-seq Sequencing Data

Anqi Ma[1,2], Zuolang Zhu[1,2], Meiqin Ye[1,2], and Fei Wang[1,2(✉)]

[1] Shanghai Key Lab of Intelligent Information Processing, Shanghai, China
[2] School of Computer Science and Technology,
Fudan University, Shanghai, China
wangfei@fudan.edu.cn

**Abstract.** Single cell RNA sequencing (scRNA-seq) provides a view of high-resolution to reveal the cellular heterogenicity. A series of analysis, such as cell-type identification, differential expression analysis, regulatory relationship detection, could uncover unprecedented biological findings. Prior to these downstream analysis, it's crucial to remove low-quality cells because they are technical noises which weaken true biological signal and mislead downstream analysis. Existing methods either require setting threshold manually or require true labels for supervised training, which is not appropriate in many cases. We present an unsupervised ensemble learning method, which could automatically identify low-quality cells from single cell RNA-seq sequencing data. This method integrates weak classifiers base on five selected features from house-keeping genes, reads mapping rate and detected genes. To avoid setting thresholds of classifiers manually, it enumerates threshold values within a reasonable range and chooses the most suitable threshold values based on a scoring function. In experiments, it exhibits high and steady accuracy on multiple datasets.

**Keywords:** scRNA-seq · Quality control · Ensemble method

## 1 Background

In recent years, single cell RNA sequencing (scRNA-seq) technology has developed rapidly. Compared with bulk RNA sequencing, it measures gene expression abundance at a single-cell resolution so that it can reveal the heterogeneity within populations of cells. This powerful technique inspires a lot of downstream analysis such as differential expression analysis, single cell mutation detection, cell-type identification and tumor evolution analysis. However, due to technical limits, it produces some low-quality cells during the cell separation and sequencing process, for example, cells may be damaged or killed when be captured during the separation process, the capture site may be empty

---

Code is available at https://github.com/mzhq/EnsembleKQC.

D.-S. Huang et al. (Eds.): ICIC 2019, LNCS 11644, pp. 493–504, 2019.
https://doi.org/10.1007/978-3-030-26969-2_47

or contains multiple cells [10]. Based on these reasons, it's crucial to perform quality control to remove low-quality cells prior to downstream analysis.

A lot of methods of quality control have been proposed, which can be divided into experimental methods and computational methods. Experimental methods, such as using microscopic inspection at capture sites to remove unqualified captures before sequencing [15], are often time-consuming and inefficient. In computational methods, the traditional way is to manually set thresholds of some features. Various features can be used, such as the expression abundance of some specific genes (e.g., housekeeping genes), the library size, the number of expressed genes, the fraction of reads mapping to mitochondria encoded genes or spike-in RNAs [1], etc. After selecting features, cells whose values of features are below or above these fixed thresholds will be removed. For example, Dr.seq [9] sets thresholds of the reads duplicate rate, the intron reads rate and the covered gene number. Scater [18] sets thresholds of the total count across all genes, the total number of expressed genes, and the percentage of counts allocated to control genes like spike-in transcripts or mitochondrial genes. SCell [5] uses a Lorenz statistical method to evaluate the background levels of cells, and sets a threshold of the q-value to remove cells whose background levels are significantly high. However, no matter which features they use, how to determine an appropriate threshold is not easy. The thresholds do not keep constant or stable in most cases, such as threshold of the number of expressed genes is set to be 1000 in a mouse retinal cells dataset [9], but becomes 5000 in another mouse embryonic stem cells dataset [11]. To alleviate this problem, some composite methods have been proposed. For example, Cellity [10] is a recent supervised learning method which uses support vector machine(SVM) for classification. Though it performs well on its own training dataset, our experiments show that when Cellity is used for prediction on other datasets, its performance drops dramatically. Since true labels for cell qualities are not known in most scRNA-seq data, the training data of supervised learning methods are often insufficient. SinQC [12] is unsupervised. It uses the gene expression pattern to classify cells into the main cluster and outliers, and assumes that gene expression outliers with poor data quality are low-quality cells. To evaluate the data quality of cells, it uses two meta-scores based on the number of mapped reads, mapping rate and library complexity. The thresholds of meta-scores are determined by user-defined FPR (false positive rate) threshold. In addition, whether a cell is an outlier depends on a p-value, which is also user-defined. Thus, it is still a fixed threshold method. What's more, SinQC can only be used in single-end sequencing data, however, most data are pair-end now.

In this article, we present EnsembleKQC, an unsupervised learning method for quality control of single-cell sequencing data. It automatically explores better thresholds for five features respectively and integrates answers from each feature to give a comprehensive evaluation about whether a single cell is low quality. To do this, it first learns five weak classifiers from each feature respectively, enumerates the threshold values within a reasonable range, integrates the classification results by K-means, and finally chooses the best classification result based on our scoring function. In experiments, EnsembleKQC shows the high recall and precision. It outperforms other quality control methods on the F1-score. Though it takes more time than other methods, the time cost is acceptable in many application cases.

The details of EnsembleKQC are introduced in the METHODS section. The datasets and experiment results are shown in the RESULTS section. Next, a discussion is provided in the DISCUSSION section. Finally, we describe our conclusion in the CONCLUSION section.

## 2 Methods

The framework of our method, EnsembleKQC, is shown in Fig. 1. First, five features which could describe the quality of single cells to some extent are selected. Then for each feature, a linear weak classifier is constructed to label each cell as high quality or low quality. Next, labels from five weak classifiers are integrated to pop out a final judgement whether a cell is high quality or not. Instead of setting thresholds of each linear classifier manually, it first pre-estimates a range of low-quality cell number, and obtains a threshold candidate list within this range for each classifier, then enumerates all possible thresholds in the candidate lists, and evaluates the goodness of thresholds in each enumeration round by a scoring function to automatically choose the best thresholds. Next, we will explain our method in detail step-by-step.

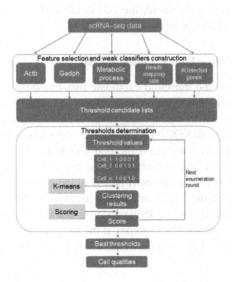

**Fig. 1.** The framework of EnsembleKQC.

### 2.1 Feature Selection and Weak Classifiers Construction

Given biological prior knowledge and sequencing data characteristics, housekeeping genes, the number of detected genes and reads mapping rate are selected as features of scRNA-seq data quality.

In molecular biology, housekeeping genes are typically constitutive genes that are required for the maintenance of basic cellular function, and are expressed in all cells of an organism under normal and pathophysiological conditions [4, 6, 24]. Due to the high conservation of housekeeping genes, cells whose expression levels of housekeeping genes are lower than others are very likely to be low-quality. Based on this phenomenon, we first select the expression of Actb, Gadph, and Metabolic process genes as features. Here Actb represents beta-actin, which is one of the non-muscle cytoskeletal actins. It is a highly conserved protein involved in cell motility, structure and integrity [7, 8]. Gadph is the abbreviation of Glyceraldehyde 3-phosphate dehydrogenase, which is an enzyme that catalyzes the glycolysis. It also functions in apoptosis induction, receptor-associated kinase, tRNA export and DNA repair [14, 21, 23]. Metabolic process is the set of life-sustaining chemical reactions in organisms, and has high similarity among different species [19].

We then select reads mapping rate and the number of detected genes as our last two features. The reason behind is that when cells are contaminated or meet other unknown errors during library preparation, in which only in a small number of genes will be detected, or only a few reads can be mapped to reference genome and most of them are discarded in the mapping step. Since the five features described above are already contained in the features of Cellity [10], our method uses Cellity to extract them.

After feature selection, we construct five weak classifiers according to each feature. Each weak classifier implements simple linear classification. If the value of this feature is above the corresponding threshold, the cell is labelled as one, which means high-quality, otherwise, the label is zero.

## 2.2  Producing the Threshold Candidate Lists

In this step, we produce the threshold candidate lists based on a range of the possible low-quality cell number. Users can provide a specific range if some characteristics of data are already known. Otherwise a fuzzy range is also acceptable, though it will take more time to run EnsembleKQC. To ensure the algorithm run correctly, the upper bound of the range should not exceed the one which will label half of the total cell number as low-quality. In our experiments, we use 10% and 20% of the total cell number as the lower bound and the upper bound of the estimated range. Then, for each feature, we obtain a threshold candidate list by extracting sorted values in this range. To reduce the size of solution size, we define a step S and take a value for every S values from the sorted feature values. For example, if the range is [40, 120] and the step is 20, the $40^{th}$, $60^{th}$, $80^{th}$, $100^{th}$, $120^{th}$ smallest values will be chosen as threshold candidates for each feature.

## 2.3  Enumerating Thresholds and Performing Clustering

We enumerate every possible combination of five thresholds from the threshold candidate lists. In each round of enumeration, we first use five weak classifiers to get labels of cells for each feature. To integrate the results of weak classifiers, we concatenate the labels to obtain a vector of size five for each cell. We then perform K-means [17] to cluster cells into two clusters based on their label vectors. We regard the cluster whose number is the majority as the cluster of high-quality cells, and the other cluster as the low-quality cells.

## 2.4   Evaluating Thresholds by Scoring Function

In each enumeration round, we evaluate the performance of thresholds by a scoring function. As for cells with high quality, most housekeeping genes express, mapping rate of reads should be high and the number of detected genes is numerous. Therefore, high-quality cells can be treated as similar to each other for the five features extracted as the signatures of cell quality. On the other side, low quality cells are dissimilar since the reasons making cells low-quality may be various. From this consideration, average similarity among high-quality cells is calculated to select good thresholds. Cosine similarity is chosen as our similarity measure to score the combination of threshold setting, which is described as follows:

$$score = \frac{2}{N_{high} \cdot (N_{high} - 1)} \sum_{i=1}^{N_{high}} \sum_{j=i+1}^{N_{high}} cosine\_similarity(cell_i, cell_j) \quad (1)$$

$$cosine\_similarity(A, B) = \frac{A \cdot B}{\|A\|\|B\|} \quad (2)$$

Here $N_{high}$ indicates the number of high-quality cells. The cosine similarity is calculated based on the raw values of the selected five feature for each cell. This scoring function evaluates the result in each enumeration round. Finally, we choose the parameter combination which achieves the largest score as our final output result. In order to reduce the number of repeated calculations, we will pre-calculate the similarity between every pair of cells before the threshold enumeration step. Furthermore, since each enumeration round is independent, parallel computing is used to speed up the process.

# 3   Results

We first demonstrate whether the five selected features could discriminate low-quality cells to some extent. Then, we explore the necessary of classifier ensemble instead of immediate clustering based on the features and compare EnsembleKQC with outlier detection algorithms and Cellity algorithm in the performance.

## 3.1   Datasets

The datasets used in experiments contains two labeled datasets and a partially labeled data set. Details of three datasets are described below.

*mES data Kolodziejczyk AA,2015, E-MTAB-2600 [13].* This dataset is a scRNA-seq dataset of 960 mouse cells under three different culture conditions, in which 175 cells are manually labeled as low-quality cells.

*mES data Buettner et al. 2015, E-MTAB-3749 [3].* This dataset is a scRNA-seq dataset of 672 mouse cells, in which 144 cells are manually labeled as low-quality cells.

*mES data Shalek 2014 E-GEOD-48968 [22].* This dataset is a scRNA-seq dataset of 1861 mouse cells from 24 different culture conditions and two different sequencing platforms. Only 1090 cells have labels, in which 345 cells are labeled as low-quality cells. Since it only provides the TPM matrix and reads mapping rate cannot be extracted, we set the mapping rate as 1 to all cells to apply EnsembleKQC in experiments.

### 3.2  Discrimination of the Features

Our method EnsembleKQC is dependent on five selected features. To show the discrimination of each feature respectively, we uniformly select 100 thresholds within the range of values for each feature. Then we use these 100 thresholds to perform classification and draw the corresponding ROC curve. Figure 2 shows the ROC curve of each classifier on two datasets with labels.

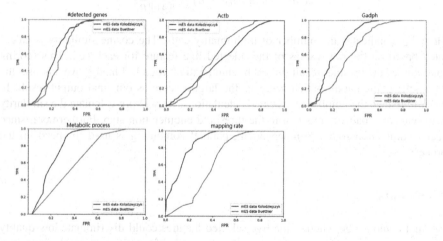

**Fig. 2.** The ROC curve of each classifier on two datasets

From Fig. 2, we can figure out that each weak classifier has its ability to distinguish low-quality cells since their ROC curves are all above the diagonal. However, the performance of each weak classifier varies on different datasets, makes it hard to determine which feature is the most effective. For example, the mapping rate has the best performance on the Kolodziejczyk data, but its performance significantly declines on the Buettner data. In this case, using the ensemble method to integrate weak classifiers can obtain more stable performance.

We then use the real number of low-quality cells to get some reference threshold values for each feature. This is to say, if the dataset has N low-quality cells, we list N-th smallest values of each feature. We scale the value of reference threshold of the Actb, Gadph and metabolic process using the formula as follows to make the expressions comparable. Mapping rate and detected genes still use the raw values.

$$scaled\_reference\_threshold_f = \frac{reference\_threshold_f - Min_f}{Max_f - Min_f} * 100 \qquad (3)$$

Here, $Min_f$ and $Max_f$ represent the minimum and the maximum value of feature $f$.
As shown in Table 1, for each feature, reference threshold varies according to different datasets. Among these features, metabolic process shows the biggest difference, which varies from 28.33 to 2.17. The reference threshold of Actb also spans a wide range, which spans from 6.09 to 19.15. Other features are more stable but still show some difference. Thus, using fixed thresholds for all datasets is not appropriate.

**Table 1.** The reference thresholds of each feature on different datasets

| Datasets | Features | | | | |
|---|---|---|---|---|---|
| | Actb | Gadph | Metabolic process | Mapping rate | Detected genes |
| Kolodziejczyk | 6.09 | 20.22 | 28.33 | 0.73 | 10140 |
| Buettner | 19.15 | 14.26 | 2.17 | 0.80 | 8957 |

We also calculate the intersection of low-quality cells detected by weak classifiers. Figure 3 shows the corresponding Wayne diagram. In order to make the classification result of each weak classifier comparable, we use the real number of low-quality cells to obtain the corresponding thresholds for each classifier (If the real number of low-quality cells is N, we use the Nth smallest values of each feature as the thresholds). The value of each grid represents the ratio of the intersection of the corresponding row and column classifier results. We find most of the intersection ratios are between 20% and 50%, indicating the results of different features are not very consistent.

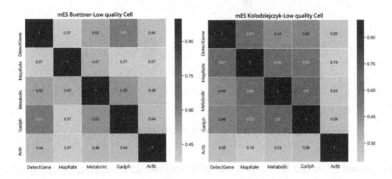

**Fig. 3.** The intersection of the low-quality cells detected by weak classifiers

To sum up, each weak classifier based on a single feature has a certain ability to detect low-quality cells, but is not strong enough. And since the performance of each feature and the proper thresholds vary according to different datasets, it's difficult to determine either which feature should be used, or how large the threshold values should be set in traditional threshold filtering methods. What's more, the consistency between the results of different features is not high.

### 3.3 Comparison with Other Methods

**Comparison with Voting**

To combine classification results of weak classifiers, we use K-means to perform clustering on concatenated label vectors. A more straightforward way is to use voting instead of K-means. In the voting method, a cell will be judged as low-quality cell if more than half of classifiers classify it to the low-quality category. Figure 4 shows the F1-score of K-means and voting on two datasets. The definition of F1-score is as follows:

$$F1 = \frac{2 \cdot Precision \cdot Recall}{Precision + Recall} \tag{4}$$

$$Precision = \frac{the\ number\ of\ true\ low - quality\ cells\ detected}{the\ total\ number\ of\ low - quality\ cells\ detected} \tag{5}$$

$$Recall = \frac{the\ number\ of\ true\ low - quality\ cells\ detected}{the\ total\ number\ of\ true\ low - quality\ cells} \tag{6}$$

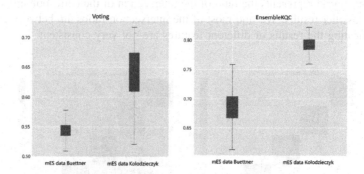

**Fig. 4.** Comparison with voting on two datasets

Here we record F1-score for each round in the threshold enumeration step to draw box plots. Since the F1-score distribution of K-means is obviously better than voting on both datasets, we think K-means is a more appropriate ensemble method.

**Comparison with Immediate Clustering Without Ensemble**

To demonstrate the effectiveness of EnsembleKQC, we also implement immediate clustering from the raw values of the five selected features. For fairness, we also employ K-means for both immediate clustering and EnsembleKQC. In immediate clustering, we perform K-means based on the raw values of features to divide cells into two clusters. The cluster with the lower number is regarded as low-quality and the other cluster is labelled as high-quality. The results are shown in Fig. 5.

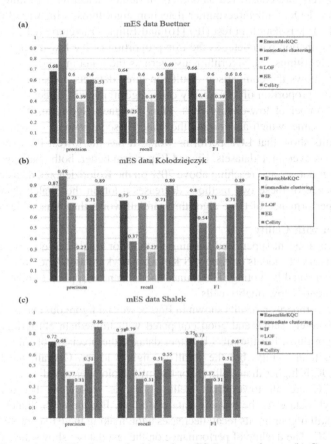

**Fig. 5.** (a) Comparison with other methods on the mES Buettner dataset (b) Comparison with other methods on the mES Kolodziejczyk dataset (c) Comparison with other methods on the mES Shalek dataset

We can figure out that the precision of immediate clustering is high. Especially on the first two datasets, it nearly reaches 100%, which means few false positives. However, its recall is only 25% and 37% on these two datasets, which means that although the precision of immediate clustering reaches the maximum, only a small number of low-quality cells are identified, and a large number of false negative samples are missed. Its performance on the mES Shalek dataset shows good, but the F1-score is still less than EnsembleKQC. In practical application, immediate clustering is not a good choice since in most data, a lot of low-quality cells cannot be filtered. Based on F1-score, EnsembleKQC performs better than immediate clustering. The reason behind this is that EnsembleKQC performs clustering for multiple rounds based on different thresholds to produce the result instead of a single immediate clustering, which makes it more robust and tolerant of noise which is rich in scRNA-seq data.

**Comparison with Outlier Detection Algorithms**

Low-quality cells are considered as outliers in outlier detection algorithms. We compare EnsembleKQC with three outlier detection algorithms, which are Local Outlier Factor (LOF) [2], Isolation Forest (IF) [16] and Elliptic Envelope (EE) [20]. Here we set the contamination parameter as the true proportion of the low-quality cells to train these three algorithms. All algorithms use zero as the random seed.

Figure 5 shows the comparison results on three datasets. Because the contamination is the true proportion of low-quality cells, algorithms will detect the same number as the true number of low-quality cells. Thus, for each algorithm, it's precision and recall are the same, which also means the same F1-score value.

The results show that LOF performs worst. It has the lowest F1-score among all seven methods over three datasets. EE and IF perform better, both reaching 60% on the mES Buettner dataset and reaching above 70% on the Kolodziejczyk dataset. However, the performance of these two methods decreases a lot on the Shalek dataset. On the whole, the performance of three algorithms are all worse than EnsembleKQC.

**Comparison with Cellity**

Cellity [10] is a recent supervised learning method for single cell quality control. This method trains SVM models on the mES Kolodziejczyk dataset. The final classification results are generated by voting. The trained model can be applied to other scRNA-seq datasets to identify low-quality cells.

From the comparison results shown in Fig. 5, we can figure out that performance of Cellity is relatively stable and good compared with immediate clustering and outlier detection algorithms. For the mES Buettner dataset, the precision of EnsembleKQC is much higher than Cellity, but the recall is slightly lower. Overall, the F1-score of EnsembleKQC is higher than Cellity. For the mES Kolodziejczyk dataset, the scores of EnsembleKQC are all lower than Cellity. This is reasonable because the mES Kolodziejczyk dataset is the training data of Cellity. For the mES Shalek dataset, its precision is still higher but its recall decreases a lot, making the F1-score 8% lower than EnsembleKQC. The decline of performance on the last dataset shows the low ability of generalization of Cellity. The reason is that Cellity is a supervised learning method and may over-fit the training data. Thus, it is not good at the datasets which are much different from the training data. In conclusion, EnsembleKQC is comparable with Cellity on its training data and performs better on other datasets.

## 4   Discussions

EnsembleKQC offers many advantages over previously used methods. It is free of setting threshold manually and does not require training data with known high-quality and low-quality cells to build a classified model. As shown on multiple real datasets, EnsembleKQC exhibited the highest accuracy compared with other methods.

The most significant advantage of EnsembleKQC is that it could automatically identify low-quality cells without prior knowledge. Traditional quality control methods remove low-quality cells by setting thresholds for some select features such as reads mapping rate, library size, the number of expressed genes, ratio of mitochondrial genes. This requires extensive experience to determine features, especially thresholds. A supervised learning model named Cellity identifies low-quality cells based on a built classifier from one mouse dataset. Its generalization performance is unsatisfactory if the experiment protocol, sequencing platform, or the species in research differ. Outlier detection algorithms treat low-quality cells as outlier, which not only regard some rare cell type as low quality, but also miss a lot of true low-quality cells. Ensemble integrates five weak classifiers, which is robust and achieves best balance between precision and recall.

Though EnsembleKQC consumes time cost, its time complexity heavily relies on the number of features and range of thresholds. Therefore, time cost of EnsembleKQC is tolerable, even if the number cells reach tens of thousands.

## 5   Conclusions

In this article, we present EnsembleKQC, an unsupervised learning ensemble method for quality control of scRNA-seq data, which automatically explores better thresholds for five features respectively and integrates answers from each feature to give a comprehensive evaluation about whether a single cell is low quality. It has good adaptability to various datasets. In experiments, EnsembleKQC shows the best performance.

## References

1. Bacher, R., Kendziorski, C.: Design and computational analysis of single-cell RNA-sequencing experiments. Genome Biol. 17(1), 63 (2016)
2. Breunig, M.M., Kriegel, H.P., Ng, R.T., et al.: LOF: identifying density-based local outliers. In: ACM Sigmod Record, vol. 29(2), pp. 93–104. ACM (2000)
3. Buettner, F., Natarajan, K.N., Casale, F.P., et al.: Computational analysis of cell-to-cell heterogeneity in single-cell RNA-sequencing data reveals hidden subpopulations of cells. Nat. Biotechnol. 33(2), 155 (2015)
4. Butte, A.J., Dzau, V.J., Glueck, S.B.: Further defining housekeeping, or "maintenance", genes Focus on "A compendium of gene expression in normal human tissues". Physiol. Genomics 7(2), 95–96 (2001)
5. Diaz, A., Liu, S.J., Sandoval, C., et al.: SCell: integrated analysis of single-cell RNA-seq data. Bioinformatics 32(14), 2219–2220 (2016)
6. Eisenberg, E., Levanon, E.Y.: Human housekeeping genes are compact. Trends Genet. 19(7), 362–365 (2003)
7. Gunning, P.W., Ghoshdastider, U., Whitaker, S., et al.: The evolution of compositionally and functionally distinct actin filaments. J. Cell Sci. 128(11), 2009–2019 (2015)
8. Hanukogle, I., Tanese, N., Fuchs, E.: Complementary DNA sequence of a human cytoplasmic actin: interspecies divergence of 3′ non-coding regions. J. Mol. Biol. 163(4), 673–678 (1983)

9. Huo, X., Hu, S., Zhao, C., et al.: Dr.seq: a quality control and analysis pipeline for droplet sequencing. Bioinformatics **32**(14), 2221–2223 (2016)
10. Ilicic, T., Kim, J.K., Kolodziejczyk, A.A., et al.: Classification of low quality cells from single-cell RNA-seq data. Genome Biol. **17**(1), 29 (2016)
11. Islam, S., Zeisel, A., Joost, S., et al.: Quantitative single-cell RNA-seq with unique molecular identifiers. Nat. Methods **11**(2), 163 (2014)
12. Jiang, P., Thomson, J.A., Stewart, R.: Quality control of single-cell RNA-seq by SinQC. Bioinformatics **32**(16), 2514–2516 (2016)
13. Kolodziejczyk, A.A., Kim, J.K., Tsang, J.C.H., et al.: Single cell RNA-sequencing of pluripotent states unlocks modular transcriptional variation. Cell Stem Cell **17**(4), 471–485 (2015)
14. Laschet, J.J., Minier, F., Kurcewicz, I., et al.: Glyceraldehyde-3-phosphate dehydrogenase is a GABAA receptor kinase linking glycolysis to neuronal inhibition. J. Neurosci. **24**(35), 7614–7622 (2004)
15. Leng, N., Chu, L.F., Barry, C., et al.: Oscope identifies oscillatory genes in unsynchronized single-cell RNA-seq experiments. Nat. Methods **12**(10), 947 (2015)
16. Liu, F.T., Ting, K.M., Zhou, Z.H.: Isolation forest. In: 2008 Eighth IEEE International Conference on Data Mining, pp. 413–422. IEEE (2008)
17. MacQueen, J.: Some methods for classification and analysis of multivariate observations. In: Proceedings of the Fifth Berkeley Symposium on Mathematical Statistics and Probability, vol. 1(14), pp. 281–297 (1967)
18. McCarthy, D.J., Campbell, K.R., Lun, A.T.L., et al.: Scater: pre-processing, quality control, normalization and visualization of single-cell RNA-seq data in R. Bioinformatics **33**(8), 1179–1186 (2017)
19. Pace, N.R.: The universal nature of biochemistry. Proc. Natl. Acad. Sci. **98**(3), 805–808 (2001)
20. Rousseeuw, P.J., Driessen, K.V.: A fast algorithm for the minimum covariance determinant estimator. Technometrics **41**(3), 212–223 (1999)
21. Sawa, A., Khan, A.A., Hester, L.D., et al.: Glyceraldehyde-3-phosphate dehydrogenase: nuclear translocation participates in neuronal and nonneuronal cell death. Proc. Natl. Acad. Sci. **94**(21), 11669–11674 (1997)
22. Shalek, A.K., Satija, R., Shuga, J., et al.: Single-cell RNA-seq reveals dynamic paracrine control of cellular variation. Nature **510**(7505), 363 (2014)
23. Sirover, M.A.: Role of the glycolytic protein, glyceraldehyde-3-phosphate dehydrogenase, in normal cell function and in cell pathology. J. Cell. Biochem. **66**(2), 133–140 (1997)
24. Zhu, J., He, F., Hu, S., et al.: On the nature of human housekeeping genes. Trends Genet. **24**(10), 481–484 (2008)

# Regulatory Sequence Architecture of Stress Responsive Genes in *Oryza Sativa*

Mohsin Ali Nasir[1], Samia Nawaz[2], Farrukh Azeem[2(✉)],
and Sajjad Haider[3]

[1] Center for Informational Biology, University of Electronic Science
and Technology of China, No. 2006, Xiyuan Ave,
West Hi-Tech Zone, Chengdu 611731, China
m.ali91@gmail.com
[2] Department of Bioinformatics and Biotechnology,
Government College University, Faisalabad 38000, Pakistan
farrukh@gcuf.edu.pk
[3] Department of Statistics, Government College University,
Faisalabad 38000, Pakistan

**Abstract.** Rice (*Oryza sativa*) is a staple food crop. Different environmental stresses critically threaten crop production in different areas. For improvement of plant yield, non-coding DNA sequences have been a focus of modern research. The focus of current study is the promoters and UTRs, which are specific regions present at the boundaries of genes and mRNA respectively, for analyzing and filtration of high required features to get improved yield. Here, a computational package "R" was used to analyze the promoters and UTRs of stress responsive genes in rice. It was observed that CACATG in promoters and TAC in 5' UTRs are globally present and scattered in the sequences under study while in 3' UTRs long chains of thymine were detected which shows that mRNA is rich to have adenine which plays role in termination of translation. In the group of salt stress responsive genes, the sequence patters CACATG, GGTTAA and ACTCAT were more frequently present and similarly, the elements TGTCTC and CACATG were over represented in promoter sequences of cold stress related genes. An analysis of similarity index, among members of a particular stress related genes, demonstrated that some of the sequences were highly similar i.e., 31.8, 30.2 and 29.3% among heat drought and cold stress related genes, respectively in promoters. Most importantly, the promoters and UTRs of housekeeping genes have no conserved pattern or sequence as other promoters and UTRs sequences. To detect the relationship among the genes of specific stress, phylogenetic analysis was also performed.

**Keywords:** Rice · UTRs · Gene prediction · Promoters · Motifs ·
Cis-regulatory elements · R coding

**Electronic supplementary material** The online version of this chapter (https://doi.org/10.1007/978-3-030-26969-2_48) contains supplementary material, which is available to authorized users.

# 1 Introduction

Rice (Oryza sativa) has been developed as a noteworthy food for over 7000 years [1]. As the population is increasing, the demand for food is also increasing every year. Different stresses i.e. biotic and abiotic like salinity, drought, heat and cold critically threaten crop production in different areas [2, 3]. Plants activate a specific and unique stress response when subjected to a combination of multiple stresses [4]. Multiple signaling pathways regulate the stress responses of plants and the patterns of expression of genes that are induced in response to different stressfactors [5, 6]. Human genome analysis and the other studies in recent era has shown that 1.5% of the genetic material codes the specific proteins or expressions while rest part of genetic material does not code any expression. Gene promoters are DNA sequences located upstream of gene coding regions and are defined by multiple features including cis-acting elements, which are specific binding sites for proteins involved in the initiation and regulation of transcription [7]. One more important non-coding part of genes is UTR (untranslated regions). In molecular genetics, an untranslated region (UTR) is more divided into two classis on both sides of a specific coding gene. These two sides are 5 (5 prime) and 3 (3 prime). For the analysis of biological data, computational biology is the major component. Computational analysis include the improvement and use of information systematic and hypothetical strategies, numerical displaying and computational reenactment procedures to the investigation of natural, behavioral, and social frameworks [8]. The name "R" alludes to the computational condition at first made by Robert Gentleman and Robert Ihaka, comparable in nature to the "S" measurable condition created at Bell Laboratories. In this study, different stresses and the genes which are responding under these stresses in rice have been focused on. UTRs and promoters of these genes have given new annotations according to their conservation and structure. Moreover, all these genes are analyzed bioinformatically and statistically depending upon their stress quality individually and in sum, to interpret the common features responsible for different stresses. Programmatically errors were occurring so to avoid it continually practice was preferred to do.

# 2 Materials and Methods

## 2.1 Selection of Representative Stresses

Stresses were varying like biotic stress and abiotic stresses e.g. salt stress, drought stress, heat stress and pathogens related stress etc. various stresses that are more common and affected the rice production were selected for this study. But in this article only salt stress and housekeeping genes are going to show all other data is present in supplementary files.

## 2.2 Selection of Stress Responsive Genes

The expressions of different genes were seen under selected stress situations. From the literature review and by the use of database named oryzabase (https://shigen.nig.ac.jp/rice/oryzab/)ase different genes were selected. These selected genes were presenting their response under different stresses.

## 2.3    Collection of Gene Promoters and UTRs Sequences

Once genes list was finalized, the promoter and UTRs sequences were retrieved of these specific genes. It was done by using different databases like phytozome (https://phytozome.jgi.doe.gov/) or NCBI (https://www.ncbi.nlm.nih.gov/). These sequences were saved in the world files. Phytozome always required a LOC number to give the gene information. Therefore different sources were used to get the different LOC number of specific genes e.g. NCBI (https://www.ncbi.nlm.nih.gov/), PubMed (https://www.ncbi.nlm.nih.gov/pubmed/), and oryzabase (https://shigen.nig.ac.jp/rice/oryzabase/).

## 2.4    Selection of Features

Hereafter, different features like to detect the specific location of the specific conserved part on the promoter and UTRs were selected. These conserved parts were TATA box site, different cis-regulatory elements and GC contents. Except to detect the location, the repetition of the conserved motifs in the promoters was also detectable. One more feature was selected that was the similarity index among all the sequences gained for a specific selected stress.

## 2.5    R Coding

There were many problems to learn R language and to draw the specific codes for the biological data which was chosen for this study. It was relatively tough to get all features by using of R programming. Different codes were made for this purpose. Than finally by the using of the StrngI program, R code was made. Which was able to implement on the data, easy to handle and understandable to reader.

## 2.6    Implementation of R Language

These codes which were made by different packages of R language like (stringI) than were implemented on the data of the promoters and UTRs of selected genes of specific stress one by one. These codes were implemented one by one on each stress; therefore each specific stress and its regarding genes have a separate result.

## 2.7    Excel Sheets

The promoter and UTRs sequences of all selected genes under specific stress were pasted into the excel sheets according to the demand of R language format which had been used during working.

## 2.8    Tree Building

The phylogenetic tree of promoter and UTRs sequences of the selected sequences was constructed using MEGA (version 7.0).

# 3 Results

## 3.1 Stress Selection

Different biotic and abiotic stresses affect the development and yield of the product of rice. In this project, the primary and major stresses which have significant impact on the yield of rice were chosen to investigate (Table 1).

**Table 1.** Selected stresses.

| Crop | Stress (situations) |
|------|---------------------|
| Rice | Drought Stress |
|      | Heat stress |
|      | Oxidative stress |
|      | Defense response |
|      | Abscisic sensitivity stress |
|      | Auxin stimulus stress |
|      | Cold stress |
|      | UV light stress |
|      | Salt stress |

## 3.2 Gene Selection

Different genes that show response in selected stress were selected. By the reading of papers about rice, regarding to different stress many genes was identified. Apart of paper reading, by the help of different websites and databases of rice, a list of different genes was made according to each specific stress. Almost nine housekeeping genes were also retrieved for the analysis.

## 3.3 R Coding

Different codes were generated to get the efficient and beneficial results. These codes are easy to understand and handle.

**Calculation of Similarity Index.** R codes were made to check the similarity index between the sequences that are all related to one stress. Here different results are available for different stresses. The result of first 10 genes of these stresses is displaying (Tables 2, 3, 4, 5, 6 and 7).

The maximum value of similarity index is highlighted by sky-blue color in all tables while the smallest value heighted by purple color.

**Table 2.** Salt stress genes similarity index. First ten genes related to this stress and the exact similarity of their promoter sequences in the form of digits. At the left side and at the top the LOC numbers of the genes are given.

| LOC Os | 05g2 6890 | 09g3 6320 | 08g3 1410 | 01g4 0950 | 06g2 1410 | 05g4 8890 | 09g3 0486 | 11g0 6820 | 02g2 1750 | 2g46 680 |
|---|---|---|---|---|---|---|---|---|---|---|
| 05g26890 | 100 | 25.1 | 27.9 | 27.4 | 26.2 | 24.1 | 25 | 28.4 | 28.1 | 24.7 |
| 09g36320 | 25.1 | 100 | 24.3 | 25.5 | 25.7 | 25.1 | 24.8 | 24.8 | 25 | 25.5 |
| 08g31410 | 27.9 | 24.3 | 100 | 24.9 | 28 | 24.3 | 28.3 | 27.7 | 26.8 | 22.7 |
| 01g40950 | 27.4 | 25.5 | 24.9 | 100 | 24.7 | 30.2 | 27.7 | 27.8 | 25.1 | 23.5 |
| 06g21410 | 26.2 | 25.7 | 28 | 24.7 | 100 | 23.7 | 25.8 | 26.7 | 23.7 | 25.8 |
| 05g48890 | 24.1 | 25.1 | 24.3 | 30.2 | 23.7 | 100 | 23.8 | 23.8 | 26.7 | 25.2 |
| 09g30486 | 25 | 24.8 | 28.3 | 27.7 | 25.8 | 23.8 | 100 | 28.3 | 25.6 | 23.6 |
| 11g06820 | 28.4 | 24.8 | 27.7 | 27.8 | 26.7 | 23.8 | 28.3 | 100 | 25.7 | 23.8 |
| 02g21750 | 28.1 | 25 | 26.8 | 25.1 | 23.7 | 26.7 | 25.6 | 25.7 | 100 | 23.6 |
| 02g46680 | 24.7 | 25.5 | 22.7 | 23.5 | 25.8 | 25.2 | 23.6 | 23.8 | 23.6 | 100 |

**Table 3.** Housekeeping genes similarity index. First ten genes related to this stress and the exact similarity of their promoter sequences in the form of digits.

| LOC Os | 03g5 0885 | 02g4 2314 | 03g0 8050 | 04g4 0950 | 01g5 9150 | 02g0 5330 | 06g4 6770 | 01g2 290 | 03g1 5020 |
|---|---|---|---|---|---|---|---|---|---|
| 03g50885 | 100 | 22.8 | 24.8 | 26.7 | 21.4 | 24.5 | 25.7 | 23.8 | 26.8 |
| 02g42314 | 22.8 | 100 | 24 | 24 | 26.9 | 24.2 | 23 | 28.3 | 25.9 |
| 03g08050 | 24.8 | 24 | 100 | 26 | 23.3 | 27 | 25.9 | 23.8 | 24.9 |
| 04g40950 | 26.7 | 24 | 26 | 100 | 25.6 | 26.5 | 29 | 25.2 | 29.9 |
| 01g59150 | 21.4 | 26.9 | 23.3 | 25.6 | 100 | 26 | 23 | 27.9 | 26.6 |
| 02g05330 | 24.5 | 24.2 | 27 | 26.5 | 26 | 100 | 28.1 | 26.9 | 26.5 |
| 06g46770 | 25.7 | 23 | 25.9 | 29 | 23 | 28.1 | 100 | 24.6 | 26.9 |
| 01g22490 | 23.8 | 28.3 | 23.8 | 25.2 | 27.9 | 26.9 | 24.6 | 100 | 25.4 |
| 03g15020 | 26.8 | 25.9 | 24.9 | 29.9 | 26.6 | 26.5 | 26.9 | 25.4 | 100 |

## 3.4   Conserved Patterns Detection

**Pattern List.** First of all, a sequence list of various conserved arrangements was developed that can be available in the promoters and UTRs.

**Table 4.** Salt stress genes similarity index. Genes related to this stress and the exact similarity of their 3' UTRs sequences in the form of digits.

| LOC Os | 11g0 6780 | 11g4 4560 | 04g4 0570 | 03g1 0370 | 09g3 6320 | 05g4 8890 | 02g4 6680 | 01g6 2244 | 12g4 4000 | 03g0 8170 |
|---|---|---|---|---|---|---|---|---|---|---|
| 11g06780 | 100 | 34.0 | 20.7 | 27.5 | 29.5 | 25.6 | 28.5 | 23.2 | 30.3 | 30.8 |
| 11g44560 | 34.0 | 100 | 22.0 | 27.2 | 29.6 | 31.5 | 30.9 | 23.1 | 23.9 | 27.2 |
| 04g40570 | 20.7 | 22.0 | 100 | 19.4 | 27.2 | 27.2 | 22.0 | 24.6 | 37.6 | 25.9 |
| 03g10370 | 27.5 | 27.2 | 19.4 | 100 | 28.6 | 29.5 | 28.5 | 25.5 | 20.9 | 27.8 |
| 09g36320 | 29.5 | 29.3 | 27.2 | 28.6 | 100 | 30.5 | 27.8 | 25.5 | 27.7 | 29.5 |
| 05g48890 | 25.6 | 31.5 | 27.2 | 29.5 | 30.5 | 100 | 25.6 | 27.0 | 30.5 | 22.1 |
| 02g46680 | 28.5 | 30.9 | 22.0 | 28.5 | 27.8 | 25.6 | 100 | 28.8 | 26.4 | 26.7 |
| 01g62244 | 23.2 | 23.1 | 24.6 | 25.5 | 25.5 | 27.0 | 28.8 | 100 | 23.9 | 26.0 |
| 12g44000 | 30.3 | 23.9 | 37.6 | 20.9 | 27.7 | 30.5 | 26.4 | 23.9 | 100 | 26.9 |
| 03g08170 | 30.8 | 27.2 | 25.9 | 27.8 | 29.5 | 22.1 | 26.7 | 26.0 | 26.9 | 100 |

**Table 5.** Salt stress genes similarity index. Genes related to this stress and the exact similarity of their 5' UTRs sequences in the form of digits.

| LOC Os | 11g0 6820 | 03g1 0370 | 09g3 6320 | 05g4 8890 | 02g4 6680 | 01g6 2244 | 12g4 4000 | 03g0 8170 | 01g0 2300 | 01g0 2390 |
|---|---|---|---|---|---|---|---|---|---|---|
| 11g06820 | 100 | 26.3 | 20.6 | 26 | 23.6 | 25.8 | 26.1 | 23.6 | 27.4 | 22.7 |
| 03g10370 | 26.3 | 100 | 25.8 | 18 | 24.3 | 29.9 | 32.8 | 23.9 | 27.4 | 27.2 |
| 09g36320 | 20.6 | 25.8 | 100 | 30 | 27.0 | 27.0 | 26.8 | 30.4 | 25.4 | 24.2 |
| 05g48890 | 26 | 18 | 30 | 100 | 28 | 22 | 24 | 20 | 26 | 28 |
| 02g46680 | 23.6 | 24.3 | 27.0 | 28 | 100 | 28.5 | 24.6 | 25.8 | 27.4 | 24.2 |
| 01g62244 | 25.8 | 29.9 | 27.0 | 22 | 28.5 | 100 | 29.1 | 27.2 | 33.3 | 22.7 |
| 12g44000 | 26.1 | 32.8 | 26.8 | 24 | 24.6 | 29.1 | 100 | 24.6 | 11.7 | 22.7 |
| 03g08170 | 23.6 | 23.9 | 30.4 | 20 | 25.8 | 27.2 | 24.6 | 100 | 17.6 | 25.7 |
| 01g02300 | 27.4 | 27.4 | 25.4 | 26 | 27.4 | 33.3 | 11.7 | 17.6 | 100 | 29.4 |
| 01g02390 | 22.7 | 27.2 | 24.2 | 28 | 24.2 | 22.7 | 22.7 | 25.7 | 29.4 | 100 |

**Table 6.** Housekeeping genes similarity index of 3' UTRs in digits.

| LOC Os | 03g5088 5 | 02g4 2314 | 04g4 0950 | 01g5 9150 | 02g0 5330 | 06g4 6770 | 01g2 2490 | 03g1 5020 |
|---|---|---|---|---|---|---|---|---|
| 03g50885 | 100 | 28.9 | 29.8 | 27.0 | 27.0 | 27.0 | 25.8 | 27.0 |
| 02g42314 | 28.9 | 100 | 25.8 | 27.5 | 25.2 | 29.0 | 28.7 | 24.6 |
| 04g40950 | 29.8 | 25.8 | 100 | 29.1 | 25.3 | 27.4 | 29.6 | 27.1 |
| 01g59150 | 27.0 | 27.5 | 29.1 | 100 | 27.8 | 28.8 | 25.5 | 26.2 |
| 02g05330 | 27.0 | 25.2 | 25.3 | 27.8 | 100 | 27.8 | 22.3 | 29.5 |
| 06g46770 | 27.0 | 29.0 | 27.4 | 28.8 | 27.8 | 100 | 28.7 | 24.5 |
| 01g22490 | 25.8 | 28.7 | 29.6 | 25.5 | 22.3 | 28.7 | 100 | 26.8 |
| 03g15020 | 27.0 | 24.6 | 27.1 | 26.2 | 29.5 | 24.5 | 26.8 | 100 |

**Table 7.** Housekeeping genes similarity index of 5' UTRs in digits.

| LOC Os | 03g5 0885 | 02g4 2314 | 04g4 0950 | 01g5 9150 | 02g0 5330 | 06g4 6770 | 01g2 2490 | 03g1 5020 |
|---|---|---|---|---|---|---|---|---|
| 03g50885 | 100 | 45.4 | 9.0 | 9.0 | 18.1 | 36.3 | 36.3 | 45.4 |
| 02g42314 | 18.1 | 19.5 | 29.2 | 21.9 | 19.5 | 26.8 | 29.2 | 31.7 |
| 04g40950 | 9.0 | 35.6 | 100 | 25 | 29.8 | 18.4 | 28.7 | 28.7 |
| 01g59150 | 14.2 | 14.2 | 28.5 | 28.5 | 14.2 | 28.5 | 28.5 | 14.2 |
| 02g05330 | 18.1 | 30.5 | 29.8 | 41.2 | 100 | 19.7 | 30 | 28.4 |
| 06g46770 | 36.3 | 19.7 | 18.4 | 36.8 | 19.7 | 100 | 28.9 | 17.1 |
| 01g22490 | 36.3 | 30 | 28.7 | 32.5 | 30 | 28.9 | 100 | 24.5 |
| 03g15020 | 45.4 | 23.8 | 28.7 | 31.2 | 28.4 | 17.1 | 24.5 | 100 |

**Color Scheme.** R codes likewise were utilized to find the diverse patterns and sequence on the promoters and UTRs (Figs. 1, 2, 3 and 4).

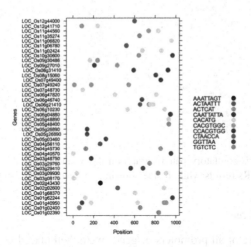

**Fig. 1.** Motifs and *cis*-regulatory pattern graph on the promoters of the salt stress genes. Here at the right side the color scheme and their adoptive patterns were given. Like AAATTAGT is demonstrated in the figure by the black color and so on. At Y axis, there is the list of the genes LOC numbers. At the X axis, there is a scale set according to average length of the promoters.

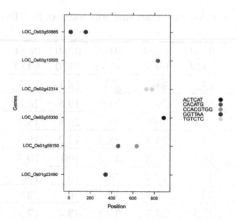

**Fig. 2.** Motifs and *cis*-regulatory pattern graph on the promoters of housekeeping genes. Here at the right side the color scheme and their adoptive patterns were given. Like ACTCAT is demonstrated in the figure by the black color and so on.

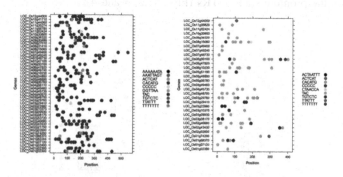

**Fig. 3.** Motifs and *cis*-regulatory pattern graph on the 3' and 5' of salt stress genes. Side a is demonstrating 3' UTRs results while b side showing the results of 5' UTRs.

## 3.5   Phylogeny Tree

The DNA sequences of all promoters of genes were then placed in MEGA 6.0 version 6.06 (Tamura et al. 2013) to multiple sequence alignments by using the ClustalW and constructed the phylogenetic tree (Figs. 5 and 6)

## 3.6   GC Contents

GC contents were also been detected by using of these codes. The GC contents ratio in all the promoter sequences of selected genes of selected stress was identified as sum. Here in the Fig. 7 the ratio in each stress has been given for promoters (Fig. 8).

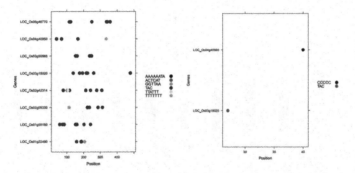

**Fig. 4.** Motifs and *cis*-regulatory pattern graph on the 3' and 5' of housekeeping genes. Side a is demonstrating 3' UTRs results while b side showing the results of 5' UTRs.

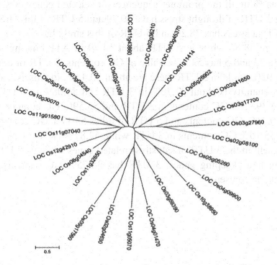

**Fig. 5.** Unrooted phylogenetic tree of promoters sequences of salt stress genes.

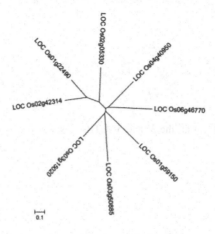

**Fig. 6.** Unrooted phylogenetic tree of promoters sequences of housekeeping genes.

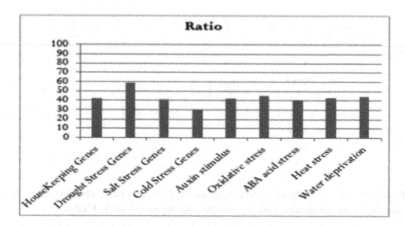

**Fig. 7.** GC contents % in all the promoter sequences of selected genes of selected stresses. As results are showing 3 UTR of drought stress have 39.79 and 5 UTR of this stress has 49.94% GC contents. 3 UTR of heat stress has 38.29 and 5 UTR of this stress has 53.55% GC contents. In 3 UTR GC contents are 38.36 while in 5 UTR about 54.79 in ATP binding stress. In oxidative stress, 3 UTR has 38.33 and 5 has 53.77% value of GC contents. 3 UTR of defense related stress GC contents are 38.92% while in 5 UTR these are about 50.42%. In Abscisic acid sensitive stress genes, GC contents quantity is about 37.72% in 3 UTR and 54.08% in 5 UTR. Genes of auxin stimulus stress have 39.93% GC contents in 3 UTR and 53.59% GC contents in 5 UTR. 3 UTR of cold stress has 40.31% GC contents and 5 UTR of it has 53.17% GC contents. UV light stress related genes have 38.16% GC contents in 3 UTR while 52.81% GC contents in 5 UTR. 36.59% GC contents are present in the 3 UTR of salt stress related genes while its 5 UTR has 50.02% GC contents. Finally, in housekeeping genes, 3 UTR has 38.75% GC contents. While in 5 UTR 56.92% GC contents are present.

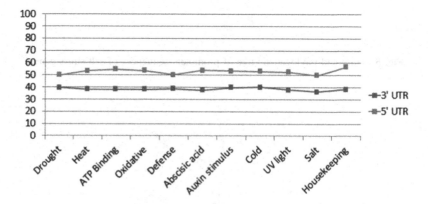

**Fig. 8.** GC contents % in all the 3' and 5' UTRs of selected genes of selected stresses.

# 4  Discussion

The current sequencing and explanation of the rice genome has provided us a chance to analyze transcriptional control of genes at a genomic level. Computational expectation of promoters, un-translated regions (UTRs) and TFs give an effective technique to disentangle the regulatory systems controlling plant gene expression. Here in this study computational analysis have been performed by the use of R language and the promoter sequences of the Rice genes related to specific environmental stresses. Promoters contains cis-regulatory components are composed of non-coding DNA having TFs binding sites, which modulate the transcription of respective genes [9]. In this study, different cis-regulatory elements and conserved motifs have been analyzed in the promoters sequence. It was observed that genes involved in different stresses contain particular sequence architecture.

Undoubtedly, the mutations in protein-coding regions could have more pleiotropic effects, on the grounds that by and large they aect all tissues in which the protein is dynamic, while the change of a tissue-specific regulatory component in the promoter ought to aect just the cells intrigued by the specific expression variations [9]. This work provides the easy handling methods of R language to get the promoter analysis and their different features in different stresses. The normal length of 5 UTRs is generally consistent over assorted ordered classes and ranges in the vicinity of 100 and 200 nucleotides, while the normal length of 3 UTRs is considerably more factor, going from around 200 nucleotides in plants and growths to 800 nucleotides in human and different vertebrates. Here in this study, we evaluated that 3 UTR is lengthier than the 5 UTR in rice. 3 UTRs sequences of different stresses of rice share may of the common and conserved patterns that are more related to Uracil chain. From this study, one can conclude that TAC which have the complement as AUG on mRNA which is starting is mostly present on the 5 UTRs which made it clear that ORFs of 5 UTRs have starting codons. And 3 UTR have long chains of thymine which have compliment of adenine in mRNA which demonstrate that 3 have long chains of adenine which plays important role in termination process. As per [10] their report is the first report that portrays the measurable investigation of the whole 5' UTR. To start with, they dissected the connection between the length of the 5' UTR and its comparing mRNA length. Their information proposed that the 5' UTR length is focused under 200 bp paying little mind to the mRNA length. The 5' UTR length could have a best range for effective interpretation. Next, they hunt down eliminator and ATG codons in the 5' UTR. Unexpectedly, 459 of 954 mRNA species contained no in-outline eliminator codon in the 5' UTR. It was likewise interesting that 278 mRNA species contained no less than one ATG upstream of the announced initiator ATG [10]. Here in this project we just picked up the complimentary of ATG that was TAC because we collected the sequences from the DNA that must be the complementary to the real UTR sequences. We concluded that many of TAC sequences are present in 5 UTRs. Infact this is the most common pattern present in whole UTRs.

## 5  Conclusion

The gene promoters and UTRs play the most important role in plant response to external stimuli. It is certainly the very first attempt to analyze genomic sequences by R coding. From current study, it is obvious that similarity index might play a role in stress response of a group of genes. It is because these genes certainly contain similar cis-elements architecture. Existence of higher similarity indexes (app.30%) in promoters and (app.40%) in UTRs surpasses the threshold of random occurrences. Therefore, a functional characterization is more likely to provide more accurate information about co-regulation of such genes. Similarly, the existence of stress specific and/or a common element and absence of in housekeeping gene promoters and UTRs, offers clear directions with higher order success rates.

## References

1. Izawa, T., Shimamoto, K.: Becoming a model plant: the importance of rice to plant science. Trends Plant Sci. **1**(3), 95–99 (1996)
2. Mantri, N., et al.: Abiotic stress responses in plants: present and future. In: Ahmad, P., Prasad, M. (eds.) Abiotic Stress Responses in Plants, pp. 1–19. Springer, New York (2012). https://doi.org/10.1007/978-1-4614-0634-1_1
3. Nauš, J., Pareek, A., Sopory, S.K., Bohnert, H.J., Govindjee (ed.): Abiotic stress adaptation in plants. Physiological, molecular and genomic foundation. Photosynthetica **48**(3), 474 (2010)
4. Reddy, I.N.B.L., et al.: Salt tolerance in rice: focus on mechanisms and approaches. Rice Sci. **24**(3), 123–144 (2017)
5. Chen, W., et al.: Expression profile matrix of Arabidopsis transcription factor genes suggests their putative functions in response to environmental stresses. Plant Cell **14**(3), 559–574 (2002)
6. Seki, M., et al.: Monitoring the expression pattern of 1300 Arabidopsis genes under drought and cold stresses by using a full-length cDNA microarray. Plant Cell **13**(1), 61–72 (2001)
7. Molina, C., Grotewold, E.: Genome wide analysis of Arabidopsis core promoters. BMC Genom. **6**(1), 25 (2005)
8. Vishnubhakat, S., Rai, A.K.: When biopharma meets software: bioinformatics at the patent office. Harv. JL Tech. **29**, 205 (2015)
9. Boccacci, P., et al.: Cultivar-specific gene modulation in Vitis vinifera: analysis of the promoters regulating the expression of WOX transcription factors. Sci. Rep. **7**, 45670 (2017)
10. Suzuki, Y., et al.: Statistical analysis of the 5′ untranslated region of human mRNA using "Oligo-Capped" cDNA libraries. Genomics **64**(3), 286–297 (2000)

# Identifying Cancer Biomarkers from High-Throughput RNA Sequencing Data by Machine Learning

Zishuang Zhang and Zhi-Ping Liu[✉]

School of Control Science and Engineering,
Shandong University, Jinan 250061, Shandong, China
zpliu@sdu.edu.cn

**Abstract.** In cancer progression, the expression level of relevant genes will change significantly in tumors comparing to their healthy counterparts. Therefore, the discovery of specific genes serving as biomarkers is of practical significance for diagnosis and prognosis. The available high-throughput '-omic' datasets provide unprecedented resources and opportunities of deriving cancer biomarkers, such as the public RNA-sequencing data generated by the Cancer Genome Atlas (TCGA) consortium. Here, we explore the identification of biomarker genes in 12 types of cancers from the classification effects in control and disease samples by machine learning. We firstly identify differentially expressed genes individually. Then, we implement feature selection by integrating recursive feature reduction and random forest classification with feature ranking. The final feature number will be determined via a parsimony principle that the features will be as few as possible, while they are still with the highest classification accuracy. In each cancer, the biomarker genes are then evaluated by tenfold cross-validations via several classification algorithms. We find extreme learning machine achieves the best classification performance when compared to the other methods. The further gene enrichment analyses indicate the dysfunctional and pathogenic mechanism in these identified biomarkers.

**Keywords:** Cancer biomarkers · TCGA RNA-sequencing data ·
Machine learning · Classification · Feature selection

## 1 Introduction

Cancer is a major threat to human life and health. Tens of millions of people are newly diagnosed with cancer and millions of people die from cancer every year [1]. It is in the urgent need of early diagnosis and treatment to reduce cancer mortality. Effective molecular biomarker is one of the efficient way of realizing the early diagnosis. Thus, it is very important to discover biomarkers of different cancers to achieve cancer screening and early detection [2, 3]. Currently, with the improvement of high-throughput technologies, the accumulating samples of omics data provide new resources for the discovery of cancer biomarkers [4, 5].

TCGA initiatives provide a huge treasure house of multiple omics data by collecting more than 30 types of human tumor samples [6]. The resources aim to provide

© Springer Nature Switzerland AG 2019
D.-S. Huang et al. (Eds.): ICIC 2019, LNCS 11644, pp. 517–528, 2019.
https://doi.org/10.1007/978-3-030-26969-2_49

free and open comprehensive cancer genomics datasets [1]. It is thus of great significance to draw biomedical conclusions from the genomic data sets generated. The public available RNA-sequencing data describes the genome-wide gene expression profiles in different phenotypic states. From them, it is expected to find out distinguishing gene patterns when compare normal gene expressions with their corresponding controls. Some feature genes can be served as biomarkers for classifying normal and disease individuals.

Generally speaking, the essence of biomarkers identification is feature selection and classification. It is to select optimal gene or gene set for achieving correct classifications in disease and normal samples. Therefore, machine-learning classification algorithms are critical in the selection of biomarkers. To date, there already some methods have been proposed to discover biomarkers from high-dimensional omics data by machine learning. For instance, a gene selection method has been proposed to discover the biomarkers for hepatocellular carcinoma from differentially expressed genes [7]. Based on machine learning, a method of SVM has been presented to identify biomarker genes for colon cancer from microarray gene expression data [8]. Similarly, a machine learning-based method has been applied to identify biomarker genes for four cancer types [9]. For proteomics data, some methods have also been built up for identifying biomarker proteins by peak clustering of mass spectrometry data [10]. Although these methods have been proposed for some specific cancers, the high false positive rate in classification determines their marginal success of generic findings, especially in an integrated framework for many cancer types simultaneously [11]. In the existing methods of biomarker discovery from TCGA data, few attentions have been paid for the concordantly combinatorial integration of the two main steps, i.e., feature selection and classification, in machine learning. The feature selection and classification need to be iteratively run for obtaining effective cancer biomarkers.

In this paper, we explore the cancer biomarkers from TCGA transcriptomic RNA-seq data by machine learning. Due to the characteristics of high dimension and high noise in RNA-seq data, simple filtering methods such as chi-square test cannot achieve good classification performance [12]. Thus, we implement the biomarker discovery based on a strategy of combining the recursive feature reduction method (feature selection) and random forest algorithm (classification). Then we take the optimal classification accuracy as the standard to select the gene set of interest. For validating the classification performance of identified biomarkers, we implement the tenfold classifications by several algorithms [13], e.g., Adaboost, extreme learning machine (ELM), naive Bayesian (NB), neural network (NN), support vector machine (SVM), as well as random forest (RF) used in the feature selection. And we find ELM performs the best classification in terms of accuracy for these identified biomarkers. The functional enrichment analysis of biomarker genes also envisions the pathogenesis of cancer.

## 2 Materials and Methods

### 2.1 Data Sets

TCGA project covers 33 types of cancer, including 10 rare cancers. In this study, we select 12 types (UCEC, KIRP, COAD, LIHC, STAD, PRAD, LUSC, HNSC, THCA, LUAD, KIRC) of them, all of which are typical sample sets in the two states of disease and control [11]. Only the datasets with the normal sample size greater than 20 are targeted. In each type of cancer, the data sets contain samples ranging from 201 to 1,218. Detailed information on the number of genes, control and disease samples for each disease can be available from our website: http://doc.aporc.org/wiki/ICIC2019. We firstly download the raw RNA-seq data from GDC (Genomic Data Commons) using TCGA-Assembler 2.0 [11]. Raw RNA-seq data is used to analyze the differential expression of genes. Then we download normalized RNA-seq data for the 12 cancers from TCGA hub of UCSC Xena [14]. The normalized RNA-seq data is used for the latter biomarker discovery.

### 2.2 Methods

Figure 1 illustrates our proposed framework of identifying cancer biomarkers from TCGA transcriptomic data. For the 12 types of cancer, we implement the same framework respectively.

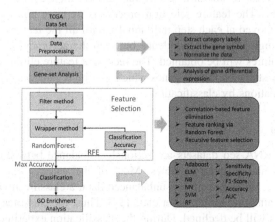

**Fig. 1.** The framework of cancer biomarker discovery from TCGA data.

After downloading the RNA-seq profiles, we implement the data preprocessing. We firstly label the samples, i.e., disease sample with '0' and normal sample with '1', and then correspond the Entrez gene IDs with their official symbols from the gene annotation files in NCBI.

We normalize the raw RNA-seq data using the median of ratios rule of DESeq2 [15]. We use DESeq2 to screen out differentially expressed genes by FDR < 0.05 (5%), fold change > 2 and *P*-value < 0.05 [11]. The differential genes provide the candidate pool for biomarker discovery. For these differential genes, we build a feature selection strategy for dimension reduction. As shown in Fig. 1, we integrate the recursive feature elimination and random forest algorithm to select the most distinctive genes in the two phenotypic states. The genes after selection with optimal classification performance are regarded as the identified cancer biomarkers. To investigate the classification perfor-mances of these identified biomarkers, we test them in the classification of disease and control samples by several machine learning algorithms. For showing their pathological implications, the functional enrichment analysis about the biological processes of gene ontology (GO) is performed in these identified biomarkers. The next section describes the main step of feature selection with machine learning in details.

### 2.3   Feature Selection with Random Forest Algorithm

We implement an integrative feature reduction process by combining a filter method and a model-based method joint with machine learning [16, 17]. The filter step briefly involves two correlation-based feature elimination processes. For two genes, if the Pearson's correlation coefficient between them is greater than 0.6, the variable with the higher mean absolute value is deleted [17]. For a gene set, the method firstly deter-mines whether the eigenmatrix of gene expression is full rank or not. If it is not, it deletes the gene that can be represented by a linear combination of the other genes. This is to ensure the subsequent matrix is full rank. In the model-based feature selection, we employ random forest to obtain a feature ranking via their contributions to the clas-sification accuracy. The feature selection process is iterative and the least important feature will be removed at each step individually until all the features will be removed [17, 18]. Accordingly, the correspondence between the accuracy of classification and the number of features will be obtained. The recursive feature elimination strategy will achieve relatively stable candidate genes which will be regarded as cancer biomarkers after further validations by classification.

### 2.4   Evaluation

Here, all the 12 RNA-seq datasets are seriously imbalance between the numbers of positive samples and negative samples. Most of the cancers have fewer negative samples (controls). When encounters unbalanced data, a classification algorithm tends to pay too much attention to the major class [19]. Thus, the classification performance in the minor class will be declined. During the classification experiments, we organize the same number of positive and negative samples by sampling. Then, we implement a tenfold cross-validation process of training and testing in the sample-balanced dataset. For each type of cancer, we conduct 10 experiments individually in the 10 datasets. We evaluate the classifications by five metrics: sensitivity (SN), specificity (SP), F1-score, accuracy (ACC) and AUC (area under curve) [19]. The mean and standard deviation values of them are then obtained.

For further investigating the classification abilities of the identified biomarkers, we employ several machine learning methods, e.g., Adaboost, ELM, NB, NN, SVM as well as the RF for a comparison study. Briefly, Adaboost continuously updates the weight of weak classifier in the training process to achieve better output results [20]. ELM is an algorithm to solve a single hidden layer neural network [21]. NB calculates the posterior probability by the prior probability of a classifying sample. Then it selects the class with the largest posterior probability as the class to which the sample belongs [22]. NN is composed by the forward propagation of signals and the back propagation of errors. The feedback errors are supervisors to correct the weights between neurons [23]. SVM is to find the optimal separation hyperplane in the feature space to maximize the margin between positive and negative samples [24, 25]. RF classifier is made up by multiple decision trees [16]. As described, we combine its effective classification ability with feature selection to identify the biomarkers.

## 3 Results

### 3.1 Identified Biomarkers

Based on differentially expressed genes, we implement the proposed feature selection strategy for further selection. The final feature number will be determined via a parsimony principle that the features will be as few as possible, while the classification is still with the highest accuracy. The number of biomarkers selected for 12 cancers is summarized in Table 1. The dimension reduction ratio of biomarkers to the number of differential genes is also shown. We can find that most of the differential genes are eliminated in our process of biomarker discovery.

**Table 1.** Feature dimension reduction process (cancers are ranked by sample size).

| Cancers | #Differential genes | #Biomarkers | Dimension reduction ratio |
|---------|--------------------|-------------|---------------------------|
| UCEC | 5177 | 31 | 99.40% |
| KIRP | 4728 | 7 | 99.85% |
| COAD | 5317 | 158 | 97.03% |
| LIHC | 4190 | 28 | 99.33% |
| STAD | 3909 | 30 | 99.23% |
| PRAD | 2972 | 22 | 99.26% |
| LUSC | 6899 | 124 | 98.20% |
| HNSC | 4510 | 32 | 99.29% |
| THCA | 3171 | 25 | 99.21% |
| LUAD | 5049 | 87 | 98.28% |
| KIRC | 5317 | 4 | 99.92% |
| BRCA | 4734 | 22 | 99.54% |

### 3.2 Classification Comparison

The cancer biomarkers selected by recursive feature reduction and random forest will be tested in the classifications by multiple algorithms to verify the effectiveness of the identified biomarkers. We derive the classification performances by Adaboost, ELM, NB, NN, SVM, as well as RF combined with feature selection.

All the six algorithms adopt 10 times of sampling and tenfold cross-validations similarly. The frequency of the maximum value of the five indices in each cancer is counted, and the statistical results are shown in Fig. 2. Note that few multiple classifiers obtain the same maximum value for some indices (i.e., ties). Table 2 lists the AUC values for 12 cancers derived from the six classifiers. From Fig. 2 and Table 2, we find that ELM achieves the most number of maximum classification values.

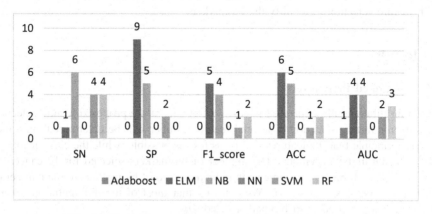

**Fig. 2.** The frequency of the maximum classification metrics.

**Table 2.** AUC values for 12 cancers.

| Cancers | Adaboost | ELM | NB | NN | SVM | RF |
|---|---|---|---|---|---|---|
| UCEC | 0.891 | 0.999 | 0.999 | 0.971 | 0.997 | 0.996 |
| KIRP | 0.861 | 0.988 | 0.987 | 0.972 | 0.993 | 0.998 |
| COAD | 0.984 | 0.991 | 0.983 | 0.983 | 0.993 | 1.000 |
| LIHC | 0.977 | 0.998 | 0.997 | 0.981 | 0.992 | 0.995 |
| STAD | 0.985 | 0.998 | 0.996 | 0.967 | 0.996 | 0.997 |
| PRAD | 0.950 | 0.965 | 0.968 | 0.918 | 0.963 | 0.964 |
| LUSC | 0.931 | 0.998 | 1.000 | 0.993 | 0.999 | 0.997 |
| HNSC | 0.985 | 0.997 | 0.990 | 0.974 | 0.998 | 0.994 |
| THCA | 0.975 | 0.989 | 0.989 | 0.968 | 0.993 | 0.992 |
| LUAD | 0.999 | 0.999 | 0.999 | 0.993 | 0.990 | 0.998 |
| KIRC | 0.966 | 0.989 | 0.990 | 0.980 | 0.987 | 0.992 |
| BRCA | 0.996 | 0.999 | 0.997 | 0.978 | 0.998 | 0.999 |

Specifically, ELM classifier has the highest SP value in the nine cancer datasets (UCEC, COAD, LIHC, LUSC, HNSC, THCA, LUAD, KIRC, BRCA), has the highest F1-score in the five types of cancer (UCEC, STAD, HNSC, THCA, BRCA), has the highest ACC value in the six cancers (UCEC, LIHC, STAD, HNSC, THCA, BRCA), and has the highest AUC in the four cancers (UCEC, LIHC, STAD, BRCA). ELM shows a good stability of classification across these different cancer types.

ELM classifies 12 types of cancer and calculates SN, SP, F1-score and ACC. Figure 3 demonstrates the ROC curves of classification in the 12 cancers. ROC curves in Fig. 3 are the average classification results in the 10 datasets constructed by 10 times of random sampling. The details of statistical indices are shown in Table 3.

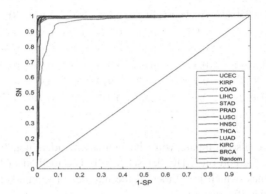

**Fig. 3.** ROC curves of 12 cancers by the identified biomarkers.

**Table 3.** The classification performance by the identified biomarkers in 12 cancer datasets via ELM.

| Cancers | #Samples | #Genes | SN | SP | F1-score | ACC | AUC |
|---|---|---|---|---|---|---|---|
| UCEC | 48 | 31 | 0.954 | 0.996 | 0.974 | 0.975 | 0.999 |
| KIRP | 64 | 7 | 0.944 | 0.994 | 0.968 | 0.969 | 0.988 |
| COAD | 82 | 158 | 0.954 | 0.993 | 0.973 | 0.973 | 0.991 |
| LIHC | 100 | 28 | 0.966 | 1.000 | 0.982 | 0.983 | 0.998 |
| STAD | 70 | 30 | 0.963 | 0.991 | 0.977 | 0.977 | 0.998 |
| PRAD | 104 | 22 | 0.944 | 0.902 | 0.925 | 0.923 | 0.965 |
| LUSC | 102 | 124 | 0.982 | 1.000 | 0.991 | 0.991 | 0.998 |
| HNSC | 88 | 32 | 0.977 | 0.970 | 0.974 | 0.974 | 0.997 |
| THCA | 118 | 25 | 0.969 | 0.983 | 0.976 | 0.976 | 0.989 |
| LUAD | 118 | 87 | 0.980 | 0.998 | 0.989 | 0.989 | 0.999 |
| KIRC | 144 | 4 | 0.972 | 0.986 | 0.979 | 0.979 | 0.989 |
| BRCA | 228 | 22 | 0.982 | 0.992 | 0.987 | 0.987 | 0.999 |

The average values of SN, SP, F1-score, ACC and AUC are 0.966, 0.984, 0.975, 0.975 and 0.993, respectively. The good classification performance of multiple classifiers proves that we have selected effective biomarkers for these cancers.

To show the generalization ability of these identified biomarkers, we implement the classification in a totally independent dataset of liver cancer as an example which is downloaded from GEO database in NCBI (ID: GSE25097). There are 268 liver tumor samples and 243 normal samples. Table 4 shows the cross-dataset validation results. From an empirical perspective, the high performance in the independent dataset proves that there is no overfitting in the classifications. The identified biomarkers can distinguish the disease samples from controls.

**Table 4.** The average classification results in an independent dataset.

| GSE250097 | Adaboost | ELM | NB | NN | SVM | RF |
|---|---|---|---|---|---|---|
| SN | 0.987 | 0.967 | 0.984 | 0.954 | 0.988 | 0.988 |
| SP | 0.988 | 0.992 | 0.963 | 0.980 | 0.984 | 0.980 |
| F1-score | 0.988 | 0.979 | 0.974 | 0.967 | 0.986 | 0.984 |
| ACC | 0.988 | 0.979 | 0.973 | 0.967 | 0.986 | 0.984 |
| AUC | 0.999 | 0.998 | 0.997 | 0.994 | 0.999 | 0.998 |

In addition, we also compare with the biomarkers identified by SVM-RFE and random-extraction method on LIHC dataset [8, 9]. The method of SVM-RFE selects 114 biomarker genes, which is 4 times of the number of biomarkers selected by our method, but the classification results are not as good as our method. The method of random extraction obtains very bad classification results. Random forest has the advantage in calculating the importance of features. Due to the space constraints, the detailed calculations and comparisons can be available from our website.

### 3.3  Case of Study

For a case study of identifying cancer biomarkers by our proposed strategy, we take Uterine Corpus Endometrial Carcinoma as a case study, which has the smallest sample size. We analyze the validity of selected biomarkers for model construction and compare the classification results of the six machine-learning algorithms. Also, we perform GO functional enrichment analysis to detect the novel pathogenesis implications.

**Feature Selection.** The original UCEC dataset contains 20,530 genes. Firstly, differential expression analysis is performed and 5,177 genes are screened out as differentially expressed. Then, we identify 31 gene biomarkers from the list of differential genes by combining the filter-based feature selection method and the random-forest machine learning method. Random forest not only implements the classification from the voting of ensemble trained decision trees, but also sorts the gene features according to their importance in the contribution of classification accuracy. Figure 4 demonstrates the correspondence between the number of genes and the accuracy of classification. In Fig. 4, we can find the left 31 genes showing in the x-axis reach the maximum value of 0.987 in the y-axis of classification accuracy. Thus, these genes are identified as the biomarkers of UCEC. They are further validated for their classification power by several machine learning methods.

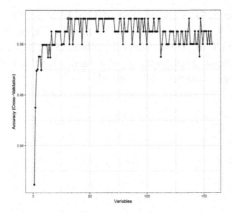

**Fig. 4.** The relationship between the number of selected features/genes and the classification accuracy in UCEC.

**Comparison of Classification Results.** To validate the biomarker genes for their classification capability, we implement the further classification experiments. The TCGA UCEC dataset contains a total of 201 samples. There are 24 control samples and thus we need handle the imbalance between positive and negative samples. Therefore, we randomly select some disease samples with the same number of control samples to construct 10 balanced datasets. We use the former six machine learning classifiers to obtain the classification performance by tenfold cross-validations. Table 5 shows the average (AVG) and variance (VAR) of evaluation metrics.

**Table 5.** The classification performance of the six classifiers in UCEC.

| TCGA-UCEC | | SN | SP | F1-score | ACC | AUC |
|---|---|---|---|---|---|---|
| AdaBoost | AVG | 0.946 | 0.879 | 0.917 | 0.913 | 0.891 |
| | VAR | 0.006 | 0.015 | 0.004 | 0.005 | 0.005 |
| ELM | AVG | 0.954 | 0.996 | 0.974 | 0.975 | 0.999 |
| | VAR | 0.002 | <0.001 | 0.001 | 0.001 | <0.001 |
| NB | AVG | 0.971 | 0.975 | 0.973 | 0.973 | 0.999 |
| | VAR | <0.001 | <0.001 | <0.001 | <0.001 | <0.001 |
| NN | AVG | 0.875 | 0.95 | 0.909 | 0.913 | 0.971 |
| | VAR | 0.003 | 0.003 | 0.001 | 0.001 | <0.001 |
| SVM | AVG | 0.983 | 0.954 | 0.969 | 0.969 | 0.997 |
| | VAR | <0.001 | <0.001 | <0.001 | <0.001 | <0.001 |
| RF | AVG | 0.975 | 0.967 | 0.971 | 0.971 | 0.996 |
| | VAR | <0.001 | 0.001 | <0.001 | <0.001 | <0.001 |

As shown in Table 5, SP, F1-score and ACC of ELM are the largest values in these classifications. The ROC curves of the six classifiers for UCEC is shown in Fig. 5. The maximum value of AUC is 0.999, which is achieved by ELM and NB. The performance of these classifications provides more evidence for the ability of the biomarkers in distinguishing disease samples with controls.

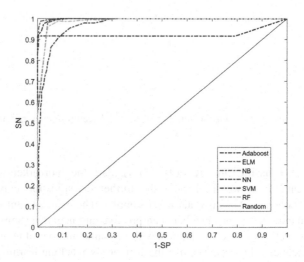

**Fig. 5.** The ROC curves of the six classifiers for UCEC.

**Functional Enrichment Analysis.** We perform the functional enrichment analysis on the 31 identified biomarker genes. And we find that the enriched GO biological processes are highly related to 'metabolism'. And we find multiple GO terms are related to ADAMTS family proteases. Existing studies have proved that ADAMTS family proteases are closely related to the development and function of reproductive organs [26]. In particular, the abnormal expression of gene 'ADAMTS5' and gene 'ADAMTSL1' is closely related to Uterine Corpus Endometrial Carcinoma [26, 27]. From the enriched functions, some new pathogenesis hypothesis of UCEC could be inferred and generated. The details can be found at our website.

## 4 Conclusion

In this paper, we identified biomarker genes for 12 cancer types from RNA-seq data by feature selection and machine learning. From the differentially expressed genes, we integrated the feature selection and random forest classification to identify the biomarkers. We firstly employed some filter methods to remove the redundant genes. Then we implemented the recursive feature reduction combined with random forest classifier to select the gene biomarkers. The random forest algorithm provides the feature rankings according to their contributions in the classification. Thus, the low-ranked genes can be eliminated iteratively. In each type of cancer, our method selected the genes as identified biomarkers with the minimal size and the highest classification performance.

For the identified cancer biomarkers, we evaluated their classification abilities by six machine learning algorithms. ELM is found to be the one with the best performance. The high accuracy in the cross-validations provides more evidence for these identified biomarkers in classifying the samples in normal and disease states. We also performed the functional enrichment analyses on these selected biomarkers. The enriched dysfunctions generate the pathogenesis implications underlying the biomarker genes. The number of biomarkers in the 12 cancer types is very diverse. Few of them only achieve several genes. For instance, there are 4 and 7 biomarker genes for KIRC and KIRP, respectively. These biomarkers need to be further justified by experimental and trial validations before being applied for clinical applications. Some of them have already been reported as cancer indicators. In the kidney cancer of KIRP and KIRC, the biomarkers identified from the two datasets have some overlaps, e.g., the selected gene 'ATP6V0A4' has been shown to be associated with kidney cancer development [28]. In conclusion, we provided a computational method of identifying cancer biomarkers from TCGA RNA-seq data.

**Acknowledgement.** This work was partially supported by the National Natural Science Foundation of China (Nos. 61572287 and 61533011), the Shandong Provincial Key Research and Development Program, China (No. 2018GSF118043), the Innovation Method Fund of China (Ministry of Science and Technology of China, No. 2018IM020200), and the Program of Qilu Young Scholars of Shandong University.

# References

1. Rodriguez, H., Pennington, S.R.: Revolutionizing precision oncology through collaborative proteogenomics and data sharing. Cell **173**, 535–539 (2018)
2. Zhu, C., Ren, C., Han, J., et al.: A five-microRNA panel in plasma was identified as potential biomarker for early detection of gastric cancer. Br. J. Cancer **110**, 2291–2299 (2014)
3. Li, M., Hong, G., Cheng, J., et al.: Identifying reproducible molecular biomarkers for gastric cancer metastasis with the aid of recurrence information. Sci. Rep. **6**, 24869 (2016)
4. Vargas, A.J., Harris, C.C.: Biomarker development in the precision medicine era: lung cancer as a case study. Nat. Rev. Cancer **16**, 525–537 (2016)
5. Bhalla, S., Chaudhary, K., Kumar, R., et al.: Gene expression-based biomarkers for discriminating early and late stage of clear cell renal cancer. Sci. Rep. **7**, 44997 (2017)
6. Chang, K., Creighton, C.J., Davis, C., et al.: The cancer genome atlas pan-cancer analysis project. Nat. Genet. **45**, 1113–1120 (2013)
7. Wei, L., Lian, B., Zhang, Y., et al.: Application of microRNA and mRNA expression profiling on prognostic biomarker discovery for hepatocellular carcinoma. BMC Genom. **15**, S13 (2014)
8. Tsai, C.-A., Chen, J.J., Baek, S.: Development of biomarker classifiers from high-dimensional data. Brief. Bioinform. **10**, 537–546 (2009)
9. Dupont, P., Helleputte, T., Abeel, T., et al.: Robust biomarker identification for cancer diagnosis with ensemble feature selection methods. Bioinformatics **26**, 392–398 (2009)
10. Swan, A.L., Mobasheri, A., Allaway, D., et al.: Application of machine learning to proteomics data: classification and biomarker identification in postgenomics biology. OMICS J. Integr. Biol. **17**, 595–610 (2013)

11. Wenric, S., Shemirani, R.: Using supervised learning methods for gene selection in RNA-Seq case-control studies. Front. Genet. **9**, 297 (2018)
12. Guyon, I., Elisseeff, A.: An introduction to variable and feature selection. J. Mach. Learn. Res. **3**, 1157–1182 (2003)
13. Wong, T.-T.: Performance evaluation of classification algorithms by k-fold and leave-one-out cross validation. Pattern Recogn. **48**, 2839–2846 (2015)
14. Goldman, M., Craft, B., Swatloski, T., et al.: The UCSC cancer genomics browser: update 2015. Nucleic Acids Res. **43**, D812–D817 (2014)
15. Anders, S., Huber, W.: Differential expression analysis for sequence count data. Genome Biol. **11**, R106 (2010)
16. Breiman, L.: Random forests. Mach. Learn. **45**, 5–32 (2001)
17. Kuhn, M.: Building predictive models in R using the caret package. J. Stat. Softw. **28**, 1–26 (2008)
18. Guyon, I., Weston, J., Barnhill, S., et al.: Gene selection for cancer classification using support vector machines. Mach. Learn. **46**, 389–422 (2002)
19. Ganganwar, V.: An overview of classification algorithms for imbalanced datasets. Int. J. Emerg. Technol. Adv. Eng. **2**, 42–47 (2012)
20. Freund, Y., Schapire, R.E.: Experiments with a new boosting algorithm. In: ICML, pp. 148–156. Citeseer (1996)
21. Huang, G.-B., Zhu, Q.-Y., Siew, C.-K.: Extreme learning machine: theory and applications. Neurocomputing **70**, 489–501 (2006)
22. Domingos, P., Pazzani, M.: On the optimality of the simple bayesian classifier under zero-one loss. Mach. Learn. **29**, 103–130 (1997)
23. Hecht-Nielsen, R.: Theory of the backpropagation neural network. In: Neural Networks for Perception, pp. 65–93. Elsevier (1992)
24. Chen, H.-L., Yang, B., Liu, J., et al.: A support vector machine classifier with rough set-based feature selection for breast cancer diagnosis. Expert Syst. Appl. **38**, 9014–9022 (2011)
25. Chang, C.-C., Lin, C.-J.: LIBSVM: a library for support vector machines. ACM Trans. Intell. Syst. Technol. **2**, 27 (2011)
26. Demircan, K., Cömertoğlu, İ., Akyol, S., et al.: A new biological marker candidate in female reproductive system diseases: Matrix metalloproteinase with thrombospondin motifs (ADAMTS). J. Turk. Ger. Gynecol. Assoc. **15**, 250–255 (2014)
27. Russell, D.L., Brown, H.M., Dunning, K.R.: ADAMTS proteases in fertility. Matrix Biol. **44–46**, 54–63 (2015)
28. Lindgren, D., Eriksson, P., Krawczyk, K., et al.: Cell-type-specific gene programs of the normal human nephron define kidney cancer subtypes. Cell Rep. **20**, 1476–1489 (2017)

# Identification of Prognostic and Heterogeneous Breast Cancer Biomarkers Based on Fusion Network and Multiple Scoring Strategies

Xingyi Li[1], Ju Xiang[1,2], Jianxin Wang[1],
Fang-Xiang Wu[3], and Min Li[1(✉)]

[1] School of Computer Science and Engineering, Central South University,
Changsha 410083, Hunan, China
limin@mail.csu.edu.cn
[2] Neuroscience Research Center and Department of Basic Medical Sciences,
Changsha Medical University, Changsha 410219, Hunan, China
[3] Department of Mechanical Engineering and Division of Biomedical
Engineering, University of Saskatchewan, Saskatoon, SK S7N5A9, Canada

**Abstract.** Breast cancer is a malignant disease that is caused by multiple factors, and the prognosis of breast cancer patients is the focus of medical research. In the present study, we have proposed a novel computational framework which identifies the prognostic biomarkers of breast cancer based on multiple network fusion and multiple scoring strategies. In order to eliminate the heterogeneity of samples, we first clustered the patient samples according to the principle components of gene expression. For each cluster, we used the fusion network to reduce the incompleteness of interactome and to take into account more disease-related information. Genes were weighted from the perspectives of biological functions, prognostic ability and correlation with known disease genes, and a network propagation model was applied to the fusion network so as to comprehensively evaluate the influence of genes on breast cancer patients. To evaluate the performance of our method, we have compared our approach with three state-of-the-art approaches. The results demonstrated that biomarkers captured by our method have both strong discriminative power in differentiating patients with different prognostic outcomes and biological significance.

**Keywords:** Prognostic biomarkers · Fusion network ·
Multiple scoring strategies

## 1 Introduction

Breast cancer is a malignant disease that is caused by complicate factors [1–3]. Therefore, the identification of prognostic and heterogeneous breast cancer biomarkers is of vital importance, they can not only guide the treatment of patients, but also help scientists to study the molecular mechanism of cancer [4].

Up to now, a number of methods identify prognostic biomarkers to predict outcomes of patients using gene expression data and single network [5, 6], but there are some problems in the network, such as incomplete. Some biomarker discovery

© Springer Nature Switzerland AG 2019
D.-S. Huang et al. (Eds.): ICIC 2019, LNCS 11644, pp. 529–534, 2019.
https://doi.org/10.1007/978-3-030-26969-2_50

methods are based on networks extracted from certain databases [7, 8], but they do not analyze network data with a variety of biological information, which may lead to some important information being ignored. Meanwhile, the network-based computational methods usually prioritize prognostic genes through the importance ranking of genes in the network [5, 9]. Those approaches are commonly influenced by the heterogeneity of cancer samples, which will lead to the poor prediction.

In this paper, we propose a novel framework based on fusion network and multiple scoring strategies to identify the prognostic and heterogeneous breast cancer biomarkers. We compare our method with some state-of-the-art methods. The results show that biomarkers identified by our method have a stronger ability to distinguish patients with different prognosis than other methods, and have the biological interpretation.

## 2   Materials and Methods

### 2.1   Data Preparation

We curated two gene expression profiles downloaded from Gene Expression Omnibus (GEO) [10], namely GSE1456 [11] and GSE2034 [12]. For GSE2034, samples that survived more than ten years were labeled as good prognosis (five years for GSE1456), and samples that survived less than five years were labeled as poor prognosis. Probes corresponding to multiple genes were discarded, and when multiple probes are mapped to the same gene, the median value was used to eliminate the influence of measurement error. Table 1 shows the details of the gene expression datasets.

**Table 1.** Summary of the gene expression datasets.

| Name | Good samples | Poor samples | Total genes | Characteristic for label |
|------|--------------|--------------|-------------|--------------------------|
| GSE1456 | 123 | 22 | 12432 | SURV_RELAPSE |
| GSE2034 | 44 | 93 | 12432 | Time to relapse or last follow-up |

The fusion network data was downloaded from [13]. There are 13,460 proteins and 141,296 interactions in the network. The data of human GO annotations was collected from Gene Ontology Consortium [14, 15] (http://www.geneontology.org/). Known disease-associated genes (DAGs) were downloaded from DisGeNET [16].

### 2.2   Prognostic and Heterogeneous Biomarker Identification

Inspired by Choi et al. [8], cancer samples were clustered by using principal component analysis (PCA) and k-means algorithm to minimize the effects of the heterogeneity of samples. For each cluster of samples, we scored the genes in the network from the perspectives of biological functions, prognostic ability and correlation with DAGs. The scores of DAGs were used as the values of these genes in the network and the prognostic ability of a gene was evaluated through the $t$ statistics of the gene. The relevance of biological functions for each gene is defined as the frequency of gene within the set of GO terms:

$$s(i) = \frac{N_i}{N} \tag{1}$$

where $N_i$ is the number of GO terms contained in gene $i$, and N denotes the number of all GO terms.

For each scoring strategies, we applied a network propagation model to rank genes in the fusion network with the initial weight of nodes [17]. The adjacency matrix is normalized by $W' = D^{-1/2}WD^{-1/2}$, i.e., $W'_{ij} = W_{ij}/\sqrt{D(i,i)D(j,j)}$, where $D$ is a diagonal matrix and $D(i,i)$ is the sum of row $i$ of W. Formally, the network propagation process can be formalized by the following iterative equation:

$$F_{t+1} = \alpha W'F_t + (1-\alpha)T \tag{2}$$

where $T$ denotes the prior information constructed by different scoring methods. $F_t$ is a vector, where the $i$-th element represents the flow of node $i$ at time step $t$ and $||F_{t+1} - F_t||^1 < 10^{-6}$. The default value of restart probability $\alpha$ is 0.6.

For each cluster, the comprehensive scores of genes are obtained by,

$$Score(i) = \frac{1}{3}\sum_{m=1}^{3} score_m(i) \tag{3}$$

where $m$ represents the three scoring strategies and $score_m(i)$ is the score of gene $i$ in the $m$-th scoring strategy.

Finally, we computed the average rank of each gene in all clusters and selected top p (such as top 1%) genes as biomarkers.

## 3    Results

### 3.1    Performance of Prognosis Prediction

To evaluate the performance of our method, three methods were compared to our method, namely NetRank [5], stSVM [6] and CPR [8]. A random forest classifier and five-fold cross-validation were used to evaluate the classification results. For unbiased evaluation, we repeated these experiments for 100 times for the entire datasets. The ROC curves and AUC were shown in Fig. 1 and Table 2. The results have shown that the performance of ROC and AUC obtained by our method is better than other methods.

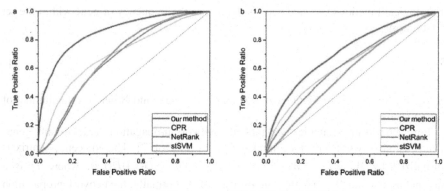

**Fig. 1.  ROC curves. a** GSE1456, **b** GSE2034.

**Table 2.** AUC of the four methods on two datasets.

| Methods | GSE1456 | GSE2034 |
|---------|---------|---------|
| My method | 0.87 | 0.73 |
| CPR | 0.70 | 0.64 |
| NetRank | 0.67 | 0.63 |
| stSVM | 0.68 | 0.58 |

## 3.2    Analysis of Functional Interpretability and Dysregulation of Biomarkers

We also analyzed the enrichment of DAGs and differential expression genes (DEGs) in biomarkers. We combined DEGs obtained by t-test and known DAGs to calculate the enrichment of these genes in the biomarkers and hypergeometric test was used to evaluate the p-value of the enrichment of the DAGs and DEGs.

$$P = 1 - \sum_{i=0}^{m-1} \frac{\binom{M}{i}\binom{N-M}{n-i}}{\binom{N}{n}} \tag{4}$$

where $N$ is the number of all genes in the gene expression data, $M$ is the number of DAGs and DEGs enriched in all genes, $n$ is the number of biomarkers, $m$ is the number of DAGs and DEGs enriched in the biomarkers.

We transformed $p - value$ to $-log_{10}(p - value)$. The significance of DAGs and DEGs ratio of four methods in two datasets were shown in Fig. 2. The biomarkers obtained by our method are most significantly enriched by DAGs and DEGs compared to other methods, showing that our method performs much better on the biological sense.

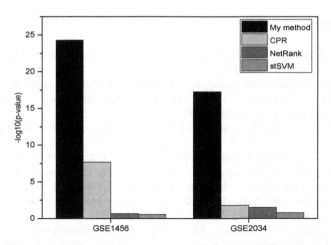

**Fig. 2. Significance of DAGs and DEGs ratio.** The significance of the ratio of DAGs and DEGs for the eight methods in two datasets was evaluated by hypergeometric test.

## 4 Conclusion and Discussion

In this paper, we proposed a novel computational framework, which identified the prognostic breast cancer biomarkers based on multiple network fusion and multiple scoring strategies. To validate the performance of this computational framework, we have compared our approach with some state-of-the-art approaches, such as CPR, NetRank and stSVM. The results have shown that the biomarkers can classify patients more accurately than other methods. Moreover, as the features should be biologically meaningful, we also analyzed the functional interpretability and dysregulation of biomarkers. We analyzed the biological sense of biomarkers identified by the four methods according to the enrichment of DAGs and DEGs. As expected, the results indicated that the biomarkers obtained by our method have high biological sense.

The limitation of our method is that considering that in the clinical application, a smaller but effective set of biomarkers are more practical. In future work, the number of biomarkers obtained by our method can be optimized so as to achieve the optimal performance with the fewest number of biomarkers.

**Acknowledgements.** This work was supported in part by the National Natural Science Foundation of China (61622213, 61732009, 61420106009, 61702054), the 111 Project (No. B18059), the Hunan Provincial Science and Technology Program (2018WK4001), the Training Program for Excellent Young Innovators of Changsha (Grant No. kq1802024), and the Hunan Provincial Natural Science Foundation of China (Grant No. 2018JJ3568).

# References

1. Weigelt, B., Peterse, J.L., Van't Veer, L.J.: Breast cancer metastasis: markers and models. Nat. Rev. Cancer **5**, 591 (2005)
2. Slamon, D.J., Clark, G.M., Wong, S.G., Levin, W.J., Ullrich, A., McGuire, W.L.: Human breast cancer: correlation of relapse and survival with amplification of the HER-2/neu oncogene. Science **235**, 177–182 (1987)
3. Key, T.J., Verkasalo, P.K., Banks, E.: Epidemiology of breast cancer. Lancet Oncol. **2**, 133–140 (2001)
4. Sawyers, C.L.: The cancer biomarker problem. Nature **452**, 548 (2008)
5. Winter, C., et al.: Google goes cancer: improving outcome prediction for cancer patients by network-based ranking of marker genes. PLoS Comput. Biol. **8**, e1002511 (2012)
6. Cun, Y., Fröhlich, H.: Network and data integration for biomarker signature discovery via network smoothed t-statistics. PLoS One **8**, e73074 (2013)
7. Wang, X., Wang, S.-S., Zhou, L., Yu, L., Zhang, L.-M.: A network-pathway based module identification for predicting the prognosis of ovarian cancer patients. J. Ovarian Res. **9**, 73 (2016)
8. Choi, J., Park, S., Yoon, Y., Ahn, J.: Improved prediction of breast cancer outcome by identifying heterogeneous biomarkers. Bioinformatics **33**, 3619–3626 (2017)
9. Liu, W., et al.: Topologically inferring risk-active pathways toward precise cancer classification by directed random walk. Bioinformatics **29**, 2169–2177 (2013)
10. Edgar, R., Domrachev, M., Lash, A.E.: Gene expression Omnibus: NCBI gene expression and hybridization array data repository. Nucleic Acids Res. **30**, 207–210 (2002)
11. Pawitan, Y., et al.: Gene expression profiling spares early breast cancer patients from adjuvant therapy: derived and validated in two population-based cohorts. Breast Cancer Res. **7**, R953 (2005)
12. Wang, Y., et al.: Gene-expression profiles to predict distant metastasis of lymph-node-negative primary breast cancer. Lancet **365**, 671–679 (2005)
13. Menche, J., et al.: Uncovering disease-disease relationships through the incomplete interactome. Science **347**, 1257601 (2015)
14. Ashburner, M., et al.: Gene Ontology: tool for the unification of biology. Nat. Genet. **25**, 25 (2000)
15. Consortium, G.O.: Expansion of the Gene Ontology knowledgebase and resources. Nucleic Acids Res. **45**, D331–D338 (2016)
16. Piñero, J., et al.: DisGeNET: a comprehensive platform integrating information on human disease-associated genes and variants. Nucleic Acids Res. **45**, D833–D839 (2016)
17. Vanunu, O., Magger, O., Ruppin, E., Shlomi, T., Sharan, R.: Associating genes and protein complexes with disease via network propagation. PLoS Comput. Biol. **6**, e1000641 (2010)

# A Novel Differential Essential Genes Prediction Method Based on Random Forests Model

Jiang Xie[1(✉)], Jiamin Sun[1], Jiaxin Li[1], Fuzhang Yang[1], Haozhe Li[3], and Jiao Wang[2(✉)]

[1] School of Computer Engineering and Science, Shanghai University,
99 Shang Da Road, Shanghai 200444, People's Republic of China
jiangx@shu.edu.cn
[2] Laboratory of Molecular Neural Biology, School of Life Sciences,
Shanghai University, 99 Shang Da Road,
Shanghai 200444, People's Republic of China
Jo717@shu.edu.cn
[3] Christian Wuhan Britain-China School, Wuhan 430022,
People's Republic of China

**Abstract.** Prediction of differential essential genes is an important field to research cell development and differentiation, drug discovery and disease causes. The goal of this work is to extract gene expression and topological changes in biomolecular networks for identifying the essential nodes or modules. Based on the random forests model, this paper proposed an essential node prediction algorithm for biomolecular networks called Differential Network Analysis method based on Random Forests (DNARF). The algorithm had two main points. First, the five-dimension eigenvector construction method was put forward to extract the differential information of nodes in networks. Second, a positive sample expansion method based on the Pearson correlation coefficient was present to solve the problem that positive and negative samples may be unbalanced. In the simulated data experiments, the DNARF algorithm was compared with three other algorithms. The results showed that the DNARF had an excellent performance on the prediction of essential genes. In the real data experiments, four gene regulatory networks were used as datasets. DNARF algorithm predicted five essential genes related to leukemia: HES1, STAT1, TAL1, SPI1 and RFXANK, which had been proved by literatures. Also, DNARF could be applied to other biological networks to identify new essential genes.

**Keywords:** Differential Network Analysis · Essential genes · Machine learning · Random Forests · Biomarker discovery

## 1 Introduction

During past decades, molecular biology has deeply investigated essential genes that affect life processes such as disease and senility. Many methods have been developed which can identify the genes with significant changes between different biological states. In the early period, scientists used some differential gene expression analysis

D.-S. Huang et al. (Eds.): ICIC 2019, LNCS 11644, pp. 535–546, 2019.
https://doi.org/10.1007/978-3-030-26969-2_51

methods [1] to complete the task, such as student's t-test, SAM, etc. However, because of the complexity and uncertainty of biological process that genes are members of strongly intertwined interactive with each other, the same method on different clinical samples may gain different results. Therefore, the results from conventional gene expression analysis methods may readily be biased. Soon afterward, network-based methods were developed, which provided a method to research multiple interactions among genes. In recent years, a popular approach named differential network analysis had been used frequently to measure the changes in networks between two different states [2, 3]. This approach could also dig out essential genes between the two states. Diffk [4] was based on Differential Degree Centrality (DDC) [2] to score all nodes in networks. Then the essential genes were gained by the ranking with the score mentioned above. DiffRank [5] ranked genes based on their contributions to the differences between the two networks and defined two new structural scoring measures.

All researches above are based on conventional methods such as statistic and graph theory. With the development of computer science, more researchers pay attention to the combination of machine learning and bionetwork analysis. Support vector machine and neural network on protein-protein networks identified whether each protein in the network is essential or inessential [6]. PREvaIL [7] was an integrative machine learning approach that using sequence, structural, and network features to determine the catalytic residues in an enzyme. Expert knowledge is used to solve machine learning problems. The 26-gene panel as a candidate biomarker set was used to classify an integrated expression data of 530 ovarian tissues [8].

However, all of these machine learning methods noticed in single network while the changes between two networks of different states are able to better reveal the law of biology process. This work presented a novel differential network analysis algorithm called Differential Network Analysis method based on Random Forests (DNARF). Through the expert knowledge, DNARF could extract the differential information of each node between two biomolecular networks. It could transform the differential information into vectors to do the learning operation. Then, the well-trained model was able to predict the essential genes which mainly cause the change between two states.

# 2 Related Methods

DNARF is based on random forests model, a kind of machine learning method. The related methods are introduced in the three following parts.

## 2.1 Random Forests Model

Random Forests (RF) is an ensemble tree-structured algorithm used for classification and regression analyses. RF consists of hundreds of decision trees, and all decision trees participate in voting the prediction. Compared with other statistical classifiers, advantages of RF includes (1) high classification accuracy, (2) a method of determining variable importance, (3) the ability to model complex interactions among predictor variables, (4) an algorithm for imputing missing values [9]. Therefore, RF is widely used in computational biology.

## 2.2   Feature Engineering

Feature Engineering is a crucial point in machine learning. It is essential to describe each node in the network using eigenvector. Here is the illustration of some features:

(1) Degree Centrality (DC) [10]: DC means how many nodes connected to someone, and it can measure one node's 'centrality' apparently. It is formalized by Eq. (1):

$$C_D(v) = deg(v) \tag{1}$$

(2) Betweenness Centrality (BC) [11]: For one node, BC is defined as the average length of the shortest path through it. If more shortest paths pass through a node in a network, it reflects that the node has a high centrality. BC's equation is as follow:

$$C_B(v) = \sum_{s \neq v \neq t \in V} \frac{\sigma_{st}(v)}{\sigma_{st}} \tag{2}$$

In which $\sigma_{st}$ means the amount of shortest path through node $s$ and node $t$, and $\sigma_{st}(v)$ means the number of ones which go through node $v$.

(3) Closeness Centrality (CC) [12]: CC indicates the degree of a node in the network that how much it communicates with other nodes. It is calculated by the total number of links for all its neighbor nodes:

$$C_c(v) = \sum_{j=1}^{g} x_{ij}(i \neq j) \tag{3}$$

CC relates to the scale of the network. A larger network may lead to a higher CC. To eliminate the impact of network size on CC, Eq. (3) is converted into Eq. (4):

$$C'_c(v) = \frac{C_c(v)}{g-1} \tag{4}$$

(4) Clustering Coefficient [13] (CCo): CCo is a coefficient indicating the level of aggregation around a node in the network.

$$C_{CCo}(v) = \frac{n}{C_k^2} = \frac{2n}{k(k-1)} \tag{5}$$

As shown above, $k$ represents the number of all neighbors of node $v$, and $n$ represents the number of edges which are connected among node $v$'s all adjacent nodes.

## 2.3   Performance Evaluation

To evaluate the performance of a prediction model, true negative (TN), false positive (FP), false negative (FN) and true positive (TP) [14] can be used to calculate several evaluation indicators to test the effect of the model.

Moreover, F1-Score is a handy indicator for measuring the accuracy of a binary classification model. The F1-Score takes *Precision* and *Recall* into account, which ranges from 0 to 1. The model is more excellent if the F1-Score is closer to 1.

## 3   Methods

DNARF aims to find the essential nodes in the differential network through the label of expert knowledge, which is based on random forests model. This part describes the two steps of DNARF algorithm. First, a novel feature extraction method built by biological network data and gene expression with expert knowledge make DNARF extract differential essential genes. Second, DNARF contained a sample expansion method to address the problem that positive and negative samples may be unbalanced.

### 3.1   Preparing the Training Set

The training set required two contrasted biomolecular networks (e.g. the disease state network and the control state network). The next step was to collect the gene expression data of each node from some databases such as Gene Expression Omnibus (GEO) (https://www.ncbi.nlm.nih.gov/geo/).

### 3.2   Generating 5-Dimension Eigenvector

Feature extraction was the key part of DNARF. A novel feature extraction method was proposed by topological structure in differential networks and gene expression data. A five-dimension vector $x_v = \{x_v^1, x_v^2, x_v^3, x_v^4, x_v^5\}$ was constructed for node $v$ to quantify the difference of $v$ in different states. The details are as follow:

(1)  $x_v^1 = \Delta DC$. The degree's change of node $v$ is the most intuitive way to describe the difference between $v$ in different states. The formula of DC is shown in (1), so the calculation method of $x_v^1$ is

$$x_v^1 = \left| C_D(v) - C_D'(v) \right| \tag{6}$$

In which, $C_D(v)$ and $C_D'(v)$ denote the two degrees of node $v$ in different states.

(2)  $x_v^2 = \Delta BC$. As shown in Eq. (2), BC means a kind of 'Centrality' of one node. The more shortest-paths go pass, the node might be more critical in the network. So, $\Delta BC$ can reflect the change of centrality in different states.

$$x_v^2 = \left| C_B(v) - C_B'(v) \right| \tag{7}$$

$C_B(v)$ and $C_B'(v)$ denote BC in different states, respectively.

(3) $x_v^3 = \Delta CC$. What is CC is shown as Eqs. (3) and (4). To estimate the impact of network size on CC, Eq. (4) is used to denote CC. $x_v^3$ is as follow.

$$x_v^3 = \left| C_C'(v) - C_C''(v) \right| \tag{8}$$

$C_C'(v)$ and $C_C''(v)$ denote BC in different states, respectively.

(4) $x_v^4 = \Delta CCo$. A node with high CCo means that it has high status in the network. $\Delta CCo$ can reflect the change of status of a node.

$$x_v^4 = \left| C_{CCo}(v) - C_{CCo}'(v) \right| \tag{9}$$

(5) $x_v^5 = p - value(E, E')$. $E$ and $E'$ respectively represents expression values of each gene in two different states. The differences between gene expression values are direct representations of genes' change. p-value can depict the difference of gene expression. It makes $x_v^5$ have statistical significance.

### 3.3 Marking the Training Data

After the vectorization step, every piece of data should be marked in training dataset. This is a binary classification problem. Each node is regarded as essential (marked as 1) or unessential (marked as 0). Commonly, the biomolecules supported by literatures are defined the essential biomolecules.

### 3.4 Expanding Positive Samples

There is an unavoidable problem that the essential biomolecules from literatures are usually not abundant, which makes positive samples and negative samples out of balance. In this condition, the model will prefer to predict the classification with more samples, and the algorithm will lack extensiveness [15].

The essential biomolecules supported by literatures are essential, but the biomolecules unmarked in literatures may also be essential. Therefore, positive samples can be extended to address the problem of unbalanced data. A positive sample extension method is presented to balance the samples of positive and negative. The experiments used the Pearson correlation coefficient and set up a threshold $\varepsilon$. The Pearson correlation coefficient is shown as follows:

$$r = \frac{N \sum x_i y_i - \sum x_i \sum y_i}{\sqrt{N \sum x_i^2 - \left(\sum x_i\right)^2} \sqrt{N \sum y_i^2 - \left(\sum y_i\right)^2}} \tag{10}$$

All biomolecules were divided to the essential biomolecules set $R$ and unmarked biomolecules set $U$. $R = \{r_0, r_1, r_2, \ldots, r_n\}$ was used to identify the essential biomolecules supported by literatures. $U = \{u_0, u_1, u_2, \ldots, u_n\}$ was used to represent the ones unmarked. Then the calculation of all Pearson correlation coefficient between $r_i \in R$ and $u_i \in U$ was completed. If the $p$-value between $r_x$ and $u_y$ was bigger than the threshold $\varepsilon$, $u_y$ could be marked as an essential biomolecule.

The setting principle is that the number of positive and negative samples is as balanced as possible after extension. Moreover, to ensure the biological sense, the threshold $\varepsilon$ is recommended to set the value over 0.8.

## 3.5    Training the Model

After preparing the training set, the random forests model was trained to predict the essential biomolecules in networks.

## 3.6    Predicting Essential Biomolecules

After training the model, the essential biomolecules in bionetworks of life process were predicted. They could provide new inspiration for the research of bioscience or medicine science.

The flow chart of DNARF is shown in Fig. 1. The parameters of $x_v^1, x_v^2, x_v^3, x_v^4$ (formula (6)–(9)) are constructed by four topological structure. The $x_v^5$ is constructed by differential expression from gene expression profile.

The 5-dimension eigenvector $x_v = \{x_v^1, x_v^2, x_v^3, x_v^4, x_v^5\}$ is built by biological network and gene expression data to train the tree models and predict essential genes. Accuracy, Precision, Recall, Specificity and F1-Score of them are calculated to evaluate the performance of prediction models.

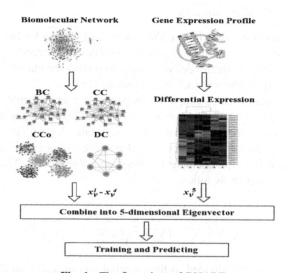

**Fig. 1.** The flow chart of DNARF.

# 4   Experiments

In this section, DNARF was verified by simulated data and explored neurocytoma and leukemia on real data.

## 4.1   Simulated Data Experiments

### (1) Data Sources

Zhang presented a simulated data generating method in differential network analyses [16]. The author claimed that a biomolecular network is a kind of scale-free network, and the network should follow the principles of scale-free networks presented by Barabási [17]. According to those principles, the author put forward the simulated data generating algorithm. The algorithm would output two networks, two sets of gene expression and a list of essential genes. This algorithm had some parameters to generate simulation data. The parameters of $n_1$ and $n_2$ meant the number of nodes in two different networks. The $m$ meant the number of essential genes in two networks, and the $\rho$ meant the proportion of differential edges driven by perturbed genes [16]. The smaller $\rho$ was, the more difficult it was to find essential genes. In the simulated experiments setting, $n_1 = n_2 = 100$, $m = 10$, $\rho = 0.1$.

### (2) Results and Analyses

In simulated experiments, DNARF was compared with three other classical algorithms which could predict essential differential genes: DCloc [18], DiffRank [5] and DEC [19]. 101 sets of data were generated here.

These three algorithms were based on the traditional numerical calculation method, so they did not need a training set. They could score all nodes but not classify whether the gene was essential or not in the differential networks. In this paper, 100 of the 101 data sets were calculated by each of the algorithms. The top 10 highest score genes in each data set were defined as essential genes. Table 1 shows the average performance of these algorithms on 100 data sets.

One of the 101 sets was used for training DNARF model, and the remaining 100 sets were used for test data. The mean value of the 100 testing is shown in Table 1.

**Table 1.** Evaluation indicators of four algorithms

| Indicators | Algorithm | | | |
|---|---|---|---|---|
| | DCloc | DiffRank | DEC | DNARF |
| Accuracy | 89.04% | 93.44% | 94.82% | **99.33%** |
| Precision | 45.20% | 67.20% | 74.10% | **98.14%** |
| Recall | 45.20% | 67.20% | 74.10% | **95.10%** |
| Specificity | 93.91% | 96.36% | 97.12% | **99.80%** |
| F1-score | 0.452 | 0.672 | 0.741 | **0.966** |

As shown in Table 1, DNARF is superior to the other three algorithms. Particularly, the F1-Score of DNARF is 0.966, which is most closed to 1. It proves that DNARF has a good performance on essential genes prediction.

Traditional numerical calculation methods are lack of generalization ability. Usually, they take few differential indicators into attention. In reality, the reason for a node to be essential can be various. DNARF combined multiple kinds of features, and it could take various kinds of differences into consideration to achieve a better prediction.

### 4.2   Real Data Experiments

#### (1)   Data Sources

In the real data experiments, two sets of networks were found in the Interactome dataset [20, 21] (http://www.regulatorynetworks.org/, January, 2019). The astrocyte gene regulatory network (NHA, 516 genes and 9296 edges) and the neuroblastoma gene regulatory network (SKNSH, 508 genes and 12761 edges) were set as training set. The microvascular endothelium, adult, blood gene regulatory network (HMVEC_dBlAd, 520 genes and 13510 edges) and the promyelocytic leukemia gene regulatory network (NB4, 525 genes and 18960 edges) were as test set.

Afterward, the paper referred to the GEO datasets for gene expression profile data. Data from GSE99051 [22] was used for NHA; data from GSE112384 [23] was used for SKNSH; data from GSE12679 [24] was used for HMVEC_dBlAd and data from GSE73157 was used for NB4.

For the training set, the data was marked and 23 essential genes for neuroblastoma were supported by 14 literatures [25–38]: TP53, BRCA1, MYCN, E2F1, FOXA1, ZFX, PRDM1, BCL6, XBP1, ASCL1, TP73, ESR1, ZBTB33, PPARA, E2F2, BACH1, BACH2, PBX1, MEIS1, GATA3, HIF1A, ZNF148 and BPTF.

#### (2)   Data Pre-processing

In the differential network analyses, significant differential changes of the common genes are focus of concern. Thus, the common genes between two differential networks should be found at first. There were 486 and 247 common genes in two sets of networks, respectively. In the training set, the SKNSH network had 12149 edges, while the NHA network had 8649 edges among 486 genes. In the test set, the HMVEC_dBlAd network had 480 edges, while the NB4 network had 663 edges among 247 genes.

Next, using the methods mentioned in Sect. 3, each group of networks was vectorized as differential eigenvectors.

Thus, there were only 23 essential genes out of 486 genes. The positive samples and negative samples were out of balance. Therefore, the extension method was used to balance the two kinds of samples. After setting the $\varepsilon = 0.9$, the number of essential genes was 198 and the number of unmarked genes was 288 after extension.

#### (3)   Results and Analyses

After the pre-processing step, SKNSH and NHA were used as training set to train the random forests model and used HMVEC_dBlAd and NB4 for test data.

In order to avoid the deviation caused by randomness, the train-test procedure was repeated for 100 times, and genes which were marked as essential are counted. The result is as follow: HES1, STAT1, RFXANK and TAL1 are marked as essential for 100 times. SPI1 is marked as essential gene for 97 times. These 5 genes can be regarded as the essential genes about leukemia predicted by DNARF.

Afterwards, these essential genes were analyzed by literatures:

HES1: HES1 plays a critical role in the development of T cells and also plays an essential role in the process of cancer. More importantly, HES1 can be treated as an essential drug target for leukemia [39].

STAT1: STAT1 is widely known as a suppressor during tumor growth. However, STAT1 can accelerate the process of leukemia [40].

TAL1: In the TAL1 gene, site-specific DNA recombination will occur in the patient of leukemia. Therefore, the TAL1 gene has a close relationship with leukemia [41].

SPI1: The Friend viruses, like the Rauscher virus, is a cause of leukemia [42]. Spi-1 gene activation is a general feature in the malignant proerythroblastic transformation which occurs in mice infected with Friend and Rauscher viruses.

RFXANK: No direct relation between RFXANK and leukemia has been found. However, RFXB [43] (another name of RFXANK) as a kind of transcription factor has positive regulation on HLA genes, which had close relationship with leukemia. According to the prediction of DNARF, a hypothesis was made that RFXANK has a direct relationship with leukemia that was not discovered. This hypothesis could be verified by biological experiments later.

In conclusion, the prediction of DNARF has biological meaning that the 5 genes are essential in the procedure of leukemia.

# 5  Conclusions

Predicting the essential genes in differential network analyses is a work full of biological meaning. It can reveal the potential laws of disease. This article presented a differential essential gene prediction algorithm called DNARF based on the random forest model with two main points. First, the information of each node was vectorized to extract differential information of nodes, which could be used to conduct the training and prediction steps. Because of essential nodes from literatures which probably led to the imbalance of positive and negative samples, a positive sample expansion method based on Pearson correlation coefficient was proposed, which balanced the data and improved the performance of the model.

In the simulated data experiments, a data generation algorithm was used to generate multiple sets of simulated networks with different parameters. The DNARF algorithm was compared with three other algorithms, DCloc, DiffRank and DEC. A series of indicators such as accuracy, recall and F1-Score proved that the DNARF had more excellent performance than other three traditional essential gene prediction algorithms. This reason was that the DNARF algorithm had a stronger generalization ability for the problem and could identify a variety of causes that made the gene to be 'essential'.

In the real data experiments, the NHA and SKNSH network were set to be the training set. The HMVEC_dBlAd and NB4 network were set to be the test set. For the training set, 23 marked genes were supported by 14 literatures, and the positive samples were extended by the method proposed in this paper. After training and predicting, the DNARF algorithm predicted five essential genes related to leukemia: HES1, STAT1, TAL1, SPI1 and RFXANK. The literatures support that the four genes HES1, STAT1, TAL1 and SPI1 have an extremely close relationship with leukemia. For, the current conclusion is that it has an indirect relationship with leukemia. However, based on the prediction of DNARF, a hypothesis could be made: There was a direct link between RFXANK and leukemia. In the future work, a series of biological experiments can be conducted to prove this hypothesis.

**Acknowledgement.** This work was partially supported by the National Natural Science Foundation of China [No. 61873156] and the Project of NSFS [No. 17ZR1409900].

# References

1. Yang, D., Parrish, R.S., Brock, G.N.: Empirical evaluation of consistency and accuracy of methods to detect differentially expressed genes based on microarray data. Comput. Biol. Med. **46**, 1–10 (2014)
2. De la Fuente, A.: From 'differential expression' to 'differential networking'–identification of dysfunctional regulatory networks in diseases. Trends Genet. **26**, 326–333 (2010)
3. Ideker, T., Krogan, N.J.: Differential network biology. Mol. Syst. Biol. **8**, 565 (2012)
4. Hudson, N.J., Reverter, A., Dalrymple, B.P.: A differential wiring analysis of expression data correctly identifies the gene containing the causal mutation. PLoS Comput. Biol. **5**, e1000382 (2009)
5. Odibat, O., Reddy, C.K.: Ranking differential hubs in gene co-expression networks. J. Bioinform. Comput. Biol. **10**, 1240002 (2012)
6. Chen, Y., Xu, D.: Understanding protein dispensability through machine-learning analysis of high-throughput data. Bioinformatics **21**, 575–581 (2004)
7. Song, J., et al.: PREvaIL, an integrative approach for inferring catalytic residues using sequence, structural, and network features in a machine-learning framework. J. Theor. Biol. **443**, 125–137 (2018)
8. Yeganeh, P.N., Mostafavi, M.T.: Use of machine learning for diagnosis of cancer in ovarian tissues with a selected mRNA panel. In: 2018 IEEE International Conference on Bioinformatics and Biomedicine (BIBM), pp. 2429–2434. IEEE, Madrid (2018)
9. Cutler, D.R., et al.: Random forests for classification in ecology. Ecology **88**, 2783–2792 (2007)
10. Seidman, S.B.: Network structure and minimum degree. Soc. Netw. **5**(3), 269–287 (1983)
11. Freeman, L.C.: A set of measures of centrality based on betweenness. Sociometry **40**, 35–41 (1977)
12. Wuchty, S., Stadler, P.F.: Centers of complex networks. J. Theor. Biol. **223**, 45–53 (2003)
13. Saramäki, J., Kivelä, M., Onnela, J.-P., Kaski, K., Kertesz, J.: Generalizations of the clustering coefficient to weighted complex networks. Phys. Rev. E: Stat. Nonlinear Soft Matter Phys. **75**, 027105 (2007)
14. Glas, A.S., Lijmer, J.G., Prins, M.H., Bonsel, G.J., Bossuyt, P.M.: The diagnostic odds ratio: a single indicator of test performance. J. Clin. Epidemiol. **56**(11), 1129–1135 (2003)
15. Chawla, N.V., Bowyer, K.W., Hall, L.O., Kegelmeyer, W.P.: SMOTE: synthetic minority over-sampling technique. J. Artif. Intell. Res. **16**, 321–357 (2002)

16. Zhang, X.F., Ou-Yang, L., Yan, H.: Incorporating prior information into differential network analysis using nonparanormal graphical models. Bioinformatics **33**, 2436 (2017)
17. Barabási, A.-L.: Scale-free networks: a decade and beyond. Science **325**, 412–413 (2009)
18. Bockmayr, M., Klauschen, F., Denkert, C., Budczies, J.: New network topology approaches reveal differential correlation patterns;in breast cancer. BMC Syst. Biol. **7**, 78 (2013)
19. Lichtblau, Y., Zimmermann, K., Haldemann, B., Lenze, D., Hummel, M., Leser, U.: Comparative assessment of differential network analysis methods. Brief. Bioinform. **18**, 837 (2016)
20. Neep, S., Stergachis, A.B., Reynolds, A., Sandstrom, R., Borenstein, E., Stamatoyannopoulos, J.A.: Circuitry and dynamics of human transcription factor regulatory networks. Cell **150**, 1274–1286 (2012)
21. Stergachis, A.B., et al.: Conservation of trans-acting circuitry during mammalian regulatory evolution. Nature **515**, 365–370 (2014)
22. Sloan, S.A., et al.: Human astrocyte maturation captured in 3D cerebral cortical spheroids derived from pluripotent stem cells. Neuron **95**, 779–790 (2017)
23. Hassannia, B., Wiernicki, B., Ingold, I., Qu, F., Berghe, T.V.: Nano-targeted induction of dual ferroptotic mechanisms eradicates high-risk neuroblastoma. J. Clin. Investig. **128**, 3341–3355 (2018)
24. Harris, L.W., et al.: The cerebral microvasculature in schizophrenia: a laser capture microdissection study. PLoS One **3**, e3964 (2008)
25. Mahdavi, M., et al.: Hereditary breast cancer; Genetic penetrance and current status with BRCA. J. Cell. Physiol. **234**, 5741–5750 (2019)
26. Aygun, N., Altungoz, O.: MYCN is amplified during S phase, and c-myb is involved in controlling MYCN expression and amplification in MYCN-amplified neuroblastoma cell lines. Mol. Med. Rep. **19**, 345–361 (2019)
27. Cai, L., et al.: ZFX mediates non-canonical oncogenic functions of the androgen receptor splice variant 7 in castrate-resistant prostate cancer. Mol. Cell **72**, 341–354 (2018)
28. Desmots, F., et al.: Pan-HDAC inhibitors restore PRDM1 response to IL21 in CREBBP-mutated follicular lymphoma. Clin. Cancer Res. **25**, 735–746 (2019)
29. Narayanan, A., et al.: The proneural gene ASCL1 governs the transcriptional subgroup affiliation in glioblastoma stem cells by directly repressing the mesenchymal gene NDRG1. Cell Death Diff. **1** (2018)
30. García-Martínez, A., et al.: DNA methylation of tumor suppressor genes in pituitary neuroendocrine tumors. J. Clin. Endocrinol. Metabol. **104**, 1272–1282 (2018)
31. Wang, L., et al.: Kaiso (ZBTB33) downregulation by Mirna-181a inhibits cell proliferation, invasion, and the epithelial-mesenchymal transition in glioma cells. Cell. Physiol. Biochem. **48**, 947–958 (2018)
32. Gao, Y., et al.: PPARα Regulates the proliferation of human glioma cells through miR-214 and E2F2. BioMed. Res. Int. **2018** (2018)
33. Davudian, S., Mansoori, B., Shajari, N., Mohammadi, A., Baradaran, B.: BACH1, the master regulator gene: a novel candidate target for cancer therapy. Gene **588**, 30–37 (2016)
34. Roychoudhuri, R., et al.: The transcription factor BACH2 promotes tumor immunosuppression. J. Clin. Investig. **126**, 599–604 (2016)
35. Blasi, F., Bruckmann, C., Penkov, D., Dardaei, L.: A tale of TALE, PREP1, PBX1, and MEIS1: interconnections and competition in cancer. Bioessays **39** (2017)
36. Lin, M., Lin, J., Hsu, C., Juan, H., Lou, P., Huang, M.: GATA3 interacts with and stabilizes HIF-1α to enhance cancer cell invasiveness. Oncogene **36**, 4243 (2017)
37. De Bustos, C., Smits, A., Strömberg, B., Collins, V.P., Nistér, M., Afink, G.: A PDGFRA promoter polymorphism, which disrupts the binding of ZNF148, is associated with primitive neuroectodermal tumours and ependymomas. J. Med. Genet. **42**, 31–37 (2005)

38. Richart, L., Real, F.X., Sanchez-Arevalo Lobo, V.J.: c-MYC partners with BPTF in human cancer. Mol. Cell. Oncol. **3**, e1152346 (2016)
39. Rani, A., Greenlaw, R., Smith, R.A., Galustian, C.: HES1 in immunity and cancer. Cytokine Growth Factor Rev. **30**, 113–117 (2016)
40. Kovacic, B., et al.: STAT1 acts as a tumor promoter for leukemia development. Cancer Cell **10**, 77–87 (2006)
41. Brown, L., et al.: Site-specific recombination of the tal-1 gene is a common occurrence in human T cell leukemia. EMBO J. **9**, 3343–3351 (1990)
42. Moreau-Gachelin, F., Ray, D., Tambourin, P., Tavitian, A.: Spi-1 oncogene activation in Rauscher and Friend murine virus-induced acute erythroleukemias. Leukemia **4**, 20–23 (1990)
43. Das, B., Majumder, D.: Information theory based analysis for understanding the regulation of HLA gene expression in human leukemia. Int. J. Infor. Sci. Tech. (2012)

# Identification of Candidate Biomarkers and Pathways Associated with Liver Cancer by Bioinformatics Analysis

Zhen-Bo Tian and Xu-Qing Tang

School of Science, Jiangnan University, Wuxi 214122, China
txq5139@jiangnan.edu.cn

**Abstract.** Liver cancer is a common malignant tumor in China, which seriously threatens people's health. The aim of this study is to explore molecular markers for early diagnosis and potential drug targets in the immunotherapy of liver cancer. The gene expression profiles of liver cancer are downloaded from TCGA database and analyzed by R package. A total of 1564 differentially expressed genes are obtained, of which 1400 are up-regulated and 164 are down-regulated. The GO and KEGG enrichment analysis of differentially expressed genes are performed by DAVID. The STRING database is used to construct a protein interaction network, which is visualized using Cytoscape software and further screened for 15 hub genes. In addition, Oncomine database and survival analysis method are used to verify the hub genes. These results shown that PLK1, CDC20, CCNB2, BUU1, MAD2L1 and CCNA2 genes are closely related to the occurrence, development and prognosis of liver cancer. These genes could be used as potential markers and drug targets of liver cancer, which are provided to assist deeper understanding the mechanism of liver cancer, and find tumor markers and drug targets.

**Keywords:** Bioinformatics analysis · Differential expressed genes · PPI · Oncomine · Survival curve analysis

## 1 Introduction

Liver cancer is a common malignant tumor in China, which seriously threatens people's health [1]. It accounts for 70–80% of primary malignant tumors of the liver, and its cancer mortality rate ranks second in tumors [2]. The occurrence and development of liver cancer are a process involving multi-gene and multi-factor synergy. It takes several pathological stages and multiple molecular event changes to realize the complex process of normal liver cells to liver cancer cells [3]. Previous studies had shown that some epidemiological factors such as aflatoxin exposure [4], chronic viral hepatitis [5], and genetic factors [6] may be associated with the risk of liver cancer. However, the underlying mechanism of liver cancer development is still unclear. Due to the high incidence, high invasiveness, high metastasis and poor prognosis of liver cancer, most patients with liver cancer are diagnosed at terminal stage. After the discovery, there are fewer effective treatments and poor efficacy. Therefore, the early diagnosis of liver cancer is a hot and difficult point in the research of liver cancer prevention and

© Springer Nature Switzerland AG 2019
D.-S. Huang et al. (Eds.): ICIC 2019, LNCS 11644, pp. 547–557, 2019.
https://doi.org/10.1007/978-3-030-26969-2_52

treatment. Targeted drugs have opened up a new path for targeted therapy of liver cancer, such as Sorafenib. But the application of Sorafenib has certain limitations, and long-term drug use will produce drug resistance [7]. Therefore, the discovery of new targets and the development of new drugs have become the research hotspots of the world medical community. Recent years, the development of high-throughput gene chip and sequencing technology has made it possible for rapidly studying the gene expression profile of liver cancer, exploring the gene expression and key gene changes of liver cancer tissues and cells under specific conditions.

The TCGA database is currently the largest database of cancer gene information, which collects clinical data, genomic variation, mRNA expression, miRNA expression, methylation and other data of various human cancers. It aims to apply high-throughput Genomic analysis techniques to help people understand the mechanism of cancer better, thereby improving the ability of prevent, diagnose and treatment. Oncomine database is the largest gene chip database and integrated data extraction platform in the world at present [8]. The integrated literature and chip data have been widely recognized by the academia for their high quality. In this study, the original data is downloaded from the TCGA database, the differentially expressed genes (DEGs) in the gene expression profiles of liver cancer tissues and para-cancerous tissues are screened by R language. The bioinformatics analysis is performed on DEGs, and Oncomine database is combined to screen out potential molecular targets which suitable for early diagnosis and immunotherapy of liver cancer. Finally, the survival analysis and literature mining are performed to verify the results, providing theoretical basis for further research.

## 2    Materials and Methods

### 2.1    Materials

The data analyzed in this paper is obtained from the TCGA database. The mRNA expression data of liver cancer tissues and adjacent tissues of liver cancer patients are downloaded from the database, including 371 samples of liver cancer patients and 50 normal samples.

### 2.2    Methods

The aim of this study is to identify molecular markers for early diagnosis and potential molecular targets in the immunotherapy of liver cancer. The methods used in the study are as follows: Screening for the DEGs, GO and KEGG enrichment analysis, PPI network construction and hub genes identification, Oncomine validation and survival analysis of hub genes. These methods are described in detail as follows:

(1)  Screening for the DEGs

Background correction, standardization and expression value calculation are performed on the original data by using the "edger" package, and screened out the DEGs. In this process, we need to use two indexes $\log FC$ and $FDR$, which the conditions for using them are generally $|\log FC| > \alpha$ and $FDR < \beta$, where: The larger the value of $\alpha$, the smaller the number of DEGs screened; The larger the value of $\beta$, the greater the number of DEGs screened out.

(2)  GO and KEGG enrichment analysis

The DAVID is a biological information database that integrates biological data and analysis tools to provide systematic and comprehensive biological function annotation information [9]. GO enrichment analysis is a common useful method for large scale functional enrichment research, gene functions can be classified into biological process (BP), molecular function (MF), and cellular component (CC). The p-value and FDR of the differential gene GO terms are calculated by statistical analysis, and the most likely related GO term is located. KEGG pathways enrichment analysis calculates the hypergeometric distribution of these differential genes and pathway based on the selected differential genes. KEGG pathways enrichment analysis will return a p-value and FDR value of each pathway in which the DEGs exist, so as to locate the most likely related pathways. In order to analyze the biological functions and enriched signaling pathways of DEGs, the DAVID 6.8 database is used to conduct GO and KEGG enrichment analysis. Then, the bar plot and ggplot2 are performed to visualize analysis of GO and KEGG, respectively.

(3)  PPI network construction and hub genes identification

The STRING database is an online analysis tool for studying protein-protein interaction (PPI) patterns that builds PPI networks and provides information on related pathways and functions [10]. Using the database to import DEGs, the network file of protein interaction is output. Then this file is carried out the visual protein interaction network analysis by Cytoscape software [11]. CytoHubba plug-in is used to define the genes of network regulatory center as hub genes [12].

(4)  Oncomine validation and survival analysis of hub genes

Oncomine database is a cancer gene chip database and integrated data mining platform. It aims to analyze cancer gene information, which can be used to compare the differential expression analysis between cancer and normal tissues, and then verify the hey genes. The KM-Plotter database (http://kmplot.com/analysis/) is applied to evaluate the prognostic values of the hub genes in liver cancer patients. Where, the KM-Plotter database contains information on 364 patients with liver cancer.

In this study, the DEGs are obtained by differential expression analysis of liver cancer gene expression profiles. Then, the GO and KEGG enrichment analysis of DEGs are performed by DAVID, and the significantly enriched functions and pathways of DEGs are obtained. Furthermore, the STRING database is used to construct a protein interaction network, and hub genes are further screened. Finally, Oncomine database and survival analysis are used to confirm the hub genes, and the genes which suitable for liver cancer markers or potential targets are obtained.

## 3   Results

### 3.1   Screening for the DEGs

According to the mRNAs matrix data of normal tissues and liver cancer tissues in the TCGA database, with $\alpha = 2$ and $\beta = 0.05$ as the screening criteria, a total of 1564 DEGs are obtained, of which 1400 are up-regulated and 164 are down-regulated (Fig. 1).

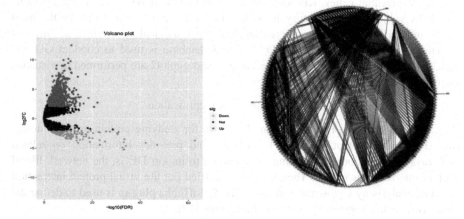

**Fig. 1.** The volcano map of the DEGs, where red indicates up-regulated genes and green indicates down-regulated genes. (Color figure online)

**Fig. 2.** PPI of 15 hub genes and other genes.

### 3.2   GO and KEGG Enrichment Analysis

GO and KEGG enrichment analysis are performed on 1564 DEGs by the DAVID. The results of GO enrichment analysis indicate that the DEGs are mainly enriched in BPs, including chemical synaptic transmission and sister chromatid cohesion. CC shows that the DEGs are significantly enriched in extracellular region, extracellular space, anchored component of membrane, plasma membrane and chromosome, and centromere region. MF shows that the DEGs are significantly enriched in sequence-specific DNA binding, hormone activity, calcium ion binding and growth factor activity (Table 1). The results of KEGG pathway enrichment analysis show that the DEGs are mainly enriched in Neuroactive ligand-receptor interaction, Cell cycle, Mineral absorption, Hypertrophic cardiomyopathy, Cardiac muscle contraction, Adrenergic signaling in cardiomyocytes and Dilated cardiomyopathy (Table 1). The screening results of GO and KEGG are visualized by the R package as shown in Fig. 3. $\beta = 0.01$ is included as inclusion criteria in GO, and p-value < 0.05 is included as inclusion criteria in KEGG.

**Table 1.** Significantly enriched GO terms and KEGG pathways of DEGs.

| Category | Term | Description | Count | FDR/p-value |
|---|---|---|---|---|
| BP term | GO:0007268 | Chemical synaptic transmission | 44 | 2.26e-04 |
| BP term | GO:0007062 | Sister chromatid cohesion | 24 | 4.38e-03 |
| CC term | GO:0005576 | Extracellular region | 198 | 2.66e-08 |
| CC term | GO:0005615 | Extracellular space | 158 | 1.08e-04 |
| CC term | GO:0031225 | Anchored component of membrane | 26 | 1.86e-03 |
| CC term | GO:0005886 | Plasma membrane | 386 | 1.16e-02 |
| CC term | GO:0000775 | Chromosome, centromeric region | 16 | 2.77e-02 |
| MF term | GO:0043565 | Sequence-specific DNA binding | 79 | 2.65e-06 |
| MF term | GO:0005179 | HORMONE activity | 24 | 4.46e-04 |
| MF term | GO:0005509 | Calcium ion binding | 88 | 6.24e-03 |
| MF term | GO:0008083 | Growth factor activity | 29 | 3.83e-02 |
| KEGG pathway | hsa04080 | Neuroactive ligand-receptor interaction | 46 | 2.33e-07 |
| KEGG pathway | hsa04110 | Cell cycle | 23 | 8.06e-05 |
| KEGG pathway | hsa04978 | Mineral absorption | 12 | 2.43e-04 |
| KEGG pathway | hsa05410 | Hypertrophic cardiomyopathy (HCM) | 16 | 4.47e-04 |
| KEGG pathway | hsa04260 | Cardiac muscle contraction | 15 | 9.36e-04 |
| KEGG pathway | hsa04261 | Adrenergic signaling in cardiomyocytes | 22 | 1.00e-03 |
| KEGG pathway | hsa05414 | Dilated cardiomyopathy | 16 | 1.01e-03 |

## 3.3    PPI Network Construction and Hub Genes Identification

In order to further understand the regulation relationship between DEGs and proteins in liver cancer tissues and normal tissues, the STING database is used to construct an interaction network diagram for protein regulation and 1498 protein interaction network nodes are obtained through analysis. The data downloaded from the STRING database is imported into the Cytoscape software for visualization, and the PPI network graph of the DEGs is obtained (Fig. 2). The Degree of the CytoHubba plugin is used to define the top 15 hub genes (GNGT1, GNG4, LPAR3, BDKRB1, PLK1, CDC20, CCNB2, ADCY8, BUB1, MAD2L1, AURKB, AGTR2, RLN3, CCNA2 and CENPE) for further analysis.

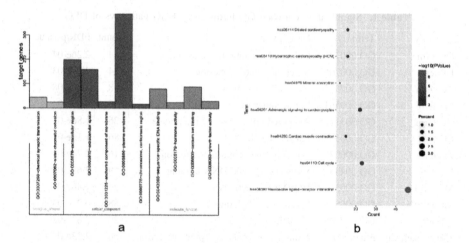

**Fig. 3.** Visualization of GO and KEGG. a: GO, the abscissa represents the function of genes and the ordinate represents the number of genes enriched in each function. b: KEGG, the ordinate indicates the name of the signal pathway and the abscissa indicates the number of genes enriched in this pathway.

### 3.4 Oncomine Validation and Survival Analysis of the Hub Genes

Oncomine database is used to analyze the differential expression of these 15 hub genes, and it is verified that the expression levels of PLK1, CDC20, CCNB2, BUB1, MAD2L1 and CCNA2 are significantly different (Fig. 4). Then the meta-analysis is performed on the 6 hub genes with significant differences (Fig. 5). In the meta-analysis, three studies are screened by significant differences and differential expression folds: the articles are published in Cancer Res [13] and Hepatology [14], respectively. According to the statistical analysis of Oncomine database, these 6 hey genes are highly expressed in liver cancer compared with normal tissues, and the differences are statistically significant. (The CDC20 gene is used as an example in both differential expression analysis and meta-analysis).

The survival curves are obtained based on the 6 hub genes using the KM-Plotter database. Furthermore, we study the prognostic value of these hub genes for the overall survival of patients with liver cancer. According to the survival analysis of KM-Plotter database, PLK1 (P = 7e-07), CDC20 (P = 5.7e-08), CCNB2 (P = 2.7e-04), BUB1 (P = 4e-05), MAD2L1 (P = 5.3e-06) and CCNA2 (P = 5.3e-06) are obtained. The analysis results of survival curve are shown in Fig. 6. According to the survival analysis results of clinical data with liver cancer, the high expression of PLK1, CDC20, CCNB2, BUB1, MAD2L1 and CCNA2 genes are closely related to the occurrence, development and prognosis of liver cancer, which is significant for the study of liver cancer.

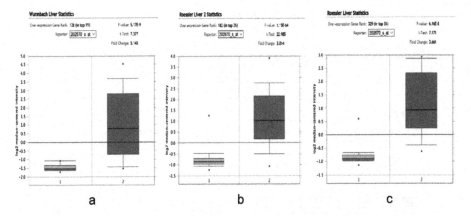

**Fig. 4.** Differential expression analysis of CDC20 gene in Oncomine database liver cancer. 1: normal tissue; 2: liver cancer tissue; a, b, c respectively represents the results of three studies.

## 4    Discussion

The occurrence and development of liver cancer is a dynamic biological process with multiple molecules, steps and factors. At present, people's understanding of its mechanism is still very limited. Most of the previous studies are limited to the influence of a single gene on tumors, but changes of multiple genes are often involved in the process of cell carcinogenesis, and these genes could interact with each other and play a role through regulatory network. Therefore, studying cancer gene expression profile at the level of multiple genes can help us better explore the pathogenesis. As a new generation of high-throughput detection technology, gene chip can detect the expression level of tens of thousands of genes at the same time, and it is a powerful tool to study the interaction between genomes and genes.

In this study, gene expression in para-cancerous tissues and liver cancer tissues is first compared, and a total of 1564 DEGs are screened, including 1400 up-regulated genes and 164 down-regulated genes. The results of GO enrichment analysis indicate that the DEGs are mainly enriched in chemical synaptic transmission, extracellular region, extracellular space, plasma membrane and calcium ion binding. The results of KEGG pathway enrichment analysis show that the DEGs are mainly enriched in Neuroactive ligand-receptor interaction, Cell cycle, Hypertrophic cardiomyopathy (HCM), Adrenergic signaling in cardiomyocytes and Dilated cardiomyopathy. Then, the STRING database and Cytoscape software are used to construct a PPI network of DEGs, and then the 15 most important genes are screened. In addition, in order to verify the expression level of these key genes, differential expression analysis and meta-analysis of 15 key genes are carried out through the Oncomine database, and six high expression genes of PLK1, CDC20, CCNB2, BUU1, MAD2L1 and CCNA2 are obtained. Finally, the KM-Plotter database is used to analyze and verify the six genes based on survival curves. It is found that the high expression of these genes is closely related to the prognosis of liver cancer, which is significant for the study of liver cancer.

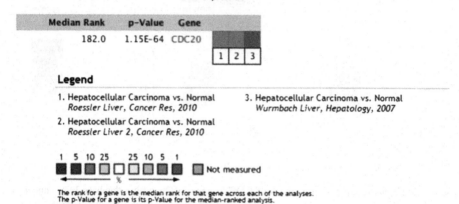

**Fig. 5.** Meta-analysis of Oncomine database of CDC20 gene in liver cancer.

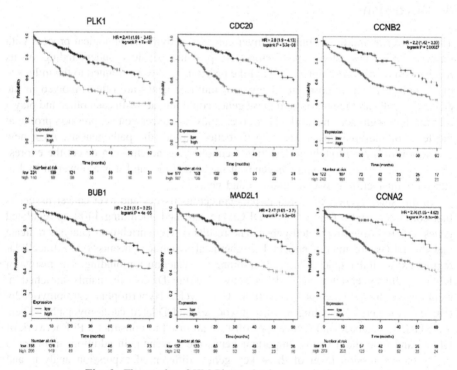

**Fig. 6.** The results of KM-Plotter database survival analysis.

As a member of the PLK family, PLK1 is a key protein of the cell cycle regulation network and plays an important role in cell mitotic regulation. In recent years, it has been found that PLK1 is highly expressed in various malignant tumors such as prostate cancer [15], neuroblastoma cells [16], acute myeloid leukemia [17] and cervical cancer [18]. PLK1 plays an important role in the initiation, maintenance and completion of mitosis, and is closely related to survival and prognosis. Inhibition of PLK1 expression by interfering with multiple stages of mitosis can lead to tumor cell death, and PLK1 is expected to be a potential target for cancer therapy [19]. CDC20 belongs to the cell cycle protein, which is the key protein to ensure the normal progression of cell mitosis and plays an important role in cell mitosis. Its abnormal expression causes normal cells to produce non-integer multiple chromosomes by disrupting cell mitosis, which in turn promotes malignant transformation of cells [20]. Previous studies have shown that abnormally high expression of CDC20 is likely to be an important factor in promoting abnormal proliferation and resistance to apoptosis in tumor cells [21]. CDC20 is highly expressed in various malignant tumor tissues such as lung cancer, liver cancer and malignant glioma [22–24]. In addition, the high expression of CDC20 is also closely related to the clinical progress of human malignant tumors. Wu *et al.* [25] found that CDC20 is significantly related to the progress and prognosis of colorectal cancer. Wang *et al.* [26] also showed that CDC20 may be a tumor treatment target. CCNA2 and CCNB2 are well-known cyclins, both of which are essential for cell cycle progression. The results of these studies show that both of them are highly expressed in HCC tissues, and patients with high expression have lower overall survival rates. Some reports [27, 28] have similar results. BUB1 is a mitotic checkpoint protein that plays an important regulatory role in cell mitosis [29]. The protein regulates the formation and activation of complexes in the late mitosis by phosphorylating some of the components of the mitotic checkpoint complex and activating the checkpoint of the spindle [30, 31]. Previous studies demonstrated mitotic kinase and mitotic checkpoint protein play an important role in cell division and maintenance of chromatin stability [32]. The abnormal expression of this gene is related to the occurrence and development of a variety of tumors, and BUB1 is also involved in the biological behavior of tumor stem cells such as breast cancer [33, 34]. MAD2L1 is a key component of the spindle assembly checkpoint, and its expression is related to the chemotherapy resistance of ovarian serous tumors [35], and reducing the expression of MAD2L1 by siRNA can reduce tumor cell growth and inhibit cell migration and invasion [36].

## 5    Conclusions

In this study, we explored the molecular function and biological process of the occurrence and development of liver cancer from the perspective of "holistic view". Based on the analysis of the liver cancer data from TCGA database, a total of 1564 DEGs are obtained, and 15 hub genes are further obtained by bioinformatics analysis. Finally, Oncomine and survival analysis confirm that the six genes of PLK1, CDC20, CCNB2, BUU1, MAD2L1 and CCNA2 are closely related to the development, the prognosis and the mechanism of liver cancer, and provides reference for finding tumor markers and drug targets.

# References

1. Chen, W.: Cancer statistics: updated cancer burden in China. Chin. J. Cancer Res. **27**(1), 1 (2015)
2. Erstad, D.J., Tanabe, K.K.: Hepatocellular carcinoma: early-stage management challenges. J. Hepatocell. Carcinoma **4**, 81–92 (2017)
3. Gong, Y.Z., Jiang, Y.A.: Significance of abnormal gene expression in the pathogenesis and development of hepatocellular carcinoma in its diagnosis and treatment. J. Gastroenterol. Hepatol. **25**(8), 848–851 (2016)
4. Moudgil, V., Redhu, D., Dhanda, S., et al.: A review of molecular mechanisms in the development of hepatocellular carcinoma by aflatoxin and hepatitis B and C viruses. J. Environ. Pathol. Toxicol. Oncol. **32**(2), 165–175 (2013)
5. Tu, T., Bühler, S., Bartenschlager, R.: Chronic viral hepatitis and its association with liver cancer. Biol. Chem. **398**(8), 817–837 (2017)
6. Tang, R., Liu, H., Yuan, Y., et al.: Genetic factors associated with risk of metabolic syndrome and hepatocellular carcinoma. Oncotarget **8**(21), 35403–35411 (2017)
7. Dong, Z.Z., Zhu, X.D., Li, Z., et al.: Advances in basic and clinical research on hepatocellular carcinoma in 2016. Chin. J. Hepatol. **25**(2), 85–93 (2017)
8. Rhodes, D.R., Kalyana-Sundaram, S., Mahavisno, V., et al.: Oncomine 3.0: genes, pathways, and networks in a collection of 18,000 cancer gene expression profiles. Neoplasia **9**(2), 166–180 (2007)
9. Huang, D.W., Sherman, B.T., Tan, Q., et al.: DAVID bioinformatics resources: expanded annotation database and novel algorithms to better extract biology from large gene lists. Nucleic Acids Res. **35**(Web Server), W169–W175 (2007)
10. Szklarczyk, D., Franceschini, A., Wyder, S., et al.: STRING v10: protein-protein interaction networks, integrated over the tree of life. Nucleic Acids Res. **43**(D1), D447–D452 (2015)
11. Shannon, P.: Cytoscape: a software environment for integrated models of biomolecular interaction networks. Genome Res. **13**(11), 2498–2504 (2003)
12. Chin, C.H., Chen, S.H., Wu, H.H., et al.: CytoHubba: Identifying hub objects and sub-networks from complex interactome. BMC Syst. Biol. **8**(Suppl 4), S11 (2014)
13. Roessler, S., Jia, H.L., Budhu, A., et al.: A unique metastasis gene signature enables prediction of tumor relapse in early-stage hepatocellular carcinoma patients. Cancer Res. **70**(24), 10202–10212 (2010)
14. Wurmbach, E., Chen, Y., Khitrov, G., et al.: Genome-wide molecular profiles of HCV-induced dysplasia and hepatocellular carcinoma. Hepatology **45**, 938–947 (2007)
15. Li, J., Wang, R., Kong, Y., et al.: Targeting Plk1 to enhance efficacy of olaparib in castration-resistant prostate cancer. Mol. Cancer Ther. **16**(3), 469–479 (2017)
16. Pajtler, K.W., Sadowski, N., Ackermann, S., et al.: The GSK461364 PLK1 inhibitor exhibits strong antitumoral activity in preclinical neuroblastoma models. Oncotarget **8**(4), 6730–6741 (2016)
17. Tao, Y.F., Li, Z.H., Du, W.W., et al.: Inhibiting PLK1 induces autophagy of acute myeloid leukemia cells via mammalian target of rapamycin pathway dephosphorylation. Oncol. Rep. **37**(3), 1419–1429 (2017)
18. Yang, X., Chen, G., Li, W., et al.: Cervical cancer growth is regulated by a c-ABL-PLK1 signaling axis. Can. Res. **77**(5), 1142–1154 (2016)
19. Liu, Z., Sun, Q., Wang, X.: PLK1, a potential target for cancer therapy. Transl. Oncol. **10**(1), 22 (2016)
20. Huang, H.C., Shi, J., Orth, J.D., et al.: Evidence that mitotic exit is a better cancer therapeutic target than spindle assembly. Cancer Cell **16**(4), 347–358 (2009)

21. Gayyed, M.F., El-Maqsoud, N.M., Tawfiek, E.R., et al.: A comprehensive analysis of CDC20 overexpression in common malignant tumors from multiple organs: its correlation with tumor grade and stage. Tumour Biol. **37**(1), 749–762 (2016)

22. Shi, R., Sun, Q., Sun, J., et al.: Cell division cycle 20 overexpression predicts poor prognosis for patients with lung adenocarcinoma. Tumor Biol. **39**(3), 101042831769223 (2017)

23. Kim, H.S., Vassilopoulos, A., Wang, R.H., et al.: SIRT2 maintains genome integrity and suppresses tumorigenesis through regulating APC/C activity. Cancer Cell **20**(4), 487–499 (2011)

24. Xie, Q., Wu, Q., Mack, S.C., et al.: CDC20 maintains tumor initiating cells. Oncotarget **6** (15), 13241–13254 (2015)

25. Wu, W.J., Hu, K.S., Wang, D.S., et al.: CDC20 overexpression predicts a poor prognosis for patients with colorectal cancer. J. Transl. Med. **11**(1), 142 (2013)

26. Wang, Z., Wan, L., Zhong, J., et al.: Cdc20: a potential novel therapeutic target for cancer treatment. Curr. Pharm. Des. **19**(18), 3210–3214 (2013)

27. Gao, C., Wang, G., Yang, G., et al.: Karyopherin subunit-α 2 expression accelerates cell cycle progression by upregulating CCNB2 and CDK1 in hepatocellular carcinoma. Oncol. Lett. **15**(3), 2815–2820 (2017)

28. Gopinathan, L., Tan, S.L.W., Padmakumar, V.C., et al.: Loss of Cdk2 and cyclin A2 impairs cell proliferation and tumorigenesis. Cancer Res. **74**(14), 3870–3879 (2014)

29. Akhoundi, F., Parvaneh, N., Modjtaba, E.B.: In silico analysis of deleterious single nucleotide polymorphisms in human BUB1 mitotic checkpoint serine/threonine kinase B gene. Meta Gene **9**(C), 142–150 (2016)

30. Ricke, R.M., Jeganathan, K.B., Van Deursen, J.M.: Bub1 overexpression induces aneuploidy and tumor formation through Aurora B kinase hyperactivation. J. Cell Biol. **193**(6), 1049–1064 (2011)

31. Jia, L., Li, B., Yu, H.: The Bub1–Plk1 kinase complex promotes spindle checkpoint signalling through Cdc20 phosphorylation. Nat. Commun. **7**, 10818 (2016)

32. Park, Y.Y., Nam, H.J., Do, M., et al.: The p90 ribosomal S6 kinase 2 specifically affects mitotic progression by regulating the basal level, distribution and stability of mitotic spindles. Exp. Mol. Med. **48**(8), e250 (2016)

33. Chang, J.T., Reiner, S.L.: Asymmetric division and stem cell renewal without a permanent niche: lessons from lymphocytes. Cold Spring Harb. Symp. Quant. Biol. **73**, 73–79 (2008)

34. Lathia, J.D., Hitomi, M., Gallagher, J., et al.: Distribution of CD133 reveals glioma stem cells self-renew through symmetric and asymmetric cell divisions. Cell Death Dis. **2**, e200 (2011)

35. Nakano, Y., Sumi, T., Teramae, M., et al.: Expression of the mitotic-arrest deficiency 2 is associated with chemotherapy resistance in ovarian serous adenocarcinoma. Oncol. Rep. **28** (4), 1200–1204 (2012)

36. Nascimento, A.V., Singh, A., Bousbaa, H., et al.: Overcoming cisplatin resistance in non-small cell lung cancer with Mad2 silencing siRNA delivered systemically using EGFR-targeted chitosan nanoparticles. Acta Biometarialia **47**, 71–80 (2017)

# Real-Time Pedestrian Detection in Monitoring Scene Based on Head Model

Panpan Lu[1], Kun Lu[1], Wenyan Wang[1], Jun Zhang[2],
Peng Chen[2(✉)], and Bing Wang[1,2(✉)]

[1] School of Electrical and Information Engineering,
Anhui University of Technology, Maanshan 243002, People's Republic of China
Wangbing@ustc.com
[2] The Institute of Health Sciences, Anhui University,
Hefei 230601, Anhui, China

**Abstract.** Pedestrian detection is an essential technology in robotics, intelligent transportation system, and intelligent video surveillance. Pedestrians in the surveillance scene have the characteristics of dense crowds and high degree of occlusion, meanwhile, it needs to meet the requirements of real-time detection. To solve this problem, the method based on head model with YOLOv3 algorithm was proposed. This work re-selects the number and dimensions of anchor boxes by k-means method on the training set, fine-tuning the network and training it to get the optimal model. The experiment results show that this work effectively improves the detection performance and real-time of pedestrian detection in the surveillance scene.

**Keywords:** Pedestrian detection · Surveillance · Head model · YOLOv3

## 1 Introduction

Pedestrian detection is the key technology for pedestrian tracking, behavior analysis, pedestrian identification and other research. It has attracted a large number of researchers in related fields [1–4]. Pedestrians have both rigid and soft characteristics, and the complex backgrounds, different lighting conditions, different camera shots, occlusions between people and pedestrian attitude all have an impact on test results. Especially in the surveillance scenario, the pedestrian traffic is large, partial occlusion is high, and real-time requirements for detection. These problems make the pedestrian real-time detection having a big challenge in the monitoring scenario [5].

In order to handle the above challenges, there are a lot of research results in this field. R Benenson et al. reviewed the pedestrian detection in the past ten years [6]. B Olga proposed a new probabilistic framework for object detection which is related to the Hough transform [7]. S Zhang et al. compares the performance of state of art methods in the field of pedestrian detection in recent years [8].

At present, the methods of pedestrian detection can be divided into method based on background modeling and method based on statistical learning [9]. Among them,

© Springer Nature Switzerland AG 2019
D.-S. Huang et al. (Eds.): ICIC 2019, LNCS 11644, pp. 558–568, 2019.
https://doi.org/10.1007/978-3-030-26969-2_53

the method based on statistical learning is divided into traditional methods and deep learning. The traditional methods of pedestrian detection is mainly two-stage methods. The first stage is extracting pedestrian features such as Haar-like [10], LBP [11] and HOG [12] from images, and the second step is training the classifier to learning features. For example, Wang et al. combined HOG features and LBP features to deal with pedestrian occlusion in 2009 [13], and improving detection accuracy. However, it is hard to be applied to actual scenes because of artificially designed features always having poor robustness, high missed detection rate, and cannot meet real-time detection due to complex calculation.

In 2012, Hinton and his students applied deep learning to image processing and achieved first place in the ILSVRC, and far exceeded the second place [14]. Since then, convolutional neural networks (CNNs) have been used for object detection and introduction of pedestrian detection studies. It has been proved that deep learning is potentially more capable than traditional method. Articles [15–17] et al. use CNNs to extract pedestrian features, which are free from the constraints of artificial design features and achieve better detection results. For example, Joint Deep adds pedestrian occlusion model and deformation model to convolutional neural network for joint training in [18]. The Deep Parts model handles occlusion by building a part pool to improve detection performance [19]. The R-CNN algorithm proposes a region proposal method of selective search to extract bounding box in the pictures based on CNNs [20]. The Fast R-CNN [21], and Faster R-CNN [22] networks, which were later improved, are based on the idea of region proposals. However, these methods have poor real-time performance due to the complex network structure and large amount of computation.

The previously proposed algorithms all solve the problem of transforming detection problems into classification problems. It was not until Redmon Joseph proposing YOLO [23] algorithm and returning the detection problem to the regression method that the real-time performance was greatly improved. But the initial version of the YOLO algorithm is not mature due to unreasonable meshing and bounding box prediction, and there are problems such as inaccurate positioning and lower recall than region proposal method. The YOLOv2 [24] algorithms are improved on the basis of YOLO algorithm, which make the detection effect further improved. But the YOLOv2 are difficult to detect small objects. The YOLOv3 [25] improves the deficiency of YOLOv2, and it is a state-of-the-art, real-time object detection system.

This work focuses on the problem of pedestrian occlusion and real-time detection in surveillance scenarios. On the basis of YOLOv3 network, a pedestrian detection method based on the head model is proposed. There are two points in this work. On the one hand, solve serious occlusion problems and improve the real-time detection in surveillance scenarios by introducing pedestrian head model. On the other hand, according to the relative width and height ratio of head model in the picture, clustering the training set by k-means method to obtain the suitable anchor boxes, fine-tuning the network parameters, and training to obtain the optimal model.

## 2  Method

### 2.1  Dataset

The deep learning network relies heavily on the training of big data. Therefore, the data is not only the experimental platform and benchmark, but also the guarantee of experimental results. This work selected a total of 1,014 images as a training set from INRIAPerson and PASCAL VOC2012 dataset, meanwhile, selected 338 images as test set from INRIAPerson and PASCAL VOC2012 dataset.

The INRIAPerson dataset is a representative pedestrian dataset because that it has a variety of shooting conditions, changes of lighting conditions, pedestrians occlude each other and the background is more complex. The dataset was collected as part of research work on detection of upright people in images and video and it is divided in four class: (a) only car (b) only people (c) include car and people (d) without car and people. The PASCALVOC dataset include VOC2007 and VOC2012 and it providing standardized image annotation and evaluation systems for performance test of detection algorithms. The PASCAL VOC2012 include 20 classes and it has 11,530 images containing 27,450 ROI annotated objects and 6,929 segmentations.

### 2.2  Network Structure

The full name of the YOLO algorithm is that you only look once, which is a state-of-the-art, real-time multi-object detection system. It has been developed to three versions now. The network structure of YOLOv1 consists of convolutional layer, pooled layer and fully connected layer. In YOLOv2, to get more spatial information, the fully connected layer is removed, and set an anchor in order to better predict the bounding box.

The model of YOLOv3 draws lessons from the multi-scale prediction idea of FPN and the basic classification network structure of ResNet. The model much more complex than YOLOv2 and it is mainly composed of a series of 1x1 and 3x3 convolution followed by a BN layer and a Leaky ReLU layer. The YOLO 2 used pass through structure to detect fine-grained features, and the YOLO 3 further uses three feature maps of different scales to detect objects. The first two layers of the feature-map are upsampled, concated with the starting feature map, add some convolutional layers. The author call it Darknet-53 because there are 53 convolutional layers in the network.

The head model detected in this work occupies a small proportion of the whole image, so that we adjusted the detection network, removed the yolo layer for detecting large targets, and only retained two small and medium targets. The yolo layer is used to detect our head model. At the same time, since this work only detects a single pedestrian category, and the original YOLOv3 is a multi-target detection algorithm, it is necessary to streamline the number of convolution kernels, so that the accuracy of detection can be ensured, and the detection speed can be improved. The network structure as shown in Fig. 1.

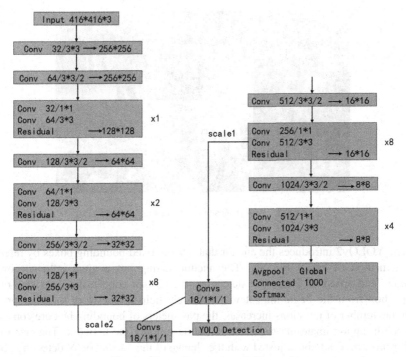

**Fig. 1.** Network structure of this work

## 2.3   Head Model

The pedestrian traffic is crowded in most cases in the surveillance scene, which makes the pedestrians having a large amount of partial occlusion. Therefore, only the detector with high robustness to partial occlusion can meet the detection needs of the scene. According to Chen's [26] statistical analysis of the pedestrian dataset SED-PD, in the surveillance scene, the visibility of the pedestrian's leg is 27%, the visible rate of the trunk is 63%, and the visibility of the head and shoulders can reach 98%. For instance, as show in the Fig. 2, when pedestrians are crowded in surveillance scenarios, this will lead to the pedestrian's torso being mostly obscured or even completely obscured. At the same time, the head model can greatly reduce the detection error caused by factors such as clothes, luggage, and body deformation compared to the whole body.

Accordingly, applying the head model to pedestrian detection is very suitable for solving pedestrian occlusion problems in the monitoring scene. Meanwhile, it also reduces the amount of calculations for detection and improves real-time performance.

## 2.4   Clustering Bounding Boxes on the Training Set

Although the YOLOv3 network has achieved the best detection results in the field of object detection, it is not fully applicable to pedestrian detection in surveillance scenarios. Consequently, this work makes some changes based on the YOLOv3 network to make it suitable for specific applications in this work.

**Fig. 2.** Pedestrians whose trunk is covered

The YOLOv2 introduces the anchor that a set of fixed bounding boxes by referring to the method of Faster R-CNN. The anchor setting has a great influence on the accuracy and speed of the object detection. If you can choose the right size of the anchor box from the beginning, it will definitely help the network to better predict. With the number of iterations increases, the parameters of bounding box are constantly adjusted to approximate the ground truth while training the network. The size of the anchor boxes needs to be adjusted with the change of the dimension of detecting object. The YOLOv3 setting three anchor boxes for each down sampling scale and clustering 9 anchor boxes on the COCO datasets.

However, the COCO dataset contains 80 classes and the scale of each category is different. Consequently, the anchor boxes of original are not applicable to the head model of this paper. After all, aspect ratio of pedestrians head in the picture is always a relatively fixed scale. Therefore, this work needs to re-select the number and size of anchor boxes.

The traditional k-means clustering method uses the Euclidean distance function, but the error is also greater if the size of the box is larger, and this work hopes that the error does not have much relationship with the size of the box. The purpose of clustering is to have a larger IOU value between the anchor box and the adjacent ground truth, which is not directly related to the size of the anchor box. So the distance metric needs to be redefined through the IOU and make the error being independent of the size of the anchor box. The distance function as Eq. 1.

$$d(groundtruth,\ centroid) = 1 - IOU(groundtruth,\ centroid) \tag{1}$$

In the above equation, the ground truth represents the true labeling box of object, centroid represents the center of the cluster, and the IOU represents the intersection ratio of the ground truth of the training sample and the cluster center. The final clustering objective function is:

$$S = min \sum_{i=0}^{k} \{1 - IOU(bbox, \ groundtruth)\} \tag{2}$$

In the above equation, bbox is a candidate box, and k is the number of anchor box selected. The distance from the bbox to the center point of the cluster should be as small as possible, and the value of the IOU should be as large as possible. Therefore, the objective function indicates that the smaller the distance between the samples in the cluster, the larger the value of the IOU.

The K-Means algorithm can automatically assign samples to different categories, but it is not possible to decide which classes to divide. So we first need to determine all the number of cluster center points, that is, the number of k, which in this work represents the number of anchor boxes. This work uses the Eq. 2 to calculate the corresponding objective function value S when k is 1–9. The result is shown in Fig. 3. According to the line graph, when the k value is increased to 6, the change of the objective function gradually becomes gentle. Therefore, it is finally determined that the number of cluster center points is 6.

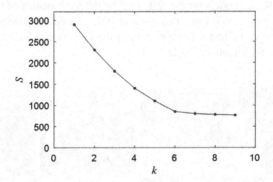

**Fig. 3.** Clustering selection

# 3  Experimental Results and Discussion

## 3.1  Network Training

Network training is an important part in our work. First of all, this work selects 1,014 images as training set from the INRIAPerson and PASCAL VOC2012 dataset. Secondly, this work labeled the training set with the labelImg tool and fine-tuned the network. This work used a small batch random gradient descent method in the training process, updating weight parameters every 64 samples. This work also used the method of data augmentation to increase the number of training samples and improve the generalization ability of the model. Finally, the network was trained 34,000 iterations. The training process of the network is shown in Fig. 4. When the network iterates to 20,000 batches, the loss value declined steadily as can be seen from Fig. 4.

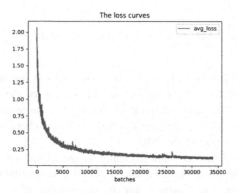

**Fig. 4.** The loss curves of training.

## 3.2 Results

Import images and labels from the training set into the model. Finally, the model are obtained after 34,000 iterations by GPU acceleration. This work tested the experimental results of these models on the test set. The detected results of the head model is shown in Fig. 5. At the same time, this work also compared the results of Yolo 2, Yolo 3 and head model on the test set, the results of the detection are shown in Fig. 6. The detailed value are presented in Table 1.

**Fig. 5.** Detected results of our model

It can be seen from the Table 1, the performance of our model is pretty swell. Compared with Faster R-CNN, YOLOv2 and YOLOv3 algorithm, the accuracy of our work was promoted 4.64%, 9.22% and 0.74% respectively, meanwhile, our work is more advantageous in terms of detection speed than Faster R-CNN and YOLOv3 since this work only detects the pedestrian head. This work tested the miss rates of different models in the case of FPPI on the test set. The miss rate was improved 6.97%, 4.69% and 1.64%. Although yolov2 algorithm has faster detection speed than our work, its accuracy and miss rate are poor.

**Fig. 6.** Comparison of results between YOLOv2, YOLOv3 and our model. The first line is the test results of YOLOv2, the second line is the test results of YOLOv3, the third line is the test results of our model.

**Table 1.** The results of the experiment.

| Algorithm model | Accuracy (%) | Test time per images (s/p) | Miss rate (%) FPPI = $10^{-1}$ |
|---|---|---|---|
| Faster R-CNN | 87.84 | 0.256 | 14.55 |
| YOLOv2 | 83.26 | 0.136 | 12.27 |
| YOLOv3 | 91.74 | 0.165 | 9.22 |
| Our work | 92.48 | 0.152 | 7.58 |

To further validate the real-time performance of our model, this work tested the fps of model on surveillance video. There are two test videos, the one is pedestrian video on foreign streets, which lasts 8 min and 45 s, and the other is pedestrian video on domestic streets, which lasts 3 min and 50 s. The size of video are 800 * 480 and 25 fps. The test results are shown in Fig. 7, and the comparison of detection frame rates between YOLOv3 and our models is shown in Table 2.

### 3.3 Discussion

This work introduces the YOLOv3 algorithm and improves it by focusing on the characteristics of pedestrians in the surveillance scene. Then the fine-tuned network is iteratively trained with sufficient training samples to get the optimal model. Finally, this work validates the trained model on the test set. It can be seen from the above results, our work has been improved in precision, test time, and fps compared with Faster R-CNN and YOLOv3 algorithm, meanwhile, the missed rate of our model is also reduced. The experimental results fully demonstrate that our work has positive

**Fig. 7.** The results of detection on the video

**Table 2.** Detection results on the video.

| Method | YOLOv3 | Our work |
|---|---|---|
| FPS (f/s) | 27 | 36 |

implications for improving the capability and real-time performance of pedestrian detection in surveillance scenarios. And it adapts to input images of different resolution and have better robustness to pedestrian detection. This model can be applied well to the field of pedestrian detection including monitoring scenarios.

## 4  Conclusion

This work proposes the head model aiming at the problem that pedestrian trunk parts are easily occluded due to the factors such as the external environment, posture and clothing in surveillance scene. Furthermore, in order to meet the real-time requirements in monitoring scenarios, the yolov3 algorithm is also introduced in this work. The experimental results show that the proposed method effectively improves the detect performance, and achieves end-to-end detection, which basically meets the requirements of pedestrian video detection in the monitoring scenario.

However, there are still some problems in the research of this work. The model has not been improved in terms of false detection. Especially character posters and buildings are prone to false detections. Moreover, the deep learning algorithm has higher requirements on hardware devices, so it needs further research to apply it to the real scene.

**Acknowledgement.** This work is supported by the National Natural Science Foundation of China (Nos. 61472282, 61672035, and 61872004), Anhui Province Funds for Excellent Youth Scholars in Colleges (gxyqZD2016068), the fund of Co-Innovation Center for Information

Supply & Assurance Technology in AHU (ADXXBZ201705), and Anhui Scientific Research Foundation for Returned Scholars.

# References

1. Barinova, O., Lempitsky, V., Kholi, P.: On detection of multiple object instances using hough transforms. IEEE Trans. Pattern Anal. Mach. Intell. **34**, 1773 (2012)
2. Ding, Y., Xiao, J.: Contextual boost for pedestrian detection. In: Computer Vision and Pattern Recognition, pp. 2895–2902
3. Dollar, P., Wojek, C., Schiele, B., Perona, P.: Pedestrian detection: an evaluation of the state of the art. IEEE Trans. Pattern Anal. Mach. Intell. **34**, 743–761 (2012)
4. Tuzel, O., Porikli, F., Meer, P.: Pedestrian detection via classification on Riemannian manifolds. IEEE Trans. Pattern Anal. Mach. Intell. **30**, 1713–1727 (2008)
5. Su, S.Z., Li, S.Z., Chen, S.Y., Cai, G.R., Wu, Y.D.: A survey on pedestrian detection. Acta Electron. Sinica **40**, 814–820 (2012)
6. Benenson, R., Omran, M., Hosang, J., Schiele, B.: Ten years of pedestrian detection, what have we learned? In: Agapito, L., Bronstein, M., Rother, C. (eds.) ECCV 2014. LNCS, vol. 8926, pp. 613–627. Springer, Cham (2015). https://doi.org/10.1007/978-3-319-16181-5_47
7. Sermanet, P., Kavukcuoglu, K., Chintala, S., Lecun, Y.: Pedestrian detection with unsupervised multi-stage feature learning. In: Computer Vision and Pattern Recognition
8. Zhang, S., Benenson, R., Omran, M., Hosang, J., Schiele, B.: How far are we from solving pedestrian detection? pp. 1259–1267 (2016)
9. Trabelsi, R., Smach, F., Jabri, I., Abdelkefi, F., Snoussi, H., Bouallegue, A.: An endeavour to detect persons using stereo cues. In: Zaman, H.B., Robinson, P., Olivier, P., Shih, Timothy K., Velastin, S. (eds.) IVIC 2013. LNCS, vol. 8237, pp. 358–370. Springer, Cham (2013). https://doi.org/10.1007/978-3-319-02958-0_33
10. Zhang, S., Bauckhage, C., Cremers, A.B.: Informed haar-like features improve pedestrian detection. In: Computer Vision and Pattern Recognition, pp. 947–954
11. Ojala, T., Pietikäinen, M., Mäenpää, T.: Gray scale and rotation invariant texture classification with local binary patterns. IEEE Trans. Pattern Anal. Mach. Intell. **24**, 971–987 (2000)
12. Dalal, N., Triggs, B.: Histograms of oriented gradients for human detection. In: IEEE Computer Society Conference on Computer Vision and Pattern Recognition
13. Wang, X., Han, T.X., Yan, S.: An HOG-LBP human detector with partial occlusion handling. In: IEEE International Conference on Computer Vision, pp. 32–39
14. Krizhevsky, A., Sutskever, I.E., Hinton, G.: ImageNet classification with deep convolutional neural networks (2012)
15. Zeng, X., Ouyang, W., Wang, X.: Multi-stage contextual deep learning for pedestrian detection. In: IEEE International Conference on Computer Vision, pp. 121–128
16. Luo, P., Tian, Y., Wang, X., Tang, X.: Switchable deep network for pedestrian detection. In: Computer Vision and Pattern Recognition, pp. 899–906
17. Hosang, J., Omran, M., Benenson, R., Schiele, B.: Taking a deeper look at pedestrians. In: Computer Vision and Pattern Recognition, pp. 4073–4082
18. Ouyang, W., Wang, X.: Joint deep learning for pedestrian detection. In: IEEE International Conference on Computer Vision, pp. 2056–2063
19. Tian, Y., Luo, P., Wang, X., Tang, X.: Deep learning strong parts for pedestrian detection. In: IEEE International Conference on Computer Vision

20. Girshick, R., Donahue, J., Darrell, T., Malik, J.: Rich feature hierarchies for accurate object detection and semantic segmentation. In: IEEE Conference on Computer Vision and Pattern Recognition, pp. 580–587
21. Girshick, R.: Fast R-CNN. Comput. Sci. (2015)
22. Ren, S., He, K., Girshick, R., Sun, J.: Faster R-CNN: towards real-time object detection with region proposal networks (2015)
23. Redmon, J., Divvala, S., Girshick, R., Farhadi, A.: You only look once: unified, real-time object detection (2015)
24. Redmon, J., Farhadi, A.: YOLO9000: better, faster, stronger, pp. 6517–6525 (2016)
25. Redmon, J., Farhadi, A.: YOLOv3: an incremental improvement (2018)
26. Chen, Q., Jiang, W., Zhao, Y., Zhao, Z.: Part-based deep network for pedestrian detection in surveillance videos. In: Visual Communications and Image Processing, pp. 1–4

# CNN and Metadata for Classification of Benign and Malignant Melanomas

José-Sergio Ruiz-Castilla$^{(\boxtimes)}$, Juan-José Rangel-Cortes,
Farid García-Lamont, and Adrián Trueba-Espinosa

Universidad Autónoma del Estado de México,
Centro Universitario UAEM Texcoco, Av. Jardín Zumpango s/n El Tejocote,
56259 Texcoco, Estado de México, Mexico
jsergioruizc@gmail.com, juan991029@gmail.com,
fglamont@gmail.com, atruebae@hotmail.com

**Abstract.** Skin cancer is detected in skin lesions. The most common skin cancer is melanoma. Skin cancer is increasing in several parts of the world. Due to the above, it is important to work on the classification of melanomas, in order to support the possible detection of malignant melanomas that cause skin cancer. We use *Convolutional Neural Networks (CNN)* for the classification of melanomas. We use images available from *International Skin Imaging Collaboration (ISIC)*. We created a repository of 1000 images and did training with a sequential *CNN* to obtain two categories: benign and malignant melanomas. In the first instance we obtained results of 94.89% accuracy and 82.25% in validation. In the second instance we created another repository of 600 images for the method that we propose that consists in adding metadata within the same pixel matrix of the image in each RGB layer. The image was shown with a band of colors at the bottom. We made training with the CNN using images with metadata and achieved the results: 98.39% of accuracy and 79% of validation. Therefore, we conclude that adding the metadata repeatedly to the pixel matrix of the image improves the results of the classification.

**Keywords:** Melanomas · Convolutional neural networks · Metadata · Classification · Prediction

## 1 Introduction

The increase in the rate of cancer is a health problem in the world. The melanoma is among the most common types of cancer in the United States of America. In Mexico, skin cancer is becoming more frequent. This is due to different factors such as prolonged exposure to ultra violet rays or exposure to water contaminated by arsenic and other chemicals. In America, cancer is the second cause of death. It is estimated that 2.8 million people are diagnosed with cancer each year and 1.3 million people die from this disease annually. Approximately, 52% of new cancer cases occur in people 65 years old or younger. In case of not taking action, an increase of more than four million new cases and 1.9 million deaths due to cancer is expected for the year 2025 [1]. Melanoma-like skin cancer can almost always be cured if discovered and treated promptly [2].

© Springer Nature Switzerland AG 2019
D.-S. Huang et al. (Eds.): ICIC 2019, LNCS 11644, pp. 569–579, 2019.
https://doi.org/10.1007/978-3-030-26969-2_54

The National Health System in Mexico has three levels for the attention to the public. In the first level, basic health services are provided. It is the main scenario for preventive health and where 80% of the ailments are treated and solved. In the second level, patients referred from the first level are attended to perform diagnostic procedures, as well as therapeutic and rehabilitation processes. At the third level, diseases of low prevalence, high risk and more complex diseases are treated. At this level, patients sent from the first and second level are attended [3].

Therefore, a proposal is made to classify melanomas to help in the pre-diagnosis of melanoma skin cancer. The proposed solution is focused on the first level of the National Health System in Mexico. A convolutional neuronal network as well as metadata will be used from the image base offered by International Skin Imaging collaboration (ISIC).

## 1.1   Skin Cancer

Skin cancer represents the most common cancer in humans, and we are currently facing increasing rates of newly diagnosed cutaneous neoplasm each year. Knowledge has been gained concerning the biology of skin cancer and risk factors; diagnosis has been improved and novel therapeutic modalities have been developed. As UV-exposure is known as the major risk factor for skin cancer, prevention can be achieved through UV-protection and regular sunscreen use. In many countries primary care physicians are performing regular skin checks and initiate early treatment, working closely with a dermatologist. Therefore, knowledge regarding diagnosis, appropriate treatments including recommended excision margins, the performance of sentinel lymph nodes biopsies, and the ability to use novel topical treatment modalities are of importance for all physicians involved [4].

## 1.2   Melanomas

Melanomas are common cancers arising from the pigment cells of the skin. The principal environmental determinant of cutaneous melanoma is sunlight, with incidence rates varying more than tenfold between ethnically similar populations residing in environments with different levels of ambient sunlight [5]. Diagnosis should be based on a full-thickness excisional biopsy, with a small side margin. Wide local excision of a cuff of normal tissue after an excision biopsy reduces local recurrence rates [6]. Because melanoma in advanced stages is still incurable, early detection is indispensable to reduce mortality. With the introduction of dermoscopy into the clinical practice, the diagnostic accuracy of pigmented skin lesions can be improved. The use of dermoscopy allows the identification of many different structures and colors, not seen by the naked eye. Colors play an important role in dermoscopy. Common colors are light brown, dark brown, black, blue, blue-gray, red, yellow, and white [7].

## 1.3   Convolutional Neural Netware

CNNs, also known as convnets, are a type of deep learning model that is used almost universally in machine vision applications. Convnets can be applied to image classification problems, particularly those involving small sets of training data. The convolution layers of a convent learn from local patterns (see Fig. 1): in the case of the images, the patterns are found in small 2D windows of the inputs.

**Fig. 1.** The images can be divided into local patterns, such as edges, textures, etc. [8]

The above feature is key and gives the convnets two interesting properties [8]:

1. The patterns they learn are invariant in translation. After learning a certain pattern in the lower right corner of an image, a *convnet* can recognize it anywhere: for example, in the upper left corner. A densely connected network would have to learn the pattern again if it appeared in a new location. This makes the *convnets* data efficient when processing images: they need fewer examples of training to learn representations that have power of generalization.

2. They can learn spatial hierarchies of patterns (see Fig. 2). A first convolution layer will learn small local patterns such as the edge; a second convolution layer will learn larger patterns made from the characteristics of the first layers, and so on. This allows social networks to efficiently learn increasingly complex and abstract visual concepts (because the visual world is fundamentally hierarchical spatial) [8].

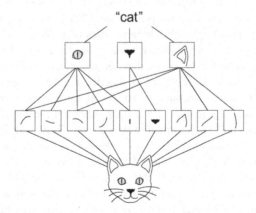

**Fig. 2.** The visual world forms a spatial hierarchy of visual modules: hyperlocal edges are combined in local objects such as eyes or ears, which are combined in high-level concepts such as "cat" [8].

## 2  Related Works

For the classification of images, it is necessary to apply techniques of image processing and extraction of characteristics such as color and borders, as well as, the segmentation of the image [9].

When looking to classify very similar images such as leaves of plants it is necessary to apply the extraction of characteristics such as texture, chromaticity and geometric shape and apply some combination of classification techniques [10, 11]. It is possible to classify fruits by extracting characteristics of fruit shape, texture and color (HSV). This classification can be very useful for supermarkets to handle fruit sales [12].

Mahbod among others [13] took available images of the ISIC in 2017. They applied three pre-processing steps in the images. First, they normalized the images by subtracting the average RGB value from the data set, since the pre-trained networks were optimized for those images. Next, the images were resized to the appropriate size using bi-cubic interpolation to feed the networks (227 × 227 and 224 × 224). Finally, the images increased to have a larger training set. Since the classifiers were trained for a problem of multiple classifications for the classes of: *melanoma, seborrheic keratosis* and classes of benign *nevi*. The performance of the classifier, in terms of accuracy (*ACC*) and area under the curve (*AUC*) of receiver operating characteristics were measured for 150 test images. The Achieving results of 79.9% to 89.2% of the value Accuracy.

Yang among others [14], *DCNN* (*Deep CNN*) multitasking was used. For the analysis of skin lesions. The *DCNN* multitasking model you implemented is based on the *GoogleNet* design architectures. The dermatoscopic training and evaluation data sets were taken from *ISIC* 2017. 2000 samples in the training data set and 150 samples in the validation data set. The results of the classification are evaluated by the area under the receiver operating characteristics curve (*AUC*). The *AUC* is a measure of how well a parameter can distinguish between two diagnostic groups (sick/normal). They compared the performance between *DCNN* multitasking and *GoogleNet*. The results obtained were 92.6% of the proposed model and 90.3% with *GoogleNet*.

Shoieb [15] among others took a pre-trained *CNN* in its last phase to classify the infected skin lesion according to the characteristics of *CNN* to train a *SVM* multiclass classifier. There are several pre-trained networks that have gained popularity. Most of these have been trained in the *ImageNet* data set, which has 1000 object categories and 1.2 million training images. Four classes of skin diseases have been selected for system validation. These classes are melanoma, basal cell carcinoma, eczema and impetigo. The data set comprises 134 images, 72 of them represent melanoma, 64 for basal cell carcinoma, 74 eczema and 31 for impetigo skin disease. The results of the classification of melanoma and non-melanoma lesions from the data obtained from the Dermatology Information System were 94% accuracy, 94% specificity and 94% sensitivity and for non-melanomas it was 94% accuracy, 94% of specificity and 94% sensitivity.

Codella among others [16] experimented with the double-cross-validation method and performed 20 times (40 total experiments) with 334 images of melanoma and 144 images of atypical nevi, as well as 2146 clearly benign lesions (2624 total) taken from *ISIC*. Two color dictionaries (*RGB*) and grayscale color spaces are built. The images are resized to dimensions of 128 × 128 pixels before the extraction of 8 × 8 patches, to learn dictionaries of 1024 elements. To train classifiers we used a nonlinear *SVM* that uses a core intersecting histogram and sigmoid. *SVM* scores were assigned to probabilities by logistic regression in the training data. They used a 50% probability as a binary classification threshold. The fusion is carried out by means of an unweighted *SVM* average score (late fusion). The results obtained among others with *Caffe CNN* were 91.9% accuracy, 90.3% specificity and 92.10% sensitivity.

Dorj among others [17] did a classification of skin cancers using a deep convolutional neuronal network. The algorithm to obtain the general methodology that they used is the following:

Acquisition of image data (collect image data, cut the images, prepare training and test sets of images).

Classification into groups that use convolutional neuronal network functions (extract training functions using a convolutional neuronal network, train and test an ECOC-SVM classifier using convolutional neural network functions).

The data set in this study originated from related Internet sites. In this study, they used *RGB* images of skin cancer with 500–1000 pixels, .jpg and .tiff. With a total of 3753 images, which are collected from the Internet (google.com and naver.com, baidu.com and bing.com). The results of the four types of melanoma classified for the accuracy factor were: 92.3% for Actinic keratosis, 91.8% for Basal cell carcinoma, 95. 1% for Squamous cell carcinoma and 92.2% for Melanoma.

Liao [18] did a classification of skin diseases. Your data set of skin diseases from two different sources: *Dermnet* and *OLE*. *Dermnet* is one of the largest sources of photographic dermatology that is publicly available. It has more than 23,000 images of skin diseases in a wide variety of skin conditions. *Dermnet* biologically organizes skin diseases in two-tier taxonomy. In their approach, they transferred the learning of *ImageNet's* pre-trained modeling with *Caffe*, a deep learning framework that supports the deep and expressive training of *CNN*. They chose *VGG16*, *VGG19* and *GoogleNet* as their pre-trained models. The results obtained were 91% with *CGG16*, 90.9% with *VGG19* and 90.7% with *GoogleNet*.

Georgakopoulos among others [19] classified dermatoscopic images with two classes as "malignant" or "non-malignant". They tried to use an improved entry to *CNN* using the Gaussian filters, Gaussian Laplacian (*LoG*), Hessian matrix and Gabor. They investigated the value of increasing *CNN* entries with filters used from computer vision. The response of each available image with these filters is calculated and used as an additional input to *CNN*. They also used Transfer learning (*TL*) using two pre-trained network architectures.

Step 1: the set of malignant skin images is enlarged by applying image rotations in multiples of 90°, as well as mirror transformations in each rotated image.

Step 2: For file1, all images are resized to 256 × 256. A pre-processing procedure extracts 224 × 224 image patches at random locations in the resized image, which are used as an input to CNN. Similarly, for CNN file2, the images are resized to 352 × 352. The pre-processing stage extracts 320 × 320 image patches in random locations of the resized image, which are used as an input to CNN.

The expanded data set is used only for training, and the produced images did not need to be stored. When performing the training with the two sets of images. The results with CNN were of, 92.3% for the first set and 92.3% for the second set. While, for the second set with LoG were 93.9% and 88.6% for the first and second sets respectively [19].

Haenssle among others [20] used and trained a modified version of Google's CNN Inception v4 architecture. They created a test set of 300 images that included 20% of melanomas from all body sites and all frequent histotypes, and 80% of benign mela-nocytic nevi from different subtypes and body sites, including so-called "simulators" Melanoma. The images of the test set-300 were retrieved from the high-quality vali-dated image library of the Department of Dermatology of the University of Heidelberg, Germany, and various camera/dermoscopy combinations were used for the acquisition of images the overlap between data sets for training/validation and testing.

In this research they focused on comparing the accuracy of the algorithm against experts in dermatology. Finding that, experts with more than five years of experience were better than the algorithm, whereas, experts with less than two years of experience were worse than the algorithm.

Moura among others [21], propose the classification of skin lesions using a descriptor formed by the combination of the characteristics of the ABCD rule (Asymmetry, Border, Color and Diameter) and pre-trained CNN. The characteristics were selected according to their gain ratios and used as inputs for the MultiLayer Perceptron (MLP) classifier.

Characteristics are properties that can be measured from an image, such as shape, color and texture. These attributes, grouped in a feature vector, are called an image descriptor. The descriptors used to extract characteristics of the lesion images of the skin and form the hybrid descriptor. The proposed ones are detailed below:

ABCD Rule: This rule is divided into four steps to describe the image and has five attributes: asymmetry (1), edge irregularity (2), color (1) and diameter (1). In the most successful tests, the proposed method achieved an accuracy rate of 94.9% using MPL.

Oliveira among others [22] comment that the extraction of characteristics to describe skin lesions is an area of research is challenging due to the difficulty to select significant characteristics. The computational diagnostic system developed for the skin lesion was applied to a set of 1104 dermoscopy images by a cross-validation procedure. The best results were obtained through an optimal trajectory forest classifier (OPF) with very promising results. The proposed system reached an accuracy of 92.3%, a sensi-tivity of 87.5% and a specificity of 97.1% when the complete set of characteristics was used. In addition, it achieved an accuracy of 91.6%, a sensitivity of 87% and a specificity of 96.2%, when 50 functions were chosen using a feature selection algo-rithm based on correlation. A total of 1104 images were selected from the original data set. Of these, 916 images were benign lesions and 188 images were malignant lesions. The images in the data set were resized at an average resolution of $400 \times 299$ pixels to simplify processing. The results of six different classifiers were: 75.8% for KNN, 68.2% for Bayes net, 86.9% for CA.5, 74.5% for MLP, 91.7% for SVM and 92.3 for OPF.

## 3 Proposed Method

### 3.1 Images and Images Metadata

For this work, we got images from ISIC a database of melanomas images and metadata able in "https://isic-archive.com/". Melanoma Project is an academia and industry partnership designed to facilitate the application of digital skin imaging to help reduce melanoma mortality. This archive serves as a public resource of images for teaching and for the development and testing of automated diagnostic systems. This is a site that store 23906 images from collaborators of area. We download the images and metadata from ISIC. We created two images packages. The characteristic of packages are shown on Table 1.

**Table 1.** Image packages from *ISIC*. 800 images for training and 400 images for validation. Also, 400 images with metdata for training and 200 images with metadata for validation

| Packages | Set | Melanoma type | Images |
|---|---|---|---|
| Images without metadata | Training | Benign melanomas | 600 |
| | | Malignant melanomas | 600 |
| | Validation | Benign melanomas | 200 |
| | | Malignant melanomas | 200 |
| Images with metadata | Training | Benign melanomas | 200 |
| | | Malignant melanomas | 200 |
| | Validation | Benign melanomas | 100 |
| | | Malignant melanomas | 100 |

### 3.2 Add Metadata to the Image

We duplicate the metadata repeatedly at the end of the matrix. The images have three channels (RGB), therefore, we duplicate the metadata in the three layers. Once the metadata is added repeatedly the image shows a band of color indicating that the metadata were added. This process was necessary to do it in the images of: training, validation and testing. See Fig. 3.

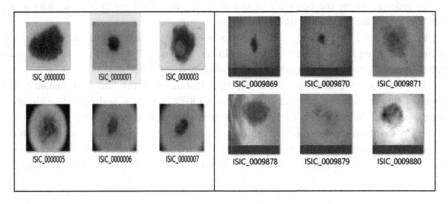

**Fig. 3.** Matrix of images (a) without metadata (b) with metadata

We choose three metadata: sex, age_approx and clin_size_long_diam_mm. We insert the gender with the values of 100 for feminine and 200 for masculine. In the same way, the size of the melanoma was inserted. Finally, the patient's age was inserted. The algorithm traverses the column from row 300 to 350 by inserting the metadata repeatedly. This process was applied to the three channels for each image of all image set.

# 4   Results

### 4.1   Instance 1 y 2. Results of Training and Validation

The first experiment was executed with CNN. With the next parameters: 8 Epochs, 350 of height, 350 of weight, 128 of batch_size, 2 classes, 1000 step_train, 300 step_-validation, 32 filter_Conv1, 64 filter_Conv2, (3, 3) of size_Filter, (2, 2) of size_Filter2 and (2,2) of size_Pool. The values obtained are sown in the Table 2.

**Table 2.**   Results of accuracy and loss of Instance 1 and 2.

| Epochs | Instance 1 | | | | Instance 2 | | | |
|---|---|---|---|---|---|---|---|---|
| | ACC | VAL_ACC | LOSS | VAL_LOSS | ACC | VAL_ACC | LOSS | VAL_LOSS |
| 1 | 0,7160 | 0,8100 | 0,6311 | 0,4821 | 0,8487 | 0,785 | 0,8487 | 0,6260 |
| 2 | 0,7701 | 0,8725 | 0,4622 | 0,3744 | 0,9393 | 0,795 | 0,1529 | 0,8093 |
| 3 | 0,8108 | 0,8650 | 0,4016 | 0,3667 | 0,9700 | 0,765 | 0,0856 | 0,8959 |
| 4 | 0,8517 | 0,8325 | 0,3301 | 0,4619 | 0,9826 | 0,795 | 0,0524 | 1,3705 |
| 5 | 0,8863 | 0,8525 | 0,2695 | 0,5041 | 0,9858 | 0,785 | 0,0421 | 1,2821 |
| 6 | 0,9152 | 0,8475 | 0,2097 | 0,6170 | 0,9830 | 0,820 | 0,0537 | 1,2562 |
| 7 | 0,9361 | 0,7825 | 0,1678 | 0,7229 | 0,9914 | 0,790 | 0,0248 | 1,3772 |
| 8 | **0,9489** | **0,8225** | 0,1313 | 0,7594 | **0,9890** | **0,790** | 0,0330 | 1,2230 |

We executed the algorithms for instance 1 and instance 2, obtaining 94.89% and 89.90% in training, while 82.25% and 79.00% were obtained in the validation. As we can see the accuracy improved in the training, however it decreased in the validation.

In order to know the behavior of the accuracy and the loss in both scenarios, we create a graph for each case that we can see in Fig. 4.

In Instance 2 with images with metadata, better results were obtained in the accuracy, although not in the validation. In the Instance 2 we can see that, the validation accuracy was not ascending, but it remained stable.

The results were obtained from the two Instances. From the first instance that contains images without metadata, results of 94.89% accuracy and 82.25% validation were obtained. While, with the set of images with metadata, 98.90% accuracy and 79.00% validation was obtained. Finally, we did tests with images not included in the training or validation and the results were 61% in Instance 1 and 85% in instance 2. We did not find results of a test with new images in the related works. See the Table 3.

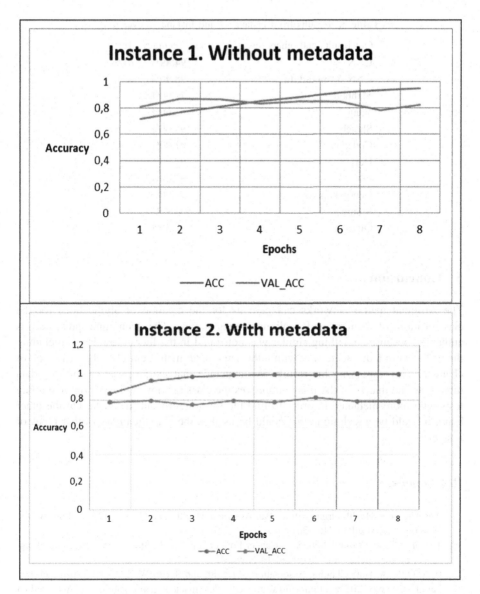

**Fig. 4.** Instance 1 y 2. Precision and loss behavior.

Table 3. Results from Instance 1 and Instance 2 ever test.

| Our results and others research works | Accuracy | Test |
|---|---|---|
| **CNN** | **94.89%** | **61%** |
| **CNN & metadata** | **98.90%** | **85%** |
| Mahbod | 79.90% | |
| Yang | 92.60% | |
| Shoieb | 94.00% | |
| Codella | 91.90% | |
| Dorj | 92.30% | |
| Liao | 91.00% | |
| Georgakopoulos | 93.90% | |
| Moura | 94.90% | |
| Oliveira | 92.30% | |

## 5   Conclusions

The Artificial neural networks can help us for the creation of applications for the classification of benign and malignant melanomas. The technique presented is authentic, because something similar was not found in the literature related. Including the metadata in a duplicate way within the same pixel matrix enriches the extraction of characteristics improving the results. We conclude that the proposal to add metadata directly to the matrix of the image improves the classification. We will continue doing tests with more metadata in order to find the most significant metadata. On the other hand, it could be a technique that could be used in the classification of other types of images.

## References

1. Pan American Health Organization, Pan American Health Organization (2017). http://www. paho.org. Accessed 14 Nov 2017
2. The Skin Cancer Foundation, Skin Cancer Foundation (2013). http://www.cancerdepiel.org/ cancer-de-piel/melanoma. Accessed 14 Nov 2017
3. P. A. Q. L. A. Burr Claudia, Association of Health and Social Welfare of Women and their Families, Mayo (2011). http://asbis.org.mx/PDF/guiapacienteparticipativo.pdf. Accessed 01 Nov 2017
4. Stockfleth, E., Rosen, T., Shumack, S.: Managing Skin Cancer. Springer, Heidelberg (2010). https://doi.org/10.1007/978-3-540-79347-2
5. Whiteman, D., Green, A.: Epidemiology of malignant melanoma. In: Dummer, R., Pittelkow, M., Iwatsuki, K., Green, A., Elwan, N. (eds.) Skin Cancer- A World-Wide Perspective, pp. 13–26. Springer, Heidelberg (2011). https://doi.org/10.1007/978-3-642-05072-5_2
6. Guggenheim, M., et al.: Melanoma. In: Dummer, R., Pittelkow, M., Iwatsuki, K., Green, A., Elwan, N. (eds.) Skin Cancer - A World-Wide Perspective, pp. 307–342. Springer, Heidelberg (2011). https://doi.org/10.1007/978-3-642-05072-5_23

7. Kolm, I., Dummer, R., Braun, R.P.: Dermoscopy. In: Dummer, R., Pittelkow, M., Iwatsuki, K., Green, A., Elwan, N. (eds.) Skin Cancer – A World-Wide Perspective, pp. 373–378. Springer, Heidelberg (2011). https://doi.org/10.1007/978-3-642-05072-5_26

8. Chollet, F.: Deep Learning with Python. Manning Publications Co., Shelter Island (2018)

9. García-Lamont, F, Cervantes, J, López, A, Rodríguez, L: Segmentation of images by color features: a survey. Neurocomputing **292**, 1–27 (2018). https://doi.org/10.1016/j.neucom.2018.01.091, http://www.sciencedirect.com/science/article/pii/S0925231218302364. ISSN 0925-2312

10. Cervantes, J., Taltempa, J., Lamont, F.G., Castilla, J.S.R., Rendon, A.Y., Jalili, L.D.: Análisis Comparativo de las técnicas utilizadas en un Sistema de Reconocimiento de Hojas de Planta. Revista Iberoamericana de Automática e Informática industrial **14**(1), 104–114 (2017). Web. 15 May 2019

11. Cervantes, J., Garcia Lamont, F., Rodriguez Mazahua, L., Zarco Hidalgo, A., Ruiz Castilla, J.S.: Complex identification of plants from leaves. In: Huang, D.-S., Gromiha, M.M., Han, K., Hussain, A. (eds.) ICIC 2018. LNCS (LNAI), vol. 10956, pp. 376–387. Springer, Cham (2018). https://doi.org/10.1007/978-3-319-95957-3_41

12. García-Lamont, F., Cervantes, J., López-Chau, A., Alvarado, M.: Fruit classification by extracting color chromaticity, shape and texture features: towards an application for supermarkets. IEEE Latin Am. Trans. **14**, 3434–3443 (2016). https://doi.org/10.1109/TLA.2016.7587652

13. Mahbod, A., Ecker, R., Ellinger, I.: Skin Lesion Classification Using Hybrid Deep Neural Networks (2017). CoRR, abs/1702.08434

14. Yang, X., et al.: Cornell University, 04 March 2017 https://arxiv.org/abs/1703.01025

15. Shoieb, D.A., Youssef, S., Aly, W.: Computer-aided model for skin diagnosis using deep learning. J. Image Graph. 116–121 (2016). https://doi.org/10.18178/joig.4.2.122-129, https://www.researchgate.net/publication/312188377_Computer-Aided_Model_for_Skin_Diagnosis_Using_Deep_Learning

16. Codella, N., Cai, J., Abedini, M., Garnavi, R., Halpern, A., Smith, J.R.: Deep learning, sparse coding, and SVM for melanoma recognition in dermoscopy images. In: Zhou, L., Wang, L., Wang, Q., Shi, Y. (eds.) MLMI 2015. LNCS, vol. 9352, pp. 118–126. Springer, Cham (2015). https://doi.org/10.1007/978-3-319-24888-2_15

17. Dorj, U.O., Lee, K.K., Choi, J.Y., et al.: Multimed. Tools Appl. **77**, 9909 (2018). https://doi.org/10.1007/s11042-018-5714-1, https://link.springer.com/article/10.1007%2Fs11042-018-5714-1

18. Liao, H.: A Deep Learning Approach to Universal Skin Disease Classification (2015). https://www.semanticscholar.org/paper/A-Deep-Learning-Approach-to-Universal-Skin-Disease-Liao/af34fc0aebff011b56ede8f46ca0787cfb1324ac

19. Georgakopoulos, S.V., Kottari, K., Delibasis, K., et al.: Neural Comput. Appl. (2018). https://doi.org/10.1007/s00521-018-3711-y, https://link.springer.com/article/10.1007/s00521-018-3711-y

20. Haenssle, H.A., et al.: Reader study level-I and level-II groups, man against machine: diagnostic performance of a deep learning convolutional neural network for dermoscopic melanoma recognition in comparison to 58 dermatologists. Ann. Oncol. **29**(8), 1836–1842 (2018). https://doi.org/10.1093/annonc/mdy166, https://academic.oup.com/annonc/article/29/8/1836/5004443

21. Moura, N., Veras, R., Aires, K., et al.: Multimed. Tools Appl. 78, 6869 (2019). https://doi.org/10.1007/s11042-018-6404-8, https://link.springer.com/article/10.1007%2Fs11042-018-6404-8

22. Oliveira, R.B., Pereira, A.S. Tavares, J.M.R.S.: Neural Comput. Appl. (2018). https://doi.org/10.1007/s00521-018-3439-8. https://link.springer.com/article/10.1007/s00521-018-3439-8

# An Application of Using Support Vector Machine Based on Classification Technique for Predicting Medical Data Sets

Mohammed Khalaf[1]([✉]), Abir Jaafar Hussain[2], Omar Alafandi[3], Dhiya Al-Jumeily[2], Mohamed Alloghani[4], Mahmood Alsaadi[1], Omar A. Dawood[5], and Dhafar Hamed Abd[1]

[1] Department of Computer Science, Al-Maarif University College, Ramadi, Anbar 31001, Iraq
M.I.Khalaf@acritt.org.uk, mahmood89.ma@gmail.com, dhafar.dhafar@gmail.com
[2] Faculty of Engineering and Technology, Liverpool John Moores University, Byrom Street, Liverpool L3 3AF, UK
{a.hussain,d.aljumeily}@ljmu.ac.uk
[3] College of Technological Innovation, Zayed University, Abu Dhabi, UAE
Omar.AlFandi@zu.ac.ae
[4] Abu Dhabi Health Services Company (SEHA), Sultan Bin Zayed Street, PO Box 109090, Abu Dhabi, United Arab Emirates
mloghani@seha.ae
[5] College of Computer Science and Information Technology, University of Anbar, Ramadi 31001, Iraq
the_lionofclub@yahoo.com

**Abstract.** This paper illustrates the utilise of various kind of machine learning approaches based on support vector machines for classifying Sickle Cell Disease data set. It has demonstrated that support vector machines generate an essential enhancement when applied for the pre-processing of clinical time-series data set. In this aspect, the objective of this study is to present discoveries for a number of classes of approaches for therapeutically associated problems in the purpose of acquiring high accuracy and performance. The primary case in this study includes classifying the dosage necessary for each patient individually. We applied a number of support vector machines to examine sickle cell data set based on the performance evaluation metrics. The result collected from a number of models have indicated that, support vector Classifier demonstrated inferior outcomes in comparison to Radial Basis Support Vector Classifier. For our Sickle cell data sets, it was found that the Parzen Kernel Support Vector Classifier produced the highest levels of performance and accuracy during training procedure accuracy 0.89733, AUC 0.94267. Where the testing set process, accuracy 0.81778, the area under the curve with 0.86556.

**Keywords:** Machine learning · Support vector machines ·
Sickle cell disorder data set · Evaluation techniques · Classification

© Springer Nature Switzerland AG 2019
D.-S. Huang et al. (Eds.): ICIC 2019, LNCS 11644, pp. 580–591, 2019.
https://doi.org/10.1007/978-3-030-26969-2_55

# 1 Introduction

Sickle Cell disease (SCD) is a chronic disease that change the red blood cells (RBC) from circle to a crescent. In this case, SCD patients face worse condition in terms of RBC moving smoothly in the blood's artery and vessel. The total amount of oxygen flow is decreased to tissues, particularly the lungs. This kind of condition causes a sever chronic pain for all SCD patient, and most important difficulty in breathing and heart not receiving sufficient amount of blood [1, 2].

In the past of few decades, The increase enhancement of biomedical information has played a crucial part in healthcare areas [3]. The objective is to raise the use of advanced technology in medical applications which can be implemented in various medical centres [4]. Artificial intelligence algorithms are implemented in a number of domain to advance the decision progress and most importantly to assist clinical professionals for correct treatment [28]. Machine Learning [5], specifically Support Vector Machines (SVM) are a sophisticated approach in scientific field that offers computers to learn from data during the training phase [6]. Then, SVM can easily evaluate the performance and accuracy of models when obtain the remaining data during testing process. The main idea of applying SVM rather than other techniques is due to uses a method known the kernel trick, which can be easily transfer the SCD data set based on these transformations. This can be assist to discover an optimal and accurate boundary between all the possible outputs. The study focused on the supervised learning as the class-label exist in our data set. The evaluation techniques comprising; Accuracy, You den's J statistic (J Score) Area under ROC Curve (AUC), F1 score, Receiver operating characteristics (ROC) curve, Precision, Sensitivity, and Specificity [3]. In this experimental study, the application of intelligent system techniques for the problem of SCD dose is highly considered.

The clinical data sets were gathered from the local hospital at the city of Liverpool, united kingdom is a supervised learning involved an input and output (classes) [7, 8]. Insufficient number of training sets makes it hard for the classifier to detect the output accurately. This matter can leads to another problem in the algorithm as known overfitting. To demonstrate that issue, high dimensional medical data could be extremely complex than low dimensional. It is so essential to decrease the random features based on a dimensionality reduction procedures for obtaining high performance and accuracy. Two significant techniques to tackle the dimensionality reduction process. The former on is called feature selection that choose the optimal subsets to produce accurate discriminative feature vectors [9, 10]. The later method is known as feature extraction that able to transfer the high-dimensional data space onto low-dimensional data [11]. We are concentrated on feature selection in this research as dealing with numerical data set.

The paper format in this study is structured as follows. In Sect. 2, we illustrate the previous work with the current algorithms used in SCD, while Sect. 3 presents the classification techniques based on a performance evaluation metrics. The support vector machines model discusses in Sect. 4, whereas the proposed methodology presented in Sect. 5. We discuss our simulation results in Sect. 6. The conclusion and future work is presented in the last section.

## 2    Current Algorithms Used in Sickle Cell Disease

Healthcare organisations have faced a number of issues in the past few years ago, to meet the demands of innovative medical fields [12]. This study is proposed to yield a an application system using different kind of SVM that is able to assist clinical centres and consequently to offer such a huge benefits for sickle patients. Many researchers around the world have implemented a scientific research based on Machine Learning approaches [13]. Several number of solutions have been proposed by researchers to deliver support for healthcare organization [14].

The SVM is applied in a number of medical applications [15]. The SVM has been used widely as a connectionist approach to the classification and determination of medical results including blood inflammations [16]. The model has been employed widely to automate the assessment of blood disorders such as SCD using morphological attributes of erythrocytes in the cell. Dalvi and Vernekar [17] proposed anaemia detection based on machine learning approaches and statistical methods to produce an optimal accuracy in RBC classification technique. Their results indicated that combined approaches received the highest accuracy. ANN provided the best outcomes in their experiments, in comparison with the proposed K- Nearest Neighbour, which acquired unsatisfactory outcomes. Dalvi and Vernekar ensemble Decision Tree classifier and KNN classifier using stacked ensembles to obtain robust results. The combination of various models is indicated as providing superior performance than that of individual models. The evaluation measures used in the study included Accuracy, Specificity, Sensitivity, and Precision, with 10-fold cross validation utlised in their experiment. Training set comprised 441 instances, while the testing set comprised the remaining 49 cases.

Advocating the resampling method, Xiong et al. [18] suggested that training a model with an imbalanced dataset and insufficient features in machine learning offers less classification accuracy. They applied an application called (SMOTE) to produce patterns for horizontal gene transfer (HGT) for detecting genome diversification. The authors received less mean error rate (MER) compare to the previous outcomes.

According to our deep investigations in the previous work, the literature review could not present a sophisticated system for classifying SCD clinical data sets in terms of supporting medical centres with correct amount of medication. Therefore, constructing the support vector machines approaches have high impact to analysis blood samples from manual methods to an intelligent system. To achieve that, the proposed research will be able to assess the SCD data set and prescribe a proper dosage. The novel application in the proposed system is how to use machine-learning algorithm to classify SCD data set based on the classification techniques.

## 3    Evaluation Metrics Techniques

Evaluation metrics techniques are considered a significant tool to estimate the accuracy and performance for the proposed approaches. Each model used in our research is calculated through a precise threshold ($0 \leq t \leq 1$) to select the final class [19, 20]. Training and out-of-sample as known testing diagnostics is used to evaluate each classifier individually. To deal with evaluation metrics outcomes, it is very crucial to utilise classification accuracy such as specificity, precision, sensitivity, F1 score, Youden's J statistic, and overall accuracy as demonstrated in Table 1. In addition to that, it is crucial to show the outcomes of True positive and False positive values of a model by using the AUC plot and ROC plot.

**Table 1.** Evualtion metric techniques calculations

| No | Metric name | Calculation |
|----|-------------|-------------|
| 1 | Sensitivity | $True\,positive/(True\,positive + False\,positive)$ |
| 2 | Specificity | $True\,negative/(True\,negative + False\,positive)$ |
| 3 | Precision | $TP/(TP + FP)$ |
| 4 | F1 score | $2 * (Precision * Recall)/(Precision + Recall)$ |
| 5 | Youden's J statistic (J Score) | $Sensitivity + Specificity - 1$ |
| 6 | Accuracy | $(TP + TN)/(TP + FN + TN + FP)$ |
| 7 | Area Under ROC Curve (AUC) | $0 <= Area\,under\,the\,ROC\,Curve <= 1$ |
| 8 | ROC | Sensitivity vs $(1 - specificity)$ |

## 4 Support Vector Machines

Support vector machines (SVM) is a supervised learning technique that have the capability to analyse large number of data set, especially with binary classes, used for classification and regression procedure [21]. In this aspect, SVM is class of models that attempts to minimise misclassification as much as possible by a training task, called as maximum margin point [22]. This model was first established and proposed by Cortes and Vapnik [23]. Providing a training sets involving an input and output, input belongs to the blood sample attributes $(x_1, x_2, x_3, \ldots, x_n)$ and the output result for the amount of medication (classes) $\{(y_1, y_2, y_3, \ldots, y_N), (x_N, y_N)\}$ where $x_i \in$ $input\,features$ and $y_i \in \{class - 1, class + 1\}$. In order to solve the optimisation, SVM can find suitable solution for that problem. There are a set of weight $w_i$ to estimate the accurate value of $(y)$. This study utilises the optimisation of maximizing the margin to determine the hyperplane and to reduce the number of weight as shown in Eq. (1) [24].

$$f(x) = w^T x_i + b$$
$$f(x) = \sum_i \lambda_i y_i (x_i^T x + b$$
$$f(x) \geq 1, \quad \forall x \in class\,1$$
$$f(x) \leq -1, \quad \forall x \in class\,2 \tag{1}$$
$$H = \frac{|g(x)|}{\|w\|} = \frac{1}{\|w\|}$$

$w^T$ belongs to the vector weight, while $f(x)$ indicates the features sets of both classes, $\lambda_i$ refers to the dual function returned after training, $x$ is the training data set, $y$ is the classes (output), $H$ refers to hyperplanes, $g(x)$ refers to number of features, $b$ bias belongs to omega 0.

Classifying clinical data set is considered a typical procedure in SVM. In this model, a data set is indicated as a p-dimensional vector to generate a model that is cable to discrete dimensional hyperplane [25]. There are a large number of hyperplanes can be implemented to separate the data set into a number of group. The idea behind using the hyperplane in our study is to deal with the largest margin in, or separation, among multi-class label. To illustrate that, it is essential to select the suitable hyperplane to maximize the total distance from the nearest point on each margin side. In order to demonstrate the complex model, which can be implemented to all data distributions. The complexity of SVM is optimal when enough number of training data sets have.

Kernel technique is considered essential and important to apply it in our proposed study as we are dealing with multi-class problems. To use and handle a appropriate kernel trick method, the model can learn without explicitly computing $\varnothing(x)$. In this aspect, this classifier tries to implement linear classifiers that can be worked into a nonlinear situation. The separation hyperplane is conducted by finding a proper solution for solving an optimization issue that chooses the paralyzed points and support vector on the incorrect side of the resulting hyperplane. The penalty parameter C is the tuning parameter for building an optimal approach that specifies well. The selected associated parameters and kernel have vital effect on how the approach able to classifies the medical data set.

## 5   The Proposed Methodology

In this section, the design part in our experimental study includes structure the proposed final model to obtain the total requirements for the accurate prototype. The aim is to evaluate the efficiency and effectiveness of using a number of SVM models on SCD data set, to predict the amount of dosage for each patient individually based on their situation. The blood samples with 14 features has been used in our empirical study. These kind of features is considered important to classify the SCD data sets [26]. To conduct our experiments using the SCD data set, Fig. 1 shows the proposed methodology. These steps involved a number of procedures, firstly to collect real raw data from local hospital, prepared data, pre-processing, split data set through training set, validation set, and testing set which can assist to build our suitable models, and presents the outcomes. Table 2 illustrates the total parameters and models that are utilised in our study.

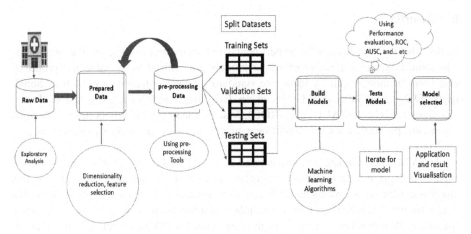

**Fig. 1.** The proposed methodology framework

**Table 2.** List of parameters

| No | Type | Number | Description |
|---|---|---|---|
| 1 | Data instances | 1896 | Data were collected from Alder Hey Children Hospital in the city of Liverpool, UK |
| 2 | Features (attributes) | 14 | Bio, Haemoglobin (Hb), Platelets (PLTS), neutrophils (white blood cell NEUT), Mean corpuscular volume (MCV), Reticulocyte Count (RETIC F), Reticulocyte Count (RETIC), Bilirubin (BILI), Alanine aminotransferase (ALT), Hb F, Lactate dehydrogenase (LDH), Weight, an aspartate aminotransferase (AST), and Mg/kg |
| 3 | Class variables | 9 | Multi class data set: [target Class 1 (250 mg)], [target Class 2 (300 mg)], [target Class 3 (500 mg)], [target Class 4 (600 mg)], [target Class 5 (700 mg)], [target Class 6 (700 mg)] [target Class 7 (1000 mg)], [target Class 8 (1200 mg)], [target Class 9 (1500 mg)] |
| 4 | Machine learning algorithms | 2 | Support Vector Classifier (SVC), Trainable classifier: Support Vector Machine, nu-algorithm (NUSVC), Parzen Kernel Support Vector Classifier (RBSVC), Radial Basis Support Vector Classifier (RBSVC), and General kernel/dissimilarity-based classification (KERNELC) |
| 5 | Evaluation metrics of classification models | 6 | Sensitivity, Specificity, Precision, F1 Score, Accuracy, and Youden's J statistic (J Score) values |
| 6 | Visualisation techniques | 2 | The area under the Curve (AUC), the Receiver Operating Characteristic (ROC) |

## 6  Results

In this chapter, we discuss the experimental outcomes and analysis of the SCD classification data set. A number of single models are utilised to examine the proposed classifiers in more depth by using the standard evaluation techniques, such as Sensitivity (SEN), Specificity (SPEC), Precision, F1 score, Youden's J 1, Confusion Matrix, Accuracy (ACC), the (AUC) and (ROC). Machine learning classifiers provide various significant properties, such as non-linear mapping, universal approximation, and parallel processing.

In SVM model, a data point is showed as a p-dimensional vector and SVM can be separated using p-1-dimensional hyperplane procedure. In fact, the main idea of this study is to identify geometrical patterns with 9 classes of the amount of medication that could be used universally across a number of models, including SVM. This study focused with a number of classifiers that are related to SVM to calculate the classification performance metrics. This thesis conducted the classification outcomes based on Support Vector Classifier (SVC), Trainable classifier: Support Vector Machine, nu-algorithm (NUSVC), Parzen Kernel Support Vector Classifier (RBSVC), Radial Basis Support Vector Classifier (RBSVC), and General kernel/dissimilarity-based classification (KERNELC). These models were used in our experiment and all of them work based on the support vector machine methodology.

Our main target is to illustrate that all these SVM models with different types of optimization setting have provided satisfactory outcomes in terms of accuracy and performance and yield by building a sophisticated model that used in medical domains. The proposed study used a single database with high dimensional data of 14 features using 9 classes. This research implemented SVM using various types of kernels, such as kernel matrix, linear and sigmoid kernel. NUSVC is dealing with linear kernel, while PKSVC works with sigmoid kernel and KERNELC compute the outcomes depending on the kernel matrix. The training results illustrated in Table 3, and the ROC and AUC histograms show in Figs. 2 and 3, respectively. During the training process to build the model, it is found that, PKSVC performed the best accuracy and AUC with 0.89733 and 0.94267, respectively. The proposed model discovered after running the simulation, the sensitivity with RBSVC outperformed all the other approaches with 0.86.

**Table 3.**  Range of SVM classifiers performance with an average of 9 classes (Train)

| Model | Sen | Spec | Prec | F1 | J1 | Acc | AUC |
|---|---|---|---|---|---|---|---|
| SVC | 0.74567 | 0.60389 | 0.20342 | 0.31444 | 0.34974 | 0.62944 | 0.68444 |
| NUSVC | 0.74344 | 0.74189 | 0.27152 | 0.389 | 0.48556 | 0.74244 | 0.79878 |
| PKSVC | 0.84844 | **0.90333** | 0.51033 | 0.62511 | 0.75189 | **0.89733** | **0.94267** |
| RBSVC | **0.86** | 0.89667 | **0.52033** | **0.63278** | **0.75644** | 0.89278 | 0.94411 |
| KERNELC | 0.819 | 0.783444 | 0.315667 | 0.449667 | 0.602556 | 0.787667 | 0.864111 |

The PKSVC and RBSVC are discovered to perform almost similar, with both ranking better outcomes for the training set. The AUC values for both models is

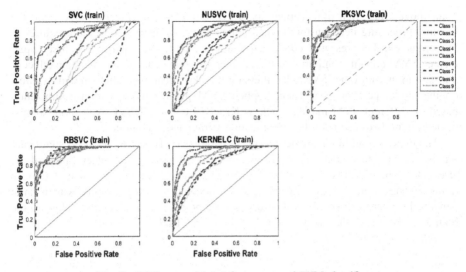

**Fig. 2.** ROC curve (Train) for a range of SVM classifiers

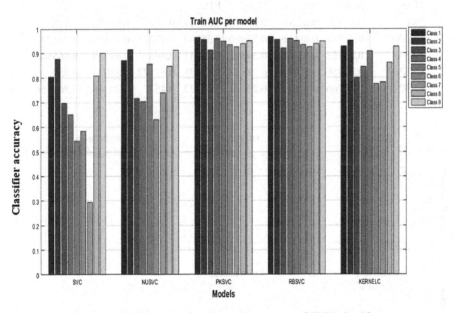

**Fig. 3.** AUC Histgram plot (Train) for a range of SVM classifiers

average with 9 classes 0.94267 and 0.94411, while obtaining 0.89733 and 0.89278 in regard to the accuracy, respectively.

In our SCD data set, the data points are considered not linearly separable due to the 9 target values (classes) with multi-class problems. To achieve high accuracy with multi-class issues, it is important to use a nonlinear mapping ($\varphi$) method within dimension space [27]. The computational complexity of the model rises, when the data point moves into high dimensional space. To build the classification algorithm, the learning procedure iteration by the data points with a number of operation requires to be completed. This research conducted a number of SVM experiments, first implemented SVM utilising random factors, and then investigated in depth the main effect of normalisation with other SVM classifiers on the classification evaluation and its effect on the model performance. Then, applied SVM parameter evaluation optimization based on different SVM models, such as KERNELC and NUSVC with more sophisticated methods to estimate the classification parameters techniques.

The linear kernel in this model has many parameters. However, the most significant one is $C$, which belongs to the cost function, and the penalty parameter values of the error rates. The cost function with each parameter comes with default value of zero. In terms of large value of cost function, it is allocated to margin errors with a large penalty. In contrast, a smaller value just ignores points that are identified close to the boundary and raises the margin side (Figs. 4, 5 and Table 4).

**Table 4.** Range of SVM classifiers performance with average of 9 classes (Test)

| Model | Sen | Spec | Prec | F1 | J1 | Acc | AUC |
|---|---|---|---|---|---|---|---|
| SVC | 0.74433 | 0.62267 | 0.20774 | 0.32111 | 0.36663 | 0.65156 | 0.675 |
| NUSVC | 0.77778 | 0.69656 | 0.23956 | 0.35944 | 0.47423 | 0.70833 | 0.78122 |
| PKSVC | **0.83122** | **0.81478** | **0.34811** | **0.48411** | **0.646** | **0.81778** | **0.86556** |
| RBSVC | 0.81356 | 0.80867 | 0.33822 | 0.47078 | 0.62222 | 0.81033 | 0.859 |
| KERNELC | 0.75766 | 0.695889 | 0.23922 | 0.35433 | 0.4537 | 0.703667 | 0.765556 |

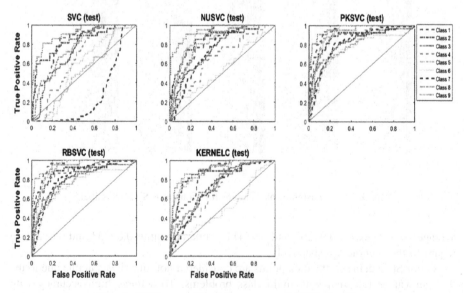

**Fig. 4.** ROC curve (Test) for a range of SVM classifiers

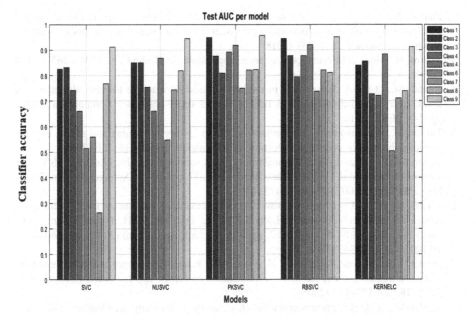

**Fig. 5.** AUC Histgram plot (Test) for a range of SVM classifiers

## 7   Conclusion

We conducted a number of Support vector machine models to enhance the clinical decision for who suffer from long-term SCD. In this aspect, to discover the accurate models that can yield high classification outcomes; this study has chosen a number of classifiers. This experimental research based on single classifier to evaluate the classification technique. The results shown that BPXNC gain the best performance and accuracy in terms of 6 categories. This model obtain result during training phase Accuracy 0.79544, AUC 0.86856. Where the testing phase, Accuracy 0.78767, AUC 0.85889. The results motivated us to implement difference artificial intelligence techniques that can provides high results. Our future work is to use the genetic algorithm and fuzzy logic, neural networks, random forest, K-nearest Neighbour to find out more comprehensively in the field of machine learning techniques.

**Acknowledgments.** The authors would like to thank Al-Maarif University College for supporting this research.

## References

1. Charache, S., et al.: Effect of hydroxyurea on the frequency of painful crises in sickle cell anemia. N. Engl. J. Med. **332**(20), 1317–1322 (1995)
2. Khalaf, M., Hussain, A.J., Al-Jumeily, D., Fergus, P., Keenan, R., Radi, N.: A framework to support ubiquitous healthcare monitoring and diagnostic for sickle cell disease. In: Huang, D.-S., Jo, K.-H., Hussain, A. (eds.) ICIC 2015. LNCS, vol. 9226, pp. 665–675. Springer, Cham (2015). https://doi.org/10.1007/978-3-319-22186-1_66

3. Zaidan, A., et al.: A review on smartphone skin cancer diagnosis apps in evaluation and benchmarking: coherent taxonomy, open issues and recommendation pathway solution. Health Technol. **8**(4), 223–238 (2018)
4. Adams, H.: Medical Informatics: Computer Applications in Health Care. JAMA **265**(4), 522 (1991)
5. Khalaf, M., et al.: A data science methodology based on machine learning algorithms for flood severity prediction. In: 2018 IEEE Congress on Evolutionary Computation (CEC), pp. 1–8. IEEE (2018)
6. Taiana, M., Nascimento, J., Bernardino, A.: On the purity of training and testing data for learning: the case of pedestrian detection. Neurocomputing **150**, 214–226 (2015)
7. Gaber, M.M., Zaslavsky, A., Krishnaswamy, S.: A survey of classification methods in data streams. In: Aggarwal, C.C. (ed.) data streams, vol. 31, pp. 39–59. Springer, Boston (2007). https://doi.org/10.1007/978-0-387-47534-9_3
8. Khalaf, M., et al.: A performance evaluation of systematic analysis for combining multi-class models for sickle cell disorder data sets. In: Huang, D.-S., Jo, K.-H., Figueroa-García, J.C. (eds.) ICIC 2017. LNCS, vol. 10362, pp. 115–121. Springer, Cham (2017). https://doi.org/10.1007/978-3-319-63312-1_10
9. Holder, L.B., Russell, I., Markov, Z., Pipe, A.G., Carse, B.: Current and future trends in feature selection and extraction for classification problems. Int. J. Pattern Recogn. Artif. Intell. **19**(02), 133–142 (2005)
10. Khalaf, M., et al.: Recurrent neural network architectures for analysing biomedical data sets. In: 2017 10th International Conference on Developments in eSystems Engineering (DeSE), pp. 232–237. IEEE (2017)
11. Williams, N., Zander, S., Armitage, G.: A preliminary performance comparison of five machine learning algorithms for practical IP traffic flow classification. ACM SIGCOMM Comput. Commun. Rev. **36**(5), 5–16 (2006)
12. Akay, M.F.: Support vector machines combined with feature selection for breast cancer diagnosis. Expert Syst. Appl. **36**(2), 3240–3247 (2009)
13. Liu, Y., Yu, X., Huang, J.X., An, A.: Combining integrated sampling with SVM ensembles for learning from imbalanced datasets. Inf. Process. Manage. **47**(4), 617–631 (2011)
14. Khalaf, M., et al.: training neural networks as experimental models: classifying biomedical datasets for sickle cell disease. In: Huang, D.-S., Bevilacqua, V., Premaratne, P. (eds.) ICIC 2016. LNCS, vol. 9771, pp. 784–795. Springer, Cham (2016). https://doi.org/10.1007/978-3-319-42291-6_78
15. Subashini, T., Ramalingam, V., Palanivel, S.: Breast mass classification based on cytological patterns using RBFNN and SVM. Expert Syst. Appl. **36**(3), 5284–5290 (2009)
16. Gil, D., Manuel, D.J.: Diagnosing Parkinson by using artificial neural networks and support vector machines. Global J. Comput. Sci. Technol. **9**(4), 63–71 (2009)
17. Dalvi, P.T., Vernekar, N.: Anemia detection using ensemble learning techniques and statistical models. In IEEE International Conference on Recent Trends in Electronics, Information & Communication Technology (RTEICT), pp. 1747–1751. IEEE (2016)
18. Tang, J., Zhang, X.: Prediction of smoothed monthly mean sunspot number based on chaos theory. Acta Phys. Sin. **61**, 169601 (2012)
19. Seliya, N., Khoshgoftaar, T.M., Van Hulse, J.: Aggregating performance metrics for classifier evaluation. In: IEEE International Conference on Information Reuse & Integration, IRI 2009,, pp. 35–40. IEEE (2009)
20. Khalaf, M., et al.: Machine learning approaches to the application of disease modifying therapy for sickle cell using classification models. Neurocomputing **228**, 154–164 (2017)

21. Wei, Z.-S., Han, K., Yang, J.-Y., Shen, H.-B., Yu, D.-J.: Protein–protein interaction sites prediction by ensembling SVM and sample-weighted random forests. Neurocomputing **193**, 201–212 (2016)
22. Sánchez A, V.D.: Advanced support vector machines and kernel methods. Neurocomputing **55**(1–2), 5–20 (2003)
23. Cortes, C., Vapnik, V.: Support-vector networks. Mach. Learn. **20**(3), 273–297 (1995)
24. Adankon, M.M., Cheriet, M.: Model selection for the LS-SVM. Application to handwriting recognition. Pattern Recogn. **42**(12), 3264–3270 (2009)
25. Gunn, S.R.: Support vector machines for classification and regression. ISIS technical report, vol. 14, no. 1, pp. 5–16 (1998)
26. Khalaf, M., et al.: The utilisation of composite machine learning models for the classification of medical datasets for sickle cell disease. In: 2016 Sixth International Conference on Digital Information Processing and Communications (ICDIPC), pp. 37–41. IEEE (2016)
27. Hric, M., Chmulík, M., Jarina, R.: Model parameters selection for SVM classification using Particle Swarm Optimization. In: Radioelektronika (RADIOELEKTRONIKA), 2011 21st International Conference, 2011, pp. 1–4. IEEE (2011)
28. Baker, T., Rana, O.F., Calinescu, R., Tolosana-Calasanz, R., Bañares, J.Á.: Towards autonomic cloud services engineering via intention workflow model. In: Altmann, J., Vanmechelen, K., Rana, Omer F. (eds.) GECON 2013. LNCS, vol. 8193, pp. 212–227. Springer, Cham (2013). https://doi.org/10.1007/978-3-319-02414-1_16

# Fully Convolutional Neural Networks for 3D Vehicle Detection Based on Point Clouds

Lihua Wen and Kang-Hyun Jo[(✉)]

The Graduate School of Electrical Engineering,
University of Ulsan, Ulsan 44610, Korea
wenlihuawlh@gmail.com, acejo@ulsan.ac.kr

**Abstract.** In this paper, a novel methodology proposed for 3D detection with the purpose of boosting the detection accuracy and keep autonomous vehicles safety. The model takes the point clouds as the input directly. Based on our modified feature pyramid networks and VGG-16 named as FFPNets, which utilizes the one-stage fully convolutional network to detect 3D cars. The experimental result shows the robustness of the model and its superiority. The average precision (AP) of the car for easy, moderate, and hard cases achieves state-of-the-art detection accuracy on KITTI datasets.

**Keywords:** 3D detection · Autonomous vehicles · Point clouds · Fully convolutional network

## 1 Introduction

Recently, Point clouds based 3D object detection is an important component of a variety of real-world applications, such as autonomous navigation. However, there is a bottleneck for autonomous vehicles commercialization, that is the high cost of sensors. Currently, the autonomous vehicle is equipped with many sensors, such as Lidar, radar, and IMU, etc. Deep learning has achieved remarkable progress on 2D computer vision tasks, including object detection [1, 2] and instance segmentation [3–5]. However, 2D detection only can get exact precision in-camera pixels. To get the exact position of objects in real 3D space, we proposed an algorithm based on point clouds.

In autonomous driving, the most commonly used 3D sensor is Lidar, which generates 3D point clouds to capture the 3D structures of objects. Point clouds shown in Fig. 1 are unordered and irregular. Some researchers [6, 7] apply an MLP directly on the sorted point set. However, the performances are poor, and the speed is slow. Inspired by VoxelNet [8], one highly efficient algorithm, named F-VFE, which based on fully convolutional networks was proposed to extract the features of point clouds directly.

To design a fast and highly efficient algorithm, I choose the one-stage architecture like YOLO [9]. Inspired by YOLO, SqueezeDet [10] and FPN [11], we proposed a one-stage fully convolutional neural network to detect 3D vehicle detection. The detection pipeline is shown in Fig. 2.

© Springer Nature Switzerland AG 2019
D.-S. Huang et al. (Eds.): ICIC 2019, LNCS 11644, pp. 592–601, 2019.
https://doi.org/10.1007/978-3-030-26969-2_56

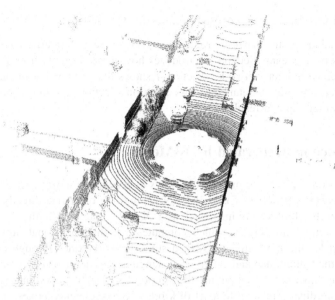

**Fig. 1.** Point clouds. Point clouds are unordered and irregular.

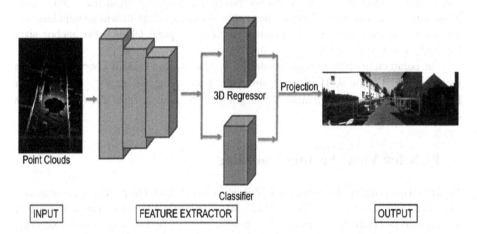

**Fig. 2.** Our pipeline. The input is point clouds. The feature extractor is the FFPNets explained above. The output is the 3D bounding boxes. In the output images, the yellow color denotes the prediction and the green color denotes the ground truth. (Color figure online)

Our contributions are summarized as the following:

1. 3D anchors are obtained by K-means.
2. A FFPNets which is highly efficient to rich features with lower PLOPs compared with the VGG [12].
3. In-class loss balance method that improves all detection accuracy at least 20%. Balancing loss can accelerate model convergence.

4. A novel loss function is proposed, it improves the model at least 10%.

The remainder of the paper is structured as follows: Sect. 2 describes the way to get 3D anchors by K-means. Section 3 introduces how to get features from point clouds with fully convolutional networks. Section 4 shows the architecture of our convolutional neural network in detail and loss function. Section 5 mainly introduces the experimental dataset and results.

## 2   3D Anchors Generated by K-Means

In 2D detection, there are commonly two ways to generate proposals, one is the region proposal networks (RPN) in Faster R-CNN [2] and the other one directly puts some anchors at each cell of feature map, such as YOLO [9]. The RPN in the Faster R-CNN starts with convolution layers, which computes a high dimensional, low-resolution feature map for the input image. Next, a small network slides through each spatial position in the feature map and generates rectangular region proposals centered around the position. Instead of computing the proposal's absolute coordinates, the RPN computes coordinates relative to a set of k pre-selected reference boxes, or anchors.

Intuitively, we want the anchors to be spatially close to the ground truth bounding boxes. In an extreme case, if an anchor box is too far away from the ground truth bounding box, learning to transform the anchor to the ground truth will be hopeless. So, we use the K-means to get 3D anchors based on original labels. One anchor size, l = 3.85, w = 1.61, h = 1.53 m for cars.

The parameters of 3D anchor besides (x, y, z, w, h, l) in LiDAR coordinate, it also includes spatial angle $\theta$. To reduce subsequent computing, we set $\theta \in \left[-90°, 90°\right]$ in LiDAR coordinate. So, we denote 3D anchor as (x, y, z, w, h, l, $\theta$). The (x, y, z) is the center of each anchor.

## 3   FCN for Voxel Feature Encoding

Existing work usually encodes 3D LIDAR point clouds into a bird's eye view map or a front view map of 2D image format [12, 14]. The bird's eye view map and the front view map representation are encoded by height, intensity and density in the voxels of a unique range first. Then, neural networks are used to the above 2D view map. While the method preserves most of the raw information of the point clouds, it lacks the 3D space information and needs more complex computation for subsequent feature extraction. We propose a more compact representation by projecting 3D point clouds to the bird's eye view with fully convolutional networks, named F-VFE. Figure 3 visualizes the pipeline of the F-VFE.

To easily understand the pipeline, here we just take one voxel as the example to show. In each voxel, we first compute the local mean as the centroid of all the points in V, denoted as $\left(v_x, v_y, v_z\right)$. Then we augment each point pi with the relative offset w.r.t. the centroid and obtain the input feature set $V_{in} = \hat{p}_i = [x_i, y_i, z_i, x_i - v_x, y_i - v_y, z_i - v_z]^T \in R^7_{i=1...t}$. Next, each pi is transformed through the 1D convolutional

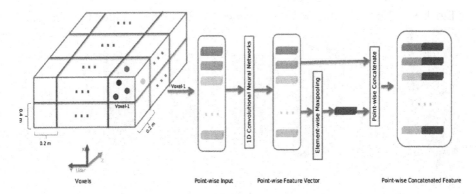

**Fig. 3.** F-VFE pipeline.

networks into a feature space. After obtaining point-wise feature representations, we use element-wise MaxPooling across all $f_i$ associated to V to get the locally aggregated feature $\widehat{f} \in R^m$ for V. Finally, we augment each $f_i$ with $\widehat{f}$ to form the point-wise concatenated feature as $f_i^{out} = \left[f_i^T, \widehat{f}_i^t\right]^T \in R^{2m}$.

In this paper, supervised learning algorithm is used to do 3D detection. So, we only focus on the range of 2D image visualization. We consider point clouds within the range of $[-2, 2] \times [-40, 40] \times [0, 70.4]$ meters along Z, Y, X axis respectively. The voxel size of $v_D = 0.4$, $v_H = 0.2$, $v_W = 0.2$ m, which leads to 3D voxel grid size D = 10, H = 400, W = 352. To balance sampled points in each non-empty voxel, we set the sampled point number as T = 35 as the maximum number.

## 4 Neural Networks and Loss Function

Neural Networks. We aim at designing fast and efficient neural networks. Now The popular neural network is VGG-16 [15]. it can extract clear features. However, the VGG-16 was designed for classifying the IMAGENET, which has 1000 classes. So, there are many redundancies of the filters for our model. First, we half the filters of VGG-16. Then we modify the VGG-16 structures like the feature pyramid networks [16], we named our feature extractor as FFPNets. The details of FFPNets is shown in Fig. 4.

From Fig. 4, we can know the Output feature includes all previous Blocks features. The detailed information shown in Table 1. Our networks mainly include 2D convolutional networks and Group normalization. In our networks, we do not use pooling to downsampling feature maps and use stride = 2 of convolution to do downsampling.

Loss Function. Let $\{a_i^{pos}\}_{i=1\cdots N_{pos}}$ be the set of $N_{pos}$ positive anchors and $\{a_i^{neg}\}_{i=1\cdots N_{neg}}$ be the set of $N_{neg}$ negative anchors. We parameterize a 3D ground truth box as $\left(x_c^g, y_c^g, z_c^g, w^g, h^g, l^g, \theta^g\right)$, where $x_c^g, y_c^g, z_c^g$ represent the center position of anchors, $w^g, h^g, l^g$ are length, width, height of the box, $\theta^g$ is the yaw rotation around Z-axis. To retrieve the ground truth box from a matching positive anchor parameterized

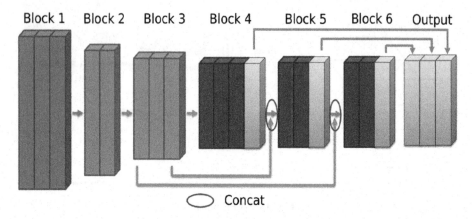

**Fig. 4.** Neural Networks. Different color denotes different size's block. The arrow denotes the propagation direction. The Concat denotes two feature maps combined by concatenation.

**Table 1.** The details of FFPNets. Our networks mainly include 2D convolutional networks and Group normalization. In our networks, we do not use pooling to downsampling feature maps and use stride = 2 of convolution to do downsampling.

|  | Layers | Property | Output size | Feature maps |
|---|---|---|---|---|
| Block 1 | Convolution group norm | ×3 | 400 × 352 | [3 × 3] ×32 |
| Block 2 | Convolution group norm | ×2 | 100 × 88 | [3 × 3] ×64 |
| Block 3 | Convolution group norm | ×4 | 100 × 88 | [3 × 3] ×128 [1 × 1] ×128 |
| Block 4 | Convolution group norm | ×4 | 100 × 88 | [3 × 3] ×128 [1 × 1] ×128 |
| Block 5 | Convolution group norm | ×3 | 100 × 88 | [3 × 3] ×256 [1 × 1] ×256 |
| Block 6 | Convolution group norm | ×3 | 100 × 88 | [3 × 3] ×256 [1 × 1] ×256 |
| Output | Concat |  | 100 × 88 | ×640 |

as $\left(x_c^a, y_c^a, z_c^a, w^a, h^a, l^a, \theta^a\right)$. To distinguish the explicit orientation when $\theta = \pm\pi$, we parameterize the regression parameters as $(x, y, z, w, h, l, cos(\theta), sin(\theta))$. We define the residual vector $\mu^* \in R^8$ containing the 8 regression targets corresponding to center location $\Delta x, \Delta y, \Delta z$ three dimensions $\Delta w, \Delta h, \Delta l$ and the rotation $\Delta\theta$ which are computed as:

$$\Delta x = \frac{x_c^g - x_c^a}{d^a}, \Delta y = \frac{y_c^g - y_c^a}{d^a}, \Delta z = \frac{z_c^g - z_c^a}{d^a} \qquad (1)$$

$$\Delta l = log\left(\frac{l^g}{l^a}\right), \Delta w = log\left(\frac{w^g}{w^a}\right), \Delta h = log\left(\frac{h^g}{h^a}\right) \qquad (2)$$

$$\Delta\theta_{cos} = cos(\theta^g) - cos(\theta^a), \Delta\theta_{sin} = sin(\theta^g) - sin(\theta^a) \qquad (3)$$

$$\Delta\theta = \Delta\theta_{cos} + \Delta\theta_{sin} \tag{4}$$

where $d^a = \sqrt{(l^a)^2 + (w^a)^2}$ is the diagonal of the base of the anchor box. We define the loss function as follows:

$$L = \frac{1}{N_{pos}}\sum_i L_{cls}(p_i^{pos}, 1) + \frac{1}{N_{neg}}\sum_j L_{cls}(p_i^{neg}, 0) + \frac{1}{N_{pos}}\sum_i L_{reg}(\mu_i, \mu_i^*) \tag{5}$$

where $N_{pos}$ and $N_{neg}$ are the number of positive and negative respectively.

We define the anchor is positive if the anchor having an IoU (>0.6) with the corresponding ground-truth, and the anchor is negative if the anchor having an IoU (<0.4) with the corresponding ground-truth. Experiments show that one image only has the positive anchor $N_{pos} \in [0, 110]$, however, the negative anchor $N_{neg} \in [15000, 17600]$. Because of the huge imbalance in positive class loss and negative class loss, the total class loss cannot decrease. To balance the positive class loss and negative class loss, we set $N_{pos} = 256$ and $N_{neg} = 768$ respectively.

After setting fixed number of $N_{pos}$ and $N_{neg}$, the experimental results improved at least 20%. To analyze the reason, we collect the data of loss, regression loss, classification loss, positive classification loss and negative classification loss.

Figure 5 shows the detailed analysis results.

## 5  Experiments and Results

The proposed model was trained using stochastic gradient descent (SGD). We used an Adam optimizer run on a GTX 1080 Ti GPU with a total of the point clouds per minibatch. All models were trained for 160 epochs (200 k iterations). The learning rate was 0.001, with an exponential decay factor of 0.8 and a decay every 20 epochs. We trained our network on the KITTI dataset [17] and evaluated our 3D object detector on the KITTI benchmarks for 3D object detection and BEV object detection. Our experimental setup is based on the LiDAR specifications of the KITTI dataset. We follow the official KITTI evaluation protocol, where the 3D IOU threshold is 0.7 for class car. We compare the methods using the average precision (AP) metric. We report the performance achieved on the KITTI validation set. Table 2 shows the performances are with loss balance and without loss balance.

Table 3. reports 3 kinds of neural networks test in our experiments. Truncated VGG-16 and Res-50 [18] means that the convolution part only be used. To get one fast model, we half the channels of VGG-16 and Res-50.

As shown in Table 4, we compare our results with previous state-of-the-art methods. For BEV detection benchmark, our method outperforms previous state-of-the-art methods with remarkable margins in all three difficulties. Figure 6 shows some images of 3D detected and BEV results.

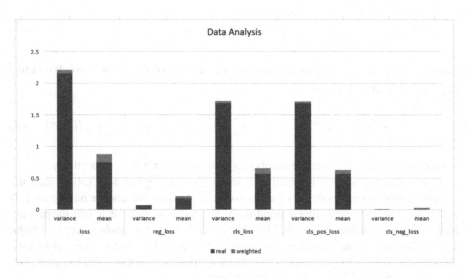

**Fig. 5.** The analysis of loss balance. The real and weighted denote the real loss without fixed number, the loss with fixed number respectively. The loss, reg-loss, cls-loss, cls-pos-loss and cls-neg-loss denote total loss, bounding box regression loss, classification loss, positive classification loss and negative classification loss respectively. We choose the mean and variance to analyze each class of loss. We can see that weighted loss is more balance than real loss.

**Table 2.** Results for loss balance. The BEV means the detection accuracy on bird's eye view. The 3D detection is the 3D accuracy on 2D images. Easy, moderate, hard represent 3 kinds of difficulties.

| Method | BEV | | | 3D detection | | |
|---|---|---|---|---|---|---|
| | Easy | Moderate | Hard | Easy | Moderate | Hard |
| Without loss balance | 82.393242 | 65.567146 | 57.947975 | 46.492203 | 34.209381 | 32.06216 |
| With loss balance | 90.175873 | 88.16745 | 79.367691 | 68.173782 | 60.368336 | 53.095936 |

**Table 3.** Performance of 3 different neural networks.

| | BEV | | | 3D detection | | |
|---|---|---|---|---|---|---|
| | Easy | Moderate | Hard | Easy | Moderate | Hard |
| Res-50 | 82.393242 | 65.567146 | 57.947975 | 46.492203 | 34.209381 | 32.06216 |
| VGG-16 | 82.896866 | 73.523773 | 66.16037 | 49.515339 | 41.532734 | 39.065273 |
| FFPNets | 90.175873 | 88.16745 | 79.367691 | 68.173782 | 60.368336 | 53.095936 |

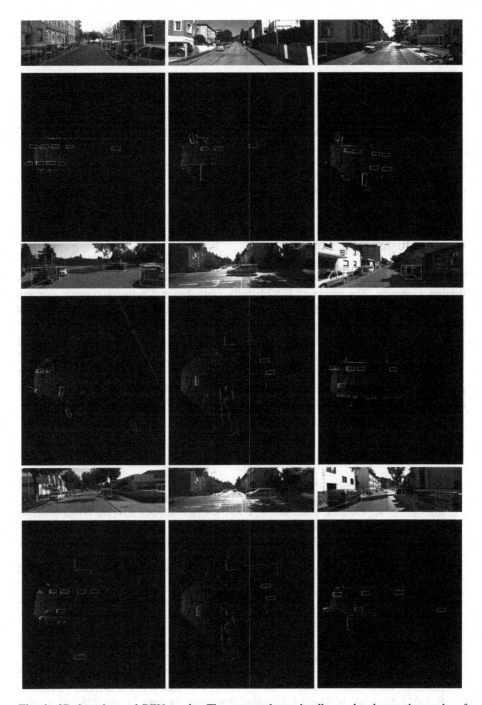

**Fig. 6.** 3D detection and BEV results. The green color and yellow color denote the results of ground-truth and prediction respectively. (Color figure online)

**Table 4.** Performance comparison of 3D object detection with previous methods on car class of KITTI test split by submitting to official test server. 3D object detection and bird's eye view detection are evaluated by Average Precision (AP) with IoU threshold 0.7.

| Method | Modality | BEV | | | 3D Detection | | |
|---|---|---|---|---|---|---|---|
| | | Easy | Moderate | Hard | Easy | Moderate | Hard |
| VoxelNet [8] | LiDAR | 89.35 | 79.26 | 77.39 | 77.47 | 65.11 | 57.73 |
| SECOND [19] | LiDAR | 88.07 | 79.37 | 77.95 | 83.13 | 73.66 | 66.20 |
| MV3D [14] | RGB+LiDAR. | 86.02 | 76.90 | 68.49 | 71.09 | 62.35 | 55.12 |
| UberATG [20] | RGB+LiDAR | 88.81 | 85.83 | 77.33 | 82.54 | 66.22 | 64.04 |
| AVOD-FPN [21] | RGB+LiDAR | 88.53 | 83.79 | 77.90 | 81.94 | 71.88 | 66.38 |
| F-PointNet [22] | RGB+LiDAR | 88.70 | 84.00 | 75.33 | 81.20 | 70.39 | 62.19 |
| Ours | LiDAR | 90.175873 | 88.16745 | 79.367691 | 68.173782 | 60.368336 | 53.095936 |

## 6  Conclusion and Future Work

The paper has presented a novel 3D object detector of point cloud for the autonomous driving scenarios. The proposed F-VFE fully convolutional network directly extracts 3D features from point cloud, which achieves significantly higher recall than previous proposal generation methods. The FFPNets shows its superiority than same Flops' truncated VGG-16 and Res-50. The classification loss balance accelerates model convergence and obviously improves detection accuracy.

Future work includes improving 3D detection accuracy and making a real time 3D detection system, which extends the current work for joint LiDAR and image based end-to-end 3D detection to further improve detection and localization accuracy.

## References

1. Girshick, R.B., Donahue, J., Darrell, T., Malik, J.: Rich feature hierarchies for accurate object detection and semantic segmentation. CoRR, vol. abs/1311.2524 (2013)
2. Ren, S., He, K., Girshick, R.B., Sun, J.: Faster R-CNN: towards real-time object detection with region proposal networks. CoRR, vol. abs/1506.01497 (2015)
3. Dai, J., He, K., Sun, J.: Instance-aware semantic segmentation via multi-task network cascades. CoRR, vol. abs/1512.04412 (2015)
4. He, K., Gkioxari, G., Dollár, P., Girshick, R.B.: Mask R-CNN. CoRR, vol. abs/1703.06870 (2017)
5. Liu, S., Jia, J., Fidler, S., Urtasun, R.: SGN: sequential grouping networks for instance segmentation. In: 2017 IEEE International Conference on Computer Vision (ICCV), pp. 3516–3524 (2017)
6. Qi, C.R., Su, H., Mo, K., Guibas, L.J.: Pointnet: deep learning on point sets for 3D classification and segmentation. CoRR, vol. abs/1612.00593 (2016)
7. Li, Y., Bu, R., Sun, M., Wu, W., Di, X., Chen, B.: PointCNN: convolution on x-transformed points. In: Bengio, S., Wallach, H., Larochelle, H., Grauman, K., Cesa-Bianchi, N., Garnett, R. (eds.) Advances in Neural Information Processing Systems 31, pp. 828–838. Curran Associates Inc., New York (2018)

8. Zhou, Y., Tuzel, O.: Voxelnet: end-to-end learning for point cloud based 3D object detection. CoRR, vol. abs/1711.06396 (2017)
9. Redmon, J., Divvala, S.K., Girshick, R.B., Farhadi, A.: You only look once: unified, real-time object detection. CoRR, vol. abs/1506.02640 (2015)
10. Wu, B., Iandola, F.N., Jin, P.H., Keutzer, K.: SqueezeDet: unified, small, low power fully convolutional neural networks for real-time object detection for autonomous driving. CoRR, vol. abs/1612.01051 (2016)
11. Lin, T., Dollár, P., Girshick, R.B., He, K., Hariharan, B., Belongie, S.J.: Feature pyramid networks for object detection. CoRR, vol. abs/1612.03144 (2016)
12. Li, B., Zhang, T., Xia, T.: Vehicle detection from 3D lidar using fully convolutional network. CoRR, vol. abs/1608.07916 (2016)
13. Caltagirone, L., Scheidegger, S., Svensson, L., Wahde, M.: Fast lidar-based road detection using fully convolutional neural networks. CoRR, vol. abs/1703.03613 (2017)
14. Chen, X., Ma, H., Wan, J., Li, B., Xia, T.: Multi-view 3D object detection network for autonomous driving. CoRR, vol. abs/1611.07759 (2016)
15. Simonyan, K., Zisserman, A.: Very deep convolutional networks for large-scale image recognition. CoRR, vol. abs/1409.1556 (2014)
16. Lin, T., Dollar, P., Girshick, R.B., He, K., Hariharan, B., Belongie, S.J.: Feature pyramid networks for object detection. CoRR, vol. abs/1612.03144 (2016)
17. Geiger, A., Lenz, P., Urtasun, R.: Are we ready for autonomous driving the kitti vision benchmark suite. In: Conference on Computer Vision and Pattern Recognition (CVPR) (2012)
18. He, K., Zhang, X., Ren, S., Sun, J.: Deep residual learning for image recognition. CoRR, vol. abs/1512.03385 (2015)
19. Yan, Y., Mao, Y., Li, B.: Second: sparsely embedded convolutional detection. Sensors **18**, 3337 (2018)
20. Liang, M., Yang, B., Wang, S., Urtasun, R.: Deep continuous fusion for multi-sensor 3D object detection. In: The European Conference on Computer Vision (ECCV), September 2018
21. Ku, J., Mozian, M., Lee, J., Harakeh, A., Waslander, S.L.: Joint 3D proposal generation and object detection from view aggregation. CoRR, vol. abs/1712.02294 (2017)
22. Qi, C.R., Liu, W., Wu, C., Su, H., Guibas, L.J.: Frustum pointnets for 3D object detection from RGB-D data. CoRR, vol. abs/1711.08488 (2017)

# Illegally Parked Vehicle Detection Based on Haar-Cascade Classifier

Aapan Mutsuddy[1], Kaushik Deb[1], Tahmina Khanam[1], and Kang-Hyun Jo[2(✉)]

[1] Department of Computer Science and Engineering,
Chittagong University of Engineering and Technology,
Chattogram 4349, Bangladesh
[2] Department of EE and Information Systems,
University of Ulsan, Ulsan 680-749, Korea
acejo@ulsan.ac.kr

**Abstract.** Detection of illegally parked vehicles in the urban traffic scene is required for handling unwanted incidents in the roads such as vehicle collision, traffic jam, accidents etc. Due to occlusion, lightning change and other factors the task becomes more laborious. A video-based proposal is presented in this paper to eliminate these problems. Initially, foreground objects are separated using Gaussian Mixture based background subtraction model. Meanwhile, median filtering is used to remove the salt-pepper noise. A simple shadow removal technique based on HSI and YCbCr color model improves the precision of the vehicle detection. Then, edge detection was performed using Canny method for precise tracking of temporarily static vehicles checking centroid co-ordinate values. Finally, for verification of the vehicles a machine learning mechanism known as 'Haar-Cascade' classifier is used. However, experimental data shows that proposed system performed well under different conditions with an accuracy of 92%.

**Keywords:** Gaussian mixture model · Median filtering ·
HSI and YCbCr model · Shadow detection · Canny edge detection ·
Haar-Cascade classifier

## 1 Introduction

The upward trend of population boom around the world has led to a dynamic rise of vehicles plying in modern day roads even though, the increase in parking spaces for these vehicles has been quite moderate. Hence, a large number of vehicles are parked illegally, thus, blocking the roads. As a consequence, traffic congestions and accidents are becoming more frequent, resulting in the compromise of public and transport safety. To tackle these issues, Intelligent Transport Systems (ITS) are expanding throughout the world to make existing infrastructures more secure and efficient.

Illegally parked vehicles not only create barriers in the lanes but also drive other vehicles to move in the adjacent lane causing congestions in the traffic scene. To monitor these irregularities in the roads traffic surveillance systems have been

D.-S. Huang et al. (Eds.): ICIC 2019, LNCS 11644, pp. 602–614, 2019.
https://doi.org/10.1007/978-3-030-26969-2_57

developed with the help of video cameras. However, with a growing number of cameras automated traffic analysis system is mandatory for providing information to the authority about traffic deformity. Several schemes [1–4] have been proposed for this purpose. The technique used in most of the system is similar. At first the vehicles are extracted from the background for foreground analysis using a background subtraction procedure and then some sort of object tracking system is used to detect the vehicle. [1–7] represent the literature history of background modeling and object tracking.

In reference [1] the foreground objects were separated using background detection techniques which can be used in normal traffic conditions. However, when the scenes become overcrowded the system fails. A system used Segmentation History Images (SHI) to detect stationary objects was introduced in [2]. Then, pixel-based approach was used on the segmented object to identify the moving pixel patches. Sometimes using single background model may fail to detect stationary objects due to lighting conditions.

Cumulative dual background subtraction was introduced in [4]. In [3] background was detected by Gaussian mixture model and the vehicles were detected by their wheels. However, one of the problems of this system is that the vehicle cannot be detected if the wheels are occluded (Fig. 1).

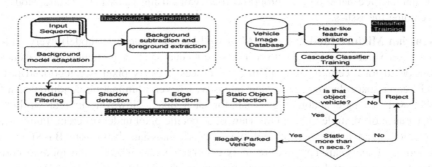

**Fig. 1.** Proposed system diagram

Shadow detection techniques after background subtraction to improve accuracy of object detection can be found in [7]. In a nutshell the contribution done in this research can be summarized as following-

- Haar feature based cascade classifier can verify different Bangladeshi vehicles including truck, CNG etc. robustly within minimal computational cost. Moreover, overall system can complete its proceedings within an average time complexity of .8 s.

The rest of the paper is organized as follows. In Sect. 2.2, the background segmentation technique is discussed. Then static object extraction is described in Sects. 2.3, 2.4 discusses about vehicle verification. In Sect. 3 experimental results are presented. Finally, Sect. 4 outlines the conclusions and possible future directions for this work.

## 2  Proposed Method

### 2.1  Overview

Our approach follows four stages as mentioned in the system diagram: background adaptation, differentiating temporarily static foreground objects and tracking as if the object is stationary or not, detecting the vehicles. First, foregrounds are extracted using GMM based background subtraction. Shadows of foreground is detected to improve detection accuracy. Then foreground objects are tracked using centroid co-ordinate value to determine if the object is in static or motion state. Finally, Haar-cascade classifier verifies the object as a vehicle.

### 2.2  Background Segmentation

In this step, background subtraction process is performed. In a traditional way the process continuously compares current frame from a predefined background model to differentiate foregrounds from background. This type of foreground extraction is affected by illumination change, noisy environment and camera vibration. Moreover, the predefined background frame may not always be available. However, all of these calculations are performed in proposed framework using Gaussian Mixture Model to overcome the issues.

**Gaussian Mixture Model:** In GMM (a well-known parametric model) every particular pixel value is modeled into a mixture of Gaussians. A particular pixel value is considered for a certain time called 'pixel process'. Every pixel is categorized in the RGB color space depending on their intensity. Uneven illumination conditions can be dealt with a single adaptive gaussian for each pixel. However, in real world the appearance of multiple plane in the view of a specific pixel under different illumination conditions lead to the use of multiple adaptive gaussian dispersion. Based on the variation of the Gaussian mixture, it is determined that whether the mixture corresponds to background color or foreground color. The probability of every pixel being in the foreground or background is estimated using the following probability,

$$p(X_t) = \sum_{i=1}^{K} \omega_{i,t}.\mathcal{N}(X_t, \mu_{i,t}, \Sigma_{i,t}) \tag{1}$$

Where, $X_t$: Current pixel in frame t, $K$: The number of distributions in the mixture, $\omega_{i,t}$: the weight of the kth distribution in frame t, $\mu_{i,t}$: the mean of the kth distribution in frame t, $\Sigma_{i,t}$: the standard deviation of the kth distribution in frame.

$\mathcal{N}(X_t, \mu_{i,t}, \Sigma_{i,t})$ is probability density function (pdf):

$$\mathcal{N}(X_t, \mu, \Sigma) = \frac{1}{(2\pi)^{\frac{n}{2}}|\Sigma|^{\frac{1}{2}}} exp^{-\frac{1}{2}(X_t-\mu)\Sigma^{-1}(X_t-\mu)} \tag{2}$$

The intensity difference follows uniform standard deviation according to [5]. RGB values maintain an uncorrelated pattern here. Co-variance matrix formula is represented in Eq. (3).

$$\Sigma_{i,t} = \sigma_{i,t}^2 I \tag{3}$$

Matched gaussians having a value greater than a designated threshold is defined as background (4) (Fig. 2).

| (a) | (b) | (c) |

**Fig. 2.** Processing example of background segmentation (a. input video, b. adapted background using gaussian mixture model, c. foreground extraction)

$$B = argmin_b(\sum_{i=1}^{b} \omega_{i,t} > T) \tag{4}$$

If a pixel matches with one of the K Gaussian, then the value of $\omega$, $\mu$ and $\sigma$ is updated with Eqs. (5), (6) and (7).

$$\omega_{i,t+1} = (1 - \alpha)\omega_{i,t} + \alpha \tag{5}$$

$$\mu_{i,t+1} = (1 - \rho)\mu_{i,t} + \rho X_{t+1} \tag{6}$$

$$\sigma_{i,t+1}^2 = (1 - \rho)\sigma_{i,t}^2 + \rho(X_{t+1} - \mu_{i,t+1})(X_{t+1} - \mu_{i,t+1})^T \tag{7}$$

If there is K number of gaussians that do not match then only $\omega$ is updated with Eq. (8).

$$\omega_{i,t+1} = (1 - \alpha)\omega_{i,t} \tag{8}$$

### 2.3   Static Object Extraction

As background subtraction often results in noisy binary images, in this step at first a nonlinear digital filtering technique called median filtering is used to remove the salt-paper noises of the extracted foreground signals.

**Shadow Detection:** The foreground objects are further put through other filtering process to detect shadows associated with the signals. As shadow always project on the other plan, so the intensity of shadow is less than any real black portion [7].

This difference in intensity can be used to remove the shadow from an image without removing any other real black portion increasing intensity. Two color models (i.e. HSI model and YCbCr model) has been used in this article to detect shadow. In order to gain intensity a projection of Y channel of YCbCr model is made into I channel of HSI model. Sometimes, pixels having less chromatic value fade due to intensity gain. In order to solve the problem chromatic gain is performed in the earlier stage of intensity gain. Some small shadow areas may still appear in the scene which are rejected using morphological operation and geometrical properties.

**Canny Edge Detection:** In this part of proposed framework edge detection is performed for further processing using popular canny method after removing shadows. First, to filter out the noises from given scene gaussian filtering was applied using Eq. (9).

$$G(x, y) = \frac{1}{2\pi\sigma^2} e^{-(x^2 + y^2)/2\sigma^2} \tag{9}$$

Gradient vector magnitude $(G)$ and direction $(\theta)$ was calculated using Eqs. (10) and (11).

$$G = \sqrt{G_x^2 + G_y^2} \tag{10}$$

$$\theta = tan^{-1} \frac{Gy}{Gx} \tag{11}$$

Next, we removed all the pixels that were not part of edge or local maximum to get thin edges from the image scene. Finally, Hysteresis thresholding is applied. The criteria use two thresholds, if a pixel gradient is higher than upper threshold it is accepted as an edge else it is rejected.

**Extracting Static Objects as ROI:** Here, static objects are extracted from the input signals as region of interest. To perform this operation first morphological closing was performed and then centroid co-ordinate position of each object is checked per frame. If the co-ordinate value remains same over the frames the object is considered static and listed as the potential candidate region for being an illegally parked vehicle, else the region is discarded for being in motion (Fig. 3).

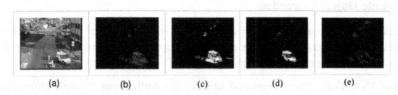

(a)          (b)          (c)          (d)          (e)

**Fig. 3.** Processing example of shadow detection and edge detection using Canny method (a. input frame, b. background subtracted frame, c. frame after median filtering, d. objects without shadow, e. detected foreground edges)

## 2.4    Vehicle Verification Using Haar Feature Based Cascade Classifier

In this sub-section the process for verifying the potential region as a vehicle is performed using well known "Haar-cascade" classifier, which was first introduced by Viola-Jones [8] for rapid face detection. Following Fig. 4. shows the steps involved in vehicle detection:

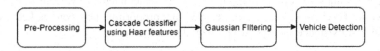

**Fig. 4.**  Steps in vehicle detection using Haar feature based cascade classifier

Firstly, captured image frames were converted into gray scale images as for image pre-processing. Training and detection are the two parts of vehicle detection schema. All procedures were performed using OpenCV. To create the Haar cascade classifier some positive and negative images are generated. Detection accuracy is better if large numbers of sample images are used. Around 5000 positive samples of vehicles and 2000 non-vehicle or related environment images as negative samples are used for the cascade classifier. Then, image noises were removed using gaussian blur technique.

**Vehicle Detection.** In this step, for detection procedure an image representation called integral image computes detectors features rapidly. Extended Haar features are used in this process. Critical visual features from a larger set are selected by a learning algorithm named adaboost. This algorithm results in with computationally effective classifiers [8]. Next, cascading process discards non-objects very quickly. Each cascaded stage is used to determine whether a sub window contains vehicle or non-vehicle. A given sub-window is immediately discarded if it fails in any of the stages.

**Counting Duration:** As static objects are extracted and verified using vehicle detection process, each stopped vehicles are tracked considering time. If the vehicles remain still after a certain time which is longer than traffic light cycles, an alarm is generated for being parked illegally (Fig. 5).

**Fig. 5.**  (a) (b) Detected illegally parked vehicles.

# 3   Experimental Result

## 3.1   Tools and Evaluation Parameters

OpenCV image processing tools were used to implement the proposed framework with python programming language on pc equipped with Core i5-3210 CPU running at 2.5 GHz and 4 GB ram. Resized video resolution was chosen 320 × 240. Precision and Recall values for different input videos are calculated to evaluate the proposed framework. Precisions are determined by the proportion between parked vehicles detected and existence of true parked vehicles in the scene.

$$\text{Prescesion} = \frac{TP}{TP + FP} \tag{12}$$

Ratio measured between truly detected parked vehicles and total number of parked vehicles represents the recall value. Recall is calculated by:

$$\text{Recall} = \frac{TP}{TP + FN} \tag{13}$$

Here, TP is true positive value when an illegal parked vehicle exists and its detected, thereafter if it is not detected then it is a false negative call. However, false detection as an illegally parked vehicle is known as false positive.

## 3.2   Dataset

Propose framework has been evaluated using two types of datasets, one Imagery Library for Intelligent Detection System (i-LIDS) and another our own dataset that consists campus CCTV footage and video captured on roads under different conditions showed in Table 1. Steps involved in detection procedure of illegally parked vehicles are described in following section.

## 3.3   Experimental Result Analysis

At first background segmentation is performed on the data sets where the foreground is extracted using Gaussian Mixture Model. Then, the output from the MOG background subtraction is filtered to remove salt-paper noise. After removing the noise from the extracted foreground, shadow removal is performed using HSI and YCbCr color models. Improvement of detection accuracy after shadow detection can be seen in Fig. 6, where it can be seen that detection box bounded the object poorly before shadow detection and sometimes considered two parallel objects as one due to shadows.

**Table 1.** Dataset description

| Video | Ground truth | Illumination condition and length |
|---|---|---|
| | i-Lids dataset | |
| Sample-1 | 1 | Daylight 00.03.31 |
| | Own dataset | |
| Sample-2 | 1 | Daylight 00.01.14 |
| Sample-3 | 1 | Daylight 00.02.43 |
| Sample-4 | 1 | Daylight 00.01.32 |
| Sample-5 | 1 | Night 00.01.32 |
| Sample-6 | 1 | Daylight 00.02.20 |
| Sample-7 | 1 | Shadowed 00.02.25 |
| Sample-8 | 1 | Night 00.01.45 |
| Sample-9 | 1 | Daylight 00.02.35 |
| Sample-10 | 1 | Daylight 00.02.32 |
| Sample-11 | 1 | Daylight 00.02.07 |

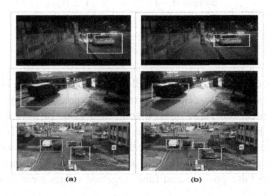

(a)          (b)

**Fig. 6.** Foreground detection accuracy improvement (Detected objects (a) before shadow detection (b) after shadow detection)

However, after shadow detection these ambiguities are seen less, hence accuracy is achieved. Then, next step is to detect the edges and reduce unwanted noise and texture using Canny Edge Detection process. Then temporarily static objects are then extracted by checking centroid pixel point co-ordinate value over frames. If the value remained same then the object is static, otherwise it is in motion. A machine learning approach "Haar-Cascade Classifier" is used to verify the vehicles and it can perform detection process very robustly. From the Fig. 7. we can see the high performance for detecting different types of vehicles regarding F1 measurement.

### 3.4  Evaluation and Comparison

All these steps mentioned above are performed with 25 video sequences at different illumination condition (day, night, un even), different vehicles. Here we presented 12 sequences which includes all different criteria mentioned above shown in Figs. 9, 10 and 11.

**Table 2.** Detection result

| Video | Ground truth | TP | FP |
|---|---|---|---|
| Sample-2 | 1 | 1 | 0 |
| Sample-3 | 1 | 1 | 0 |
| Sample-4 | 1 | 1 | 0 |
| Sample-5 | 1 | 1 | 0 |
| Sample-6 | 1 | 1 | 0 |
| Sample-7 | 1 | 1 | 0 |
| Sample-8 | 1 | 1 | 0 |
| Sample-9 | 1 | 1 | 0 |
| Sample-10 | 1 | 1 | 0 |
| Sample-11 | 1 | 1 | 1 |

Table 2 shows the results of detecting illegally parked vehicles. Detection accuracy of the samples was tested by calculating recall and precision. The proposed system achieved average precision value of 86% and recall value of 89% for i-LIDS datasets shown in Table 3. Here, TP = True Positive; FP = False Positive; FN = False Negative. While testing our own datasets average precision and recall values we yielded are 90% and 92%. Due to vehicle misclassification and poor background modeling in crowded scenes, most of the false positive and false negative values were produced.

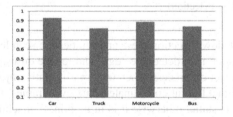

**Fig. 7.** F1 score measured for different types of detected vehicles.

**Table 3.** Detection accuracy of tested samples

| Frames | Precision (TP/(TP+FP)) | Recall (TP/(TP+FN)) | Average accuracy |
|---|---|---|---|
| Own dataset | .90 | .92 | 92.3% |
| i-LIDS dataset | .866 | .898 | |

A graphical representation between True positive rate and False positive rate for 3500 frames regarding daylight scene, i-Lids dataset and night scene is shown ROC curve in Fig. 8.

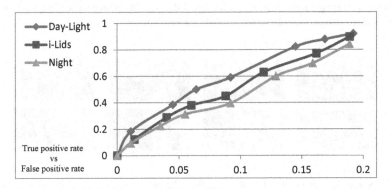

**Fig. 8.** ROC curve

Average time complexity of the proposed framework is shown in Table 4, where it can be seen that proposed method can perform its steps within very short time limit.

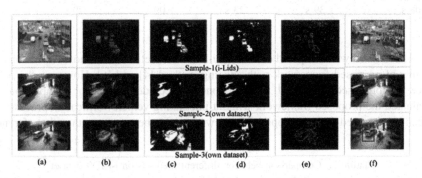

**Fig. 9.** Processing example of illegally parked vehicle detection (a) Input RGB frame (b) Extracted Foregrounds with shadows (c) Filtering (d) Foregrounds without shadows (e) Canny edge detection (f) Detected illegally parked vehicle

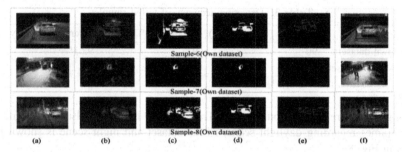

Fig. 10. Illegally parked vehicle detection (different lightning) (a) Input RGB frame (b) Extracted Foregrounds with shadows (c) Filtering (d) Foregrounds without shadows (e) Canny edge detection (f) Detected illegally parked vehicle

Fig. 11. Some sequential results of detecting illegally parked vehicles (a) (b) vehicles approaching to park illegally (c) counting time after parking (d) detected illegally parked vehicles

**Table 4.** Time complexity analysis

| Background subtraction | Shadow detection | Vehicle detection | Classification | Average computation time |
|---|---|---|---|---|
| 0.08 s | .39 s | .23 s | .09 s | .79 s |

The method was compared evaluating i-LIDS dataset. Table 5 shows the compared results of proposed framework with existing method [4] and [6]. However, the reference [6] proposed a method for abandoned object detection. For comparison we evaluated the proposed framework without integrating vehicle verification process and it can be observed that the proposed framework yields better output than the reference.

**Table 5.** Comparison between different methods on i-LIDS dataset

| Evaluation dataset (i-LIDS) | Precision (TP/(TP+FP)) | Recall (TP/(TP+FN)) |
|---|---|---|
| Proposed framework | .866 | .898 |
| [4] | .89 | .89 |
| Proposed framework (without vehicle verification) | .74 | .81 |
| [6] | .40 | .67 |

Though strong illumination changes and overcrowded scene caused background modeling ineffective in some cases resulting in producing false positives, from Fig. 10 we can see the proposed framework can handle decent gradual illumination change and uneven lightning conditions. Sample-7 of Fig. 10. shows a parked bike occluded by a tree is also detected by the proposed system even though it was in an uneven illumination area. However, when it was parked in a darker condition under shadow of a tree, system was unable to detect it. Figure 12(a) and (b) shows the result with a better view.

**Fig. 12.** (a) Illegally parked vehicle detection at occluded and uneven illumination condition. (b) undetected vehicle in a darker region

## 4   Conclusion

In this study we represent a framework for detecting illegally parked vehicles to improve existing traffic surveillance system. The main contribution of this dissertation work is to detect illegally parked vehicles for real time processing. Initially, the background subtraction is performed to extract the foreground objects using Gaussian Mixture Model which is adaptive with illumination change and avoids false foreground extraction. Median filtering is applied to remove the salt pepper noise from the foreground objects. After that shadow detection process is used to remove the shadows from the foreground signals to avoid miss-classification of the objects. Then, Canny edge detection is performed to find the edges of objects. Using Morphological operation and centroid co-ordinate value calculation static foreground objects are classified. For vehicle verification Haar feature based cascade classifier is trained which is well known for its robustness, and applied in the region of interest to identify the object as vehicle. Finally, static vehicles are tracked using time, if the vehicles remain static in the road region for a certain period of time then the alert is generated.

The proposed framework starts to model the background using Gaussian Mixture model and initially it assumes all static objects as a background. If an illegally parked vehicle stays in the scene before the modeling of background the system cannot identify that. Again, if vehicles stayed stationary for long period of time system considers the parked vehicles as a background object. Cumulative background modeling can be used to overcome this issue. However, in overcrowded places consisting many people, vehicle detection using Convolutional Neural Network may provide higher accuracy.

**Acknowledgements.** This research was supported by the MSIT (Ministry of Science and ICT), Korea, under the ICT Consilience Creative program (IITP-2019-2016-0-00318) supervised by the IITP (Institute for Information & communications Technology Planning & Evaluation).

# References

1. Filonenko, W.A., Jo, K.H.: Illegally parked vehicle detection using adaptive dual background model. In: 41st Annual Conference of the IEEE Industrial Electronics Society, Yokohama, pp. 2225–2228, 09–12 November 2015
2. Hassan, W., Birch, P., Young, R., Chatwin, C.: Real-time occlusion tolerant detection of illegally parked vehicles. Int. J. Control Autom. Syst. 10(5), 972–981 (2012)
3. Bevilacqua, A., Vaccari, S.: Real time detection of stopped vehicles in traffic scenes. In: IEEE Conference on Advanced Video and Signal Based Surveillance, London, UK, pp. 266–270, 05–07 September 2007
4. Wahyono, W., Jo, K.H.: Cumulative dual foreground differences for illegally parked vehicles detection. IEEE Trans. Ind. Inform. 13(5), 2464–2473 (2017)
5. Stauffer, C., Grimson, W.: Adaptive background mixture models for real-time tracking. In: IEEE Computer Society Conference on Computer Vision and Pattern Recognition, Fort Collins, USA, pp. 246–252, 23–25 June 1999
6. Fan, Q., Pankanti, S.: Modeling of temporarily static objects for robust abandoned object detection in urban surveillance. In: IEEE International Conference on Advanced Video and Signal Based Surveillance (AVSS), Austria, pp. 36–41, 30 August–02 September 2011
7. Khanam, T., Deb, K.: Baggage recognition in occluded environment using boosting technique. KSII Trans. Internet Inf. Syst. 11(11), 5436–5458 (2017)
8. Viola, P., Jones, M.: Rapid object detection using a boosted cascade of simple features. In: IEEE Computer Society Conference on Computer Vision and Pattern Recognition, Kauai, HI, USA, pp. 511–518, 08 December 2001

# Graph-SLAM Based Hardware-in-the-Loop-Simulation for Unmanned Aerial Vehicles Using Gazebo and PX4 Open Source

Khoa Dang Nguyen[1,2(✉)], Trong-Thang Nguyen[3],
and Cheolkeun Ha[4]

[1] Faculty of Electrical and Electronic Engineering, Phenikaa Institute for
Advanced Study (PIAS), Phenikaa University, Hanoi 100000, Vietnam
khoa.nguyendang@phenikaa-uni.edu.vn
[2] Phenikaa Research and Technology Institute (PRATI),
A&A Green Phoenix Group, 167 Hoang Ngan, Hanoi 100000, Vietnam
[3] Faculty of Electrical-Electronic Engineering, Vietnam Maritime University,
484 Lach Tray, Le Chan, Haiphong, Vietnam
dr.nguyentrongthang@gmail.com
[4] School of Mechanical and Automotive Engineering, University of Ulsan,
Ulsan 680-749, South Korea
cheolkeun@gmail.com

**Abstract.** This paper presents a method to simulate the graph simultaneous localization and mapping (Graph-SLAM) for a Unmanned Aerial vehicle (UAV) by using the hard-in-the-loop-simulation (HILS). This method uses the Gazebo software to render six-degree-of-freedom (6 DOF) UAV model, the virtual sensor model and virtual RGB-D camera model. To drive the UAV in Gazebo, the flight control based on PX4 open source code is performed on the Pixhawk board hardware. A Graph-SLAM algorithm open source named RTAB-MAP which is modified and installed on the Raspberry board, is used to estimate the 3D mapping of the environment and localization of UAV in map. A control application software (CAS) is developed to connect all parts of HILS such as the Gazebo, Pixhawk and Raspberry by using the multithread architecture. Numerical simulation has been performed to demonstrate the effectiveness of the HILS configuration approach.

**Keywords:** Graph-SLAM · HILS · UAV

## 1 Introduction

Hardware-in-the-loop simulation (HILS) have been well-known as a tool to simulate the operation of robot in the real time condition, which is combined by the software and hardware board. Herein, the software is used to presented the dynamic model of robot and environment working. Meanwhile, the hardware board is used to install the algorithm to control the robot. Nowadays, HILS is often used in the testing unmanned aerial vehicle (UAV) to avoid the high risk of property damage and dangerous for humans. It is very necessary in the development cycles of UAV system.

© Springer Nature Switzerland AG 2019
D.-S. Huang et al. (Eds.): ICIC 2019, LNCS 11644, pp. 615–627, 2019.
https://doi.org/10.1007/978-3-030-26969-2_58

For some testing of UAV based on HILS, in flight control design approach, some algorithms [1, 2] were designed, which show some interesting results. In UAV applications approach, the vision was attracted significant interests of researchers such as tracking landing pad and tracking target. Herein, the real camera is used to capture the virtual image which is shown on the desktop monitor screen [3, 4], projector screen [5, 6]. Although they showed some good configuration in HILS development, their applicability is limited following reasons. First, the cost of real camera is the trouble with the developer system. Second, the monitor screen is difficult to present the 3D environment which is RGB image and depth information.

To overcome this limitation, the Gazebo software [7, 8] is used in the HILS configuration, which can support the dynamic model of UAV, sensor model, camera model and 3D environment for the simulation. In HILS configuration, Gazebo was combined with Odroid board [9] and Pixhawk [10, 11]. Although their researches have presented the vision simulation based on HILS, but their real flight testing is very difficult to perform. As in first case, the reason is the lack of sensor (IMU sensor, GPS sensor) in Odroid board and the overload of operation system by running the flight controller algorithm and vision algorithm in one microchip. In second case, the vision algorithm is worked on the desktop computer. Therefore, it is a challenge when the UAV flights in real environment.

Despite the Gazebo simulation has more advantage for the vision algorithm development. Nevertheless, the SLAM (specially Graph-SLAM) with HILS based on the RGB-D virtual camera have not been still performed.

The main contribution of this paper is to create full HILS configuration for testing the flight control algorithm and Graph-SLAM algorithm on the real-time condition. Specifically, a quadrotor UAV is used to demonstrate the HILS configuration. First, the Gazebo software is used to make the dynamic model, sensor model and RGB-D camera model. Second, the flight control algorithm is applied to drive the operation of the quadrotor UAV, which is installed on Pixhawk hardware board. And a Graph-SLAM algorithm is modified from RTAB-MAP open source code for making the 3D mapping and localization of the quadrotor in the map. SLAM algorithm is installed on the Raspberry hardware board. Third, the middleware software named CAS in our previous research [11, 12] is used to ensure the communication between the parts of HILS.

The remainder of this paper is organized as follows: Sect. 2 includes the system modeling of the quad-rotor UAV while the HILS configuration is presented in Sect. 3. Numerical simulation is performed in Sect. 4 to demonstrate the effectiveness of our HILS for the Graph-SLAM. Finally, some concluding remarks are given in Sect. 5.

## 2   System Modeling

The UAV named the quadrotor is used to present the operation flights in the HILS configuration, which has the configuration as in Fig. 1. It contains four propellers and four motors connected to a body frame. The motors are fixed with a rotating direction.

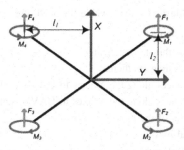

**Fig. 1.** Description of the quad-rotor UAV

In order to design the flight controller for quadrotor UAV, the mathematical model and the dynamic model of the quadrotor UAV need to be considered as: [13, 14]

$$
\begin{bmatrix} U_1 \\ U_2 \\ U_3 \\ U_4 \end{bmatrix} = \begin{bmatrix} K_f & K_f & K_f & K_f \\ -l_1 K_f & -l_1 K_f & l_1 K_f & l_1 K_f \\ l_2 K_f & -l_2 K_f & -l_2 K_f & l_2 K_f \\ -K_m & K_m & -K_m & K_m \end{bmatrix} \begin{bmatrix} \omega_1^2 \\ \omega_2^2 \\ \omega_3^2 \\ \omega_4^2 \end{bmatrix} \tag{1}
$$

$$
\ddot{\phi} = \frac{1}{I_{xx}} \left( U_2 - J_r \dot{\theta} \omega_r + \left( I_{yy} - I_{zz} \right) \dot{\Psi} \dot{\theta} \right) \tag{2}
$$

$$
\ddot{\theta} = \frac{1}{I_{yy}} \left( U_3 - J_r \dot{\phi} \omega_r + \left( I_{zz} - I_{xx} \right) \dot{\phi} \dot{\Psi} \right) \tag{3}
$$

$$
\ddot{\Psi} = \frac{1}{I_{zz}} \left( U_4 + \left( I_{xx} - I_{yy} \right) \dot{\theta} \dot{\phi} \right) \tag{4}
$$

$$
\ddot{x} = \left( cos(\phi) cos(\psi) sin(\theta) + sin(\phi) sin(\psi) \right) \left( \frac{U_1}{m} \right) \tag{5}
$$

$$
\ddot{y} = \left( cos(\phi) sin(\psi) sin(\theta) - sin(\phi) cos(\psi) \right) \left( \frac{U_1}{m} \right) \tag{6}
$$

$$
\ddot{z} = \left( cos(\phi) cos(\theta) \right) \left( \frac{U_1}{m} \right) - g \tag{7}
$$

$$
\omega_r = \omega_1 - \omega_2 + \omega_3 - \omega_4 \tag{8}
$$

where $U_1$, $U_2$, $U_3$ and $U_4$ are the throttle, roll, pitch and yaw control input, respectively. $\omega_i$ is angular velocity of the $i^{th}$ motor; Roll ($\phi$), pitch ($\theta$), yaw ($\psi$) are rotational and (x, y, z) are translational system. $I_{xx}, I_{yy}, I_{zz}$ are the moments of inertia about the principle axes in the body frame; Some constant values are shown in Table 1.

**Table 1.** Constant value for quadrotor UAV.

| Symbol | Description |
|--------|-------------|
| $K_f$ | Rotor thrust coefficient |
| $K_m$ | Rotor drag coefficient |
| $l_1$ | Distances along the X axes to UAV's center of gravity |
| $l_2$ | Distances along the Y axes to UAV's center of gravity |
| $J_r$ | Inertia of the motor |
| $m$ | Total mass |
| $g$ | Gravitational |

## 3 HILS Development

In this section, the HILS is developed for testing the flight control algorithm and Graph-SLAM for the UAV in the real-time platform. Four parts is developed to make a full HILS. First, Gazebo is present the dynamic model, sensor model, 3D environment. Specially, the virtual RDB-D is made for the Graph-SLAM. Second, the flight control is build by using the PX4 open source, which is compiled and uploaded to the Pixhawk board. Third, the Raspberry board is used to install the Graph-SLAM algorithm which has the camera signal from Gazebo. Four, the CAS is used to connect all informations between the Gazebo, Pixhawk and Raspberry of HILS. The HILS configuration is proposed as shown in Fig. 2.

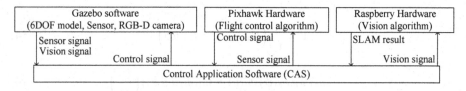

**Fig. 2.** Proposed HILS system

### 3.1 Gazebo Development

To present the operations of the quadrotor UAV, the Gazebo is used, which includes the dynamic model, sensor model, 3D environment and 3D visualization. First, the 3D CAD of the quad-rotor UAV is drawn in the Solidworks software with all real sizes. Then it is inserted to Gazebo to make the dynamic model of UAV. The virtual quadrotor UAV is presented in Fig. 3.

Second, the input and output signals are defined for controlling the quadrotor UAV model which are written in the code of the plugins. Herein, the input is the angular velocity for four motors which come from Pixhawk board. Therefore, The force and moment are calculated to drive the quad-rotor UAV. The outputs often are the IMU (for measuring the orientation, linear acceleration, angular velocity) and the camera signals (RGB image and depth information), which can be developed based on the libraries in Gazebo such as physics (*physics.hh*), sensor (*sensor.hh*), camera (*cameraSensor.hh*).

**Fig. 3.** The virtual quadrotor UAV in Gazebo

**Fig. 4.** A visualization in the Gazebo simulation software

Third, two above parts are defined in a world file to show the 3D environment as well as the operation of the quadrotor UAV. A display of simulation in the Gazebo is shown in Fig. 4.

### 3.2 Flight Control Algorithm

To control the quad-rotor UAV, a flight control algorithm is modifying based on the PX4 open source code which is compiled and uploaded to Pixhawk board hardware. For demonstrating HILS configuration, a tracking controller is developed as in Fig. 5.

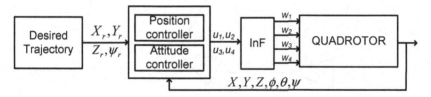

**Fig. 5.** Tracking controller of the quad-rotor UAV

Herein, two controllers: the position control and attitude control based on PID controller are used to track a desired trajectory. The inputs of the controllers are $(X, Y, Z, \psi_r)$. And the outputs are $(U_1, U_2, U_3, U_4)$ which are used to calculate the angular velocity for four motors by using the inverse function $(InF)$ of Eq. (1). The feedback of controller as roll $(\phi)$, pitch $(\theta)$, yaw $(\psi)$, X, Y, Z is estimated from the IMU sensor which comes from gazebo [15, 16].

## 4 Graph-SLAM Development

The Gazebo and Pixhawk are developed to demonstrate the performance of the flight controller. To show the better point in HILS configuration, the SLAM algorithm (Graph-SLAM) is presented as in Fig. 6, which can make a 3D mapping and localization of UAV in map.

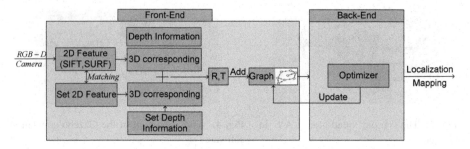

**Fig. 6.** Schema of Graph-SLAM process

In Fig. 6, Graph-SLAM contains two processes: *Front-End and Back-End.*

*Front-End(Graph construction)*: Let $x = (x_1, ... x_N)$ be a vector of parameters, where $x_i$ describes the pose of node $i$. Let's define $T_{i,j}$ and $R_{i,j}$ are the transformation and the rotation between two poses $x_i$ and $x_j$, respectively. A graphical of Graph-SLAM is shown as in Fig. 7.

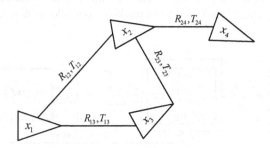

**Fig. 7.** Graphical of Graph-SLAM

In this paper, a the RGB-D virtual camera is used for the input of the Graph-SLAM algorithm, which provides the depth information as well as the RGB image information.

Based on the RGB image information, the 2D image features are extracted and its matched with previous features by using speeded-up robust features (SURF) [17]. Based the features, the random sample consensus (RANSAC) [18] algorithm is applied to remove outliers in image matching. The feature matching is called the loop closure detection. Therefore, two set 2D features are detected. Each feature can be located to the 3D point information $(X, Y, Z)$ by using the depth information. The set 3D points of features are used to calculate the new pose of the graphical by using the comparison between the current 3D point features and previous 3D point features. Then the iterative closest point (ICP) [19] algorithm are used to get the transformation T and the rotation R among the frames are used to add to the edge between two poses. Therefore, the new pose can be calculated by

$$P_{new} = T + R * P_{previous} \tag{9}$$

*Back-End(Graph optimization):* by using the maximum-likelihood estimation method [20], as

$$x^* = \arg \min_x \frac{1}{2} \sum_{(i,j) \in E} e_{i,j}^T(x) \Omega_{i,j} e_{i,j}(x) \tag{10}$$

where $e_{i,j}(x)$ is the error between the predicted and observed relative poses between the $i^{th}$ and $j^{th}$ nodes. $\Omega_{i,j}$ is the measurement information matrix.

The $E$ is the set of edges between two nodes. The $e_{i,j}(x)$ can be presented as

$$e_{i,j}(x) = h_{i,j}(x) - z_{i,j}(x) \tag{11}$$

where $z_{i,j}$, $h_{i,j}$ are the measurement value obtained from sensor and the prediction model between two nodes, respectively. As above part, the 3D corresponding between two frames (denoted $F_1$ and $F_2$) obtained. Apply the transformation $T$ and the $R$ rotation for each 3D feature in $F_1$, the 3D feature corresponding in $F_2$ can be calculated from prediction model. However, all sensors are impacted by the noise. Therefore, the error $e_{i,j}(x)$ can be computed by Eq. (11).

The optimization Graph-SLAM process is the minimization the error $e_{i,j}(x)$. By using the Levenberg-Marquardt [20], the pose of Graph-SLAM can be followed as

$$x = x + \Delta x \tag{12}$$

where $\Delta x$ is calculated by [20]

$$
\begin{aligned}
H &= \sum_{(i,j) \in E} J_{i,j}^T \Omega_{i,j} J_{i,j} \\
b &= \sum_{(i,j) \in E} J_{i,j}^T \Omega_{i,j} e_{i,j}(x) \\
J_{ij} &= \left. \frac{\partial e_{i,j}(\breve{x})}{\partial \breve{x}} \right|_{\breve{x}=x}
\end{aligned}
\tag{13}
$$

The $J_{i,j}$ is Jacobian of the error $e_{i,j}(x)$ with respect to the pose $x$.

By the optimization graph process, all poses in the graph are updated based on the updated translation and rotation. Therefore, the localization and mapping of the quad-rotor UAV is constructed. For implementation the Graph-SLAM, the RTAB-Map open source software [21, 22] is modified and installed on the Raspberry hardware board.

## 5 CAS Development

By using CAS, described in our previous publication [11, 12], the connection between Pixhawk hardware (flight controller), Raspberry (Graph-SLAM) and Gazebo is established. Based on the multithread architecture, it can ensure the data transfer among the components in HILS configuration with low loss ratio and incremental synchronization. The detailed structure of the CAS proposed in this paper is presented in Fig. 8.

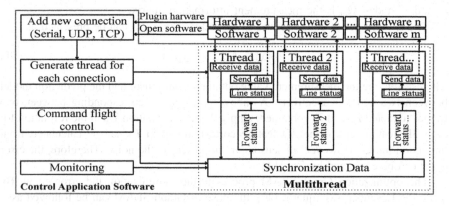

**Fig. 8.** Structure of the CAS

For implementing our HILS, the hardware 1, hardware 2 and software 3 are defined as pixhawk board, Raspberry board and Gazebo simulation software, respectively. The layout of the overall HILS is shown in Fig. 9.

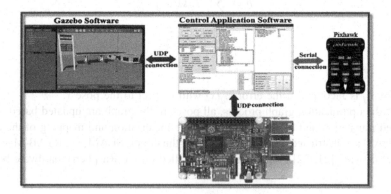

**Fig. 9.** Connection in HILS configuration

## 6  Implementation of HILS and Results

In this section, simulation is carried out to demonstrate the effectiveness of HILS in the test Graph-SLAM algorithm. The HILS configuration is built as in Fig. 2. A scenarios in Gazebo is created for HILS, which includes a room with size as in Fig. 10 and the quad-rotor UAV model contained the RGB-D camera. The 3D environment in the Gazebo is shown in Fig. 11.

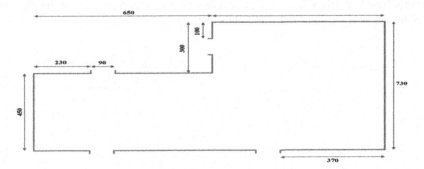

**Fig. 10.** Cross section of the map under 1[m]

**Fig. 11.** 3D environment in gazebo software for HILS

The Graph-SLAM is developed and installed on the Raspberry board to make the 3D mapping of the environment and the localization the pose of the robot which is based on the RGB-D camera signal from the Gazebo.

By using the above configuration, a desired trajectory is defined for the quad-rotor UAV as in Fig. 12. The quad-rotor UAV is take-off and flight around the environment which is set in the Gazebo. Two simulations are performed to demonstrate the effectiveness of the Graph-SLAM as well as HILS. In first simulation, the Graph-SLAM is applied without the Back-End process. As a result in Fig. 13, the map is not really good to present the 3D environment which is defined as in Fig. 11. Because of the optimization to correct the pose of the quad-rotor UAV is not performed. In second simulation, the Back-End process is applied. The result in Fig. 14 is shown that a 3D mapping is created and the pose of the quad-rotor UAV is corrected in the map.

The results are shown very clearly by using 2D view of the 3D map as in Figs. 15 and 16. The Graph-SLAM with Back-End really provides the estimation mapping better than Graph-SLAM without Back-End. Therefore, the position of the quad-rotor UAV is improved. Although, the quad-rotor UAV does not really track high accurately the desired trajectory because the noise comes from the environment. But the similarly between the position from GPS data response and the position from the Graph-SLAM showed that the Graph-SLAM can provide the high performance in both mapping and localization. The result proved convincingly that the Graph-SLAM algorithm could perform based on HILS setup which provides high performance for test the algorithm of both vision algorithms and flight algorithms.

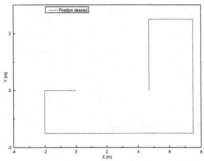

**Fig. 12.** Desired trajectory of quadrotor UAV

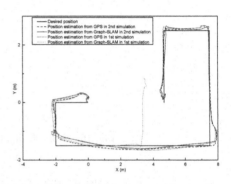

**Fig. 16.** 2D position of the quad-rotor UAV simulate in HILS system

**Fig. 13.** 3D mapping of the quad-rotor UAV simulate in HILS system without the optimization

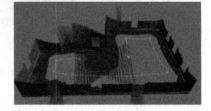

**Fig. 14.** 3D mapping of the quad-rotor UAV simulate in HILS system with the optimization

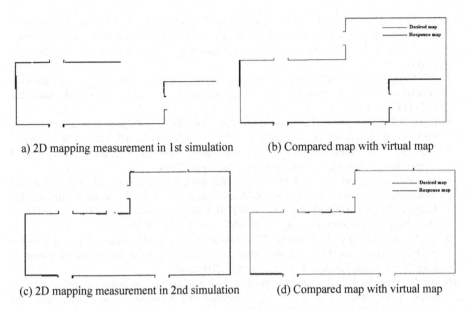

a) 2D mapping measurement in 1st simulation     (b) Compared map with virtual map

(c) 2D mapping measurement in 2nd simulation     (d) Compared map with virtual map

**Fig. 15.** Cross section of the quad-rotor UAV in two modes

## 7 Conclusion

This paper presents a HILS configuration for testing the Graph-SLAM algorithm which is performed on a quadrotor UAV. Herein, the Gazebo is used to present the dynamic model, sensor model, RGB-D camera model for the quad-rotor UAV. Meanwhile, the PX4 source code is developed for a tracking position control of the quadrotor UAV. Specially, the Graph-SLAM is developed to make the 3D mapping of environment and localization of UAV in map. As a result, the Graph-SLAM algorithm had high performance with the small error. HILS could provide a comprehensive to test any algorithms for a quadrotor UAV.

**Acknowledgments.** This work was supported by the KHNP (Korea Hydro & Nuclear Power Co., Ltd) Research Fund Haeorum Alliance Nuclear Innovation Center of Ulsan University.

## References

1. Al-Zentani, D.M., Zerek, A., Elmelhi, A.M., Academy, I.: Design of a flight control system based on HILS test platform. In: International Conference on Control Engineering & Information Technology, Tunisia (2017)
2. Jeonghoon, K., Yunsick, S.: Autonomous UAV flight control for GPS-based navigation. IEEE Access **6**, 37947–37955 (2018)
3. Duan, H., Zhang, Q.: Visual measurement in simulation environment for vision based UAV autonomous aerial refueling. IEEE Trans. Instrum. Meas. **64**, 2468–2480 (2015)

4. Mingu, K., Daewon, L., Jaemann, P., Chulwoo, P., Hyoun, J.K., Youdan, K.: Vision-based hardware-in-the loop simulation test of vision based net recovery for fixed wing unmanned aerial vehicle. In: Third Asia-Pacific International Symposium on Aero-space Technology (2011)
5. Trilaksono, B.R., Triadhitama, R., Adiprawita, W., Wibowo, A.: Hardware-in-the-loop simulation for visual target tracking of octorotor UAV. Aircr. Eng. Aerosp. Technol. **83**, 407–419 (2011)
6. Gans, N.R., Dixon, W.E., Lind, R., Kurdila, A.: A hardware in the loop simulation platform for vision-based control of unmanned air vehicles. Mechatronics **19**, 1043–1056 (2009)
7. Wei, Q., et al.: Manipulation task simulation using ROS and Gazebo. In: International Conference on Robotics and Biomimetics, Indonesia (2014)
8. Meyer, J., Sendobry, A., Kohlbrecher, S., Klingauf, U., von Stryk, O.: Comprehensive simulation of quadrotor UAVs using ROS and Gazebo. In: Noda, I., Ando, N., Brugali, D., Kuffner, J.J. (eds.) SIMPAR 2012. LNCS (LNAI), vol. 7628, pp. 400–411. Springer, Heidelberg (2012). https://doi.org/10.1007/978-3-642-34327-8_36
9. Odelga, M., Stegagno, P., Bülthoff, H.H., Ahmad, A.: A setup for multi-UAV hard-ware-in-the-loop simulations. In: 2015 Workshop on Research, Education and Development of Unmanned Aerial Systems, Mexico, pp. 204–210 (2015)
10. Bu, Q., Wan, F., Xie, Z., Ren, Q., Zhang, J., Liu, S.: General simulation platform for vision based UAV testing. In: 2015 IEEE International Conference on Information and Automation, China, pp. 2512–2516 (2015)
11. Khoa, D.N., Cheolkeun, H.: Development of hardware-in-the-loop simulation based on Gazebo and Pixhawk for unmanned aerial vehicles. Int. J. Aeronaut. Space Sci. **19**(1), 238–249 (2018)
12. Nguyen, K.D., Ha, C.: Vision-based hardware-in-the-loop-simulation for unmanned aerial vehicles. In: Huang, D.-S., Bevilacqua, V., Premaratne, P., Gupta, P. (eds.) ICIC 2018. LNCS, vol. 10954, pp. 72–83. Springer, Cham (2018). https://doi.org/10.1007/978-3-319-95930-6_8
13. Jithu, G., Jayasree, P.R.: Quadrotor modelling and control. In: International Conference on Electrical, Electronics, and Optimization Techniques, India (2016)
14. Fernando, H.C.T.E., De Silva, A.T.A., De Zoysa, M.D.C., Dilshan, K.A.D.C., Munasinghe, S.R.: Modelling, simulation and implementation of a quadrotor UAV. In: International Conference on Industrial and Information Systems, Sri Lanka (2013)
15. Christian, B.A., Ruth, P.J.L., Diogenes, A.D.P.: Position estimation using inertial measurement unit (IMU) on a quadcopter in an enclosed environment. Int. J. Comput. Commun. Instrum. Eng. **3**, 1477–2349 (2016)
16. Seong-Hoon, W., William, M., Farid, G.: Position and orientation estimation using Kalman filtering and particle diltering with one IMU and one position sensor. In: 34th Annual Conference of IEEE Industrial Electronics, USA (2008)
17. Bay, H., Tuytelaars, T., Van Gool, L.: SURF: speeded up robust features. In: Leonardis, A., Bischof, H., Pinz, A. (eds.) ECCV 2006. LNCS, vol. 3951, pp. 404–417. Springer, Heidelberg (2006). https://doi.org/10.1007/11744023_32
18. Lan-Rong, D., Chang-Min, H., Yin-Yi, W.: Implementation of RANSAC algorithm for feature-based image registration. J. Comput. Commun. **1**, 46–50 (2013)
19. Ying, H., Bin, L., Jun, Y., Shunzhi, L., Jin, H.: An iterative closest points algorithm for registration of 3D laser scanner point clouds with geometric features. Sensors **17**, 1862 (2017)
20. Grisetti, G., Kummerle, R., Stachniss, C., Burgard, W.: A tutorial on graph-based SLAM. IEEE Trans. Intell. Transp. Syst. Mag. **2**(4), 31–43 (2010)

21. Labbé, M., Michaud, F.: Long-term online multi-session graph-based SPLAM with memory management. Auton. Robots **42**(6), 1133–1150 (2018)
22. Labbé, M., Michaud, F.: Appearance-based loop closure detection for online large-scale and long-term operation. IEEE Trans. Rob. **29**(3), 734–745 (2013)

# Evaluating CNNs for Military Target Recognition

Xiao Ding$^{(\boxtimes)}$, Lei Xing, Tao Lin, Jiabao Wang,
Yang Li, and Zhuang Miao

Army Engineering University of PLA, Nanjing 210007, China
768519341@qq.com, 462201660@qq.com, 756920886@qq.com,
jiabao_1108@163.com, solarleeon@outlook.com,
emiao_beyond@163.com

**Abstract.** Convolutional neural network (CNN) is an efficient algorithm in deep learning. Aiming at the field of military target recognition, this paper constructs a dataset for military target recognition in battlefield, which contains ten kinds of targets. The characteristics of the dataset are described and analyzed. Three classical CNN models (AlexNet, VGGNet and ResNet) and two learning strategies (dropout and data augmentation) are evaluated on the dataset. Under the same condition of the dataset and the same super-parameter setting, the effects of different models are presented and analyzed. The experimental results show that the mean average precisions of ResNet and VGGNet are better than AlexNet, and the accuracies of both ResNet and VGGNet are over 90% with only thousands of training images. At the same time, the dropout and data augmentation strategies have a strong effect for improving the performance.

**Keywords:** Military target recognition · Convolutional neural network · Dropout · Data augmentation

## 1 Introduction

With the rapid development of science and technology, the form of war is also undergoing a profound revolution. In the future, a war will gradually become unmanned. So target recognition is very important for unmanned combat. Unmanned systems, such as UAV reconnaissance, vehicle-borne, air-borne and ship-borne, will play an increasingly important role. Military target recognition becomes the key technique in early warning detection, precision guidance, battlefield reconnaissance, identification of enemy or friend. High maneuverability, strong confrontation and harsh environment in modern warfare will result in blurring, occlusion, background interference and other additional factors, such as target distortion, and similar background caused by interference camouflage. It is very important to study how to identify military targets with the above problems.

In recent years, Convolutional Neural Network (CNN) [1] has become an effective technical model for computer vision target recognition. The end-to-end structure and its performance exceed that of traditional algorithms. However, the technology of convolutional neural network for target recognition is currently mainly used in the civil

© Springer Nature Switzerland AG 2019
D.-S. Huang et al. (Eds.): ICIC 2019, LNCS 11644, pp. 628–638, 2019.
https://doi.org/10.1007/978-3-030-26969-2_59

field. In this paper, we apply CNN for classification of tank, artillery and low altitude helicopter on battlefield.

## 2   Methods

### 2.1   Convolutional Neural Networks

**AlexNet.** AlexNet [2] is the first large-scale neural network designed by Alex for the ImageNet Image Classification Competition in 2012 [3]. It won the Champion ship with 57.1% accuracy. The model has five convolution layers and three pooling layers. After the fifth convolution layer, two full connection layers are connected, and finally input to the end of the softmax classifier for classification. In this paper, we fine-tune AlexNet on the new constructed military dataset. As the two full-connection layers in AlexNet has too many parameters and are easy to over-fitting. So we adopt the dropout strategy, which can suppress over-fitting problem.

**VGGNet.** VGGNet [4] is the representative network of deep learning in 2014. In order to solve the problem of parameter initialization, VGGNet adopts a pre-training mode, which is often seen in classical neural networks. It trains a part of the small network first, and then increase the network based on the small network. VGGNet is also composed of convolution layer and full connection layer. In the convolution part, VGGNet uses a convolution of size 3 × 3 and stride 1. The full connection layers and the classifier are similar to AlexNet. VGG has several variants for different demands of application. In this paper, VGG-16 is used.

**ResNet.** ResNet [5] has made great progress in depth and accuracy compared with AlexNet and VGGNet. Based on the classical convolution operation, ResNet introduces skip-connections, which convert convolution neural network into residual network. When the input and output have the same dimension, the skip-connection is used. ResNet-50 [5] has a depth of 175 layers, including 52 convolution layers and 2 pooling layers. The two pooling layers are located behind the first convolution layer and before the full connection layer respectively. Different from AlexNet and VGGNet, ResNet has batch normalization (BN) [6, 7], which can improve the gradient vanishing problem and speed up the convergence of optimization for learning.

### 2.2   Fine-Tune and Learning Strategy

**Fine-tune.** The CNN networks always have lots of parameters, which is hard to learn. In practice, to learning a CNN network from scratch needs large-scale dataset and high-efficient computing resource. However, the military target dataset used in this paper only has nearly 10,000 images, and training is easy over-fitting, so we use the fine-tune technology [8] based on the ImageNet pre-trained models. The military target dataset has different number of classes, compared with ImageNet dataset. When fine-tune is applied, the full-connection layers and softmax layer of the original network are removed, and a new convolution layer and softmax layer are replaced.

**Dropout.** Dropout [9] refers to the random drop of the nodes of a layer in the network in training, and the dropped nodes may temporarily be considered not to be part of the network. When dropout is used to train a network, instead of using L2 norm as penalty, an upper limit is set for each hidden node's L2 norm. In the training process, if the node does not satisfy the upper limit constraints, the upper limit is used to normalize the weights, so that the initial update of the weights has a larger learning rate.

**Data Augmentation.** Data augmentation [10, 11] aims to suppress over-fitting by increasing the amount of data. Through geometric transformation of image, one or more combinational transforms are used to increase the amount of input data. Typical data conversion includes region cropping, scale transformation, stretching or scaling to a fixed size and horizontal flip. It should be noted that the data augmentation is suitable for training and testing images. In the training phase, data augmentation can generate additional training samples, which can reduce the over-fitting effect caused by less training data. In the testing phase, data augmentation helps to improve classification accuracy. The augmented samples can be regarded as independent images, or as a separate representation formed by pooling or overlapping operations.

In this experiment, we randomly crop a sub-region from the image in accordance with the input size of the network model at each iteration, so that each training sample remains random, so as to achieve the effect of data augmentation.

## 3   Data Set

The military target recognition dataset is constructed by collecting images from the Internet. The images are collected from the popular image search engines, such as Google, Bing and Baidu. There are 10 types of military targets including jeep (JPC), military truck (JYKC), armored vehicle (ZJC), early warning aircraft (YJJ), fighter aircraft (ZDJ), helicopter (ZSJ), cannon howitzer (JNLDP), rocket gun (HJP), mortar (PJP), battle tank (ZZTK). Figure 1 shows some examples of 10 types of military targets in the dataset.

**Fig. 1.** Examples of 10 types of military targets

The dataset contains 9055 images. For each category, the number of examples varies from 250 to 1200, and most of the categories have about 1000 images. The number of images in each class is shown in Fig. 2.

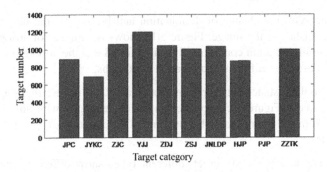

**Fig. 2.** The numbers of each class

## 3.1 Characteristics

CNN needs a large number of examples to ensure the convergence of learning, and requires the model to have good adaptability to various complex variations. When we collect the images of different types of military targets, we collect them with different viewpoints, illuminations, scales and complex background. Figure 3. shows various examples with different viewpoints, illumination, scale, type, background interference and so on.

(a)

(b)

(c)

(d)

(e)

**Fig. 3.** Images with different variations

**Viewpoints.** With different captured viewpoints, the same target shows different characteristics in the appearance. Figure 3(a) shows the images of military trucks taken from different viewpoints. It can be seen from the images that they show great differences in different perspectives.

**Illumination.** With the change of illumination and brightness, the target shows different characteristics in the image. Figure 3(b) shows the images of armored vehicles under different illumination conditions. It can be found that the armored vehicles show great differences under different illumination conditions.

**Scales.** Due to the distance between camera and target, the same target shows different scales in the image. Figure 3(c) shows examples of helicopter with different scales. It can be seen from the figure that helicopter and other targets show great differences in different scales.

**Types.** For the same class of target, different types show different appearance and characteristics. Figure 3(d) shows the images of jeeps of different types. It can be found that jeeps show great differences with different types.

**Complex Background.** Military targets in complex background show different characteristics than those in normal circumstances. Figure 3(e) shows the image of battle tank and helicopter in complex background environment. It can be seen that the battle tank and helicopter show great differences in complex background.

### 3.2 Application of Data

In order to compare the effect of the models effectively, we divide the dataset into three parts: training set (train), verification set (val) and test set (test). The specific division of various types of data is shown in Table 1.

**Table 1.** Division of dataset

| Category | #train | #val | #test | #Total |
|----------|--------|------|-------|--------|
| JPC | 374 | 90 | 425 | 889 |
| JYKC | 275 | 57 | 359 | 691 |
| ZJC | 431 | 118 | 512 | 1061 |
| YJJ | 481 | 121 | 600 | 1202 |
| ZDJ | 449 | 106 | 489 | 1044 |
| ZSJ | 404 | 105 | 497 | 1006 |
| JNLDP | 438 | 96 | 502 | 1036 |
| HJP | 360 | 74 | 434 | 868 |
| PJP | 99 | 26 | 132 | 257 |
| ZZTK | 392 | 99 | 510 | 1001 |
| Total | 3703 | 892 | 4460 | 9055 |

The division mainly takes into account that the parameter learning of convolutional neural networks requires a large number of training samples to ensure the robustness of the model, so the proportion of training sets is relatively large.

As the input image of convolution neural network requires a fixed scale and depth, all images are preprocessed, and the scale of the image is normalized to a uniform size. Each image is resized to $256 \times 256$ in the experiment. Meanwhile, the gray image is transformed into an image of three channels.

## 4  Experiment

### 4.1  Implementation

Experiments are conducted based on MatConvNet[1]. We use a workstation of Lenovo D30, with 2.6 GHz CPU. We use NVIDIA Tesla K20c GPU to accelerate training.

The parameters of the networks are initialized by random numbers distributed evenly on [-0.01, 0.01]. The learning process has 25 epochs, and the learning rate is initialized to 0.001, and decreases to 0.0001 at 20 epochs. This makes the gradient bigger in the early stages and the convergence speed faster in the later stages. In experiments, the accuracy and mean average precision (MAP) [12] is used to evaluate the performance of different methods.

### 4.2  Experimental Results and Analysis

With fine-tune technology, the results of AlexNet, VGGNet and ResNet are shown in Table 2, where the accuracy and MAP performance of 10 types of military targets.

**Table 2.**  Experimental results

| Model | JPC | JYKC | ZJC | YJJ | ZDJ | ZSJ | HJP | JNLDP | PJP | ZZTK | AVG |
|-------|-----|------|-----|-----|-----|-----|-----|-------|-----|------|-----|
| AlexNet | 0.87 | 0.80 | 0.82 | 0.93 | 0.94 | 0.91 | 0.86 | 0.84 | 0.84 | 0.77 | 0.86 |
| VGGNet | 0.93 | 0.80 | 0.87 | 0.96 | 0.96 | 0.93 | 0.91 | 0.90 | 0.89 | 0.84 | 0.90 |
| ResNet | 0.96 | 0.87 | 0.86 | 0.95 | 0.98 | 0.95 | 0.91 | 0.89 | 0.89 | 0.84 | 0.91 |

Table 2 gives the experimental results of AlexNet, VGGNet and ResNet for classification of ten kinds of targets. From Table 2, we can find that:

(1) ResNet achieves the highest MAP with 0.91, while AlexNet has the worst performance. VGGNet has similar performance with ResNet.
(2) For VGGNet and ResNet, the accuracy of each class in the dataset is more than 0.8. The MAPs of JPC, YJJ, ZDJ, ZSJ and HJP are more than 0.9. For AlexNet, the accuracy of ZZTK is only 76.9%, which has a large gap from ZDJ.

---

[1]  http://www.vlfeat.org/matconvnet/.

To further analyze the performance of three models, we present the classification confusion matrices of AlexNet, VGGNet and ResNet in Figs. 4, 5 and 6 respectively. The results are presented in percent.

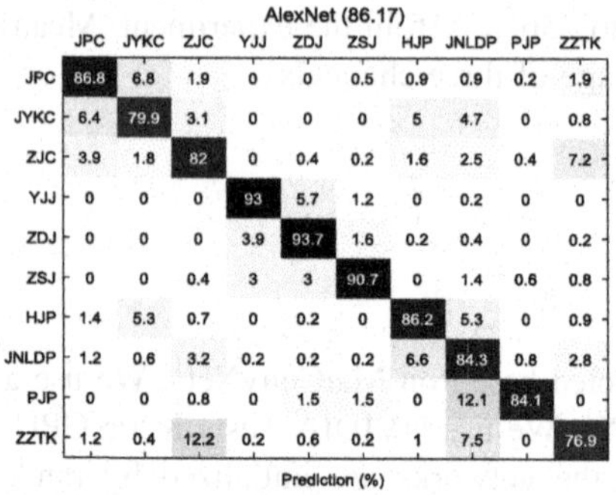

**AlexNet (86.17)**

|       | JPC  | JYKC | ZJC  | YJJ  | ZDJ  | ZSJ  | HJP  | JNLDP | PJP  | ZZTK |
|-------|------|------|------|------|------|------|------|-------|------|------|
| JPC   | 86.8 | 6.8  | 1.9  | 0    | 0    | 0.5  | 0.9  | 0.9   | 0.2  | 1.9  |
| JYKC  | 6.4  | 79.9 | 3.1  | 0    | 0    | 0    | 5    | 4.7   | 0    | 0.8  |
| ZJC   | 3.9  | 1.8  | 82   | 0    | 0.4  | 0.2  | 1.6  | 2.5   | 0.4  | 7.2  |
| YJJ   | 0    | 0    | 0    | 93   | 5.7  | 1.2  | 0    | 0.2   | 0    | 0    |
| ZDJ   | 0    | 0    | 0    | 3.9  | 93.7 | 1.6  | 0.2  | 0.4   | 0    | 0.2  |
| ZSJ   | 0    | 0    | 0.4  | 3    | 3    | 90.7 | 0    | 1.4   | 0.6  | 0.8  |
| HJP   | 1.4  | 5.3  | 0.7  | 0    | 0.2  | 0    | 86.2 | 5.3   | 0    | 0.9  |
| JNLDP | 1.2  | 0.6  | 3.2  | 0.2  | 0.2  | 0.2  | 6.6  | 84.3  | 0.8  | 2.8  |
| PJP   | 0    | 0    | 0.8  | 0    | 1.5  | 1.5  | 0    | 12.1  | 84.1 | 0    |
| ZZTK  | 1.2  | 0.4  | 12.2 | 0.2  | 0.6  | 0.2  | 1    | 7.5   | 0    | 76.9 |

Prediction (%)

**Fig. 4.** Confusion matrix of AlexNet

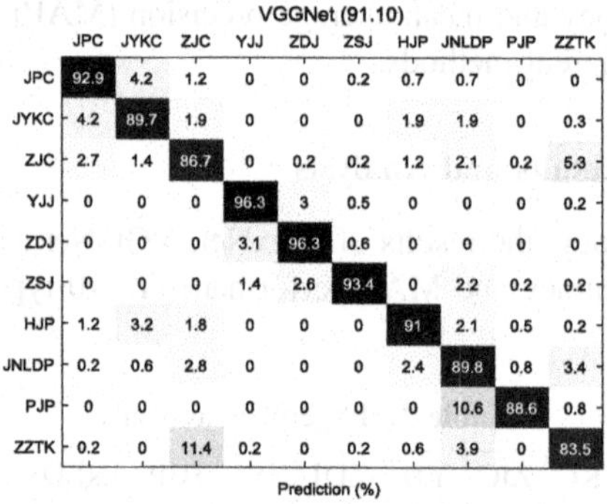

**VGGNet (91.10)**

|       | JPC  | JYKC | ZJC  | YJJ  | ZDJ  | ZSJ  | HJP  | JNLDP | PJP  | ZZTK |
|-------|------|------|------|------|------|------|------|-------|------|------|
| JPC   | 92.9 | 4.2  | 1.2  | 0    | 0    | 0.2  | 0.7  | 0.7   | 0    | 0    |
| JYKC  | 4.2  | 89.7 | 1.9  | 0    | 0    | 0    | 1.9  | 1.9   | 0    | 0.3  |
| ZJC   | 2.7  | 1.4  | 86.7 | 0    | 0.2  | 0.2  | 1.2  | 2.1   | 0.2  | 5.3  |
| YJJ   | 0    | 0    | 0    | 96.3 | 3    | 0.5  | 0    | 0     | 0    | 0.2  |
| ZDJ   | 0    | 0    | 0    | 3.1  | 96.3 | 0.6  | 0    | 0     | 0    | 0    |
| ZSJ   | 0    | 0    | 0    | 1.4  | 2.6  | 93.4 | 0    | 2.2   | 0.2  | 0.2  |
| HJP   | 1.2  | 3.2  | 1.8  | 0    | 0    | 0    | 91   | 2.1   | 0.5  | 0.2  |
| JNLDP | 0.2  | 0.6  | 2.8  | 0    | 0    | 0    | 2.4  | 89.8  | 0.8  | 3.4  |
| PJP   | 0    | 0    | 0    | 0    | 0    | 0    | 0    | 10.6  | 88.6 | 0.8  |
| ZZTK  | 0.2  | 0    | 11.4 | 0.2  | 0    | 0.2  | 0.6  | 3.9   | 0    | 83.5 |

Prediction (%)

**Fig. 5.** Confusion matrix of VGGNet

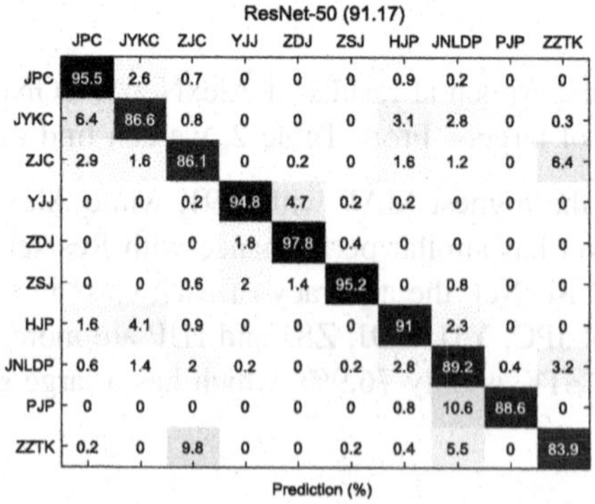

**ResNet-50 (91.17)**

|       | JPC  | JYKC | ZJC  | YJJ  | ZDJ  | ZSJ  | HJP  | JNLDP | PJP  | ZZTK |
|-------|------|------|------|------|------|------|------|-------|------|------|
| JPC   | 95.5 | 2.6  | 0.7  | 0    | 0    | 0    | 0.9  | 0.2   | 0    | 0    |
| JYKC  | 6.4  | 86.6 | 0.8  | 0    | 0    | 0    | 3.1  | 2.8   | 0    | 0.3  |
| ZJC   | 2.9  | 1.8  | 86.1 | 0    | 0.2  | 0    | 1.6  | 1.2   | 0    | 6.4  |
| YJJ   | 0    | 0    | 0.2  | 94.8 | 4.7  | 0.2  | 0.2  | 0     | 0    | 0    |
| ZDJ   | 0    | 0    | 0    | 1.8  | 97.8 | 0.4  | 0    | 0     | 0    | 0    |
| ZSJ   | 0    | 0    | 0.6  | 2    | 1.4  | 95.2 | 0    | 0.8   | 0    | 0    |
| HJP   | 1.8  | 4.1  | 0.9  | 0    | 0    | 0    | 91   | 2.3   | 0    | 0    |
| JNLDP | 0.6  | 1.4  | 2    | 0.2  | 0    | 0.2  | 2.8  | 89.2  | 0.4  | 3.2  |
| PJP   | 0    | 0    | 0    | 0    | 0    | 0    | 0.8  | 10.6  | 88.6 | 0    |
| ZZTK  | 0.2  | 0    | 9.8  | 0    | 0    | 0.2  | 0.4  | 5.5   | 0    | 83.9 |

Prediction (%)

**Fig. 6.** Confusion matrix of ResNet

Confusion matrices show the detailed classification accuracy of ten types of military target in the test set. The value of 0 in the matrix indicates that there is no confusion between the corresponding target classes. Non-zero (except for black area) indicates that there is a misclassification between the corresponding categories, and black diagonal represents the accuracy of classification. From the figures, it can be seen that each class of military target has been misclassified, and the misclassification rates of different types of targets are different. The results of three models consistently show that battle tank (ZZTK) is always misclassified as armored vehicle (ZJC). The reason is that different types of military targets with high similarity of features can easily be misclassified, and the predicted results directly affect the accuracy.

Figure 7 gives the misclassified examples of ResNet. In Fig. 7, the truth category is in bracket and the prediction category is outside bracket. From these misclassified examples, it can be found that the main reason of misclassification lies in the similarity of features between categories of military targets. For example, the image of the early warning aircraft (YJJ) is very similar to that of the fighter aircraft in some angles and complex background. The images taken by the armored vehicle (ZJC) and the battle tank (ZZTK) from different angles also have similar characteristics and are easy to be misclassified. In the process of ResNet model prediction, a total of 394 examples are misclassified.

**Fig. 7.** Misclassified examples of ResNet

### 4.3    Analysis of Learning Strategies

In order to reduce the over-fitting problem caused by fewer training examples, the dropout and data augmentation are adopted in experiments.

The experiment compares the classification accuracy of ten kinds of targets and the MAP. In the experiment, drop rate is set to 0 and 0.7, and the random cropping is used for data augmentation. The accuracy of target classification is measured in percentage (%).

Figures 8, 9 and 10 shows the impact of dropout and data augmentation based on AlexNet, VGGNet and ResNet. The classification accuracy of ten classes and the overall MAP are presented.

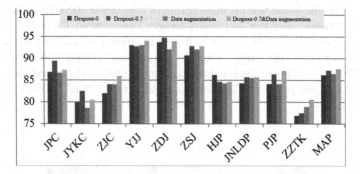

**Fig. 8.**  Comparison of dropout and data augmentation strategy by AlexNet

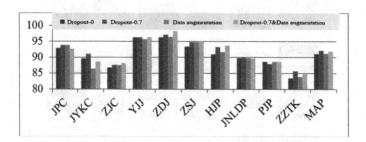

**Fig. 9.**  Comparison of dropout and data augmentation strategy by VGGNet

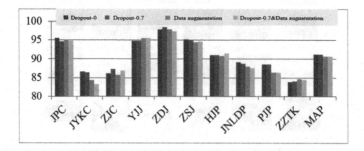

**Fig. 10.**  Comparison of dropout and data augmentation strategy by ResNet

From Figs. 8, 9 and 10, it can be found that:

(1) The MAP of AlexNet, VGGNet and ResNet achieves 87.26%, 92.11% and 91.21% respectively when dropout rate was 0.7, which have 1.09%, 1.01% and 0.04% higher than that without dropout strategy. After data augmentation, the MAPs of AlexNet, VGGNet and ResNet models achieves 86.41%, 91.21% and 90.76% respectively. The first two results have 0.24% and 1.01% higher than those of AlexNet, VGGNet, while the MAP of ResNet reduces by 0.41%.

(2) Compared with the data augmentation strategy, dropout has better effect on suppressing over-fitting when the scale parameter is 0.7. However, when both dropout and data augmentation technology are used to suppress over-fitting, only AlexNet has better accuracy than that of dropout or data augmentation.

(3) With data augmentation, the MAPs of AlexNet and VGGNet increased by 1.43% and 0.83%, respectively, while that of ResNet decreased by 0.41%. The reason is that the data augmentation used in this paper is relatively simple (only random cropping). Although the number of examples is increased, the prediction accuracy can't be improved.

## 5 Conclusion

This paper mainly focuses on the evaluation of military target recognition by convolutional neural networks. A dataset for military target recognition is constructed. The characteristics of the dataset are described and analyzed. Three classical CNN models (AlexNet, VGGNet and ResNet) and two learning strategies (dropout and data augmentation) are evaluated on the dataset. Under the same condition of the dataset and the same super-parameter setting, the effects of different models are presented and analyzed. The experimental results show that the mean average precisions of ResNet and VGGNet are better than AlexNet, and the accuracies of both ResNet and VGGNet are over 90% with only thousands of training images. Besides, the dropout and data augmentation strategies have a strong effect for improving the performance.

Based on above research, it can be concluded that CNN achieves great performance for military target recognition. When the number of images is not large-scale, fine-tuning is a good choice. Besides, the dropout and data augmentation can improve the classification accuracy by suppressing over-fitting.

## References

1. LeCun, Y., Bengio, Y., Hinton, G.: Deep learning. Nature **521**(7553), 436–444 (2015)
2. Krizhevsky, A., Sutskever, I., Hinton, G.: Imagenet classification with deep convolutional neural networks. Adv. Neural. Inf. Process. Syst. **25**, 1097–1105 (2012)
3. Simonyan, K., Zisserman, A.: Very deep convolutional networks for large-scale image recognition. CoRR abs/1409.1556 (2014)
4. Russakovsky, O., Deng, J., Su, H., et al.: Imagenet large scale visual recognition challenge. Int. J. Comput. Vis. **115**(3), 211–252 (2015)

5. He, K., Zhang, X., Ren, S., et al.: Deep residual learning for image recognition. CoRR abs/1512.03385 (2015)
6. Ioffe, S., Szegedy, C.: Batch normalization: accelerating deep network training by reducing internal covariate shift. In: Proceedings of 32th International Conference on Machine Learning, pp. 448–456 (2015)
7. Li, Y., Wang, N., Shi, J., et al.: Adaptive batch normalization for practical domain adaptation. Pattern Recogn. **80**, 109–117 (2016)
8. Yosinski, J., Clune, J., Bengo, Y., et al.: How tansferable are features in deep neural networks? Adv. Neural. Inf. Process. Syst. **27**, 3320–3328 (2014)
9. Srivastava, N., Hinton, G., Krizhevsky, A., et al.: Dropout: a simple way to prevent neural networks from overfitting. J. Mach. Learn. Res. **15**(1), 1929–1958 (2014)
10. Perez, L., Wang, J.: The effectiveness of data augmentation in image classification using deep learning. CoRR abs/1712.04621 (2017)
11. Sajjad, M., Khan, S., Muhammad, K., et al.: Multi-grade brain tumor classification using deep CNN with extensive data augmentation. J. Comput. Sci. **30**, 174–182 (2019)
12. Zheng, L., Shen, L., Tian, L., et al.: Scalable person re-identification: a benchmark. In: IEEE International Conference on Computer Vision, pp. 1116–1124 (2015)

# Detection and Segmentation of Kidneys from Magnetic Resonance Images in Patients with Autosomal Dominant Polycystic Kidney Disease

Antonio Brunetti[1], Giacomo Donato Cascarano[1], Irio De Feudis[1],
Marco Moschetta[2], Loreto Gesualdo[2],
and Vitoantonio Bevilacqua[1(✉)]

[1] Department of Electrical and Information Engineering,
Polytechnic University of Bari, Bari, Italy
vitoantonio.bevilacqua@poliba.it
[2] Department of Emergency and Organ Transplants,
University of Bari Medical School, Bari, Italy

**Abstract.** The segmentation of kidneys from Magnetic Resonance (MR) images of subjects affected by Autosomal Dominant Polycystic Kidney Disease (ADPKD) is a task of fundamental importance as it allows a non-invasive assessment and monitoring of the disease over time. In this work, a fully automated procedure based on Convolutional Neural Networks (CNNs) is proposed for detecting images containing the kidney and subsequently segment them classifying each pixel. Specifically, a mono-objective genetic algorithm was designed for optimising the CNN architecture, modelling the number of encoders, the structure of each encoder and the final fully-connected layers. The input dataset for the classification task included 526 MR images: 366 containing kidney and 160 did not include any pixel of the kidney. All the images containing the kidney were split into left and right side and used as input dataset for the segmentation procedure. For both the tasks, the training set, validation set and test set were generated according to the 60-20-20 percentages. Accuracy higher than 95% for the classification task and 90% for the segmentation algorithm show the reliability of the proposed approach, also improving the results of previous works.

**Keywords:** Convolutional Neural Networks · Classification · Segmentation · Genetic Algorithm · ADPKD · Polycystic Kidney

## 1 Introduction

The Autosomal Dominant Polycystic Kidney Disease (ADPKD) is one of the most common genetic disorders primarily affecting the kidneys. This disease is characterised by the onset of cysts into the kidneys, slowly increasing in both volume and number, causing a progressive increase of the Total Kidney Volume (TKV) [1, 2]. Also, it is a systemic disease affecting other organs near the kidney itself. Thus, cysts might grow in

© Springer Nature Switzerland AG 2019
D.-S. Huang et al. (Eds.): ICIC 2019, LNCS 11644, pp. 639–650, 2019.
https://doi.org/10.1007/978-3-030-26969-2_60

other organs, such as liver and pancreas, and the cardiovascular system may suffer from hypertension and aneurysms [3, 4]. As the genetic test for diagnosing ADPKD is demanding and expensive, the screening of risky subjects includes diagnostics imaging acquisitions, such as Ultra-sonography (US), Computed Tomography (CT) or Magnetic Resonance (MR). Choosing the correct imaging technique for investigating the disease is essential since, in some cases, it can avoid the diagnostic genetic test [5]. Although the screening and early detection are crucial phases for risky subjects, assessing the disease progression over time is fundamental for people affected by ADPKD. The Total Kidney Volume (TKV) estimation is the most used indicator for assessing the evolution of the disease.

According to the literature, several methods allow performing the TKV estimation [6–8]. The commonly used approaches include the ellipsoid-equations based method, the stereology and the manual segmentation [9]. The first method approximates the kidney to an ellipsoid and computes its volume based on the length, width and depth obtained analysing the acquired images [10]. Stereology consists in the superimposition of a grid, with specific textures, on each slice of the volume; the area of the kidney is obtained merely counting all the minimal regions of the texture containing parts of the kidney to get the bi-dimensional area on each slice; the final volume is then obtained considering the thickness of the slices by means of interpolation. Manual segmentation, instead, requires the manual contouring of the kidney regions contained in each slice; like the previous approach, the final volume is obtained by interpolating all the contoured areas considering the thickness of the slices. Although the manual segmentation approach allows obtaining a more accurate estimation of the TKV than the other methods, it requires a longer and more tedious work. On the contrary, the other methods require a minimum interaction to obtain the volume estimation [10–13]. Like stereology and manual segmentation, all the recent approaches for the semi-automatic or automatic TKV computation perform image processing and segmentation algorithms operating on each slice; the final model of the kidney is then reconstructed for the estimation of its volume [14–18]. Despite all the reported works dealing with this task, the segmentation of polycystic kidneys with the support of automatic systems based on image processing strategies is still a tricky task, especially using MR imaging [19].

In recent years, the introduction of strategies based on Deep Learning (DL) for the classification and segmentation tasks allowed designing Computer-Aided Diagnosis (CAD) systems overcoming the limitations due to previous technologies, leading to higher performance [20–23]. These advantages make DL approaches useful in several fields, from computer vision tasks to medical imaging [24, 25]. In fact, Deep Neural Networks (DNNs) and Convolutional Neural Networks (CNNs) have been employed for different aims in medical imaging procedures, including image classification, automatic detection of Regions Of Interest (ROIs) and semantic segmentation.

In a previous work, the authors investigated several DL approaches for segmenting the kidney from MR images [26]. In details, the authors firstly evaluated different CNN architectures for segmenting the whole MR series of images; then, they explored the Regions with CNN (R-CNN) classifiers to prior detect ROIs containing the kidney, to which apply a subsequent segmentation algorithm.

In this work, a fully-automated pipeline for the detection and segmentation of areas containing kidney in MR images from subjects affected by ADPKD has been designed and developed. It includes a former classification step for the detection of images containing the kidney and a subsequent segmentation phase identifying the areas containing the pixels belonging to the kidney (Fig. 1). Specifically, since the detection of slices containing kidney is crucial for improving the segmentation performance in the subsequent phase, the design of the CNN architecture to perform the classification task has been optimised through a mono-objective Genetic Algorithm finding the optimal subset of parameters for defining the different layers of the CNN classifier. Based on the architecture obtained for the classification task, a CNN for the segmentation step have been designed opportunely creating the respective decoding part of the network for reconstructing the input image.

This work is organized as follows: Sect. 2 reports the materials, in terms of patients, protocol for acquiring the images and their characteristics; Sect. 3 describes the classification approaches, including the image classification, the semantic segmentation, and the genetic algorithm designed for optimising the classification architecture; Sect. 4 reports the results obtained in classifying and segmenting the images, whereas Sect. 5 contains the final discussion of the work.

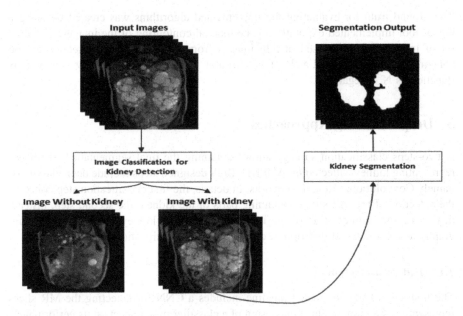

**Fig. 1.** Representation of the implemented automatic pipeline for the segmentation of images containing the kidney.

## 2   Materials

From February 2017 to July 2017, 18 patients affected by ADPKD (mean age: 33.4 years; range: 19–47) underwent Magnetic Resonance examinations for assessing the TVK. The acquisition protocol was carried out by the physicians from the Department of Emergency and Organ Transplantations (DETO) of the Bari University Hospital.

A 1.5 TMR device (Achieva, Philips Medical Systems, Best, The Netherlands) was used to perform MR examinations, by using a four-channel breast coil. The protocol did not use contrast material intravenous injection and consisted of:

- Transverse and coronal short TI inversion recovery (STIR) turbo-spin-echo (TSE) sequences (TR/TE/TI = 3.800/60/165 ms, field of view (FOV) = 250 × 450 mm (AP × RL), matrix 168 × 300, 50 slices with 3-mm slice thickness and without gaps, 3 averages, turbo factor 23, resulting in a voxel size of 1.5 × 1.5 × 3.0 mm$^3$; sequence duration of 4.03 min);
- Transverse and coronal T2-weighted TSE (TR/TE = 6.300/130 ms, FOV = 250 × 450 mm (AP x RL), matrix 336 × 600, 50 slices with 3-mm slice thickness and without gaps, 3 averages, turbo factor 59, SENSE factor 1.7, resulting in a voxel size of 0.75 × 0.75 × 3.0 mm$^3$; sequence duration of 3.09 min);
- Three-dimensional (3D) T1-weighted high resolution isotropic volume (THRIVE) sequence (TR/TE = 4.4/2.0 ms, FOV = 250 450 150 mm (AP × RL × FH), matrix 168 × 300, 100 slices with 1.5 mm slice thickness, turbo factor 50, SENSE factor 1.6, data acquisition time of 1 min 30 s).

The ground truth for evaluating the implemented algorithms was created by using a digital tool implemented explicitly for the manual contouring of the images. The final set of images was composed of 526 images from T2-weighted TSE sequence: 366 containing kidney (labelled as P - Positive) and 160 did not include any kidney parts (labelled as N - Negative).

## 3   Deep Learning Approaches

A two-steps classification strategy allows obtaining the final segmentation of images representing kidneys affected by ADPKD. Both designed steps include deep classifiers, namely Convolutional Neural Networks. In details, the first classification step concerns the detection of the MR slices containing parts of the kidney; the second step, instead, deals with the segmentation of the images showing the kidney. The following paragraphs report a detailed description of the designed classification approaches.

### 3.1   Kidney Detection

The first step of the automated pipeline includes a CNN for detecting the MR slices representing the kidney. Since the design of a classifier may impact on its performance, there are several strategies based on optimisation algorithms allowing to numerically search for the optimal set of parameters for its designing, configuration and tuning [27, 28]. A mono-objective genetic algorithm (GA) allowed designing an optimised topology of the CNN for the classification task [21, 29]. Each genotype described convolutional architectures composed of at least one encoder, up to three; each encoder included by up to three groups of the following operators: Bi-dimensional Convolution Layer (conv2d layer), Batch Normalisation Layer, Rectified Linear Unit (ReLU) layer.

The chromosome also modelled the number of kernels and the filter size for each convolutional layer. Finally, up to 2 fully connected layers (with a maximum of 512 neurons per layer) may precede the softmax layer for the final classification between the two classes. The training algorithm for all the CNNs was ADAM [30], showing little memory consumptions respect to other algorithms. In addition, there is a max pooling layer (filter size [4 4], stride [4 4]) to perform subsampling from one encoder to the subsequent. The training procedure also included the dataset augmentation for improving the overall classification performance and the generalisation capabilities of the classifier [20]. The evolutionary optimisation started from an initial random population of 100 individuals. The crossover probability was 0.8 whereas the mutation operator was an adaptive feasible mutation function randomly generating directions adaptive with respect to the last successful or unsuccessful generation (the mutation chooses a direction and step length that satisfies bounds and linear constraints). Figure 2 shows a complete representation of a candidate solution of the GA; Table 1 reports the representation of the CNN parameters included in the GA chromosome.

**Table 1.** Representation of the GA parameters for designing the CNN performing the image classification task.

| Network Configuration | | | |
|---|---|---|---|
| *Layers* | *GA Parameters* | | |
| | *Number of Layers* | *Number of Filters* | *Filter Dimension* |
| Image Input | – | – | – |
| Encoder 1 | [1–3] | [8–256] | [1–7] |
| Max-pooling | – | – | – |
| Encoder 2 | [0–3] | [8–256] | [1–7] |
| Max-pooling | – | – | – |
| Encoder 3 | [0–3] | [8–256] | [1–7] |
| Max-pooling | – | – | – |
| | *Number of Neurons* | | |
| Fully-Connected 1 | [0–512] | | |
| Fully-Connected 2 | [0–512] | | |
| Softmax | 2 | | |

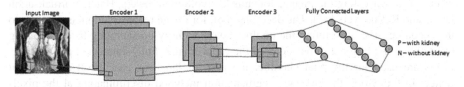

**Fig. 2.** The representation of a Convolutional Neural Network candidate solution of the Genetic Algorithm. Each encoder can have up to 3 sequences of bi-dimensional convolutional layers, batch normalisation layers and Rectified Linear Unit (ReLU) layers. Between two consecutive encoders, there is max pooling layer with kernel size [4 4].

Every individual into the search space was trained, validated and tested on a random permutation of the dataset dividing it into the training set, validation set and test set according to the 60-20-20 percentages. Each input image was then classified and labelled as Positive (P) if it contained at least one pixel belonging to the kidney, Negative (N) otherwise. The fitness function for evaluating the genotypes was the classification Accuracy (Eq. 1) reached considering the performance on the test set. Considering Eq. 1, TP, TN, FP and FN were the number of Positive images correctly classified as Positive, the number of Negative images correctly classified as Negative, the number of Negative images classified as Positive and the number of Positive images classified as Negative, respectively.

$$Accuracy = \frac{TP + TN}{TP + TN + FP + FN} \tag{1}$$

### 3.2 Kidney Segmentation

Like the classification step, a Convolutional Neural Network performed the segmentation of the images showing the kidneys, labelling each pixel of the input image [28, 31, 32].

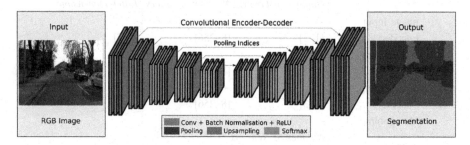

**Fig. 3.** Encoder-Decoder representation of a CNN for the semantic segmentation task (image from [33]).

As represented in Fig. 3, a CNN designed to perform semantic segmentation has an encoder-decoder structure, where each encoder, connected to the others by pooling layers for downscaling the image, contains Convolutional layers, Batch Normalisation layers and ReLu layers [33]. The decoding part, for image reconstruction, is specular to the encoding one, substituting the pooling layers with by up-sampling layers allowing to re-obtain the original dimensions of the input image. Fully-connected and softmax layers are used for the final classification and for labelling each pixel of the input image. In this work, the semantic segmentation included discriminating al the pixels between two classes: Kidney and Background (Fig. 4).

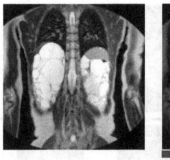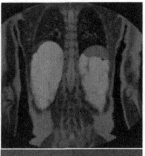

Kidney        Background

**Fig. 4.** The Ground Truth for image segmentation. (left) An MR image from subjects affected by ADPKD; histogram equalisation was performed for better visualisation. (right) The labelling of the considered slice.

For designing the CNN for segmentation purpose, the optimised CNN topology obtained in the previous step was replicated into the encoding section of the semantic classifier. The decoding part was also generated from it. All the images containing at least one pixel of the kidney class constituted the dataset of the semantic segmentation. Specifically, each image was split into a left and right part, increasing both the number and variety of the sample size. As for the image classification task, the input dataset was randomly divided into a training set, a validation set and a test set according to the 60-20-20 percentages. Figure 5 shows an example of the kidney segmentation result; the following sections, instead, discuss the final results of both the classification and segmentation algorithms.

## 4   Results

Regarding the first step of the automated pipeline, the input dataset included 526 images: 366 containing kidney (labelled as Positive - P), whereas 160 containing the background only (labelled as Negative - N). Although the fitness function of the GA considered only the Accuracy (Eq. 1), the overall performance evaluation took into account also the Sensitivity (Eq. 2) and Specificity (Eq. 3) measured on the test set.

The optimised individual found by the GA showed a topology constituted by three encoders: the first one included two convolutional layers (11 kernels with size [2 2]); the second encoder included two convolutional layers (182 kernels with size [5 5]); the last encoder included one convolutional (32 kernels with size [5 5]). The two fully-connected layers included 56 and 232 neurons, respectively. Table 2 reports the confusion matrix of the classification results on the test set for the best genotype. As represented, 95.15% of Accuracy is obtained, whereas Sensitivity and Specificity reach 95.83% and 93.55%, respectively.

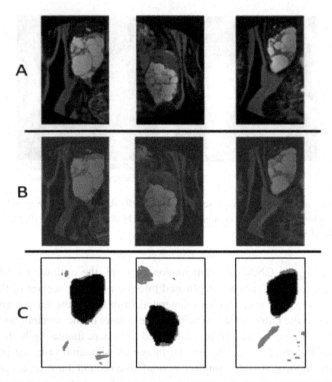

**Fig. 5.** The output of the kidney segmentation by CNN. (A) Some input images. (B) Super-imposition of the input image to the output of the segmentation: the red pixels are "kidney", whereas the others belong to the background. (C) Superimposition of the segmentation output (purple) and the ground truth obtained by manual segmentation of the kidney (green); the black pixels are the True Positives. (Color figure online)

$$Sensitivity(or\,Recall) = \frac{TP}{TP+FN} \tag{2}$$

$$Specificity = \frac{TN}{TN+FP} \tag{3}$$

**Table 2.** Confusion matrix computed on the test set for the best individual found by the genetic algorithm.

|  |  | True condition | |
|---|---|---|---|
|  |  | *Positive* | *Negative* |
| Predicted condition | *Positive* | 69 | 2 |
|  | *Negative* | 3 | 29 |

The input dataset for the segmentation task included 732 images (every image of the Positive class was split into left and right parts). As in [26], Accuracy (Eq. 1), Boundary F1 Score (Eq. 4), or BF Score, and the Jaccard Similarity Coefficient (Eq. 6), or Intersection over Union – IoU, were considered for evaluating the performance of the CNN on the test set. In detail, the BF score measures how close the predicted boundary of an object matches the ground truth boundary; it is defined as the harmonic mean of the Precision (Eq. 5) and Recall (Eq. 2) values [34]. The resulting score spreads in the range [0, 1]; the more BF score is close to 1, the more the boundaries of the segmented object and the ground truth match. The Jaccard Similarity Index, instead, is the ratio between the correctly classified pixels (TP) and the total number of pixels belonging to the positive class and the pixels wrongly predicted as positive (FP).

$$Boundary\ F1\ Score = \frac{2 * Precision * Recall}{Precision + Recall} \qquad (4)$$

$$Precision = \frac{TP}{TP + FP} \qquad (5)$$

$$Jaccard\ Similarity\ Index = \frac{TP}{TP + FP + FN} \qquad (6)$$

Table 3 reports the normalised confusion matrix of the classification performance for the segmentation task. Moreover, the overall performance, based on the previous metrics shows Accuracy reaching 91.06%, whereas IoU and BF Score reach 0.8296 and 0.5234, respectively.

**Table 3.** Normalised confusion matrix computed on the test set for the semantic segmentation.

|  |  | True condition | |
|---|---|---|---|
|  |  | Positive | Negative |
| Predicted Condition | Positive | 0.9441 | 0.1229 |
|  | Negative | 0.0559 | 0.8771 |

## 5  Discussion and Conclusion

In this work, a fully-automated procedure was designed in order to detect and segment MR images from people affected by ADPKD. Since the design of classifiers strongly influences their performance, an evolutionary approach allowed designing an optimised CNN topology, in terms of number of encoders, number of convolutional layers per encoder, number of filters and their dimension, for classification and segmentation purposes.

Regarding the classification strategy, the developed CNN obtained an Accuracy higher than 95% in detecting images containing the kidney. The segmentation approach, instead, reached an Accuracy higher than 90%, improving the overall performance obtained in previous work [28].

The obtained results show that employing classifiers based on Deep Learning strategies is a reliable approach for both automatic classification and segmentation of diagnostic images. Moreover, assessing ADPKD through MR reduces the invasiveness of the examination (no ionising radiation or contrast medium), maintaining high reliability of the analysis.

In the future, procedures working on the three-dimensional space of the MR series may also improve the classification and segmentation capabilities of the designed pipeline, by analysing the input at a voxel level rather than slice per slice. Performing even more accurate assessments of the TKV through imaging investigations, including MR not involving ionising radiation, could help the screening and monitoring over time of the disease, reducing both time and costs of more invasive diagnosis methodologies.

**Acknowledgements.** This study has been partially funded from the PON MISE 2014-2020 "HORIZON2020" program, project PRE.MED.

# References

1. Emamian, S.A., Nielsen, M.B., Pedersen, J.F., Ytte, L.: Kidney dimensions at sonography: correlation with age, sex, and habitus in 665 adult volunteers. AJR Am. J. Roentgenol. **160**, 83–86 (1993)
2. Grantham, J.J.: Autosomal dominant polycystic kidney disease. N. Engl. J. Med. **359**, 1477–1485 (2008)
3. Irazabal, M.V., et al.: Short-term effects of Tolvaptan on renal function and volume in patients with autosomal dominant polycystic kidney disease. Kidney Int. **80**, 295–301 (2011)
4. Bergmann, C., Guay-Woodford, L.M., Harris, P.C., Horie, S., Peters, D.J.M., Torres, V.E.: Polycystic kidney disease. Nat. Rev. Dis. Prim. **4**, 50 (2018)
5. Pei, Y., et al.: Imaging-based diagnosis of autosomal dominant polycystic kidney disease. J. Am. Soc. Nephrol. **26**, 746–753 (2015)
6. King, B.F., Reed, J.E., Bergstralh, E.J., Sheedy, P.F., Torres, V.E.: Quantification and longitudinal trends of kidney, renal cyst, and renal parenchyma volumes in autosomal dominant polycystic kidney disease. J. Am. Soc. Nephrol. **11**, 1505–1511 (2000)
7. Vauthey, J.-N., et al.: Body surface area and body weight predict total liver volume in Western adults. Liver Transplant. **8**, 233–240 (2002)
8. Hoy, W.E., Douglas-Denton, R.N., Hughson, M.D., Cass, A., Johnson, K., Bertram, J.F.: A stereological study of glomerular number and volume: preliminary findings in a multiracial study of kidneys at autopsy. Kidney Int. **63**, S31–S37 (2003)
9. Bae, K.T., Commean, P.K., Lee, J.: Volumetric measurement of renal cysts and parenchyma using MRI: phantoms and patients with polycystic kidney disease. J. Comput. Assist. Tomogr. **24**, 614–619 (2000)
10. Irazabal, M.V., et al.: Imaging classification of autosomal dominant polycystic kidney disease: a simple model for selecting patients for clinical trials. J. Am. Soc. Nephrol. **26**, 160–172 (2015)
11. Higashihara, E., et al.: Kidney volume estimations with ellipsoid equations by magnetic resonance imaging in autosomal dominant polycystic kidney disease. Nephron **129**, 253–262 (2015)

12. Bae, K.T., et al.: Novel approach to estimate kidney and cyst volumes using mid-slice magnetic resonance images in polycystic kidney disease. Am. J. Nephrol. **38**, 333–341 (2013)
13. Grantham, J.J., Torres, V.E.: The importance of total kidney volume in evaluating progression of polycystic kidney disease. Nat. Rev. Nephrol. **12**, 667 (2016)
14. Turco, D., Severi, S., Mignani, R., Aiello, V., Magistroni, R., Corsi, C.: Reliability of total renal volume computation in polycystic kidney disease from magnetic resonance imaging. Acad. Radiol. **22**, 1376–1384 (2015)
15. Kim, Y., et al.: Automated segmentation of kidneys from MR images in patients with autosomal dominant polycystic kidney disease. Clin. J. Am. Soc. Nephrol. **11**, 576–584 (2016)
16. Kline, T.L., Edwards, M.E., Korfiatis, P., Akkus, Z., Torres, V.E., Erickson, B.J.: Semiautomated segmentation of polycystic kidneys in T2-weighted MR images. Am. J. Roentgenol. **207**, 605–613 (2016)
17. Kline, T.L., et al.: Automatic total kidney volume measurement on follow-up magnetic resonance images to facilitate monitoring of autosomal dominant polycystic kidney disease progression. Nephrol. Dial. Transplant. **31**, 241–248 (2016). https://doi.org/10.1093/ndt/gfv314
18. Sharma, K., et al.: Automatic segmentation of kidneys using deep learning for total kidney volume quantification in autosomal dominant polycystic kidney disease. Sci. Rep. **7**, 2049 (2017)
19. Magistroni, R., Corsi, C., Martí, T., Torra, R.: A review of the imaging techniques for measuring kidney and cyst volume in establishing autosomal dominant polycystic kidney disease progression. Am. J. Nephrol. (2018). https://doi.org/10.1159/000491022
20. Bevilacqua, V., Brunetti, A., Guerriero, A., Trotta, G.F., Telegrafo, M., Moschetta, M.: A performance comparison between shallow and deeper neural networks supervised classification of Tomosynthesis breast lesions images. Cogn. Syst. Res. **53**, 3–19 (2019)
21. Brunetti, A., Carnimeo, L., Trotta, G.F., Bevilacqua, V.: Computer-assisted frameworks for classification of liver, breast and blood neoplasias via neural networks: a survey based on medical images. Neurocomputing **335**, 274–298 (2019). https://doi.org/10.1016/J.NEUCOM.2018.06.080
22. Lee, J.-G., et al.: Deep learning in medical imaging: general overview. Korean J. Radiol. **18**, 570–584 (2017)
23. Bevilacqua, V., et al.: A novel deep learning approach in haematology for classification of leucocytes. In: Smart Innovation, Systems and Technologies (2019). https://doi.org/10.1007/978-3-319-95095-2_25
24. Litjens, G., et al.: A survey on deep learning in medical image analysis (2017). https://doi.org/10.1016/j.media.2017.07.005
25. Schmidhuber, J.: Deep Learning in neural networks: an overview (2015). https://doi.org/10.1016/j.neunet.2014.09.003
26. Bevilacqua, V., Brunetti, A., Cascarano, G.D., Palmieri, F., Guerriero, A., Moschetta, M.: A deep learning approach for the automatic detection and segmentation in autosomal dominant polycystic kidney disease based on magnetic resonance images. In: Huang, D.-S., Jo, K.-H., Zhang, X.-L. (eds.) ICIC 2018. LNCS, vol. 10955, pp. 643–649. Springer, Cham (2018). https://doi.org/10.1007/978-3-319-95933-7_73
27. Bevilacqua, V., Mastronardi, G., Menolascina, F., Pannarale, P., Pedone, A.: A novel multi-objective genetic algorithm approach to artificial neural network topology optimisation: the breast cancer classification problem. In: IEEE International Conference on Neural Networks - Conference Proceedings, pp. 1958–1965 (2008). https://doi.org/10.1109/ijcnn.2006.246940

28. Bevilacqua, V., et al.: A novel approach to evaluate blood parameters using computer vision techniques. In: 2016 IEEE International Symposium on Medical Measurements and Applications, MeMeA 2016 - Proceedings (2016). https://doi.org/10.1109/MeMeA.2016.7533760

29. Bevilacqua, V., et al.: Computer assisted detection of breast lesions in magnetic resonance images (2016). https://doi.org/10.1007/978-3-319-42291-6_30

30. Kingma, D.P., Ba, J.: Adam: A Method for Stochastic Optimization (2014) https://arxiv.org/abs/1412.6980

31. Bevilacqua, V.: An innovative neural network framework to classify blood vessels and tubules based on Haralick features evaluated in histological images of kidney biopsy. Neurocomputing **228**, 143–153 (2017). https://doi.org/10.1016/j.neucom.2016.09.091

32. Garcia-Garcia, A., Orts-Escolano, S., Oprea, S., Villena-Martinez, V., Garcia-Rodriguez, J.: A review on deep learning techniques applied to semantic segmentation. arXiv Prepr. arXiv: 1704.06857 (2017)

33. Badrinarayanan, V., Kendall, A., Cipolla, R.: SegNet: a deep convolutional encoder-decoder architecture for image segmentation. IEEE Trans. Pattern Anal. Mach. Intell. **39**, 2481–2495 (2017). https://doi.org/10.1109/TPAMI.2016.2644615

34. Csurka, G., Larlus, D., Perronnin, F.: What is a good evaluation measure for semantic segmentation? In: Burghardt, T., Damen, D., Mayol-Cuevas, W.W., Mirmehdi, M. (eds.) British Machine Vision Conference, BMVC 2013, Bristol, UK, 9–13 September 2013. BMVA Press (2013). https://doi.org/10.5244/C.27.32

# Research of Formal Analysis Based on Extended Strand Space Theories

Meng-meng Yao$^{(\boxtimes)}$ ⓘ, Jun Zhang, and Xi Weng ⓘ

Jiangnan Institute of Computing Technology, Wuxi, China
wellstudy@163.com

**Abstract.** In recent years, research based on strand space theories is a hotspot in the field of formal analysis method of security protocols. However, some extended theories based on strand space still has defects, and the analysis of security protocols is not accurate enough. And the model based on improved strand space theories is complex and not easy to develop automated verification tool. Based on this, the same execution bundle, matching strand, private item, etc. are proposed in this paper. In strand space, the matching strand of a given regular strand can be determined by the private item and the security key of the message item to prove the consistency of the security protocols. This paper proves the validity and accuracy of the extended strand space theories by analyzing Woo-Lam and Kerberos protocols.

**Keywords:** Strand space · Formal analysis · Security protocols

## 1 Introduction

With the rapid development of network security, security protocols such as IKEV2, TLS, and block-chain fair protocols play an important role in the field of network security. Formal analysis of security protocols is an effective way to ensure security of protocols. The strand space is a formal analysis method based on theorem proving.

In 1983, Dolev and Yao firstly proposed the hypothesis and model of formal analysis. Febreag, Herzog, Guttman proposed strand space and authentication tests based on the Dolev-Yao model [1–3]. Strand space is based on string operations. In the early years, based on strand space and authentication tests, the theories of strand space were extended, and security protocols were successfully analyzed [4–8]. In [9] the ISO/IEC9798-3 protocol is analyzed by authentication tests and a defect is discovered for the protocol. A electronic voting protocol is analyzed and a attack model based on strand space is designed, which improves the security of electronic voting protocol [10]. In [11], the semantics of the relationship between message items in the strand space model is improved, so that the nested encryption component is applied in the authentication tests. In [12], the message algebra and penetrator strands are extended and some defects of the remote authentication protocol are discovered. The theory of strand space is extended so that security protocols can largely defend against Type Defect Attacks(TFAs) [13]. Based on the authentication tests theorems, a improved formal analysis method. is proposed in [14].

Because of the feature of strand space, the design of automated analysis tools has been facilitated. The model SHAPE is proposed based on the strand space and a

© Springer Nature Switzerland AG 2019
D.-S. Huang et al. (Eds.): ICIC 2019, LNCS 11644, pp. 651–661, 2019.
https://doi.org/10.1007/978-3-030-26969-2_61

automated analysis tool CPSA(Cryptographic Protocol shapes Analyzer) is implemented. In [15], the fair exchange protocol is analyzed by CPSA tool, which expands the scope of strand space applications. The relevant security protocols are analyzed by the CPSA tool, and defects in the protocols are found [16, 17]. A framework for describing the protocol state model and a hybrid analysis method are proposed based on SHAPE in [18]. The consistency of the parameters of the cryptographic protocol and the relevance of the subject are studied, so that the strand space formal analysis method is easy to implement automatically [19, 20]. An algorithm is proposed for automated analysis in [21].

The automatic verification tool based on strand space is a goal of formal analysis of security protocols [22]. However, the following problems exist in the automated analysis model: 1. The model is too complicated and the versatility is poor. 2. Most of them are semi-automatic implementations, subjective intervention is required. 3. The algorithm is too complicated. 4. The attack path cannot be given.

Based on the research of above literatures, theory of strand space is extended in this paper, which is easy to develop automate analysis tools. Improved definitions and theorems are proposed in this paper, which extends knowledge set of the strand space and improves accuracy and practicality of analysis protocols. The main contributions of this paper are as follows:

(1) The definitions such as the same execution bundle, matching strand, matching node, etc. are proposed.
(2) The defect is pointed out of the authentication tests theorem;
(3) A formal analysis method based on extended strand space theory is proposed.
(4) The improved formal analysis method is used to analyze protocols such as Woo-Lam and Kerberos protocols, which is easy to implement automatically.

## 2   Basic Notions

### 2.1   Symbols

See Table 1.

**Table 1.** Symbols and definitions.

| Symbols | Definitions |
| --- | --- |
| $M$ | The set of unencrypted message item |
| $P$ | The set of penetrator strands |
| $KP$ | The set of compromised keys |
| $\{M\}_K$ | Message M is encrypted with K |
| $KAB$ | The shared key with principal A and B |
| $KA$ | The private key of principal A |
| $PKA$ | The public key of principal A |
| $Na$ | The nonce of principal A |
| $Nb$ | The nonce of principal B |
| $\sum$ | The set of strand space |

## 2.2  Notions

The basic concepts and propositions used in this paper are as follows [2, 3], and the proof of the propositions is in reference [2].

**Proposition 1(Incoming test theorem).** Let $C$ be a bundle with $n$, $n' \in C$ and let $n \Rightarrow^+ n'$ is an incoming test for $a$ in $t = \{h\}_K$. Then there must be regular nodes $m$, $m' \in C$ such that $t \subset term(m')$ and edge $m \Rightarrow^+ m'$ is the transforming edge for $a$.

**Proposition 2 (Unsolicited test theorem).** Let $C$ be a bundle with $n \in C$ and let $n \Rightarrow^+ n'$ is an unsolicited test for $a$ in $t = \{h\}_K$. Then there is a regular positive node $m$ with $t \subset term(m)$.

# 3  The Defect of Incoming Test Theorem

The Woo-Lam protocol is a authentication protocol, and the honest principal B authenticates the honest principal A through the server S. Figure 1 is bundle of the Woo-Lam protocol, which contains the following four types of strand sets:

(1) The init[A, B, S, Nb] is a set of initiator strands, and the message trace is <+A, −Nb, +{Nb}KAS>;
(2) The resp[A, B, S, Nb, H = {Nb}KAS] is a set of responder strands, and the message trace is <−A, +Nb, −H, +{A, B, {Nb}KAS}, −{A, Nb}KBS>;
(3) The serv[A, B, S, Nb] is a set of server strands, and the message trace is <−{A, B, {Nb}KAS}, +{A, Nb}KBS>;
(4) The P is the set of penetrator strands.

In the set of responder strands, H indicates an encrypted item which the honest principal cannot recognize the content.

**Assumption 1.** Suppose $C$ is a bundle of $\sum$. If Sr $\in$ resp[A, B, S, Nb, H = {Nb}KAS] is on bundle $C$ with $C\text{-}height = 5$, and $Nb$ is only originates in $\sum$ and KAS, KBS $\notin$ KP. Then there are must be a server strand Ss $\in$ serv[A, B, S, Nb] in the bundle $C$ with $C\text{-}height = 2$, and a responder strand Si $\in$ init[A, B, S, Nb] with its $C\text{-}height$ is 3 at least.

Proof. As shown in Fig. 1, $Nb$ is sent to <Sr, 2> where KBS $\notin$ KP, then the edge <Sr, 2> $\Rightarrow^+$ <Sr, 5> construct the incoming test for $Nb$, and $\{Nb, A\}_{KBS}$ is test component. By proposition 1, there must be regular nodes $m$ and $m'$ in the $C$, and $m \Rightarrow^+ m'$ is the transformation edge for $Nb$, and $\{Nb, A\}_{KBS}$ is the component of $m'$. $m$ and $m'$ are in the $C$.

The node $m$ is a negative node, $term(m) = \{A, \{Nb\}_{KBS}\}_{KBS}$. $m$ is the first node of a server strand Ss* = serv[A*, B*, S*, Nb*]. Since $term(m) = \{A, \{Nb\}_{KBS}\}_{KBS}$, $Nb$ is uniquely originating at <Sr, 2> and KAS, KBS $\notin$ KP, so A* = A, B* = B, S* = S, Nb* = Nb. Since (Ss*, 1) $\Rightarrow^+$ (Ss*, 2) is the transformation edge in the $C$, so $C\text{-}height = 2$.

As is apparent from Fig. 1, the node <Ss, 1> constructs an unsolicited test for $Nb$ and KAS $\notin$ KP. By Proposition 2, there must be a regular positive node denoted as $m$, and the $\{Nb\}_{KAS}$ is component of $m$. $m$ is the third regular node of some initiator strand

$Si^* = serv[A^*, B^*, S^*, Nb^*]$. Because $term(m) = \{Nb\}_{KAS}$, $S^* = S$, $Nb^* = Nb$. That is, there must be a initiator strand $Si \in init[A, B, S, Nb]$ with *C-height* is 3 at least.

By the above process of proof, B can authenticate A. Assumption 1 is true.

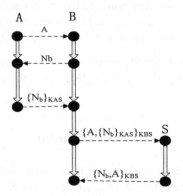

**Fig. 1.** Bundle of Woo-Lam protocol

In fact, there is a attack of Woo-Lam protocol in Fig. 2. When principal A sends a message to principal Z, the attacker intercepts the message and forwards it to the principal B. The attacker intercepts B's second message $\{Nb\}$ and forwards it to A. After receiving $\{Nb\}$, A generates $\{Nb\}_{KAS}$. Finally, the attacker successfully breaks the security of the Woo-Lam protocol. The main reason for success of the attack in Fig. 2 is that the message item $\{Nb\}_{KAS}$ does not identify which principal it is sent to, so it can be forwarded to another principal by the attacker. Analysis of this protocol by incoming test illustrates the incompleteness of authentication test theorem. There are regular nodes, but not necessarily the associated regular nodes.

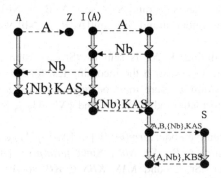

**Fig. 2.** Protocol attack message of Woo-Lam

# 4  Extend Strand Space Theories

## 4.1  Notions

**Definition 1.** Let $t \in C$. $t \neq term(n)$ for any node $n \in C$. If $t$ is only originates at node $n' \in C$, $t$ is said to be a private item of node $n'$, and $t = pri\_term(n')$ is used to represent the private item of the node n'. The set of private items is represented by **PTS**.

**Definition 2.** Any $k$ which is a key satisfies that $k \notin KP$ and $k \notin M$, then $k$ is called security key. A private key is security key. The set of security keys is represented by **SKS**.

**Definition 3.** If $k$ is a security key, $\{t\}_k$ is called security message, $\{t\}_k \notin KP$ and $\{t\}_k \notin M$. The set of security messages is represented by **SMS**.

**Definition 4.** If the key $k = f(x, y, z)$ is a private item of a regular node n in the bundle $C$, and at least one of x, y, and z is a component of $term(n)$, then $k$ is called negotiation key, the negotiation key of a node n is represented by $k = ke\_term(n)$. The negotiation key set is represented by **KES**.

**Definition 5.** $r$ which is nonce(number once) satisfies the following conditions: (1) $r$ is a random number; (2) there is a node $m$ in the bundle $C$, $r \in term(m)$ and $r$ is only originates at the node $m$. The set of nonce is represented by **NRS**.

**Definition 6.** There is a node $n$ in the bundle $C$, $t \in term(n)$ and $t \in M$, $t \notin NRS$, then $t$ is called negotiation message. The set of negotiated message is represented by **KMS**.

**Definition 7.** If $A$ is a component of a node in bundle $C$ and identifies the identity of a honest principal, $A$ is called identity message. The set of identity message is represented by **IS**.

**Definition 8.** The set of components consists of all components of the message items in a bundle $C$, which is represented by **CS**.

**Definition 9.** There are regular strands s1 and s2 in bundle $C$, and *C-height* = $x$ of s1. If the following conditions are true:

(1)  *C-height* = $y$ of s2 where $y \leq x$;
(2)  $term((s1,i)) = term((s2,i))$ where $i \leq y$;
(3)  If the symbol of (s1, i) is +, the symbol of (s2, i) is −. if the symbol of (s1, i) is −, the symbol of (s2, i) is +;
(4)  The honest principal of s1 is A and the principal of s2 is B. All interactive message items generated by A and B follow the protocol specification.

Then s2 is called the matching strand of s1, which is represented by s2 $\propto$ s1. If $x = y$, then s1 and s2 are said to be matching strand each other, which is represented by the s1 $\infty$ s2. A regular strand may have several matching strands.

**Definition 10.** In bundle C, s2 $\propto$ s1 and *C-height* = $x$ of s2. If $i \leq x$, (s2, i) is the matching node of (s1, i).

**Definition 11.** There are regular strands s1 and s2 in a bundle $C$. If s2 $\propto$ s1, the bundle $C$ is the same execution bundle of s1 and s2. If the bundle $C$ is the same execution bundle of regular strands S1,..., Sn, and S is a matching strand of $Si(0 \leq i \leq n)$, then $C$ is the same execution bundle of S1,..., Sn, S.

**Definition 12.** The action of receiving message is defined as receiving events denoted as $recv(X, Y, Msg, Z, S)$ and the action of sending message is defined as sending events denoted as $send(X, Y, Msg, Z, S)$. $X$ represents principal like the receiver or sender. If represents sender, $Y$ represents receiver. And if $X$ represents receiver, $Y$ represents sender. $Msg$ represents transporting message item. $Z$ represents the principal that generated the message item. $S$ represents that whether the interactive strands are in the same execution bundle.

**Definition 13.** In bundle C, n, $n' \in C$, $t = \{h\}_K$, and $a \subset t$. Regarding the addition, subtraction, multiplication and division of $a$ and a natural number is called simple transformation denoted as $\partial a$ and $\partial a \subset t$.

**Proposition 3 (Incoming test theorem).** Let $C$ be a bundle with $n$, $n' \in C$ and let $n \Rightarrow^+ n'$ is an incoming test for $\partial a$ in $t = \{h\}_K$. Then there must be regular nodes $m$, $m' \in C'$ such that $t \subset term(m')$ and edge $m \Rightarrow^+ m'$ is the transforming edge for $\partial a$. If $C = C'$, $n$, $n'$, $m$, $m'$ are in the same execution bundle.

**Proposition 4 (Unsolicited test theorem).** Let $C$ be a bundle with $n \in C$ and let $n \Rightarrow^+ n'$ is an unsolicited test for $\partial a$ in $t = \{h\}_K$. Then there is a regular positive node $m$ with $t \subset term(m)$.

### 4.2    Message Item and Events

The parameters in the $recv()$ and $send()$ are determined by the message item. The Kerberos protocol is represented by strand space as shown in Fig. 3 [23].

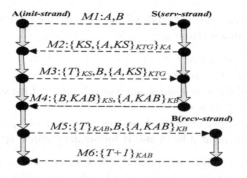

**Fig. 3.** Bundle of Kerberos protocol

The Kerberos protocol is interactive with the server S, the principal A and principal B, which includes the *init-strand* of the principal A, the *serv-strand* of the server S and the *recv-strand* of the principal B. The ticket service is on the server S, and the *KTG* is the key of the ticket service. The Kerberos protocol executes by six messages: *M1-M6*.

Suppose that there is an initiator strand init1 in the *init-strand* set, and *term* (init1,6) = $\{T + 1\}_{KAB}$, KAB is the long-term key of the principal A and the principal B. The honest principals A, B, and S can use this security key to encrypt or decrypt message. By KAB, it can be judged that the message item $\{T + 1\}_{KAB}$ must be generated at a regular node. Since $\{T\}_{KAB} \subset term(\text{init1,5})$ and $\{T + 1\} \subset term(\text{init1,6})$, by Proposition 3, there must be a responder strand denoted as recv1 and *C-height* of recv1 is 2. recv1 includes nodes (recv1,1) and (recv1, 2). recv1 and init1 are in the same execution bundle. $term(\text{recv1},1) = \{T\}_{KAB}$, B, $\{A, KAB\}_{KB}$, because the key KB is a security key, $\{A, KAB\}_{KB} \notin P$. Component B is in *P*. If B is altered by a attacker, it conflicts with $\{T\}_{KAB}$(KAB is the shared key between principal A and principal B), so B is not been altered and $B \in IS$.

The analysis in above shows that the process is objective analysis and there is no subjective intervention. Therefore, the message item and the component of the message item can determine the values such as the sending principal, the receiving principal, and the same execution bundle of the message item.

### 4.3    Analysis of Message Item

According to the authentication tests, as long as the key is secure, the message item encrypted by the key is also secure. A new analysis method is proposed based on the authentication tests [24]. The security of message item is judged based on whether the key is secure in the method. When a regular node receives a message, it judges from the following three points:

(1)  Which principal generated the message item?
(2)  Which principal sent the message?
(3)  Whether the communicational strands are in the same execution bundle?

If these three points are met, the node is matching node of the given regular node by the Definitions 9 and 10.

## 5    Analysis by Extended Strand Space Theory Model

### 5.1    Analysis Method

According to the consistency definition of the protocol in strand space [3], the authentication attribute of the relevant protocol can be proved as long as the matching strand of the given regular strand is in the same execution bundle. A new method based on the extended strand space theory in this paper is proposed. The method is described as follows:

(1)  Describing the *recv()* and *send()* events for each node of a given regular strand.
(2)  Starting from the last negative regular node of the given regular strand, detecting each parameter in the *recv()* event, and sequentially searching for the matching nodes.

658    M. Yao et al.

(3) It is finish after all matching nodes are found or the attack patch is given.

This method has the following advantages:

(1) The analysis model and the attack model are easy to implement automatically.
(2) This method is universal and concise.
(3) The result of analysis of protocol is accurate by detecting parameters of *recv()* and *send()*.
(4) If a defect in the protocol is found, an attack path can be given.

### 5.2 Analysis of Woo-Lam Protocol

**Assumption 2.** Suppose $C$ is a bundle of $\sum$. If Sr $\in$ resp[$A$, $B$, $S$, $Nb$, $H = \{Nb\}_{KAS}$] is in bundle $C$ with $C$-*height* = 5, and $Nb$ is only originates in $\sum$ and $KAS$, $KBS \notin P$. Then there are must be a matching strand of Sr denoted as Ss $\in$ serv[$A$, $B$, $S$, $Nb$] in the bundle $C$ with $C$-*height* = 2, and a matching strand of Sr denoted as Si $\in$ init[$A$, $B$, $S$, $Nb$] with its $C$-*height* is 3 at least.
Proof. The sequence of events of Sr are as follows:

Event1: *recv*($B$, $X1$, $A$, $Y1$, $Z1$);

Event2: *send*($B$, $A$, $Nb$, $B$, 1);

Event3: *recv*($B$, $X2$, $\{Nb\}_{KAS}$, $Y2$, $Z2$);

Event4: *send*($B$, $S$, $\{A, \{Nb\}_{KAS}\}_{KBS}$, $B$, 1);

Event5: *recv* ($B$, $X3$, $\{A, Nb\}_{KBS}$, $Y3$, $Z3$).

$X1$, $X2$, $X3$, $Y1$, $Y2$, $Y3$, $Z1$, $Z2$, $Z3$ represent that parameters are currently undeterminable.

The received message item is $\{A, Nb\}_{KBS}$ in Event5 and $KBS$ is a private item of honest principal B and principal S. According to the protocol specification, it is determined that $Y3 = S=X3$. Therefore, the message item is sent by the regular node <Ss, 2> of a some server strand denoted as Ss. $A$ and $Nb$ are components of the message item $\{A, Nb\}_{KBS}$, it is determined that the message is actually sent to the principal A, and $Nb$ is sent in Event2. By Proposition 5, Ss and Sr are in the same execution bundle, that is, $Z3 = 1$. And the matching node of <Sr, 5> is <Ss, 2>. Since <Ss, 2> sent the message item $\{A, Nb\}_{KBS}$, according to the protocol specification, *term*(<Ss, 1>) = $\{A, \{Nb\}_{KAS}\}_{KBS}$. It is known from Event4 that the matching node of <Sr, 4> is <Ss, 1>. That is, Ss is a matching strand of Sr and $C$-*height* = 2. Event5 is *recv* ($B$, $S$, $\{A, Nb\}_{KBS}$, $S$, 1).

Because KAS is private item of principal A and principal S, principal B can not know the content of $\{Nb\}_{KAS}$ in Event3. According to the protocol specification, $\{Nb\}_{KAS}$ is generated by principal A and $Y2 = A$, but the message item cannot be determined that principal A is sent to principal B. There must be a regular initiator strand denoted as Si. The regular node <Si, 3> is at Si, which transmits the message item $\{Nb\}_{KAS}$, and the values of $X2$ and $Z2$ cannot be determined. According to the protocol specification, <Si, 2> must have received the message item $Nb$ and <Si, 1> must have sent a message denoted as XM, but it cannot be determined which principal sent. It is also not possible to determine the value of $X1$. That is, Si and Sr may be not in the same execution bundle. If $XM = Z$($Z$ is another honest principal), then there is the attack in Fig. 2.

From the above, the assumption 2 is not true.

### 5.3   Comparison with Existing Formal Analysis Methods

Theories in this paper are strong versatility. The model of theories is not complicated, and it is easy to implement automated verification tools. Comparison with existing formal analysis methods are shown in Tables 2 and 3 below.

**Table 2.**  Abilities of discovering flaws

| Protocol | Known flaws | Authentication tests | Theories in this paper |
|---|---|---|---|
| Woo-Lam | Yes | Undiscover known flaws | Discover known flaws |
| IKEv2 | Yes | Undiscover known flaws | Discover known flaws |
| Kerberos | No | Undiscover known flaws | Undiscover known flaws |
| BAN-Yaholom | Yes | Undiscover known flaws | Discover known flaws |
| TLS handshake protocol | No | Undiscover known flaws | Undiscover known flaws |
| TKE | No | Undiscover known flaws | Discover flaws |

**Table 3.**  Features of formal analysis methods

| Formal analysis methods | Complexity | Universality | Development of automated tools |
|---|---|---|---|
| Authentication tests | Normal | Strong | Difficult |
| CPSA [17] | Difficult | Difficult | Simple |
| Strong authentication tests | Difficult | Difficult | Difficult |
| New authentication tests | Normal | Normal | Difficult |
| IVAP algorithm | Difficult | Difficult | Normal |
| Automatic verification algorithm [20] | Normal | Difficult | Simple |
| Theories in this paper | Normal | Strong | Simple |

## 6   Conclusion

The extended theories based on strand space are proposed in this paper, which improved the set of knowledge of strand space. Compared with the original tests theorem, it increases the correctness of analysis protocol and expands the scope of using of strand space theory. Because the analysis process relies entirely on the defined theorems and safety principles, there is no subjective judgment. So it is easy to develop automated verification tools based on the extended theories proposed in this paper. The future work is that algorithms of automated analysis will be implemented and algorithms of message item recognition will be studied. And automated verification tools for security protocols analysis will be developed.

# References

1. Fabrega, F.J.T., Herzog, J.C., Guttman, J.D.: Strand space: Why is a security protocol correct. In: The 18th IEEE Symposium 1998, Oakland, USA, pp. 160–171. IEEE, Oakland (1998)
2. Fabrega, F.J.T., Herzog, J.C., Guttman, J.D.: Strand space: proving security protocols correct. J. Comput. Secur. **7**(2,3), 191–230 (1999)
3. Guttman, J.D., Thayer, F.J.: Authentication tests and the structure of bundles. Theoret. Comput. Sci. **283**(2), 333–380 (2002)
4. Li, L., Wang, L., Chen, J., Wang, R., Zhang, Z.: Fairness analysis for multiparty nonrepudiation protocols based on improved strand space. Discrete Dyn. Nat. Soc. **2014**(1), 1–7 (2014)
5. Chang, Y., Zhang, Z., Wang, J.: A security protocol for trusted access to cloud environment. Recent Adv. Electr. Electron. Eng. **8**(2), 135–144 (2015)
6. Lei, Y.U., Wei, S., Zhuo, Z.: Research on consistence of strand parameters for protocol principals in authentication test theory. Comput. Eng. Appl. **51**(13), 86–90 (2015)
7. Fan, Y., Escobar, S., Meadows, C., Meseguer, J., Santiago, S.: Strand spaces with choice via a process algebra semantics. In: International Symposium on Principles & Practice of Declarative Programming (2016)
8. Dong, X., Yang, C., Sheng, L., Wang, C., Ma, J.: A new method to deduce counterexamples in secure routing protocols based on strand space model. Secur. Commun. Netw. **9**(18), 5834–5848 (2017)
9. Xiong, L., Peng, D.Y.: An improved authentication test for security protocol analysis. Commun. Technol. **47**(08), 951–954 (2014)
10. Sureshkumar, V., Anitha, R.: Analysis of electronic voting protocol using strand space model. In: Martínez Pérez, G., Thampi, Sabu M., Ko, R., Shu, L. (eds.) SNDS 2014. CCIS, vol. 420, pp. 416–427. Springer, Heidelberg (2014). https://doi.org/10.1007/978-3-642-54525-2_37
11. Song, W.T., Bin, H.U.: One strong authentication test suitable for analysis of nested encryption protocols. Comput. Sci. **42**(1), 149–169 (2015)
12. Wei, F., Feng, D.G.: Analyzing trusted computing protocol based on the strand spaces model. Chin. J. Comput. **38**(4), 701–716 (2015)
13. Song, W.T., Hu, B.: Approach to detecting type-flaw attacks based on extended strand spaces. Comput. J. **58**(4), 572–587 (2018)
14. Yao, M.M., Zhu, Z.C., Liu, M.D.: An improved formal analysis method based on authentication tests. Netinfo Secur. **19**(1), 27–33 (2019)
15. Doghmi, S.F., Guttman, J.D., Thayer, F.J.: Searching for shapes in cryptographic protocols// Proceedings of the 13th TACAS. LNCS4424. Berlin: Springer. 523–537(2007)
16. Guttman, J.D.: State and progress in strand spaces: Proving fair exchange. J. Autom. Reason. **48**(2), 159–195 (2012)
17. Ramsdell, J.D., Guttman, J.D.: CPSA Primer. The MITER Corporation, Bedford (2012)
18. Ramsedll, J.D.: Proving Security Goals with Shape Analysis Sentences. The MITER Corporation, Bedford (2013)
19. Ramsdell, J.D., Dougerty, D.J., et al.: A Hybrid Analysis for Security Protocols with State. The MITER Corporation, Bedford (2014)

20. Liu, J.: Automatic verification of security protocols with strand space theory. J. Comput. Appl. **35**(7), 1870–1876 (2015)
21. Yu, L., Wei, S.M., Jiang, M.M.: Formal analysis method of security protocol based on correlation degree of principals. Netinfo Secur. (2018)
22. Yuan, B.A., Liu, J., Zhou, H.G.: Study and development on cryptographic protocol. J. Mil. Commun. Technol. **38**(1), 90–94 (2018)
23. Wei, H., Xie, Z.L., Yi, G.Z.: Analysis of kerberos protocol based on strand spacd theory. Comput. Technol. Dev. **2013**(12), 109–112 (2013)

# The Application of Artificial Intelligence in Music Education

Mingli Shang[✉]

Zhengzhou Institute of Technology, Zhengzhou, China
253370721@qq.com

**Abstract.** This article comprehensively reviewed the development history and evolution of the Artificial Intelligence Technology (AIT), systematically demonstrated the convenience, practicality and limitations of Artificial Intelligence (AI) in music education, compared and studied the advantages and disadvantages of both traditional and AI-applied music education mode, statistically analyzed results of the study through questionnaire survey, and deduced that the combination of music education and AI will become the new trend of future music education.

**Keywords:** Artificial Intelligence · Music education · Practicality · Convenience · Limitations

## 1 Introduction

In July 2015, "Artificial Intelligence" was included into the "Guiding Opinions of the State Council on Actively Promoting Internet Plus Action"; in May 2016, the National Development and Reform Commission and other four departments jointly published the "Implementation Program of the Three-Year Action on Artificial Intelligence in Internet Plus" [1]. On March 5, 2017, Premier Keqiang Li published "report on the work of the government", pointing out that it is necessary to accelerate the cultivation, development and expansion of emerging industries, in which AIT is one of the most important parts. This is the first time that AI has been included in the Report on the Work of Government in China, fully reflecting the government's emphasis on the research and development of AIT.

In terms of the current development of AI in music education, the integration of music education and AI has achieved some accomplishments. For example, Wolfie, a music education platform of Tonara and an APP with more focus on classroom teaching, provides teachers and students with teaching and evaluation tools, helps music educators guide students to correct their errors in practice during the process of teaching-learning-practice, and provides teachers and students with a completely different way of music education-learning interaction. In terms of the integration of AI and music education, the arising of The ONE, a smart piano, also brings a new teaching mode in which piano learning and user experience are integrated. It shall be particularly emphasized that The ONE piano advocates the music education concept of convenience, efficiency,

D.-S. Huang et al. (Eds.): ICIC 2019, LNCS 11644, pp. 662–668, 2019.
https://doi.org/10.1007/978-3-030-26969-2_62

happiness and simpleness, which makes learners experience more fun in learning. Such teaching mode is a model of the integration of AI and music education.

The author took AI and music education as the research direction, applied literature research method, interdisciplinary approach, survey study and other research methods for the purpose of providing a detailed and scientific reference for the study of AI and music education. The author hopes that through the research can make some academic contribution to the combination of AI and music education, so that high-tech achievements can better serve music education.

## 2   The Emergence and Development of AI

AI emerged in the 1950s. It is a high-tech multidisciplinary and comprehensive science on the research of intelligent machines and machine intelligence. It has achieved world-renowned research results in fields including natural language processing, knowledge processing, pattern recognition, automatic programming, knowledge base, and intelligence robot and so on [2].

The concept of AI was formally proposed by John Maccarthy at the Dartmouth Conference in 1956. It is one of the important branches of the computer science and is known as one of the top three technologies in the world. Professor Nelson, an expert at Stanford University in the United States, once defined AI as "Artificial Intelligence is a discipline of knowledge – about how to express, acquire and use knowledge." Dr. Weston of the Massachusetts Institute of Technology thinks that "Artificial Intelligence means the study of how to make computers do intelligent things that only people can do in the past." Therefore, AI can be regarded as a discipline rule that studies human intelligent activities and can build artificial systems with certain intelligent behaviors and contents.

Although AI is more and more influential in the society today, its development process is not as effective as Moore's Law. Research bottlenecks in the study of machine translation in 1965 and the feedback facts from governments and universities have proved that the development of AI is not so easy, nor as miraculous as experts' expectation. In the 1980s, AI researchers began to realize that the reasonable operation of knowledge engineering and expert systems requires a large amount of knowledge manually input other than by the system. It is self-evident that such knowledge input is quite costly. However, machine learning can be figured out easily with the design of a bottom-up heuristic approach.

## 3   The Convenience of AI in Music Education

The famous American music educator Michael L. Mark once wrote in his book "CONTEMPORARY MUSIC EDUCATION": Electronic music was initially regarded as a form of music learning. People at that time just took it as a new form of music learning, failing to realize its potential in school music education later [3]. It can be inferred that AI had appeared in music education in the 1960s as a high-tech means. At that time, developed countries in the world applied such technology to electronic keyboard instrumental music and developed an electronic music synthesizer with certain intelligent functions. It stores, plays various tones of instruments, changes tones at any time, and it is small for easy carrying and handling. In the subsequent teaching application, music educators realized and experienced the advantages and significance of electronic synthesizers in music teaching and they applied them to music education for the convenience of collective teaching mode in the classroom. The music teaching mode provides greater convenience. The intelligent functions can be used to change the single teaching mode through the ages, the study atmosphere in the classroom can be better and the students' learning initiative can be improved. The educators at that time applied such instrument in teaching for exploration. This is the early tentative application of intelligent technology in music education.

In the late 20th century, with the rapid development of science and technology, the music teaching market and classes gradually accepted the more humanized and intelligent music synthesizers. These new music synthesizers are more convenient and intelligent. Compared with those of the old generation, they can be used for musical creation of homophony, harmony and polyphony by means of programming or even algorithm and can store in the system a variety of tones. The tones of Chinese national musical instruments and Western musical instruments are all available and even sounds of animals and other sound effects can be simulated. Learners can choose the tones they need for creation. It should be noted that the instrument system has humanized and powerful phonetic sequence function. The phonetic sequence functions and editing functions are unmatched by any traditional instrument [4]. The instrument system can also have intelligent "behaviors" such as procedural collection and editing based on instructions and information input given by the creator and the application user. Basically, the learners' needs can be satisfied by the instrument eventually. Therefore, the research, development and promotion of such new instruments has urged music educators and learners to seek new teaching methods and new teaching concepts so as to adapt to the impacts and challenges of science and technology. Now, the performance requires cooperation of many people is to be replaced by the musical creation or performance of one person, and the efficiency of learning and practicing will be greatly improved. Students can play songs they created and works they were practicing at any time. Key points in teaching can also be better evaluated and feedback in music practice. The in-depth research and development and extensive application of such instrument provides a practical basis and platform for the presentation and establishment of teaching methods such as "experiential learning", "experiential teaching" and "comprehensive evaluation".

With the rapid development and updating of electronic musical instruments, intelligent electronic instruments are more powerful and have been accepted by more and more music lovers with rapid application and popularization. Meanwhile, electronic musical instrument manufacturers also developed a variety of intelligent pianos with various models and complete functions. These pianos can be equipped with a recording program to automatically interpret different styles of instrumental music according to the settings of different programs. Compared to previous electronic instruments, these latest music synthesizers have been updated in terms of tone quality, performance effects, and phonetic sequence functions. More convenient and efficient programming function replaces complicated operation, and the new phonetic sequence function replaced manual mode with automatic mode and even voice mode. The instrumental music can be found in classrooms of primary and secondary schools, or music educational universities and even social music education institutions. Intelligent instruments help students understand music elements, consolidate their theoretical knowledge and enhance students' ability in classroom learning, comprehension and adaptability. It can be inferred that they have great advantages and values indeed.

In recent years, music technology has been fully applied in the school curriculums. With different teaching and learning content in class, it is possible to adjust the class scene at any time accordingly. With the development and popularization of computers and mobile clients, people can download relevant intelligent software at any time to assist musicians to learn. For example, the application of music software such as notation software, automatic page turning software, music education evaluation software, etc. In addition, music science and technology can effectively utilize and integrate music creation materials of students to lay the foundation for new works. This provides a new model for music classroom teaching and a new direction, teaching ideas and thinking space for music educators leading them to a new world.

## 4 Practicality of AI in Music Education

The Transits into an Abyss played by the London Symphony Orchestra, was staged in July 2012. This was the first time that a top orchestra played a completely machine-created work. The song was created by a computer cluster called "Iamus" that uses intelligent algorithms. Iamos is a figure in Greek mythology and is said to understand the language of birds. The system started with minimal information, such as the instrument for playing the music, and created an extremely complex piece of music without human intervention. The music resonates with the audience within minutes. Iamos has created millions of unique pieces of modern classical style and it is likely to adjust its style to try other musical genres in the future.

In addition, the computer systems now can write music similar to master's work with algorithm. For example, David Cope, a computer scientist and composer and the author of "Music Intelligence Experiment", designed the humanoid robot EMMY who created a lot of convincing music. From Bach's Hymns, Mozart's Sonatas, Chopin's Mazuka, Beethoven's Tenth Symphony and Mahler's Opera in Five Acts, etc., the application of AIT in the field of music has become more and more extensive.

The AI-ED application has produced the Intelligent Tutoring System (ITS) in the early days. The intelligent teaching system is the main form of integration of AIT and education, and it is the main research and development direction of teaching in the future. The rapid development of information technology and the development and improvement of development models for new teaching system have promoted the use of AIT to develop new teaching systems in which ITS is a typical model [5]. ITS contains the field model, the learner model, and the teacher model and embodies the entire content of the teaching system development with incomparable advantages and great appeal [6].

The application of AIT helps teachers understand students' learning behaviors and habits in a more quick, effective and in-depth way and teach with appropriate methods so as to better teach in accordance with students' aptitude and help them do better. The human-computer interactive learning model can help learners track and observe their learning process and progress, learn to summarize and reflect, and thereby stimulate their enthusiasm and initiative. Now there are some intelligent learning robots like "Zhixiaole" and "Xiaoai Speaker" on the market [7]. The intelligent robots can naturally lead kids into a dialogue scene and topic that interests them according to their physical and mental ages. By questioning and communicating with kids through games or intelligent behaviors, the robots can teach and inform in interactive dialogues. They also analyze the status, emotions and interests, hobbies, etc. of kids while recording the kids' mastery of knowledge and their intellectual and mental development, etc. on the cloud big data platform. The robot can obtain relevant information in interaction through various algorithms, and can continuously update and renew the information to better serve kids like an adaptive tutor.

## 5    Limitations of AIT in Music Education

Science and technology has both advantages like convenience and efficiency and challenges to our way of life. The application of AI to music education certainly has its drawbacks and limitations [8].

i.  Particularity of Music Education

Genetics, nanotechnology, and Artificial Intelligence/Robotics are three fundamental singularity revolutions among which the Artificial Intelligence/Robotics is the most profound technology. The creation in non-biological intelligence of Artificial Intelligence/Robotics has surpassed that of humans in certain fields. It is believed that in the near future, artificial intelligence will surpass human intelligence in more aspects [9]. Despite this, there are still query, criticisms and even disavowal from various fields and professionals, such as concerns from the government, arguments among various schools of AI theories. The author mainly discussed the query from the music industry.

The author believes that application of AI in music teaching is comparatively lagging behind other disciplines. When AIT becomes as commonplace and widespread as modern computers, traditional music teaching mode will inevitably be challenged and overturned by it [10].

Music can purify the mind and improve the aesthetic ability. In the process of learning music, AIT or even computer technology can be used to assist teaching on related theoretical knowledge. Most of the theoretical knowledge is established concepts and knowledge, such as scales and intervals scale, interval, chord, pitch shift, major or minor key in the Western classical music knowledge system. These are concepts and frameworks must be mastered in music learning and they are in a knowledge system that has remained almost unchanged for many years. It is similar to mathematics in the one-answer-only fact without any other possibility. In terms of complex perceptual aspects such as tone, emotion and music content expression, everyone has their own opinions. And the question is, can AIT correctly understand the music content? This question a matter of great concern and has long been arguing over.

And many people deny that AIT can be of great use in music teaching. But the author believes that AIT can make a big difference in music education.

ii. Recognition of the Industry

A questionnaire survey I conducted on AI and music education in March 2019 can be taken as an example. Although 72% of respondents in the survey believed that the integration of music education and AI is bound to be the general trend of music education development in the future, the rest 28% still insisted that music is to be perceived and only human brain can experience and perceive complex emotions implied by music while machine does better in regular and procedural work. To do better in music learning or piano playing, face-to-face teaching for one student at a time is essential as the foundation of music education. Some believe that AI and the Internet can be convenient for music learners in their learning but it only plays a minor role as an auxiliary. This argument has certain reason and basis in the current development status of AIT, but if we pay more attention to the development of AIT, we will find that many engineers and computer experts in the world are working on exploring AIT for the fact whether it can have consciousness and emotion [11]. Through a series of algorithms, the desired result can ultimately be achieved with the support of people in the field of music in jointly solving the setbacks in music teaching, so as to fasten the pace of science and technology in serving music teaching.

# 6 Conclusion

AI can offer great convenience for music learning and enhance learning interest and efficiency in music education and teaching. It takes great advantage of the superior resource database to function as an excellent music teacher in a distinctive form in underdeveloped areas lacking outstanding music teachers to generally improve students' enthusiasm for music learning. Surely, the AIT now still needs further improvement to cover the gap between it and music emotional teaching. With the rapid development of technology, though, AI will certainly break through the technical bottleneck of music emotional teaching, and it will be extensively integrated into music education to form a new trend for music education in the future.

# References

1. Artificial Intelligence Officially included in Government Reports March 11, 2017 Public platform of Sohu. http://mt.sohu.com/20170311/n483042891.shtml
2. Jiang, Z.: Superlative Science: on Artificial Intelligence. Posts and Telecom Press, Beijing (2016)
3. Mark, M.L.: Contemporary Music Education, 2nd edn. Schirmer Books, New York (1986)
4. Yong, G.: Modern music education and AI\IT. In: Explorations in Music. 2月 (2007). 1004-2172(2007)02-0104-04
5. Taoye, X.: On the application of artificial intelligence based on expert system in education. Sci. Technol. Inf. (11) (2011)

6. Compiled by Swarma Club: Superlative Science: on Artificial Intelligence. Posts and Telecom Press, Beijing (2015)
7. Zhao, D.-S., Wang, L.-Y.: An empirical study on influencing factors of online consumer brand choice behavior. In: Proceedings of the 2nd International Conference on Politics, Economics and Law (ICPEL 2017) (2017)
8. Yonghong, G., Hengyu, G.: On the relationship between education and technology in the study of educational technology: take the application of AI in education as an example. Glob. Market Inf. Guide (1) (2015)
9. Kaplan, J.: Humans Need not Apply. Zhejiang People's Publishing House, Hangzhou (2016). (Trans. Pan, L.)
10. Kurzweil, R.: On Intelligence. Zhejiang Music Press, Hangzhou (2016). (Trans. Yangyan, S.)
11. Harari, Y.N.: Homo Deus. Citic Publishing House, Beijing (2017). (Trans. Junhong, L.)

# Single-Machine Green Scheduling Problem of Multi-speed Machine

Ai Yang[1], Bin Qian[1,2(✉)], Rong Hu[1,2],
Ling Wang[3], and Shang-Han Li[1]

[1] School of Information Engineering and Automation, Kunming University
of Science and Technology, Kunming 650500, China
bin.qian@vip.163.com
[2] School of Mechanical and Electrical Engineering, Kunming University
of Science and Technology, Kunming 650500, China
[3] Department of Automation, Tsinghua University, Beijing 100084, China

**Abstract.** This paper proposes an accurate dynamic programming algorithm (EDPA) to solve the single-machine scheduling problem with release time (SMSP_RTs), which targets total carbon emissions. First of all, considering the constraints of each workpiece with different release time and expected completion time, and the different speeds of the machine consume different energy, the length of time the workpiece is processed at different speeds is different, and the problem is established on this basis. Sorting model. The model can be expressed as a ternary method. Then, construct a state recursive equation and design an exact dynamic programming algorithm (EDPA). EDPA has the complexity of pseudo-polynomial time to obtain the optimal solution to the problem. Finally, through simulation experiments, the effectiveness of the proposed EDPA is verified, and the influence of machine speed on energy saving and emission reduction is analyzed.

**Keywords:** Single machine scheduling problem · Multi-speed ·
Total carbon emissions · Dynamic programming algorithm

## 1 Introduction

"Made in China 2025" puts green manufacturing into a key strategic task, which will require companies to consider environmental indicators while taking into account environmental indicators. Therefore, energy saving and emission reduction is particularly important in modern industrial production. On the one hand, energy saving and emission reduction can save production costs and reduce energy consumption. On the other hand, carbon emissions as an important reference for environmental indicators, companies have to limit carbon emissions under the constraints of laws and regulations. The unreasonable scheduling strategy of enterprises in the production process is likely to cause energy waste, which makes the carbon emissions remain high. Therefore, improving the utilization rate of energy in the production scheduling is the key to energy conservation and emission reduction. Green Workshop Scheduling Problem (GSSP) is a workshop scheduling problem for green manufacturing. Efficient optimization

© Springer Nature Switzerland AG 2019
D.-S. Huang et al. (Eds.): ICIC 2019, LNCS 11644, pp. 669–677, 2019.
https://doi.org/10.1007/978-3-030-26969-2_63

scheduling method can effectively improve economic efficiency, achieve energy saving, emission reduction, consumption reduction, cost reduction, reduce environmental impact, and obtain economic indicators and green. Collaborative optimization of indicators [1]. Therefore, the green shop scheduling problem has become a hot topic in academic research, and has been widely used in the fields of power, logistics, cloud computing and so on. In the traditional workshop scheduling research, the introduction of green indicators makes the problem more complicated. While increasing the difficulty of solving, it also makes the research problem more theoretical and practical.

In the research on GSSP problem, Che et al. [2] established a mixed integer linear programming model to solve the single machine scheduling problem with the goal of limiting delivery time and minimum energy consumption; Mouzon et al. [3] to minimize the total completion time and Energy consumption is an optimization index, and energy efficiency scheduling rules are proposed to solve the single machine scheduling problem. Wang et al. [4] established the integer programming model to solve parallel machine scheduling under the condition that the maximum power limit is the constraint and the completion time is the optimization goal. Problem; Bin [5] designed the branch and bound algorithm to solve the parallel machine scheduling problem with the goal of minimizing the weighted completion time and cost; Wu et al. [6] used delivery time satisfaction, time, equipment utilization and Cost is the optimization goal, and a hybrid genetic algorithm is proposed to solve the flexible flow shop scheduling problem. Ding JY et al. [7] minimize the completion time and total carbon emissions, and use the greedy algorithm based on the non-dominated solution structure. Flow shop scheduling problem; Zhang et al. [8] proposed a hybrid genetic algorithm with local search to minimize the total weighted delay and total energy consumption. Workshop scheduling problem; Zhang et al. [9] aimed at completion time and total carbon emissions, using genetic algorithm to solve the completion time, and then using the method of adjusting non-critical processes to reduce carbon emissions, which is a multi-optimization model of shop scheduling problem. According to the appeal literature, GSSP has become a research hotspot of production scheduling, but the related literature is still very limited. Therefore, the research on such problems is of great significance.

The single machine scheduling problem is a typical production scheduling problem, and many documents have been studied [10–14]. And the solution of the single machine scheduling problem is relatively simple, which is regarded as the prototype and foundation of the multi-machine scheduling problem. Paulo et al. [10] proposed a new cultural genetic algorithm to solve the single-machine scheduling problem with set-related setup time, and the optimization target is the total delay time. Guo et al. [11] proposed a mixed integer programming model to solve the machining conditions. In the case of gradual deterioration, the single-machine scheduling problem is optimized for the total weighted delay time; Akar et al. [12] used the neural network rule obtained by artificial neural network to solve the single-machine scheduling problem with release time, and the optimization target is the total weighted delay time. Zhong Tao et al. [13] proposed a hybrid evolutionary algorithm (HEA) for solving single-machine scheduling problems with sequence-dependent dependencies and preparation time. The optimization goal is to minimize the total delay time Ye et al. [14]. The ant colony algorithm solves the single machine scheduling problem, and the optimization target is the total weighted delay time;

In this paper, we focus on a class of green SMSP_RTs that are widely used in practice, and minimize the total carbon emissions as the optimization target to study problem modeling, scheduling algorithms and energy saving strategies. First of all, considering the constraints of each workpiece with different release time and expected completion time, and the different speeds of the machine consume different energy, the length of time the workpiece is processed at different speeds is different, and the problem is established on this basis. Sorting model. The model can be expressed as a ternary method. Then, construct a state recursive equation and design an exact dynamic programming algorithm (EDPA). EDPA has the complexity of pseudo-polynomial time to obtain the optimal solution to the problem. Finally, through simulation experiments, the effectiveness of the proposed EDPA is verified, and the influence of machine speed on energy saving and emission reduction is analyzed.

## 2   Problem Description

The green SMSP_RTs are described as follows: n workpieces are machined on the same machine, the machine has several gear positions, each gear position of the machine corresponds to different speed and power, the workpiece has arrival time and delivery time, and is in different gear positions. The processing time is not the same. The higher the speed, the higher the energy consumption, but the shorter the processing time. The goal of the scheduling is to select the machining sequence of the workpiece and the gear position of the machining, while optimizing the total carbon emissions of the machining, while satisfying the start time and due date."

### 2.1   Symbol Definition

$n$: number of jobs;
$a_j$: release time of job $j$;
$d_j$: due date of job $j$;
$x_j$: starting processing time of the job $j$;
$k$: machine position, $k = 1..\eta$;
$k_j$: machine position of job $j$;
$P_k$: the power of the machine in the $k$ position;
$T_{jk}$: processing time of the job $j$ in the $k$ position;
$c_{jk}$: completion time of the job $j$ in the $k$ position;
$\gamma$: the conversion coefficient of electric and carbon;
$Q_{jk}$: energy consumed to machine the job $j$ in the $k$ position;
$Q_{total}^{j}$: total energy consumed by machining to job $j$;
$Q_{total}$: total energy consumption for processing all workpieces;
$TCE$: total carbon emission for processing all workpieces.

## 2.2  Problem Hypothesis

(1) The machine is ready at time $t = 0$.
(2) The machine can only process one job at a time.
(3) Once the machining starts, it cannot be interrupted until the workpiece is processed. (4) The machine is in a normal state throughout the machining process, that is, regardless of the failure of the machine, sudden power failure, etc.

## 2.3  Scheduling Optimization Model

The optimization goal is the total amount of carbon emissions.

$$f = \min \text{TCE} \quad f = \min TCE \tag{1}$$

s.t

The starting processing time of the workpiece must be greater than the arrival time, which satisfies the formula (2).

$$x_j \geq a_j \tag{2}$$

The completion time of the workpiece is equal to the sum of the machining time corresponding to the starting gear and the selected gear,which satisfies the formula (3).

$$c_{jk} = a_j + p_{ik} \tag{3}$$

The completion time of the job $j$ must be less than the due date of the job $j$, which satisfies the formula (4).

$$c_{jk} \leq d_j \tag{4}$$

The due date of the job $j$ should meet the constraint of formula (5)

$$a_j + \min(T_k) \leq d_j \tag{5}$$

The arrival time and delivery time of the workpiece meet certain rules, which satisfies the formula (6–7).
Hypothesis:

$$d_j \leq d_i \tag{6}$$

Already:

$$r_j \leq r_i \tag{7}$$

The energy consumption of the machined workpiece is equal to the product of the machining time and machine power corresponding to the selected gear position,which satisfies the formula (8).

$$Q_j = T_{jk} * P_k \tag{8}$$

The processing time for selecting the high gear will be less than the processing time for the low gear, which satisfies the formula (9).

$$T_{jk} < T_{jk-1} \tag{9}$$

The processing energy consumption for selecting the high gear should be greater than the processing energy for the low gear. which satisfies the formula (10).

$$Q_{jk} > Q_{jk-1} \tag{10}$$

Total energy consumed by machining to workpiece $j$, which satisfies the formula (11).

$$Q_{total}^j = \sum_{i=1}^{j} Q_i \tag{11}$$

The total processing energy consumption is equal to the sum of the processing energy of all machines, which satisfies the formula (12).

$$Q_{total} = \sum_{j=1}^{n} Q_j \tag{12}$$

The total carbon emissions of all workpieces are equal to the total energy consumption and the carbon conversion factor, which satisfies the formula (13).

$$TCE = \gamma Q_{total} \tag{13}$$

# 3  Dynamic Programming Algorithm and Strategy Analysis

## 3.1  Lemma

There is one most sorted order so that the machined workpiece must follow the EDD (estimated due date) allocation rule.

Prove: That there is a best order $\Pi = (j_1 \ldots j_i, j_j \ldots j_n)$ in this sorting $\Pi$, job $i$ and $j$ are interchanged, that is, the $i$-th and $j$-th positions in the original sorting are interchanged to obtain a new sort $\Pi' = (j_1 \ldots j_i, j_j \ldots j_n)$. The sorting $\Pi$ is delayed by $d_j - d_i$ from the sorting $\Pi'$. This will probably cause the artifact $j$ not to satisfy the constraint of $c_j \geq d_j$.

## 3.2  Lemma

For the job $j$, the high gear position is preferentially selected for machining, $k = \eta$, If the gear position n − 1 satisfies the condition of $c_{jk} \leq d_j$, then $k = \eta - 1$, and so on, knowing that the condition of $c_{jk} \leq d_j$ is not satisfied.

Prove: If $k = \eta$ and $k = \eta - 1$ boss satisfied the condition of $c_{jk} \leq d_j$, the former will consume more $P_k * T_{jk} - P_{k-1} * T_{jk-1}$ energy than the latter. This will cause the target function to increase by $\gamma * (P_k * T_{jk} - P_{k-1} * T_{jk-1})$. According to the constraint 9, $\gamma * (P_k * T_{jk} - P_{k-1} * T_{jk-1})$ is greater than 0.

## 3.3  Dynamic Programming Algorithm

The production tasks are divided into stages according to the production sequence of the products. Each stage needs to determine the time when the current product starts processing, the gears that should be selected for the current product machine, and the current total carbon emissions need to be calculated. The state produced by step $j$ is

$$v_j = \{x_j, k_j, c_{jk}, Q_{jk}, \sum_{i=1}^{j} Q_{ik}\}.$$ The variables in state 1 are explained below.

$x_j$: starting processing time of the job $j$;

$k_j$: machine position of job $j$;

$c_{jk}$: completion time of the job $j$ in the $k$ position;

$Q_{jk}$: energy consumed to machine the job $j$ in the $k$ position;

$\sum_{i=1}^{j} Q_{ik}$: total energy consumed by machining to job $j$;

Given a state $v_{j-1} = \{x_{j-1}, k_{j-1}, c_{j-1k}, Q_{j-1k}, \sum_{i=1}^{j-1} Q_{ik}\}$, suppose there are 3 gears, $\eta = 3$. The following states may have the following conditions:

(1) The workpiece can only be processed at the highest position, $k = 3$;
(2) The workpiece can be processed at the $k = \eta - 1$ position, $k = 2$;
(3) The workpiece can be processed at the $k = \eta - 2$ position, $k = 1$.

The dynamic programming algorithm is given below:

Step 1: Sort the workpieces according to EDD rules;
Step 2: Initialization, $v_0 = \{0, 0, 0, 0, 0\}$;
Step 3: Generated $v_j$ by $v_{j-1}$;

For $j = 1$ to $n$ do
Set $\upsilon_j = \varnothing$.

For $(x_{j-1}, k_{j-1}, c_{j-1k}, \sum_{i=1}^{j-1} Q_{j-1k}) \in \upsilon_{j-1}$

$k_j = 3$, $x_j = \max(c_{j-1}, a_j)$;
  If $x_j + T_{jk-1} \le d_j$ then
    If $x_j + T_{jk-2} \le d_j$ then

$$\upsilon_j = \{x_j, 1, x_j + T_{j1}, \sum_{i=1}^{j-1} Q_{j-11} + P_1 T_{j1}\}$$

    Else

$$\upsilon_j = \{x_j, 2, x_j + T_{j1}, \sum_{i=1}^{j-1} Q_{j-12} + P_2 T_{j2}\}$$

  Else

$$\upsilon_j = \{x_j, 3, x_j + T_{j3}, \sum_{i=1}^{j-1} Q_{j-13} + P_3 T_{j3}\}$$

    EndIf // If $0 < x_j + T_{jk-1} \le d_j$
  EndIf // If $0 < x_j + T_{jk-2} \le d_j$
EndFor. // For $j = 1$ to $n$ do

Step 4: [optimize] Total carbon emissions $TCE = \gamma \sum_{j=1}^{n} Q_{jk}$, the optimal solution can be obtained by backtracking through the state tree constructed in step 3.

### 3.4   Time Complexity Analysis

In the dynamic programming algorithm mentioned in Sect. 3.3, the time complexity of the EDD sorting in step 1 is $O(n \log n)$, that there may be three cases of workpieces, so there are $O(3^n)$ a total of possible states, so the time complexity of the algorithm is Less time complexity than the exhaustive method is $O(n!)$.

## 4   Simulation Experiment Analysis

In order to verify the validity of the algorithm, Select the parameters of document (6) and document (15), and made some modifications (Tables 1 and 2).

The traditional algorithm is processed according to the fastest gear. From this case, it can be seen that the dynamic programming algorithm can significantly reduce the total carbon emissions, and has a significant effect on the energy saving and emission reduction of enterprises (Tables 3 and 4).

**Table 1.** Processing parameters of the machine

| Machine position | 1 | 2 | 3 |
|---|---|---|---|
| Power/W | | 210 | 1230 | 1510 |

**Table 2.** Order parameter

| Parameter | 1 | 2 | 3 | 4 | 5 | 6 | 7 |
|---|---|---|---|---|---|---|---|
| Release time | 20 | 120 | 280 | 400 | 580 | 750 | 840 |
| Processing time | 70/60/50 | 90/80/70 | 160/120/100 | 110/90/70 | 210/180/150 | 140/120/100 | 110/90/70 |
| Due date | 100 | 300 | 450 | 500 | 750 | 850 | 940 |

**Table 3.** Experimental results of the case

| Parameter | 1 | 2 | 3 | 4 | 5 | 6 | 7 |
|---|---|---|---|---|---|---|---|
| starting time | 20 | 120 | 280 | 400 | 580 | 750 | 850 |
| position | 1 | 1 | 1 | 2 | 1 | 3 | 2 |

**Table 4.** Optimize the target results of the case

| Algorithm | Total carbon emission/$kgCO_2$ |
|---|---|
| Dynamic programming algorithm | 362775 |
| Traditional algorithm | 921100 |

## 5    Conclusion

Aiming at the green SMSP_RTs with multiple speeds of the machine, this paper proposes an accurate dynamic programming algorithm to solve. This is the first time that a dynamic programming algorithm has been used to solve this problem. The specific conclusions are as follows:

(1) Considering that different gears of the machine have different powers, the machining time of the workpieces in different gear positions is different, and the sorting model of the problem is established.
(2) Using the nature of the sorting model problem, a rule based on workpiece sorting and machine gear selection to ensure that the optimal solution is not lost is proposed.
(3) Based on the proposed rules to construct the state recursive equation, an accurate dynamic programming algorithm with pseudo-polynomial time is designed to obtain the optimal solution of the problem.

The future research work is aimed at the green multi-machine scheduling problem, mining the nature of valuable problems, designing the fusion problem decomposition strategy and the effective solution algorithm of dynamic programming.

**Acknowledgements.** This research is partially supported by National Natural Science Foundation of China (51665025), Applied Basic Research Key Project of Yunnan, China, and National Natural Science Fund for Distinguished Young Scholars of China (61525304).

# References

1. Wang, L., Wang, J., Wu, C.: Research progress in green shop scheduling optimization. Control Decis. (2018). Gong, H., Tang, L., Duin, C.W.: A two-stage flow shop scheduling problem on a batching machine and a discrete machine with blocking and shared setup times. Comput. Oper. Res. **37**(5), 960–969 (2010)
2. Che, A., Lv, K., Levner, E., et al.: Energy consumption minimization for single machine scheduling with bounded maximum tardiness. In: IEEE, International Conference on Networking, Sensing and Control, pp. 146–150. IEEE (2015). Fang, K., Uhan, N.A., Zhao, F., Sutherland, J.W.: Flow shop scheduling with peak power consumption constraints. Ann. Oper. Res. **206**(1), 115–145 (2013)
3. Mouzon, G., Yildirim, M.: A framework to minimise total energy consumption and total tardiness on a single machine. Int. J. Sustain. Eng. **1**(2), 105–116 (2008). Liu, C.H., Huang, D.H.: Reduction of power consumption and carbon footprints by applying multi-objective optimisation via genetic algorithms. Int. J. Prod. Res. **52**(2), 337–352 (2014)
4. Wang, Y.C., Wang, M.J., Lin, S.C.: Selection of cutting conditions for power constrained parallel machine scheduling. Robot. Comput. Integr. Manuf. **43**, 105–110 (2017)
5. Bin, H.: Research on parallel machine scheduling considering machine switch. Ind. Eng. Manage. **16**(2), 60–64 (2011)
6. Wu, X., Sun, Y.: Green scheduling problem of flexible job shop with multiple speeds of machine. Comput. Integr. Manuf. Syst. **24**(4) (2018)
7. Ding, J.Y., Song, S., Wu, C.: Carbon-efficient scheduling of flow shops by multi-objective optimization. Eur. J. Oper. Res. **248**(3), 758–771 (2015)
8. Zhang, R., Chiong, R.: Solving the energy-efficient job shop scheduling problem: a multi-objective genetic algorithm with enhanced local search for minimizing the total weighted tardiness and total energy consumption. J. Clean. Prod. **112**, 3361–3375 (2016)
9. Guohui, Z., Shijie, D.: Research on low carbon flexible job shop scheduling problem considering machine speed. J. Comput. Appl. **34**(4), 1072–1075 (2017)
10. Guo, P., Cheng, W., Wang, Y.: Scheduling step-deteriorating jobs to minimise the total weighted tardiness on a single machine. Int. J. Syst. Sci. Oper. Logistics **4**(2), 16 (2017)
11. Çakar, T.: Single machine scheduling with unequal release date using neuro-dominance rule. J. Intell. Manuf. **22**(4), 481–490 (2011)
12. Chen, C.L., Bulfin, R.L.: Complexity of single machine, multi-criteria scheduling problems. Eur. J. Oper. Res. **70**(1), 115–125 (1993)
13. Zhong, T., Xiao, W., Xu, H., et al.: Hybrid evolutionary algorithm for single-machine scheduling problem with preparation time. J. Comput. Appl. **30**(11), 3248–3252 (2013)
14. Ye, Q.: Research on a class of single machine scheduling problem based on improved ant colony algorithm. Hefei University of Technology (2008)
15. Wang, J.: Sustainable machine scheduling problem with minimizing carbon emissions. Control Decis. **32**(6), 1063–1068 (2017)

# Whale Optimization Algorithm with Local Search for Open Shop Scheduling Problem to Minimize Makespan

Hui-Min Gu[1], Rong Hu[1,2(✉)], Bin Qian[1,2], Huai-Ping Jin[1],
and Ling Wang[3]

[1] School of Information Engineering and Automation,
Kunming University of Science and Technology, Kunming 650500, China
ronghu@vip.163.com
[2] School of Mechanical and Electronic Engineering,
Kunming University of Science and Technology, Kunming 650500, China
[3] Department of Automation, Tsinghua University, Beijing 100084, China

**Abstract.** In this paper, a hybrid whale optimization algorithm (HWOA) with local search strategy is proposed to minimize the makespan for the Open-shop Scheduling Problem (OSP), which is one of the most important scheduling types in practical applications. Firstly, the large-order-value (LOV) encoding rule is presented to transform HWOA's individuals from continuous vectors into job permutations, which makes HWOA suitable for dealing with the OSP and performing global search in the solution space. Secondly, a local search mechanism guided by different neighborhoods is designed to enhance the search depth in the promising regions of excellent solutions. Computational experiments and comparisons show that HWOA performs well on random generation problems. This is the first time that HWOA has been used to address the OSP.

**Keywords:** Open-shop Scheduling Problem · Makespan ·
Whale optimization algorithm · Local search

## 1 Introduction

With the introduction of "Industry 4.0" and "Made in China 2025", the manufacturing industry will face severe challenges of transformation and upgrading and scheduling is the core and key technology of the manufacturing process. Reasonable scheduling can shorten the manufacturing period, reduce resource waste, and improve economic and management benefits. Open-shop Scheduling Problem (OSP) is a kind of special production scheduling problem that exists in practice. OSP can be applied not only to enterprise production management, industrial process modeling, but also to inspection and service industries, such as: inspection and diagnosis of machine network operations in hospitals [1], inspection and maintenance of large equipment such as automobiles and aircraft [2], logistics and road construction [3], medical examination [4], teacher or student examination arrangement [5] and so on, the common feature of the above examples is that there is no order constraint between the operations, so this type of problem can be simplified to OSP, it can be seen that it has a strong application

© Springer Nature Switzerland AG 2019
D.-S. Huang et al. (Eds.): ICIC 2019, LNCS 11644, pp. 678–687, 2019.
https://doi.org/10.1007/978-3-030-26969-2_64

background. Therefore, this paper studies the effective algorithm for OSP with the minimum makepsan as the optimization goal. In terms of computational complexity, OSP is an optimal combination of complexity and difficulty. Gonzalez and Sahni [6] have shown that when there are more than two machines, OSP is an NP-Hard problem, meaning that with a few exceptions, it is basically impossible to find the optimal solution in polynomial time. In summary, the study of the algorithm for solving the OSP that minimizes the makepsan is not only important for the development of operations research and combinatorial optimization theory, but also has important practical significance.

The OSP is more complex and it is more difficult to find the optimal solution in a limited time. In the past few decades, domestic and foreign scholars have conducted extensive research on OSP, and mainly focused on minimizing the makespan. Gonzalez and Sahni [6] first studied the OSP, gave the definition and a linear time solving algorithm for the case of two machines. Brucker and Hurink [7] used the branch and bound method based on Disjunctive Graph (DG) to minimize the optimal solution of some examples of Taillard [8]. Liaw [9] proposed a local search algorithm based on tabu search for non-preemptive OSP to minimizing makespan. At the same time, using the neighborhood structure by defining the block structure on the critical path to improve the current solution. The optimal solution of most examples and the calculation time is reasonable. Research on OSP focuses more on artificial heuristics based on artificial intelligence. Prins [10] aimed at minimizing the makespan of OSP, pointing out that the OSP's no-process constraint is the reason that the performance of the meta-heuristic algorithm based on DG-based branch and bound method and tabu search is weak in solving OSP. Several genetics are proposed to solve OSP, not only obtained the optimal solution of the Taillard [8] example, but also improved the optimal solution of an example. Sha [11] proposed a new particle swarm optimization algorithm for minimizing makespan. It can solve the discrete OSP by using the priority and interpolation operators to modify the particle position and motion representation respectively, it obtains many new optimal solution of the Taillard [8] example. Gao [12] proposed the PSO-OSP algorithm for minimizing makespan. By analyzing the defects of particle swarm optimization (PSO) in information sharing mechanism, a new information sharing mechanism based on swarm intelligence was proposed. The neighborhood knowledge guides the local search and overcomes the blind search caused by the randomness of the meta heuristic algorithm.

The whale optimization algorithm (WOA) is a novel biomimetic group intelligence proposed by Mirjalili [13] in 2016. WOA has been proved to be superior to PSO, DE, gravitational search algorithm (GSA) and GWO algorithm in solving accuracy and convergence speed [13]. At present, the research on whale optimization algorithm is still in its preliminary stage. In the field of shop scheduling: Yan [14] aimed at minimizing the maximum completion time of the job shop, a quantum whale optimization algorithm is proposed. The algorithm makes use of the quantum computing and optimization ideas to some extent to make up for the shortcomings of WOA with low convergence precision and easy to fall into local optimum. Abdel-Basset et al. [15] proposed a simple hybrid whale optimization algorithm for the replacement flow shop problem. It can be seen that there is no literature report on solving the open shop scheduling problem, so it is very important to carry out relevant research.

This paper proposes a hybrid whale optimization algorithm to minimize makespan of OSP. Firstly, the Large-order-value (LOV) coding rule is presented to transform HWOA's individuals from continuous vectors into job permutations, which can make OSP solve the problem solution space effectively. Then, a local search mechanism guided by perturbation and exploitation is designed to enhance the optimization accuracy of the algorithm. Finally, the experimental results show that HWOA has good performance on random generation problems.

In Sect. 2, the mixed integer linear programming (MILP) model of OSP is briefly introduced. In Sect. 3, WOA is described briefly. In Sect. 4, HWOA is proposed and described in detail. In Sect. 5, simulation results and comparisons are provided. Finally, we end the paper with some verdicts and future work in Sect. 6.

## 2    Problem Description of OSP

Firstly, define the symbols required in this article, including parameters, indicators and variables as shown in Table 1.

**Table 1.** Symbols and definitions

| Notation | Description | Notation | Description |
|---|---|---|---|
| $n$ | Number of jobs | $O_{ij}$ | The operation of job $i$ on machine $j$ |
| $m$ | Number of machines | $p_{ij}$ | The processing time of $O_{ij}$ |
| $J_i$ | Job $i$ | $F_{ij}$ | The complete time of $O_{ij}$ |
| $M_j$ | Machine $j$ | $C_i$ | The complete time of job $i$ |
| $\pi$ | Job permutations | $S_{ij}$ | The start time of $O_{ij}$ |
| $\theta$ | Set of all $O_{ij}$ | | |

### 2.1    Problem Description

Given a set $J = \{J_i | i = 1, 2, \ldots, n\}$ of $n$ jobs and a set $M = \{M_j | j = 1, 2, \ldots, m\}$ of $m$ machines. Each job $J_i$ consists of $m$ operations $(O_{ij}, \forall i, \forall j)$, where $O_{ij}$ has to be processed on machine $M_j$ for $(p_{ij} > 0, \forall i, \forall j)$ time units. The operations of each job can be processed in any order; each machine can process at most one operation at a time. The goal of OSP scheduling is to determine the sequential combination of all jobs and machines while satisfying the above assumptions and constraints, so that the scheduling performance metrics are optimal.

### 2.2    Mathematical Models of OSP

Before describing the model, we need to define the decision variables

$$Y_{ikj} = \begin{cases} 1, & \text{for machine } j, \text{ if process } O_{kj} \text{ when finish } O_{ij}, \ 0 \leq j \leq m, \ 0 \leq i \neq k \leq n \\ 0, & \text{or else} \end{cases}$$

$$X_{iff} = \begin{cases} 1, \textit{for job } i, \textit{ if process } O_{if} \textit{ when finish } O_{ij}, \ 0 \le i \le n, \ 0 \le j \ne f \le m \\ 0, \textit{ or else} \end{cases}$$

Under the symbols and decision variables defined above, let $C_{max}$ be the objective function, which represents makespan. The mixed integer linear programming model of OSP can be given as follows

$$Min \ C_{max} \tag{1}$$

$$s.t. \ \sum_{i=0,i\ne k}^{n} Y_{ikj} = 1, \forall k,j \tag{2}$$

$$\sum_{k=1,k\ne i}^{n} Y_{ikj} \le 1, \forall i,j \tag{3}$$

$$\sum_{k}^{n} Y_{0kj} = 1, \forall j \tag{4}$$

$$Y_{ikj} + Y_{kij} \le 1, \forall j, 0 < i \ne k < n \tag{5}$$

$$S_{kj} - S_{ij} \ge P_{ij}, \forall j, 0 < i \ne k < n \tag{6}$$

$$S_{if} - S_{ij} \ge P_{ij}, \forall i, 0 < j \ne f < m \tag{7}$$

$$C_i \ge F_{ij} > 0, \forall i,j \tag{8}$$

Among them, formula (2) means that each job can be processed once on any machine; formula (3) means that only one job can be processed after machining one job per machine; formula (4) means that the 0 process can be a pre-process of any one of the job; formula (5) means that one process cannot be the former or the latter of the other process at the same time; formula (6) indicates that the same machine cannot process two job at the same time; formula (7) indicates that for the same process, it cannot be processed on two machines at the same time; formula (8) means that the completion time of one job is not less than the processing time of any one of the processes.

## 3   Whale Optimization Algorithm

The WOA is a novel meta-heuristic that simulate the social behavior of humpback whales in chasing their prey. The modeling process can be described as follows

### 3.1   Bubble-Net Attacking Method (Exploitation Phase)

The humpback whale's bubble-net attack method swims around the prey along a spiral path within a gradually constricted circle, the strategy of encircling the prey can be represented by the following equations

$$D = |C \times X^*(t) - X(t)| \tag{9}$$

$$X(t+1) = X^*(t) - A \times D \tag{10}$$

$$A = 2 \times a \times r - a \tag{11}$$

$$C = 2 \times r \tag{12}$$

where $t$ indicates the current iteration, $A$ and $C$ are coefficient vectors, $X^*$ is the position vector of the best solution obtained so far, $X$ is the position vector; $X^*$ would be updated in each iteration if there is a better solution. $a$ is linearly decreased from 2 to 0 over the process of iterations and $r$ is a random vector in [0, 1].

The behavior of the spiral-shaped path can be formulated and assuming a 50% probability of selecting the position of the whale in the shrink-wrap or spiral model. Set $p$ to a random value in [0, 1], the bubble network attack process can be fully described as follows

$$X(t+1) = \begin{cases} X^*(t) - A \times D & p < 0.5 \\ D' \times e^{bl} \times \cos(2 \times \pi \times l) + X^*(t) & p \geq 0.5 \end{cases} \tag{13}$$

Where $b$ is a constant for defining the shape of the logarithmic spiral, $l$ is a random number in [−1, 1].

### 3.2 Search for Prey (Exploration Phase)

In fact, humpback whales search randomly according to the position of each other, the mathematical model is as follows:

$$D = |C \times X_{rand} - X| \tag{14}$$

$$X(t+1) = X_{rand} - A \times D \tag{15}$$

where $X_{rand}$ is a random position vector (a random whale) chosen from the current population.

## 4   HWOA for OSP with Makespan

Since traditional WOA is suitable for solving continuous problems, it is usually combined with several methods to solve discrete problems. This paper proposes a hybrid whale optimization algorithm, which combines WOA with local search strategy to improve the performance of WOA. HWOA consists of four main steps: encoding, decoding, population initialization, and local search design. All four steps will be discussed in the following sections.

## 4.1    Encoding and Solution Expression

The important problem in applying WOA to OSP is to find a suitable mapping between job sequence and individuals (continuous vectors) in WOA. For the $n$ job and $m$ machine problem, each vector contains $n$ number of dimensions corresponding to $n$ operations. In this paper, we adopt a largest-order-value (LOV) rule [16] to convert the $i$th individual of WOA $X_i = [x_{i,1}, x_{i,2}, \ldots, x_{i,n}]$ to the job solution $\pi_i = [\pi_{i,1}, \pi_{i,2}, \ldots, \pi_{i,n}]$. According to LOV rule, $X_i = [x_{i,1}, x_{i,2}, \ldots, x_{i,n}]$ are firstly ranked by descending order to get the sequence $\varphi_i = [\varphi_{i,1}, \varphi_{i,2}, \ldots, \varphi_{i,n}]$. Then the job permutation $\pi_i$ is calculated by the following formula:

$$\pi_{i,\varphi_{i,k}} = k \tag{16}$$

To better understand the LOV rule, a simple example is provided in Table 2. In this instance, when $k = 1$, $\varphi_{i,1} = 6$, $\pi_{i,\varphi_{i,1}} = \pi_{i,6} = 1$; when $k = 2$, $\varphi_{i,2} = 3$, $\pi_{i,\varphi_{i,2}} = \pi_{i,3} = 2$, and so on. This representation is unique and simple in terms of finding new permutations (Table 2).

**Table 2.**  Solution representation

| k | 1 | 2 | 3 | 4 | 5 | 6 |
|---|---|---|---|---|---|---|
| $x_{i,k}$ | 0.11 | 2.86 | 1.25 | 3.9 | 3.93 | 2.73 |
| $\varphi_{i,k}$ | 6 | 3 | 5 | 2 | 1 | 4 |
| $\pi_{i,k}$ | 5 | 4 | 2 | 6 | 3 | 1 |

Permutation list is one of the widely used encoding schemes. It is purely the relative order of all the $g$ operations in which they are processed ($g = n \times m$). For example, consider a problem of two machines and three jobs. It sums up to six operations: $O_{12}$, $O_{22}$, $O_{32}$, $O_{11}$, $O_{21}$, $O_{31}$. Every permutation of these six operations, like $\theta = \{O_{12}, O_{22}, O_{32}, O_{11}, O_{31}, O_{21}\}$, corresponds to one solution.

## 4.2    Decoding

It is known that in terms of makespan, the search space is reduced in such a way that it does not eliminate the optimal solution by this decoding. In this paper, we mean to examine the performance of decoding by the set of non-delay schedule. We apply the procedure proposed by Giffler and Thompson [17]: all operations are put into a set including unscheduled ($U$) operations. $S_{ij}$ means the earliest possible at which the operation $O_{ij}$ could be started. We count $y$ which equals the minimum of $S_{ij}$ of operations in $U$. All the operations whose starting time is equal to $y$ are put into a set called $R$. Among all operations in $R$, the operation $O^*$ with the earliest relative position in permutation is scheduled and extracted from $U$.

### 4.3   Population Initialization

For population initialization, because the solution space of OSP is large and there is no prior knowledge of the global optimal solution to the problem, we will generate the initial population randomly, that is, each individual is randomly generated $g$ $(g = n \times m)$ non-repeating continuous real numbers.

### 4.4   Local Search

It is known that the neighborhood can lead the search to a more various region and then enhance the search ability of HWOA in a way, so, we employ *insert* and *cut here* as the neighborhood for local search. Two operators can be described as follows.

**Insert:** randomly select two different locations of the sequence and insert the back one before the front one.

**Cut:** randomly select two different locations of the sequence and insert the job between the two locations before the front of the sequence.

These operators are executed in the given order to generate another solution, and then the new solution replaces the old one if it precedes the old. The above procedure is designed as follows:

**Step 1:** Convert individual $X_i(t)$ to a job permutation $\pi_{i\_0}$ according to the LOV rule.

**Step 2:** Perturbation phase.
　　　Randomly select $u$ and $v$, where $u \neq v$; $\pi_i = insert(\pi_{i\_0}, u, v)$.

**Step 3:** Exploitation phase.
　　　Set $k = 1$;
　　　Do
　　　Randomly select $u$ and $v$, where $u \neq v$;
　　　$\pi_{i\_1} = cut(\pi_i, u, v)$;
　　　If $f_{i\_1} < f_i$, then $\pi_i = \pi_{i-1}$;
　　　$k = k + 1$;
　　　While $k < (n \times (n - 1))$;

**Step 4:** If $f_i < f_{i\_0}$, then $\pi_{i\_0} = \pi_i$.

**Step 5:** Convert $\pi_{i\_0}$ back to $X_i(t)$.

With the above design, the procedure of HWOA for solving the OSP with makespan is illustrated in Fig. 1.

## 5   Simulation Result and Comparisons

### 5.1   Experimental Setup

To test the performances of WOA, we will test the effectiveness of algorithm on randomly generated test questions. For OSP, $n$ jobs are randomly machined on $m$ machines, and the processing time of each process is the integers generated in [1, 30], we compare the effective algorithms proposed by Naderi et al. [18].

HWOA and HGA are coded in Delphi 2010 and run on a PC with Intel 7700HQ 2.80 GHz. The proposed HWOA's parameters are set as follows: Since the number of

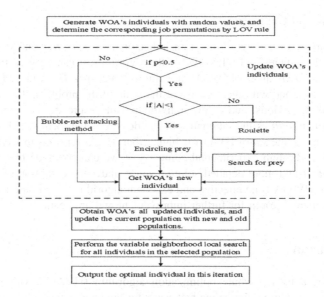

**Fig. 1.** Framework of HWOA

instances of different scales measured by the algorithm is huge, and the running time of each generation is quite different, the running generation (*gen*) and the population of the corresponding problem (*pop*) will be respectively given. The comparison algorithm runs 20 times independently for each test problem at the same runtime as HWOA, where *min*, *max*, *avg*, and *std* are the minimum, maximum, average, and standard deviation, respectively. The optimal results corresponding to each problem are shown in bold, and the test results are shown in Table 3.

**Table 3.** Comparisons of HWOA and HGA

| n_m | HWOA | | | | HGA | | | | gen | pop |
|---|---|---|---|---|---|---|---|---|---|---|
| | *min* | *max* | *avg* | *std* | *min* | *max* | *avg* | *std* | | |
| 10_5 | **175** | 217 | **214.9** | 9.15 | 217 | 217 | 217 | 0 | 50 | 20 |
| 30_5 | **482** | 493 | **492.45** | 2.39 | 493 | 493 | 493 | 0 | 50 | 20 |
| 20_10 | **334** | 339 | **338.75** | 1.08 | 339 | 339 | 339 | 0 | 50 | 20 |
| 50_5 | **764** | 788 | **786.8** | 5.23 | 788 | 788 | 788 | 0 | 50 | 20 |
| 30_10 | **488** | 511 | **509.85** | 5.01 | 511 | 511 | 511 | 0 | 50 | 20 |
| 50_10 | **882** | 888 | **887.7** | 1.30 | 888 | 888 | 888 | 0 | 10 | 20 |
| 40_20 | **730** | 736 | **735.7** | 1.30 | 736 | 736 | 736 | 0 | 5 | 20 |
| 60_10 | **991** | 1007 | **1006.2** | 3.49 | 1007 | 1007 | 1007 | 0 | 5 | 20 |

## 5.2  Results and Comparison

In the comparison of the algorithms, four indicators *min*, *max*, *avg* and *std* are used to evaluate the advantages and inferiority of the algorithm. It can be seen from Table 3 that the *avg* and *min* values of HWOA are almost better than HGA in each instance of various problems, and with the increase of the scale of the problem, the superiority of HWOA is more obvious, and a few individuals can ensure the performance of the algorithm, which verifies the rationality of the design of the local search mechanism. On the other hand, the *std* of HWOA is bigger than HGA, it means that HWOA is not stable enough. Through experimental comparison, we can conclude that HWOA can search for larger and more detailed solution space under the condition of running the same algebra. HWOA is an effective and excellent algorithm for solving the makespan of OSP, and there is room for further improvement and application.

## 6  Conclusion

In this paper, a hybrid whale optimization algorithm (HWOA) with local search strategy is proposed to minimize the makespan for the Open-shop Scheduling Problem (OSP), which is one of the most important scheduling types in practical applications. Firstly, the large-order-value (LOV) encoding rule is presented to transform HWOA's individuals from continuous vectors into job permutations, which makes HWOA suitable for dealing with the OSP and performing global search in the solution space. Secondly, a local search mechanism guided by *inter*-based perturbation and *cut*-based exploitation is designed to enhance the search depth in the promising regions of excellent solutions and improve the ability of WOA. Computational experiments and comparisons show that HWOA performs well on random generation problems. This is the first time that HWOA has been used to address the OSP. The future research work is to solve the multi-objective OSP and design a more efficient WOA.

**Acknowledgements.**  This research is partially supported by National Natural Science Foundation of China (51665025), Applied Basic Research Key Project of Yunnan, China and National Natural Science Fund for Distinguished Young Scholars of China (61525304).

## References

1. Kubiak, W., Sriskandarajah, C., Zaras, K.: A note on the complexity of open shop scheduling problems. Inform. Syst. Oper. Res. **29**(4), 284–294 (1991)
2. Low, C., Yeh, Y.: Genetic algorithm-based heuristics for an open shop scheduling problem with setup, processing, and removal times separated. Robot. Comput.-Integr. Manuf. **25**(2), 314–322 (2009)
3. Matta, M.E.: A genetic algorithm for the proportionate multiprocessor open shop. Comput. Oper. Res. **36**(9), 2601–2618 (2009)
4. Prins, C.: An overview of scheduling problems arising in satellite communications. J. Oper. Res. Soc. **45**(6), 611–623 (1994)

5.  Liaw, C.F.: An efficient tabu search approach for the two-machine preemptive open shop scheduling problem. Comput. Oper. Res. **30**(14), 2081–2095 (2003)
6.  Gonzalez, T.F., Sahni, S.: Open shop scheduling to minimize finish time. J. ACM **23**(4), 665–679 (1976)
7.  Brucker, P., Hurink, J., Jurisch, B., Wostmann, B.: A branch and bound algorithm for the open shop problem. Discrete Appl. Math. **76**(1–3), 43–59 (1997)
8.  Taillard, E.: Benchmarks for basic scheduling problems. Eur. J. Oper. Res. **64**(2), 278–285 (1993)
9.  Liaw, C.F.: A tabu search algorithm for the open shop scheduling problem. Comput. Oper. Res. **26**(2), 109–126 (1999)
10. Prins, C.: Competitive genetic algorithms for the open-shop scheduling problem. Math. Methods Oper. Res. **52**(3), 389–411 (2000)
11. Sha, D.Y., Hsu, C.Y.: A new particle swarm optimization for the open shop scheduling problem. Comput. Oper. Res. **35**(10), 3243–3261 (2008)
12. Gao, L., Gao, H.B., Zhou, C.: Open shop scheduling based on particle swarm optimization. J. Mech. Eng. **42**(2), 129–134 (2006)
13. Mirjalili, S., Lewis, A.: The whale optimization algorithm. Adv. Eng. Softw. **95**, 51–67 (2016)
14. Yan, X., Ye, C.M., Yao, Y.Y.: Solving job-shop scheduling problem by quantum whale optimization algorithm. Appl. Res. Comput. **36**(04), 1–2 (2019)
15. Abdel-Basset, M., Gunasekaran, M., El-Shahat, D., et al.: A hybrid whale optimization algorithm based on local search strategy for the permutation flow shop scheduling problem. Future Gener. Comput. Syst. **85**, 129–145 (2018)
16. Qian, B., Zhou, H.-B., Hu, R., Xiang, F.-H.: Hybrid Differential evolution optimization for no-wait flow-shop scheduling with sequence-dependent setup times and release dates. In: Huang, D.-S., Gan, Y., Bevilacqua, V., Figueroa, J.C. (eds.) ICIC 2011. LNCS, vol. 6838, pp. 600–611. Springer, Heidelberg (2011). https://doi.org/10.1007/978-3-642-24728-6_81
17. Giffler, B., Thompson, G.L.: Algorithms for solving production scheduling problems. Oper. Res. **8**(4), 487–503 (1960)
18. Naderi, B., Ghomi, S.M.T.F., Aminnayeri, M., et al.: A study on open shop scheduling to minimise total tardiness. Int. J. Prod. Res. **49**(15), 4657–4678 (2011)

# Water Wave Optimization
# for the Multidimensional Knapsack Problem

Hong-Fang Yan, Ci-Yun Cai, De-Huai Liu, and Min-Xia Zhang[✉]

College of Computer Science and Technology,
Zhejiang University of Technology, Hangzhou 310023, China
minxia_zhang@yeah.net

**Abstract.** Water wave optimization (WWO) is a novel evolutionary algorithm
that draws inspiration from shallow water wave model for continuous optimization problems. Multidimensional knapsack problem (MKP) is a well-
known NP-hard combinatorial optimization problem which has a wide range of
applications in practice. This paper proposes a discrete WWO algorithm for
MKP, which adapts the key evolutionary operators including propagation and
breaking to effectively search the discrete solution space of MKP. We also
propose an adaptive WWO algorithm for MKP by integrating four different
breaking operators and adaptively selecting among them according to their
historical performance in the evolution. Experimental results show that the
proposed discrete WWO algorithms have a competitive performance compared
to a number of well-known evolutionary algorithms including genetic algorithm,
particle swarm optimization, fruit fly optimization, ant colony optimization, and
some state-of-the-art hybrid evolutionary algorithms.

**Keywords:** Water wave optimization (WWO) ·
Multidimensional knapsack problem (MKP) ·
Combinatorial optimization · Self-Adaptive

## 1 Introduction

The multidimensional knapsack problem (MKP) [1] is a classic combinatorial optimization problem. It can be stated as follows. There are $n$ items to be selected for
inclusion in $m$ knapsacks. Each knapsack $j$ has a capacity $c_j$, and each item $i$ requires $a_{ij}$
units of resource consumption in the knapsack $j$ and yields $p_j$ units of profit upon
inclusion ($i = 1, 2, \ldots, n; j = 1, 2, \ldots, m$). The goal is to find a subset of items that
yields maximum profit without exceeding the knapsack capacities:

$$\max f(\mathbf{x}) = \sum_{i=1}^{n} p_i x_i \tag{1}$$

Supported by grants from National Natural Science Foundation of China under Grant No. 61662036.

D.-S. Huang et al. (Eds.): ICIC 2019, LNCS 11644, pp. 688–699, 2019.
https://doi.org/10.1007/978-3-030-26969-2_65

$$\text{s.t.} \ \sum_{i=1}^{n} a_{ij}x_i \le c_j, \ j = 1, \ldots, m \tag{2}$$

$$x_i \in \{0, 1\}, \ i = 1, \ldots, n \tag{3}$$

where $x_i = 1$ denotes that item $i$ is selected and $x_i = 0$ otherwise, and $p_i$, $a_{ij}$ and $c_j$ are all non-negative numbers.

MKP has a wide range of applications in manufacturing, transportation, finance, etc. There are many practical problems, such as packing, capital budgeting, cutting stock, project selection, resource allocation, and so on, which can be attributed to MKP. Consequently, it is of great theoretical and practical significance to design efficient algorithms for solving MKP.

Nevertheless, MKP has been proved to be NP-hard, which indicates that there are no known polynomial time algorithms for the problem. Consequently, the computational time of classical exact algorithms, such as exhaustive search, branch-and-bound, integer programming, etc., will become unaffordable when the MKP instance size is relatively large. In recent years, evolutionary algorithms have been a popular approach for MKP, because they can efficiently solve large-size MKP instances with an acceptable computational time. A number of popular evolutionary algorithms, such as genetic algorithm (GA) [2], particle swarm optimization (PSO) [3], ant colony optimization (ACO) [4], etc., has shown their effectiveness and efficiency compared to traditional approaches for solving MKP.

In this study, we propose a new evolutionary algorithm for MKP based on water wave optimization (WWO) [5]. WWO is a nature-inspired algorithm based on shallow water wave model that regards each solution to the problem as a wave. The core idea of WWO is to assign each solution a wavelength that is inversely proportional to its fitness, uses a propagation operator to change each solution based on its wavelength, uses a refraction operator to remove stagnant solutions, and uses a breaking operator to perform local search. Since its proposal, WWO have gained great research interest and has been applied or adapted to many combinatorial optimization problems [6–13]. However, to our knowledge, the studies on WWO for knapsack problems are still few. We propose a discrete WWO for MKP by adapting the key operators of the basic WWO to effectively search the discrete solution space of MKP. We also propose an adaptive WWO algorithm for MKP by integrating four different local search operators into the algorithm. The experimental results show that the proposed discrete WWO algorithms have a competitive performance compared to a number of popular evolutionary algorithms.

The remainder of this paper is structured as follows. Section 2 briefly introduces the basic WWO algorithm. Section 3 proposes the discrete WWO algorithm for MKP. Section 4 conducts computational experiments. Finally, Sect. 5 concludes with a discussion.

## 2   Water Wave Optimization

WWO [5] is a novel metaheuristic evolutionary algorithm that draws inspiration from shallow water wave model to solve optimization problems. In WWO, each solution is analogous to a wave, the solution space is analogous to the seabed area, and the fitness of a solution is measured inversely by its seabed depth: the higher the energy (i.e., fitness), the smaller the wavelength, and the smaller range the wave propagates, as shown in Fig. 1. It should be noted that, for a high-dimensional optimization problem, the 2-D or 3-D space of the seabed should be extended to a hyperspace space.

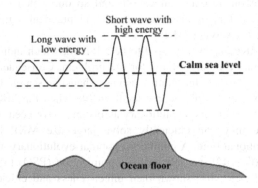

**Fig. 1.** Illustration of the relationship between wavelengths and wave energy (fitness) in WWO.

As most evolutionary algorithms, WWO maintains a population of solutions (waves), which are evolved by three operators named propagation, refraction, and breaking.

**Propagation.** At each generation, each wave **x** propagates once to creates a new wave **x′** by shifting each dimension $d$ in a range proportional to its wavelength $\lambda_x$ as follows:

$$x'(d) = x(d) + \lambda_x \cdot rand(-1, 1) \cdot L(d) \tag{4}$$

where function $rand(-1, 1)$ produces a uniformly distributed random number within the range $[-1, 1]$, and $L(d)$ is the $d$-th dimension of the search space.

If the new wave **x′** is better than the original **x**, then **x** is replaced by **x′**.

After each generation, each wave **x** updates its wavelength according to its fitness $f(\mathbf{x})$ as follows:

$$\lambda_x = \lambda_x \cdot \alpha^{-(f(x)-f_{min}+\epsilon)/(f_{max}-f_{min}+\epsilon)} \tag{5}$$

where $f_{max}$ and $f_{min}$ are the maximum and minimum fitness among the population, respectively, $\alpha$ is the wavelength reduction coefficient which is set up 1.0026, and $\epsilon$ is a very small number to avoid division by zero.

**Refraction.** If a wave **x** does not improve after a predefined number of generations, it moves to a newly generated wave **x′** by refraction as follows:

$$x'(d) = \text{norm}\left(\frac{x^*(d) + x(d)}{2}, \frac{|x^*(d) - x(d)|}{2}\right) \tag{6}$$

where $x^*$ is the current best wave in the population, and $norm(\mu, \sigma)$ produces a Gaussian random number with mean $\mu$ and standard deviation $\sigma$.

After a refraction operation, the wavelength of $x$ is updated as follows:

$$\lambda_{x'} = \lambda_x \frac{f(x)}{f(x')} \tag{7}$$

**Breaking.** Whenever a new best wave $x^*$ is found, it breaks into several solitary waves:

$$x'(d) = x^*(d) + \text{norm}(0, 1) \cdot \beta \cdot L(d) \tag{8}$$

where $\beta$ is the breaking coefficient. If none of the solitary waves is better than $x^*$, then $x^*$ keeps in the population; otherwise $x^*$ is replaced by the fittest one among the solitary waves.

**Framework.** Algorithm 1 presents the framework of the basic WWO.

---

**Algorithm 1:** The general framework of WWO

1  Randomly initialize a population $P$ of solutions to the problem;
2  Let $x^*$ be the best among the solutions;
3  **while** *the stopping condition does not match* **do**
4      **foreach** $x \in P$ **do**
5          Propagate $x$ to a new $x'$ based on Eq. (4);
6          **if** $f(x') > f(x)$ **then**
7              Replace $x$ with $x'$;
8              **if** $f(x') > f(x^*)$ **then**
9                  Replace $x^*$ with $x'$;
10                 Break $x^*$ based on Eq. (8);
11         **else**
12             $h_x = h_x - 1$;
13             **if** $h_x = 0$ **then**
14                 Refract $x$ to a new $x'$ based on Eq. (6) and Eq. (7);
15     Update the wavelengths based on Eq. (5);
16 **return** $x^*$.

---

## 3  A Discrete WWO Algorithm for MKP

In this section, we adapt WWO to solve the MKP. As we all know, the basic WWO is proposed for continuous optimization problems, while the MKP is a combinatorial optimization problem. Thus, we adapt the operators of WWO to evolve MKP solutions in a discrete search space.

### 3.1   Solution Representation and Fitness Evaluation

We employ the basic 0–1 vector representation for MKP solutions. That is, given an MKP instance with $n$ items, each solution $\mathbf{x}$ is encoded as an $n$-dimensional solution vector, where each component $x_i = 1$ denotes that item $i$ is included in the knapsack and $x_i = 0$ otherwise.

As MKP is also a constrained optimization problem, we evaluate the fitness value of each solution $\mathbf{x}$ by incorporating constraint violations into the objective function:

$$f(\mathbf{x}) = \left( \sum_{i=1}^{n} p_i x_i \right) - M \sum_{j=1}^{m} \max \left( \sum_{i=i}^{n} a_{ij} x_i - c_j, 0 \right) \tag{9}$$

where $M$ is a large constant.

### 3.2   Propagation for MKP

The basic principle of WWO propagation is that the better (the worse) the solution, the more (the less) the solution changes. For MKP, we redefine the propagation operator as performing local search steps. As each solution $\mathbf{x}$ is an $n$-dimensional solution vector, we consider changing a component $x_i$ from 0 to 1 or from 1 to 0 as one local search step. A propagation operation on $\mathbf{x}$ performs a number of $k$ local search steps: the better (the worse) the solution, the smaller (the larger) the number $k$. Here $k$ is set to a random integer in the range of $[1, \lambda_{\mathbf{x}}]$.

Obviously, different from the basic WWO, wavelengths in the discrete WWO should be integers. We predefine a range $[\lambda_{min}, \lambda_{max}]$, and initialize the wavelength of each solution as a random integer in this range. After each generation, Eq. (10) replaces Eq. (5) in the basic WWO for wavelength updating:

$$\lambda_{\mathbf{x}} = \lambda_{min} + (\lambda_{max} - \lambda_{min}) \frac{f_{max} - f(\mathbf{x}) + \in}{f_{max} - f_{min} + \in} \tag{10}$$

As we can observe from Eq. (10), the best solution in the population has a wavelength of $\lambda_{min}$, the worst solution has a wavelength of $\lambda_{max}$, and the wavelengths of the other solutions are within $[\lambda_{min}, \lambda_{max}]$ and are inversely proportional to their fitness values.

Another difference between Eqs. (5) and (10) is that, based on Eq. (5), the average wavelength of the solutions gradually decreases with the number of generations, and thus makes the algorithm explore larger areas in early stages and exploit small regions in later stages. Equation (10) does not have such an effect, because the discrete search space of MKP cannot be decreased infinitely as in continuous search spaces.

It should be noted that, when a solution $\mathbf{x}$ propagates to a new solution $\mathbf{x}'$, in most cases, the objective function of the new $\mathbf{x}'$ does not need to be evaluated based on Eqs. (1) and (9). Instead, $f(\mathbf{x}')$ can be evaluated based on the original solution $\mathbf{x}$ and the local search steps between $\mathbf{x}$ and $\mathbf{x}'$:

- For each local search step that changes $x_i$ from 1 to 0, we have $f(\mathbf{x}') = f(\mathbf{x}) - p_i$;
- For each local search step that changes $x_i$ from 0 to 1, if the constraints are not violated, we have $f(\mathbf{x}') = f(\mathbf{x}) - p_i$; otherwise we have $f(\mathbf{x}') = f(\mathbf{x}) + p_i - M \sum_{j=1}^{m} \max\left(\sum_{i=i}^{n} a_{ij}x_i - c_j, 0\right)$;

If there are $k$ local search steps from $\mathbf{x}$ to $\mathbf{x}'$, the time complexity for evaluating the objective function (1) from scratch is $O(n)$, while the time complexity of the above increment evaluation is $O(k)$, and in most cases $k$ is much smaller than $n$.

### 3.3  Breaking for MKP

As in the basic WWO, the breaking operator breaks a best wave $\mathbf{x}^*$ newly found into a set of $n_b$ solitary waves (where $n_b$ is a control parameter). For the MKP, each solitary wave is considered as a neighbor solution to $\mathbf{x}^*$, which is produced by the following procedure:

(1) Sort the items in the knapsacks in increasing order of profit;
(2) Remove the item of the smallest profit from the knapsacks;
(3) Add other items in decreasing order of profit until there is no item can be added.

### 3.4  Reduction of Population Size and Removal of Refraction

The discrete WWO inherits a strategy of the simplified WWO [14] that removes the refraction operator from the algorithm. In order to remove stagnant or low-quality solutions from the population, it iteratively reduces the population size $NP$ from an upper limit $NP_{max}$ to a lower limit $NP_{min}$:

$$NP = NP_{min} + (NP_{max} - NP_{min}) \frac{g}{g_{max}} \tag{11}$$

where $g$ is the current generation number, and $g_{max}$ is the maximum allowable number of the generations. Whenever the size in the current generation is less than that in the previous generation, the worst solution in the population is removed.

### 3.5  An Improved Adaptive WWO Algorithm

The breaking operator of WWO is essentially a local search operator. We also propose an adaptive WWO algorithm for MKP by integrating the four different breaking operators, including the one (denoted by LS1) described in Sect. 3.3 and the following three local search operators

**LS1** Which sorts the items in the knapsacks in increasing order of profit, remove the item of the $k$-th smallest profit from the knapsacks ($2 \leq k \leq 6$), and add other items in de-creasing order of profit until there is no item can be added;
**LS2** Which sorts the items in the knapsacks in decreasing order of average consumption $\sum_{j=1}^{m} a_{ij}$, remove the item of the largest consumption from the

knapsacks, and add other items in decreasing order of profit until there is no item can be added;

**LS3** Which sorts the items in the knapsacks in increasing order of the ratio of profit to consumption $p_i / \left( \sum_{j=1}^{m} a_{ij} \right)$, remove the item of the smallest ratio from the knapsacks, and add other items in decreasing order of the ratio of profit to consumption until there is no item can be added.

Initially, the four local search operators have the same probability of 0.25 of being selected for breaking a newly found best wave. After the first *LP* generations (where *LP* is a parameter for controlling the learning period), the probability of each operator is adjusted at each generation based on its performance during the previous *LP* generations:

$$\rho_l = \frac{n_l^I / n_l}{\sum_{l=1}^{4} n_l^I / n_l} \tag{12}$$

where $n_l$ is the number of invocations of the *l*-th operator, and $n_l^I$ is the number of invocations that produce better solutions ($1 \leq l \leq 4$).

Algorithm 2 presents the framework of the discrete WWO algorithm for MKP, where Lines 11 and 17 are attributed to the adaptive WWO version.

---
**Algorithm 2:** The discrete WWO algorithm for MKP.
---
1   Randomly initialize a population $P$ of solutions to the problem;
2   Let $x^*$ be the best among the solutions;
3   **while** *the stopping condition does not match* **do**
4      **foreach** $x \in P$ **do**
5          Let $k = rand(1, \lambda_x)$;
6          Propagate $x$ to a new $x'$ by randomly reversing $k$ components of $x$;
7          **if** $f(x') > f(x)$ **then**
8              Replace $x$ with $x'$ in the population;
9              **if** $f(x) > f(x^*)$ **then**
10                  Replace $x^*$ with $x$ in the population;
11                  Select a breaking operator according to its probability $\rho_l$;
12                  Perform the breaking operation on $x^*$;
13                  **if** *the best neighboring solution is better than* $x^*$ **then**
14                      Replace $x^*$ with the best neighboring solution;
15      Update the wavelengths based on Eq. (10);
16      Update the population size based on Eq. (11);
17      Remove the worst solution if the population size is decreased;
18      Update the selection probabilities of the breaking operators based on Eq. (12);
19 **return** $x^*$.

---

# 4   Computational Experiment

To test the performance of the proposed discrete WWO algorithm for MKP, we use 10 test instances from OR-Library [2] and [15], the sizes ($m \times n$) of which range from $2 \times 28$ to $30 \times 500$. The following seven popular evolutionary algorithms for MKP are used for comparison:

- A basic GA [2].
- A binary PSO algorithm [16].

- A binary fruit fly optimization (FFO) algorithm [17].
- An ACO algorithm combined with local search [18].
- A hybrid harmony search (HHS) algorithm with local search [19].
- A binary artificial algae algorithm (BAAA) with elite local search [20].
- A hybrid GA (HGA) with local search based on simulated annealing [21].

For WWO, we set $NP_{max} = 5m \ln(n/2)$, $NP_{min} = 12$, $\lambda_{min} = 1$, $\lambda_{max} = 0.9m$, and $n_b = 6$. For the adaptive version (denoted by WWO-M), we set the learning period $LP = 30$. The control parameters of the other algorithms are set as suggested in the literature. All algorithms use the same stopping condition that the number of function evaluations reaches $50\ mn$. Each algorithm is run 30 times on each test instance, and the algorithm performance is evaluated in terms of the relative percentage deviation (RPD) of the obtained objective function value from the best known one.

The comparison can be divided into two groups. In the first group, we compare the basic WWO version (that uses the single breaking operator) with the first three algorithms (GA, PSO, and FFO), as they all do not use dedicated local search operators. In the second group, we compare WWO-M with the other four algorithms (ACO, HHS, BAAA, and HGA), as they are equipped with dedicated local search operators.

Table 1 presents the medians and standard deviations of the RPD values of the algorithms in the first group. On each instance, the minimum median value among the four algorithms is shown in bold. In addition, we conduct the non-parametric Wilcoxon rank sum test on the results of the algorithm, and mark a superscript $^+$ (or $^-$) before a median value if the result of the corresponding algorithm is significantly worse (or better) than that of WWO (at a confidence level of 95%). On instance 1, PSO, FFO, and WWO always obtain the exact optimal solution. On the remaining nine instances, FFO and WWO both obtain the best median value on instances 2 and 5, PSO uniquely obtains the best median value on instance 6, FFO uniquely obtains the best median value on instances 3 and 4, and WWO uniquely obtains the best median value on instances 7, 8, 9 and 10. That is, FFO performs better on small-size and medium-size instances, WWO performs better on large-size instances, and the overall performance of WWO is the best among the four algorithm on the test set. According to the statistical tests, the result of WWO is significantly better than GA on all ten instances, better than PSO on six instances, and better than FFO on four instances. On the contrary, there is only one case that the result of WWO is significantly worse than PSO on instance 6. The results demonstrate the adapted evolutionary operators in the discrete WWO are efficient in exploring the search space of MKP.

Table 2 presents the medians and standard deviations of the RPD values of the algorithms in the second group. Similarly, the minimum median value among the five algorithms on each instance is shown in bold, and a superscript $^+$ ($^-$) before a median value indicates that the result of the corresponding algorithm is significantly worse (better) than that of WWO-M (at a confidence level of 95%). On instances 1 and 6, all five algorithms always obtain the exact optimal solutions. ACO exhibits the worst performance, which is significantly worse than WWO-M on the remaining eight instances. On instance 2, the four algorithms (except ACO) always obtain the exact optimal solution. On instance 3, BAAA, HGA, and WWO-M always obtain the exact optimal solution. WWO-M performs significantly better than HGA on all the remaining

**Table 1.** Results of GA, PSO, FFO, and WWO for MKP on the test instances.

| Instance | Size ($m \times n$) | Metrics | GA | PSO | FFO | WWO |
|---|---|---|---|---|---|---|
| 1 | $2 \times 28$ | median | $^+$0.39 | 0 | **0** | **0** |
| | | std | 0.24 | 0 | 0 | 0 |
| 2 | $2 \times 105$ | median | $^+$1.50 | 0.36 | **0.32** | **0.32** |
| | | std | 0.39 | 0.25 | 0.30 | 0.44 |
| 3 | $5 \times 100$ | median | $^+$0.85 | $^+$0.74 | **0.37** | 0.55 |
| | | std | 0.53 | 0.17 | 0.26 | 0.33 |
| 4 | $10 \times 100$ | median | $^+$1.93 | $^+$1.33 | **1.27** | 1.31 |
| | | std | 0.37 | 0.27 | 0.45 | 0.60 |
| 5 | $5 \times 250$ | median | $^+$0.86 | $^+$0.95 | **0.60** | **0.60** |
| | | std | 0.31 | 0.57 | 0.51 | 0.26 |
| 6 | $30 \times 60$ | median | $^+$5.14 | $^-$**0.65** | $^+$1.83 | 1.55 |
| | | std | 1.25 | 0.68 | 0.60 | 0.71 |
| 7 | $10 \times 250$ | median | $^+$1.57 | $^+$1.89 | $^+$1.15 | **1.01** |
| | | std | 0.82 | 0.70 | 0.72 | 0.51 |
| 8 | $30 \times 100$ | median | $^+$3.61 | 2.07 | $^+$2.29 | **1.90** |
| | | std | 2.21 | 1.57 | 1.43 | 0.99 |
| 9 | $10 \times 500$ | median | $^+$3.73 | $^+$3.46 | 3.31 | **3.20** |
| | | std | 1.44 | 1.35 | 1.31 | 1.29 |
| 10 | $30 \times 500$ | median | $^+$5.27 | $^+$4.66 | $^+$3.96 | **3.83** |
| | | std | 1.98 | 1.03 | 0.89 | 0.70 |

six instances and better than HHS on three instances. There is only one case that WWO-M performs significantly worse than HHS on instance 5. The overall performances of BAAA and WWO-M are the best among all comparative algorithms, and there is no significant difference between the results of BAAA and WWO-M.

**Table 2.** Results of ACO, HHS, BAAA, HGA, and WWO-M for MKP on the test instances.

| Instance | Size ($m \times n$) | Metrics | ACO | HHS | BAAA | HGA | WWO-M |
|---|---|---|---|---|---|---|---|
| 1 | $2 \times 28$ | median | **0** | **0** | **0** | **0** | **0** |
| | | std | 0 | 0 | 0 | 0 | 0 |
| 2 | $2 \times 105$ | median | $^+$0.25 | **0** | **0** | **0** | **0** |
| | | std | 0.22 | 0.10 | 0 | 0 | 0 |
| 3 | $5 \times 100$ | median | $^+$0.02 | **0** | **0** | **0** | **0** |
| | | std | 0.14 | 0.12 | 0 | 0 | 0 |
| 4 | $10 \times 100$ | median | $^+$1.18 | 0.37 | **0.29** | $^+$1.05 | 0.39 |
| | | std | 0.52 | 0.38 | 0.39 | 0.51 | 0.22 |
| 5 | $5 \times 250$ | median | $^+$0.42 | $^-$**0.23** | 0.33 | $^+$0.55 | 0.33 |
| | | std | 0.31 | 0.23 | 0.15 | 0.20 | 0.18 |
| 6 | $30 \times 60$ | median | **0** | **0** | **0** | **0** | **0** |
| | | std | 0.12 | 0 | 0 | 0.26 | 0 |

*(continued)*

**Table 2.** (*continued*)

| Instance | Size ($m \times n$) | Metrics | ACO | HHS | BAAA | HGA | WWO-M |
|---|---|---|---|---|---|---|---|
| 7 | $10 \times 250$ | median | [+]1.08 | [+]0.61 | 0.48 | [+]1.14 | **0.45** |
|  |  | std | 0.36 | 0.33 | 0.28 | 0.35 | 0.25 |
| 8 | $30 \times 100$ | median | [+]1.88 | [+]1.23 | **1.03** | [+]1.67 | **1.03** |
|  |  | std | 1.02 | 0.35 | 0.24 | 0.40 | 0.21 |
| 9 | $10 \times 500$ | median | [+]3.09 | 1.41 | 1.25 | [+]1.73 | **1.07** |
|  |  | std | 0.98 | 0.41 | 0.36 | 0.95 | 0.41 |
| 10 | $30 \times 500$ | median | [+]3.71 | [+]2.69 | 2.43 | [+]2.80 | **2.35** |
|  |  | std | 0.69 | 0.60 | 0.46 | 0.68 | 0.48 |

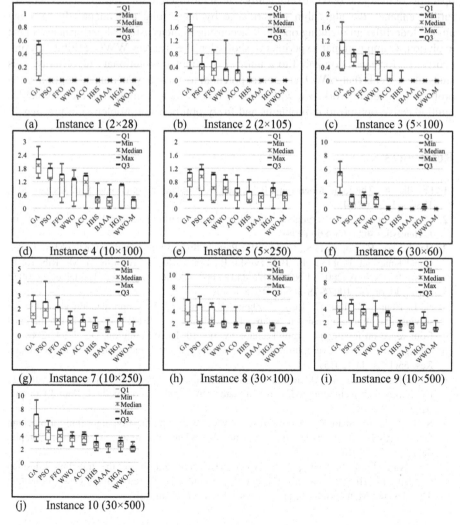

(a)    Instance 1 (2×28)

(b)    Instance 2 (2×105)

(c)    Instance 3 (5×100)

(d)    Instance 4 (10×100)

(e)    Instance 5 (5×250)

(f)    Instance 6 (30×60)

(g)    Instance 7 (10×250)

(h)    Instance 8 (30×100)

(i)    Instance 9 (10×500)

(j)    Instance 10 (30×500)

**Fig. 2.** Box plots of the results of the comparative algorithms on MKP test instances.

In addition, we present the box plots of the results in Fig. 2 to show the median, maximum, minimum, the first quartile (Q1) and the third quartile (Q3) values obtained by the algorithms. In general, the algorithms in the second group perform much better than those in the first group, because the dedicated local search operators can effectively improve the solution accuracies of the algorithms. In summary, the experiments demonstrate that the proposed discrete WWO is efficient in solving the MKP instances, and the WWO-M exhibits competitive performance compared to the existing superior evolutionary algorithms for MKP.

## 5   Conclusion

In this paper, we propose a discrete WWO algorithm for MKP by adapting the key operators of the basic WWO to efficiently search the discrete solution space of MKP. We further enhance the algorithm performance by integrating and adaptively selecting among four different breaking operators according to their historical performance in the evolution. Computational experiments show that the discrete WWO exhibits a competitive performance compared to a number of popular evolutionary algorithms. This study also reveals that WWO has a great potential to solve a wide range of optimization problems. In future work, we will adapt WWO to more complex problems (such as multiobjective optimization problems and fuzzy optimization problems) in practice.

## References

1. Fréville, A.: The multidimensional 0-1 knapsack problem: an overview. Eur. J. Oper. Res. 155(1), 1–21 (2004)
2. Chu, P., Beasley, J.: A genetic algorithm for the multidimensional knapsack problem. J. Heuristics 4(1), 63–86 (1998)
3. Hembecker, F., Lopes, H.S., Godoy, W.: Particle Swarm Optimization for the Multidimensional Knapsack Problem. In: Beliczynski, B., Dzielinski, A., Iwanowski, M., Ribeiro, B. (eds.) ICANNGA 2007. LNCS, vol. 4431, pp. 358–365. Springer, Heidelberg (2007). https://doi.org/10.1007/978-3-540-71618-1_40
4. Ji, J., Huang, Z., Liu, C., Liu, X., Zhong, N.: An ant colony optimization algorithm for solving the multidimensional knapsack problems. In: IEEE/WIC/ACM International Conference on Intelligent Agent Technology, pp. 10–16, Washington, DC, USA (2007)
5. Zheng, Y.J.: Water wave optimization: a new nature-inspired metaheuristic. Comput. Oper. Res. 55(1), 1–11 (2015)
6. Azadi Hematabadi, A., Akbari Foroud, A.: Optimizing the multi-objective bidding strategy using min–max technique and modified water wave optimization method. Neural Comput. Appl. 1–19 (2018)
7. Shao, Z., Pi, D., Shao, W.: A novel discrete water wave optimization algorithm for blocking flow-shop scheduling problem with sequence-dependent setup times. Swarm Evol. Comput. 40(1), 53–75 (2018)
8. Wu, X.B., Liao, J., Wang, Z.C.: Water wave optimization for the traveling salesman problem. In: Huang, D.S., Bevilacqua, V., Premaratne, P. (eds.) Intelligent Computing Theories and Methodologies, pp. 137–146. Springer, Cham (2015)

9. Wu, X., Zhou, Y., Lu, Y.: Elite opposition-based water wave optimization algorithm for global optimization. Math. Probl. Eng. **2017**, 25 (2017)
10. Zhang, J., Zhou, Y., Luo, Q.: Nature-inspired approach: a wind-driven water wave optimization algorithm. Appl. Intell. **49**(1), 233–252 (2019)
11. Zhang, J., Zhou, Y., Luo, Q.: An improved sine cosine water wave optimization algorithm for global optimization. J. Intell. Fuzzy Syst. **34**(4), 2129–2141 (2018)
12. Zhao, F., Liu, H., Zhang, Y., Ma, W., Zhang, C.: A discrete water wave optimization algorithm for no-wait flow shop scheduling problem. Expert Syst. Appl. **91**, 347–363 (2018)
13. Zhou, X.-H., Xu, Z.-G., Zhang, M.-X., Zheng, Y.-J.: Water wave optimization for artificial neural network parameter and structure optimization. In: Qiao, J., Zhao, X., Pan, L., Zuo, X., Zhang, X., Zhang, Q., Huang, S. (eds.) BIC-TA 2018. CCIS, vol. 951, pp. 343–354. Springer, Singapore (2018). https://doi.org/10.1007/978-981-13-2826-8_30
14. Zheng, Y.J., Zhang, B.: A simplified water wave optimization algorithm. In: 2015 IEEE Congress on Evolutionary Computation (CEC), pp. 807–813. IEEE, Sendai, Japan (2015)
15. Beasley, J.E.: Or-library: Distributing test problems by electronic mail. J. Oper. Res. Soc. **41**(11), 1069–1072 (1990)
16. Bansal, J.C., Deep, K.: A modified binary particle swarm optimization for knapsack problems. Appl. Math. Comput. **218**(22), 11042–11061 (2012)
17. Wang, L., Zheng, X.L., Wang, S.Y.: A novel binary fruit fly optimization algorithm for solving the multidimensional knapsack problem. Knowl.-Based Syst. **48**, 17–23 (2013)
18. Ke, L., Feng, Z., Ren, Z., Wei, X.: An ant colony optimization approach for the multidimensional knapsack problem. J. Heuristics **16**(1), 65–83 (2010)
19. Zhang, B., Pan, Q.K., Zhang, X.L., Duan, P.Y.: An effective hybrid harmony search-based algorithm for solving multidimensional knapsack problems. Appl. Soft Comput. **29**, 288–297 (2015)
20. Zhang, X., Wu, C., Li, J., Wang, X., Yang, Z., Lee, J.M., Jung, K.H.: Binary artificial algae algorithm for multidimensional knapsack problems. Appl. Soft Comput. **43**, 583–595 (2016)
21. Rezoug, A., Bader-El-Den, M., Boughaci, D.: Hybrid genetic algorithms to solve the multidimensional knapsack problem. In: Talbi, E.-G., Nakib, A. (eds.) Bioinspired Heuristics for Optimization. SCI, vol. 774, pp. 235–250. Springer, Cham (2019). https://doi.org/10.1007/978-3-319-95104-1_15

# A Complex-Valued Firefly Algorithm

Chuandong Song[✉]

School of Information Science and Engineering, Zaozhuang University,
Zaozhuang, China
63575490@qq.com

**Abstract.** In order to improve the diversity of population and convergence rate further, a firefly algorithm (FA) based on the complex-valued encoding (CFA) is proposed. In CFA, each variable could be encoding by complex number (the real part and the imaginary part). These parts could be optimized in parallel. The amplitude and angle of complex number are calculated for individual evaluation. Three real-valued benchmark functions are utilized to test the performance of CFA. The experiment results show that CFA has better convergence speed and accuracy than FA because it increases the diversity of population.

**Keywords:** Complex encoding · Firefly algorithm · Real encoding ·
Optimal solution

## 1 Introduction

Swarm intelligence optimization algorithm is a new evolutionary computing technology in recent years. Its mechanism originates from its simulation of various group behaviors of animals such as nature and human society [1]. It uses the information exchange and cooperation among individuals in the swarm to achieve the goal of optimization. Compared with the traditional optimization methods, swarm intelligence optimization algorithm owns the simple principle, convenient implementation and high efficiency, so swarm intelligence optimization algorithm has been widely used in many areas, such as engineering technology, network communication, finance, automatic control, resource allocation [2–4].

Real-valued firefly algorithm (FA) is a novel evolutionary algorithm based on group search, which was proposed by Yang in 2010 [5]. The basic idea of this algorithm is to move firefly with the low glow to the firefly with the high glow within a certain range, so as to search the optimal solution effectively. Because the principle of FA is relatively simple and easy to implement, and it has good global optimization ability and could converge quickly, it has been applied to industrial optimization, dynamic path planning, image processing, economic dispatch and other fields [6, 7]. Mohanty proposed FA for design optimization of a shell and tube heat exchanger with total cost as fitness function [8]. Sekhar et al. proposed a hybrid fuzzy PID controller based on FA and derivative filter for load frequency control of power system [9]. Lei et al. proposed a new markov clustering method based on the FA to infer protein complexes from dynamic protein–protein interaction networks [10].

© Springer Nature Switzerland AG 2019
D.-S. Huang et al. (Eds.): ICIC 2019, LNCS 11644, pp. 700–707, 2019.
https://doi.org/10.1007/978-3-030-26969-2_66

Basic FA relies extremely on initial population distribution and convergent slow in the later stage of optimization. The real-valued population contains a limited amount of information, so the diversity of individuals is limited and the algorithm may be trapped in local solutions [11]. In order to improve the diversity of population further, this paper proposed a firefly algorithm based on the complex-valued encoding (CFA). In CFA, each variable could be encoding by complex number, which contains two parts: the real part and the imaginary part. These parts could be optimized in parallel. The amplitude and angle of complex number are calculated for individual evaluation. Thus the dimension of each individual increases two times, which can improve the diversity of the population and prevent being trapped in local optimal solutions.

## 2  Proposed Method

### 2.1  Real-Valued Firefly Algorithm

Firefly algorithm is a group optimization method, which simulates the luminescence behavior of firefly in the nature. The firefly could search the partners and move to the position of better firefly according to brightness property. Each individual is regarded as a firefly. The fireflies with low brightness are attracted by the fireflies with high brightness. For each individual, the attraction and luminance of other individuals vary according to distance. Suppose that firefly vector is $[x_1, x_2, \ldots, x_n]$ ($n$ is the number of fireflies). The brightness of firefly $i$ is computed as

$$B_i = B_{i0} * e^{-\gamma r_{ij}} \tag{1}$$

Where $B_{i0}$ represents maximum brightness of firefly $i$ by the fitness function as $B_{i0} = f(x_i)$. $\gamma$ is coefficient of light absorption, and $r_{ij}$ is the distance factor between the two corresponding fireflies $i$ and $j$.

The movement of the less bright firefly toward the brighter firefly is computed by

$$x_i(t+1) = x_i(t) + \beta_i(x_j(t) - x_i(t)) + \alpha \varepsilon_i. \tag{2}$$

Where $\alpha$ is step size randomly created in the range [0, 1], and $\varepsilon_i$ is Gaussian distribution random number.

The flowchart of FA is described as follows:

(1) Initialize firefly population, fitness function $f(\cdot)$ and parameters in FA, which contain $\beta$, $\gamma$ and $\alpha$.
(2) The distance of any two fireflies ($r_{ij}$) in the populations is calculated.
(3) According to fitness function, the brightness and attractiveness of each firefly are calculated according to Eq. (1).
(4) For each firefly, search the most attractive individual around it, and update its location according to Eq. (2).
(5) If the iteration is completed or the accuracy is reached, algorithm will be stopped, otherwise go to step (2).

## 2.2 Complex-Valued Firefly Algorithm

Complex-valued firefly algorithm (CFA) uses diploid to express each firefly and greatly expands the information capacity of population. The flowchart of CFA is depicted in Fig. 1. The detailed process is descried as follows.

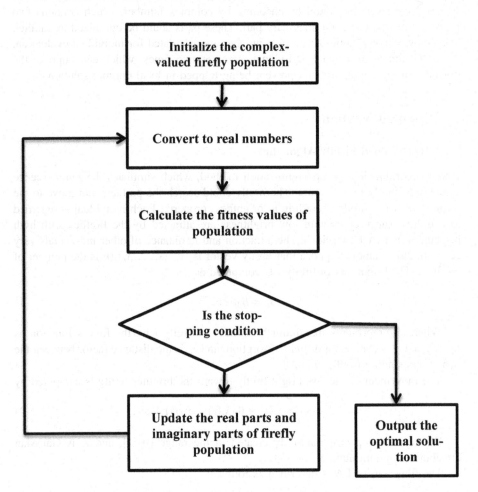

**Fig. 1.** The flowchart of complex-valued firefly algorithm.

**Complex-Valued Population Initialization.** For a problem with $m$ independent variables, the interval of the problem is set as $[a_k, b_k]$ $(k = 1, 2, \ldots, m)$. $m$ moduli and phase angles are created randomly. The selected intervals of moduli and phase angle are defined as follows [11].

$$\rho_k = [0, \frac{b_k - a_k}{2}], \quad k = 1, 2, \ldots, m. \tag{3}$$

$$\theta_k = [-2\pi, \ 2\pi], \quad k = 1, 2, \ldots, m. \tag{4}$$

According to modulus vector $[\rho_1, \rho_2, \cdots, \rho_m]$ and phase angle vector $[\theta_1, \theta_2, \cdots, \theta_m]$, $m$ complex numbers are created as follows.

$$x_k + i y_k = \rho_k(\cos \theta_k + i \sin \theta_k), \quad k = 1, 2, \ldots, m. \tag{5}$$

**Position Updating.** The real parts and imaginary parts of the positions of firefly population are updated in parallel.

(1) Real part updating

$$B_{i,R} = B_{i,0} * e^{-\gamma r_{ij,R}}. \tag{6}$$

$$x_{i,R}(t+1) = x_{i,R}(t) + \beta_{i,R}(x_{j,R}(t) - x_{i,R}(t)) + \alpha\varepsilon_{i,R}. \tag{7}$$

Where $x_{i,R}(t+1)$ and $x_{i,R}(t)$ are the real parts of firefly $i$ at $t - th$ time point, respectively. $B_{i,R}$ is the real part of the brightness of firefly $i$. $r_{ij,R}$ is the real part of the distance between two fireflies $i$ and $j$. $\varepsilon_{i,R}$ is the real part of complex-valued Gaussian distribution random number.

(2) Imaginary part updating

$$B_{i,I} = B_{i,0} * e^{-\gamma r_{ij,I}}. \tag{8}$$

$$x_{i,I}(t+1) = x_{i,I}(t) + \beta_{i,I}(x_{j,I}(t) - x_{i,I}(t)) + \alpha\varepsilon_{i,I}. \tag{9}$$

Where $x_{i,I}(t+1)$ and $x_{i,I}(t)$ are the imaginary parts of firefly $i$ at $t - th$ time point, respectively. $B_{i,I}$ is the imaginary part of the brightness of firefly $i$. $r_{ij,I}$ is the imaginary part of the distance between two fireflies $i$ and $j$. $\varepsilon_{i,I}$ is the imaginary part of complex-valued Gaussian distribution random number.

**Fitness Value Calculation.** When complex-valued firefly is utilized to resolve real-valued problems, complex-valued firefly population need to be converted into real number in order to calculate the fitness values, which could be determined with modulus and phase angle as follows.

$$\rho_i = \sqrt{x_{i,R}^2 + x_{i,I}^2}, \quad i = 1, 2, \ldots, m. \tag{10}$$

$$x_i = \rho_i \operatorname{sgn}\left(\sin\left(\frac{x_{i,I}}{\rho_i}\right)\right) + \frac{a_i + b_i}{2}, \quad i = 1, 2, \ldots, m. \tag{11}$$

Where $x_i$ denotes the converted real number.

# 3 Experiments

In order to compare the search ability of FA and CFA, three benchmark functions are utilized, which are described in Table 1. The parameters of FA and CFA are the same. The population size is set to 60 and the maximum generation is set as 200.

**Table 1.** Benchmark functions.

| No. | Testing functions | Dimension | Range |
|-----|-------------------|-----------|-------|
| $f_1$ | $f(x) = \min \sum_{i=1}^{n} x_i^2$ | 30 | $[-100, 100]$ |
| $f_2$ | $f(x) = \min \sum_{i=1}^{n} (x_i^2 - 10\cos(2\pi x_i) + 10)$ | 30 | $[-5.12, 5.12]$ |
| $f_3$ | $f(x) = \min \ -20\exp(-0.2\sqrt{\sum_{i=1}^{n} x_i^2}) - \exp(\frac{1}{n}\sum_{i=1}^{n}\cos(2\pi x_i)) + 20 + e$ | 30 | $[-32, 32]$ |

Through 20 runs, the testing results by FA and CFA with the same parameters are listed in Table 2. From Table 2, it can be clearly seen that CFA has the better fitness values than its real-valued version (FA). For functions $f_1$ and $f_2$, CFA could achieve theoretical optimal solution. For function $f_3$, CFA could sharply improve the accuracy for Best, Mean, Worse and Std results.

**Table 2.** Testing results of functions ($f_1$, $f_2$ and $f_3$) with FA and CFA.

| Benchmark functions | Method | Best | Mean | Worse | Std |
|---------------------|--------|------|------|-------|-----|
| $f_1$ | FA | 0.0189 | 0.1639 | 0.3721 | 0.0492 |
|       | CFA | 0 | 0 | 0 | 0 |
| $f_2$ | FA | 7.654E-8 | 20.371 | 158.048 | 12.376 |
|       | CFA | 0 | 0 | 0 | 0 |
| $f_3$ | FA | 3.291 | 4.514 | 6.9182 | 2.1458 |
|       | CFA | 8.88E-16 | 1.5609E-15 | 3.3819E-15 | 1.281E-15 |

The evolution curves of fitness value for $f_1$, $f_2$ and $f_3$ are depicted in Figs. 2, 3 and 4, respectively. From the curve results, we can see that CFA has better convergence speed and accuracy than FA. FA is easy to make population to obtain the local optimal solution. Due to complex-valued factor, CFA has better diversity of population and is easier to search the optimal solution.

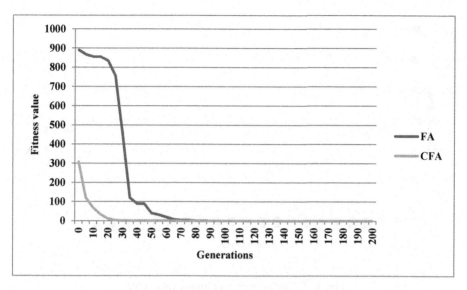

**Fig. 2.** Evolution curves of fitness value for $f_1$.

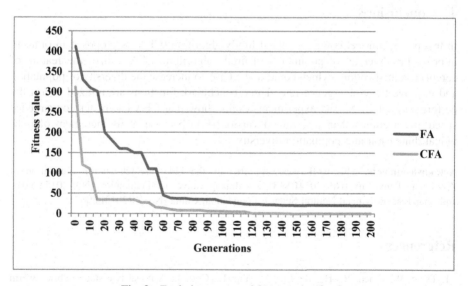

**Fig. 3.** Evolution curves of fitness value for $f_2$.

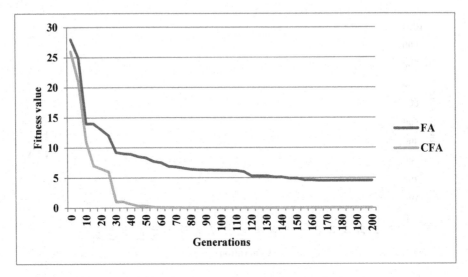

**Fig. 4.** Evolution curves of fitness value for $f_3$.

## 4   Conclusions

In this paper, a novel complex-valued firefly algorithm (CFA) is proposed in order to improve the diversity of population of firefly algorithm. CFA utilizes the feature of complex-valued coding, which could make CFA to increase the diversity of population and improve the convergence rate. Three benchmark functions are utilized to test the performance of CFA. The experiment results show that CFA has better convergence speed and accuracy than real-valued version (FA) because of its increment of individual dimension and population diversity.

**Acknowledgments.** This work was supported by the PhD research startup foundation of Zaozhuang University (No. 2014BS13), Zaozhuang University Foundation (No. 2015YY02), and Shandong Provincial Natural Science Foundation, China (No. ZR2015PF007).

## References

1. Deng, W., Chen, R., He, B., Liu, Y., Yin, L., Guo, J.: A novel two-stage hybrid swarm intelligence optimization algorithm and application. Soft Comput. **16**(10), 1707–1722 (2012)
2. Alomari, A., Phillips, W., Aslam, N., Comeau, F.: Swarm intelligence optimization techniques for obstacle-avoidance mobility-assisted localization in wireless sensor networks. IEEE Access **6**, 22368–22385 (2017)
3. Khoshahval, F., Zolfaghari, A., Minuchehr, H., Sadighi, M., Norouzi, A.: PWR fuel management optimization using continuous particle swarm intelligence. Ann. Nucl. Energy **37**(10), 1263–1271 (2010)

4. Srinivas, J., Giri, R., Yang, S.H.: Optimization of multi-pass turning using particle swarm intelligence. Int. J. Adv. Manuf. Technol. **40**(1–2), 56–66 (2009)
5. Yang, S.X.: Firefly algorithm, stochastic test functions and design optimization. Int. J. Bio-Inspir. Comput. **2**(2), 78–84 (2010)
6. Horng, M.H.: Vector quantization using the firefly algorithm for image compression. Expert Syst. Appl. **39**(1), 1078–1091 (2012)
7. Yang, X.S., Hosseini, S.S.S., Gandomi, A.H.: Firefly algorithm for solving non-convex economic dispatch problems with valve loading effect. Appl. Soft Comput. **12**(3), 1180–1186 (2012)
8. Mohanty, D.K.: Application of firefly algorithm for design optimization of a shell and tube heat exchanger from economic point of view. Int. J. Therm. Sci. **102**(2), 228–238 (2016)
9. Chandra Sekhar, G.T., Sahu, R.K., Baliarsingh, A.K., Panda, S.: Load frequency control of power system under deregulated environment using optimal firefly algorithm. Int. J. Electr. Power Energy Syst. **74**, 195–211 (2016)
10. Lei, X., Wang, F., Wu, F.X., Zhang, A., Pedrycz, W.: Protein complex identification through Markov clustering with firefly algorithm on dynamic protein–protein interaction networks. Inf. Sci. **329**(6), 303–316 (2016)
11. Li, L., Zhou, Y.: A novel complex-valued bat algorithm. Neural Comput. Appl. **25**(6), 1369–1381 (2014)

# Gene Expression Prediction Based on Hybrid Evolutionary Algorithm

Haifeng Wang and Sanrong Liu[(✉)]

School of Information Science and Engineering,
Zaozhuang University, Zaozhuang, China
23384722@qq.com

**Abstract.** Gene expression profiles are essential to reverse-engineering gene regulatory network. In this paper, a Hill function-based ordinary differential equation (ODE) model is proposed to predict time-series gene expression data. A novel hybrid evolutionary algorithm based on binary particle swarm optimization (BPSO) and grey wolf optimization (GWO) is proposed to identify the structure and parameters of the ODE model. Gene expression dataset from SOS DNA repair network is used to test the performance of our method. Results show that ODE based on Hill function could forecast gene expression data more accurately than general ODE model and recurrent neural network.

**Keywords:** Gene expression · Hill function · Grey wolf optimization · Hybrid evolutionary · Ordinary differential equation

## 1 Introduction

Gene expression regulation is a complex process that regulates gene expression in organisms, makes the process of gene expression in an orderly state, and responds to changes in environmental conditions [1]. The regulation of gene expression can be carried out at many levels, including gene level, transcription level, post-transcription level, translation level and post-translation level [2]. Gene expression regulation is the molecular basis of cell differentiation, morphogenesis and individual development in organisms.

DNA microarray and high-throughput sequencing technology could supply amount of gene expression data, which could intuitively reflect the gene regulation process [3]. But sequencing experiments are expensive, and the sequencing data contain a lot of noise [4]. And current technology must obtain enough gene expression signals at a time point for a cell sample, so cell cycle of the sample cells must synchronize at the beginning of the experiment. However, the inherent randomness of cell operations makes cells more and more asynchronous in cell cycle. Therefore, with the growth of time points, the expression data obtained become more and more blurred. These factors lead to that the expression levels of many genes are missing. Thus it is very important to predict the unknown expression level of genes according to the expression data of other genes.

Ao et al. proposed an ensemble of recurrent Elman neural networks (ENNs) and support vector machines (SVMs) for modeling microarray continuous time series data

© Springer Nature Switzerland AG 2019
D.-S. Huang et al. (Eds.): ICIC 2019, LNCS 11644, pp. 708–716, 2019.
https://doi.org/10.1007/978-3-030-26969-2_67

in order to further improve the prediction accuracy of the individual models [5]. Maraziotis et al. presented a novel neural fuzzy recurrent network for concerning gene expression time course prediction [6]. Yang et al. proposed flexible neural tree (FNT) model for gene regulatory network inference and time-series prediction from gene expression profiling [7]. Ordinary differential equation (ODE) model could model the known observation data and predict the future development. Because its characteristics of forward fitting, backward prediction with little error and high accuracy, ODE model has been applied for gene regulation network inference [8].

In order to model gene regulatory network (GRN) accurately, Hill and sigmoidal functions, which could well identify biochemical mechanism, have been added into the ODE model. Santillán analyzed the biochemical reactions and discussed the feasibility and constrains of gene regulation process identification with Hill function [9]. Baralla et al. presented Michaelis Menten-type Models, Mendes model based on Hill function and S-system model to describe GRN and particle swarm optimization (PSO) was presented to evolve the parameters of models [10]. These past proposed methods have proved that Hill functions are very suitable for describing gene regulatory process. In this paper, ODE model based on Hill function is proposed to predict gene expression profiles. A novel hybrid evolutionary algorithm is proposed to identify the ODE model based on Hill functions. In the hybrid evolutionary algorithm, binary particle swarm optimization (BPSO) is proposed to select automatically the excitation and suppression genes in Hill functions, and grey wolf optimization (GWO) is utilized to optimize the parameters of ODE model.

## 2   Proposed Method

### 2.1   Ordinary Differential Equation Based on Hill Functions

The regulations of each target gene are identified by one ODE based on Hill function. The number of ODEs is equal to the size of GRN. The $i - th$ Hill function-based ODE-Hill is described with Eq. (1) [11].

$$\frac{dx_i}{dt} = p_i \prod_{j \in \Omega_i} h^-(x_j, Q_{ij}, R_{ij}) \prod_{k \in \Upsilon_i} h^+(x_k, Q_{ik}, R_{ik}) - \lambda_i x_i. \tag{1}$$

Where

$$h^-(x_j, Q_{ij}, R_{ij}) = \frac{Q_{ij}^{R_{ij}}}{x_j^{R_{ij}} + Q_{ij}^{R_{ij}}}, \tag{2}$$

and

$$h^+(x_k, Q_{ik}, R_{ik}) = 1 + \frac{x_k^{R_{ij}}}{x_k^{R_{ik}} + Q_{ik}^{R_{ik}}}. \tag{3}$$

Where Eqs. (2) and (3) are inhibiting and activating functions, $x_i$ is the expression level of $i-th$ gene, $p_i$ represents a rate constant, $\Omega_i$ is a subset of all inhibiting genes of $i-th$ gene, $\Upsilon_i$ is a subset of all activating genes of $i-th$ gene, $\lambda_i$ denotes degradation coefficient, $R_{ij}$, $Q_{ij}$, $R_{ik}$ and $Q_{ik}$ are parameters of Hill function.

## 2.2  Hybrid Evolutionary Algorithm

In order to optimize the structure and parameters of ODE model based on Hill function, a novel hybrid evolutionary algorithm is proposed. In the hybrid evolutionary algorithm, binary-coding particle swarm optimization (BPSO) algorithm is utilized to search the best structure of model and grey wolf optimization (GWO) algorithm is utilized to optimize the parameters of model.

**Binary-Coding Particle Swarm Optimization.** Binary-coding particle swarm optimization algorithm is a binary version of particle swarm optimization (PSO), which is an evolutionary computation technique used for optimization of parameters of model, such as neural network [12].

Each particle $x_i$ represents a potential solution. A swarm of particles moves through space, with the moving velocity of each particle represented by a velocity vector $v_i$. At each step, each particle is evaluated and keeps track of its own best position, which is associated with the best fitness it has achieved so far in a vector $Pbest_i$. The best position among all the particles is kept as $Gbest$. In BPSO, the moving trajectory and velocity of each particle is defined in term of probability. The moving trajectory represents changes of probabilities of a certain value. The moving velocity is defined as probability of a state or another state. Thus each bit $x_{id}$ of one particle is restricted to 0 or 1. Each $v_{id}$ represents the probability of bit $x_{id}$ taking the value 1. A new velocity $v_{id}$ for particle $i$ is updated as follows.

$$v_i(t+1) = w * v_i(t) + c_1 r_1 (Pbest_i - x_i(t)) + c_2 r_2 (Gbest(t) - x_i(t)). \qquad (4)$$

Where $w$ is the inertia weight, $c_1$ and $c_2$ are positive constant, and $r_1$ and $r_2$ are uniformly distributed random number in [0,1].

$x_{id}$ is calculated as follows.

$$x_{id} = \begin{cases} 1, & r < Sig(v_{id}) \\ 0, & other \end{cases}. \qquad (5)$$

Where $r$ is created randomly from range [0.1, 1.0], and the function $Sig(\cdot)$ is defined as follows.

$$Sig(v_{id}) = \frac{1}{1 + e^{-v_{id}}}. \qquad (6)$$

**Grey Wolf Optimization.** Grey wolf optimization (GWO) algorithm is a novel swarm intelligence optimization algorithm based on wolf hunting behaviour, which was proposed by Mirjalili et al. in year 2014 [13]. GWO algorithm has the characteristics of fewer parameters, fast convergence, strong global search ability and simple implementation. GWO algorithm was proposed by imitating the hierarchy and predation mechanism of grey wolf populations in nature. The grey wolf populations have the strict hierarchy, which are divided into four classes ($\alpha$, $\beta$, $\delta$ and $\omega$) according to the leadership level. In wolf populations, the wolf with $\alpha$ level is called the dominant one, which has the best fitness value in general. The wolf with $\beta$ level has the second best fitness value, which is the best successor of the wolf with $\alpha$ level. The wolf with $\delta$ level mainly carries out the orders of the wolves with $\alpha$ and $\beta$ levels, but it could order the wolves with $\omega$ level. The wolves with $\omega$ level have the lowest level.

Grey wolves could encircle the prey when hunting. The mathematical model of encircling behaviour is described as follows.

$$D = \left| C \cdot X_p(t) - X(t) \right|. \tag{7}$$

$$X_p(t+1) = X_p(t) - A \cdot D. \tag{8}$$

$$A = 2\alpha \cdot r_1 - \alpha. \tag{9}$$

$$C = 2 \cdot r_2. \tag{10}$$

Where $D$ denotes the distance between grey wolf and prey, $X_p(t)$ is the current position of prey, $X(t)$ denotes the current position of grey wolf, $C$ and $A$ are algorithm coefficients, and $\alpha$ is a convergence factor, which decreases linearly from 2 to 0. $r_1$ and $r_2$ are random variables in the interval [0, 1].

Due to that the position of the prey always changes, in order to search the better prey, suppose the positions of the wolves with $\alpha$, $\beta$ and $\omega$ levels as the ones closer to the prey. Therefore, during each iteration, the positions of three best wolves ($\alpha$, $\beta$ and $\delta$) are saved, which are utilized to update the positions of other wolves with Eq. (11).

$$X(t+1) = \frac{X_1(t) + X_2(t) + X_3(t)}{3}. \tag{11}$$

Where $X_1(t)$, $X_2(t)$ and $X_3(t)$ are the calculated position results when the positions of the wolves with $\alpha$, $\beta$ and $\delta$ levels are considered as the prey positions ($X_p$), respectively.

**Optimization of ODE Based on Hill Functions.** In this paper, a hybrid evolutionary algorithm is proposed, which can automatically select inhibiting and activation gene sets, and optimize parameters. Suppose that gene regulatory network contains $n$ genes. The corresponding Hill functions-based ODE of each gene could be derived independently. For the $i-th$ gene, the optimization algorithm of ODE based on Hill functions is described as follows.

(1) Binary particle swarm optimization is utilized to automatically select inhibiting and activation gene sets. In BPSO, initialize the binary populations. Each particle contains two parts: inhibiting gene set and activating gene set, whose structure is depicted in Fig. 1. For GRN with $n$ genes, the dimension of each particle is set to $2n$. Each bit of particle ($p_i$, $i = 1, 2, \ldots, 2n$) is binary. If $p_j$ is set to 1, the $j-th$ gene could

inhibit the expression of the $i-th$ gene. In order to infer the regulations of the $i-th$ gene, inhibiting gene set and activating gene set in Hill functions could not contain the $i-th$ gene. Therefore, in the chromosome of each particle, $p_i$ and $p_{n+i}$ are set as 0 during the optimization process. Then the fitness values of all particles are calculated. Update the positions of particles with Eq. (5). Repeat this process until that stopping criterion is met.

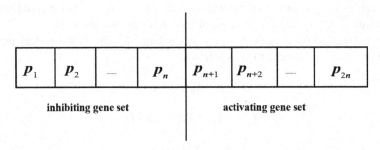

**Fig. 1.** Chromosome structure of binary grey wolf optimization algorithm.

(2) In the optimization process of BPSO, grey wolf algorithm is utilized to optimize the parameters of ODE based on Hill functions. During this process, the structure of ODE is fixed. According to the structure, the number of parameters is counted. If $p_j$ is 1 in the structure, two kinds of parameters ($R_{ij}$ and $Q_{ij}$) are given.

(3) Fitness function. In order to search the optimal model, mean square error (MSE) is utilized as fitness function, which is defined as follows.

$$f_i = \sum_{k=1}^{T} (y_{ik} - y'_{ik})^2. \tag{12}$$

Where $T$ denotes the number of gene expression sample points, $y_{ik}$ is the expression level of the $i-th$ gene at $k-th$ sample point, $y'_{ik}$ is the predicted expression level of the $i-th$ gene at $k-th$ sample point, which could be obtained by solving ODE model using Four-order Runge-Kutta method. For differential equation $\frac{dy}{dt} = f(x, y)$, the solution is as follows.

$$
\begin{aligned}
k_1 &= f(x(t), y(t)), \\
k_2 &= f(x(t) + \frac{h}{2}, y(t) + h * \frac{k_1}{2}), \\
k_3 &= f(x(t) + \frac{h}{2}, y(t) + h * \frac{k_2}{2}), \\
k_4 &= f(x(t) + h, y(t) + h * k_3), \\
y(t+1) &= y(t) + h * \frac{k_1 + 2k_2 + 2k_3 + k_4}{6}.
\end{aligned}
\tag{13}
$$

Where $h$ is the step size.

## 3  Experiments

SOS DNA repair network is used to evaluate the performance of our proposed algorithm. SOS repair is that cells give the stress response when DNAs are seriously injured. SOS DNA repair network includes 8 genes (**uvrD**, **lexA**, **umuD**, **recA**, **uvrA**, **uvrY**, **ruvA** and **polB**) and 9 regulatory relationships [14]. The dataset contains four experiments and each experiment has 50 sample points. The gene expression levels are normalized with Eq. (14).

$$g' = \frac{g - g_{min}}{g_{max} - g_{min}} \tag{14}$$

Our method is implemented using C programming language on windows7 platform, and executed on Intel 2.60 Ghz processor and 4 GB of RAM. The parameters of our method are listed in Table 1. The datasets from three experiments are utilized to train the ODE model and other dataset is utilized to test. Recurrent Neural Network (RNN) model is also used to predict gene expression dataset from SOS network. Particle swarm optimization is utilized to optimize the parameters of RNN model [15]. The predicted results of eight genes are depicted in Fig. 2. The predicted results reveal that the predicted curves by our proposed method are closer to the actual data than RNN model. Figure 3 shows the predicted errors of eight genes by RNN and our method, which reveal that our predicted errors mainly concentrate near zero.

Table 2 shows the predicting MSE comparison of SOS network among RNN, ODE (general ODE model) [16] and ODE based on Hill functions. From Table 2, it can be

**Table 1.** Parameters in our proposed method.

| Parameters | Values |
|---|---|
| Population size in BPSO | 50 |
| Maximum iteration in BPSO | 200 |
| $c_1$, $c_2$ in BPSO | 2.0 |
| Population size in GWO | 100 |
| Maximum iteration in GWO | 100 |
| $\alpha$ in GWO | 2.0 |

seen that the prediction of our proposed method is more accurate than RNN and ODE models.

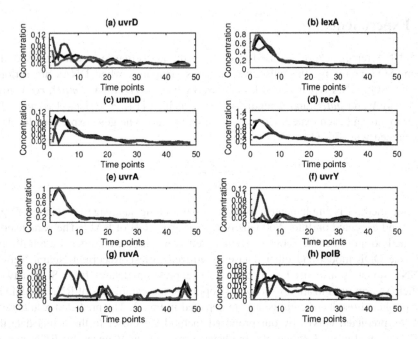

**Fig. 2.** The predicted results of eight genes in SOS network. The black line denotes the actual gene expression data, the blue line represents the predicted data by RNN model and the red line denotes the predicted results by our method. (Color figure online)

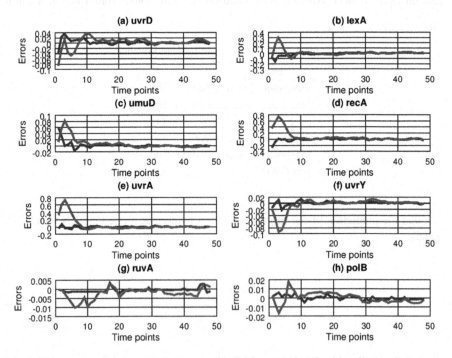

**Fig. 3.** The predicted errors of eight genes in SOS network. The blue line represents the predicted errors by our method and the red line denotes the predicted errors by RNN model. (Color figure online)

**Table 2.** Predicted performances of eight genes in SOS network.

| Methods | Predicted MSE |
|---|---|
| ODE | $2.11 \times 10^{-4}$ |
| RNN | $7.39 \times 10^{-4}$ |
| ODE based on Hill functions | $9.64 \times 10^{-5}$ |

## 4  Conclusions

In order to predict time-series gene expression data accurately, this paper proposes a novel hybrid evolutionary algorithm to identify the ODE based on Hill functions. In the hybrid evolutionary algorithm, binary particle swarm optimization is proposed to select automatically the excitation and suppression genes in Hill functions, and grey wolf optimization is used to optimize the parameters of ODE model. The prediction performance of SOS repair network reveals that our method could predict gene expression data more accurately than general ODE and RNN models.

## References

1. Wingender, E.: TRANSFAC: an integrated system for gene expression regulation. Nucleic Acids Res. **28**(1), 316–319 (2000)
2. Duechler, M., Leszczyńska, G., Sochacka, E., Nawrot, B.: Nucleoside modifications in the regulation of gene expression: focus on tRNA. Cell. Mol. Life Sci. **73**(16), 3075–3095 (2016)
3. Li, M., Belmonte, J.C.: Ground rules of the pluripotency gene regulatory network. Nat. Rev. Genet. **18**(3), 180 (2017)
4. Chan, T.E., Stumpf, M.P.H., Babtie, A.C.: Gene regulatory network inference from single-cell data using multivariate information measures. Cell Syst. **5**(3), 251–267 (2017)
5. Ao, S.I., Palade, V.: Ensemble of elman neural networks and support vector machines for reverse engineering of gene regulatory networks. Appl. Soft Comput. **11**(2), 1718–1726 (2011)
6. Maraziotis, I.A., Dragomir, A., Bezerianos, A.: Gene networks reconstruction and time-series prediction from microarray data using recurrent neural fuzzy networks. IET Syst. Biol. **1**(1), 41–50 (2007)
7. Yang, B., Chen, Y., Jiang, M.: Reverse engineering of gene regulatory networks using flexible neural tree models. Neurocomputing **99**(1), 458–466 (2013)
8. Polynikis, A., Hogan, S.J., Bernardo, M.D.: Comparing different ode modelling approaches for gene regulatory networks. J. Theor. Biol. **261**(4), 511–530 (2009)
9. Santillán, M.: On the use of the hill functions in mathematical models of gene regulatory networks. Math. Model. Nat. Pheno. **3**(2), 85–97 (2008)
10. Baralla, A., Cavaliere, M., de la Fuente, A.: Modeling and parameter estimation of the SOS response network in E.coli, MS thesis, University of Trento, Trento, Italy (2008)
11. Elahi, F.E., Hasan, A.: A method for estimating hill function-based dynamic models of gene regulatory networks. Royal Soc. Open Sci. **5**(2), 171226 (2018)

12. Yang, B., Zhang, W., Wang, H., Song, C., Chen, Y.: TDSDMI: inference of time-delayed gene regulatory network using S-system model with delayed mutual information. Comput. Biol. Med. **72**, 218–225 (2016)
13. Mirjalili, S., Mirjalili, S.M., Lewis, A.: Grey wolf optimization. Adv. Eng. Softw. **69**(7), 46–61 (2014)
14. Zhang, Y., Pu, Y., Zhang, H., Cong, Y., Zhou, J.: An extended fractional kalman filter for inferring gene regulatory networks using time-series data. Chemometr. Intell. Lab. Syst. **138**, 57–63 (2014)
15. Xu, R., Donald Wunsch, I.I., Frank, R.: Inference of genetic regulatory networks with recurrent neural network models using particle swarm optimization. IEEE/ACM Trans. Comput. Biol. Bioinform. **4**(4), 681–692 (2007)
16. Chen, Y., Yang, B., Meng, Q., Zhao, Y.: Time-series forecasting using a system of ordinary differential equations. Inf. Sci. **181**(1), 106–114 (2011)

# MPdeep: Medical Procession
# with Deep Learning

Qi Liu[1], Wenzheng Bao[2(✉)], and Zhuo Wang[2]

[1] The Affiliated Hospital of Xuzhou Medical University, Xuzhou,
People's Republic of China
[2] The School of Information and Electrical Engineering,
Xuzhou University of Technology, Xuzhou, People's Republic of China
baowz55555@126.com

**Abstract.** The thoracic diseases, can be regarded as one of the most serious and dangerous diseases, threat to the health of human being. Among these diseases, some of them may cause functional impairment and sequela in some degree and some dangerous ones may lead to death and organ failure or death. With the development of the artificial intelligence, the deep learning method can be utilized to deal with such issue. In this work, we proposed the MPdeep to demonstrate the medical imaging procession in the field. With such method, we can find out that the identical image quality of contrast enhanced chest CT, application of 40% DR reduced 48.5% radiation dose, and combination of 40% DR and 100 kV further reduce 58.9% radiation dose.

**Keywords:** Medical image procession · Deep learning · Diseases

## 1 Introduction

Currently, thoracic diseases, can be regarded as one of the most serious and dangerous diseases, threat to the health of human being. Among these diseases, some of them may cause functional impairment and sequela in some degree and some dangerous ones may lead to death and organ failure or death [1–3]. Therefore, the World Health Organization (WHO) has taken into account thoracic diseases several decades. Meanwhile, it was pointed that effective methods to detecting these diseases can make great contributions in prevention and treatment them. In other words, chest CT examination plays the significant roles in this field [4–7].

When it comes to chest CT examination, such approach has been widely utilized in the thoracic diseases examination. With several years' efforts, such approach has achieved high potential for growth and advancement. A great deal of hospitals and medical institutions have made use of such method in China [8]. According to some related researches reporting, more than 120 million and cases have been utilized such method every years and the account of cases will increase by years. It was noted that the CT techniques optimized for small bowel imaging are playing an increasing role in the evaluation of small bowel disorders. Several related studies and researches have shown the advantages of such methods over tradition barium fluoroscopic examinations. The preference of CT has been geographical and based on expertise and public policy.

© Springer Nature Switzerland AG 2019
D.-S. Huang et al. (Eds.): ICIC 2019, LNCS 11644, pp. 717–725, 2019.
https://doi.org/10.1007/978-3-030-26969-2_68

On the other hand, some side effects of such method can't to be ignored in this field. The ionizing radiation of CT may increase the risks of cancer and lead to the pollution. With the increasing awareness of radiation exposure, there has been a more global interest in implementing techniques that either reduce or eliminate radiation exposure. Therefore, the novel technologies are urgent to develop and design. Therefore, how to ensure the quality of the images meet the requirements of the clinical diagnosis at the same time reduce patient radiation dose, has become an important direction of the current image in this field. The typically and traditional methods deal with such issue forcing on optimizing parameters. Such method may lead to increase the image noises in some degree. Iterative Reconstruction (IR), which is force on selective optimization to reduce image noise. In this work, we utilize the IR method to deal with some traditional noise image. And then, the IR method images compared with the FBP (filtered back projection) images. The performances show that the IR method has better performances than FBP ones.

## 2  Methods and Materials

### 2.1  Dataset

In this work, we have achieved the 60 cases during April in 2014. These cases have been divided into two groups, which include the IR method and the FBP ones. The Body Mass Index (BMI) of 60 cases range from 18.5 kg/m to 27 kg/m in this work. And the ages of the two groups range from 44 to 82. In group A, we make use of the parameters, which are 120 kV, mA (NI = 12), DR = 0. In the other group, we utilize the parameters, which are 120 kV, mA (NI = 12), DR = 40%. At the same time, it was pointed that GE Optima CT660 has been employed as the scanner in this research. All the patients take the supine position, closed the iv line right in the middle of the elbow vein puncture, utilized the German production of Ulrich automatic double cylinder of high pressure syring. And the iohexol, whose parameter is 1.2 ml/kg, is employed as the contrast agent in this work. Meanwhile, its injection rate is 3 ml/s and 20 ml normal saline with this injection. And the parameters of scanner are 0.5 s racking time, 1 pitch and large body style. And the detailed information can be shown in Fig. 1.

On the other hand, the whole employed samples are satisfied the basic needs of medical image process in this work. At the same time, all the employed images have been tested by two imaging physicians in this field and the value of Kappa is 0.613 that means the consistency maintain in a good level. It was pointed that the doses of calculation of radiation can be treated as one of the most significant elements in this issue. The construction of group A is [(240.41 ± 77.81) mGy/cm, (4.09 ± 1.32)mSv] and the ones of group B is [(126.24 ± 37.12) mGy/cm, (2.11 ± 0.51)mSv]. From the construction of two groups, we can easily find out that the group B is merely the 51.5% in group A. Therefore, we want to find the effective and low-harmful doses to be utilized in the medical imaging procession.

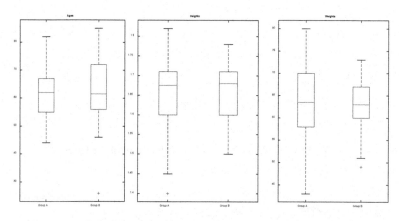

**Fig. 1.** The detailed information of two groups

## 2.2   Algorithms

How to find effective features is the core issue in image classification and pattern recognition [9–12]. Humans have an amazing skill in extracting meaningful features, and a lot of research projects have been undertaken to build an FE system as smart as human in the last several decades. Deep learning is a newly developed approach aiming for artificial intelligence. Deep learning-based methods build a network with several layers, typically deeper than three layers. Deep neural network (DNN) can represent complicated data. However, it is very difficult to train the network. Due to the lack of a proper training algorithm, it was difficult to harness this powerful model until Hinton and Salakhutdinov proposed a deep learning idea [13–18].

Deep learning involves a class of models that try to learn multiple levels of data representation, which helps to take advantage of input data such as image, speech, and text. Deep learning model is usually initialized via unsupervised learning and followed by fine-tuning in a supervised manner. The highlevel features can be learnt from the low-level features. This kind of learning leads to the extraction of abstract and invariant features, which is beneficial for a wide variety of tasks such as classification and target detection [19–24].

The human visual system can tackle classification, detection, and recognition issues very effectively. Therefore, machine learning researchers have developed advanced data processing methods in recent years based on the inspirations from biological visual systems.

CNN is a special type of DNN that is inspired by neuroscience. From Hubel's earlier work, we know that the cells in the cortex of the human vision system are sensitive to small regions. The responses of cells within receptive fields have a strong capability to exploit the local spatial correlation in images [25, 26].

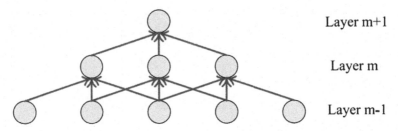

**Fig. 2.** Local connections in the architecture of the CNN

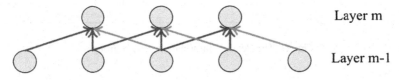

**Fig. 3.** Shared weights in the architecture of the CNN

Additionally, there are two types of cells within the visual cortex, i.e., simple cells and complex cells. While simple cells detect local features, complex cells "pool" the outputs of simple cells within a neighborhood. In other words, simple cells are sensitive to specific edge-like patterns within their receptive field, whereas complex cells have large receptive fields and they are locally invariant. The architecture of CNN is different from other deep learning models. There are two special aspects in the architecture of CNN, i.e., local connections and shared weights. CNN exploits the local correlation using local connectivity between the neurons of near layers. We illustrate this in Fig. 2, where the neurons in the $m$th layer are connected to three adjacent neurons in the $(m\text{-}1)$ th layer, as an example. In CNN, some connections between neurons are replicated across the entire layer, which share the same weights and biases [27–29]. In Fig. 3, the same color indicates the same weight. Using a specific architecture like local connections and shared weights, CNN tends to provide better generalization when facing computer vision problems. A complete CNN stage contains a convolution layer and a pooling layer. Deep CNN is constructed by stacking several convolution layers and pooling layers to form deep architecture. The convolutional layer is introduced first. The value of a neuron $v_{ij}^x$ at position $x$ of the $j$th feature map in the $i$th layer is denoted as follows:

$$v_{ij}^x = g\left(b_{ij} + \sum_m \sum_{p=0}^{P_i-1} w_{ijm}^p v_{(i-1)m}^{x+p}\right) \qquad (1)$$

$$g(x) = \tanh(x) = \frac{e^x - e^{-x}}{e^x + e^{-x}} \qquad (2)$$

Where, $m$ indexes the feature map in the previous layer $((i-1)$th layer) connected to the current feature map, $w_{ijm}^p$ is the weight of position p connected to the mth feature map, $P_i$ is the width of the kernel toward the spectral dimension, and $b_{ij}$ is the bias of $j$th feature map in the $i$th layer.

Pooling can offer invariance by reducing the resolution of the feature maps. Each pooling layer corresponds to the previous convolutional layer. The neuron in the pooling layer combines a small N × 1 patch of the convolution layer. The most common pooling operation is max pooling, which is used throughout this paper. The max pooling is as follows:

$$a_j = \max_{N \times 1}(a_i^{n \times 1} u(n, 1)) \tag{3}$$

Where, $u(n, 1)$ is a window function to the patch of the convolution layer, and $a_j$ is the maximum in the neighborhood. All layers, including the convolutional layers and pooling layers of the deep CNN model, are trained using a backpropagation algorithm.

**Spectral FE Framework for HSI Classification**

In this section, we present a 1-D FE method considering only spectral information. This method stacks several CNNs to develop a deep CNN model with L2 regularization. Generally, the classification of HSI includes two procedures, including FE and classification. In the FE procedure, LR is taken into account to adjust the weights and biases in the back-propagation. After the training, the learned features can be used in conjunction with classifiers such as LR, K-nearest neighbor (KNN), and SVMs.

**Fig. 4.** Architecture of deep CNN with spectral FE of HSI

The proposed architecture is shown in Fig. 4. The input of the system is a pixel vector of hyperspectral data, and the output of the system is the label of the pixel vector. It consists of several convolutional and pooling layers and an LR layer. In Fig. 4, as an example, the flexible CNN model includes two convolution layers and two pooling layers. There are three feature maps in the first convolution layer and six feature maps in the second convolution layer. After several layers of convolution and pooling, the input pixel vector can be converted into a feature vector, which captures the spectral information in the input pixel vector. Finally, we use LR or other classifiers to fulfill the classification step. The power of CNN depends on the connections (weights) of the network; hence, it is very important to find a set of proper weights. Gradient back-propagation is the core fundamental algorithm for all kinds of neural networks. In this paper, the model parameters are initialized randomly and trained by an error back-propagation algorithm. Before setting an updating rule for the weights, one needs to properly set an "error" measure, i.e., a cost function. There are several ways to define such a cost function. In our implementation, a mini-batch update strategy is adopted, which is suitable for large data set processing, and the cost is computed on a mini-batch of inputs. Here $m$, denotes the mini-batch size. Two variables $xi$ and $zi$ denote the $i$th predicted label and the label in the minibatch, respectively. The $i$ summation is done over the whole mini-batch. Our hope turns to optimize (4) using mini-batch stochastic gradient descent.

$$c_0 = -\frac{1}{m}\sum_{i=1}^{m}[x_i\log(z_i)+(1-x_i)\log(1-z_i)] \tag{4}$$

LR is a type of probabilistic statistical classification model. It measures the relation between a categorical variable and the input variables using probability scores as the predicted values of the input variables. To perform classification by utilizing the learned features from the CNN, we employ an LR classifier, which uses softmax as its output-layer activation. Softmax ensures that the activation of each output unit sums to 1 so that we can deem the output as a set of conditional probabilities. For given input vector R, the probability that the input belongs to category $i$ can be estimated as follows:

$$P(Y=i|R,W,b) = s(WR+b) = \frac{e^{W_iR+b_i}}{\sum e^{W_iR+b_i}} \tag{5}$$

Where, W and b are the weights and biases of the LR layer, and the summation is done over all the output units. In the LR, the size of the output layer is set to be the same as the total number of classes defined, and the size of the input layer is set to be the same as the size of the output layer of the CNN. Since the LR is implemented as a single-layer neural network, it can be merged with the former layers of networks to form a deep classifier.

## 3   Results and Discussions

In recent years, low dose CT examination technology has become the focus of medical imaging. Medical and health industry management and supervision organization introduced the relevant laws and regulations to guide manufacturers and research institutions to develop more advanced and more effective innovation design. The results have been shown in the following tables (Table 1).

With technology of deep learning and machine learning, At the same time, also regulated health care practitioners in the practical work of the rational use of low dose, abide by the radiation of the reasonable and lowest (as low as german-russian co-operation achievable, ALARA) principle, as far as possible on the premise of meet the requirements of clinical diagnosis of patients with lower doses. Such as the lower tube voltage and improve noise index, reduce the tube current and methods of shortening the exposure time, but these methods or due to the limited radiation dose to reduce, or due

Table 1. The performances of Group A and Group B

| Performances | Group A | Group B | F value | P value |
|---|---|---|---|---|
| Pulmonary artery | | | | |
| CT(HU) | 250.84 ± 68.60 | 233.51 ± 60.63 | 6.391 | 0.34 |
| SNR | 40.82 ± 13.66 | 37.34 ± 10.63 | 1.021 | – |
| CNR | 30.67 ± 13.30 | 27.32 ± 10.32 | 1.648 | – |
| Thoracic aorta | | | | |
| CT(HU) | 266.34 ± 34.20 | 270.81 ± 42.39 | 47.501 | 0.667 |
| SNR | 43.26 ± 8.45 | 43.55 ± 9.25 | 5.534 | 0.998 |
| CNR | 33.11 ± 7.39 | 33.52 ± 8.37 | 9.645 | 0.853 |
| Muscle | | | | |
| CT(HU) | 62.25 ± 5.81 | 62.18 ± 4.60 | 7.791 | 0.801 |

to the increase of noise limits the clinical application of selective recognition to reduce the iterative reconstruction algorithm and remove the image noise, so scan parameters optimization method combined with the iterative reconstruction be low dose research hotspot in recent years. DR technology is a kind of based on adaptive statistical iterative reconstruction Technology of CT scanner built-in dose lower technology, the technology based on noise reduction rate and the percentage of ASiR automatically choose matching mA value, the linear relationship between the dose to reduce rate and ASiR numerical basic consistent, in the blood vessels of the ROI CT value, SNR and CNR, no statistically significant difference were observed in the background noise, group A and group B radiation dose is reduced by about 48.5%, namely pure application of DR technology in line with the principle of ALARA, but reduce the limited of radiation dose.

# 4   Conclusions

In order to achieve further reduce the radiation dose, the purpose of this study from the X-ray intensity formula $I = KiZU2$, the lower kV is relatively lower tube current can more effectively reduce radiation dose. Huda and found that as the kV reduced, X-ray photon energy closer to contain a high ordinal element structure, such as tissue or blood vessels containing iodine K electron binding energy, the photoelectric effect, X-ray attenuation degree increase, CT value increases, blood vessels and the surrounding higher contrast of substance, but with lower tube voltage reduce X-ray photons, although increases image noise, but for the influence of the blood vessels could be increased CT value compensated by blood vessels, thus lower tube voltage can be applied to enhance CT examination and vascular imaging. This study, according to the results of group C (100 kV) from group A and group B (120 kV) radiation dose was reduced by 58.9% and 20.4% respectively, B group of thoracic aorta and pulmonary artery CT value and image Background noise than in group A were higher, but the SNR of the group B thoracic aorta ROI and CNR, compared with A, B two groups also appeared to rise, makes the aortic blood vessels and normal tissue contrast, two groups patients, there was no statistically significant difference subjective image quality can meet the diagnostic requirements, therefore, DR technology combined with low kV, 100 kV) technology is still in line with the ALARA principle.

**Acknowledgments.** This work was supported by the grants of the National Science Foundation of China, Nos. 61133011.

# References

1. Benediktsson, J.A., Ghamisi, P.: Spectral-spatial classification of hyperspectral remote sensing images. Artech House, Boston (2015)
2. Hughes, G.: On the mean accuracy of statistical pattern recognizers. IEEE Trans. Inf. Theory **14**(1), 55–63 (1968)
3. Dias, J.B., et al.: Hyperspectral remote sensing data analysis and future challenges. IEEE Geosci. Sens. Mag. **1**(2), 6–36 (2013)
4. Jia, X., Kuo, B., Crawford, M.M.: Feature mining for hyperspectral image classification. Proc. IEEE **101**(3), 676–679 (2013)
5. Licciardi, G., Marpu, P.R., Chanussot, J., Benediktsson, J.A.: Linear versus nonlinear PCA for the classification of hyperspectral data based on the extended morphological profiles. IEEE Geosci. Remote Sens. Lett. **9**(3), 447–451 (2011)
6. Villa, A., Benediktsson, J.A., Chanussot, J., Jutten, C.: Hyperspectral image classification with independent component discriminant analysis. IEEE Trans. Geosci. Remote Sens. **49**(12), 4865–4876 (2011)
7. Bandos, T.V., Bruzzone, L., Camps-Valls, G.: Classification of hyperspectral images with regularized linear discriminant analysis. IEEE Trans. Geosci. Remote Sens. **47**(3), 862–873 (2009)
8. Bruce, L.M., Koger, C.H., Li, J.: Dimensionality reduction of hyperspectral data using discrete wavelet transform feature extraction. IEEE Trans. Geosci. Remote Sens. **40**(10), 2331–2338 (2002)
9. Jimenez, L.O., Landgrebe, D.A.: Hyperspectral data analysis and supervised feature reduction via projection pursuit. IEEE Trans. Geosci. Remote Sens. **37**(6), 2653–2667 (1999)
10. Lunga, D., Prasad, S., Crawford, M.M., Ersoy, O.: Manifold-learningbased feature extraction for classification of hyperspectral data: a review of advances in manifold learning. IEEE Signal Process. Mag. **31**(1), 55–66 (2014)
11. Han, T., Goodenough, D.: Investigation of nonlinearity in hyperspectral imagery using surrogate data methods. IEEE Trans. Geosci. Remote Sens. **46**(10), 2840–2847 (2008)
12. Tenenbaum, B., Silva, V., Langford, C.: A global geometric framework for nonlinear dimensionality reduction. Science **290**(5500), 2319–2323 (2000)
13. Roweis, S., Saul, L.K.: Nonlinear dimensionality reduction by locally linear embedding. Science **290**(5500), 2323–2326 (2000)
14. Bachmann, C.M., Ainsworth, T.L., Fusina, R.A.: Improved manifold coordinate representations of large-scale hyperspectral scenes. IEEE Trans. Geosci. Remote Sens. **44**(10), 2786–2803 (2006)
15. Scholkopf, B., Smola, A.J.: Learning With Kernels. MIT Press, Cambridge (2002)
16. Kuo, B.C., Li, C.H., Yang, J.M.: Kernel nonparametric weighted feature extraction for hyperspectral image classification. IEEE Trans. Geosci. Remote Sens. **47**(4), 1139–1155 (2009)

17. Plaza, A., Plaza, J., Martin, G.: Incorporation of spatial constraints into spectral mixture analysis of remotely sensed hyperspectral data. In: Proceedings IEEE International Workshop Machine Learning Signal Processing, Grenoble, France, 2009, pp. 1–6

18. Fauvel, M., Tarabalka, Y., Benediktsson, J.A., Chanussot, J., Tilton, J.C.: Advances in spectral–spatial classification of hyperspectral images. Proc. IEEE **101**(3), 652–675 (2013)

19. Tarabalka, Y., Fauvel, M., Chanussot, J., Benediktsson, J.A.: SVM- and MRF-based method for accurate classification of hyperspectral images. IEEE Geosci. Remote Sens. Lett. **7**(4), 736–740 (2010)

20. Fauvel, M., Benediktsson, J.A., Chanussot, J., Sveinsson, J.: Spectral and spatial classification of hyperspectral data using SVMs and morphological profiles. IEEE Trans. Geosci. Remote Sens. **46**(11), 3804–3814 (2008)

21. Li, J., Bioucas-Dias, J.M., Plaza, A.: Spectral–spatial classification of hyperspectral data using loopy belief propagation and active learning. IEEE Trans. Geosci. Remote Sens. **51**(2), 844–856 (2013)

22. Chen, Y., Nasrabadi, N.M., Tran, T.D.: Hyperspectral image classification using dictionary-based sparse representation. IEEE Trans. Geosci. Remote Sens. **49**(10), 3973–3985 (2011)

23. Song, B., Li, J., Bioucas-Dias, J.M., Benediktsson, J.A.: Remotely sensed image classification using sparse representations of morphological attribute profiles. IEEE Trans. Geosci. Remote Sens. **52**(8), 5122–5136 (2013)

24. Bengio, Y., Courville, A., Vincent, P.: Representation learning. a review and new perspectives. IEEE Trans. Pattern Anal. Mach. Intell. **35**(8), 1798–1828 (2013)

25. Kruger, N., et al.: Deep hierarchies in primate visual cortex what can we learn for computer vision? IEEE Trans. Pattern Anal. Mach. Intell. **35**(8), 1847–1871 (2013)

26. Krizhevsky, A., Sutskever, I., Hinton, G.: ImageNet classification with deep convolutional neural networks. In: Proceedings Neural Information Processing System, pp. 1106–1114, Lake Tahoe, NV, USA (2012)

27. Hinton, G., Salakhutdinov, R.: Reducing the dimensionality of data with neural networks. Science **313**(5786), 504–507 (2006)

28. LeCun, Y., Bottou, L., Bengio, Y., Haffner, P.: Gradient-based learning applied to document recognition. Proc. IEEE **86**(11), 2278–2324 (1998)

29. Chen, Y., Lin, Z., Zhao, X., Wang, G.: Deep learning-based classification of hyperspectral data. IEEE J. Sel. Topics Appl. Earth Observ. Remote Sens. **7**(6), 2094–2107 (2014)

# Predicting Potential Drug-Target Interactions with Multi-label Learning and Ensemble Learning

Lida Zhu[1(✉)] and Jun Yuan[2]

[1] College of Informatics, Huazhong Agricultural University, Wuhan, China
zhulinda@hotmail.com
[2] School of Computer Science and Technology,
Hankou University, Wuhan, China

**Abstract.** Identifying Drug-Target Interactions (DTIs) is an important process in drug discovery. Wet experimental methods are expensive and time-consuming for detecting DTIs. Therefore, computational approaches were provided to deal with this task and have many effective strategies. In this paper, by using a machine-learning models for identifying drug targets, we provide a novel tool for DTIs prediction. In recent years, most of computational methods only are used to find the drug-drug similarity or target-target similarity, which cannot perfectly capture all characteristics to identify DTIs. To improve the performance of prediction, we focus on critical drug-related features and ignore irrelevant features to represent drugs, targets and relationship between them. Moreover, we further develop the ensemble learning model by integrating individual feature-based multi-label models for predicting DTIs. Experiments of evaluation show that the proposed approach achieves better results than other outstanding methods on benchmark datasets.

**Keywords:** Drug discovery · Drug targets · Drug-Target Interactions ·
Association features · Linear neighborhood similarity · Multi-label learning

## 1 Introduction

Drug discovery is one of the most complex and important scientific research topics. The identification of Drug-Target Interactions (DTIs) has been a successful strategy in the target-based drug discovery process and enabled a great expansion of compounds available for the medical treatment. Due to the fast speed of development of modern molecular biology methods and the accumulation of the knowledge of the human genome, pharmaceutical industry has been further developed during the past few years. However, the wet experiment for identification of DTI is still a time-consuming process and expensive with low success rate [1]. Thus, many researchers spend great attentions by using computational technique to promote ligand research [2].

Since many biological signaling pathways involve ligand-receptor interactions. Extensive studies have been proved that drugs act on the cell membrane by physical and chemical interactions. In recent years, machine-learning methods were applied to the DTIs prediction, extending works from case study to more data. These methods

© Springer Nature Switzerland AG 2019
D.-S. Huang et al. (Eds.): ICIC 2019, LNCS 11644, pp. 726–735, 2019.
https://doi.org/10.1007/978-3-030-26969-2_69

used known information from given set of drugs, target proteins and known interaction between targets and drugs, to predict novel drug-target interactions [3]. Generally, computational methods could roughly be classified into two categories: feature-based methods and similarity-based methods. For feature-based methods, many studies were summarized as structure-based, molecular activity-based, side-effect-based and multi-omics-based predictions according to the used data for inference [4]. These computational methods for predicting DTIs also provides an overview with available web-servers and databases for useful DTI features [5]. Lei Wang et al. developed a rotation forest-based computational model RFDT for predicting DTI using the structural properties from drugs and protein sequence information [6]. For similarity-based method, they are based on a hypothesis that similar drugs interact with the same targets, and similar targets interact with the same drugs [7]. Recently, a lot of attentions were paid on polypharmacology and chemogenomics, about several agents exert their biological effects with binding multiple target proteins [8, 9].

The drug-target interaction data includes drug-target association relationship, drug-drug similarity (e.g. chemical structure similarity between compounds using SIM-COMP [10]) and target-target similarity (e.g. sequence similarity between proteins using a normalized version of Smith-Waterman score [11]). And many exists methods have provided a theoretical basis by including the chemical and biological features of drugs (i.e. chemical structures, target protein, gene expression profiles, gene ontology annotations, etc.). Pauwels took into account chemical substructures of drug candidate molecules, and respectively adopted four machine learning methods (KNN, SVM, ordinary canonical correlation analysis and sparse canonical correlation analysis) to construct prediction models [12]. Liu integrated the phenotypic information, chemical information and biological information about drugs, and then built the prediction models by using different machine learning classifiers (LR, NB, KNN, random forest and SVM) [13]. And furthermore, Xiao-Ying Yan et al. proposed an effective strategy via decision template to integrate multiple classifiers designed with multiple similarity metrics [14]. These methods provide a promising solution to build the relationship between drug-related features and DTIs.

Inspired by these pioneering works, we attempt to predict the potential targets of approved drugs by using known targets. Based on intuition, DTI prediction is often formulated as a binary classification task, which aims to predict whether a DTI is present or not. However, DTIs prediction could be considered as a multi-label learning problem actually based on polypharmacology [9]. Thus, we adopt the issue in the frame of multi-label learning. In this paper, we formulate approved drugs, targets and DTIs as a feature-based multi-label models, and combine several feature-based model as base predictor by adopting ensemble learning model to improve performance for predicting new potential DTIs.

## 2   Method

### 2.1   Datasets

To the best of our knowledge, there are several databases which describe drug features.

**Drug Information.** DrugBank [15], TTD [16] and DGIdb [17] are the three major databases we collected drugs and drug-target association.

**Feature Information**:

**Target Information:** 606 drugs covers 1,067 targets.

**Indication Information:** By using the disease classed provided by Pharmaprojects (similarity threshold: 0.75) [18], the collection of the drug indication information for the agents comes from DrugBank, TTD, and ClinicalTrials [19] and involves 642 types of diseases.

**Structure Information:** The structural similarity with substructure key-based 2D Tanimoto similarity score was calculated from the PubChem database.

**Gene Expression Information:** The gene expression profiles was from CMAP [20].

**Module Information:** The drug module information was used bio-clustering method based on drug expression profile data.

Until recently, several datasets, i.e. Mizutani's dataset [21], and Liu's dataset [13], were used in the previous studies to develop the state-of-art methods. More importantly, these datasets are available to the public. Hence, we adopt these datasets as the benchmark datasets to compare and evaluate various methods. Table 1 displays the details of three datasets, covering the number of approved drugs, number of targets and the number of targets of the approved drugs (DTIs). As our work is to infer the potential targets from known targets, the utilization of known targets of approved drugs will help to build prediction models.

**Table 1.** The details about benchmark datasets.

| Datasets | Drug | Target | Indications | Substructure | Transporter | Enzyme | Pathway | Gene expression |
|----------|------|--------|-------------|--------------|-------------|--------|---------|-----------------|
| Mizutani | 658 | 1368 | NA | 881 | NA | NA | NA | NA |
| Liu | 832 | 786 | 869 | 881 | 72 | 111 | 173 | NA |
| Our data | 606 | 1067 | 642 | 881 | NA | NA | NA | 435 |

## 2.2 Drug Similarity Measurement

To assess the correlations of drugs and their target genes, we characterized the similarity between drugs and drugs, target and target, by using the Tanimoto coefficient. The Tanimoto coefficient (TC) is calculated using the following equation:

$$TC = \frac{N_{AB}}{N_A + N_B - N_{AB}} \tag{1.1}$$

Where $N_A$ is the number of drug A-related genes or other features, $N_B$ is the number of drug B-related genes or other features, and $N_{AB}$ is common gene or common features of drug A and drug B.

## 2.3    Problem Definition and Solution Strategy

Since a drug is usually targeting on multiple genes while a gene could also be targeted by multiple drugs, DTIs prediction problem can be treated as a multi-label classification task. In this paper, we adopted a method of multi-label k-nearest neighbor (MLKNN) which is capable to determine critical features and construct high-accuracy multi-label prediction models for DTIs prediction simultaneously.

Representation of drug features and drug labels is a crucial step in multi-label learning. Both positive and negative samples from the 15,367 collected agent-target pairs covering 606 agents and 1,067 targets.

Given a dataset of n drugs denoted as $\{(x_i, y_i)\}(i = 1)^n$, $x_i$ and $y_i$ are the p-dimensional feature vector and q-dimensional disease vector for the $i$th drug, respectively. Our goal is to build the functional relationship $Y = F(X):2^P \rightarrow 2^q$ between exploratory variables (feature vector) and target values (agent-target vector) for multi-label learning purpose.

Firstly, four MLKNN models were constructed based on four features. Then, each model was evaluated by the internal five-fold cross validation on the training data. As a result, five MLKNN models were built based on five internal folds and selected features. The prediction result is the average scores of the outputs by five MLKNN models. Finally, we used the ensemble learning methods to combine four features and generated high-accuracy prediction models.

## 2.4    Logistic Regression

Logistic regression (LR) is a statistical method used to estimate the probability of a response variable based on the given variables using a logistic function [22]. We built an LR model to predict the DTIs for performance comparison by using the R function "glm" (R, version 3.2.2, 2015).

## 2.5    Multi-label Learning Method

Formally, the multi-label learning is to build a model that maps inputs to binary vectors, rather than scalar outputs of the ordinary classification.

Given the training set $\{(x_i, y_i)\}_{i=1,...,n}$, $x_i$ is the $i$th instance (drug), and $y_i$ is the target gene vector. $y_i(l) = 1$. If the $i$th instance can target the $l$th gene, otherwise $y_i(l) = 0$, $l = 1, 2, ..., q$. The $k$ nearest neighbors (in training set) of instance $x_i$ are denoted by $N(x_i)$, $i = 1, 2, ..., n$. Thus, based on $l$th target of these neighbors, a membership counting vector can be denoted as:

$$C_{x_i}(l) = \sum_{a \in N(x_i)} y_a(l), l = 1, 2, ..., q \qquad (1.2)$$

where $C_{xi}(l)$ counts the number of neighbors of $x_i$ targeting the $l$th gene, and $0 \leq C(x_i)(l) \leq k$.

For a test drug $t$, MLKNN identifies its $k$ nearest neighbors in the training set and calculate $C_t(l)$. Let $H_1^l$ be the event that a drug targets $l$th gene and $H_0^l$ be the event that a drug does not target $l$th gene. Let $E_j^l$ be the event that a drug just has $j$ neighbors with

$l$th target gene in its $k$ nearest neighbors. For the instance $t$, its label for $l$th target gene $y_t(l)$ is determined by the following principle:

$$y_t(l) = \arg\ \max{}_{b \in \{0,1\}} P(H_b^l | E_{Ct(l)}^l), l = 1, 2, \ldots, q \tag{1.3}$$

Using the Bayesian rule, above equation can be rewritten as:

$$y_t(l) = \arg\ \max{}_{b \in \{0,1\}} \frac{P(H_b^l)P(E_{Ct(l)}^l | H_b^l)}{P(E_{Ct(l)}^l)}$$
$$= \arg\ \max{}_{b \in \{0,1\}} P(H_b^l)P(E_{Ct(l)}^l | H_b^l) \tag{1.4}$$

In the prediction model, $P(H_b^l)$ and $P(E_{Ct(l)}^l | H_b^l)$ are calculated based on the training set. So, the prior probabilities are calculated.
$P(H_1^l) = (s + \sum_{i=1}^{n} y_i(l))/(s \times 2 + n)$ and $P(H_0^l) = 1 - P(H_1^l)$. Then, the posterior probabilities $P(E_{Ct(l)}^l | H_0^l)$, $P(E_{Ct(l)}^l | H_1^l)$ are calculated by following equations:

$$P(E_j^l | H_1^l) = \frac{(s + c[j])}{s \times (k+1) + \sum_{i=0}^{k} c_i[i]}$$

$$P(E_j^l | H_0^l) = \frac{(s + c'[j])}{s \times (k+1) + \sum_{i=0}^{k} c_i'[i]} \tag{1.5}$$

$$l = 1, 2, \ldots, q; j = 1, 2, \ldots, k$$

where $s$ is the smooth factor. $c_l[i]$ is the number of instances which just has $i$ neighbors with $l$th target gene in their $k$ nearest neighbors; $c_l'[i]$ is the number of instances which just has $i$ neighbors without $l$th target gene in their $k$ nearest neighbors.

To facilitate the use of the machine-learning prediction models, we developed a tool that allows a quick and intuitive access to the background information and predicted new DTIs.

## 2.6   Ensemble Learning Strategy

In this paper, an ensemble learning method was designed to combine various features and develop high-accuracy prediction models. In this study, an ensemble classifier was generated using the linear weighted sum of outputs from classifiers based on four features.

Given m features, we build m individual feature-based MLKNN models, and use them as base predictors. Since features may make different contributes, it is natural to adopt weighted scoring ensemble strategy, which assigns m base predictors with m weights $\{w_1, w_2, \ldots, w_m\}$. For a testing instance, the $i$th predictor will give scores for q target genes, denoted as $S_i = \{s_i^1, s_i^2, \ldots, s_i^q\}$, $i = 1, 2, \ldots, m$ The final prediction produced by the ensemble model is the linear weighted sum of outputs from base predictors.

$$EnsembleScore = [w_1, w_2, \ldots, w_m] \times \begin{bmatrix} s_1 \\ s_2 \\ \vdots \\ s_m \end{bmatrix}$$

$$= [w_1, w_2, \ldots, w_m] \times \begin{bmatrix} s_1^1 & s_1^2 & \cdots & s_1^q \\ s_2^1 & s_2^2 & \cdots & s_2^q \\ \vdots & \vdots & \ddots & \vdots \\ s_m^1 & s_m^2 & \cdots & s_m^q \end{bmatrix} \tag{1.6}$$

Tuning weights for base predictors are critical for the ensemble models. The weights are non-negative real values between 0 and 1, and the sum of weights equals 1. We adopt the internal 5-CV AUPR on training data is used as the fitness score.

## 3  Experiments Results and Discussions

### 3.1  Evaluation Metrics

In the DTIs prediction, the predicted scores for activities were usually merged for evaluation, and the metrics for ordinary binary classification were often adopted. The area under ROC curve (AUC) and the area under the precision-recall curve (AUPR) can be used to evaluate models regardless of any threshold. However, there are much more negative labels than positive labels in the agent-activities prediction, and machine-learning methods are likely to produce overestimated AUC scores. Since AUPR takes into account recall as well as precision, it is used as the most important metric.

We used the following evaluation metrics to evaluate the performance of machine-learning models: Precision, Accuracy (ACC), Recall, Specificity, Mathew's correlation coefficient (MCC). These metrics can be calculated by the number of true positives (TP), false positives (FP), true negatives (TN), and false negatives (FN).

$$Precision = TP/(TP + FP) \tag{1.7}$$

$$ACC = (TP + TN)/(TP + FN + TN + FP) \tag{1.8}$$

$$Recall = TP/(TP + FN) \tag{1.9}$$

$$Specificity = TN/(TN + FP) \tag{1.10}$$

$$MCC = \frac{(TP \times TN - FP \times FN)}{\sqrt{((TP + FP) \times (TP + FN) \times (TN + FP) \times (TN + FN))}} \tag{1.11}$$

Several metrics were designed for multi-label classification, i.e. Hamming loss, one-error, coverage, ranking loss and average precision. Hamming loss is the fraction of the wrong labels to the total number of labels. The one-error evaluates the fraction of examples whose top-ranked label is not in the relevant label set. The coverage evaluates how many steps are needed, on average, to move down the ranked label list so as to cover all the relevant labels of the example. The average precision evaluates whether the average fraction of relevant labels ranked higher than a particular label. Therefore, we adopt AUPR, average precision, one-error, coverage, ranking loss and hamming loss for the agent-activities prediction.

## 3.2   Performances of Prediction Models

We divided the drugs with individual features, and then respectively adopt MLKNN as the multi-label learning engines to construct prediction models. To provide a comprehensive evaluation of the method, we randomly repeat the data separation, and implement 1000 runs of 5-cross validation for Logistic Regression (LR), MLKNN models and ensemble model based on MLKNN. The comparison results of metric scores of different measures are respectively demonstrated in Table 2.

**Table 2.** Comparison of different measures

| Methods | Auc | Aupr | Precision | Recall | Acc | Specificity | Mcc |
|---|---|---|---|---|---|---|---|
| LR | 0.7969 | 0.3218 | 0.0042 | 0.0664 | 0 | 0.9920 | 0.1501 |
| MLKNN | 0.7488 | 0.1341 | 0.0106 | 0.2825 | 0.7443 | 0.9356 | 0.2572 |
| MLKNN+ensemble | 0.8090 | 0.2514 | 0.0187 | 0.1310 | 0.5499 | 0.9920 | 0.4026 |

According to the result shows in Table 2, the ensemble model of MLKNN has better performance than MLKNN based on the same features. The index of AUC, AUPR, Precision and MCC reveals the ensemble model of MLKNN has the best performance. Meanwhile, the index of Recall, ACC and specificity indicates the ensemble model does not have its best performance. However, due to the imbalanced dataset, the index of AUC, AUPR, Precision and MCC is considered more important.

## 3.3   Comparison with Different Split of Imbalanced Dataset

It is observed that the models produce overestimated AUC scores for the imbalanced data in the experiment, and AUC is not a suitable metric for the problem. Since LR is robust to the data split of the cross validation, we make analysis and comparison based on the same data split in the following result (Table 3). The data was split by 1:1, 1:3 and 1:5 of positive: negative for performance comparison.

**Table 3.** Comparison of different imbalanced measure

| Positive:negative | Auc | Aupr | Recall | Specificity | Mcc |
|---|---|---|---|---|---|
| 1:1 | 0.7961 | 0.7674 | 0.7467 | 0.7232 | 0.4707 |
| 1:3 | 0.7979 | 0.5520 | 0.3310 | 0.9355 | 0.3419 |
| 1:5 | 0.7998 | 0.4431 | 0.1775 | 0.9748 | 0.2595 |

According to the result shows in Table 3, as a typical binary classifier, LR can produce good result only when the positive/negative of dataset split of 1:1, which indicates the robustness of LR's performance is low. In addition, the AUC is unable to characterize LR's performance in various positive/negative ratios. Meanwhile, the AUPR value presents better performance than AUC.

### 3.4   Comparison with Benchmark Datasets

To the best of our knowledge, some state-of-the-art methods, provided useful public available datasets i.e. Mizutani's method [17] and Liu's method [10], could be used for DTIs prediction.

Since our work is to predict the potential DTIs, Mizutani's and Liu's datasets are adopted as the benchmark datasets for the fair comparison. Here, we obtained various metric scores. We implement Liu's dataset and only utilize the drug-related biological features and chemical features to predict new DTIs. Since benchmark datasets construct the prediction models based on specific datasets and specific features, we construct our models based on the same datasets and the same features for the fair comparison. In this way, our method can be compared with benchmark datasets under the same conditions. The performances of different methods are demonstrated in Table 4. In terms of AUPR, our method makes obvious improvements over Mizutani's datasets and Liu's datasets. Moreover, our method produces better performances in terms of multi-label learning metrics (Hamming Loss, ranking loss, one error, coverage and average precision). In conclusion, our method can produce better performances than state-of-the-art methods.

**Table 4.** Comparison of different benchmark datasets

| Dataset | Auc | Aupr | Hamming loss | Ranking loss | One error | Coverage | Average precision |
|---------|------|-------|--------------|--------------|-----------|----------|-------------------|
| Mizutani | 0.6563 | 0.1013 | 0.0055 | 0.2294 | 0.6910 | 498.34 | 0.2864 |
| Liu | 0.6982 | 0.1655 | 0.0042 | 0.2825 | 0.7443 | 294.80 | 0.2572 |
| Our data | 0.8090 | 0.2514 | 0.0187 | 0.1310 | 0.5499 | 206.36 | 0.4026 |

## 4   Conclusion

In this paper, based on the idea that the relationship among drugs and targets is largely influenced by the chemical structure of the drug and the sequence information of the protein, we propose a novel computational method to infer potential unknown drug-target interactions by integrating chemical structure similarity, sequence similarity, gene expression similarity and indication similarity information. Rather than considered DTI prediction as a binary classification task, which aims to predict whether a DTI is present or not. This study transforms the DTIs prediction as a multi-label learning task. We propose a novel multi-label learning approach for DTIs prediction, which can produce high-accuracy performances. In order to combine various features effectively, we construct individual feature-based MLKNN models and use them as base predictors. Then, we combined base predictors by using the weighted scoring ensemble strategy, and developed the final prediction models for DTIs prediction. Compared with the state-of-the-art methods, the ensemble method produces much better performances on the benchmark datasets. In conclusion, the proposed MLKNN and the ensemble method are promising tools for predicting DTIs. In the future, we plan to integrate more biology knowledge to extract representative features, using more advanced machine learning method to improve the prediction performance.

**Acknowledgment.** This research was partially supported by the Fundamental Research Funds for the Central Universities (Grant 2662017PY115).

# References

1. Chen, X., Yan, C.C., Zhang, X., et al.: Drugtarget interaction prediction: databases, web servers and computational models. Brief. Bioinform. **17**(4), 696–712 (2016)
2. Han, J., Kamber, M., Pei, J.: Data Mining: Concepts and Techniques. Morgan Kaufmann, Burlington (2006)
3. Cheng, T., Hao, M., Takeda, T., Bryant, S.H., Wang, Y.: Large-scale prediction of drug-target interaction: a data-centric review. AAPS J. **19**, 1–12 (2017)
4. Huang, G., Yan, F., Tan, D.: A review of computational methods for predicting drug targets. Curr. Protein Pept. Sci. **19**(6), 562–572 (2018)
5. Anusuya, S., et al.: Drug-target interaction prediction methods and applications. Curr. Protein Pept. Sci. **19**(6), 537–561 (2018)
6. Wang, L., You, Z.-H., Chen, X., Yan, X., Liu, G., Zhang, W.: RFDT: a rotation forest-based predictor for predicting drug-target interactions using drug structure and protein sequence information. Curr. Protein Pept. Sci. **19**(5), 445–454 (2018)
7. Palma, G., Vidal, M.-E., Raschid, L.: Drug-target interaction prediction using semantic similarity and edge partitioning. In: Mika, P., et al. (eds.) ISWC 2014. LNCS, vol. 8796, pp. 131–146. Springer, Cham (2014). https://doi.org/10.1007/978-3-319-11964-9_9
8. Medina-Franco, J.L., Giulianotti, M.A., Welmaker, G.S., Houghten, R.A.: Shifting from the single to the multitarget paradigm in drug discovery. Drug Discovery Today **18**(9–10), 495–501 (2013)
9. Anighoro, A., Bajorath, J., Rastelli, G.: Polypharmacology: challenges and opportunities in drug discovery: miniperspective. J. Med. Chem. **57**(19), 7874–7887 (2014)
10. Hattori, M., Okuno, Y., Goto, S., Kanehisa, M.: Development of a chemical structure comparison method for integrated analysis of chemical and genomic information in the metabolic pathways. J. Am. Chem. Soc. **125**(39), 11853–11865 (2003)
11. Smith, T.F., Waterman, M.: Identification of common molecular subsequences. J. Mol. Biol. **147**(1), 195–197 (1981)
12. Pauwels, E., Stoven, V., Yamanishi, Y.: Predicting drug side-effect profiles: a chemical fragment-based approach. BMC Bioinformatics **12**, 169 (2011)
13. Liu, M., et al.: Large-scale prediction of adverse drug reactions using chemical, biological, and phenotypic properties of drugs. J. Am. Med. Inf. Assoc. **19**(E1), E28–E35 (2012)
14. Yan, X.-Y., Zhang, S.-W.: Identifying drug-target interactions with decision templates. Curr. Protein Pept. Sci. **19**(5), 498–506 (2018)
15. Law, V., et al.: DrugBank 4.0: shedding new light on drug metabolism. Nucleic Acids Res. **42**, D1091–D1097 (2014). http://www.drugbank.ca/. Accessed 30 November 2015
16. Qin, C., et al.: Therapeutic target database update 2014: a resource for targeted therapeutics. Nucleic Acids Res. **42**, D1118–D1123 (2014). http://bidd.nus.edu.sg/group/cjttd/. Accessed 30 November 2015

17. Wagner, A.H., et al.: DGIdb 2.0: mining clinically relevant drug-gene interactions. Nucleic Acids Res. **44**, D1036–D1044 (2015). http://dgidb.genome.wustl.edu/. Accessed 30 November 2015
18. Mcinnes, B.T., Pedersen, T., Pakhomov, S.V.: UMLS-interface and UMLS-similarity: open source software for measuring paths and semantic similarity. AMIA Annu. Symp. Proc. **2009**, 431–435 (2009)
19. Ligeti, B., Pénzváltó, Z., Vera, R., Győrffy, B., Pongor, S.: A network-based target overlap score for characterizing drug combinations: high correlation with cancer clinical trial results. PLoS One **10**(6), e0129267 (2015)
20. Lamb, J., et al.: The connectivity map: using gene-expression signatures to connect small molecules, genes, and disease. Science **313**(5795), 1929–1935 (2006)
21. Mizutani, S., Pauwels, E., Stoven, V., Goto, S., Yamanishi, Y.: Relating drug-protein interaction network with drug side effects. Bioinformatics **28**(18), i522–i528 (2012)
22. Mccallum, A., Nigam, K.: A comparison of event models for Naïve Bayes text classification. In: AAAI-98 Workshop on Learning for Text Categorization, vol. 62, pp. 41–48 (1998)

# Robust Circulating Tumor Cells Detection in Scanned Microscopic Images with Cascaded Morphological and Faster R-CNN Deep Detectors

Yun-Xia Liu[1,2(✉)], An-Jie Zhang[1], Qing-Fang Meng[1,2],
Ying-Jie Chen[3], Yang Yang[4], and Yue-Hui Chen[1,2]

[1] School of Information Science and Engineering, University of Jinan,
Jinan, China
ise_liuyx@ujn.edu.cn
[2] School of Control Science and Engineering,
Shandong University, Jinan, China
[3] Department of Clinical Laboratory, The Second Hospital
of Shandong University, Jinan, China
[4] School of Information Science and Engineering, Shandong University,
Jinan, China

**Abstract.** Robust detection and numeration of circulating tumor cells (CTCs) in scanned microscopic images of peripheral blood is essential to clinical diagnosis, individualized treatment, and prognosis judgement evaluation. Automated detection algorithm based on machine learning methods helps to reduce the subjectivity and labor intensity of cytologists in their clinical practice. In this paper, a robust CTCs detection algorithm based on cascading of two stages of detectors is proposed. Firstly, a morphological rule based detector is applied to screen out most normal cells. Based on the detection results of the first stage detection, hard negative sample selection is performed according to confidence score values of an integrated deep classifier. Finally, a second stage faster R-CNN detector is trained on positive and negative samples to obtain the detection results. Experimental results carried out on a self-established CTCs database show that the proposed algorithm achieves robust and quasi-realtime CTCs detection.

**Keywords:** Image cytometry · Circulating tumor cells · Cell detection · Morphological operation · Deep learning · Faster R-CNN

## 1 Introduction

Circulating tumor cells (CTCs) refers to individual cell or clusters of cells that shed from a primary tumor into vasculature or lymphatics circulation, which will cause subsequent growth of additional metastases tumors in distant organs [1]. Research reveals that accurate detection and numeration of CTCs plays an important role in clinical diagnosis and staging of tumors, individualized treatment design, as well as prognosis judgement evaluation [2]. Effective automated CTCs detection method based

© Springer Nature Switzerland AG 2019
D.-S. Huang et al. (Eds.): ICIC 2019, LNCS 11644, pp. 736–746, 2019.
https://doi.org/10.1007/978-3-030-26969-2_70

on image processing and machine learning methods can help to alleviate the subjectivity of different cytologists. Detected suspicious CTCs regions serving as cytologists' prompt have successfully reduced their labor intensity in clinical practice.

In the past few years, capture and detection of CTCs from peripheral blood has been widely explored by researchers and many methods have been proposed [3]. Immunological reaction-based detection methods, such as the flow cytometry [4], the CellSearch System [5] rely on immunological reaction of specific tumor associated markers. Subsequent image analysis for these methods is relative easy as detected CTCs are dyed with different colors. Svensson et al. [6] developed a naive Bayesian classifier (NBC) based on Red-Green-Blue (RGB) color features. However, these methods suffer from the disadvantage of limited generality that the surface markers are usually cancer type specific.

An effective way to counter this challenge is to employ cell filtration-based isolation method, also known as Isolation-by-Size-of-Epithelial-Tumor (ISET), for CTCs detection [7, 8]. ISET enables CTCs detection of all types of cancer, which utilizes the physical characteristic of CTCs that they are large in size as compared with normal blood cells. However, the rareness of CTCs in peripheral blood (in frequency of 1–10' per mL), the morphological similarity with normal blood cells and complex visual information (varying dying conditions, clusters of cells, occlusion, and so on) in scanned microscopic images impose great challenge to robust CTCs detection. In our previous work [9], an automatic CTCs detection system is proposed based on morphological digital image processing algorithm. However, the performance is far from satisfactory. Inspired by the recent breakthroughs of deep learning algorithms in medical image processing field [10, 11], a faster R-CNN detection network is proposed in [12], which demonstrates state of the art CTCs detection performance.

In order to further improve the robustness of CTCs detection algorithm based on scanned microscopic images of peripheral blood samples, a two stage cascaded detection algorithm is proposed in this paper. A first stage morphological detection is carried out based on prior knowledge that CTCs are larger in cell size to locate suspicious CTCs position, considering the intrinsic high throughput characteristic of the problem. Detected results of the first stage are applied to a Faster R-CNN network trained only on positive samples (CTCs) to generate high quality negative samples (normal blood cells) based on their confidence scores. The second stage detection is based on Faster R-CNN deep detector trained on positive and negative samples for reliable CTCs detection. Experimental results are presented to demonstrate the validity of the proposed algorithm.

The remainder of this paper is organized as follows. Part 2 introduces related works of microscopic images acquisition, CTCs database, as well as previously proposed morphological and faster R-CNN detectors. The overall architecture and details of the proposed two stage cascade CTCs detection algorithm is discussed in Part 3. Part 4 presents experimental results. Part 5 concludes the paper and suggest possible directions for future research.

## 2 Related Works

This section provides an introduction of the microscopic images acquisition methods and the CTCs sample database established for algorithm evaluation. Our previous work about rule-based morphological CTCs detector and faster R-CNN detectors are also discussed in this section.

### 2.1 Microscopic Images Acquisition and CTCs Database Method

Standard and repeatable high quality microscopic images are essential for robust CTCs detection in practical applications. All microscopic images in this paper are Scanned at ×40 magnification with the rapid CTCs microscopic imaging system developed by Wuhan YZY Medical technology Co. Ltd. Considering the liquid nature of blood, series of images are scanned at 5 different focal lengths (denoted as $L = \{-2, -1, 0, 1, 2\}$) simultaneously to provide multiple cues for cytologists' subsequent examination purpose. In this way, a peripheral blood sample results in whole slide image (WSI) with typical resolution of $40000 \times 40000 \times 5$, which usually occupies 2 GB storage space. The high throughput characteristic imposes challenge on computation efficiency of CTCs detection algorithms. Figure 1 illustrates several typical microscopic CTCs images. We see that CTCs exists in two types: single CTC (Last column of Fig. 1(a)) and CTC clusters of three or more individual CTCs (circulating tumor microemboli, CTM, shown in first three columns of Fig. 1(a)). Meanwhile, there are strong visual similarities between CTCs and normal blood cells.

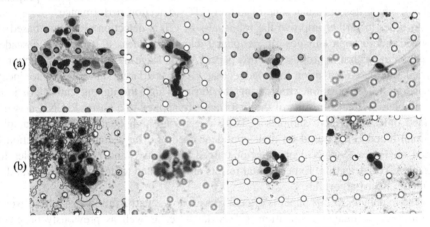

**Fig. 1.** Illustration of typical (a) CTCs and (b) Normal blood cells.

We have collected 88 WSI from 35patients (9 females and 26 males, aging from 36 to 65) attending the second hospital of Shandong university from June to December 2017. They all suffer from esophageal squamous carcinoma. Written informed consent was obtained from all patients. The study was granted approval by the ethics committee of the second hospital of Shandong university. We invite three senior experienced cytologists to manually annotate all 88 WSI according to the Hoffman rule [13] which is the clinical golden standard for clinical CTCs decision. Final annotation results are obtained based on major voting between the three cytologists. It is a labor intensive work that lasts for three months. With the prompt suspicious CTC/CTM candidates by faster R-CNN detection network proposed by [12], a CTCs dataset contains 73 CTCs and 281 CTM, 354 positive training events in total was established. It is a median size database for CTCs detection, where the range of CTCs/CTM per WSI is between 0 and 47, with a median number of 6.

## 2.2 Morphological Image Processing Based CTCs Detector

In [9], an automatic detection method based on morphological image processing is proposed to improve the objectiveness and efficiency of CTCs detection. However, the precision performance of the morphologic detector is very low (about 0.27%), that many normal blood cells are mis-claimed as CTCs candidates.

The contribution of the morphologic detector is to screen out most normal blood cells in a computational effective manner (typical detection time around 65 s/WSI) so that fine operations with high computation complexity could be applied for further examination. As discussed in Sect. 2.1, to locate 0–5 CTCs visually similar to normal blood cells among millions of blood cells in scanned microscopic images is a very difficult and time demanding task. Meanwhile, we shall later see how the detection result of the morphologic detector could be utilized for representative negative sample selection in Sect. 3.

## 2.3 Faster R-CNN CTCs Detector

A faster R-CNN based CTCs detection method is proposed in our previous work [12] (Flowchart shown in Fig. 2), where computation time reduction compared with convolutional neural networks (CNN) based methods is achieved by shared CNN features reuse. Nearly cost-free region proposals is obtained via the region proposal network. A CTCs classification layer (highlighted in blue) is designed to identify whether current proposal is CTCs while the position is obtained by a bounding box regression layer. With the end to end self-learning deep structure, the proposed faster R-CNN CTCs detector report superior precision and recall performance as compared to [6] with a much more difficult application settings. Furthermore, it is worth mentioning that the faster R-CNN CTCs detector is capable of detect small clusters of CTM that has escaped all of the three cytologists' eyes. In this way, it contributes 13 CTC and 41CTM in the CTCs database discussed in Sect. 2.1, which is a strong evidence of the effectiveness of the deep CTCs detector.

However, the faster R-CNN network is only trained on positive samples in [12], where the integrated classifier in the network is blind to negative samples, resulting in high false positive rate. A deep learning detector trained on both positive and negative samples are expected to result in more reliable CTCs detection performance.

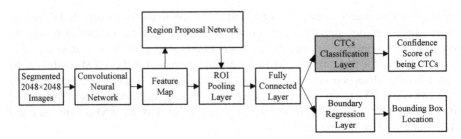

**Fig. 2.** Flowchart of the Faster R-CNN CTCs detector.

## 3    The Proposed Two Stage Cascaded CTCs Detection Algorithm

### 3.1    Motivation and System Overview

As discussed in Sect. 2, CTCs detection is a time-demanding task considering its high throughput characteristic. Real time application is a prerequisite in CTCs detection algorithm design. Meanwhile, identification of CTCs is a highly unbalanced classification problem. The rareness of CTCs leads to the fact that it is difficult to accumulate large amount of positive samples, while inclusion of all negative samples is impossible. How to select representative negative samples plays an important role in the severe unbalanced CTCs detection scenario.

In the proposed two stage cascaded CTCs detection algorithm as shown in Fig. 3, the WSI is firstly divided into 2048 × 2048 sub-images before they are fed into the first stage morphological detector. Then, hard negative sample selection is performed based on the integrated CTCs classifier in the faster R-CNN detector trained only on positive samples based on their confidence scores. Finally, the second stage faster R-CNN detector is trained on both annotated positive and selected representative negative samples to give the final detection result.

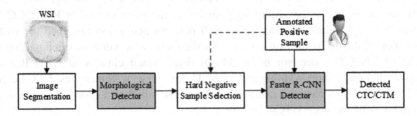

**Fig. 3.** Overview of the proposed cascaded CTCs detection algorithm.

### 3.2    The First Stage Morphological CTCs Detector

The first stage morphological CTCs detector is based on the fact that CTCs is usually large in size as compared with normal blood cells. Firstly, in order to eliminate the uncertainty of dyeing operations and the influence of filter holes on the detection results, color image is first transformed into gray image by

$$I = 0.2989 \times R + 0.587 \times G + 0.114 \times B \tag{1}$$

where R, G and B are red, green and blue channels of color images, and $I$ are gray images. Taking a typical microscopic image as example (shown in Fig. 4(a)), the converted gray image is shown in Fig. 4(b). It can be seen that the gray image retains the structural information of the original image while effectively reducing the image gray difference between different seal sample microscopic images caused by different experimental conditions.

Secondly, Otsu threshold [14] is adopted for image binarization due to its dynamic threshold selection adaptability and computational simplicity. Based on $K$-order histogram $H = \{H_k\}_{k=1,...K}$ of $I$, the binarized image distinguishing foreground CTCs from background normal blood cell can be obtained as

$$BI(m,n) = \begin{cases} 0, & I(m,n) \geq T_h \\ 1, & I(m,n) < T_h \end{cases}, \tag{2}$$

where $T_h$ is the optimal threshold that maximizes the standard deviation of foreground and background areas

$$T_h = \underset{t=0,...,K-1}{\arg\max} \sigma^2 = \underset{t=0,...,K-1}{\arg\max} \left[ \sum_{k=t}^{K-1} H_k(\mu_0 - \mu)^2 + \sum_{k=0}^{t-1} H_k(\mu_1 - \mu)^2 \right], \tag{3}$$

with $\mu_0 = \sum_{k=t}^{K-1} kH_k$, $\mu_1 = \sum_{k=0}^{t-1} kH_k$ and $\mu = \sum_{k=0}^{K} kH_k$ denoting the average grayscale value of the foreground, background and total areas, respectively. Resulting binary image is shown in Fig. 4(c).

Finally, the morphological closing operation is performed on the binary images $BI$ with a disk-shaped structuring element with radius set to be 56 pixels. All boundary pixels together with the detected region are recorded for further processing as shown in Fig. 4(d). Detection results of all $L = \{-2, -1, 0, 1, 2\}$ five layers are fused together by uniting any overlapping regions as the merged ROI. Detected regions are enlarged by half their size to make sure cells are in center of the detected ROIs.

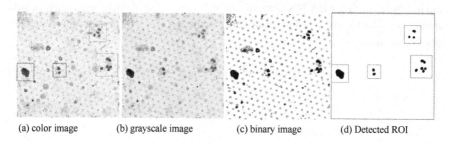

(a) color image          (b) grayscale image          (c) binary image          (d) Detected ROI

**Fig. 4.** Intermediate results of the first stage morphological detector. (Color figure online)

As we see in Fig. 4(d) that the suspicious large cell and cell clusters are successfully detected among various small sized cells. Image patches of corresponding regions $\Omega = \{R_1, R_2, \ldots, R_N\}$ (shown in dashed rectangles in Fig. 4(a)) are saved and serve as input for hard negative sample selection in the next step.

### 3.3   Hard Negative Sample Selection

Further examination of detected results of the first stage of morphological detector reveals that some normal blood cells and dyeing impurity structures (Shown in blue dashed rectangles in Fig. 4(a), while the ground truth positive samples are shown in red dashed rectangles) are mis-claimed as CTCs candidates. This is in line with our intuition that simple size based rule is unlikely to result in reliable detection results.

Meanwhile, we have observed that the detected false positives are not equally contributed to the success distinguishing of CTCs from normal cells. As illustrated in Fig. 5 that carefully selected hard samples (close to CTCs in visual appearance) will help to guide the classification boundary between CTCs and normal cells in the CTCs identification classifier training procedure (Fig. 5).

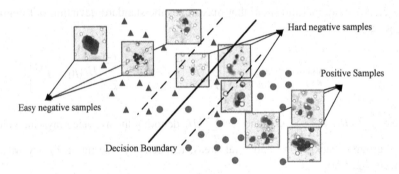

**Fig. 5.** Illustration of easy and hard negative samples.

Considering the faster R-CNN detector [12] provides state-of-the-art CTCs detection, we employ the output confidence score of its integrated classifier as criterion for easy and hard negative sample selection. All detected ROIs $\Omega = \{R_1, R_2, \ldots, R_N\}$ from the first stage morphological detector are classified with the classifier (highlighted in blue in Fig. 2) in faster R-CNN detector trained only on cytologists' annotated positive samples to obtain corresponding confidence score $\{p_i\}_{i=1,\ldots,N}$, then the hard negative samples are selected as

$$\Omega_{HardN} = \bigcup_{R_i} \{\tau_1 \le p_i \le \tau_2\}, \tag{4}$$

where $\tau_1$ and $\tau_2$ are parameters to be determined for hard sample selection that easy negative samples and positive samples can be denoted accordingly as $\Omega_{EasyN} = \bigcup_{R_i} \{p_i < \tau_1\}$ and $\Omega_P = \bigcup_{R_i} \{p_i > \tau_2\}$.

## 3.4    The Second Stage Faster R-CNN CTCs Detector

We followed the paradigm in [12] to construct the second stage faster R-CNN CTCs detector that all $L = \{-2, -1, 0, 1, 2\}$ five layers of segmented $2048 \times 2048$ sub-images are utilized as training images. Experimental results reveals performance improvement from five layers data augmentation that intrinsic relationships between layers can be utilized.

The VGG-16 network is adopted for convolutional feature extraction. Guided by the principle of transfer learning, we use VGG-16 model pre-trained on cytologists' annotated positive samples with Caffe32 framework. The four step alternating optimization training strategy is adopted to optimize the loss as

$$L(\{p_i\}, \{t_i\}) = \frac{1}{N_{cls}} \sum_i L_{cls}(p_i, p_i^*) + \frac{1}{\lambda} \frac{1}{N_{reg}} \sum_i p_i^* L_{reg}(t_i, t_i^*), \tag{5}$$

where $i$ denotes index of CTCs candidate regions, $p_i$ is the CTCs confidence score by the classifier with the ground truth confidence score denoted as $p_i^*$, $t_i$ and $t_i^*$ are predicted and ground truth bounding box parameters, and $\lambda$ is a hyper parameter to balance the classification and regression errors. Softmax loss is adopted for $L_{cls}$ while smooth $l_1$ loss is adopted for bounding box prediction.

## 4    Experimental Results and Discussion

To validate the effectiveness of the proposed CTCs detector, we carried out abundant experiments on the CTCs database established in Sect. 2.1. The confidence score thresholds were set to $\tau_1 = 0.3$ and $\tau_2 = 0.99$ empirically according to our experimental results, leading to 648 hard negative samples selected in total, which is comparable to the amount of total positive samples in the CTCs database.

For faster R-CNN network training, the initial learning rate was set to 0.001 with a rate of decrease of 0.1. Confidence score for final CTCs identification was fixed to 0.8. All experiments are conducted on a PC with Intel(R) Core(TM) i7-6800k CPU @ 3.40 GHz, equipped with a NIVIDIA TITAN XP GPU on Ubuntu 16.04 operating system. The average training time is about 105 min on the established CTCs database while the detection time is within 8 min per WSI, which can meet the real-time clinical requirement.

We randomly select 75% of segmented images as training images and test detection recall $R = \frac{TP}{TP+FN}$ and Precision $P = \frac{TP}{TP+FP}$ on the remaining 25% images, where TP, FN, and FP denote number of true positive, false negative and false positive samples, respectively. There is no overlapping between the training images and testing images. The $F_1$ score defined as the harmonic mean of precision and recall

$$F_1 = \frac{2RP}{R+P} \qquad (6)$$

is also evaluated for the overall performance of a detector. The experiments are carried out five times where results are given in Table 1.

**Table 1.** CTCs detection performance of the proposed cascaded algorithm.

| # | No. CTCs | No. Neg. samples | TP | FN | FP | Precision | Recall | $F_1$ |
|---|----------|------------------|-----|-----|-----|-----------|--------|-------|
| 1 | 119 | 152 | 110 | 9 | 7 | 94.02% | 92.44% | 92.77% |
| 2 | 114 | 147 | 98 | 16 | 11 | 89.91% | 85.96% | 87.89% |
| 3 | 120 | 156 | 105 | 15 | 15 | 87.50% | 87.50% | 87.50% |
| 4 | 114 | 128 | 97 | 17 | 12 | 85.09% | 89.00% | 87.00% |
| 5 | 116 | 176 | 104 | 12 | 12 | 89.66% | 89.66% | 89.66% |
| Averaged | | | | | | **89.24%** | **88.91%** | **88.96%** |

As we can observe in Table 1 that irrespective of the numbers of positive CTCs and negative normal cells varies according to different training and testing images settings, reliable CTCs detection results are obtained for all 5 experiments. Averaged $F_1$ of 88.96% is obtained. From the performance comparison of the proposed algorithm in Table 2, we can see that consistent performance improvement is achieved by cascading the first stage morphological detector with hard negative sample selection. Comparison results with the NBC algorithm reveals that the proposed algorithm demonstrates more robustness and competitive advantages.

**Table 2.** Performance comparison of different CTCs detection algorithms.

| Method | Precision | Recall | $F_1$ |
|--------|-----------|--------|-------|
| Morphological detector | 0.27% | 70.2% | 53.8% |
| Faster R-CNN deep detector | 82.0% | 89.3% | 85.5% |
| NBC (Semi-supervised) [6] | 86.0% | 88.0% | 86.9% |
| Proposed cascaded detector | **89.24%** | **88.91%** | **88.96%** |

Figure 6 illustrates some CTCs detection results for visual examination, where TP, TN, FP and FN labels are marked as green, blue, orange and red, respectively. Ground truth labeled positive and negative samples are marked as black and yellow. We can observe that CTCs can be correctly detected in most cases. Some FN samples with lower confidence scores will also be presented to cytologists for final check.

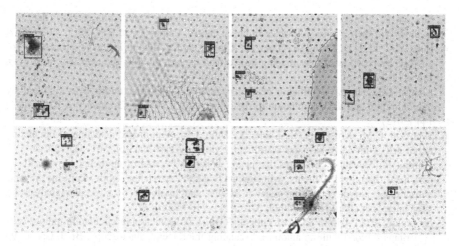

**Fig. 6.** Samples of CTCs detection results based on the cascaded detector. (Color figure online)

## 5 Conclusions

In this study, we propose a two stage cascaded detection algorithm for robust CTCs detection based on scanned microscopic images. The first stage morphological detector screens out most normal cells and provide high quality false positive samples. Hard negative sample selection is performed based on the confidence score of the integrated classifier with a faster R-CNN deep detector trained only on annotated positive samples. The final detection results are obtained by faster R-CNN detector fine-tuned on both positive and negative samples. The effectiveness of the proposed algorithm is validated via abundant experimental results.

Inspired by the performance improvement brought by involving high quality negative samples, we plan to further study the iterative negative sample selection scheme to polish the decision boundary. In addition, we plan to study effective model updating methods so that huge amount of label-free data could also contributes.

**Acknowledgement.** This work was supported by the National Nature Science Foundation of China (No. 61305015, No. 61203269), the Shandong Province Key Research and Development Program, China (Grant No. 2016GGX101022), the National Natural Science Foundation of Shandong Province (Grant No. ZR2017BF031) and the National Key Research And Development Plan (No. 2018YFC0831105).

## References

1. Plaks, V., Koopman, C.D., Werb, Z.: Circulating tumor cells. Cancer Biol. Ther. **341**(6151), 1186–1188 (2013)
2. Paterlini-Brechot, P., Benali, N.L.: Circulating tumor cells (CTC) detection: clinical impact and future directions. Cancer Lett. **253**(2), 180–204 (2007)

3. Mostert, B., Sleijfer, S., Foekens, J.A., et al.: Circulating tumor cells (CTCs): detection methods and their clinical relevance in breast cancer. Cancer Treat. Rev. **35**(5), 463 (2009)

4. Heo, Y.J., Lee, D., Kang, J., Lee, K., Chung, W.K.: Real-time image processing for microscopy-based label-free imaging flow cytometry in a microfluidic chip. Sci. Rep. **7**(1), 11651 (2017)

5. Adams, D.L., et al.: Cytometric characterization of circulating tumor cells captured by microfiltration and their correlation to the cellsearch® CTC test. Cytometry A **87**(2), 137–144 (2015)

6. Svensson, C.M., Krusekopf, S., Lücke, J., Thilo, F.M.: Automated detection of circulating tumor cells with naive bayesian classifiers. Cytometry Part A **85**(6), 501–511 (2014)

7. Pinzani, P., Salvadori, B., Simi, L., et al.: Isolation by size of epithelial tumor cells in peripheral blood of patients with breast cancer: correlation with real-time reverse transcriptase–polymerase chain reaction results and feasibility of molecular analysis by laser microdissection. Hum. Pathol. **37**(6), 711–718 (2006)

8. Ntouroupi, T.G., Ashraf, S.Q., McGregor, S.B., et al.: Detection of circulating tumour cells in peripheral blood with an automated scanning fluorescence microscope. Br. J. Cancer **99**(5), 789–795 (2008)

9. Liu, Y.X., Yang, Y., Chen, Y.H.: Automatic detection of circulating tumor cells based on microscopic images. In: 2017 Asia-Pacific Signal and Information Processing Association Annual Summit and Conference (APSIPA ASC), pp. 769–773. IEEE (2017)

10. Xu, J., Xiang, L., Liu, Q., et al.: Stacked sparse autoencoder (SSAE) for nuclei detection on breast cancer histopathology images. IEEE Trans. Med. Imaging **35**(1), 119–130 (2016)

11. Xie, Y., Xing, F., Shi, X., et al.: Efficient and robust cell detection: a structured regression approach. Med. Image Anal. **44**, 245–254 (2018)

12. Zhang, A., Zou, Z., Liu, Y., et al.: Automated detection of circulating tumor cells using faster region convolution neural network. J. Med. Imaging Health Inform. **9**(1), 167–174 (2019)

13. Hofman, V., Long, E., Ilie, M., Llie, M., Vignaud, J.M., Fléjou, J.F.: Morphological analysis of circulating tumour cells in patients undergoing surgery for non-small cell lung carcinoma using the isolation by size of epithelial tumour cell (ISET) method. Cytopathology Off. J. Br. Soc. Clin. Cytol. **23**(1), 30–38 (2012)

14. Gonzalez, R.C., Woods, R.E.: Digital Image Processing. Addison Wesley, San Francisco (2002)

# Electrocardiogram Diagnosis Based on SMOTE+ENN and Random Forest

Li Sun[1], Ziwei Shang[1], Qing Cao[2], Kang Chen[2(✉)], and Jiyun Li[1(✉)]

[1] School of Computer Science and Technology,
Dong Hua University, Shanghai 201620, China
jyli@dhu.edu.cn
[2] Rui Jin Hospital, Shanghai 200000, China
chenkang1978@163.com

**Abstract.** Many Electrocardiogram (ECG) classification algorithms have been successfully performed on standard dataset. Yet when faced with real world data, due to the issues of imbalanced data distribution, inconsistent data label formats, the performance of these algorithms are not ideal. In this paper, we propose an improved random forest algorithm, in which SMOTE+ENN is used to solve the data imbalance problem, while ECG medical knowledge including MIT-BIH arrhythmia database expert annotations are adopted to align and create the real-world data label. Experiments on ECG data from both standard dataset and real world dataset of a famous hospital showed the efficacy of the algorithm: the out-of-bag data (OOB) accuracy rate on the public data from MIT-BIH arrhythmia database (MITDB) reached 99.22% and 96.62% on real-world data.

**Keywords:** Electrocardiogram (ECG) · SMOTE+ENN · Data imbalance · Advanced random forest

## 1 Introduction

The Electrocardiogram (ECG) signal measures electrochemical signal conduction within heart muscle, it can discover disorders in our heart. The function of ECG is to access the electrical recording and muscular function such as conduction time, peak voltage etc. Physicians use ECG to observe the anomalies, but this process is usually time-consuming, so adapting machine learning algorithm in ECG diagnosis is essential. Luckily, several efficient classification methods have been put forward such as Neural Network [1, 2], K-means [3], k-Nearest Neighbor (KNN) [4], support vector machine (SVM) [5, 6]. However, these researches focus on improving algorithms but lack of combination with medical field. Also ignored the imbalance distribution of data, which cannot provide safe and effective auxiliary treatment for doctors.

In this study, SMOTE+ENN algorithm is used to solve the data imbalance problem. In the preprocessing stage, the data format is unified and the required features are extracted with the guidance of related medical knowledge. Then aligns labels with the help of MIT-BIH arrhythmia database expert annotation and ECG medical knowledge, also create data label with the assist of electrocardiologists from a famous hospital in

© Springer Nature Switzerland AG 2019
D.-S. Huang et al. (Eds.): ICIC 2019, LNCS 11644, pp. 747–757, 2019.
https://doi.org/10.1007/978-3-030-26969-2_71

Shanghai. During classification stage, an improved random forest is adopted to classify ECG into 17 classes.

The improvement of random forest is by adjusting the parameters. The improved random forest obtained the OOB accuracy rate of 96% on real world data set. In this paper, when verifying the classification effect, the evaluation criteria for the imbalance problem are used like confusion matrix, accuracy, recall rate, F1 score and ROC curve for evaluation and comparison.

## 2   Related Work

### 2.1   ECG Medical Background

**ECG Leads.** In ECG diagnosis, different leads produce different amplitude and interval of the signal. 12-lead is the most widely used for ECG recording technology in hospital, which contains 6 Precordial Leads (V1, V2, V3, V4, V5, V6), 3 Limb leads (I, II, III) and 3 augmented Limb leads (aVR, aVL, aVF). Each lead views the heart from a different angle [7].

**ECG Characteristic Points.** In medical ECG recording, a normal ECG signal consist of 5 segments namely P, Q, R, S and T. P is the first electrical positive signal in ECG, formed by atrial complex, the duration of the normal P wave does not exceed 110 ms. Additionally, a QRS is the result of contraction of the ventricles, it is also the largest and significant part of the ECG signal. Q wave is the first downward deflection, followed by R wave, which is the first upward deflection, S wave is the negative deflection. The normal health person's QRS duration is no more than 80 ms, which varies from 60 to 100 beats/min is sinus rhythm [8]. Finally, T wave is the last recognized wave, represents the repolarization or relaxation of the ventricles [9].

### 2.2   Study Background of ECG Classification

In the state of machine learning in ECG classification have many achievements. In [10], Masetic et al. proposed a ECG heartbeat classification system to diagnosing heart failure. They use autoregressive Burg method for feature extraction. C4.5 decision tree, KNN, SVM, ANN and random forest algorithms are applied for ECG classification phase. They conclude random forest classifiers gives the best classification accuracy. In [11], Pandit et al. presents the improvement on feature to increase the accuracy of ECG classification. KNN, ANN, SVM, Decision Tree, Random Forest, AdaBoostM1 etc. are used in classify the ECG.

However, data quality and the medical interpretability still lack in this field. In our study, we pay special attention to data quality, also focus on the application and combination with related medical knowledge to accomplish medical interpretability. The background and professional medical knowledge are applied as guidance in feature extraction and data label creating process.

# 3 Approach

## 3.1 Dataset

The real world data set is derived from the ECG data of more than 10,000 patients in a famous hospital in Shanghai. There are 12425 electrocardiograms in binary class data, including 2236 positive cases (normal beat) and 10189 negative cases (abnormal beat). The ratio of positive and negative samples is 1:5, which shows serious data imbalance problems. Multi-classification data contained above 5900 samples, which exists sample imbalance distribution.

## 3.2 Data Imbalance Processing Based on SMOTE+ENN

Over-sampling and under-sampling are solutions for imbalanced data. Oversampling is to replicate samples of rare categories, balancing the data set by increasing the number of rare sample class. SMOTE (synthetic minority oversampling technique) is a classical over-sampling method, it not simply copies existing data, but generates new data through algorithms based on original data. Table 1 illustrates the SMOTE algorithm procedure.

**Table 1.** SMOTE (synthetic minority oversampling technique) algorithm.

---

**Input:** Minority sample i, and it's feature vector $x_i, i \in (1, 2, ..., T)$:
**Procedure:**
1. Find k neighbors around the sample $x_i$ ,mark these k neighbors as $x_{i(near)}$, near $\in$ $(1, 2, ..., k)$;
2. Pick a sample $x_{i(nn)}$ randomly from k neighbors, the generate a random number $\xi_1$ between 0~1, get a new sample $x_{i1}$:
$$x_{i1} = x_i + \xi_1 \cdot (x_{i(nn)} - x_i)$$
3. Repeat step 2 for N times, generate N new samples $x_{i(new)}$, new $\in (1, 2, ..., N)$;
**Output:** New samples $x_{i1}, x_{i2}, x_{i3}, ..., x_{in}$

---

However, when SMOTE generated minority samples are easy to overlap with the surrounding majority samples, which will lead to useless information or noise. So we suggest to combine with under-sampling method.

Under-sampling refers to balance the data sets by reducing the sample size of the rich classes. Wilson et al. proposed edited nearest neighbor (ENN), the mainly idea is: a sample belongs to majority class, if more than half of its K neighbors do not belong to the same class, the sample will be erased. Another is Tomek Link algorithm, when two samples are nearest neighbors but they belong to different categories, then they form a Tomek Link, which means either sample is noise or both samples are near the boundary. By removing Tomek Link can "clean up" the overlapping samples.

In Fig. 1, data distribution on MITDB and a famous hospital shows SMOTE+ENN (bottom left) produce the least overlap data, indicating this method can remove more overlapping samples, form more uniform samples, so SMOTE+ENN is the superior solution among others for sample imbalance problem.

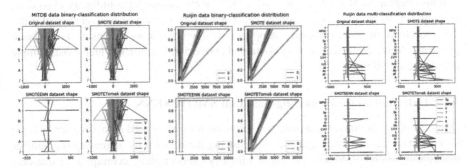

**Fig. 1.** Data imbalance processing comparison based on MITDB (left), a famous hospital's binary-class data (middle) and multi-class data (right). There are four pictures in each sub-figures stand for original dataset shape, dataset shape after SMOTE, SMOTE+ENN and SMOTE+Tomek Link algorithms.

### 3.3 Feature Extraction

We perform discrete wavelet transform DWT based on quadratic spline wavelet to identify the waveform [12]. Extracted features are in Tables 2 and 3. Feature in Table 3 was derived from a famous hospital in Shanghai. They use the average value of ECG. Except RV5 uses V5 lead, SV1 uses V1 lead, other features use the standard twelve lead.

**Table 2.** Feature extraction on MITDB ECG data.

| Name | Example | Description |
|---|---|---|
| PT | PT(i) = abs(locatedP(i) − locatedT(i)) | Interval between P, T wave |
| RR | RR(i) = abs(locatedR(i+1) − locatedR(i)) | Interval between R, R wave |
| QRS | QRS(i) = abs(locatedS(i)−locatedQ(i)) | Wavelength of QRS wave |
| TT | TT(i) = abs(locatedTBegin(i) − locatedTEnd(i)) | Wavelength of T wave |
| ampP | ampP(i) = ecgdata(locatedP(i)) | Amplitude of P wave |
| ampRS | ampRS(i) = ecgdata(locatedR(i)) − ecgdata(locatedS(i)) | Wave width between QRS |
| ampT | ampT(i) = ecgdata(locatedT(i)) | Amplitude of T wave |
| ampTS | ampTS(i) = ecgdata(locatedT(i)) − ecgdata(locatedS(i)) | Wave width between T wave and S wave |
| HR | HR(i) = 60*HZ(360)/R_R(i) | Heart rate |

**Table 3.** Feature extraction on a famous hospital's ECG data.

| No. | Feature name | Description | Measurement |
|---|---|---|---|
| 1 | P | Time interval of P wave | millisecond (ms) |
| 2 | QRS_Width | Time interval of QRS wave | millisecond (ms) |
| 3 | PR | Time interval between P wave and R wave | millisecond (ms) |
| 4 | QRSDZ | QRS wave discrete value | millisecond (ms) |
| 5 | QT | Time interval between Q wave and T wave | millisecond (ms) |
| 6 | QTC | QT interval corrected by heart rate | millisecond (ms) |
| 7 | RV5 | Amplitude of R wave in V5 lead | millivolt (mv) |
| 8 | SV1 | Amplitude of S wave in V1 lead | millivolt (mv) |
| 9 | HEART_RATE | Heart rate | millivolt (mv) |

## 3.4  Real-World Data Label Creating

In order to get a clear label category, we need to deal with the messy data label format. First, extract the key words of heart disease diagnosis from the "diagnosis result" field. Secondly, based on the extracted type, combine some sub-class into mainly types. Additionally, use expert annotation in MITDB and ECG medical knowledge to align these types, based on the number of sample, essentiality and guidance of electrocardiologists from this famous hospitals, we chose 17 types to form the real-world data label library.

The MIT-BIH arrhythmia database (MITDB) [13] from PhysioNet website consists 48 patient record, 45 records have a modified limb lead II (MLII), obtained by placing the electrodes on the chest [14]. Lead is the key element of ECG diagnose, since different types of lead may affect the amplitude and duration of each ECG wave, we choose MLII exists in 45 records to unify the lead. Each ECG signal in MITDB are labeled by expert cardiologist, according to the annotation of arrhythmia in Physionet [15].

For data label creation, we divided the of ECG Abnormality into 5 main groups: arrhythmia, ischemic heart disease, paced rhythm, pre-excitation syndrome and unclassified heartbeat. Several sub-divisions exist in this 5 types, some of the classes labels are named from MITDB: like Left ventricular premature beats (V), right bundle branch block (R) and so on. While some others labels don't exist in MITDB, use the Abbreviation of disease's name such as myocardial infarction (Myocardial infarction: MI) instead. Table 6 gives full description labels of this famous hospital in detail.

## 3.5  Model

**Random Forest (RF).** Random Forest originally proposed by Leo Breiman and Adele Cutler, is an extension of Bagging, which contains many CART trees as weak learners for building Bagging collections, and adds random attribute selection [16]. The principle of random forest is to vote for the best score among all these CART base learners and get the best classification. Since Random forests is a resemble model by tree, it can reduce variance and overfitting, also improve machine learning stability and accuracy

[17]. In addition, according to Juyoung Park, random forest is robust to outliers and noise, saves time from training process [18]. So we chose RF as an ideal classifier.

**Improved Random Forest.** Parameters in random forests can enhance the predictive power, speed and performance. The improvement is to adjust the parameters of Bagging framework and decision tree. There are 7 parameters includes two Bagging framework parameters: "n_estimators" and "oob_score". Five other decision tree parameters are in Table 4. Each parameter is performed in turn to see which value obtain the highest score. Table 4 show the best parameters and the correspondent score of RF model on binary-classification and multi-classification data from a famous hospital in Shanghai. We use OOB as an evaluate criteria later, so set "oob_score" parameter "true".

**Table 4.** Parameter adjustment on a famous hospital's binary-classification data.

| No. | Dataset | Binary-classification | | Multi-classification | |
|-----|---------|-----------------------|--|----------------------|--|
| | Parameters | Best_params_ | Best_score_ | Best_params_ | Best_score_ |
| 1 | n_estimators | 120 | 0.8886 | 80 | 0.7377 |
| 2 | max_depth | 17 | 0.8936 | 26 | 0.8908 |
| 3 | min_samples_leaf | 11 | 0.8939 | 1 | 0.9762 |
| 4 | min_samples_split | 2 | 0.8939 | 2 | 0.9762 |
| 5 | max_features | 3 | 0.8939 | 1 | 0.9798 |
| 6 | oob_score | true | \ | true | \ |
| 7 | random_state | 10 | \ | 10 | \ |

## 4 Experiment

### 4.1 Evaluating Criteria

For sample imbalance problem, the accuracy no longer has reference value. This paper uses confusion matrix, precision (P), recall (R), F1 score (F1), ROC curve (ROC), Area Under Curve (AUC) and OOB (out of bag) as evaluating criteria. The prediction rate given by RF isn't quite accurate, in self-sampling process of RF Bagging, each base learner uses only about 2/3 of the sample in the initial training set, the remaining 1/3 does not appear in the collection of samples collected by Bootstrap. This 1/3 of data is out of bag OOB (out of bag), which can be used as a verification set to perform "out-of-package estimation" of generalization performance [16]. In [19, 20] authors suggest using OOB error estimate as an integral part of generalization error estimate, and Breiman gives empirical examples to show estimate error of test set is the same with OOB, which means OOB error estimate can replace the test set.

### 4.2 Results and Analysis

Tables 5, 6 and 7 depict the number comparison of original samples and processed samples after SMOTE+ENN ensemble algorithm. In Table 6, there are 12 types: N, V,

A, /, L, R, s, T, AF, Af, WPW, Q using label migration, the label names are named from expert annotations in MITDB. As shown in Table 7, the time span of the MITDB data for 45 subjects' ECG heartbeats is 1-minute long.

**Table 5.** Data distribution after SMOTE+ENN on a famous hospital's binary classification ECG data.

| Data | Positive (P, normal beat) | Negative (N, abnormal beat) | Ratio (P: N) |
|------|---------------------------|------------------------------|--------------|
| Original | 2236 | 10189 | 1:5 |
| SMOTE+ENN | 9269 | 6611 | 1.4:1 |

**Table 6.** Data distribution after SMOTE+ENN on a famous hospital's multi-classification ECG data.

| Type | Label | Description | Original | SMOTE+ENN |
|------|-------|-------------|----------|-----------|
| Normal | N | Normal beat | 2203 | 189 |
| Normal | LVH | High voltage of left ventricular | 33 | 2129 |
| Arrhythmia | AF | Atrial fibrillation | 212 | 2108 |
| Arrhythmia | V | Premature ventricular contraction | 123 | 1938 |
| Arrhythmia | A | Atrial premature beats | 99 | 1984 |
| Arrhythmia | Af | Atrial flutter | 62 | 2101 |
| Arrhythmia | Sa | Sinus arrhythmia | 612 | 1963 |
| Arrhythmia | Ab | Atrioventricular block | 149 | 2157 |
| Arrhythmia | L | Left bundle branch block beat | 92 | 2164 |
| Arrhythmia | R | Right bundle branch block beat | 121 | 2070 |
| Arrhythmia | IVB | Intraventricular block | 60 | 2161 |
| WPW syndrome | WPW | Pre-excitation syndrome | 36 | 2177 |
| Heart disease | MI | Myocardial infarction | 350 | 1792 |
| Heart disease | s | ST-T change | 606 | 1472 |
| Heart disease | T | T-wave change | 791 | 1189 |
| Pacing rhythm | / | Paced beat | 228 | 2096 |
| Unclassified | Q | Unclassifiable beat | 131 | 1925 |

**Table 7.** Data distribution after SMOTE+ENN on MITDB multi-classification ECG data.

| No. | Label | Description | Original | SMOTE+ENN |
|-----|-------|-------------|----------|-----------|
| 1 | N | Normal beat | 1812 | 1146 |
| 2 | V | Premature ventricular contraction | 103 | 1640 |
| 3 | A | Atrial premature beats | 36 | 1629 |
| 4 | L | Left bundle branch block beat | 195 | 1751 |
| 5 | R | Right bundle branch block beat | 250 | 1665 |
| 6 | / | Paced beat | 119 | 1797 |

**Table 8.** OOB accuracy comparison based on SMOTE, SMOTE+Tomek and SMOTE+ENN.

| Database | Original | SMOTE | SMOTE+Tomek | SMOTE+ENN |
|---|---|---|---|---|
| MITDB multi-classification | 90.54% | 95.96% | 96.01% | **97.49%** |
| Hospitals' binary classification | 80.47% | 87.09% | 87.20% | **94.67%** |
| Hospitals' multi-classification | 54.15% | 86.61% | 86.97% | **92.76%** |

By comparing OOB accuracy in Table 8, SMOTE+ENN algorithm gain the highest score among them, which means has the best performance and OOB accuracy. Hospitals' binary classification dataset gains the OOB accuracy 94%, multi-classification is 92%, MITDB data is up to 97%. Therefore, we can conclude that conduct SMOTE+ENN algorithm is a robust way to solve data imbalance distribution.

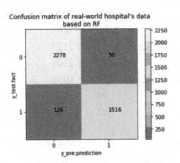
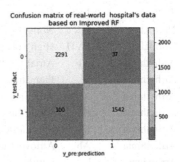

**Fig. 2.** The confusion matrix of a real-world hospitals' binary classification data based on RF and Improved RF. The RF classifier is in the left, and the Improved RF classifier is in the right.

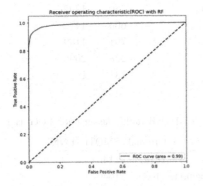
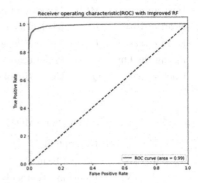

**Fig. 3.** The ROC curve of a real-world hospitals' binary classification data based on RF and Improved RF. The RF classifier is in the left, and the Improved RF classifier is in the right.

**Table 9.** OOB accuracy comparison between RF and improved RF on hospital's data.

| Classifier | Precision | Recall | F1 score | AUC |
|---|---|---|---|---|
| RF | 96.81% | 92.33% | 94.51% | 98.63% |
| **Improved RF** | **97.66%** | **93.91%** | **95.75%** | **99.22%** |

Figures 2 and 3 are the comparison of RF and Improved RF classifier on hospital's binary classification data. Table 9 shows precision, recall, F1 and AUC among them, the improved RF classifier gains the superior performance than RF classifier, precision rate reached by 97.66%, recall rate increased 1.58% from 92.33% to 93.91%.

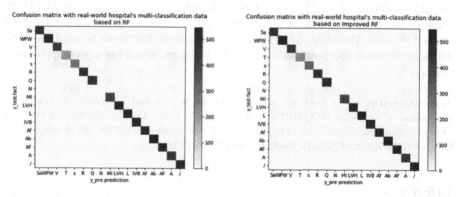

|  | precision | recall | f1-score |  |  |  | precision | recall | f1-score |
|---|---|---|---|---|---|---|---|---|---|
| Sa | 0.983 | 0.990 | 0.987 |  | Sa | 0.996 | 0.994 | 0.995 |
| WPW | 0.902 | 0.983 | 0.941 |  | WPW | 0.945 | 0.987 | 0.966 |
| V | 0.983 | 0.987 | 0.985 |  | V | 0.994 | 0.994 | 0.994 |
| T | 0.986 | 0.984 | 0.985 |  | T | 0.993 | 0.987 | 0.990 |
| s | 0.990 | 0.979 | 0.984 |  | s | 0.994 | 0.992 | 0.993 |
| R | 0.998 | 0.996 | 0.997 |  | R | 0.998 | 0.994 | 0.996 |
| Q | 0.995 | 0.996 | 0.995 |  | Q | 0.991 | 0.998 | 0.995 |
| N | 0.996 | 0.996 | 0.996 |  | N | 0.995 | 1.000 | 0.997 |
| MI | 0.908 | 0.956 | 0.932 |  | MI | 0.950 | 0.963 | 0.957 |
| LVH | 0.333 | 0.070 | 0.115 |  | LVH | 0.889 | 0.186 | 0.308 |
| L | 0.957 | 0.957 | 0.957 |  | L | 0.967 | 0.981 | 0.974 |
| IVB | 0.977 | 0.992 | 0.984 |  | IVB | 0.990 | 0.992 | 0.991 |
| Af | 0.990 | 0.986 | 0.988 |  | Af | 0.984 | 0.990 | 0.987 |
| Ab | 0.893 | 0.766 | 0.825 |  | Ab | 0.953 | 0.806 | 0.873 |
| AF | 0.909 | 0.965 | 0.936 |  | AF | 0.905 | 0.982 | 0.942 |
| A | 0.989 | 0.985 | 0.987 |  | A | 0.991 | 1.000 | 0.995 |
| / | 0.870 | 0.820 | 0.844 |  | / | 0.893 | 0.885 | 0.889 |
| micro avg | 0.962 | 0.962 | 0.962 |  | micro avg | 0.974 | 0.974 | 0.974 |
| macro avg | 0.921 | 0.906 | 0.908 |  | macro avg | 0.966 | 0.925 | 0.932 |
| weighted avg | 0.959 | 0.962 | 0.960 |  | weighted avg | 0.974 | 0.974 | 0.972 |

**Fig. 4.** The evaluating criteria of a real-world hospital's multi-classification data based on RF (left) and Improved RF (right).

**Fig. 5.** The confusion matrix of a real-world hospital's multi-classification data based on RF (left) and Improved RF (right).

Figures 4 and 5 show the performance between random forest (RF) and improved random forest on hospital's multi-classification data. The micro and weighed average performance of precision, recall and F1 Score have increased from 96% to 97% after using improved RF when compared with RF classifier.

**Table 10.** OOB accuracy comparison between RF and Improved RF.

| Classifier | MITDB multi-class | Hospitals' binary class | Hospitals' multi-class |
|---|---|---|---|
| RF | 96.94% | 94.44% | 91.06% |
| **Improved RF** | **99.22%** | **96.45%** | **96.62%** |

Table 10 gives the OOB accuracy comparison between RF classifier and the Improved RF classifier. By contrasting 3 datasets, the Improved RF classifier obtains higher OOB score. Hospitals' multi-classification increased 5.56% from 91.06% to 96.62%, on MITDB dataset possesses the highest OOB accuracy of 99.22%.

## 5   Conclusion

This paper presents an application research on Electrocardiogram diagnosis based on SMOTE+ENN and random forest. From the experiment result, SMOTE+ENN can generate ideal data with minimum noise, this method also obtains the highest OOB accuracy among other data balance methods, prove that SMOTE+ENN is an ideal way to solve the sample imbalance distribution produced by real-world data. In classification phase an improved random forest model was proposed by adjusting parameters. By comparing three data sets, Improved RF classifier obtained the highest OOB score on MITDB dataset of 99.22%, on a famous hospital's multi-classification data OOB accuracy increased 5.56% from 91.06% to 96.62%, on binary classification data also reached 96.45%. Therefore, SMOTE+ENN and random forest can be used for ECG diagnosis on real-world hospitals' data sets in the future, which has a good application prospect.

**Acknowledgment.**   This work was supported by the Science and Technology Development Foundation of Shanghai (18511102703, 16JC1400802, 16JC1400803), the Special Fund of Shanghai Municipal Commission of Economy and Informatization (RX-RJJC-08-16-0483, 2017-RGZN-01004, XX-XXFZ-02-18-2666, XX-XXFZ-01-18-2604).

## References

1. Rao, I., Rao, T.S.: Performance identification of different heart diseases based on neural network classification. Int. J. Appl. Eng. Res. **11**(6), 3859–3864 (2016)
2. Gautam, M.K., Giri, V.K.: An approach of neural network for electrocardiogram classification. APTIKOM J. Comput. Sci. Inf. Technol. **1**(3), 115–123 (2016)

3. Kaur, M., Arora, A.S.: Unsupervised analysis of arrhythmias using K-means clustering. Int. J. Comput. Sci. Inf. Technol. 1(5), 417–419 (2010)
4. Faziludeen, S., Sankaran, P.: ECG beat classification using evidential K-nearest neighbours. In: Twelfth International Conference on Communication Networks, ICCN (2016), Twelfth International Conference on Data Mining and Warehousing, ICDMW (2016), Twelfth International Conference on Image and Signal Processing, ICISP, vol. 89, pp. 499–505 (2016)
5. Tabassum, T., Islam, M.: An approach of cardiac disease prediction by analyzing ECG signal. In: 3rd International Conference on Electrical Engineering and Information Communication Technology, pp. 1–5. IEEE (2016)
6. Karpagachelvi, S., Arthanari, M., Sivakumar, M.: Classification of electrocardiogram signals with support vector machines and extreme learning machine. Neural Comput. Appl. 21(6), 1331–1339 (2012)
7. Hadjem, M., Salem, O., Nait-Abdesselam, F.: An ECG monitoring system for prediction of cardiac anomalies using WBAN. In: 2014 IEEE 16th International Conference on E-Health Networking, Applications and Services (Healthcom), pp. 441–446 (2014)
8. Malgina, O., Milenkovic, J., Plesnik, E., Zajc, M., Tasic, J.F.: ECG signal feature extraction and classification based on R peaks detection in the phase space. In: IEEE in GCC Conference and Exhibition, pp. 381–384 (2011)
9. Spodick, D.H.: Normal sinus heart rate: sinus tachycardia and sinus bradycardia redefined. Am. Heart J. 124(4), 1119–1121 (1992)
10. Masetic, Z., Subasi, A.: Congestive heart failure detection using random forest classifier. Comput. Methods Programs Biomed. 130, 54–64 (2016)
11. Pandit, D., Zhang, L., Aslam, N., Liu, C., Chattopadhyay, S.: Improved abnormality detection from raw ECG signals using feature enhancement. In: 2016 12th International Conference on Natural Computation, Fuzzy Systems and Knowledge Discovery (ICNC-FSKD), pp. 1402–1406 (2016)
12. Sahambi, J.S., Tandon, S.N., Bhatt, R.K.P.: Using wavelet transforms for ECG character-ization. An on-line digital signal processing system. IEEE Eng. Med. Biol. Mag. Q. Mag. Eng. Med. Biol. Soc. 16(1), 77–83 (1997)
13. Moody, G.B., Mark, R.G.: The impact of the MIT-BIH arrhythmia database. IEEE Eng. Med. Biol. Mag. 20(3), 45–50 (2001)
14. MIT-BIH Arrhythmia Database Directory. https://physionet.org/physiobank/database/html/mitdbdir/intro.htm#symbols. Accessed 28 March 2019
15. PhysioBank Annotations. https://physionet.org/physiobank/annotations.shtml. Accessed 28 March 2019
16. Zhihua, Z.: Machine Learning, 1st edn. Tsinghua University Press, Beijing (2016)
17. Pan, G., Xin, Z., Shi, S., et al.: Arrhythmia classification based on wavelet transformation and random forests. Multimed. Tools Appl. 77(17), 21905–21922 (2018)
18. Park, J., Lee, S., Kang, K.: Arrhythmia detection using amplitude difference features based on random forest. In: International Conference of the IEEE Engineering in Medicine & Biology Society. IEEE (2015)
19. Wolpert, D.H., Macready, W.G.: An efficient method to estimate bagging's generalization error. Mach. Learn. 35(1), 41–55 (1999)
20. Breiman, L.: Bagging predictors. Mach. Learn. 24(2), 123–140 (1996)

# Complex Leaves Classification with Features Extractor

Daniel Ayala Niño[(⊠)], José S. Ruíz Castilla,
Ma. Dolores Arévalo Zenteno, and Laura D. Jalili

UAEMEX (Universidad Autónoma del Estado de México),
Jardín Zumpango s/n, Fraccionamiento El Tejocote, 56259 Texcoco, Mexico
danielayalanio@yahoo.com.mx, jsergioruizc@gmail.com,
darevaloz@gmail.com, lauradojali2@gmail.com

**Abstract.** Currently the recognition of plants from leaves has been a field of research very studied, the current algorithms can perfectly classify the leaves of different families. However, for the current algorithms it is difficult to classify leaves belonging to the same family but different species, because they are very similar to each other that even experts have difficulties to make this classification. The next step in the leaf's classification problem, is to classify the leaves that belong to the same family. Being this a problem that has not been targeted yet by the current literature. In this paper, we propose a method to extract the best features that describe the leaf, using a genetic algorithm. And then, testing these features with different classifiers and comparing its accuracy with two Convolutional Neural Networks models. The results demonstrated that the accuracy can be improved depending on the selected model and features.

**Keywords:** Complex leaves identification · Vision system · Features selection

## 1 Introduction

Nowadays, technology has proven to be of vital importance for the development of diverse areas. The most important upturns have been those applied by machine learning, since it can effectively imitate "human tasks". The areas in which it has been developed are diverse, such as agriculture where big data and machine learning are applied in the protection of crops against pests [1], in medicine [2, 3] and botany [4], being of vital importance for the development of the economy, of the environment, it can help us in the research and creation of new pharmaceuticals, and improve agricultural production.

Whereby, it is necessary to be able to classify them. For example, botanists use the leaf's characteristics as a comparative tool for the study of plants. This is because it is collected in several months, whereas flowers and fruits do not last that much [4]. Features like texture, color, and morphological structures, play an important role in distinguishing plant species [21]. Most of these features are hand-crafted, which requires expertise and time to classify great varieties of plants. This brought the first problem, an expert on one species or family may be unfamiliar with another, known as the "taxonomic impediment" [6].

D.-S. Huang et al. (Eds.): ICIC 2019, LNCS 11644, pp. 758–769, 2019.
https://doi.org/10.1007/978-3-030-26969-2_72

This has been a problem that literature has addressed over the years. Various vision systems have been created to classify these leaves. Having a very small margin of error. But the data set on which they worked were classes of leaves very different from each other. This kind of leaves were easy to classify, given its difference, that would be appreciated by eye. Thus, motivating the main objective of this research, the development of a system for the identification of complex leaves. For this research the complex leaves will be composed of leaves from the same family, but different species, something that has not been addressed in the literature. In this research, image processing techniques will be applied in order to extract descriptors from it, which will allow us to identify the class to which it belongs, creating a vector of characteristics, using genetic algorithms we select the best characteristics that describe the leaves. We work over 8 families of leaves, with their respective subclasses (species).

The first section described above is the introduction, the second section describes the state of the art. In the third section the method used for the extraction of characteristics is shown. In the fourth section the experimental results will be described. And finally, in the fifth section the conclusions are shown.

## 2 State of the Art

In this section we will discuss the different approaches related to the classification of plants through the analysis of the leaf. Throughout the years, several studies have been developed in the literature related to the classification of plants by leaves. And these have proved to be the best option for their classification.

Different approaches have been developed, such as leaf morphology. There are several characteristics that can be used on the leaf to describe its morphology, such as elongation size, ellipse [8, 9]. Zhao et al. [10] proposed the independent-IDSC (I-DSC) features, a new form descriptor based on counting, this can recognize simple and compound leaves. Aakif and Khan [11] extracted morphological characteristics, Fourier descriptors [12] and a new feature called shape-defining. All these characteristics were introduced in an RNA obtaining an accuracy of 96%. Another investigation where they make use of morphological characteristics of the leaf is the work done by Du et al. [13], who proposed a classifier of 20 kinds of plants by the extraction of morphological characteristics of the leaf, this would include geometric characteristics of the leaf and its invariant moments, resulting in a total of 15 characteristics with which the Move Median Centers (MMC) classifier would train.

The leaf could also be classified by means of its texture. This procedure focuses on extracting information about the texture of the leaf, using the pixels of the image. One way to do this is by applying fractal theory, such as Backes et al. [14], thus achieving a precision of 90% on 10 types of leaves. Rashad et al. [15] He sorted the leaves by texture, combining classifiers such as the Learning Vector Quantization (LVQ) together with the Radial Basis Function (RBF). Naresh and Nagendraswamy [16] used Modified Local Binary Patterns to extract characteristics of the texture of the leaf, to later be classified by close neighbors. Olsen et al. [17] used Histograms of Oriented Gradients (HOGs) characteristics to be able to represent the textures in the image of a leaf.

The leaves can also be classified by its color, this can be done from the comparison of its histograms [18]. This technique has a problem, since the leaf's chromaticity is not static, it is variable. Factors such as time, weather and others affect the state of the leaf. An alternative is to use textural characteristics of the leaf since they provide information about the intensities of the colors in the image, their geometric shape, size and the shape of the previously segmented region and chromatic that refer to the intensity of the color from a segmented region [19].

As you have seen there are several techniques to classify the leaves. There are even investigations that use all the techniques mentioned above, creating a great vector of characteristics that describes the leaf. This being a disadvantage for the classifier that is used, since there may be several variables that create noise causing the classifier to be confused. A more recent work is the proposed method by Jair et al. [7], he proposed a method to solve the performance of machine learning algorithms in the classification of complex leaves (for this research, complex leaves were from different family and species, but with great similarities), using different feature selection techniques to improve the results, his work envisaged a problem for the classification of complex leaves, until now. A different approach is settled in this research, since in the previous work the data set of complex leaves was composed of leaves that are similar but all of them were from different families and species. Given that, is important to also know which kind of subclass a specific leaf belongs to. Only an expertise on this family could classify it. Being this the main objective of this paper. We propose a novel method, using feature selection with genetic algorithms, and this is tested with diverse classifiers.

## 3   Methodology

### 3.1   Model

The reliability and efficiency of an object classification system depend directly on each of the steps carried out, from the preprocessing techniques used, the segmentation methods, the extracted characteristics and the classifiers used for the recognition. In this section, the methodology carried out is discussed in detail, Fig. 1 shows the general process carried out.

### 3.2   Segmentation Techniques

The segmentation algorithm used was the Otsu algorithm. This is because the images used during the experiment were images taken in controlled environments [7]. In order to obtain a good segmentation even when there are changes in global brightness conditions, the region of the leave in each image was segmented using the following steps: (1) Computation of high-contrast gray scale from optimal linear combination of RGB color components; (2) Estimate optimal border using cumulative moments of zero order and first order (Otsu method). (3) Morphological operations to fill possible gaps in the segmented image. By segmenting the image, the proposed system can use only the region of the leaf, determine its edges and calculate properties by extracting features.

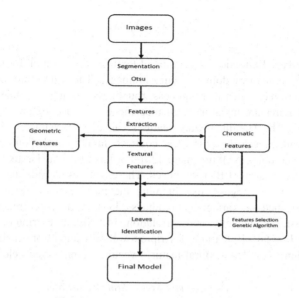

**Fig. 1.** Proposed methodology diagram

## 3.3 Feature Extraction

Based on the results presented in [7], 3 types of characteristics were extracted, which are: (1) Chromatic characteristics; (2) Textural Characteristics; (3) Geometric characteristics.

**Geometric Characteristics.** Geometric features allow extracting information about the size and shape of the segmented region. For this case, it would be the roundness of the leaf, the area, the length of the edge of the leaf, the elongation, defined by the length and width of the leaf, the $x$ and $y$ coordinates of the center of gravity of the sheet, rectangularity, and the rest. A total of 28 elemental features were considered.

For a system to be able to classify a leaf regardless of its position, and if it is different from the others, this system must be invariant to the rotation, translation or inversion of the image. Hu's moments fulfill these characteristics. Also, Fourier descriptors were used, since it consider the set of points that make up the contour of an object as a sequence of complex numbers, with which a one-dimensional periodic function $f$ that models the contour of the object is constructed; the coefficients of the Fourier transform of the function $f$ characterize the contour of an object in the frequency domain. Only its first 8 descriptors were used. The Flusser moments are a group of statistical characteristics derived from the invariant moments of Hu, and like these, they are invariant to the translation, rotation and scale. Flusser moments are also invariant under similar general transformations, so they can be used for the recognition of deformed objects. The four moments were used for this research. As the Flusser moments the R moments are defined based on Hu's moments. All ten R moments were considered. Having at the end a geometric vector of 57 features, with the distribution showed below in Eq. 1.

$$X_g = \left[X_{gb}, X_{Hu}, X_f, X_R, X_{DF}\right] \tag{1}$$

**Textural Features.** Extraction algorithms of textural features look for basic repetitive patterns with periodic or random structures in images. These structures are obtained by properties in the image such as roughness, roughness, granulation, fineness, softness, etc. Textural features are invariant to displacements, because texture repeats a pattern along a surface. The Haralik texture features can extract characteristics like homogeneity, gray-tone linear dependencies, contrast, number and nature of boundaries present, and the complexity of the image [20]. For this research Haralik texture features and Local Binary Patterns (LBP) were used. These features consider the distribution of intensity values in the region, by obtaining the mean and range of the following variables: mean, median, variance, smoothness, bias, Kurtosis, correlation, energy or entropy, contrast, homogeneity, and correlation. 14 textural descriptors of each image were obtained. In total 219 textural descriptors were obtained from each image. 73 for each color channel [7]. The textural features vector is represented below in Eq. 2:

$$X_t = \left[X_{Rlbp}, X_{RH}, X_{Glbp}, X_{GH}, X_{Blbp}, X_{BH}\right] \tag{2}$$

**Chromatic Features.** This kind of features provide information about the color intensity of a segmented region. For this research, these characteristics were calculated for each intensity channel (Red, Green, Blue). The standard intensity features describe the mean, standard deviation of intensity, first and second derivative in segmented region. The Hu's moments also are used to extract intensity information and Gabor features based on 2D Gabor functions. A total of 122 characteristics were obtained for each channel (RGB), that is 366 chromatic features. The chromatic feature vector is represented below in Eq. 3.

$$X_c = \left[X_{Re}, X_{RHu}, X_{Ge}, X_{GHu}, X_{Be}, X_{BHu}\right] \tag{3}$$

## 4   Experimental Results

The proposed recognition system was implemented using MATLAB. The experiments were carried out to test the efficiency and precision of the proposed system. The metric used to evaluate the performance of the classifier was accuracy and this is obtained from the classifier hits divided by the total of data set. Other methods were used to measure the performance of the model, like ROC, the F-measure, true positive rate (TP rate) and false positive rate (FP rate). Also, we compare the performance of the proposed method with different models of Convolutional Neural Networks (CNN), and again, their performance was evaluated by accuracy. The CNN models were implemented using Python 3.6.8, using TensorFlow and Keras, with the Spyder 3.3.3 environment.

## 4.1  Data Set

For the experiments, we worked with a data set provided by the Fruit Department of the Postgraduate School. The data set was obtained under specific conditions in a completely controlled environment. The leaves taken belong to 8 different families, these being very similar to each other, as can be seen in Fig. 2, each family has approximately 2 to 6 different species. All species are not shown in Fig. 2, this is just to illustrate the similarities between their respective species.

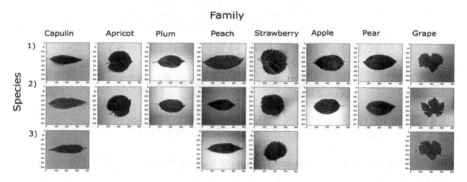

**Fig. 2.** 8 different families that compose the dataset. If one family shows only two species, is because the dataset only has two species of that family, the other families might have more species that are not showed in Fig. 2.

The distribution of the data set is the next: Capulin 4 species (296 samples), Apricot 2 species (155 samples), Plum 2 species (96 samples), Peach 6 species (498 samples), Strawberry 3 species (194 samples), Apple 2 species (105 samples), Pear 2 species (112 samples) and Grape 3 species (166 samples). Adding up 1622 samples.

A different dataset was created, this to compare the performance of the proposed method in contrast with different models of CNN. The created subset is composed only of Peach leaves, this has 6 different species.

## 4.2  Pre-processing

The images were resized to a size of 300 by 400 and some imperfections were removed, like black spots in the images. A Gaussian filter of 3 by 3 was used. From previous research results it is known that the results in the segmentation do not vary significantly on images where only the leaf of the plant with white background than those with different background.

It is known that in some cases, for Convolutional Neural Networks it is not necessary to apply advanced filters to the image, since the CNN can learn which features to extract to make a good classification. The images of this subset were resized to a size of 416 by 416, with the 3 channels (R, G and B). Data Augmentation was applied to all the training images. Image transformations like flips in the $y$ and $x$ axes and both (flip in the $x$ plus a flip in the $y$ axes), as seen in the Fig. 4. Other transformations or filters were not applied, this because we want to maintain the original form of the leaf. With the data augmentation we obtained a dataset of 1592 samples (Fig. 3).

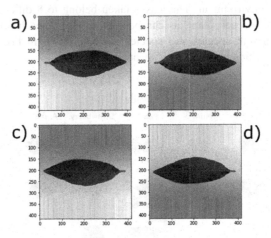

**Fig. 3.** (a) Original leaf, (b) Horizontal flip, (c) Vertical flip, (d) Vertical + Horizontal flip

### 4.3    Convolutional Models

The most important part of a CNN is the convolution layer, unlike an Artificial Neural Network (ANN), each pixel is not connected to a neuron, but a pixel of its receptive field. This type of architecture allows to concentrate on low-level characteristics in the first convolution layer and to assemble high-level characteristics in the following layers. This allows a CNN the ability to classify and extract the characteristics from images without needing to tell it how to do it. After each convolution layer comes a pooling layer, this kernel is creating subsamples from the inputted image, in order to reduce the computational cost and space. At the end of these convolution and pooling layers, the full connected layer follow, this last layer works like a hidden layer of an ANN.

A convolutional network is constituted by several convolutional layers, combined with layers of batch normalization, ReLU activations, and pooling layers so that afterward a vector of the penultimate layer is created to pass to the FC layer. An example is shown in Fig. 4, a model of two convolutional layers and one full connected layer. You can join as many layers as you want, making the performance of the convolutional network improve, theoretically.

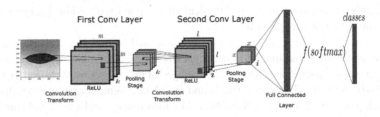

**Fig. 4.**  Example of a two convolutional layers model and one FC layer

Two different models were created to compare the performance of the proposed model. The first model as shown in Table 1, has 3 convolutional layers (ConvL) and one full connected layer (FCL). And the second model (Table 2), has 5 convolutional layers and one full connected layer. More details about the models are detailed in Tables 1 and 2.

**Table 1.** Convolutional model 1 summary

| ConvL 1 | MaxPool 1 | ConvL 2 | MaxPool 2 | ConvL 3 | MaxPool 3 | FCL 1 |
|---------|-----------|---------|-----------|---------|-----------|-------|
| 10X10X6 | 2X2 | 4X4X10 | 4X4 | 4X4X20 | 3X3 | 500 |

**Table 2.** Convolutional model 2 summary

| ConvL 1 | MaxPool 1 | ConvL 2 | MaxPool 2 | ConvL 3 | MaxPool 3 | ConvL 4 | MaxPool 4 | ConvL 5 | MaxPool 5 | FCL 1 |
|---------|-----------|---------|-----------|---------|-----------|---------|-----------|---------|-----------|-------|
| 10X10X6 | 3X3 | 5X5X10 | 3X3 | 5X5X20 | 3X3 | 5X5X25 | 2X2 | 5X5X30 | 2X2 | 1920 |

## 4.4   Validation

In order to validate the results, k-fold cross validation was used, with k = 10. Cross validation is a technique used to validate prediction models. The technique guarantees that the results obtained from a statistical analysis can be used to generalize in an independent data set.

That is, the technique allows us to estimate how accurate a classification model will play in practice. In classification problems the classifier is usually given a known dataset (training dataset) on which it is trained and once the model is obtained it is tested with an unknown dataset (test dataset) or not seen by the classifier. The objective of cross-validation is to test the model's ability to test new data that was not used during training in order to avoid over training.

It consists of repeating and computing the arithmetic average obtained from the evaluation measures on different partitions. It is a technique widely used in artificial intelligence projects to validate generated models. In the cross-validation of k iterations or k-fold cross-validation the sample data are divided into k subsets. One of the subsets is used as test data and the rest (k − 1) as training data. The cross-validation process is repeated during k iterations, with each of the possible subsets of test data. Finally, the arithmetic mean of the results of each iteration is performed to obtain a single result.

In the case of the CNN experiment, no cross-validation was applied. This because of the complexity of repeating the training process of the CNN k times. The dataset created was separated in the test set and the training set before the data augmentation, having a total of 398 samples for the training set and 100 samples for the test set, approximately 20% of the dataset was destined to the test set. Then data augmentation was applied as explained above.

## 4.5   Results

The experimental results from the proposed model selecting features with a genetic algorithm are summarized in Table 3. From the table we can summarize that the algorithm with the best accuracy rate is the Support Vector Machine algorithm (SVM). It has the highest rates for the next leaves: Capulin 88.4354%, Apricot 71.7949%, Plum 84%, Strawberry 83.6735% and Peach 83.9378%. In the case of the Pear leaves the SVM algorithm has the same accuracy as the Logistic Regression algorithm, with a 94% accuracy approximately. A 94.33% accuracy was obtained for the Apple leaves with the Decision Tree algorithm. And for the Grape leaves Logistic regression was the best algorithm with an accuracy of 95.29%. This was the highest accuracy rate obtained in all experiments. The lowest accuracy rate obtained was 58%, for the Logistic Regression with Apricot leaves, something to highlight of this family is that all the classifiers had a bad performance, i.e. the highest accuracy was SVM with a 71.79% accuracy.

**Table 3.**   Classification results with proposed approach.

|  |  | Accuracy | ROC | F-measure | TP rate | FP rate |
|---|---|---|---|---|---|---|
| Capulin leaves | Naive Bayes | 75.510% | 0.892 | 0.754 | 0.755 | 0.083 |
|  | SVM | 88.435% | 0.961 | 0.884 | 0.884 | 0.039 |
|  | Logistic R. | 78.231% | 0.931 | 0.781 | 0.782 | 0.073 |
|  | Decision Tree | 74.829% | 0.873 | 0.748 | 0.748 | 0.084 |
| Apricot leaves | Naive Bayes | 69.230% | 0.724 | 0.689 | 0.692 | 0.317 |
|  | SVM | 71.794% | 0.711 | 0.711 | 0.718 | 0.297 |
|  | Logistic R. | 58.974% | 0.574 | 0.588 | 0.590 | 0.418 |
|  | Decision Tree | 65.384% | 0.674 | 0.654 | 0.654 | 0.347 |
| Grape leaves | Naive Bayes | 72.941% | 0.788 | 0.718 | 0.729 | 0.147 |
|  | SVM | 91.764% | 0.948 | 0.917 | 0.918 | 0.041 |
|  | Logistic R. | 95.294% | 0.979 | 0.953 | 0.953 | 0.025 |
|  | Decision Tree | 88.235% | 0.915 | 0.882 | 0.882 | 0.058 |
| Plum leaves | Naive Bayes | 70% | 0.721 | 0.685 | 0.700 | 0.333 |
|  | SVM | 84% | 0.839 | 0.840 | 0.840 | 0.162 |
|  | Logistic R. | 76% | 0.828 | 0.760 | 0.760 | 0.230 |
|  | Decision Tree | 68% | 0.710 | 0.680 | 0.680 | 0.311 |
| Strawberry leaves | Naive Bayes | 63.265% | 0.744 | 0.627 | 0.633 | 0.186 |
|  | SVM | 83.673% | 0.897 | 0.836 | 0.837 | 0.082 |
|  | Logistic R. | 70.408% | 0.871 | 0.702 | 0.704 | 0.148 |
|  | Decision Tree | 71.428% | 0.794 | 0.705 | 0.714 | 0.143 |
| Apple leaves | Naive Bayes | 90.566% | 0.929 | 0.906 | 0.906 | 0.097 |
|  | SVM | 90.566% | 0.939 | 0.906 | 0.906 | 0.097 |
|  | Logistic R. | 86.792% | 0.866 | 0.868 | 0.868 | 0.135 |
|  | Decision Tree | 94.339% | 0.974 | 0.943 | 0.943 | 0.059 |

(*continued*)

**Table 3.** (*continued*)

|  |  | Accuracy | ROC | F-measure | TP rate | FP rate |
|---|---|---|---|---|---|---|
| Pear leaves | Naive Bayes | 80.769% | 0.837 | 0.808 | 0.808 | 0.197 |
|  | SVM | 94.230% | 0.939 | 0.942 | 0.942 | 0.064 |
|  | Logistic R. | 94.230% | 0.973 | 0.942 | 0.942 | 0.064 |
|  | Decision Tree | 92.307% | 0.903 | 0.923 | 0.923 | 0.079 |
| Peach leaves | Naive Bayes | 76.165% | 0.922 | 0.761 | 0.762 | 0.050 |
|  | SVM | 83.937% | 0.961 | 0.839 | 0.839 | 0.032 |
|  | Logistic R. | 73.057% | 0.942 | 0.728 | 0.731 | 0.055 |
|  | Decision Tree | 72.020% | 0.852 | 0.718 | 0.720 | 0.058 |

For the Convolutional models, the results are summarized in Table 4. Both models did not generalize the data well. The fist model has the highest accuracy with 62%, something unexpected because it is not as deeper as the second model with 5 convolutional layers. One of the reasons for what these two models gave poor results is because its simple architecture. That is, it is necessary to create more sophisticated models to increase the accuracy of the convolutional model and it has been demonstrated that increasing the number of samples in the training set could increase the accuracy of the model.

**Table 4.** Classification results for convolutional models with peach leaves.

|  | Accuracy |
|---|---|
| Convolutional model 1 | 62% |
| Convolutional model 2 | 55% |

## 5  Conclusions

In this paper different models were tested to classify complex leaves from 8 different families. A genetic algorithm was used to select the best features that describes the leaves, creating a vector of geometric, textural and chromatic characteristics of the leave. The results demonstrated that the best classifier was SVM, given that it had the higher accuracy score in 5 families out of 8, and a shared accuracy with Pear leaves, using Logistic Regression. The Apricot leaves had the lowest accuracy scores. By now nothing can be inferred about the reasons of this poor results, only that the leave's complexity is very high. The proposed model was compared with two convolutional models, these models did not generalize well the data, as results: the first model an accuracy of 62% and the second model an accuracy of 55%. Convolutional Neural Networks cannot be discarded for solving the complex leaves classification problem, CNN has demonstrated being a key element for image classification. Given these results, a more sophisticated convolutional model needs to be created. Being this a problem that will be targeted in future work.

# References

1. Ip Ryan, H.L., Ang, L.-M., Seng, K.P., Broster, J.C., Pratley, J.E.: Big data and machine learning for crop protection. Comput. Electron. Agric. **151**, 376–383 (2018)
2. Zhang, Z., Sejdić, E.: Radiological images and machine learning: trends, perspectives, and prospects. Comput. Biol. Med. **108**, 354–370 (2019)
3. Amador, J.D.J., Espejel Cabrera, J., Cervantes, J., Jalili, L.D., Ruiz Castilla, J.S.: Automatic calculation of body mass index using digital image processing. In: Figueroa-García, J.C., Villegas, J.G., Orozco-Arroyave, J.R., Maya Duque, P.A. (eds.) WEA 2018. CCIS, vol. 916, pp. 309–319. Springer, Cham (2018). https://doi.org/10.1007/978-3-030-00353-1_28
4. Wang, Z., Sun, X., Zhang, Y., Ying, Z., Ma, Y.: Leaf recognition based on PCNN. Neural Comput. Appl. **27**(4), 899–908 (2016)
5. Lopez Chau, A., Rojas Hernandez, R., Trujillo Mora, V., Cervantes Canales, J., Rodriguez Mazahua, L., Garcia Lamont, F.: Detection of compound leaves for plant identification. IEEE Lat. Am. Trans. **15**(11), 2185–2190 (2017)
6. Carvalho, M.R., et al.: Taxonomic impediment or impediment to taxonomy? A commentary on systematics and the cybertaxonomic-automation paradigm. Evol. Biol. **34**, 140–143 (2007)
7. Cervantes, J., Garcia Lamont, F., Rodriguez Mazahua, L., Zarco Hidalgo, A., Ruiz Castilla, J.S.: Complex identification of plants from leaves. In: Huang, D.-S., Gromiha, M.M., Han, K., Hussain, A. (eds.) ICIC 2018. LNCS (LNAI), vol. 10956, pp. 376–387. Springer, Cham (2018). https://doi.org/10.1007/978-3-319-95957-3_41
8. Jalili, L.D., Morales, A., Cervantes, J., Ruiz-Castilla, J.S.: Improving the performance of leaves identification by features selection with genetic algorithms. In: Figueroa-García, J.C., López-Santana, E.R., Ferro-Escobar, R. (eds.) WEA 2016. CCIS, vol. 657, pp. 103–114. Springer, Cham (2016). https://doi.org/10.1007/978-3-319-50880-1_10
9. Cervantes, J., García-Lamont, F., Rodríguez-Mazahua, L., Rendon, A.Y., Chau, A.L.: Recognition of Mexican sign language from frames in video sequences. In: Huang, D.-S., Jo, K.-H. (eds.) ICIC 2016. LNCS, vol. 9772, pp. 353–362. Springer, Cham (2016). https://doi.org/10.1007/978-3-319-42294-7_31
10. Zhao, C., Chan, S.S.F., Cham, W.-K., Chu, L.M.: Plant identification using leaf shapes—a pattern counting approach. Pattern Recogn. **48**(10), 3203–3215 (2015)
11. AaKif, A., Khan, M.F.: Automatic classification of plants based on their leaves. Biosyst. Eng. **139**, 66–75 (2015)
12. Neto, J.C., Meyer, G.E., Jones, D.D., Samal, A.K.: Plant species identification using elliptic fourier leaf shape analysis. Comput. Electron. Agric. **50**(2), 121–134 (2006)
13. Du, J.-X., Wang, X.-F., Zhang, G.-J.: Leaf shape based plant species recognition. Appl. Math. Comput. **185**(2), 883–893 (2007)
14. Backes, A.R., Bruno, O.M.: Plant leaf identification using multi-scale fractal dimension. In: Foggia, P., Sansone, C., Vento, M. (eds.) ICIAP 2009. LNCS, vol. 5716, pp. 143–150. Springer, Heidelberg (2009). https://doi.org/10.1007/978-3-642-04146-4_17
15. Rashad, M.Z., El-Desouky, B.S., Khawasik, M.S.: Plants images classification based on textural. Int. J. Comput. Sci. Inf. Technol. **3**(4), 93–100 (2011)
16. Naresh, Y.G., Nagendraswamy, H.S.: Classification of medicinal plants: an approach using modified LBP with symbolic representation. Neurocomputing **173**(3), 1789–1797 (2016)
17. Olsen, A., Han, S., Calvert, B., Ridd, P., Kenny, O.: In situ leaf classification using histograms of oriented gradients. In: International Conference on Digital Image Computing: Techniques and Applications (DICTA), pp. 1–8 (2015)

18. Tico, M., Haverinen, T., Kuosmanen, P.: A method of color histogram creation for image retrieval. In: Proceedings of the Nordic Signal Processing Symposium (NORSIG-2000), Kolmarden, Sweden, p. 157–160 (2000)
19. Cervantes, J., et al.: Análisis Comparativo de las técnicas utilizadas en un Sistema de Reconocimiento de Hojas de Planta. Revista Iberoamericana de Automática e Informática industrial **14**(1), 104–114 (2017). [S.l.], ISSN 1697-7920
20. Ma, L.-H., Zhao, Z.-Q., Wang, J.: ApLeafis: an android-based plant leaf identification system. In: Huang, D.-S., Bevilacqua, V., Figueroa, J.C., Premaratne, P. (eds.) ICIC 2013. LNCS, vol. 7995, pp. 106–111. Springer, Heidelberg (2013). https://doi.org/10.1007/978-3-642-39479-9_13
21. Garcia, F., Cervantes, J., Lopez, A., Alvarado, M.: Fruit classification by extracting color chromaticity, shape and texture features: towards an application for supermarkets. IEEE Lat. Am. Trans. **14**(7), 3434–3443 (2016)

# IDP - OPTICS: Improvement of Differential Privacy Algorithm in Data Histogram Publishing Based on Density Clustering

Lina Ge[1,2,3(✉)], Yugu Hu[1], Hong Wang[1], Zhonghua He[1],
Huazhi Meng[1], Xiong Tang[1], and Liyan Wu[1,2,3]

[1] College of Information Science and Engineering,
Guangxi University for Nationalities, Nanning, China
66436539@qq.com
[2] Key Laboratory of Network Communication Engineering,
Guangxi University for Nationalities, Nanning, China
[3] Key Laboratory of Guangxi High Schools Complex System
and Computational Intelligence, Nanning, China

**Abstract.** In recent years, incidents of privacy leaks have frequently occurred. How to protect the group and personal privacy has become a focus issue in the field of information security. Differential privacy technology can protect the privacy of published data by adding noise. In this paper, we propose the IDP-OPTICS algorithm to solve the problem of excessive noise accumulation and low data availability in the histogram data release under the static data set. First, we preprocess the data with a density-based clustering algorithm. Then, we apply differential privacy to the data and finally add noise to the processed data by adding heteroscedasticity. The experimental data shows that IDP-OPTICS achieves a reduction in the problem of excessive noise accumulation and a balance of noise distribution as a whole.

**Keywords:** Differential privacy · Histogram publishing · Laplace transform · Density clustering

## 1 Introduction

Differential privacy [1] is a privacy protection technology that Dwork first proposed in 2006. Dwork [2] publishes data with added Laplace noise in a histogram of equal width. McSherry [3] applies the exponential mechanism to differential privacy to make non-numeric data applicable. For differential privacy of the tree structure histogram, through the continuous improvement of papers [4–9], the method of dividing the original data set is proposed and the accuracy of count query is improved. Wang [10] also contributed to the related work. Nowadays, differential privacy is widely used in data mining, network data analysis and other fields.

© Springer Nature Switzerland AG 2019
D.-S. Huang et al. (Eds.): ICIC 2019, LNCS 11644, pp. 770–781, 2019.
https://doi.org/10.1007/978-3-030-26969-2_73

Blum et al. [11] applied the k-means algorithm based on the SULQ framework to differential privacy protection for the first time. Furthermore, Dwork et al. [12] improved the k-means algorithm for differential privacy protection from the perspective of budget allocation for Blum's algorithm. However, the availability of clustering results is still poor. Wu et al. [13] proposed a DP-DBSCAN algorithm. However, the algorithm itself has problems with the input parameters. André et al. [14] proposed a DiPCoDing method which reduces the sensitivity of differential privacy and improves data utility. Based on the DP-DBSCAN algorithm, Ni et al. [15] proposed a DP-MCDBSCAN model which effectively enhance data clustering.

In this paper, based on the existing clustering algorithm, we optimized the algorithm and the way of adding noise, realized the balance of privacy budget parameters, further reduced the accumulation of noise and improved the availability of data.

# 2  Preliminaries

## 2.1  Background of Differential Privacy

The differential privacy is a privacy protection technology based on data loss. By adding random noise to interfere with sensitive data to ensure the properties in the data are unchanged. The algorithm defines an extremely strict attack model: even if the attacker already knows all the records except one record, it is still impossible to infer whether the record exists in the database. A simple example is as follows:

The raw data of a credit-history table is given in Table 1. The table contains three attributes: Name, Age, and Credit History. The statistics are shown in Table 2.

**Table 1.** Credit-history raw data set.

| Name | Age | Credit history |
|------|-----|----------------|
| Dave | 25 | Good |
| Lucy | 20 | Bad |
| Cart | 23 | Good |
| Bob | 28 | Bad |
| Jim | 30 | Good |
| Gog | 34 | Excellent |
| Hey | 27 | Excellent |

**Table 2.** Credit-history statistics.

| Credit history | Counts |
|----------------|--------|
| Excellent | 2 |
| Good | 3 |
| Bad | 2 |

**Table 3.** Statistics after adding noise.

| Credit history | Counts |
|----------------|--------|
| Excellent | 2.65 |
| Good | 2.49 |
| Bad | 2.23 |

If the attacker already knows that there are 7 people in this data set and the credit history of any 6 people except Hey. Then, the attacker can guess the credit history of Hey based on the counts known in Table 2. If add Laplace noise to the counts, the queried by the attacker will be shown in Table 3.

Through the count set {2.65, 2.44, 2.23}, the attacker can neither calculate that Hey belongs to the excellent, nor judge which credit history records Hey belongs to.

## 2.2   Theories of Differential Privacy

**Definition 1 (Differential Privacy [1]).** A randomized algorithm $\mathcal{M}$ with domain $\mathbb{N}^{|x|}$ is $(\varepsilon, 0)$-differentially private if for all $S \subseteq Range(M)$ and for all data sets $D_1, D_2 \in N^{|x|}$ such that $\|D_1 - D_2\|_1 \leq 1$:

$$\Pr[\mathcal{M}(D_1) \in S] \leq \exp(\varepsilon) \Pr[\mathcal{M}(D_2) \in S] \tag{1}$$

we say that $\mathcal{M}$ is $(\varepsilon, 0)$-differentially private, where $D_1$ and $D_2$ are two data sets with at most one different record, $\varepsilon$ is a privacy parameter that indicates the degree of privacy protection. The smaller the $\varepsilon$, the higher the degree of privacy protection.

**Definition 2 (Global Sensitivity [1]).** Given a function set $F$, an input data set, where any difference is at most one data set $D_1$ and $D_2$ with different records, the result set is $R$, and any function $f$ in the function set exists $f(D) \in R$, then the global sensitivity of $F$ is defined as:

$$G(\mathcal{M}) = \mathop{max}_{D_1, D_2} \left( \sum_{f \in F} |f(D_1) - f(D_2)| \right) \tag{2}$$

**Definition 3 (Local Sensitivity [16]).** Gives a set of functions $F$, an input data set, where given a difference of at most one data set $D_1$ and $D_2$ with different records, the result set is $R$, and any function $f$ in the function set exists $f(D) \in R$, then the local sensitivity of $F$ is defined as:

$$S(\mathcal{M}) = \mathop{max}_{D_1, D_2} \left( \sum_{f \in F} |f(D_1) - f(D_2)| \right) \tag{3}$$

**Definition 4 (Laplace Mechanism).** Given any function $f : \mathbb{N}^{|x|} \to \mathbb{R}^{\kappa}$, the Laplace mechanism is defined as:

$$\mathcal{M}_L(x, f(\cdot), \varepsilon) = f(x) + (Y_1, \ldots, Y_k) \tag{4}$$

where $Y_i$ are i.i.d. random variables drawn from $Lap(\Delta f/\varepsilon)$.

$\Delta f$ is the sensitivity parameter of Laplace, and $Lap(\Delta f/\varepsilon)$ is a symmetric exponential distribution with an expected value of 0 and a standard deviation of $(\sqrt{2}\Delta f)/\varepsilon$. The probability density function is $Lap(x|b) = 1/2b \, e^{\wedge}((-|x|)/b)$, where $b = \Delta f/\varepsilon$.

### 2.3    Differential Privacy Method Based on Histogram Publishing

The differential privacy based on histogram publishing proposed by Dwork in [1] is the main privacy data publishing method currently used. By adding noise to the interval of each histogram and publishing, the user is provided with a count query while protecting sensitive information. In general, depending on the width of histogram, it can be divided into three basic types: the uniform histogram [17], V-optimized histogram [17], and compressed histogram [18].

For differential privacy based on histogram publishing, the shortcomings of the histogram will directly affect the data release. Therefore, how to improve the accuracy and data availability of interval counting query is an important issue in the research of differential privacy histogram publishing under static data sets.

## 3    Density-Based Differential Privacy Algorithm

The OPTICS algorithm based on density clustering can cluster according to the shape of the data set covering dense areas, and is not sensitive to the input of parameters. Due to the query probability of the data is different; there may be a large difference in access to the data. The feature of OPTICS just fits the problem of differential privacy data coverage. Through pre-processes the dataset with clustering method, the reconstruction error can be reduced. Therefore, we can add noise based on the results of OPTICS clustering to achieve $(\varepsilon, 0)$-differential privacy.

Before introducing algorithm 1, we will first explain some relative information about it. The input $D$: the given data set, $E$: the neighborhood radius, $MinPts$: the number of core objects in $E$. The ordered queue $L_1$: is used to store core objects and their direct density reachable objects and is sorted in ascending order of reachable distance. The result queue $L_2$: is used to store the output order of the sample points. The output queue $L$: is used to store data in the clustering result to which Laplace noise is added.

The algorithm is described as Algorithm 1.

---

**Algorithm 1.**Density-based differential privacy algorithm

---
1: **Input:** sample set $D, E, MinPts, Lap(\Delta f/\varepsilon)$;
2: **Output:** $L, L_1, L_2$;
3: Create $L, L_1, L_2$
4: **if** points in $D$ have been processed **then**
5:   **end**
6:   **return** $L, L_1, L_2$
7: **else**
8:   select an unprocessed (not in $L_2$) core object sample point
9:   find out all its direct density up to the sample point
10:   **If** sample point is not in $L_2$ **then**
11:    Put point in $L_1$; sorted by reachable distance
12:   **end if**
13: **end if**
14: **if** $L_1 = $ empty **then**
15:**goto** Line 4.
16: **else**
17:   take the first sample point from the $L_1$; save the taken sample point to $L_2$
18:   **if** sample point not in $L_2$ **then**
19:    **if** sample point is not a core point **then**
20:     goto Line 14
21:    **else**
22:     find all direct density reachable points of the extension point
23:    **end if**
24:    **if** direct-reached density sample point already in $L_2$ **then**
25:     pass
26:    **else**
27:     goto Line 38
28:    **end if**
29:    **if** direct-reached density sample point already exists in $L_1$ **then**
30:     **if** new reachable distance $<$ old reachable distance **then**
31:      old reachable distance = new reachable distance; resorted $L_1$
32:     **else**
33:      insert the point to $L_1$; reorder $L_1$
34:     **end if**
35:    **end if**
36:   **end if**
37: **end if**
38: Add Laplace noise to the clustering cluster; output result set and store in $L$

---

The algorithm preserves differential privacy.

Although DP-DBSCAN is relatively sensitive to parameter input, clustering results are better than OPTICS-based clustering. Considering the clustering itself of OPTICS, it is mainly because the low-frequency data accumulates at the end of the queue to result in poor clustering. Therefore, further improvements of OPTICS are needed.

# 4  Improved Differential Privacy Algorithm Design Based on Density Clustering

For density-based clustering algorithms, the closer the distance between two data points, the greater the probability of belonging to the same cluster. Our work is based on the OPTICS-based differential privacy algorithm, adding the adjacency list *refresh* for each object to store the nearest neighbor data points, and use the pointer field in the adjacency list to put the low-frequency point into the cluster where the high frequency area adjacent to it is located. If there are no adjacent high-frequency regions near the object, then gather the object near the adjacent low-frequency data points.

## 4.1  Basic Design Ideas

We create an adjacency list for each core object, a non-sorted seed queue $L1$ to store the points to be expanded and an $r$ pointer used to point to the point in $L1$ where the reachable distance is the smallest. When the point in $L1$ is the core object, its neighbor point is directly updated to $L1$. After all the points have been updated, $L1$ is searched again and $r$ points to the point in $L1$ where the reachable distance is the smallest. When processing the next point in $L1$, simply remove the point pointed to by $r$. After processing each point in $L1$ in this way, $L1$ is traversed once to find the point with the smallest reach. The time complexity of searching $L1$ is $O(n)$. The low frequency data is stored in a compressed form. The specific compression method is within the same neighborhood radius, not the point of the core object and the number satisfies $\lfloor 50\%MinPts \rfloor$, then we can think of this as a cluster, or the point is the secondary core point. Otherwise, it is a discrete point, and it is necessary to add a large noise.

## 4.2  Improved Algorithm Design

Before introducing algorithm 2, we will first explain some relative information about it. The input $D$: the given data set, $E$: the neighborhood radius, *MinPts*: the number of core objects in $E$. The seed queue $L1$: is used to store core objects and their direct density reachable objects and is sorted in ascending order of reachable distance. The result queue $L2$: is used to store the output order of the sample points.

The algorithm is described as algorithm 2.

---

**Algorithm 2.** Improved algorithm: IDP-OPTICS

---

1: **Input:** $D$, $E$, $MinPts$, $Lap(\Delta f/\varepsilon_i)$, $Lap(\Delta f/\varepsilon_j)$
2: **Output:** Data set with reachable distance information with added Laplace noise
3: Create L1, L2, adjacency list refresh, pointer r
4: **for** E neighborhood of each object
5:   Traverse $D$
6:   **if** object is a core object **then**
7:     create an adjacency table
8:   **end if**
8: Initialize L1, L2, r is empty; tag data object = "unprocessed"
9: **if** all points in data set $D$ have been processed **then**
10:   **goto** Line 29
11: **else**
12:   select an unhandled object to join the L1; point the r to the object
13: **end if**
14: **if** L1 is empty **then**
15:**goto** Line 9
16: **else**
17:   take the first sample point p from L1 to expand
18:   **if** p is not a core point **then**
19:     reach distance = "Undefined"; **goto** Line 26
20:   **elseif** q in L1 && (new reachable distance value of q$<$old value of q) **then**
21:     update the reachable distance value of q
22:   **else**
23:       add p directly to the end of the L1; **goto** Line 26
24:   **end if**
25: **end if**
26: **delete** p from the L1; **write** p and its reach to L2
27: Traverse the L1; Let the r point to the object with the smallest reach
28: **goto** Line 14
29: Extract L2 data with the steeply descending and steeply rising regions
30: Adding $Lap(\Delta f/\varepsilon_i)$ to the high frequency data cluster
31: Add $Lap(\Delta f/\varepsilon_j)$ to the low frequency cluster

---

The improved algorithm satisfies differential privacy.

Proof. Data sets $D_1$ and $D_2$ add or delete a record in 1 dimension with a sensitivity of 1. If on the n-dimensional space $[0,1]^n$, the sensitivity of the entire query sequence is n. We assume $D_1$ and $D_2$ are based on the IDP-OPTICS clustering algorithm and the results are $S_1$ and $S_2$ respectively, where S represents any clustering. To prove that the differential privacy algorithm based on IDP-OPTICS clustering satisfies $\varepsilon$-differential privacy, the theoretical derivation is provided as follows:

(1) According to the main idea of the above algorithm, the distance between two points after adding Laplace noise is $dis' = \sum_{i=1}^{n}(x_i, y_i)^2 + Lap(a)$, thus $P[S_1 \in S] \propto \exp(-\varepsilon \times |S - |S_1||)$ and $P[S_2 \in S] \propto \exp(-\varepsilon \times |S - |S_2||)$.

(2) According to the three-sided relationship of the triangle, there is $|S - |S_1|| > |S - |S_2|| - ||S_1| - |S_2||$, if $D_1$ and $D_2$ are equal, then $|S - |S_1|| = |S - |S_2|| - ||S_1| - |S_2||$, so $|S - |S_1|| \geq |S - |S_2|| - ||S_1| - |S_2||$.

(3) According to the sum of the two sides is smaller than the third, $||S_1| - |S_1|| < |S_1 \oplus S_2|$, if $D_1$ and $D_2$ are equal, then $||S_1| - |S_1|| = |S_1 \oplus S_2|$, so $||S_1| - |S_1|| \leq |S_1 \oplus S_2|$. Thus it is proved that $P[S_1 \in S]/P[S_2 \in S] \leq \exp(\varepsilon)$.

It is proved that the IDP-OPTICS algorithm satisfies ($\varepsilon$, 0)-differential privacy. □

## 5  Experimental Evaluation

### 5.1  Example Simulation

Let us illustrate with a concrete example:

We assume that the neighborhood radius E = 2, the point within the minimum neighborhood radius MinPts = 5, and there are 11 points A(10,11), B(10,10), C(11,11), D(10,12), E(10,13), F(11,12), G(12,12), H(100,200), I(100,201), J(101,200), K (200,300) in the data set and calculate the Euclidean distance between points. It can be known that A, C, D, and F points are core objects; B, E, and G belong to the points in their clusters; H, I, and J are not core objects, but the compression condition is satisfied at a radius of E = 2. Therefore, H, I, and J can be classified into one cluster, but K is a discrete point because it does not satisfy the compression condition. We use adjacency tables to store. From the adjacency list shown in Fig. 1 we can see that A, B, C, D, E, F, G are high frequency points and H, I, J, K are low frequency points.

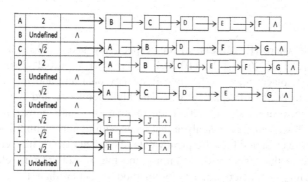

**Fig. 1.** Example of using IDP-OPTICS clustering.

### 5.2  Analysis of Experimental Data for Improved Algorithms

The configuration of computer is Windows 7, AMD Athlon(tm) II X4 640 processor and 4 GB of memory. We ran the experiment on MATLAB R2012 (a), Weka 3.6 version, Eclipse 3.7.0 and the data set is from the credit-g.arff data set in the data folder of Weka 3.6 version, Relation: german_credit; Instances: 1000; Attributes: 21.

We use the DP-DBSCAN, the IDP-OPTICS and the OPTICS-based differential privacy algorithm to perform multiple experiments on 1000 records in the credit-g.arff data set. We take values for E and MinPts {(0.1,5), (1,10), (5,10), (10,50), (0.5,8), (2,10), (10,20), (20, 50)} and take the average of 150 experiments. Through the noise processing of high-frequency data and low-frequency data, by experience, we allocation the $\varepsilon$ as $\varepsilon_i:\varepsilon_j = 20:1$. So $\varepsilon_i = 0.475$, $\varepsilon_j = 0.025$. Comparative analysis is shown in Fig. 2. When IDP-OPTICS achieves relatively small values in E and MinPts, the clustering is better in terms of locality; overall, it is better than differential privacy and DP-DBSCAN algorithm based on OPTICS clustering.

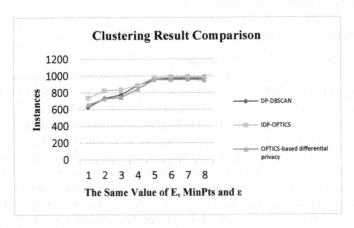

**Fig. 2.** OPTICS-based differential privacy and IDP-OPTICS clustering results under the same E, MinPts and $\varepsilon$.

In the case where (E, MinPts, $\varepsilon$) are respectively taken {(0.1,5,0.1), (1,10,0.3), (2,15,0.5), (4,18,1.0), (6,20,1.2), (10,20,1.3), (15,25,1.5), (20,25,20)}, we add the same-variance Laplace noise to the high-frequency data and add the heteroscedastic Laplace noise to the low-frequency data. The high-frequency and the low-frequency data still add noise in a ratio of 1:20. Their relative errors are shown in Fig. 3.

From Fig. 3, we can clearly analyze that when adding noise with the same variance method, the more clustered high-frequency data, the greater the relative error; while the heteroscedastic method is used to add noise and the overall error is relatively small. Therefore, adding the heteroscedasticity method is better than the same variance method and has better performance in processing data.

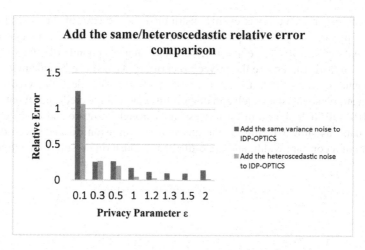

**Fig. 3.** Comparison of IDP-OPTICS relative errors for adding two different modes of noise.

In the case where ε = 0.5 and (E, MinPts) are respectively taken {(0.1,5), (1,10), (3,13), (5,15), (10,15), (15,20), (20, 25)}, we use IDP-OPTICS, DP-DBSCAN and OPTICS-based differential privacy algorithm. After several experiments, 150 experimental averages were obtained, and the time comparison of the three was as shown in Fig. 4. The OPTICS-based differential privacy algorithm is time-consuming compared to DP-DBSCAN and IDP-OPTICS. Since IDP-OPTICS adopts the adjacency list form, it is superior to the DP-DBSCAN algorithm in time complexity. Overall, the IDP-OPTICS algorithm is still feasible.

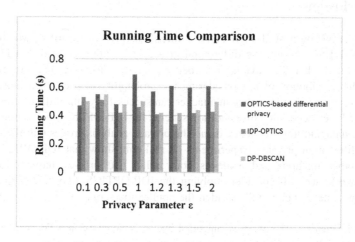

**Fig. 4.** Three algorithms time consumption.

The credit-g.arff data set is evenly distributed. In the case of the central position parameter $\mu = 0$, $L = 0.5$, the probability of the attacker's success in querying the result data set is $\rho = 1 - e^{\wedge}(-\varepsilon/2)/2$. The upper bound of the $\varepsilon$ satisfies $\varepsilon \leq \ln2(1 - \rho)\Delta f/L$ can ensure the privacy security of data: It can be calculated that the upper bound of $\varepsilon$ is 0.42. In the case of $\varepsilon = 0.42$, we select the credit_history, employment, personal_status, job and other information in the credit-g.arff data set, and use the IDP-OPTICS algorithm for processing. Through compare and analyze with the original data, as shown in Fig. 5, the abscissa is a division distance according to a histogram of a constant width, and the ordinate is a clustering result of one thousandth.

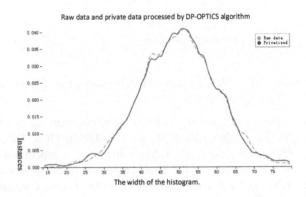

**Fig. 5.** Comparison of the original data with the privacy protection data processed by the IDP-OPTICS algorithm.

## 6   Conclusions

In this paper, we propose IDP-OPTICS, a differential privacy algorithm which is based on the DP-DBSCAN and the differential privacy algorithm based on OPTICS clustering. It has higher data availability, better data set clustering and can effectively balance the distribution of $\varepsilon$. Considering the probability of users querying data and determining the maximum $\varepsilon$ that an attacker can successfully attack dataset records, we designed a heteroscedastic noise addition scheme for IDP-OPTICS to add noise to high frequency data and low frequency data. The privacy analysis proves that IDP-OPTICS satisfies differential privacy. Experiments on real data show that IDP-OPTICS is outperform other methods and achieves both utility and privacy required. In future research work, we will consider applying the IDP-OPTICS algorithm to the recommendation system, which will broaden the application scope of differential privacy.

**Acknowledgements.**   This work is supported by the National Science Foundation of China under Grant No. 61862007, Guangxi Natural Science Foundation under Grant No. 2018GXNSFAA138147, Key Laboratory of Guangxi High Schools Complex System and Computational Intelligence under Grant No. 2016CSCI09, and Teaching Reform Project of Guangxi University for Nationalities under Grant No. 2016XJGY33 and No. 2016XJGY34.

# References

1. Dwork, C.: Differential privacy. In: Proceedings of the 33rd International colloquium on Automata, Languages and Programming, pp. 1–12 (2006)
2. Dwork, C., McSherry, F., Nissim, K., Smith, A.: Calibrating noise to sensitivity in private data analysis. In: Halevi, S., Rabin, T. (eds.) TCC 2006. LNCS, vol. 3876, pp. 265–284. Springer, Heidelberg (2006). https://doi.org/10.1007/11681878_14
3. Mcsherry, F., Talwar, K.: Mechanism design via differential privacy. In: IEEE Symposium on Foundations of Computer Science, Providence, RI, USA, pp. 94–103 (2007)
4. Hay, M., Rastogi, V., Miklau, G., et al.: Boosting the accuracy of differentially private histograms through consistency. Proc. VLDB Endow. 3(1–2), 1021–1032 (2010)
5. Peng, S., Yang, Y., Zhang, Z., et al.: DP-tree: indexing multi-dimensional data under differential privacy (abstract only), pp. 864–864. ACM (2012)
6. Xiao, Y., Xiong, L., Yuan, C.: Differentially private data release through multidimensional partitioning. In: Jonker, W., Petković, M. (eds.) SDM 2010. LNCS, vol. 6358, pp. 150–168. Springer, Heidelberg (2010). https://doi.org/10.1007/978-3-642-15546-8_11
7. Xiao, X., Wang, G., Gehrke, J.: Differential privacy via wavelet transforms. IEEE Trans. Knowl. Data Eng. 23(8), 1200–1214 (2010)
8. Xiong, P., Zhu, T.Q., Jin, D.W., et al.: A differential privacy protection algorithm for decision tree construction. 31(10), 3108–3112 (2014)
9. Wu, Y.J.: Privacy Protection Data Release: Models and Algorithms, pp. 258–262. Tsinghua University Press, China (2015)
10. Wang, H., Ge, L.N., Wang, L.Y., et al.: Survey of differential privacy for data histogram publication. Appl. Res. Comput. 34(6), 1609–1612 (2017)
11. Blum, A., Dwork, C., Mcsherry, F., et al.: Practical privacy: the SuLQ framework. In: Twenty-Fourth ACM Sigact-Sigmod-Sigart Symposium on Principles of Database Systems, DBLP, Baltimore, Maryland, USA, vol. 6, no. 13–15, pp. 128–138 (2005)
12. Dwork, C.: A firm foundation for private data analysis. Commun. ACM 54(1), 86–95 (2011)
13. Wu, W.M., Huang, H.K.: Research on DP-DBScan clustering algorithm based on differential privacy protection. Comput. Eng. Sci. 37(4), 830–834 (2015)
14. André, L.C.M., Brito, F.T., Linhares, L.S., et al.: DiPCoDing: a differentially private approach for correlated data with clustering. In: International Database Engineering & Applications Symposium. ACM (2017)
15. Ni, L., Li, C.: DP-MCDBSCAN: differential privacy preserving multi-core DBSCAN clustering for network user data. IEEE Access 6, 21053–21063 (2018)
16. Nissim, K., Raskhodnikova, S., Smith, A.: Smooth sensitivity and sampling in private data analysis. In: Proceedings of the Thirty-Ninth Annual ACM Symposium on Theory of Computing, pp. 75–84. ACM (2007)
17. Jagadish, H.V., Poosala, V., Koudas, N., et al.: Optimal histograms with quality guarantees. In: Gupta, A., Shmueli, O., Widom, J. (eds.) VLDB 1998, Proceedings of the 24th International Conference on Very Large Data Bases, pp. 275–286. Morgan Kaufmann, New York (1998)
18. Poosala, V., Haas, P.J., Ioannidis, Y.E., et al.: Improved histograms for selectivity estimation of range predicates. ACM Sigmod Rec. 25(2), 294–305 (1999)

# Author Index

Printed in the United States
By Bookmasters